DATE DUE FOR RETURN

SOIL and WATER CHEMISTRY

An Integrative Approach

SOIL and WATER CHEMISTRY

An Integrative Approach

MICHAEL E. ESSINGTON

CRC PRESS

Boca Raton London New York Washington, D.C.

Library of Congress Cataloging-in-Publication Data

Essington, Michael E.
  Soil and water chemistry : an integrative approach / by Michael E. Essington.
      p.   cm.
  Includes bibliographical references and index.
  ISBN 0-8493-1258-2 (alk. paper)
  1. Soil chemistry. 2. Water chemistry. I. Title.

S592.5.E88 2003
631.4′1—dc22   2003058468

Visit the CRC Press Web site at www.crcpress.com

© 2004 by CRC Press LLC

No claim to original U.S. Government works
International Standard Book Number 0-8493-1258-2
Library of Congress Card Number 2003058468
Printed in the United States of America 1 2 3 4 5 6 7 8 9 0
Printed on acid-free paper

Preface

Soil and Water Chemistry: An Integrative Approach was written to meet the needs of undergraduate and first-year master's students in soil and environmental chemistry courses. The book may also serve as a reference for professionals in the soil sciences and allied disciplines. The discipline of soil chemistry, or its contemporary counterpart, environmental soil chemistry, examines the chemical and mineralogical characteristics of the soil environment and the chemical processes that distribute matter between the soil solid, solution, and gaseous phases. Essentially, a soil chemistry course and this book offer a basic understanding of the complexity of the natural system that occupies an exceedingly thin layer at the Earth's surface. Traditionally, the application of chemical principles to the study of soils has been limited to agronomic systems, and primarily to the behavior of agrichemicals. However, it is well established that as a discipline soil chemistry is not limited to describing the processes that control the availability of nutrients to plants. Indeed, the chemical properties and processes that control the behavior of nutrients and pesticides in soils are the same as those that operate on a vast array of inorganic and organic substances that are outside the purview of production agriculture.

Recent texts in soil chemistry, those published in the last decade, have attempted to embrace the environmental aspects of the discipline, as evidenced by their various titles (e.g., *Environmental Soil Chemistry*, *Environmental Chemistry of Soils*, and *Environmental Soil and Water Chemistry: Principles and Applications*). Topically, *Soil and Water Chemistry: An Integrative Approach* continues the "environmental" trend established by its predecessors. However, my intent is to focus on the needs of undergraduate students in soil chemistry and allied disciplines and offer a balanced presentation of the chemical processes operating in soils.

This book contains more information and topic coverage than an instructor might cover in a single semester, and introduces some topics that may be too advanced for an undergraduate course. This extensive coverage is by design. I envision that it will allow instructors the latitude to choose their own "essential" topics, while providing additional information or a more advanced treatment for others. This book also contains more than 300 original figures and approximately 90 tables to help make the material more accessible. I have also reviewed some of the more common methodologies and analytical techniques used to characterize soil chemical properties. In addition, each chapter contains several sample examples that illustrate problem solving techniques.

Each chapter concludes with a section containing numerous exercises. The problems are not esoteric, nor do they require advanced training in soil chemistry to begin to formulate a solution. Each problem has at one time or another appeared in one of my problem sets. They are tested, relevant, and doable, but ask more of students than to simply (or not so simply) generate a number. It may be inevitable that the path to complete a computation or series of computations is tortuous, but this tortuosity should not blind the student to the purpose for determining a numerical answer. It is important for students to understand concepts, and to recognize that the answers to their computations have physical meaning. I have attempted to include exercises that embody both traits; compute an answer and discuss its significance.

This book has its roots in the undergraduate course I offer in environmental soil chemistry. This is a required course for the environmental and soil sciences major, and is perhaps the last technically oriented course these students will take during their undergraduate experience. I have high expectations of students, particularly with respect to the amount of information retained from prerequisite courses. However, I am also a pragmatic person and recognize that materials introduced in an organic chemistry course taken as a sophomore may have long since been relegated to the "recycle bin" of the mind. The information is there, it just needs to be restored. As with my course in environmental soil chemistry, this textbook begins with an overview of the soil environment and the chemical processes that operate to distribute matter between the soil solid, solution, and atmosphere. Students are then introduced to the concept of speciation, and they are presented with a list of common oxidation states and species for nearly every element as it might exist in the soil solution (with the exception of aqueous complexes). I do not specifically discuss units, unit conversions, or mass transfer computations in lecture (e.g., if an X g mass of soil is extracted with Y mL of water and the water contains Z mg L^{-1} of element A, what was the concentration of A in the soil, in mg kg^{-1}?). Instead, I rely on problem sets to refresh and restore this information. However, I am annually queried, "Where was I supposed to have learned how to do this?" I now recognize that 3 years or more of college has not prepared students for this important rudimentary capability. Therefore, these topics are also covered in Chapter 1. Also introduced in Chapter 1 is the concept of spatial variability and spatial statistics. We often discuss the elemental composition of soils in soil chemistry courses, indicating that for every element there is a mean and median value and a range of concentrations observed in soils of the world. Soil chemical properties are spatially variable on a local scale as well; they change with location on a landscape and depth in a profile.

Chapters 2, 3, and 4 are devoted to the soil solids. Chapter 2, "Soil Minerals," begins by discussing the "glue" that bonds atoms together in mineral structures and the rules that describe how these atoms are arranged in three-dimensional space (Pauling's rules). The remainder of the chapter describes the silicates, emphasizing the phyllosilicates, and the hydrous metal oxides. Finally, x-ray diffraction and its application to identifying clay minerals are discussed. Chapter 3, "Chemical Weathering," focuses on clay mineral transformations. This chapter also (re)introduces a very important capability that must be mastered by any individual in a chemistry-based course or discipline: balancing chemical reactions. Chapter 4, "Organic Matter in Soil," examines the organic component of the soil solid phase. The reader is (re)introduced to the organic functional groups and structural components that occur in soil organic matter. The distinction between non-humic and humic substances is drawn, as well as the mechanisms for isolating humic substances. The nonhumic substances are described, as are their transformations from biomolecules to humic substances. The chemical and (pseudo)structural characteristics of the humic substances are also discussed.

One of the largest chapters in this book is Chapter 5, "Soil Water Chemistry." The chapter begins by discussing chemical characteristics of water, the universal solvent, and ends by examining some important analytical methods used to determine the concentrations of dissolved substances in soil solutions. These topics, and those in between, constitute a course in water chemistry and reflect my belief that the aqueous chemistry of a substance dictates its fate and behavior. The nonideality of soil solutions, hydration-hydrolysis, Lowry-Brønsted and Lewis acidity and basicity, aqueous complexation, geochemical modeling, and soil solution sampling methods are topics that are addressed with detail and rigor. Chapters 6, 7, and 8 examine the processes that distribute matter between the soil solid and solution phases. In Chapter 6, "Mineral Solubility," the soil solid and solution characteristics that control the precipitation and dissolution of soil minerals are examined. This chapter also examines the influences of temperature and impurities in soil minerals on mineral stability and the compositions of soil solutions. Chapter 7, "Surface Chemistry and Adsorption Reactions," rivals Chapter 5 in size and scope. Adsorption and partitioning reactions are the principal mechanisms by which all organic solutes and many inorganic substances are retained in soils (the other mechanism is precipitation). The chapter describes the soil surfaces and identifies the inorganic and organic functional groups that react with solutes to form surface species. The chapter also examines those factors that influence the reactivity of soil surface functional groups and applies surface- and solute-specific information to predict adsorption behavior (surface complexation modeling). The descriptive models that are commonly employed to provide an empirical characterization of adsorption are also examined (e.g., Langmuir and Freundlich isotherm models). Although ion exchange is also an adsorption process (or is it—all adsorption processes are ion exchange processes?), it is standard practice to discuss exchange phenomena separately from adsorption. This is done in Chapter 8, "Cation Exchange." This chapter focuses on the history, methods of characterizing the soil's capacity to exchange cations, the qualitative characteristics of cation exchange, and the techniques to quantify exchange behavior.

Oxidation-reduction processes in soils are examined in Chapter 9, "Oxidation-Reduction Reactions in Soils." Although this topic is introduced in a later chapter, this should not be taken to imply that the redox behavior of an element is of minor importance. Quite the contrary; the redox status of an environment is a master chemical variable (along with pH) that directly dictates the fate and behavior of redox-sensitive elements, which in turn may influence the chemistry of other soil constituents. Methods for determining soil redox status, reduction-oxidation sequences in soils, and the redox chemistry of chromium, selenium, and arsenic are discussed. The final two chapters (Chapters 10 and 11) are devoted to topics of regional interest: "Acidity in Soil Materials" and "Soil Salinity and Sodicity." The genesis, characterization, management, and chemical properties of these differing soil systems are discussed. I have also included case studies that examine the reclamation of pyritic acid mine spoils and sodic mine spoils.

I am deeply appreciative to the many individuals who donated their time and expertise to the preparation of this book. John Sulzycki at CRC Press planted the "textbook bug" and gave me the opportunity to bring this project forward. He also provided encouragement and continued to demonstrate confidence that I would complete this project, even after I missed several deadlines. Julia Nelson critically reviewed every chapter with a keen eye. Her critiques were thorough and immeasurably improved the clarity of the manuscript and caught my many typos, misspellings, and grammatical errors. I am deeply indebted to her for her efforts. I am also indebted to Gary Pierzynski, George Vance, April Ulery, Dean Hesterberg, and Malcolm Sumner for their highly constructive reviews and suggestions. I applaud their selfless contribution to the discipline of soil and water chemistry and to the education of future "Earth" scientists and technicians by giving of their time and expertise. I was buoyed by their positive feedback and by their desire to see a book that has utility for students and professionals. Finally, I have made every effort to produce a text that is complete for the intended audience and conceptually sound. I also recognize that no book is without errors. For the errors that remain, I hope they are few in number and minor in magnitude. If errors are discovered, or if you as the reader have comments and suggestions that would improve future editions of this book, please being them to my attention. I welcome your input.

In addition to the individuals cited above who helped shepherd this book from manuscript to reality, there are many people who have mentored me and provided me with the tools, insight, and drive necessary to complete this book. First and foremost is a fellow soil scientist who steered me away from a major in biology and toward the soil science curriculum at New Mexico State University. The late Edward Essington uttered these words during a job fair while I was still a senior in high school: "Get a degree that will allow you to do more that wait tables when you graduate." I took his advice. I should thank George O'Connor and Al Page for their continual encouragement and support throughout my career, and Shas Mattigod and Garrison Sposito, who set the bar for me many years ago. Finally, without the students whom I have directed or who have taken my soil chemistry courses, this book would not have come to fruition. Their research has provided an abundance of material for this text, and their response to the materials in the lecture notes has been instrumental in the production of this book. You all have my gratitude.

Michael E. Essington
Knoxville, Tennessee

The Author

Michael E. Essington is professor of soil and water chemistry in the Institute of Agriculture at The University of Tennessee in Knoxville. In addition to teaching courses in soil chemistry and clay mineralogy, Dr. Essington's special research interests center on the role of aqueous speciation in environmental chemistry, with particular emphasis on trace element adsorption and precipitation phenomena. These interests have resulted in more than 120 publications and technical reports. Dr. Essington received his B.S. in agriculture from New Mexico State University in 1980 and his Ph.D. in soil science from the University of California, Riverside, in 1985. He was a research scientist at the Western Research Institute in Laramie, Wyoming from 1985 until 1990 and has been at The University of Tennessee since then. He is a member of the Soil Science Society of America, the American Society of Agronomy, Sigma Xi, and Gamma Sigma Delta. Dr. Essington's professional activities include serving as an associate editor for the *Soil Science Society of America Journal*; soil chemistry division chair for the Soil Science Society of America; and USDA-NRI panel member for the Soils and Soil Biology Program.

Table of Contents

Dedicated to my girls

———————

Nina

Erin

Meghan

Chelsea

and

Deanna

1 The Soil Chemical Environment: An Overview

Soil, the thin layer of unconsolidated material that covers the Earth's surface, is a natural resource that is perhaps uniquely responsible for human development and continued existence on this planet. Humans require food, water, shelter (including clothing and housing), and a means to separate themselves from their own waste products. Soil is the main reservoir that supplies water and essential nutrients for the growth of plants for food, fiber for clothing, and forests for building materials. At the same time, soil is the principal receptacle for waste material, including human, animal, and industrial wastes. Soil is expected to filter and process human and animal waste products, and to tie-up or degrade industrial waste substances to prevent toxic materials from being transferred back into the food chain and drinking water supplies. However, land development and population growth continue to consume and overtax soil resources such that productivity requirements and environmental demands require an increasingly precise level of management to optimize the performance of soils.

Soil chemistry is central to providing the understanding needed to predict the fate and behavior of substances in the terrestrial environment, whether involving the controlled release of plant nutrients required for sustainable crop production or the strong binding and immobilization of potentially toxic elements to protect water quality, human health, and the environment. Soil chemistry aims to understand fundamental processes that regulate the transfer of substances between various phases in soils (solid, aqueous, and gaseous), while integrating these individual processes into a comprehensive set of rules that can be used to predict and manage how substances are retained in or removed from soils. This chapter provides a general overview of the soil environment, the chemical processes that occur in soil, and the chemical nature (concentration and speciation) of elements in soil.

1.1 PHASES AND CHEMICAL PROCESSES IN SOIL

If soils did not have the inherent ability to bind chemical elements, the plant nutrients and waste products applied to soils would be transported vertically (leached) by percolating rainwater deep into the earth where they would be unavailable for plant growth or degradation by soil microorganisms and they would accumulate in groundwater. Alternatively, chemical elements could be transported laterally through subsoil and deeper geologic materials and ultimately discharged to surface waters. The resulting pollution problems would make groundwater unusable for drinking water, and rivers and streams unfit for human consumption and recreation.

Before the mid-1800s the ability of soils to actually react with chemicals was unknown. Indeed, soil was simply considered to be a support medium for plants, only retarding the flow of water and plant nutrients by physical filtration. It was during the 1850s that an agricultural scientist named J.T. Way, spurred on by the preliminary findings of an agriculturalist named H.S. Thompson, conducted a series of studies that illustrated the innate reactivity of the soil solid phase. His results were not easily accepted by the scientific establishment of the time; indeed, some of the most respected agricultural scientists (such as J. von Liebig) urged the scientific community to "oppose" the experiments of J.T. Way.

There is no doubt that mobility of chemical substances in soils may be retarded by the solid phase. It is also known that the ability of soil solids to regulate the movement of chemicals is

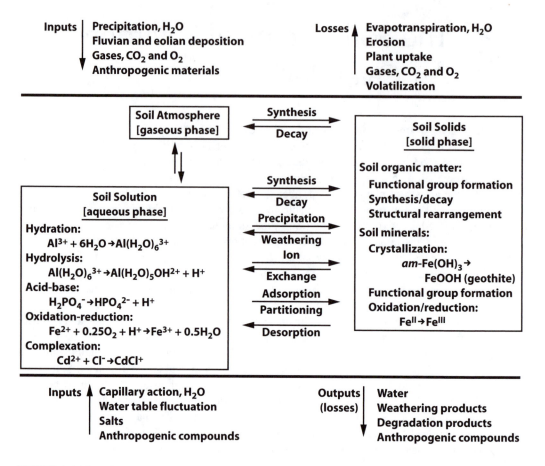

FIGURE 1.1 This diagram presents an overview of the components of and the processes that occur in a soil environment.

dictated by the processes that distribute chemicals between the immobile soil solids and the mobile soil water and gaseous phases. In chemistry, a phase is a part of a system that is uniform (homogeneous) throughout in chemical composition and physical properties, and that is separated from other uniform parts by a boundary. The soil environment may be described as consisting of three phases: the soil atmosphere or gaseous phase, the soil solution or aqueous phase, and the soil solids or solid phase (Figure 1.1). The soil gaseous and aqueous phases are true phases; however, the soil solid phase is not a single phase, but rather a composite of several phases. This distinction will be discussed in later chapters; however, for the purpose of this general discussion, the soil solid phases may be collectively divided into compartments. These solid phase compartments are mineral (inorganic compounds), humus (soil organic matter), and biotic (living organisms). Distinctly different from the soil solid, aqueous, and gaseous phases are the boundary phases, which are comprised of substances that accumulate at the interface between soil minerals or organic matter and the soil solution. For example, ions that are exchangeable may be identified as residing in the soil exchanger phase. Ions and molecules that are retained by processes other than exchange or precipitation (the formation of a true solid phase) may be identified as residing in the adsorbed phase. These boundary phases may also be included in the soil solid phase.

The soil environment is bounded by the soil surface (at the top of the soil profile) and the soil parent material (beneath the soil profile). In addition to these boundaries, arbitrary lateral boundaries may also be imposed. Soils are open systems, so energy (thermal) and matter can enter and exit the soil environment at these boundaries. At the soil surface, matter can enter the soil environment via rainfall (H_2O), atmospheric particulates, fertilizers and other anthropogenic inputs (waste disposal and utilization), the diffusion of gases, such as carbon dioxide (CO_2), oxygen (O_2), and nitrogen (N_2), and deposition (fluvial and eolian). Matter may also exit the soil environment via evaporation and transpiration, diffusion of gases (CO_2, O_2, NH_3), plant uptake of essential nutrients and nonessential elements followed by biomass removal, direct ingestion by grazing animals, volatilization (organic compounds), and erosion. At the lower boundary, matter may be leached out of the soil profile, ultimately appearing in ground and surface waters. Substances may also enter the soil by lateral flow or at the lower boundary by fluctuating water tables or in water drawn up by capillary action.

1.1.1 INTRAPHASE SOIL PROCESSES

Within each of the three soil phases, a single element may exist in several different chemical forms. Each form of an element has unique characteristics that impact the fate and behavior of the element. We are surrounded by soils that contain very high levels of aluminum (71,000 mg Al kg^{-1}, on average), a nonessential element that can be toxic to plants, aquatic life, and humans. Yet, humans show no concern for the fact that they are surrounded by large quantities of this potentially toxic substance; lakes, rivers, and streams teem with aquatic life, even though a portion of the water they hold may have percolated through soil; and soils naturally sustain a diverse and plentiful array of microbial and plant life. However, not all environments are unaffected by Al. Aluminum toxicity in plants can occur in humid regions when poor crop management practices are employed. Aquatic organisms in streams that are impacted by highly acidic runoff and leachates from pyritic mine spoils will show the effects of Al toxicity. Research by soil scientists has shown that when properly managed the toxic effects of Al can be reversed by returning Al to solid phases that are sparingly soluble. For example, Al phytotoxicity and high mine spoil leachate concentrations can be easily remedied by raising the pH of the soil or the mine spoil by liming with calcium carbonate (calcite, $CaCO_3$), altering the solid form of Al and returning the element to mineral phases that are sparingly soluble and unavailable to plants. Potentially toxic metals such as lead (Pb), cadmium (Cd), mercury (Hg), arsenic (As), and chromium (Cr) are ever-present constituents of soils, albeit in much lower concentrations than Al. As with Al, these metals are of little concern in a natural setting. However, land disposal of industrial wastes and the utilization of municipal wastes (e.g., sewage sludge) often elevate the soil concentrations of these potentially toxic metals to levels that could have a detrimental environmental impact. Again, research has illustrated that the solid phases in which these elements reside dictate their availability and toxicity, and that soils play a key role in the transformation of potentially toxic elements from soluble and available forms to those that are relatively innocuous.

As the discussion above suggests, elements that reside in the soil solid phase may change form. Indeed, within each of the three soil phases, substances may change chemical form by participating in chemical reactions (Figure 1.1). The soil solid phase is comprised of organic substances (biotic and abiotic matter) and soil minerals (inorganic compounds). Within the soil organic fraction, organic compounds are synthesized and degraded, the structure of the organic compounds may rearrange into different configurations, and functional groups (carboxyl, phenolic-hydroxyl, carbonyl, and amino) may form, transform, or ionize (develop charge). In the soil mineral phase, amorphous solids (minerals that have no long-range structure) crystallize, surface functional groups form and ionize ($\equiv AlOH^0 \rightarrow \equiv AlO^- + H^+$; or $\equiv AlOH^0 + H^+ \rightarrow \equiv AlOH_2^+$, where $\equiv Al$ represents an aluminum atom in a mineral structure), and redox-sensitive elements in mineral structures undergo oxidation ($Fe^{II} \rightarrow Fe^{III}$) or reduction ($Mn^{IV} \rightarrow Mn^{II}$).

A soil's reactivity is mostly regulated by the aqueous phase. In the soil solution, oxidation-reduction, hydration-hydrolysis, acid-base, and complexation reactions occur. These reactions impact the speciation (chemical form) of substances in the soil solution, which in turn impacts bioavailability, toxicity, the interactions that occur between the soil phases, and mobility. For example, chromate (CrO_4^{2-}), a highly toxic, mobile, anionic form of Cr, may be reduced to Cr^{3+}, a less toxic, relatively immobile, cationic form of Cr. Metal cations, such as Pb^{2+}, are surrounded by a sphere of water molecules (they are hydrated, as are anions). Depending on the metal and the pH of the soil solution, a water of hydration can decompose (deprotonate or ionize) and produce a proton and a hydrolysis product. For example, when soil solution pH is greater than 7.7, Pb predominantly exists as the hydrolysis product $PbOH^+$, formed by the reaction: $Pb^{2+} + H_2O \rightarrow PbOH^+ + H^+$. The $PbOH^+$ is a chemically unique aqueous species that reacts quite differently in soil than Pb^{2+}. Hydrolysis detoxifies soluble Al by removing the element from its toxic form, Al^{3+}, and placing it into $AlOH^{2+}$, $Al(OH)_2^+$, $Al(OH)_3^-$, and $Al(OH)_4^-$ forms.

Nearly all dissolved substances in the soil solution are capable of accepting or donating protons. Species that accept protons are called Lowry-Brønsted bases, and those that donate protons are Lowry-Brønsted acids. For example, an organic amino group is a base because it may accept a proton: $R-NH_2^0 + H^+ \rightarrow R-NH_3^+$. Not only does this process help buffer (control) the pH of the soil solution, the ionization of the organic moiety alters its reactivity and behavior in the soil environment. Acidic species, those that donate protons to the soil solution, also aid in the buffering of the soil solution, and the change in chemical form alters their environmental behavior as well. The hydrolysis of Pb^{2+} described above is an example of an acid-producing reaction that alters the chemical form of Pb. Similarly, the dissociation reaction, $H_2PO_4^- \rightarrow HPO_4^{2-} + H^+$, is also an acid-producing reaction that alters the chemical form of phosphorus.

Substances that are dissolved in the soil solution may interact to form soluble species that display chemically different behavior than they did before the interaction. For example, the free form of Cd in soil solution is the divalent cation, Cd^{2+}. In the presence of the chloride ion, Cl^-, Cd will exist as the free cation and in association with chloride, as $CdCl^+$. The $CdCl^+$ is a unique soluble species (it is not a solid) that behaves quite differently from the divalent ion, Cd^{2+}. The process of forming the ion association, or adduct for an addition product, is called aqueous complexation or ion pair formation: $Cd^{2+} + Cl^- \rightarrow CdCl^+$. Aqueous complexation is not limited to inorganic substances. Naturally occurring and synthetic organic compounds, particularly those than contain carboxyl and amino groups, form strong complexes with metal cations, which drastically increase their mobility and bioavailability. Inorganic and organic species that participate in these types of reactions—and nearly all soluble substances in the soil solution do participate in these reactions—are called Lewis acids and bases. A Lewis acid is a cation, often a metal cation, such as Cd^{2+}, that initiates a chemical reaction by employing an unoccupied electronic orbital. A Lewis base is a ligand, a substance that initiates a chemical reaction by employing a doubly occupied electronic orbital. Ligands may be charge neutral (such as H_2O or an organic amine, $R-NH_2^0$) or an anion (such as Cl^-). Lowry-Brønsted acidity is a special case of Lewis acidity, where H^+ is the cation.

1.1.2 INTERPHASE SOIL PROCESSES

Industrial activities in modern society are supported by energy derived from the burning of fossil fuels such as petroleum and coal. As a result of this combustion, the concentration of carbon dioxide (CO_2) (and other gases) in the Earth's atmosphere has steadily increased since the mid- to late 1800s. At present, the atmospheric content of CO_2 is considered to be increasing at a rate of 7.2 Pg C yr^{-1} (1 Pg = 10^{12} kg), principally from fossil fuel combustion (5.5 Pg C yr^{-1}) and land use changes (1.6 Pg C yr^{-1}) (Swift, 2001). This increase in global atmospheric CO_2 has triggered substantial international concerns over its potential effects on global warming and global weather patterns. Soil plays an important role in the regulation of atmospheric CO_2 levels and in carbon sequestration. Current estimates indicate that the global reservoir of soil-bound carbon is about 3300 Pg (found in carbonates and organic carbon), which is greater than four times the amount of carbon in the atmosphere (720 Pg). Current estimates also indicate that CO_2 emissions from soils by microbial respiration, 60 Pg C yr^{-1}, are more than 10 times greater than that from the combustion of fossil fuels (5.5 Pg C yr^{-1}). Additionally, carbon emission by respiration from plants (60 Pg C yr^{-1}) and carbon lost through the net destruction of vegetation (2 Pg C yr^{-1}) indicate that there is a net loss of carbon (2 Pg C yr^{-1}) from soils and associated ecosystems.

Capturing more CO_2 into plant biomass and transferring this carbon into soil organic matter is an important mechanism that might be employed through the manipulation of land management practices to offset fossil fuel emissions. On a smaller and more general scale, the transfer of matter between phases in the soil environment directly influences the chemical processes that impact the fate and behavior of soil components (Figure 1.1). In the soil atmosphere, the chemically significant components are CO_2 (as it impacts soil solution pH) and O_2 (as it impacts soil redox status). These soil atmosphere gases readily diffuse through the soil atmosphere under water-unsaturated conditions, and they can dissolve into the soil solution. However, gases have differing water solubilities. For example, CO_2 has a water solubility of 33.8 mmol L^{-1} at 25°C and 1 atm CO_2 pressure; whereas, O_2 has a water solubility of 1.28 mmol L^{-1} at 25°C and 1 atm O_2 pressure. Carbonic acid ($H_2CO_3^0$) is created when CO_2 dissolves and is hydrated ($CO_2 + H_2O \rightarrow CO_2 \bullet H_2O$). Carbonic acid makes rainfall naturally acidic and hastens the weathering of soil minerals. Oxygen gas is an oxidizing agent (electron acceptor) and the presence of oxygen results in an oxidizing (aerobic) environment, while the absence of O_2 leads to reducing conditions. The low water solubility of O_2 can cause the development of reduced conditions in water-saturated soil. Whether the environment is oxidizing or reducing greatly impacts both biotic (biological) and abiotic (nonbiological) processes in soil.

During the degradation of biopolymers in soil, organic matter such as lignin, proteins, and carbohydrates, CO_2, inorganic substances, and low-molecular-mass organic compounds such as acetic acid and citric acid are released to the soil solution. Inorganic elements are also released during the dissolution, or weathering, of soil minerals. Many of the released elements reprecipitate to form new and relatively stable mineral phases, such as the hydrous metal oxides. However, some substances may remain soluble and quite mobile, as if the soil had no retention capacity. The ecological disaster that befell the Kesterson National Wildlife Refuge located on the west side of the San Joaquin Valley in California was two decades in the making. During the 1950s and 1960s large tracts of land were brought under irrigated agriculture. However, facilities to remove drainage from the region were not available. Without adequate drainage, salinity levels in the soils of the region and in the shallow groundwater gradually increased during the 1960s and 1970s. Increasing soil salinity prompted even more intense irrigation practices to maintain productivity, resulting in the additional build-up of salinity, water-logging, and finally the abandonment and loss of arable land. Beginning in 1981, the San Luis Drain was opened to discharge the subsurface drainage to a reservoir at the Refuge. In 1983 it was discovered that high levels of selenium (Se) in the reservoir caused a high incidence of deformity and mortality in waterfowl hatchlings at the Wildlife Refuge. The introduction of irrigated agriculture into the valley led to the solubilization of Se from the seleniferous soils that formed on alluvium derived from the sedimentary rocks that border the valley. Selenium concentrations in the shallow groundwater continued to build during the two decades prior to the opening of the San Luis Drain. When the drain was finally opened, the Se-rich water was transported to the Wildlife Refuge. In this case, Se in the soil was solubilized because the Se-bearing minerals were easily dissolved by the irrigation waters. The Se remained soluble and mobile because the chemical form of this element in the soil solution was not effectively retained by the reactivity of the soil (the soil had no ability to regulate the particular solution form of Se).

The Kesterson example above illustrates that not all substances are effectively retained in soil. In reality, all substances are mobile to some degree in soils (no substance is truly immobile). In addition to precipitation (the formation of a solid phase), substances may be retained in soils by mechanisms that are generally described as sorption processes. Organic substances and numerous clay-sized (<2μm) minerals (such as hydrous metal oxides) in the soil environment have very reactive surfaces. These surfaces have functional groups that are ionizable (charge is created). Soil minerals may also contain structural imperfections (the minor substitution of one element for another) because they form in chemically complex environments. These minor substitutions may result in a structural charge imbalance that causes surface charge (as in the clay minerals). Ions in solution are attracted to these charged inorganic and organic soil surfaces by relatively weak electrostatic forces. These ions are called exchangeable. Ions are more strongly attached to inorganic and organic surfaces by forming direct chemical bonds, which results in very stable structures. Soluble substances that form direct chemical bonds with soil particle surfaces are called adsorbed. Adsorbed ions can be released back to the soil solution during a process termed desorption. Nonionic and nonpolar molecules, such as pesticides and organic solvents, are retained in soils by various mechanisms, depending on the molecular properties of the compounds. One important mechanism for the retention of these compounds is hydrophobic partitioning, in which a substance is essentially repelled from the polar soil solution and partitions into the nonpolar molecular framework of soil organic matter.

It might appear from the above general discussion that a complex, and potentially chaotic, array of processes occur within the soil environment. Adding to this complexity are the facts that soil contains nearly all the elements in the periodic table, and that these elements reside in an innumerable amount of organic and inorganic substances, although on average only seven elements (oxygen, carbon, silicon, aluminum, iron, calcium, and potassium) constitute 98% of the mass of soils. Furthermore, the chemical properties of a soil are both spatially and temporally variable. Yet, for all their apparent complexity, there exists considerable order and distinction among the various soil chemical processes. Moreover, there is commonality in the behavior of chemical constituents in the environment, allowing for the grouping of elements and functional groups on the basis of chemical behavior. Finally, the chemical processes that occur in the environment are uniquely quantifiable, allowing for the description and prediction of substance behavior.

1.2 ELEMENTS IN THE SOIL ENVIRONMENT: THEIR CONCENTRATIONS AND IMPORTANT SPECIES

Life on this planet has developed and adapted to use the chemical resources available for structure and metabolic processes. Thus, elements in abundance at the Earth's surface and those that could be utilized effectively in biochemical processes became necessary (essential) for life (Table 1.1). For example, carbon is the basic building block of life. Carbon is a versatile element having several stable oxidation states (−IV, II, IV) and is unique among the elements for the vast number of compounds it can form. Carbon compounds serve as structural components and as a reservoir for much of the energy (ultimately derived from sunlight during photosynthesis) that is transferred from plants to animals, humans, and microorganisms. Silicon (Si), the second most abundant element in the Earth's crust (behind oxygen), is a nonessential element for plants, and only required in small amounts in animals. Unlike C, Si exists in only one valence state (IV) and cannot form complex chain structures such as those formed by C. Therefore, although Si is more abundant than C, it does not have the chemical versatility required for life-sustaining biochemical processes, and it was supplanted by C as the building block of life. In general, elements that concentrated at the Earth's surface during its creation (the more volatile and lighter elements) are essential or beneficial for plants and animals. Conversely, heavier and less volatile elements that concentrated at the Earth's core are nonessential. With the exception of molybdenum (Mo, 95.94 g mol^{-1}), tin (Sn, 118.71 g mol^{-1}), and iodine (I, 126.9044 g mol^{-1}), no element with an atomic mass greater than that for selenium (Se, 78.96 g mol^{-1}) is essential for life.

TABLE 1.1
The Elemental Content (Median and Range) of Uncontaminated Soils Collected from around the World and the Mean Elemental Content of the Earth's Crust (the Valences and Aqueous Speciation of the Elements in Soil Environments Are Also Shown)

Element	Atomic Mass[b]	mg kg^{-1a} Soil	Earth's Crust	ER[c]	Important Chemical Species and Oxidation States[d]
		The Most Abundant Constituents of Organic Soils and Soil Organic Matter			
O*	15.9994	490,000	474,000	1.0	O^0 [$O_2(g)$], O^{-II} [H_2O] (oxidant: $O_2(g) + 4e^- + 4H^+ = 2H_2O$)
C*	12.011	20,000 (7,000–500,000)	480	42	organic, C^{IV} [CO_3^{2-}, HCO_3^-, $H_2CO_3^0$, $CO_2(g)$]
N*	14.00674	2,000 (200–5,000) ($\sim^1/_{10}$ C)	25	80	organic, N^V [NO_3^-], N^{-III} [NH_4^+, $NH_3(g)$]
P*	30.97376	800 (35–5,300) ($\sim^1/_5$ N)	1,000	0.80	organic, P^V [HPO_4^{2-}, $H_2PO_4^-$]
S*	32.006	700 (30–1,600) ($\sim^1/_5$ N)	260	2.7	organic, S^{VI} [SO_4^{2-}], S^{-II} [$H_2S(g)$, HS^-, S^{2-}]
		The Most Abundant Elements in Mineral Soils			
Si*	28.0855	330,000 (250,000–410,000)	277,000	1.2	Si^{IV} [$H_4SiO_4^0$]
Al	26.98153	71,000 (10,000–300,000)	82,000	0.87	Al^{III} [Al^{3+}, $AlOH^{2+}$, $Al(OH)_2^+$, $Al(OH)_3^0$, and $Al(OH)_4^-$]
Fe*	55.845	40,000 (2,000–550,000)	41,000	0.96	Fe^{II} [Fe^{2+}], Fe^{III} [Fe^{3+}, $FeOH^{2+}$, $Fe(OH)_2^+$, $Fe(OH)_3^0$, and $Fe(OH)_4^-$]
		Other Major Elements			
Ca*	40.078	15,000 (700–500,000)	41,000	0.37	Ca^{2+}
K*	39.0983	14,000 (80–37,000)	21,000	0.67	K^+
Mg*	24.305	5,000 (400–9,000)	23,000	0.22	Mg^{2+}
Na*	22.98977	5,000 (150–25,000)	23,000	0.22	Na^+
Ti	47.867	5,000 (150–25,000)	5,600	0.89	Ti^{IV}
Mn*	54.938	1,000 (20–10,000)	950	1.1	Mn^{2+}
		Micro and Trace Elements			
Ba	137.327	500 (100–3,000)	500	1.0	Ba^{2+}
Zr	91.224	400 (60–2,000)	190	2.1	Zr^{IV}
Sr	87.62	250 (4–2,000)	370	0.68	Sr^{2+}
F*	18.9984	200 (20–700)	950	0.21	F^-
Cl*	35.453	100 (8–1,800)	130	0.77	Cl^-
Zn*	65.39	90 (1–900)	75	1.2	Zn^{2+}
V*	50.9415	90 (3–500)	160	0.57	V^{IV} [VO^{2+}], V^V [VO_2^+, $VO_2(OH)_2^-$, $VO_3(OH)^{2-}$]
Cr*	51.9961	70 (5–1,500)	100	0.70	Cr^{III} [Cr^{3+}], Cr^{VI} [$HCrO_4^-$, CrO_4^{2-}, $Cr_2O_7^{2-}$]
Ni*	58.6934	50 (2–750)	80	0.63	Ni^{2+}
Pb	207.2	35 (2–300)	14	2.5	Pb^{2+}
Cu*	58.9332	30 (2–250)	50	0.60	Cu^{2+}
Li	6.941	25 (3–350)	20	1.3	Li^+
B*	10.811	20 (2–270)	10	0.50	B^{III} [$B(OH)_3^0$, $B(OH)_4^-$]
Br	79.904	10 (1–110)	0.37	27	Br^-
Co*	58.9332	8 (0.05–65)	20	0.4	Co^{2+}

(continued)

TABLE 1.1

The Elemental Content (Median and Range) of Uncontaminated Soils Collected from around the World and the Mean Elemental Content of the Earth's Crust (the Valences and Aqueous Speciation of the Elements in Soil Environments Are Also Shown) (Continued)

Element	Atomic Mass[b]	mg kg^{-1a} Soil	Earth's Crust	ER[c]	Important Chemical Species and Oxidation States[d]
		Micro and Trace Elements			
As*	74.9216	6 (0.1–40)	1.5	4.0	As^{III} [HAsO$_3^{2-}$], As^V [HAsO$_4^{2-}$, H$_2$AsO$_4^-$]
Mo*	95.94	1.2 (0.1–40)	1.5	0.80	Mo^{VI} [MoO$_4^{2-}$]
Se*	78.96	0.4 (0.1–2)	0.05	8.0	Se^{IV} [HSeO$_3^-$, SeO$_3^{2-}$], Se^{VI} [SeO$_4^{2-}$], Se^{-II} [Se^{2-}]
Cd	112.411	0.35 (0.01–2)	0.11	3.2	Cd^{2+}
Hg	200.59	0.06 (0.01–0.5)	0.05	1.2	Hg^{II} [Hg(OH)$_2^0$], Hg^I [Hg$_2^{2+}$], Hg^0 [Hg(l), Hg(g)]

^{a} Soil elemental concentrations represent the median elemental content and range (in parentheses). Elemental content values for the Earth's crust represent the mean. Data are from Bowen (1979). Additional tabulations of the elemental content of soils have been compiled by Helmke (2000).

^{b} Units are g mol^{-1}.

^{c} ER is the enrichment ratio and is equal to the median soil content of an element divided by the mean Earth's crust content.

^{d} Soluble complexes are not included.

* Denotes an essential or beneficial element for plants or animals.

Although only a relatively small number of elements are essential for life, all elements that have atomic numbers of 92 (uranium, U) and less, with the exception of technetium (Tc, atomic number 43) and promethium (Pm, atomic number 61), occur naturally (Figure 1.2). A small number of elements are spatially localized in their occurrence. For example, polonium (Po) and astatine (At) are U decay products found only in association with U-bearing minerals. However, with few exceptions, the elements are widely dispersed in soils. Figure 1.3 illustrates the concentration ranges and median concentrations of the elements (on a log scale) as they occur in uncontaminated soils collected from around the world (data for selected elements are tabulated in Table 1.1). There are more than six orders of magnitude difference between the median soil content of one of the least abundant elements (Hg, 0.06 mg kg^{-1}) and one of the most abundant (Si, 330,000 mg kg^{-1}). It is important to recognize that the elemental content of soils is highly variable relative to location on the landscape and depth within the profile, ranging over several orders of magnitude for any given element. For example, the median zinc (Zn) content of soil is 90 mg kg^{-1} with a range of 1 to 900 mg kg^{-1}, which is nearly three orders of magnitude. The highly variable nature of the elemental composition data indicates that soils are chemically heterogeneous, their composition being influenced by the soil forming factors, particularly parent material and intensity of weathering (climate and time).

The median elemental content of soils closely mirrors the mean values for the Earth's crust (Figure 1.3, Table 1.1). With few exceptions, the elemental composition of the crust falls within the ranges observed in soils. The enrichment or depletion of an element in soils relative to the Earth's crust may be expressed by the enrichment ratio (ER), the median element concentration in soil divided by its mean concentration in the Earth's crust content (Table 1.1). According to Sposito (1989), ER values that fall between 0.5 and 2 should be interpreted as indicating that there is no

The Periodic Table of the elements

Key to each cell: element name / atomic number / **symbol** / atomic weight

IA	IIA	IIIA	IVA	VA	VIA	VIIA	VIIIA			IB	IIB	IIIB	IVB	VB	VIB	VIIB	VIII
H 1 (1.00794)																	**He** 2 (4.0026)
Li 3 (6.941)	**Be** 4 (9.01218)											**B** 5 (10.811)	**C** 6 (12.011)	**N** 7 (14.0067)	**O** 8 (15.9994)	**F** 9 (18.99840)	**Ne** 10 (20.179)
Na 11 (22.98977)	**Mg** 12 (24.305)											**Al** 13 (26.98153)	**Si** 14 (28.0855)	**P** 15 (30.97376)	**S** 16 (32.006)	**Cl** 17 (35.453)	**Ar** 18 (39.948)
K 19 (39.0983)	**Ca** 20 (40.078)	**Sc** 21 (44.9559)	**Ti** 22 (47.867)	**V** 23 (50.9415)	**Cr** 24 (51.9961)	**Mn** 25 (54.9380)	**Fe** 26 (55.845)	**Co** 27 (58.9332)	**Ni** 28 (58.6934)	**Cu** 29 (63.546)	**Zn** 30 (65.39)	**Ga** 31 (69.723)	**Ge** 32 (72.61)	**As** 33 (74.9216)	**Se** 34 (78.96)	**Br** 35 (79.904)	**Kr** 36 (83.80)
Rb 37 (85.4678)	**Sr** 38 (87.62)	**Y** 39 (88.90585)	**Zr** 40 (91.224)	**Nb** 41 (92.9064)	**Mo** 42 (95.94)	**Tc** 43 (97.907)	**Ru** 44 (101.07)	**Rh** 45 (102.9055)	**Pd** 46 (106.42)	**Ag** 47 (107.8682)	**Cd** 48 (112.411)	**In** 49 (114.82)	**Sn** 50 (118.71)	**Sb** 51 (121.76)	**Te** 52 (127.60)	**I** 53 (126.9044)	**Xe** 54 (131.29)
Cs 55 (132.9054)	**Ba** 56 (137.327)	**La*** 57 (138.9055)	**Hf** 72 (178.49)	**Ta** 73 (180.95)	**W** 74 (183.84)	**Re** 75 (186.207)	**Os** 76 (190.23)	**Ir** 77 (192.22)	**Pt** 78 (195.08)	**Au** 79 (196.9665)	**Hg** 80 (200.59)	**Tl** 81 (204.383)	**Pb** 82 (207.2)	**Bi** 83 (208.98)	**Po** 84 (208.98)	**At** 85 (209.99)	**Rn** 86 (222.02)
Fr 87 (223.02)	**Ra** 88 (226.0254)	**Ac**** 89 (227.0278)	**Rf** 104 (263.11)	**Db** 105 (262.11)	**Sg** 106 (266.12)	**Bh** 107 (264.12)	**Hs** 108 (269.13)	**Mt** 109 (268.14)									

* Lanthanides

Ce 58 (140.116)	**Pr** 59 (140.9077)	**Nd** 60 (144.24)	**Pm** 61 (144.91)	**Sm** 62 (150.36)	**Eu** 63 (151.96)	**Gd** 64 (157.25)	**Tb** 65 (158.9254)	**Dy** 66 (162.50)	**Ho** 67 (164.93)	**Er** 68 (167.26)	**Tm** 69 (168.9342)	**Yb** 70 (173.04)	**Lu** 71 (174.967)

** Actinides

Th 90 (232.038)	**Pa** 91 (231.0359)	**U** 92 (238.029)	**Np** 93 (237.05)	**Pu** 94 (244.06)	**Am** 95 (243.06)	**Cm** 96 (247.07)	**Bk** 97 (247.07)	**Cf** 98 (251.08)	**Es** 99 (252.08)	**Fm** 100 (257.10)	**Md** 101 (258.10)	**No** 102 (259.10)	**Lr** 103 (262.11)

FIGURE 1.2 The Periodic Table of the elements.

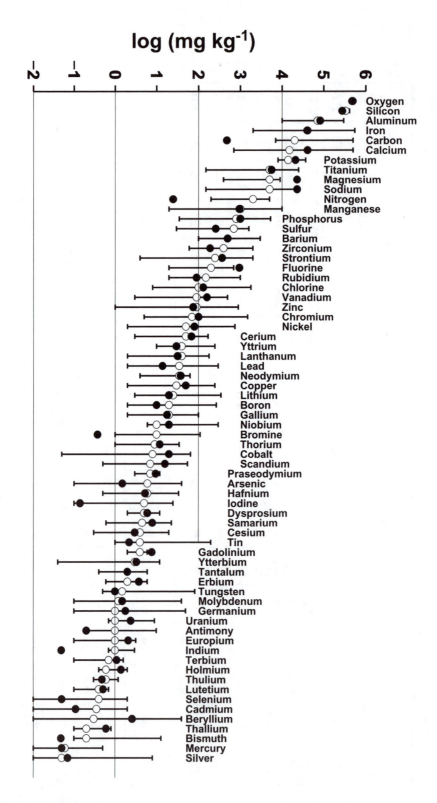

FIGURE 1.3 The median elemental content (○) and range (error bars) observed in uncontaminated surface soils and the mean elemental content of the Earth's crust (●). (Data from Bowen, 1979.)

depletion or enrichment (values that lie in this range are not different from ER = 1). Values of ER that are less than 0.5 indicate depletion. Values between 2 and 10 indicate some enrichment, and ER values greater than 10 indicate strong enrichment. Carbon and nitrogen (N) are strongly enriched in soils due to their accumulation in organic compounds. Bromine (Br), iodine (I), antimony (Sb), and indium (In) are also strongly enriched in soils relative to crustal materials. Moderate enrichment is observed for sulfur (S), zirconium (Zr), lead (Pb), arsenic (As), selenium (Se), and cadmium (Cd). Calcium (Ca), magnesium (Mg), and sodium (Na) are depleted in soils to the greatest degree, due to their occurrence in minerals that are easily weathered (physically and chemically transformed) when exposed at the surface (they are readily leached from soils during weathering). Other elements that are depleted in soils include fluorine (F), boron (B), and cobalt (Co).

The abundance of an element in the soil environment is an easily measured property. However, the elemental content of a soil is not a useful parameter for assessing fate and behavior. A high concentration of a potentially problematic element, relative to the uncontaminated soil concentration range (Table 1.1), does not indicate that a problem or risk exists. Conversely, an elemental concentration that is within the uncontaminated soil concentration range does not necessarily indicate that a problem does not exist. Instead, the behavior of an element in the environment, its mobility, phytoavailability, and toxicity are dictated by its valence and speciation in both the solid and solution phases. For a given valence, which is a property controlled by the redox status of the soil (if the element is redox sensitive), the free species of an element is the building block for the formation of the more complex chemical species that are produced through hydrolysis, acid-base, and aqueous complexation reactions. Further, chemical reactions in the soil environment are commonly written using the free species as a product or a reactant. Therefore, knowledge of the free ionic species of the elements, as they are observed in the environment, is a necessary starting point from which to examine the environmental chemistry of an element. The oxidation state, free species, acid-base, and hydrolytic species of some elements are provided in Table 1.1.

In organic soils and soil organic compounds, carbon (C), oxygen (O), nitrogen (N), phosphorus (P), and sulfur (S) are the most abundant elements. In organic compounds, C can range from the C^{IV} (as in the carboxyl functional group: $R-COOH$) to the C^{-IV} (as in the methyl group: $R-CH_3$) oxidation states. The C^0 oxidation state, as carbon occurs in carbohydrates, is also important. Organic carbon in soil occurs in an enormous assortment of substances. Nearly all classes of biochemical compounds are represented in soil. Biomolecules present in living organisms, exuded by living organisms, and released upon their death are found in soil and include monosaccharides, oligosaccharides, and polysaccharides; amino acids, peptides, and proteins; nucleic acids and antibiotics; fats, waxes, phospholipids, and other lipids; and lignins, cutins, and suberins. The decomposition products of these biomolecules are also present and recombine as microbial synthates to form substances that are truly unique to soils. These are the humic substances: humins, humic acids, and fulvic acids. The behavior of organic compounds and their interactions with other substances in the aqueous and solid phase are controlled by the type and abundance of polar or ionic functional groups and nonpolar structural components (aromatic ring structures and aliphatic chains).

Inorganic carbon is found principally in the C^{IV} valence state. Inorganic species in which C^{IV} occur include carbon dioxide gas ($CO_2(g)$), dissolved carbon dioxide ($CO_2 \cdot H_2O$), and carbonic acid ($H_2CO_3^0$). The latter two species are not discernable from one another in solutions, and the transition between the two is kinetically restricted. Therefore, both are represented in a combined term, $H_2CO_3^*$. Carbonic acid is an example of a polyprotic acid; it can donate more than one proton (H^+ ion) to a solution via deprotonation. Carbonic acid donates one proton to produce the bicarbonate anion: $H_2CO_3^0 \rightarrow H^+ + HCO_3^-$, which in turn donates a proton to produce the carbonate ion: $HCO_3^- \rightarrow H^+ + CO_3^{2-}$. As will be shown in later chapters, the particular species of inorganic carbon (or any other polyprotic acid) found in the soil solution ($H_2CO_3^0$, HCO_3^-, or CO_3^{2-}) is dependent on solution pH.

Oxygen is the most abundant element in the soil environment, found predominantly in the O^{-II} valence state. In the O^0 valence state (as in $O_2(g)$), oxygen is an important electron acceptor, $O_2(g) + 4e^- + 4H^+ \rightarrow 2H_2O$, serving as an oxidant in oxidation–reduction reactions. In the preceding reaction, the oxygen in $O_2(g)$ is reduced from a valence of O^0 to O^{-II} (as found in H_2O and mineral structures). An important and helpful rule of thumb is that one electron (e^-) will be consumed by $\frac{1}{4}O_2(g)$, when $O_2(g)$ is the oxidant. The oxidation (or aerobic decomposition) of soil organic matter can provide an example of the oxidation–reduction process. Chemically, soil organic matter can be expressed by the average formula for a carbohydrate, CH_2O, where the average oxidation state of carbon is C^0. The oxidation half-reaction for the complete oxidation of soil organic matter to produce $CO_2(g)$ is:

$$CH_2O(org) + H_2O(l) \rightarrow CO_2(g) + 4H^+(aq) + 4e^- \tag{1.1}$$

There are four electrons produced in Equation 1.1, electrons that must be consumed through a reduction half-reaction, producing a complete oxidation–reduction reaction. Since four electrons are produced, and $\frac{1}{4}O_2(g)$ consumes one e^-, one $O_2(g)$ molecule must be reduced:

$$O_2(g) + 4e^-(aq) + 4H^+(aq) \rightarrow 2H_2O(l) \tag{1.2}$$

Although Equations 1.1 and 1.2 represent chemical reactions, they may also be treated as mathematical expressions. Thus, adding Equations 1.1 and 1.2 leads to the complete oxidation–reduction reaction:

$$CH_2O(org) + O_2(g) \rightarrow CO_2(g) + H_2O(l) \tag{1.3}$$

Nitrogen, an important macronutrient for plants, animals, and microbes, occurs in a number of valence states in the soil environment. Two common valence states for soil nitrogen are N^{-III} (as in ammonia, NH_4^+, and amino acids, $R-NH_2$) and N^V (as in nitrate, NO_3^-). In solution, ammonium ($NH_3(g)$) is a strong base and hydrolyzes water, producing hydroxide and the ammonia ion:

$$NH_3(g) + H_2O(l) \rightarrow NH_4^+(aq) + H^+(aq) \tag{1.4}$$

In aerated systems, NH_4^+ is oxidized to NO_3^-, a microbially mediated process that is illustrated in the reaction:

$$NH_4^+(aq) + 2O_2(g) = NO_3^-(aq) + H_2O(l) + 2H^+(aq) \tag{1.5}$$

Note that the oxidation of N^{-III} in ammonia to N^V in nitrate requires the transfer of eight electrons. Because $\frac{1}{4}O_2(g)$ accepts one e^-, the oxidation reaction requires two $O_2(g)$ molecules.

Phosphorus occurs entirely in the P^V valence state in soils. The basic inorganic P species is the orthophosphate ion, PO_4^{3-}. This oxyanion occurs in solutions in various states of protonation, depending on solution pH. In neutral to acidic environments (pH < 7.2), orthophosphate exists as the $H_2PO_4^-$ species. As pH increases into the alkaline range, a proton is released and the dominant species is HPO_4^{2-}. The polyprotic phosphoric acid ($H_3PO_4^0$) and the completely dissociated PO_4^{3-} species dominate only at the extremes of the acidic and basic pH conditions; thus, they are not species that actually occur in measurable concentrations in soil solutions under typical environmental conditions.

Sulfur can exist in several valance states, depending on the redox status of the soil. In highly reduced soil systems, inorganic S occurs in the S^{-II} oxidation state (in hydrogen sulfide gas, $H_2S(g)$; or in the hydrogen sulfide ion, HS^-). In systems other than highly reducing environments, S in soils occurs as S^{VI} in the sulfate ion, SO_4^{2-}. In organic matter, S can occur in both the S^{-II} oxidation state (as the thiol, R–SH, in amino acid) and the S^{VI} oxidation state (such as in sulfate esters).

In addition to oxygen, silicon (Si), aluminum (Al), and iron (Fe) are the most abundant elements in mineral soils. Silicon is always in the Si^{IV} valence state in soils. In the mineral phase, Si exists as the silicate species, SiO_4^{4-}. In soil solutions, silicic acid is the dominant species, $H_4SiO_4^0$. Note that silicic acid is a neutral-charged aqueous species, and not a solid. Only in very basic solutions (pH > 9.8) will silicic acid deprotonate. Aluminum exists in only the Al^{III} valence state. Depending on solution pH, Al may be present in soil solutions as the free Al^{3+} species (note that Al^{III} denotes the oxidation state of aluminum, while Al^{3+} is a species that contains aluminum as Al^{III}), or as a hydrolysis product: $AlOH^{2+}$, $Al(OH)_2^+$, $Al(OH)_3^0$, or $Al(OH)_4^-$. Iron is found in two oxidation states, Fe^{II} and Fe^{III}. In moderately reduced (anaerobic) systems and in primary minerals, Fe^{2+} is the dominant form of iron. In oxidized systems and in secondary minerals, Fe^{3+} dominates. In soil solutions, ferric iron occurs in several hydrolysis products: $FeOH^{2+}$, $Fe(OH)_2^+$, $Fe(OH)_3^0$, or $Fe(OH)_4^-$ (depending on pH). Both aluminum and iron(III) are amphoteric because they may exist as a cationic or an anionic species in solution.

All alkali metals (Periodic group IA elements), lithium (Li), sodium (Na), potassium (K), rubidium (Rb), and cesium (Cs), exist only in the +I oxidation state, and occur as monovalent cations: Li^+, Na^+, K^+, Rb^+, and Cs^+. Similarly, the alkaline earth metals (Periodic group IIA elements), beryllium (Be), magnesium (Mg), calcium (Ca), strontium (Sr), and barium (Ba), exist only in the +II oxidation state, and occur as divalent cations: Be^{2+}, Mg^{2+}, Ca^{2+}, Sr^{2+}, and Ba^{2+}. Another group of elements that are found in only one oxidation state are the halides (Periodic group VIIB elements). These elements are found in the –I state in soil as the monovalent anions: fluoride (F^-), chloride (Cl^-), bromide (Br^-), and iodide (I^-). Elements in each of these Periodic groups tend to display very similar environmental behavior, that is, the behavior of Sr^{2+} in the soil is similar to that of Ca^{2+}. Titanium (Ti) and zircon (Zr) exist in the +IV valence state; however, their occurrence in the soil environment is restricted to the solid phase where they are surrounded by oxygen atoms.

Many of the environmentally problematic elements occur in uncontaminated soils in very low concentrations. These elements are commonly referred to as trace elements or heavy metals. The use of these terms, however, relates very little information concerning environmental behavior or biological toxicity. Indeed, elemental classification according to geochemical abundance or specific gravity does not imply biological hazard, even though trace elements and heavy metals are terms that are used to denote potentially toxic elements. The least descriptive term is heavy metals. Elements from beryllium (9.01 g mol^{-1}) to uranium (238.03 g mol^{-1}) have been identified as heavy metals, although the first row transition elements (vanadium, 50.94 g mol^{-1}, to zinc, 65.39g mol^{-1}), as well as arsenic (74.92 g mol^{-1}), selenium (78.96 g mol^{-1}), molybdenum (95.94 g mol^{-1}), cadmium (112.41 g mol^{-1}), mercury (200.59g mol^{-1}), and lead (207.2 g mol^{-1}), are typically implied by this ambiguous moniker. Unlike a heavy metal, which is a vaguely defined term, a trace element may be defined as an element whose soil concentration is less than 100 mg kg^{-1} or whose soil solution concentration is significantly below 10^{-4} mol L^{-1} (and typically below 10^{-6} mol L^{-1}). Because of their low concentrations, trace elements are not found in soils in discrete mineral phases. Instead, trace elements occur as minor substituents in silicates and aluminosilicates (olivines, pyroxenes, amphiboles, micas, and feldspars); hydrous metal (iron, aluminum, and manganese) oxides; iron sulfides; calcium and magnesium carbonates; and calcium, iron, and aluminum phosphates. A significant percentage of the total soil content of a trace element may also reside in the adsorbed phase.

Minor and trace elements that are typically found in only one oxidation state in soils include boron (B^{III} in boric acid, $B(OH)_3^0$, and borate, $B(OH)_4^-$), cobalt(II) (Co^{2+}), nickel(II) (Ni^{2+}), copper(II) (Cu^{2+}), zinc(II) (Zn^{2+}), molybdenum (Mo^{VI} in molybdate, MoO_4^{2-}), cadmium(II) (Cd^{2+}), and lead(II) (Pb^{2+}). However, many minor and trace elements are redox sensitive within the range of redox conditions observed in the environment. The two oxidation states of chromium found in the soil environment are Cr^{III} (in the relatively inert Cr^{3+} form) and Cr^{VI} (principally in the toxic chromate forms, $HCrO_4^-$ and CrO_4^{2-}). Manganese is found in three valence states: Mn^{II}, Mn^{III}, and Mn^{IV}. In soil solutions, the Mn^{2+} species dominates; whereas, all three Mn oxidation states are found in soil minerals. Both arsenic and selenium exist as oxyanions in the environment. Arsenic can occur in the As^{III} and As^V oxidation states as the arsenite ($HAsO_3^{2-}$) and arsenate ($H_2AsO_4^-$ and $HAsO_4^{2-}$) species (arsenate speciation is similar to that of phosphate). Selenium can occur in the Se^{-II}, Se^{IV}, and Se^{VI} oxidation states as the selenide (H_2Se^0, HSe^-, and Se^{2-}), selenite ($HSeO_3^-$ and SeO_3^{2-}), and selenate (SeO_4^{2-}) species (selenate speciation is similar to that of sulfate). Several oxidation states are possible for vanadium; however, V^{IV} (as VO^{2+}) and V^V (as VO_2^+ and the hydrolysis products, $VO_2(OH)_2^-$ and $VO_3(OH)^{2-}$) are stable throughout the range of soil solution pH and redox conditions. In primary minerals, V^{3+} (V^{III}) is common; whereas, vanadate (V^V as VO_4^{3-}) is common in secondary V-bearing minerals. Mercury, which is one of the least abundant elements in soils, occurs in the Hg^{II} oxidation state in aerated solutions. The $Hg(OH)_2^0$ species dominates throughout a wide range of soil pH conditions. In anaerobic conditions, Hg^I (in Hg_2^{2+}) and Hg^0 (elemental vapor or liquid Hg) may occur.

1.3   UNITS AND CONVERSIONS

Launched December 11, 1998, the mission of the Mars Climate Orbiter was to provide information on the Martian climate. As the orbiter neared the Martian atmosphere on September 23, 1999, after spending 286 days (9 ½ months) in transit, controllers lost contact. Loss of the orbiter was not the result of a collision with a cosmic particle or the failure of a critical component at a crucial time. Instead, the failure of the $655.2 million project was the result of a miscommunication between the spacecraft team in Colorado and the mission navigation team in California. Specifically, one team was using English units (yards), while the other was using metric (meters) as they worked to guide the orbiter to Mars. The problem was not that each team was using different units, but that each team assumed the other was using comparable units. The 214.6 billion yard distance (196.2 billion meters) to Mars is significantly less than the 214.6 billion meter distance assumed by one of the teams. Thus, the Orbiter rather abruptly intersected with Mars at a speed of 12,300 miles per hour (5.5 km sec^{-1}), still programmed to travel an additional 18.4 billion meters. While several lessons were learned from the Mars Climate Orbiter debacle, there are two points to be gleaned from this illustration. First, there are often several ways in which to express a physical quantity. For example, distance can be expressed in an English unit (foot), a centimeter-gram-second (cgs) unit (centimeter), or a meter-kilogram-second (mks) unit (meter). Second, if the system used to express a physical quantity is not agreed upon, or standardized, the consequences of the miscommunication may be expensive and embarrassing, and potentially detrimental to human health and the environment.

In order to remove ambiguities in reported measurements and to bring uniformity of style and terminology to communicating measurements, the SI system (Système International d'Unités, or International System of Units) of reporting measurements has been adopted by scientific societies and countries the world over. It is also the system employed in this text. There are two classes of SI units: base units and derived units. Base units are dimensionally independent (e.g., meters, kilograms, and seconds) (Table 1.2); whereas derived units are expressed as algebraic terms of base units (e.g., area, m^2, and volume, m^{-3}). The derived units can also be given special names (e.g., hectares and liters) which themselves may be used in expressing other derived units. For example, the SI unit for pressure (force per unit area) is the pascal (Pa). It is a derived unit expressed as N m^{-2}

TABLE 1.2
Base Units of the International System of Units

Quantity	Unit	Symbol
Length	meter	m
Mass	kilogram	kg
Time	second	s
Electric current	ampere	A
Thermodynamic temperature	kelvin	K
Amount of substance	mole	mol
Luminous intensity	candela	cd

TABLE 1.3
Examples of Derived Units in the International System of Units (SI)

Quantity	Name	Symbol	Expression in Terms of SI Base or Other Derived Units
Acceleration	meter per second squared	—	$m\ s^{-2}$
Area	square meter	—	m^2
	hectare	ha	m^2
Capacitance	farad	F	$C\ V^{-1}$, $m^{-2}\ kg^{-1}\ s^4\ A^2$
Celsius temperature	degree Celsius	°C	K
Concentration	mole per cubic meter	—	$mol\ m^{-3}$
Density	kilogram per cubic meter	—	$kg\ m^{-3}$
Electric charge, quantity of electricity	coulomb	C	$s\ A$
Electrical conductance	siemen	S	$A\ V^{-1}$, $m^{-2}\ kg^{-1}\ s^3\ A^2$
Electric potential, potential difference, electromotive force	volt	V	$W\ A^{-1}$, $m^2\ kg\ s^{-3}\ A^{-1}$
Electric resistance	ohm	ω	$V\ A^{-1}$, $m^2\ kg\ s^{-3}\ A^{-2}$
Energy, work, quantity of heat	joule	J	$N\ m$, $m^2\ kg\ s^{-2}$
Force	newton	N	$m\ kg\ s^{-2}$
Frequency	hertz	Hz	s^{-1}
Heat capacity, entropy	—	$J\ K^{-1}$	$m^2\ kg\ s^{-2}\ K^{-1}$
Pressure	pascal	Pa	$N\ m^{-2}$, $kg\ s^{-2}$
Power	watt	W	$J\ s^{-1}$, $m^2\ kg\ s^{-3}$
Specific energy		$J\ kg^{-1}$	$m^2\ s^{-2}$
Specific heat capacity, specific entropy	—	$J\ kg^{-1}\ K^{-1}$	$m^2\ s^{-2}\ K^{-1}$
Specific surface area	—	—	$m^2\ kg^{-1}$
Velocity	meter per second	—	$m\ s^{-1}$
Volume	liter	L	m^3

(newtons per square meter). The SI unit for force is the newton, which is derived from the base units, $m\ kg\ s^{-2}$. The derived units having special interest to environmental soil chemistry are identified in Table 1.3. The SI system also employs prefixes to indicate orders of magnitude of SI units (Table 1.4). The objective in employing prefixes is to reduce the use of nonsignificant digits or leading zeros in decimal fractions. Preferably, the prefix should be selected so that the numerical

TABLE 1.4
Prefixes Employed to Indicate Orders of Magnitude

Order of Magnitude	Prefix	Symbol	Order of Magnitude	Prefix	Symbol
10^{18}	exa	E	10^{-1}	deci	d
10^{15}	peta	P	10^{-2}	centi	c
10^{12}	tera	T	10^{-3}	milli	m
10^{9}	giga	G	10^{-6}	micro	μ
10^{6}	mega	M	10^{-9}	nano	n
10^{3}	kilo	k	10^{-12}	pico	p
10^{2}	hecto	h	10^{-15}	femto	f
10^{1}	deka	da	10^{-18}	atto	a

TABLE 1.5
Selected Units Used in Environmental Soil Chemistry

Quantity	Application	Unit	Symbol
Angle of diffraction	X-ray diffraction	degrees two-theta	$°2\Theta$
Bulk density	soil bulk density	megagram per cubic meter	$Mg\ m^{-3}$
Cation exchange capacity	ion retention	centimole of charge per kilogram	$cmol_c\ kg^{-1}$
Concentration	mass basis	mole per kilogram	$mol\ kg^{-1}\ (m)^a$
		milligram per kilogram	$mg\ kg^{-1}$
		gram per kilogram	$g\ kg^{-1}$
	volume basis	moles per cubic meter	$mol\ m^{-3}$
		mole per liter	$mol\ L^{-1}\ (M)^a$
		millimole per liter	$mmol\ L^{-1}$
		milligram per liter	$mg\ L^{-1}$
		microgram per liter	$μg\ L^{-1}$
Electrical conductivity	soil salinity	decisiemen per meter	$dS\ m^{-1}$
Interatomic spacing	crystallography,	nanometer	nm
	clay mineralogy	Angstrom	Å

^{a} m, molal or mole per unit mass; M, molar or mole per unit volume.

value of the measurement lies between 0.1 and 1000. In addition to the units presented in Tables 1.2 and 1.3, there are preferred units for the expression of select soil properties (Table 1.5).

One of the more vexing of problems encountered in the presentation and assimilation of information is unit conversion. The mechanics of performing unit conversions are not dependent on the particular property that is measured, although the appropriate application and manipulation of units is best examined in measurement-specific discussions. The concentration of a substance in a system is a recognizable characteristic. Yet, there are many ways in which to express composition. X-ray fluorescence spectrometry (XRF) is an analytical technique employed in the geological sciences to determine the elemental composition of geologic materials. The results of an XRF analysis of a mine spoil material are presented in Table 1.6. Note that the instrument is calibrated to generate compositional data on an oxide basis for some substances and on an elemental basis for others, in percentage (%) units (also known as parts per hundred). For example, 22.3% of the mine spoil is composed of Al_2O_3, and 0.49% is composed of S. It is important to recognize that the spoil material probably does not contain the compound Al_2O_3 or elemental S, but that the total concentrations of Al and S are expressed as if they were present in these forms. As is indicated in Table 1.5, the acceptable units for expressing composition data on a mass basis are $mol\ kg^{-1}$, $g\ kg^{-1}$,

---

TABLE 1.6

The Composition of a Mine Spoil Material as Determined by X-Ray Fluorescence Spectrometry (XRF) and Presented using the International System of Units

Oxide Presentation			Elemental Presentation		
Compound	%	g kg^{-1}	Element	g kg^{-1}	mol kg^{-1}
Al_2O_3	22.3	223	Al	118	4.37
Cl	0.40	4.0	Cl	4.0	0.11
Fe_2O_3	7.78	77.8	Fe	54.4	0.974
K_2O	3.46	34.6	K	28.7	0.734
MgO	1.6	16	Mg	9.6	0.39
P_2O_5	0.133	13.3	P	7.49	0.242
S	0.49	4.9	S	4.9	0.15
SiO_2	55.7	557	Si	260	9.26
TiO_2	1.01	10.1	Ti	6.06	0.127

---

or mg kg^{-1}. The conversion of percentage units to SI units is necessitated because percentages are unacceptable units (they are ambiguous). The first step in performing the conversion of units is to recognize that a percentage indeed represents parts per hundred. Another way to express parts per hundred is in the units of gram per hectogram (g hg^{-1}). In order to convert g hg^{-1} (a non-SI unit) to grams per kilogram (g kg^{-1}, an SI unit), the conversion factor must have units of hectograms per kilogram. Since there are 10 hectograms in a kilogram, the conversion factor is 10 hg kg^{-1}. The conversion process is illustrated for Al_2O_3:

$$22.3\% \, Al_2O_3 = 22.3 \, \frac{g \, Al_2O_3}{hg \, spoil} \tag{1.6}$$

$$22.3 \, \frac{g \, Al_2O_3}{hg \, spoil} \times \frac{10 \, hg}{kg} = 223 \, \frac{g \, Al_2O_3}{kg \, spoil} \tag{1.7}$$

Note that the units are simply treated as fractions. In Equation 1.7, the hg unit in the denominator of the Al_2O_3 concentration units is cancelled by the hg unit in numerator of the conversion factor, leaving the units of g kg^{-1}.

Expressing compositional data on an oxide basis is appropriate; however, it is more common to express the chemical composition of a material on an elemental basis. Again using Al_2O_3 as an example, the oxide composition of Al in the mine spoil is 223 g kg^{-1}. To convert the units of g of Al_2O_3 per kg of spoil to the units of g of Al per kg of spoil, a conversion factor that has units of g Al per g Al_2O_3 must be employed. In order to determine the mass of Al in a given mass of Al_2O_3, we will use the molecular masses of each substance (which can be determined using the molecular masses of Al and O in Figure 1.2). The molecular mass of Al is 26.98 g mol^{-1}, and of Al_2O_3 is 101.96 g mol^{-1} (2×26.98 g Al mol$^{-1} + 3 \times 16$ g O mol^{-1}). There are two moles of Al in every mole of Al_2O_3, or:

$$26.98 \, \frac{g \, Al}{mol \, Al} \times \frac{2 \, mol \, Al}{mol \, Al_2O_3} = 53.96 \, \frac{g \, Al}{mol \, Al_2O_3} \tag{1.8}$$

Multiplying this value by the reciprocal of the molecular weight of Al_2O_3:

$$53.96 \, \frac{g \, Al}{mol \, Al_2O_3} \times \frac{mol \, Al_2O_3}{101.96 \, g \, Al_2O_3} = 0.529 \, \frac{g \, Al}{g \, Al_2O_3} \tag{1.9}$$

This is the conversion factor needed to convert from the g Al_2O_3 kg^{-1} basis to g Al kg^{-1}. The elemental composition of Al in the spoil material is computed as:

$$223\frac{\text{g Al}_2\text{O}_3}{\text{kg spoil}} \times \frac{0.529 \text{ g Al}}{\text{g Al}_2\text{O}_3} = 118\frac{\text{g Al}}{\text{kg spoil}} \tag{1.10}$$

The expression of the elemental composition data in Table 1.6 in the units of g kg^{-1} is acceptable, as the values lie between 0.1 and 1000. The elemental composition may also be expressed on a mole basis. To convert from g kg^{-1} to mol kg^{-1} requires a conversion factor that is the reciprocal of the molecular mass of Al:

$$118\frac{\text{g Al}}{\text{kg spoil}} \times \frac{\text{mol Al}}{26.98 \text{ g Al}} = 4.37\frac{\text{mol Al}}{\text{kg spoil}} \tag{1.11}$$

or 4.37 mol kg^{-1}.

Another common unit transformation is the conversion of an extractable concentration, in volume-based units, back to a soil mass–based concentration. Consider the nitric acid digestion of a soil sample that has potentially received waste materials containing cadmium. A 2.5-g sample of soil (dry weight basis) is digested in 25 mL of 4 M HNO_3. After digestion, the solution is separated from the solid by filtration. A 1-mL aliquot of the extract is placed in a 100-mL volumetric flask, which is brought to volume with metal-free, deionized-distilled water. The diluted aliquot is then analyzed for Cd and found to contain 0.53 mg Cd L^{-1}. A question that is commonly asked is: based upon this analysis, has this soil received cadmium wastes? As mentioned previously, the knowledge of elemental concentrations provides little, if any, information about risk. There must be some basis for establishing contamination, such as the Cd content of the same soil that is known to have not received waste (background levels). Again, reality does not always conform to the optimal, and in this case the question must be answered without knowledge of background levels.

The first step in addressing the question is to compute the concentration of Cd on a soil basis. There are two conversion factors that must be employed in order to accomplish this. Dilutions are often required to bring the solution concentration of a substance into the analytical range of an instrument. In this case, a 100-fold dilution was used (1 mL of soil extract into 100 mL total volume), resulting in a measured Cd concentration of 0.53 mg L^{-1}. To convert back to the Cd concentration in the original soil digestate, the following expression is used:

$$c_{\text{aliquot}} \times V_{\text{aliquot}} = c_{\text{dilution}} \times V_{\text{total}} \tag{1.12}$$

In this expression, c_{aliquot} is the concentration of the analyte in the undiluted extract, V_{aliquot} is the volume of the undiluted sample used to prepare the dilution, c_{dilution} is the concentration of the analyte in the diluted sample, and V_{total} is the total volume of dilution. In our example, c_{aliquot} is the quantity we wish to compute, V_{aliquot} is 0.001 L (1 mL), V_{total} is 0.1 L (100 mL), and c_{dilution} is 0.53 mg L^{-1}. The concentration of Cd in the original digestate may then be computed:

$$c_{\text{aliquot}} = 0.53\frac{\text{mg Cd}}{\text{L}} \times \frac{0.1 \text{ L}}{0.001 \text{ L}} = 53 \text{ mg Cd L}^{-1} \tag{1.13}$$

The second conversion will result in the expression of Cd concentration on a soil basis. For this conversion we note that 2.5 g of soil was extracted with 25 mL of nitric acid; or 0.0025 kg of soil was extracted with 0.025 L of nitric acid. Because we want to convert mg L^{-1} to mg kg^{-1}, the conversion factor must have units of L kg^{-1}. Therefore, our conversion to obtain the concentration of Cd extracted from the soil on a mass basis is:

$$Cd_{soil} = 53\frac{mg\ Cd}{L} \times \frac{0.025\ L}{0.0025\ kg} = 530\ mg\ Cd\ kg^{-1} \tag{1.14}$$

According to the information presented in Table 1.1, the median Cd content of uncontaminated soil is 0.35 mg kg^{-1}. Further, the upper limit of uncontaminated soil Cd is 2 mg kg^{-1}. The concentration of Cd extracted from the potentially contaminated soil was 530 mg kg^{-1}. Although the 4 M HNO$_3$ digestion is a rather harsh extractant, the soil Cd content computed in Equation 1.14 is not a total concentration, it is an extractable concentration (total Cd would be greater than extractable). Irrespective, the extractable Cd level in the soil far exceeds the median and upper limit values of total Cd in uncontaminated soil. Thus, the probability that this soil has received Cd-bearing waste is quite high.

1.4 HETEROGENEITY OF SOIL CHEMICAL CHARACTERISTICS

It was indicated in Section 1.2 that the elemental content of soils is highly variable relative to location on the landscape and depth within the profile. A compilation of the elemental compositions of surface soil samples collected from around the world indicates a composition range of several orders of magnitude for any given element (Table 1.1). However, one need not venture far to observe the spatially heterogeneity of soil chemical and physical properties. It is a basic tenet of soil science that soil properties differ with depth due to the process of horizonation. This process, which is under the influence of the soil forming factors (parent material, vegetation, climate, topography, and time), is responsible for the elution (mobilization) and illution (accumulation) of soil components within a pedon. For example, the elution and illution of soil organic carbon (SOC) is a distinguishing feature of a Spodosol (relative to other soil orders, such as that illustrated for an Alfisol) (Figure 1.4a), just as the illution and neoformation of clay-sized material in the subsoil is

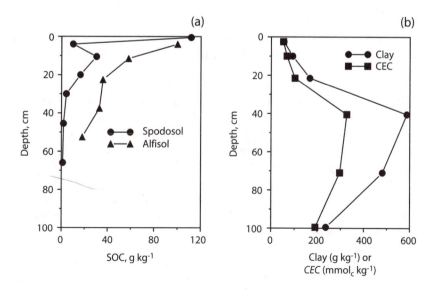

FIGURE 1.4 Soil properties as a function of depth in the soil profile: (a) soil organic carbon (SOC) in a Spodosol and an Alfisol; (b) clay content and cation exchange capacity (CEC) in an Aridisol.

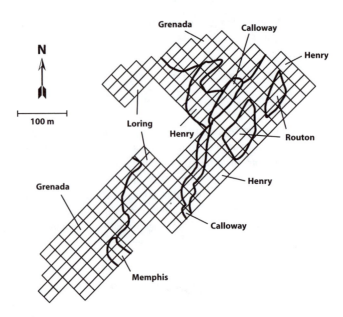

FIGURE 1.5 Soil map of a 10.5 ha production cotton field located at the Milan Agricultural Experiment Station in Milan, TN. Each rectangle represents a 27.4 m × 16.2 m monitoring unit. The classifications of the soil series at the subgroup level are: Calloway, Aquic Fraglossudalf; Grenada, Oxyaquic Fraglossudalf; Henry, Typic Fragiaqualf; Loring, Oxyaquic Fragiudalf; Memphis, Typic Hapludalf; and Routon, Typic Epiaqualf.

an indicator of an argillic horizon as seen in the Aridisol (Figure 1.4b). However, and aside from soil classification criteria, the clay and organic matter content of soil are properties that greatly influence compound fate and behavior. The observed behavior of an ionic substance in a 0- to 15-cm surface soil sample of the Aridisol described in Figure 1.4b (73 g kg^{-1} clay with a cation exchange capacity of 6.0 cmol$_c$ kg^{-1}) may be quite different from that observed in a 30- to 50-cm subsoil sample (587 g kg^{-1} clay with a cation exchange capacity of 32.6 cmol$_c$ kg^{-1}). Similarly, the behavior of a nonpolar and hydrophobic organic compound in a 0- to 8-cm surface sample of an Alfisol (10.0 g kg^{-1} SOC) will differ from that in a 45- to 60-cm subsoil sample (1.8 g kg^{-1} SOC) (Figure 1.4a). Thus, it is important to recognize that the behavior of a substance in the soil environment, which is often inferred from *ex situ* studies, is dependent upon the location in the soil profile from which the soil sample originated.

The variability of surface soil characteristics on the landscape may also be extensive, even within soil mapping units. Consider the 10.5-ha production cotton field located at the Milan Agricultural Experiment Station in Milan, TN. The field consists of several soil series (Figure 1.5); however, the soils are loessal and generally differ from one another only with respect to the depth to fragipan (ranging from 60 to 150+ cm) and the degree to which the fragipans have developed. Thus, the chemical characteristics of surface 15-cm samples are not expected to be influenced by slight changes in the soil classification (the chemical properties in the field soil are expected to be relatively homogeneous). Discrete (grid cell) and intensive sampling of this field has resulted in 235 composite surface (0 to 15 cm) soil samples, each composite sample representing a grid cell defined by an imaginary grid (Figure 1.5). The samples were then subjected to standard chemical characterization, including pH, Mehlich-3-extractable elements (Mehlich-3 is a pH 2 extracting solution that contains ammonium fluoride, the synthetic chelate EDTA, acetic acid, and nitric acid and is commonly employed to predict plant response to applied fertilizers and plant available trace elements), SOC, and effective cation exchange capacity (*ECEC* = sum of exchangeable Ca^{2+}, Mg^{2+}, K$^+$, Na$^+$, Mn^{2+}, and Al^{3+}). These data will be examined in the following sections.

TABLE 1.7
Mean ($\bar{x}$), Median (m), Standard Deviation (s), Coefficient of Variation (CV), Range, and Chi Square Statistic (χ^2) for Selected Chemical Characteristics of Soil Samples Collected from a Production Cotton Field at the Milan Experiment Station in Milan, TN[a]

Property[b]	$\bar{x}$	m	s	CV,%	Range	χ^{2c}
			Mehlich-3 extractable, mg kg^{-1}			
K	152	158	36.3	23.8	76–257	34.09
Mg	121	117	56.2	46.5	34–264	116.4
Mn	238	223	71.4	30.0	116–486	48.56
P	52	50	19.9	38.3	16–123	17.12[d]
SOC, g kg^{-1}	8.96	8.88	1.27	13.8	6.28–13.46	16.26[d]
pH[e]	5.57	5.52	0.39	7.0	4.77–7.00	28.84
$ECEC$ cmol$_c$ kg^{-1}	74.6	73.0	17.8	23.9	31.6–125	13.85[d]

[a] The statistical descriptions represent soil samples collected and composited from 235 monitoring units (Figure 1.4).
[b] SOC, soil organic carbon; $ECEC$, effective cation exchange capacity.
[c] Chi square test statistic for the combined effects of skewness and kurtosis.
[d] Data are normally distributed ($\alpha = 0.05$).
[e] Statistical data represent 236 monitoring units.

1.4.1 DESCRIPTIVE STATISTICAL PROPERTIES

The data obtained from soil analyses can be described in several ways. One analysis mechanism is to describe the data relative to its central tendency and variability (Table 1.7). In classical statistics, data that describe a property are characterized by a frequency distribution. A frequency distribution is a histogram that illustrates the number of observations that occur within specified increments of sample values. For example, in the frequency distribution that describes the variability of the soil $ECEC$ in the Milan field (Figure 1.6a), eight samples have values that range between 102.5 and 107.4 mmol$_c$ kg^{-1} and 29 samples have values that range between 68.3 and 73.2 mmol$_c$ kg^{-1}.

The frequency distribution can be described by three parameters: the sample mean, median, and standard deviation. The mean ($\bar{x}$) and median (m) are estimates of the central tendency of a data set, while the standard deviation (s) is a measure of the spread or variability in the sample population about the mean:

$$\bar{x} = \frac{1}{n} \sum_{i=1}^{n} x_i \tag{1.15}$$

$$s^2 = \frac{1}{n} \sum_{i=1}^{n} [x_i - \bar{x}]^2 \tag{1.16}$$

where x_i is the numerical value of the ith observation and n is the total number of samples. The square of the sample standard deviation (s) (Equation 1.16) is called the sample variance (s^2). The median value for a property is determined by first ranking the observations from low to high.

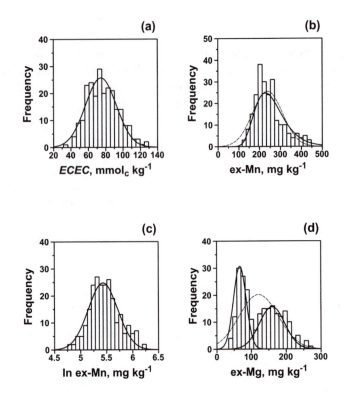

FIGURE 1.6 Frequency distributions for selected chemical properties of surface (0 to 15 cm depth) soil samples collected from a 10.5 ha production cotton field in Milan, TN ($n = 235$): (a) effective cation exchange capacity (ECEC), (b) Mehlich-3-extractable Mn, (c) transformed (natural logarithmic transformation) of Mehlich-3-extractable Mn, and (d) Mehlich-3-extractable Mg. Lines in (a) and (c) represent the normal distribution curve for the data. In (b) the solid line represents a normal distribution curve applied to transformed data (natural logarithm transformation); whereas the dashed line is the normal distribution curve applied to the untransformed data. In (d) the dashed line is the normal distribution curve applied to all the data; while the solid curves are the normal distribution models for a bimodal distribution.

The observation that is at the midpoint of the ranking is the median value. For the *ECEC* data, $\bar{x} = 74.6$ mmol$_c$ kg^{-1}, $m = 73.0$ mmol$_c$ kg^{-1}, and $s = 17.8$ mmol$_c$ kg^{-1}. In general, good agreement between the mean and median values for the *ECEC* suggests that the data are normally distributed about the mean, although specific statistical tests must be employed to substantiate the normalcy of a sample population. For example, the chi square test statistic indicates that the *ECEC* data are indeed normally distributed, with a confidence level of 95%.

When the data for a given parameter are normally distributed about the mean, the distribution is called binomial. In this case, $\bar{x}$ and s may be employed to generate the normal distribution curve that models the actual frequency distribution:

$$Y = \frac{ni}{s\sqrt{2\pi}} \exp\left[-\frac{1}{2}\left(\frac{x - \bar{x}}{s} \right)^2 \right]$$ (1.17)

where Y is the frequency and i is the class interval. Approximately 95% of the sample observations are found within plus or minus two standard deviations of the mean value when the data are described by a normal distribution (Figure 1.6a). Another indicator of the variability in a sample population and the homogeneity of the data is the coefficient of variation (CV). Values for CV are computed with the expression:

$$CV = 100 \times \frac{s}{\bar{x}}$$ (1.18)

When CV values for a property are below 15%, there is little variability and the data are relatively homogeneous. Coefficient of variation values that are between 16 and 35% indicate moderate variability; while CV values that are greater than 36% indicate high variability and a relatively heterogeneous data set. The CV value for *ECEC* (CV = 23.9%) suggests that this soil property has moderate variability in the field, while the CV value for pH (CV = 7%) indicates that this property has low variability in the field.

Figure 1.6b illustrates the distribution of Mehlich-3-extractable manganese (ex-Mn) data collected for samples obtained from the Milan field. This frequency distribution is skewed and the data are not normally distributed (also called an asymmetrical distribution). In this case, the median ex-Mn value ($m = 223$ mg kg^{-1}) is less than the mean value ($\bar{x} = 238$ mg kg^{-1}), and the normal distribution curve does not model the actual distribution (this is also indicated by the chi square test statistic in Table 1.7). Furthermore, the range of data encompassed by $\bar{x} \pm 2s$ is not a true indicator of the 95% data range. However, the asymmetrical distribution may be transformed to a normal distribution by performing a logarithmic transformation of the data (Figure 1.6c). The mean, median, standard deviation values of the transformed data are $\bar{x} = 5.431$, $m = 5.407$, and $s = 0.2837$ in units of ln (mg kg^{-1}). Using $\bar{x}$ and s of the transformed data and Equation 1.17 yields a normal distribution curve that very closely models the transformed frequency distribution. The transformed ex-Mn data may be evaluated by classical statistics. For example, 95% of the observations lie between exp(5.431 ± 0.2837), or 130 and 403 mg kg^{-1}.

The Mehlich-3-extractable magnesium (ex-Mg) data are described by $\bar{x} = 121$ mg kg^{-1}, $m = 117$ mg kg^{-1}, $s = 56.2$ mg kg^{-1}, and CV = 46.5% (indicating that the data are heterogeneous). The mean and median are relatively consistent, suggesting a normal distribution, and variability in the data is high, but this is not particularly unusual in field data. However, the chi square statistic indicates that the distribution of the ex-Mg data does not fit a normal distribution curve. Indeed, the frequency distribution of the ex-Mg data is an example of a bimodal distribution (Figure 1.6d). This type of distribution indicates that the soil samples from the Milan field do not represent a single population, with respect to this parameter. Deconvolution of the ex-Mg frequency distribution indicates two unique sets of data (two populations). Approximately one half of the data (117 samples) are described by $\bar{x} = 66.9$ mg kg^{-1}, $s = 18.4$ mg kg^{-1}, and CV = 27.5%, while the remaining 118 samples are described by $\bar{x} = 159.7$ mg kg^{-1}, $s = 36.3$ mg kg^{-1}, and CV = 22.7%. It appears from the frequency distribution of ex-Mg that two populations have been sampled from the Milan field. Indeed, the two mean values, 66.9 mg kg^{-1} and 159.7 mg kg^{-1}, appear to be very different. However, an apparent difference is not proof of difference. In order to make a definitive statement concerning the uniqueness of the two ex-Mg means (and the two populations), the Student's *t*-test can be employed. In this case, the means are significantly different within a very high degree of confidence ($\alpha < 0.001$).

1.4.2 Geostatistics

The classical statistical approach for describing the spatial distribution of soil properties assumes that a measured property at any location within a sampling area is expected to approximate the mean value of the property. This approach expects variability or dispersion of measured values and a normal distribution of values about the mean (as shown in Figure 1.6). The expected variability is described by the sample variance (s^2), which is assumed to be random and spatially uncorrelated within the sampled area. Statistical methods that describe soils data relative to central tendency and variability are useful but somewhat limited relative to describing spatial relationships. For example, the *ECEC* data discussed above are described by statistical parameters that indicate the data are normally distributed but relatively heterogeneous. However, these descriptive statistics provide no information about the spatial relationships that exist on the landscape. The ex-Mg data suggest that there are two unique populations of soil types in the field relative to this parameter. It may be inferred that one soil type is located in one part of the field, and the other type in another part; however, there is no way to validate this inference from the descriptive statistics.

Despite an assumption to the contrary when using classical statistics, the chemical and physical properties of soils on a landscape are not randomly distributed (variability is neither random nor uncorrelated to spatial location). In general, samples taken within close proximity of one another are spatially related (spatially correlated), while samples obtained from locations that are a large distance apart are uncorrelated. Geostatistics offers a mechanism to quantify the spatial structure and dependence of a measured soil property on landscape position, and to predict values for the property at unsampled locations. Mulla and McBratney (1999) and Trangmar et al. (1985) should be consulted for detailed descriptions of geostatistical techniques and applications to soils.

1.4.2.1 Variograms

The spatial dependence and structure of a measured property is described by a variogram. The variogram is a plot of sample semivariance (variance of a limited sampling, y-axis) vs. the separation distance (distance between sampling locations or *lag* distance, x-axis). The semivariance values to be plotted on a variogram are computed using the expression:

$$\gamma(h) = \frac{1}{2N(h)} \sum_{i=1}^{n(h)} [z_i - z_{i+h}]^2 \tag{1.19}$$

where $\gamma(h)$ is the semivariance at separation distance h (lag distance), $N(h)$ is the total number of sample pairs at separation distance h, z_i is a measured parameter at location i on the landscape, and z_{i+h} is the measured parameter at location $i + h$. The values obtained for z_i may represent single and discrete locations on the landscape, or they may represent the mean value of a sampling area (monitoring unit). In the case of the latter, z_i is called a regionalized parameter. In general, $\gamma(h)$ will be relatively small when h is small, and will increase to the sample variance (s^2, Equation 1.16) when h becomes large. This trend in $\gamma(h)$ is also an indication that samples located close to one another (small h) are statistically related, while those separated by large h values are not. Variograms may be omnidirectional (using all spatial data that radiate from location i to $i + h$ on a circle of radius h) or directional, i.e., constructed to represent the spatial dependence of a measured property along directional vectors (using spatial data that radiate from location i to $i + h$ on an arc of radius h), such as $0° \pm 22.5°$, $45° \pm 22.5°$, $90° \pm 22.5°$, and $135° \pm 22.5°$. Directional variograms are called anisotropic variograms. If the anisotropic variograms display a directional dependence, that is, if the variograms constructed for the different direction vectors are different, the spatial data are *anisotropic*. Isotropic variograms are omnidirectional and are employed to describe spatial data that are not directionally dependent (spatial data are *isotropic*)

In practice, a variogram is constructed using one of the many commercially available computer software packages that perform geostatistical computations. The isotropic variogram for the *ECEC* data collected from the Milan field is illustrated in Figure 1.7a. The semivariance as a function of

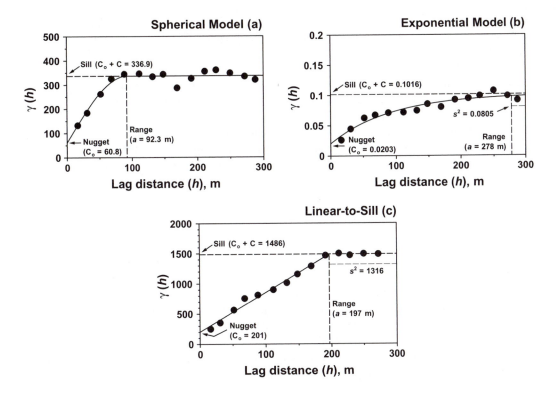

FIGURE 1.7 Examples of variograms and commonly applied semivariogram models: (a) ECEC (mmol$_c$ kg^{-1}), spherical model; (b) ln ex-Mn (mg kg^{-1}), exponential model; and (c) ex-K (mg kg^{-1}), linear-to-sill model.

lag distance (h) can be fit to a semivariogram model using nonlinear least-squares regression analysis. The *ECEC* semivariance data are modeled using the spherical model:

$$\gamma(h) = C_o + C\left[1.5\left(\frac{h}{a}\right) + 0.5\left(\frac{h}{a}\right)^3\right] \quad \text{for } h < a \tag{1.20a}$$

$$\gamma(h) = C_o + C \quad \text{for } h \geq a \tag{1.20b}$$

where the parameters C_o, C, and a provide quantitative information on the spatial structure of the measured property. The *nugget* parameter (C_o) is the predicted semivariance at a lag of zero and represents random or field variability that cannot be detected within the minimum sample spacing as well as the variability associated with experimental sampling or measurement errors. The parameter (C) is the *structural variance* or *scale* of the semivariogram. The sum of C_o and C is the *sill*, which represents the sample variance when the lag distance is very large (random variance), assuming the sample mean and variance are constant or stationary throughout the sampling area. The *range of spatial dependence* (a) is the lag distance at which the semivariance becomes constant. The range is also the distance at which samples become spatially independent and uncorrelated to one another.

For the *ECEC* data (in mmol$_c$ kg^{-1}), the nugget semivariance is 60.8 and the sill semivariance is 336.9. The ratio of the nugget semivariance to the sill semivariance is a measure of the spatial dependency of the measured property. The data show strong spatial dependency if the percent nugget is <25%; moderate spatial dependency if percent nugget is between 25 and 75%; and weak spatial dependency if percent nugget is >75%. The percent nugget for the *ECEC* data is 18%, and there is strong spatial dependency associated with the property. The sill semivariance (336.9) is also greater than the sample variance (317.9); however, a chi square test indicates that these two variances are not statistically different at $\alpha = 0.05$. This chi square test result suggests that the mean and standard deviation of the *ECEC* data are constant throughout the field and these data are stationary. The range of spatial dependence of *ECEC* is 92 m. At sampling distances greater than 92 m, the *ECEC* of soil samples are not spatially correlated.

In addition to the spherical semivariogram model, the exponential and linear-to-sill models are commonly used to describe semivariance as a function of lag distance. The exponential model is used to describe the transformed ex-Mn semivariance in the Milan field (Figure 1.7b):

$$\gamma(h) = C_o + C\left[1 - \exp\left(-\frac{3h}{a}\right)\right] \tag{1.21}$$

For ex-Mn, $C_o = 0.0203$, $C_o + C = 0.1016$, % nugget = 20% (strong spatial dependency), and $a = 278$ m. In this example, the nugget and sill semivariance values are from the ln-transformed ex-Mn data, as this transformation was necessary to obtain a normal distribution (see the previous section). The sill semivariance (0.1016) and the sample variance ($s^2 = 0.08048$) are not significantly different at $\alpha = 0.05$. Again, this result suggests that the mean and standard deviation of the ex-Mn data are constant throughout the field.

The linear-to-sill model describes the special dependency of Mehlich-3-extractable potassium (ex-K) data in the Milan field (Figure 1.7c). The linear-to-sill model is:

$$\gamma(h) = C_o + C\left(\frac{h}{a}\right) \quad \text{for } h < a \tag{1.22a}$$

$$\gamma(h) = C_o + C \quad \text{for } h \geq a \tag{1.22b}$$

For ex-K, $C_o = 201$, $C_o + C = 1486$, percent nugget = 14% (strong spatial dependency), and $a = 197$ m. The sill semivariance (1486) and the sample variance ($s^2 = 1315$) are not significantly different at $\alpha = 0.05$, suggesting that the mean and standard deviation of the ex-K data are constant throughout the field.

The variogram parameters that describe the spatial variability of a select number of soil chemical properties in the Milan field are shown in Table 1.8. Perhaps the most important variogram parameter for establishing the minimum spacing between soil sampling locations which is necessary to describe the spatial variability of a soil property is the range of spatial dependence (a). If the sampling distances are greater than the range, sampling points are not spatially related, and there is no mechanism to predict the soil properties at unsampled locations. The range values for the various properties measured in the Milan field show considerable variability. In general, soil properties that have large range values are those that are not greatly affected by soil type. For example, soil pH has a range value of 257 m. This large a value is due to the fact that pH values tend to be similar in the southeastern United States regardless of soil type, and it can also be a managed property (and thus more uniform in agronomic systems). Conversely, the *ECEC* values for the soils in the Milan field are closely related to soil type and particularly to the degree of topsoil erosion (which affects soil type). Consequently, this property has a relatively short range (92 m). Unfortunately, the determination of the variogram parameters necessary to establish the spatial variability of a property requires an intensive and extensive soil sampling regime. As a rule-of-thumb, sample spacing should be no more than ¼ to ½ the value of the range in order to

TABLE 1.8
Variogram Parameters and Goodness-of-Prediction Parameters for Selected Chemical Characteristics of Soil Samples Collected from a Production Cotton Field at the Milan Experiment Station in Milan, TN[a]

Property	Model[b]	C_o	$C_o + C$	a, m	Model r^2	% Nugget[c]	Spatial Class[d]	G 22 m	G 43 m	G 87 m	r^2 22 m	r^2 43 m	r^2 87 m
Mehlich-3 extractable, mg kg^{-1}													
K	LS	201	1486	197	0.990	14	S	75.0	78.3	43.5	0.75	0.79	0.44
Mg	S	10	3385	106	0.976	0	S	79.9	69.4	8.9	0.80	0.71	ns
Mn[f]	E	0.0203	0.1016	278	0.929	20	S	62.0	39.2	—[g]	0.62	0.40	ns
P[f]	S	0.022	0.162	96	0.884	14	S	59.3	22.2	—	0.59	0.22	ns
SOC, g kg^{-1}	E	0.001	1.807	85	0.990	0	S	37.1	—	—	0.37	ns	ns
pH, units	E	0.041	0.121	255	0.920	34	M	44.5	38.0	15.0	0.45	0.38	0.17
ECEC cmol$_c$ kg^{-1}	S	60.8	337	92	0.929	18	S	53.0	39.1	—	0.53	0.40	ns

[a] The spatial descriptions represent soil samples collected and composited from 235 monitoring units that measure 90 m × 53.3 m (Figure 1.4). SOC, soil organic carbon; ECEC, effective cation exchange capacity; C_o, nugget semivariance; $C_o + C$, sill semivariance; a, range of spatial dependence in meters; model r^2, nonlinear regression coefficient for the semivariogram model; G, goodness-of-prediction measure; r^2 linear regression coefficient for cross-validated data.

[b] Semivariogram models: LS, Linear-to-sill model (Equation 1.22); E, exponential model (Equation 1.21); spherical model (Equation 1.20). The computed model parameters apply only to the 27.4 m × 16.2 m sampling grid (22 m average distance between sampling locations). Semivariogram models and associated nugget, sill, and range values for the larger sampling distances (43 and 87 m) differ from the 22 m models, but are not illustrated.

[c] $\frac{C_o}{C_o + C} \times 100$.

[d] S = strong spatial dependency; M = moderate spatial dependency.

[e] G and r^2 values are shown as a function of the average distance between the center points of monitoring units. Unless indicated by ns, all r^2 values are significant at α < 0.001.

[f] Transformed property, ln (x).

[g] The distance between sampling locations exceeds the range of spatial dependency.

generate a variogram that adequately describes spatial variability. However, in the absence of preexisting information on the spatial variability of a soil property at a specific site, ancillary information from a nearby site or a preliminary sampling survey of the site could be employed to establish a first-approximation spatial sampling scheme.

1.4.2.2 Interpolation by Kriging

In 1966, D.G. Krige applied a statistical interpolation technique to predicting the location of ore-bodies for the mining industry. This technique, known as kriging, uses the structural characteristics of a variogram (nugget, sill, and range) and measured spatial data for a property to predict values at unsampled locations on the landscape. Kriging is essentially an error-minimization technique, where the value for a soil property at an unsampled location is predicted such that the variance of the estimated value is minimized. There are several kriging methodologies including punctual, block, universal, disjunctive, and cokriging. A detailed discussion of these techniques is beyond the scope of this text. Again, Mulla and McBratney (1999) and Trangmar et al. (1985) should be consulted for detailed information.

The Milan data are amenable to the block kriging technique. This method uses average values of a property in small regions or monitoring units (or blocks) to estimate the average value of the property in an unsampled block. Block kriging was used to create the spatial distribution maps for *ECEC*, ex-Mn, and soil organic carbon (SOC) in the Milan field (Figure 1.8). Because the spatial

FIGURE 1.8 The spatial distribution of (a) effective cation exchange capacity (ECEC, $cmol_c$ kg^{-1}), (b) Mehlich-3-extractable Mn (mg kg^{-1}), and (c) soil organic carbon content (g kg^{-1}) in the surface soil (0–15 cm) on the 10.5 ha production cotton field located at the Milan Agricultural Experiment Station in Milan, TN.

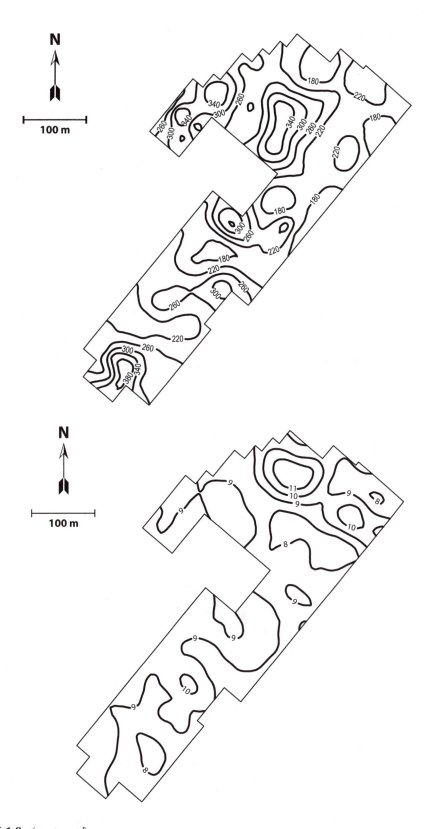

FIGURE 1.8 (*contunued*).

variability of the soil chemical parameters in the Milan field are known (all blocks were sampled), the maps in Figure 1.8 essentially represent the true spatial distribution of the chemical parameters. However, the ability of the block kriging technique to estimate the value of a parameter at an unsampled location may be evaluated. One method to evaluate the suitability of an interpolation technique is called cross-validation (the technique is also called jack-knifing). In this method each measured point in the spatial data is individually removed and its value estimated by kriging as though it were an unsampled point. The predicted values may then be compared to the actual values.

Measured and estimated values obtained by cross-validation for $ECEC$, ex-Mn, and SOC are compared in Figure 1.9. A measure of how well the block kriging technique estimates values at unsampled locations is the linear regression coefficient, r^2 (also called the linear coefficient of

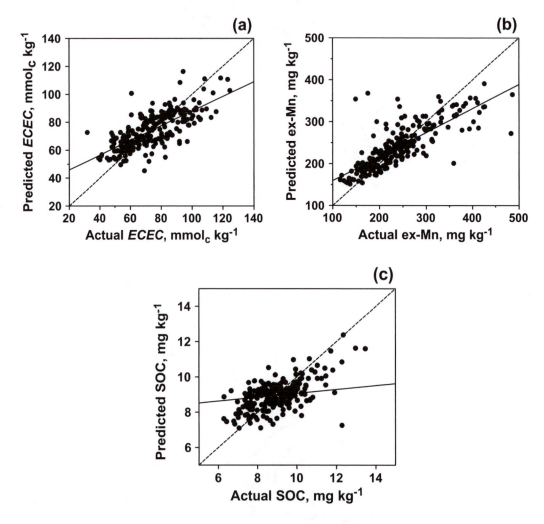

FIGURE 1.9 Cross-validation results for (a) effective cation exchange capacity (ECEC, mmol$_c$ kg^{-1}), (b) Mehlich-3-extractable Mn (mg kg^{-1}), and (c) soil organic carbon content (g kg^{-1}). The predicted values are computed by individually removing each measured point in the spatial data and its value estimated by kriging as though it were an unsampled point. The dashed lines represent a one-to-one correspondence between predicted and actual values and the solid lines were obtained by regressing predicted on actual using least-squares linear regression analysis (r^2 values are given in Table 1.8 for the 22 m spacing).

determination), that results from regressing the actual values on the predicted values. The r^2 value is related to G, a goodness-of-prediction parameter, by:

$$G = \left(1 - \left[\frac{\sum_i \{A(i) - A'(i)\}^2}{\sum_i \{A(i) - \bar{x}\}^2}\right]\right) \times 100 \cong r^2 \times 100 \tag{1.23}$$

where $A(i)$ is the measured value at location i, $A'(i)$ is the predicted value, and $\bar{x}$ is the mean value for the parameter. A large positive value for G or r^2 indicates that the property is accurately predicted by kriging ($G = 100\%$ or $r^2 = 1$ corresponds to prefect prediction); whereas, a small r^2 value ($r^2 = 0$ indicates that no correlation exists) and a small or negative G value indicates that a property is poorly predicted. For example, the G and r^2 values for SOC are 37.1% and 0.374, and for ex-Mn the G and r^2 values are 62.0% and 0.624 (Table 1.8). Thus, ex-Mn is predicted to a higher level of confidence than SOC at unsampled locations using block kriging.

Increasing the distance between sampling locations will degrade the ability of an interpolation technique to predict soil properties at unsampled locations. This effect is illustrated in Table 1.8. For example, a twofold increase in the average distance between monitoring units reduces the G for ex-P from 59.3% at 22 m distance to 22.2% at an average distance of 43 m. At a sampling distance of 87 m the range of spatial dependence is exceeded (semivariogram nugget equals the sill) and kriging cannot be employed to estimate ex-P values at unsampled locations. However, ex-K values may be predicted at unsampled locations using grid spacings 43 m and 87 m, although confidence in the estimated values decreases when the distance between sampling locations is increased to 87 m.

1.5 EXERCISES

1. A 50-g sample of soil (dry weight basis) and 100 mL of methanol are placed in a 250-mL centrifuge bottle. The mixture is shaken for 1 h, centrifuged to separate the solid and extract, and the supernatant solution removed for analysis. A 1-mL aliquot of the extract is placed in an auto-sampler vial with 4 mL H_2O. The solution is analyzed using HPLC and found to contain 26.28 µg L^{-1} of the herbicide fluometuron. What is the concentration of fluometuron in the soil, assuming the extraction procedure is 95% efficient?

2. The following extraction procedure is used as part of a scheme to determine the organic phosphorus content of a soil. A 2-g sample of soil (dry weight basis) is placed in a 50-mL volumetric flask. A 3-mL volume of concentrated H_2SO_4 is added and the flask swirled. A 4-mL volume of H_2O is added in 1-mL increments, and the flask swirled between each incremental addition. The sides of the flask are washed down with H_2O and allowed to cool before being brought to volume with H_2O. After thoroughly mixing the contents of the volumetric flask, they are poured through filter paper, and the filtrate retained for analysis. The solids remaining in the volumetric are washed onto the filter paper and washed with copious amounts of H_2O. The filter paper and soil cake (assumed to contain 2-mL of H_2O) are placed in a 250 mL Erlenmeyer flask with 98 mL of 0.5 M NaOH and shaken for 2 h. Again, the contents of the flask are filtered and the filtrate retained for analysis. A 2-mL aliquot of the acid extract is combined with a 4-mL aliquot of the base extract and a 1-mL subsample of the mixture is diluted to 10 mL. The diluted sample is analyzed for P using inductively coupled argon plasma spectro-photometry (ICAP) and found to contain 4.27 mg L^{-1} P. On a soil basis, what is the concentration of P extracted by this procedure (in units of mg kg^{-1})?

3. The following passage was prominently displayed on alkaline battery packages by a well-known battery manufacturer: "No mercury added for a safer and cleaner environment—contains only naturally occurring trace elements." Comment on this statement.

4. The saturation extract of a soil has the following chemical composition:

Component	mmol L^{-1}
SO$_4$	2.512
Na	1.000
K	0.603
CO$_3$	1.585
Cl	1.000
Mg	0.794
Ca	2.512

Assume you have access to the following salts: NaCl, CaCl$_2$•2H$_2$O, CaSO$_4$•2H$_2$O, MgSO$_4$•7H$_2$O, KHCO$_3$, KCl, NaHCO$_3$, Na$_2$SO$_4$. Determine the types and amounts (in mg) of salts necessary to synthesize 20 L of the soil extract, matching as closely as possible the composition listed above.

5. A 30-mL volume of a K$_2$HPO$_4$ solution containing 50 mg P L^{-1} is placed in contact with 40 g of soil (dry weight basis) in a centrifuge tube. After a 7-d equilibration period, the suspension is centrifuged to separate the equilibrium solution from the soil. A 5-mL aliquot of this equilibrium solution is diluted with 45 mL of a 10 mM CaCl$_2$ solution. The diluted solution is analyzed for P and found to contain 1.0 mg P L^{-1}. Answer the following:
   • What is the concentration of P retained by the soil (in mg P kg^{-1})?
   • What is the concentration of PO$_4$ retained by the soil (in mg PO$_4$ kg^{-1})?
   • What was the molar concentration of P in the initial solution (before placing in contact with the soil)?
   • What was the molar concentration of K$_2$HPO$_4$ in the initial solution?

6. Using O$_2$(g) as the oxidant, write balanced oxidation-reduction reactions for the following:
   • Arsenite to arsenate
   • Cr(OH)$_3$(s) to chromate
   • Hydrogen selenide (H$_2$Se) to selenate

7. Using sulfide as the reductant (with SO$_4^{2-}$ as a product), write balanced oxidation-reduction reactions for the following:
   • Arsenate to arsenite
   • Chromate to Cr(OH)$_3$(s)
   • Selenate to hydrogen selenide (H$_2$Se)

8. Using nitrate as the oxidant (with nitrogen gas as the reduction product), write balanced oxidation-reduction reactions for the following:
   • Methane to carbon dioxide
   • Hydrogen sulfide to sulfate
   • Ferrous to ferric iron

9. After reading the book, *Fateful Harvest: The True Story of a Small Town, a Global Industry, and a Toxic Secret* by Duff Wilson (Harperperennial Library, 2002), which details the plight of a farmer who applied arsenic-, cadmium-, lead-, and beryllium-containing industrial waste material to his land, a Tennessee state legislator introduced a bill to the General Assembly that contained the following sections:
   • It is unlawful to use toxic materials or substances as fertilizer or in any other land preparation application if such materials or substances were ever classified as toxic in any other circumstances.

- The prohibition shall also apply to the use of any substance containing heavy metals as fertilizer or any other land preparation application.

You are asked to comment on the bill by a lobbyist for the Farm Bureau. What are your comments?

10. The units parts per million (ppm) and parts per billion (ppb) are commonly used in the popular press, in some scientific literature, and by many regulatory agencies to express trace concentrations of elements. However, these units are ambiguous and not acceptable in the SI system or in the majority of the scientific literature. Why are these terms ambiguous?

11. An investigative reporter with a local newspaper attempts to link elevated nitrate levels in city groundwater supplies to a local dairy operation. Independent analysis of a ground-water sample obtained near the dairy indicates a nitrate concentration of 33.4 mg L^{-1}. The reporter, armed with the Part 141 National Interim Primary Drinking Water Regulations which establish a maximum contaminant level for NO_3 (as N) of 10 mg L^{-1}, lobbies his editor to run his article which he says clearly establishes that the city water supply is polluted and contains nitrate concentrations that are more than three times the drinking water standard. Should the editor run the article?

12. Soil that is in the vicinity of a zinc mine may be contaminated with various trace metals. The potential contamination is evaluated by subjecting a sample of the soil to an extraction procedure that provides a measure of the total elemental concentration. A 3.5-g sample of soil (dry weight basis) is placed in a glass digestion tube with 25 mL of 4 M HNO_3 and the tube is covered with a glass funnel. The assemblage is placed in a heating block and the sample digested for 16 hours at 80°C. After digestion, the sample is cooled and the total volume brought to 35 mL with deionized-distilled water. After filtering, a 5-mL aliquot of the extract is placed in a 10-mL volumetric flask that is brought to volume with deionized-distilled water. The diluted sample is then analyzed for Zn and Pb by inductively coupled plasma spectrometry and found to contain 7.6 mg Zn L^{-1} and 0.79 mg Pb L^{-1}. Has this soil been contaminated by mining if uncontaminated soil in the region contains ($\bar{x} \pm s$) 146 ± 5 mg Zn kg^{-1} and 21 ± 8 mg Pb kg^{-1}? Defend your answer.

13. Precision agriculture, or site-specific management, is currently practiced by numerous production agriculture operations in the United States. In general, a producer will contract a firm to perform the site-specific assessment and apply fertilizer nutrients and other agrochemicals where needed. However, commercial operations use a 1-ha sampling grid when characterizing the spatial variability of soil properties. Is it feasible for a producer to employ a commercial site-specific management firm to determine fertilizer P application rates on the 10.5-ha Milan Experiment Station field described in this chapter (see Table 1.8)? Defend your response.

14. A 500-mg sample of soil (dry weight basis) that has been subjected to long-term sewage sludge amendments is dissolved in 3.5 mL of a HNO_3-HCl-HF solution. Following dissolution, the mixture is placed in a 50-mL volumetric flask and brought to volume with deionized water. The solution is then analyzed for Cd, Cr, Ni, and Pb using atomic absorbance spectrophotometry. Absorbance readings of 0.0043 for Cd, 0.563 for Cr, 0.079 for Ni, and 0.123 for Pb are obtained. You have determined the standard curves relating absorbance to mg L^{-1} of metal in solution to be as follows:

   Cd:mg L^{-1} = 2.778 × Abs.
   Cr:mg L^{-1} = 6.803 × Abs.
   Ni:mg L^{-1} = 14.493 × Abs.
   Pb:mg L^{-1} = 20.408 × Abs.

What are the mg kg^{-1} concentrations of Cd, Cr, Ni, and Pb in the sludge-amended soil sample? Compare your concentration values to the range and median values for these elements in uncontaminated soils (Table 1.1). Based on this limited data set, what preliminary conclusions can be drawn concerning the potential accumulation of these elements in the soil as a result of long-term sewage sludge applications?

15. A 10-g soil sample (dry weight basis) is combined with 20 mL of a solution containing 163.8 μg L^{-1} of the pesticide norflurazon in a Teflon centrifuge tube. After a 1-month equilibration, the tube is centrifuged to separate soil and solution phases. A 10-mL aliquot of the solution is removed, analyzed by HPLC, and found to contain 57.1 μg L^{-1} norflurazon. To the soil and solution remaining in the centrifuge tube is added 20 mL of acetonitrile to extract the norflurazon retained by the soil. The tube is shaken for 24 h, centrifuged, and an aliquot of the extracting solution removed for HPLC analysis. The extracting solution is found to contain 88.7 μg L^{-1} norflurazon. What is the extraction efficiency of the acetonitrile (what percentage of the pesticide retained by the soil is removed by the acetonitrile)?

REFERENCES

Bowen, H.J.M. *Environmental Chemistry of the Elements*. Academic Press, London, 1979.

Helmke, P.A. The chemical composition of soils. In *Handbook of Soil Science*. M.E. Sumner (Ed.) CRC Press, Boca Raton, FL, 2000, pp. B3-B24.

Mulla, D.J. and A.B. McBratney. Soil spatial variability. In *Handbook of Soil Science*. M.E. Sumner (Ed.) CRC Press, Boca Raton, FL, 1999, pp. A321–A352.

Sposito, G. *The Chemistry of Soils*. Oxford University Press, New York, 1989.

Swift, R.S. Sequestration of carbon by soil. *Soil Sci.* 166:858–871, 2001.

Trangmar, B.B., R.S. Yost, and G. Uehara. Application of geostatistics to spatial studies of soil properties. *Adv. Agron.* 38:45–94, 1985.

2 Soil Minerals

The solid phase of soil is composed of two distinct materials: inorganic and organic. Of these, the inorganic materials, or minerals, predominate in virtually all soils except Histosols. A mineral is defined as any naturally occurring inorganic substance. In the soil sciences, minerals are grouped into two idealized categories: crystalline and amorphous. Crystalline minerals are naturally occurring inorganic compounds composed of atoms arranged in a three-dimensional and periodic (repeating) pattern. Amorphous minerals are naturally occurring inorganic compounds that have no periodic three-dimensional arrangement of atoms. Minerals exert a pronounced influence on the chemistry of soil solutions. Through precipitation and dissolution (weathering), they regulate the chemical composition of the soil solution (see Chapter 6). Mineral surfaces are also inherently reactive, potentially forming strong or weak chemical bonds with soluble substances and further regulating the composition of the soil solution (see Chapters 7 and 8). A majority of the discussion of minerals in this chapter is devoted to the silicates, the most common class of minerals in the soil environment. Aside from their abundance, a subclass of the silicates, the phyllosilicates or clay minerals, have a substantial impact on the physical and chemical characteristics of soil. With few exceptions, the phyllosilicates and associated clay materials exert the greatest impact on the fate and behavior of substances in the soil environment.

2.1 CHEMICAL BONDS

The elements can be divided into two groups based on their tendency to give up or to attract electrons. Those elements on the left side of the periodic chart (alkali metals, Na and K, and alkaline earth metals, Ca and Mg) possess electronic orbitals that are largely empty. These elements will donate valence electrons to achieve a noble gas configuration and become more stable. Those elements on the right side of the periodic chart (halogens, F and Cl) are electron acceptors because they possess an electronic orbital that is nearly full. These elements accept an electron to achieve a noble gas configuration and become more stable. The elements between the alkali metals and the halogens in the periodic chart vary in their tendencies to be electron acceptors or donors, depending on how close they are to filling or depleting their orbitals with electrons. Elements that are electron donors are termed metals; whereas, elements that are electron acceptors are termed nonmetals. Metals and nonmetals are complementary: an electron donated by a metal is accepted by a nonmetal. When electron transfer is achieved, the metal (electron donor) becomes a cation (a positively charged species) and the nonmetal (electron acceptor) becomes an anion (a negatively charged species).

Cations and anions are attracted to one another by Coulombic forces. This electrostatic attraction may result in the formation of a chemical bond that is principally ionic in character. In this type of bond, electrons in the anion are not shared with the cation, and the angles at which bonds radiate from the individual atoms are not tied to the geometric configuration of the electronic orbitals: the bonds are undirected. The predominance of the ionic bond is common between elements having vastly different metal and nonmetal character, such as Na^+ and Cl^- in halite, which is known as common table salt. When two elements of similar metal and nonmetal character interact, such as N and O in NO_3^-, electrons may be shared (but not transferred) by the atoms forming a bond having a predominance of covalent character. Ionic and covalent bonds are conceptual idealizations that approximate real chemical bonds. The covalent character of a bond depends on the difference in the tendency of the atoms to attract electrons. The covalent character of a bond increases as atoms become more equal in their ability to attract electrons. This leads to the following two points: (1) no bond is 100% ionic and (2) the only 100% covalent bonds are found in molecules formed from two atoms of the same element (e.g., H_2, O_2, N_2).

There are semiquantitative measures available that allow one to predict the character of a bond between two elements. A measure of the ability of an atom in a molecule to attract electrons is termed the electronegativity (*EN*). The *EN* of elements can provide a measure of the ionic character of chemical bonds and can indicate the extent to which two atoms in a molecule actually share their valence electrons. Elements with low *EN* act as electron donors. For example cesium, with a [Xe] $6s^1$ electronic configuration, has the lowest *EN* (*EN* = 0.7) and the strongest tendency to be an electron donor by donating the single $6s$ valence electron to achieve the atomic configuration of xenon (as Cs^+). Elements with high *EN* act as electron acceptors. Fluorine, with a [He] $2s^2 2p^5$ electronic configuration, has the highest *EN* (*EN* = 4.0) and the strongest tendency to accept an electron and achieve the atomic configuration of neon (as F^-). Thus, a chemical bond between these two atoms is strongly ionic, as F^- has little tendency to share the acquired electron and Cs^+ has little tendency to gain an additional electron.

The actual character of the bond formed between two elements is predicted by the difference in their *EN* values (ΔEN). Elements with similar *EN* form principally covalent bonds with little ionic bond character. Conversely, two elements with dissimilar *EN* values form a bond whose ionic character is proportional to the magnitude of ΔEN. The relationship of bond character to ΔEN values is displayed in Figure 2.1 using the *EN* and ΔEN values presented in Table 2.1 and the relationship:

$$\% \text{ ionic character} = 100 \times [1 - \exp(-0.25 \times \Delta EN^2)] \tag{2.1}$$

developed by Linus Pauling to describe the nature of the chemical bond. When the ΔEN value is greater than approximately 1.7 (~50% ionic character), the bond is identified as an ionic bond. Notable examples of ionic bonds are the base cation-oxygen bonds (e.g., K—O, Na—O, Ca—O, and Mg—O), the Al—O bond, and the Si—O bond. Values of ΔEN that range between approximately 0.6 and 1.7 are indicative of a polar covalent bond. In this bond type, the atom with the greater *EN* has a greater proportion of the shared electron density, becoming slightly negative. The atom

---

TABLE 2.1
The Ionic Character of Several Metal–Oxygen Bonds in Relation to the Difference in Electronegativity (ΔEN) between Metals and Oxygen

Element	EN	Bond	Compound	ΔEN	% Ionic
O	3.5				
N	3.0	N—O	NO_3^-	0.5	6
C	2.5	C—O	CO_3^{2-}	1.0	22
S	2.5	S—O	SO_4^{2-}	1.0	22
P	2.2	P—O	PO_4^{3-}	1.3	35
H	2.1	H—O	H_2O	1.4	39
Si	1.8	Si—O	SiO_2	1.7	51
Al	1.5	Al—O	$Al(OH)_3$	2.0	63
Mg	1.2	Mg—O	$MgSO_4$	2.3	74
Ca	1.0	Ca—O	$CaCO_3$	2.5	79
Na	0.9	Na—O	$NaNO_3$	2.6	82
K	0.8	K—O	KNO_3	2.7	84

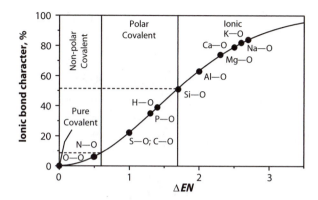

FIGURE 2.1 The ionic and covalent character of metal-oxygen bonds as a function of the absolute difference between the metal and oxygen electronegativities (ΔEN). The curve represents the equation: % ionic character = $[1 - \exp(-0.25 \times \Delta EN^2)] \times 100$, which was developed by Linus Pauling to describe bond character.

with the smaller EN has a smaller proportion of the shared electron density, and becomes slightly positive. The C—O, S—O, P—O, and H—O bonds are examples of the polar covalent bond type (the bond type that leads to the unique properties of water and the hydrogen bond). When ΔEN is less than approximately 0.7, as in the N—O bond, the bond is nonpolar covalent. In nonpolar covalent bonds, the valence electrons almost completely reside with the more electronegative species (e.g., oxygen in the N—O bond). Diatomic molecules, such as H_2, O_2, and N_2, and materials consisting of a single element, such as diamond, contain pure covalent bonds ($\Delta EN = 0$) with equal sharing of the valence electrons.

Covalent and ionic bonds are two of the most common bond types in soil minerals. But they are by no means the only important bond types. Another type of bond that occurs in minerals is an electrostatic interaction called a hydrogen bond. Water and hydroxide are common structural components of soil minerals and the O—H bond is polar covalent, with an ΔEN value of 1.4 (approximately 61% covalent). Thus, O attracts electrons from hydrogen resulting in a partial negative charge on the oxygen and a partial positive charge on the proton in structural H_2O and in OH. Moreover, oxygen has a greater supply of electrons than hydrogen, further enhancing this separation of charge. The separation of charge leads to a weak, undirected attraction between a water or hydroxyl proton of one molecule and a structural oxygen of another. This weak attraction is a special case of a dipole-dipole interaction, termed a hydrogen bond. In water, the hydrogen bond strength is approximately $\frac{1}{20}$ the strength of the polar covalent O—H bond within the water molecule (23.3 kJ mol^{-1} vs. 492 kJ mol^{-1}). Hydrogen bonds are not unique to proton-oxygen interactions; they may also form when a proton interacts with other electronegative atoms, such as F and N.

The hydrogen bond is responsible for the unique properties of liquid water (discussed in Chapter 5), and is a significant bond type in soil minerals. Kaolinite is a common clay mineral composed of alternating sheets of aluminum hydroxide and silicate (Figure 2.2). Adjacent aluminum hydroxide and silicate sheets in kaolinite are linked by principally ionic bonds having strong covalent character (37% for the Al—O bond and 49% for the Si—O bond) to form a cohesive unit. These units are then bound to adjacent units by hydrogen bonds between Si-bound oxygen atoms and protons on Al-bound hydroxide molecules. Although the hydrogen bond is weak relative to other bonding mechanisms (as indicated above), this weakness does not render the kaolinite structure unstable. Indeed, kaolinite is common in soil environments, and particularly abundant in the clay- and silt-sized fractions of soils that are in advanced stages of weathering.

Neutral surfaces or molecules that are nonpolar can also be attracted to one another by very weak electrostatic interactions called van der Waals forces. Van der Waals bonding arises from permanent or induced dipole interactions, or from interactions that are caused by the dynamic

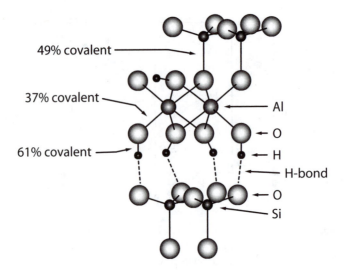

FIGURE 2.2 The bonding structure in kaolinite. The Si—O bonds in the Si tetrahedra are 49% covalent. The apical Si tetrahedral oxygen atoms are shared with Al in octahedral coordination (37% covalent bond). The resulting unit, the planar Si tetrahedral sheet and the planar Al octahedral sheet, is the basic 1:1 structural unit of kaolinite. The 1:1 units are connected through hydrogen bonds between basal Si oxygen atoms and Al-bound hydroxyls.

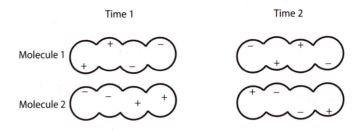

FIGURE 2.3 The electron correlation effect that results in van der Waals' bonding of two uncharged molecules at time 1 and time 2.

polarization (also called dispersive forces) of neutral molecules (Figure 2.3). Although neutral, these molecules have electrons that move in an electronic cloud. For this type of bonding to occur, the movement of electrons in one neutral molecule must be correlated or paired with the movement of protons in another neutral molecule at every instant. This type of bond is weak and undirected, and the bond strength depends on the approximate area of contact. Thus, the greater the area over which two neutral molecules interact (the larger the interacting molecules), the greater the attractive force. This type of bond occurs in the clay minerals talc and pyrophyllite, where a neutral-charged layer of silicon-bound oxygen atoms is bonded to an adjacent neutral-charged layer of silicon-bound oxygen atoms.

2.2 PAULING'S RULES

The types of bonds that occur in minerals are important for a variety of reasons, but two are principal: (1) the physical and chemical properties of all minerals depend on the character of the bonds that hold them together, and (2) weathering reactions and the stability of soil minerals are functions of the nature of the crystal chemical bonds. In effect, the weakest bond type in a mineral, to a large extent, will dictate the mineral's physical and chemical properties. As was stated previously, the most important bond types in soil minerals are the ionic and covalent bonds. Chemical bonds that are predominately covalent in character are stronger than those that are predominately ionic in character. Thus, the relative stability of a mineral in a weathering environment is determined by the ionic bonds (the weaker linkage) that are inherent to the mineral.

The ionic interactions also influence the structure of soil minerals. The O^{2-} anion dominates in soil minerals (46.6% by weight and 93.8% by volume). With the exception of the Si—O bond (which is approximately 50% covalent), metal bonds with the O^{2-} anion are mainly ionic (Table 2.1 and Figure 2.1). Because ionic interactions dominate, and because cations and anions do not share electrons in ionic bonding, the ions in a mineral structure can be treated as hard or rigid spheres. Thus, the arrangement of atoms in a crystal structure may be elucidated by examining the ways in which rigid spheres can be packed together in three-dimensional space. In addition to the geometric packing constraints, the electrostatic interactions between atoms in a mineral structure must also be considered in predicting their likely arrangement; in effect, the observed structural arrangement will be that which minimizes the electrostatic repulsive forces and maximizes the attractive forces between atoms. A set of rules that describes the likely geometric arrangement of atoms in an ionic structure, while minimizing the total electrostatic energy, was formulated by Linus Pauling (Pauling, 1960). These rules are known as Pauling's rules of crystal configuration:

The First Rule: A coordinated polyhedron of anions is formed about each cation, the cation-anion distance being determined by the radius sum and the coordination number of the cation by the radius ratio.

This rule states that each cation in a crystal structure is surrounded by anions. The distance from the center of the cation to the center of an adjacent anion is the radius of the cation plus the radius of the anion (because ionic bonding dominates and ions are then assumed to be hard spheres). The number of anions that can pack around and coordinate a cation is the coordination number (CN). The CN of a cation is a function of cation and anion size (radius). For example, the radius of Si^{4+} is 0.042 nm and that of O^{2-} is 0.140 nm. The radius ratio (cation radius/anion radius) is 0.042/0.140 = 0.30. From Table 2.2, a radius ratio of 0.30 gives Si^{4+} a CN of 4. This means that a Si^{4+} ion in a stable structure with O^{2-} anions will be surrounded by 4 O^{2-} ions, and Si^{4+} is said to be in tetrahedral coordination (see Table 2.3 for predicted and observed metal coordination numbers). For each CN, there is a limiting radius ratio that describes the condition of "closest packing." For example, for CN = 3 the limiting radius ratio is 0.155. Similarly, for CN = 4 the limiting radius ratio is 0.255. The limiting radius ratio is the radius ratio at which the coordinating anions just touch. If the central cation were any smaller, the anions would overlap and set up a strong repulsion to one another. The structure would then be unstable and a lower CN would be needed to maintain stability. This rule is also called the *no rattle rule*; if the central cation has room to move inside a given polyhedron, the configuration is unstable. Some of the more common coordinated structures observed in minerals are illustrated in Figure 2.4. As described in this first rule, as the size of a

TABLE 2.2
The Radius Ratio, the Predicted Coordination Number, and an Example Polyhedron Predicted through the Application of Pauling Rule 1

Radius Ratio	Coordination Number	Polyhedron Type
<0.155	2	Linear
0.155–0.225	3	Equilateral triangle
0.225–0.414	4	Tetrahedral
0.414–0.732	6	Octahedral
0.732–1.00	8	Cubic
>1.00	12	Dodecahedral

TABLE 2.3
The Relationship between Ionic Size (Pauling Radius, r_{ion}), Radius Ratio (r_{ion}/r_O), and Coordination Number (CN), with Oxygen ($r_O = 0.140$ nm) as the Coordinating Anion

Ion	r_{ion} (nm)	r_{ion}/r_O	Predicted CN	Observed CN	Ion	r_{ion} (nm)	r_{ion}/r_O	Predicted CN	Observed CN
Cs^+	0.167	1.19	12	12	Li^+	0.068	0.49	6	6
Rb^+	0.147	1.05	12	8–12	Ti^{4+}	0.068	0.49	6	6
Ba^{2+}	0.134	0.96	8	8–12	Mg^{2+}	0.066	0.47	6	6
K^+	0.133	0.95	8	8–12	Fe^{3+}	0.064	0.46	6	6
Sr^{2+}	0.112	0.80	8	8	Cr^{3+}	0.063	0.45	6	6
Ca^{2+}	0.099	0.71	6	6,8	Al^{3+}	0.051	0.36	4	4,6
Na^+	0.097	0.69	6	6,8	Si^{4+}	0.042	0.30	4	4
Mn^{2+}	0.080	0.57	6	6	P^{5+}	0.035	0.25	4	4
Fe^{2+}	0.074	0.53	6	6	Be^{2+}	0.035	0.25	4	4
V^{3+}	0.074	0.53	6	6	S^{6+}	0.030	0.21	4	4
					B^{3+}	0.023	0.16	3	3,4

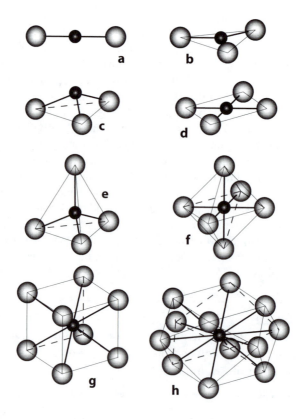

FIGURE 2.4 Examples of coordinated structures that occur in soil minerals. The dark circles represent the central metal cations and the light circles are the coordinating anions: (a) linear two-coordination, (b) planar triangular three-coordination, (c) trigonal pyramidal three-coordination, (d) planar square four-coordination, (e) tetrahedral four-coordination, (f) octahedral six-coordination, (g) cubic eight-coordination, and (h) cubooctahedral 12-coordination.

central metal cation in a coordinated structure increases, the number of anions that can fit around the cation also increases. For example, the radii of the following cations increase in the order B^{3+} (0.023 nm) < P^{5+} (0.035 nm) < Fe^{3+} (0.064 nm) < K^+ (0.133 nm) < Cs^+ (0.167 nm). When in a coordinated structure with O^{2-}, the radius ratios of the cations also increases: B^{3+} (0.164) < P^{5+} (0.250) < Fe^{3+} (0.457) < K^+ (0.950) < Cs^+ (1.19). Thus, the coordination number and coordinated structure possible for each cation are (Table 2.2 and Figure 2.4): B^{3+}, CN = 3 and planar triangular; P^{5+}, CN = 4 and tetrahedral; Fe^{3+}, CN = 6 and octahedral; K^+, CN = 8 and cubic; and Cs^+, CN = 12 and cubooctahedral.

> **The Second Rule (The Electrostatic Valence Principle):** In a stable structure, the valence of each anion, with changed sign, is equal or nearly equal to the sum of the strengths of the electrostatic bonds to it from adjacent cations.

In a stable coordinated structure, total strength of the valence bonds that reach an anion from all the neighboring cations is equal to the charge on the anion. This rule simply states that minerals do not bear charge, they are electronically neutral. Thus, the negative charge of an anion must be exactly neutralized by positive charge from surrounding cations in a stable structure. For example, in silicate structures, Si^{4+} is in tetrahedral (CN = 4) coordination, surrounded by 4 O^{2-} ions. The strength of each ionic Si—O bond is $+\frac{4}{4} = +1$ (where strength = cation valence/CN). This means that each bond that radiates from Si^{4+} in tetrahedral coordination has strength of +1. It follows that an O^{2-} ion in the structure can only be bound to two Si^{4+} ions (Figure 2.5a). In the side view of

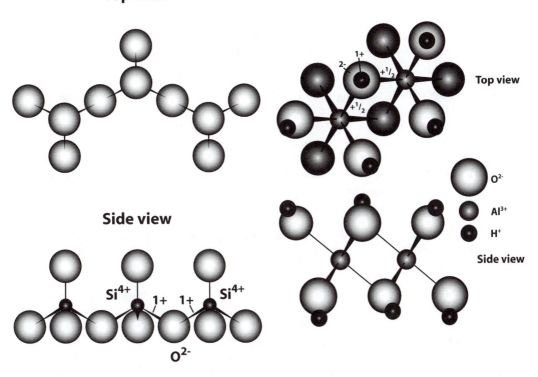

FIGURE 2.5 An illustration of Pauling Rule 2: (a) The linkage of silica tetrahedra through basal oxygen atoms. Each bond radiating from a Si^{4+} atom has strength of +1. Therefore, each O^{2-} anion can only bond to two Si^{4+} atoms. (b) The linkage of aluminum octahedra through hydroxide anions. Each bond radiating from an Al^{3+} atom has strength of $+\frac{1}{2}$. The single negative charge of the hydroxide anion can be satisfied by two Al^{3+} atoms (alternatively, the –2 charge of the hydroxyl oxygen is satisfied by the +1 of a proton and the two Al^{3+} atoms).

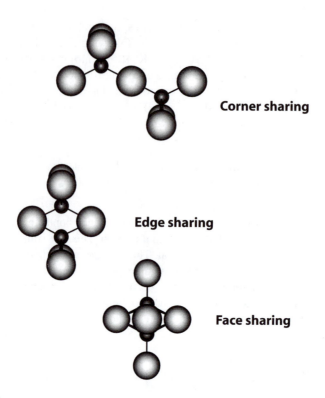

FIGURE 2.6 Corner, edge, and face sharing tetrahedral polyhedron linkages. Note that the central metal cations move closer together as the number of shared elements increases.

two linked Si tetrahedra, the 2– charge of a shared corner (basal) oxygen is exactly neutralized by the two 1+ strength bonds radiating from the two surrounding Si cations. It is also important to recognize that this basal oxygen cannot be bound to any other cation. To do so would violate this Pauling rule. In the gibbsite $[Al(OH)_3]$ structure (Figure 2.5b), Al^{3+} is in octahedral (CN = 6) coordination (surrounded by 6 OH^- ions). The strength of the ionic Al—OH bond is $+\frac{3}{6} = +\frac{1}{2}$ (s = cation valence/CN). Each bond that radiates from Al^{3+} has a strength of $+\frac{1}{2}$. It then follows that an OH^- ion in the aluminum octahedron can only be bound by two Al^{3+} ions. Correspondingly, the 2– charge of a shared basal oxygen of an Al-octahedra is exactly neutralized by the two $+\frac{1}{2}$ strength bonds radiating from two Al^{3+} cations, plus the +1 bond from the hydrogen atom.

> **The Third Rule:** The presence of shared edges and especially shared faces in a coordinated structure decreases its stability; the effect is large for cations with large valence and small coordination number (ligancy).

The closer the positive-charge centers (cations) get to one another, the less stable the mineral structure becomes. Structures that contain polyhedrons that share a corner anion are more stable than structures that contain polyhedrons that share two edge anions (Figure 2.6). The least stable structure arises when polyhedrons share a face (share three or more anions). This rule explains why linked silica tetrahedra, which only share corners, are stable. The highly charged Si^{4+} cation is surrounded by only four O^{2-} anions (CN = 4), with a bond strength of $s = +1$. However, aluminum octahedra, Al^{3+} surrounded by six O^{2-} anions (CN = 6, s = +0.5), form stable linked structures by sharing polyhedron edges.

> **The Fourth Rule:** In a crystal containing different cations, those with large valence and small coordination number tend not to share polyhedron elements with each other.

This rule is similar to rule 3, in that structural cations require separation and shielding. In a stable structure, the electrostatic interaction of high-valence cations is minimized through the shielding provided by the coordinating anions. A structure containing only small (low coordination number), highly charged cations (high charge density) that share anions will be less stable than a structure in which the high valence cation polyhedra are linked through polyhedra containing low-valence cations with higher coordination numbers. An expression of this rule can be found in phosphorus-bearing minerals and compounds. In nature, inorganic phosphorus occurs in the ortho-phosphate minerals. In these minerals, the highly charged P^{5+} cation is in tetrahedral configuration with O^{2-}, forming the orthophosphate polyhedron (PO_4^{3-}). The phosphate polyhedra are linked through mono-, di-, and trivalent cations in octahedral or higher coordination. Example minerals include the calcium phosphates [brushite, $CaHPO_4{\bullet}2H_2O$; octocalcium phosphate, $Ca_8H_2(PO_4)_6{\bullet}5H_2O$; and apatite, $Ca_5(PO_4)_3(OH, F)$], strengite [$Fe^{III}PO_4{\bullet}2H_2O$], and variscite [$AlPO_4{\bullet}2H_2O$]. Polyphosphate compounds, or polymerized orthophosphates (polymerized by sharing corner O^{2-} anions in a manner similar to that illustrated for silica tetrahedra in Figure 2.5), such as pyrophosphates ($P_2O_7^{4-}$) and triphosphates ($P_3O_{10}^{5-}$), do not form in nature. However, these compounds can be produced through various industrial processes (such as in the ignition of orthophosphates). Indeed, the polyphosphates are common to phosphate fertilizers, but rapidly decompose to orthophosphates when incorporated into the soil environment.

The Fifth Rule (The Principle of Parsimony): The number of essentially different kinds of atoms or coordinated polyhedron in a crystal tends to be small.

This rule is a natural consequence of the first four rules. The chemical composition of minerals is generally not (very) complex. Pauling Rule 1 states that the radius ratio dictates the presence of a cation in a given polyhedral configuration. It is also evident from Table 2.3 that several different cations can reside in any given polyhedral configuration. For example, Mn^{2+}, Fe^{2+}, V^{3+}, Li^+, Ti^{4+}, Mg^{2+}, Fe^{3+}, and Cr^{6+} are predicted and found to occur in sixfold coordination when oxygen is the coordinating anion. Yet, the Principle of Parsimony states that a stable structure will contain only a small number of essentially different kinds of atoms. There are two readily apparent reasons that support the validity of this rule. First, cations do vary in size, and although several different cations can reside in a particular coordination, generally only one particular cation fits best in the structure (is best suited for the location). Other cations are not as well suited to the location, causing stress and instability. Second, a +4 or a +6 cation residing in a location normally occupied by a +2 cation will result in electrostatic imbalances, destabilizing the structure.

2.3 CRYSTAL STRUCTURE

All crystalline minerals consist of a three-dimensional (x-, y-, and z-dimensions), repeating array of atoms. A three-dimensional group of atoms in a crystal that can be isolated, such that these atoms represent the smallest group of atoms that can be translocated throughout the crystal is the unit cell. A unit cell is the smallest repeating three-dimensional array of a crystal and is defined by the parameters a, b, and c (the x-, y-, and z-dimensions of the unit cell) and the angles α, β, and γ where α is the angle between the b and c axes; β is the angle between the a and c axes; and γ is the angle between the a and b axes. The imaginary translocation of any arbitrarily selected unit cell of atoms a distance of a (or some multiple of a) along the x-axis of the crystal will exactly overlay an identical group of atoms (the atoms will be superimposed). Similarly, translocation of the unit cell at multiples of b along the y-axis, or of c along the z-axis, or of any combination of a, b, or c distances along the x, y, or z axes will also overlap the identical three-dimensional array of atoms. For example, halite, which is the mineral form of NaCl, has a cubic structure: $a = b = c = 0.56402$ nm and $\alpha = \beta = \gamma = 90°$ (Figure 2.7). The chemical composition of the halite unit cell, termed the unit cell formula, is Na_4Cl_4.

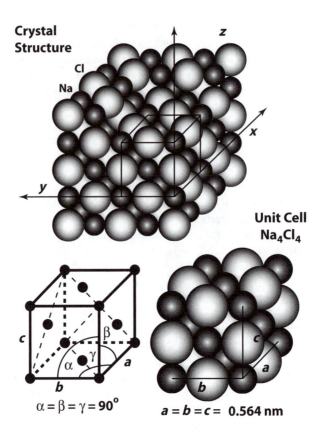

FIGURE 2.7 The halite (NaCl) lattice and unit cell.

In addition to the cubic structure illustrated for halite, there are several other structures (or crystal systems) found in mineralogy. However, in mineralogy, crystal systems are not described by the locations of atoms in the structure, or even the chemical composition of the structure. For clarity in visualizing the structural pattern embodied in a three-dimensional structure, it is convenient to replace the structural pattern (the atoms or groups of atoms within the structure) with a set of points, such that each point represents an atom or group of atoms that has identical surroundings. In so doing, a point lattice (or space lattice) is created, with each lattice point representing the complete chemical pattern of the crystal. Thus, a lattice point in the halite crystal contains one Na^+ ion and $\frac{1}{6}$ of each of the six Cl^- ions in the coordinated polyhedron, because each Cl^- is surrounded and shared by six Na^+ ions (or simply one Na^+ and one Cl^-, the halite chemical formula) (Figure 2.7).

Crystal structure and crystal lattice are not equivalent terms. A crystal structural pattern is real and tangible; it describes the actual physical assemblage of atoms in the structure. A point lattice is conceptual and reflects an infinite and uniform distribution of points in space. It might seem that there could be an endless number of point lattice patterns; however, there are only 14 unique ways in which lattice points can be arranged in three-dimensional space. These are called the Bravais lattices (Figure 2.8). The Bravais lattices are divided into six crystal systems depending on the symmetry of the three-dimension structure. For example, the cubic (or isometric) lattice system has the greatest symmetry ($a = b = c$; $\alpha = \beta = \gamma = 90°$; four threefold axes along the space diagonal), while the triclinic lattice system has no external symmetry ($a \neq b \neq c$; $\alpha \neq \beta \neq \gamma = 90°$). Within each crystal system are the Bravais types (Table 2.4). The Bravais types are identified by P for the primitive lattice; C for a primitive lattice that includes a lattice point centered in the ab plane of the monoclinic system ($a \neq b \neq c$; $\alpha = \gamma = 90°$, $\beta > 90°$); I (from German innenzentriert) for a primitive lattice that includes a lattice point that is centered within the body of the primitive lattice (termed body centered) of the

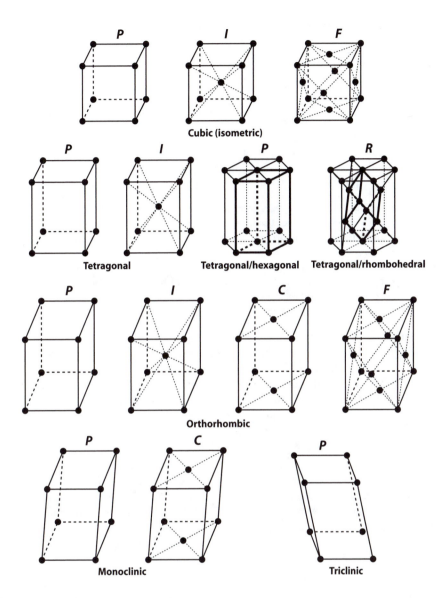

FIGURE 2.8 The 14 Bravais lattice types. The lattice parameters and structural characteristics of each lattice type are described in Table 2.4.

cubic, tetragonal ($a = b \neq c$; $\alpha = \beta = \gamma = 90°$), and orthorhombic ($a \neq b \neq c$; $\alpha = \beta = \gamma = 90°$) systems; F for a primitive lattice that includes lattice points centered on all faces of the primitive lattice of the cubic and orthorhombic systems; and R for the rhombohedral lattice type ($a = b = c$; $\alpha = \beta = \gamma < 120°$) of the hexagonal system ($a = b \neq c$; $\alpha = \beta = 90°$, $\gamma = 120°$). The halite crystal illustrated in Figure 2.7 belongs to the face-centered cubic Bravais lattice system.

2.4 SILICATE CLASSES

Silicate minerals dominate the soil mineral phase. The primary silicates, those silicate minerals formed during the cooling of liquid magma, dominate the sand- and silt-sized fraction of soils. The secondary silicates, those formed through simple or complex weathering reactions involving primary and other secondary silicates, are principally found in the clay fraction and are responsible

TABLE 2.4
Crystal Systems and Bravais Lattice Types

Crystal System[a]	Bravais Type(s)[b]	Dimensional Relationships[c]	Angular Relationships[d]
Cubic (isometric)	P, F, I	$a = b = c$	$\alpha = \beta = \gamma = 90°$
Tetragonal	P, I	$a = b \neq c$	$\alpha = \beta = \gamma = 90°$
Orthorhombic	P, F, I	$a \neq b \neq c$	$\alpha = \beta = \gamma = 90°$
Hexagonal-trigonal	P, R	$a = b \neq c$	$\alpha = \beta = 90°; \gamma = 120°$
Monoclinic	P, C	$a \neq b \neq c$	$\alpha = \gamma = 90°; \beta > 90°$
Triclinic	P	$a \neq b \neq c$	$\alpha \neq \gamma \neq \gamma = 90°$

[a] Structural representations of each crystal system are shown in Figure 2.8.
[b] P, primitive lattice; C, primitive lattice that includes a lattice point centered in the ab plane; I, primitive lattice that includes a lattice point that is centered within the body (body centered); F, primitive lattice that includes lattice points centered on all faces.
[c] a, b, c represent the X, Y, and Z dimensions of the unit cell.
[d] α is the angle between the b and c unit cell axes; β is the angle between the a and c unit cell axes; γ is the angle between the a and b unit cell axes.

for the reactivity of soils (phyllosilicate and other clay minerals). The easiest way in which to view the primary silicate minerals, and to understand their structure and weatherability, is through the mineral classification system.

At the highest and least detailed level of the mineral classification scheme is the *Class* category. The class category employs the anion or anion group as the distinguishing characteristic. In Dana's mineral classification scheme (Gaines et al. 1997), there are nine classes: native elements; sulfides; oxides and hydroxides; halides; carbonates, nitrates, and borates (includes iodates); sulfates, chromates, and molybdates (includes selenates, selenites, tellurates, tellurites, and sulfites); phosphates, arsenates, and vanadates (includes antimonates and tungstates); silicates; and organic materials. The silicates are divided further into the *Silicate Classes* based on the degree of polymerization of the silica tetrahedral unit $[SiO_4^{4-}]$. A silica tetrahedron consists of four O^{2-} ions coordinated to and surrounding one Si^{4+} ion (which is in fourfold coordination, CN = 4). These individual silica units may then be linked to one another by sharing apical and basal O^{2-} ions. The number of oxygen atoms that each tetrahedral shares in a silicate structure is the basis for the classification.

2.4.1 NESOSILICATES

The nesosilicates (also termed orthosilicates) are a class of silicate minerals that are composed of independent silica tetrahedral units that do not share oxygen atoms with neighboring tetrahedral units (Figure 2.9). The basic Si unit is the SiO_4^{4-} polyhedron. The charge per Si unit is −4, and the Si-to-O ratio is 0.25. The individual silica tetrahedra are linked to one another by metal cations, such as Mg^{2+}, Fe^{2+}, Ca^{2+}, and Mn^{2+} in CN = 6 (octahedral) coordination. The olivines [(Mg, $Fe^{II})_2SiO_4$, where Fe^{II} represents the oxidation state of Fe in the mineral] are a group of minerals that typify the nesosilicates. In the olivine formula, note that Mg and Fe^{II} are grouped together by parentheses and separated by a comma. This is common practice. When a mineral formula contains an expression of this nature, it is stating that the atoms within the parentheses reside in the same polyhedron location within the crystal (the position contains either Mg^{2+} or Fe^{2+} in octahedral coordination). The formula is also a generalization; the exact number of Mg^{2+} and Fe^{2+} atoms, considered separately, is not defined. However, the total number of Mg^{2+} and Fe^{2+} atoms is constrained, and in the case of olivine must equal 2. Two minerals in the olivine group are forsterite $[Mg_2SiO_4]$ and fayalite $[Fe_2^{II}SiO_4]$. These two minerals are end-members of a solid solution or

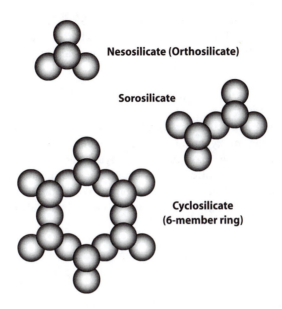

FIGURE 2.9 The basic silica tetrahedral units of the nesosilicate, sorosilicate, and cyclosilicate minerals.

isomorphic series: $[(Mg_x Fe^{II}_{2-x})SiO_4]$, where x varies between 0 and 2, and where Mg^{2+} and Fe^{2+} may freely substitute for one another in the crystal structure:

forsterite end member

Mg_2SiO_4

increasing Mg^{2+} ↑ ↓ increasing Fe^{2+}

$Fe^{II}_2SiO_4$

fayalite end member

The formation of a solid solution is regulated by the Pauling rules (principally rule 1) and results from a process termed isomorphic substitution. By definition, isomorphic substitution is the replacement of one atom by another of similar size in a crystal structure, a process that occurs when the mineral is formed, and also does not change the structure of the crystal. For example, Al^{3+} will substitute for Si^{4+} in tetrahedral coordination, and Mg^{2+}, Fe^{2+}, and Fe^{3+} will substitute for Al^{3+} in octahedral coordination. Isomorphism (a result of isomorphic substitution) is two or more minerals having different chemical compositions but nearly identical structures. Particular compositions of solid solutions may also have specific mineral names. For instance, chrysolite is an olivine with a chemical formula of $Mg_{1.8}Fe^{II}_{0.2}SiO_4$, which may also be written as $0.9[Mg_2SiO_4] \bullet 0.1[Fe^{II}_2SiO_4]$. Chrysolite is 90% forsterite and 10% fayalite and the Mg^{2+} and Fe^{2+} ions are distributed randomly throughout the mineral structure.

The nesosilicates have the smallest Si-to-O ratio among the primary silicates. This means that these minerals have the smallest number of Si—O bonds (which are 50% covalent and provide stability against weathering) relative to the number of base cation–oxygen bonds (which are greater than 70% ionic and less resistant to weathering). As a result, the nesosilicates generally weather rapidly in soils. Other nesosilicates common in the soil environment are the garnets $[(Ca,Mg,Fe^{II},Mn^{II})(Al,Fe^{III},Cr^{III})(SiO_4)_3]$, sphene $[CaTiO(SiO_4)]$ (the TiO^{2+} unit is a divalent cation in the structure), and topaz $[Al_2SiO_4(F,OH)_2]$, where the isomorphic substitution of Al^{3+} for Si^{4+} results in excess positive charge that is neutralized by F^- and OH^-]. An example of a highly stable nesosilicate is zircon [(Zr,Hf)SiO$_4$]. The Zr—O and Hf—O bonds are approximately 33% covalent and Zr and Hf are highly charged cations (4+), compared to the 2+ charge of base cations common to the other nesosilicates. High cation charge and a more compact polyhedron structure yield a stable mineral.

2.4.2 SOROSILICATES

Sorosilicates are composed of double tetrahedral units. The basic Si unit is Si$_2$O$_7^{6-}$, with a charge per Si unit of –3 and a silicon-to-oxygen ratio of 0.286 ($2/7$). In this class of minerals, each individual silica tetrahedral unit shares a single oxygen atom (Figure 2.9). The double tetrahedral units are bound to neighboring units by metal cations in octahedral coordination. Generally, sorosilicates are not common in soils due to their high weatherability. Akermanite [Ca$_2$MgSi$_2$O$_7$] and gehlenite [Ca$_2$Al(AlSi)O$_7$] are example sorosilicates. These two minerals are also end-members of a solid solution:

$$akermanite\ [Ca_2MgSi_2O_7]$$

$$\uparrow\downarrow$$

$$solid\ solution\ [Ca_2(Mg_{1-x}Al_x)(Al_xSi_{2-x})O_7\ \{x < 1\}]$$

$$\downarrow\uparrow$$

$$gehlenite\ [Ca_2Al(AlSi)O_7]$$

In this solid solution, Al^{3+} isomorphically substitutes for Si^{4+} in tetrahedral coordination. The Al^{3+} also acts as a bridging cation, substituting for Mg^{2+} in octahedral coordination. A common soil sorosilicate (highly resistant to weathering) is epidote, [Ca$_2$(Al,Fe^{II})Al$_2$O(SiO$_4$)(Si$_2$O$_7$)(OH)].

2.4.3 CYCLOSILICATES

The cyclosilicates are composed of three- or six-membered silica tetrahedra rings. The basic Si units are Si$_3$O$_9^{6-}$ and Si$_6$O$_{18}^{12-}$. The charge per Si unit is –2 and the silicon-to-oxygen ratio is 0.33 ($1/3$). Each individual silica tetrahedra shares two oxygen atoms (Figure 2.9). Each three- or six-member silicate ring is bound to a neighboring ring by metal cations. Example cyclosilicates are beryl [Be$_3$Al$_2$(SiO$_3$)$_6$] and tourmaline [(Na,Ca)(Li,Mg,Al)(Al,Fe^{II},Mn^{II})$_6$(BO$_3$)$_3$(Si$_6$O$_{18}$)(OH)$_4$]. Note that in tourmaline, isomorphic substitution can be extensive. The Na$^+$ and Ca^{2+} atoms reside in the same coordination, and when summed, must equal 1. The Li$^+$, Mg^{2+}, and Al^{3+} atoms reside in the same coordination and also must sum to 1. Finally, Al^{3+}, Fe^{2+}, and Mn^{2+} atoms reside in the same coordination, but must sum to 6.

2.4.4 INOSILICATES

The inosilicates are the chain silicates. They are divided into two groups: the pyroxenes and the amphiboles. The pyroxenes are composed of single chains of silica tetrahedra linked through basal oxygen atoms (Figure 2.10). The basic Si unit is SiO$_3^{2-}$, the charge per Si unit is –2, and the silicon-to-oxygen ratio is 0.33 ($1/3$). Each silica tetrahedron shares two basal oxygen atoms. Again, the chains are bound to neighboring chains by divalent metal cations in octahedral coordination. Example pyroxenes are augite [Ca(Mg,Fe^{II})Si$_2$O$_6$], enstatite [MgSiO$_3$], hypersthene [(Mg,Fe^{II})SiO$_3$], diopside [CaMgSi$_2$O$_6$], and hedenbergite [CaFe^{II}Si$_2$O$_6$].

The amphiboles are composed of double chains of silica tetrahedra linked through basal oxygen atoms. The basic Si unit is Si$_4$O$_{11}^{6-}$, the charge per Si unit –1.5, and the silicon-to-oxygen ratio is 0.364 ($4/11$). Each silica tetrahedron on the outside of the chain shares two basal oxygen atoms (Figure 2.10).

Pyroxene

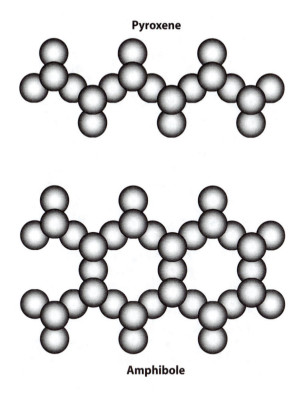

Amphibole

FIGURE 2.10 The basic silica tetrahedral units of the single-chain (pyroxenes) and double-chain (amphiboles) inosilicate minerals.

The silica tetrahedra on the inside of the chain share three basal oxygen atoms. Similar to the pyroxenes, the double chain structures are bound to neighboring chains primarily by divalent metal cations in octahedral coordination. Example amphiboles are hornblende [$NaCa_2Mg_5Fe_2^{II}(AlSi_7)O_{22}(OH)$], tremolite [$Ca_2Mg_5Si_8O_{22}(OH)_2$], and actinolite [$Ca_2Mg_4Fe^{II}Si_8O_{22}(OH)_2$].

2.4.5 PHYLLOSILICATES

The phyllosilicates are also termed the layer or sheet silicates. The basic Si unit is $Si_2O_5^{2-}$, having a charge per Si unit of −1 and a silicon-to-oxygen ratio of 0.4 ($\frac{2}{5}$). Each individual silica tetrahedra shares three basal oxygen atoms, resulting in a sheet structure (Figure 2.11). An example group of the primary phyllosilicates is the micas. The micas are composed of sheets of silica tetrahedron that are held together by cations in octahedral coordination with the apical oxygen atoms of the Si tetrahedra. The configuration that results from this sharing of apical oxygen atoms is a 2:1 layer structural unit (2 Si tetrahedral layers sandwiching an octahedral layer) (Figure 2.12). However, unlike the ideal hexagonal structure illustrated in Figure 2.11 for the silica sheet, the Si-tetrahedra must rotate slightly for the apical oxygen atoms to align with the coordination of the octahedral layer, resulting in a structure that is ditrigonal, rather than hexagonal (Figure 2.13).

The mica minerals are divided into two groups, depending on the occupation of the octahedral layer. The chemical formulae for the mica minerals and other phyllosilicate minerals are commonly expressed as $\frac{1}{2}$-unit cell formulae. The $\frac{1}{2}$-unit cell formula is also employed in the classification of the phyllosilicates. In the octahedral layer, there are three possible positions for a cation to reside in each $\frac{1}{2}$-unit cell. If two of these three positions are occupied by cations, the mineral is dioctahedral.

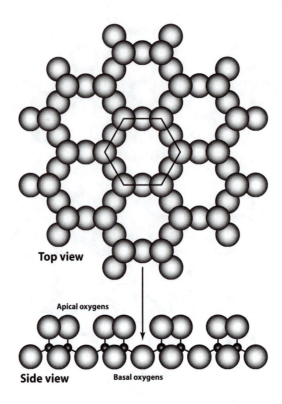

Top view

Apical oxygens

Side view   **Basal oxygens**

FIGURE 2.11 Top and side views of the idealized silica tetrahedral unit (silica sheet) of the phyllosilicate minerals. The arrow correlates a basal oxygen atom in the top and side views.

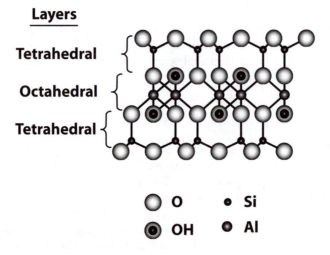

Layers

Tetrahedral

Octahedral

Tetrahedral

○ O   • Si
◉ OH ◉ Al

FIGURE 2.12 A side view of the 2:1 unit structure of a dioctahedral mica mineral. Note that the apical oxygen atoms of the silica (tetrahedral) layer are also coordinating aluminum in the octahedral layer.

Further, in order to maintain charge balance, cations in the octahedral layer of a dioctahedral mineral must be trivalent. An example of a dioctahedral mica mineral is muscovite. The idealized $\frac{1}{2}$-unit cell chemical formula for muscovite is $KAl_2(Si_3Al)O_{10}(OH)_2$. The way in which this chemical formula is written provides key information about the location of the atoms in the muscovite structure (Figure 2.14). The octahedral layers contain two Al^{3+} atoms per $\frac{1}{2}$-unit cell. This is

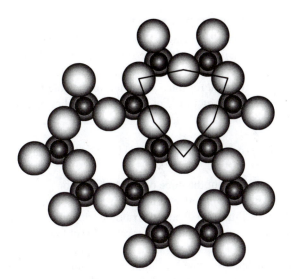

FIGURE 2.13 Top view of the tetrahedral sheet of phyllosilicate minerals illustrating the rotation of the individual silica tetrahedra and the ditrigonal symmetry.

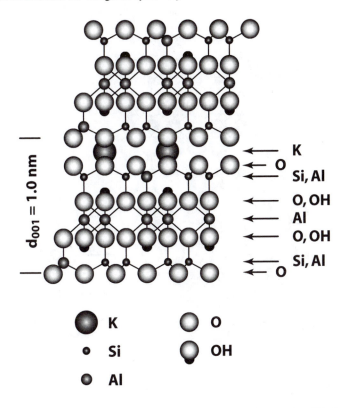

FIGURE 2.14 The structure of the dioctahedral mica muscovite $[KAl_2(Si_3Al)O_{10}(OH)_2]$.

appropriate, as muscovite is dioctahedral and only needs two trivalent cations to satisfy charge of the coordinating O^{2-} and OH^- anions. The tetrahedral layer contains three Si^{4+} atoms and one Al^{3+} atom per $\frac{1}{2}$-unit cell. In this tetrahedral layer, one Al^{3+} atom has isomorphically substituting for one Si^{4+} atom, which results in a charge deficit of -1 per $\frac{1}{2}$-unit cell. Note that the ratio of four tetrahedrally coordinated atoms (three Si^{4+} + one Al^{3+}) to the ten oxygen atoms is $2/5$, which is the silicon-to-oxygen ratio for the phyllosilicates. Since minerals are electrically neutral, additional cation charge is required to satisfy the imbalance created by the substitution of Al^{3+} for Si^{4+}. In this mineral, the charge deficit is balanced by K^+, which is the interlayer cation. From this discussion, we can see that the chemical formula for mica is written to indicate the locations (or coordination numbers) of atoms in the mineral structure. For the 2:1 layer silicates the general form of the $1/2$-unit cell chemical formula is:

$$\{\text{interlayer cation}\}\{\text{octahedral occupation}\}\{\text{tetrahedral occupation}\}O_{10}(OH)_2$$

In the second mica group, three of the three octahedral positions are occupied by cations. These minerals are trioctahedral. Further, in order to maintain charge balance, cations in the octahedral layer of a trioctahedral mineral must be divalent. Example trioctahedral mica minerals are phlogo-pite and biotite. The idealized $1/2$-unit cell chemical formula for phlogopite is $KMg_3(Si_3Al)O_{10}(OH)_2$, and that for biotite is $K(Fe^{II}Mg_2)(Si_3Al)O_{10}(OH)_2$. In both phlogopite and biotite, the octahedral layers contain three divalent cations: three Mg^{2+} atoms per $1/2$-unit cell in phlogopite and two Mg^{2+} and one Fe^{2+} in biotite. The remaining structural components are identical to muscovite: three Si^{4+} and one Al^{3+} atoms per $1/2$-unit cell in the tetrahedral layer and 1 charge-neutralizing interlayer cation (K^+) per $1/2$-unit cell (Figure 2.15).

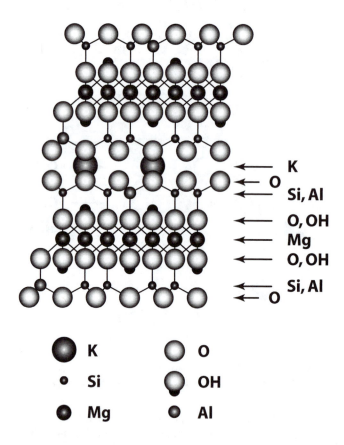

FIGURE 2.15 The structure of the trioctahedral mica phlogopite [$KMg_3(Si_3Al)O_{10}(OH)_2$].

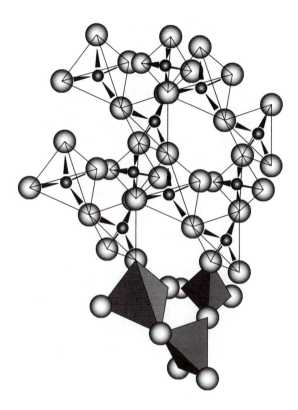

FIGURE 2.16 The tectosilicate framework structure of quartz [SiO_2]. The dark circles represent Si atoms and the light circles are oxygen atoms.

2.4.6 TECTOSILICATES

In the tectosilicate minerals there is complete sharing of the oxygen atoms in the silicate tetrahedra (and as in all previous cases, each shared oxygen can only bond to two Si^{4+} ions). This results in a framework structure (Figure 2.16). The basic silica unit is SiO_2^0. The charge per silica unit is 0, and the silicon-to-oxygen ratio is 0.5 ($\frac{1}{2}$). Of the silicates, the tectosilicate minerals contain the greatest amount of covalent bonding. As a result, they tend to be resistant to weathering and common in soils, particularly in the sand- and silt-sized fractions. A tectosilicate with no isomorphic substitution will have the chemical formula SiO_2. A common mineral with this chemical formula is quartz. Less common polymorphs of quartz include the minerals tridymite and cristobalite. Polymorphism is the condition in which elements crystallize in more than one crystalline form with no change in chemical composition. Quartz, tridymite, and cristobalite have the same chemical formula (SiO_2), but their structures differ. For example, the crystal system of quartz is hexagonal-trigonal ($a = b = 0.4913$ nm, $c = 0.5405$ nm) with a unit cell formula of Si_3O_6. Tridymite has a unit cell composition of $Si_{330}O_{660}$ that is triclinic ($a = 0.9932$ nm, $b = 1.7216$ nm, and $c = 8.1864$ nm; $\alpha = \beta = \gamma = 90°$); while, cristobalite has a unit cell composition of Si_4O_8 that is tetragonal ($a = b = 0.4971$ nm, $c = 0.6918$ nm). There are numerous examples of polymorphism in mineralogy, including vaterite, aragonite, and calcite [$CaCO_3$], and diaspore and boehmite [$AlOOH$].

A group of tectosilicate minerals with Al^{3+} isomorphically substituted for Si^{4+} is the feldspars. If Al^{3+} occupies 1 out of every 4 Si tetrahedral positions, then a -1 charge per unit formula is created in the structure [$AlSi_3O_8^-$], which in turn requires a monovalent cation (Na^+ or K^+) for charge balance. If Al^{3+} occupies 2 out of every 4 Si^{4+} tetrahedral positions, then a -2 charge is created in the structure [$Al_2Si_2O_8^{2-}$], which in turn requires a divalent cation (Ca^{2+}) for charge balance.

These mono- and divalent charge-balancing cations are in cubic coordination (CN = 8) and occupy cavities in the silica framework. There are two main feldspar groups: potassium feldspars and plagioclase (or alkali) feldspars. The potassium feldspars [$KAlSi_3O_8$] include the polymorphs: orthoclase, microcline, and sanidine. The two end members of the discontinuous solid solution that comprise the plagioclase feldspars are albite [$NaAlSi_3O_8$] and anorthite [$CaAl_2Si_2O_8$]. This solid solution is discontinuous because the coupled, free substitution of Ca^{2+} for Na^+ and Al^{3+} for Si^{4+} only occurs over small compositional ranges. Further, albite and anorthite may also exist as discrete phases in a crystal matrix, existing as a mixture rather than a true solid solution. Classically, the albite-anorthite solid solution series is illustrated as:

albite [$NaAlSi_3O_8$] end member

↑

more Na and Si

↑

oligoclase [$(Na_{0.8}Ca_{0.2})(Al_{1.2}Si_{2.8})O_8$], 80% albite and 20% anorthite

andesine [$(Na_{0.6}Ca_{0.4})(Al_{1.4}Si_{2.6})O_8$], 60% albite and 40% anorthite

labradorite [$(Na_{0.4}Ca_{0.6})(Al_{1.6}Si_{2.4})O_8$], 40% albite and 60% anorthite

bytownite [$(Na_{0.2}Ca_{0.8})(Al_{1.8}Si_{2.2})O_8$], 20% albite and 80% anorthite

↓

more Ca and Al

↓

anorthite [$CaAl_2Si_2O_8$] end member

A unique group of tectosilicate minerals is the zeolites. Unlike most other tectosilicate minerals, the zeolites are characterized by a porous structure with cages and large interconnected spaces and channels. The framework of a zeolite mineral consists of linked SiO_4^{4-} and AlO_4^{5-} tetrahedra, such that the (Si + Al) to O ratio is $\frac{1}{2}$ (and the oxygen atoms are completely shared, as required for classification as a tectosilicate mineral). Structurally, the basic unit is [$(Al_xSi_{1-x})O_2$]$^{x-}$. The isomorphic substitution of Al^{3+} for Si^{4+} results in the development of a negative charge on the structural framework that is exactly equal to the number of Al^{3+} ions in the structure. This charge deficit is satisfied (neutralized) by highly hydrated monovalent and divalent cations, typically Na^+, K^+, Ca^{2+}, and Ba^{2+}. Along with water molecules, these cations reside in the vacuoles and channels that are framed by the [$(Al_xSi_{1-x})O_2$]$^{x-}$ tetrahedra. Unlike the feldspars, which also exhibit Al^{3+} substitution for Si^{4+} and the neutralization of the resulting structural charge by Na^+, K^+, and Ca^{2+}, the base cations in zeolite are not fixed in the mineral structure. Indeed, the porous nature of many zeolites allows for the diffusion of charge-satisfying cations and water molecules into and out of the structure.

In general, zeolites are not common, but they are found in soils of volcanic origin, such as tuffaceous soils (developing on pyroclastic rock), in soils subjected to the eolian or fluvial deposition of tuffaceous materials, and in saline-sodic soils. Zeolite minerals that have been identified in sedimentary environments include analcite, chabazite, erionite, heulandite (clinoptilolite), laumonite, mordenite, and phillipsite (Table 2.5). Clinoptilolite, a variety of heulandite, is the most common naturally occurring zeolite. The chemical formula for a zeolite has the form:

$$(N_x^+M_y^{2+})[Al_{(x+2y)}Si_{n-(x+2y)}]O_{2n} \bullet mH_2O$$

TABLE 2.5
Selected Characteristics of Natural Zeolites that Have Been Identified in Sedimentary Environments[a]

Mineral	Structural Formula (Unit Cell)	Void Volume, %	Channel Dimensions, nm	CEC, cmol$_c$ kg^{-1}
Analcite	Na$_{16}$[Al$_{16}$Si$_{32}$]O$_{96}$•16H$_2$O	18	0.16 × 0.42	460
Chabazite	Ca$_6$[Al$_{12}$Si$_{24}$]O$_{72}$•40H$_2$O	47	0.38 × 0.38	420
Erionite	(Ca$_{3.5}$Na$_7$K$_2$)[Al$_9$Si$_{27}$]O$_{72}$•27H$_2$O	35	0.36 × 0.51	320
Heulandite	Ca$_4$[Al$_8$Si$_{28}$]O$_{72}$•24H$_2$O	39	0.31 × 0.75	290
(Clinoptilolite)			0.36 × 0.46	
			0.28 × 0.47	
Laumonite	Ca$_4$[Al$_8$Si$_{16}$]O$_{48}$•16H$_2$O	34	0.40 × 0.53	420
Mordenite	Na$_8$[Al$_8$Si$_{40}$]O$_{96}$•24H$_2$O	28	0.65 × 0.70	220
			0.26 × 0.57	
Phillipsite	(Ca$_2$Na$_4$K$_2$)[Al$_6$Si$_{10}$]O$_{32}$•12H$_2$O	31	0.38 × 0.38	380
			0.30 × 0.43	
			0.32 × 0.33	

[a] Structural formula and channel dimensions were obtained from Baerlocher et al. (2001). Void volume and *CEC* obtained from Ming and Mumpton (1989).

where N$^+$ and M^{2+} are the monovalent and divalent cations that neutralize the charge that is generated on the framework by the isomorphic substitution of Al^{3+} for Si^{4+}. Because these charge-neutralizing cations are highly hydrated, they are not coordinated by the structural oxygen atoms; water molecules reside between the adsorbed cations and the negative surface. Instead of being coordinated, the cations are held by weak electrostatic interactions (Coulombic forces), and as a result they can be displaced by other cations. This process, known as cation exchange, is an exploitable characteristic of the zeolite minerals. The ability of zeolites to retain cations is indicated by the cation exchange capacity (or *CEC*). For the natural zeolites shown in Table 2.5, computed *CEC* values range from 220 cmol$_c$ kg^{-1} for mordenite to 460 cmol$_c$ kg^{-1} for analcite, which compares well to the range of 229 to 568 cmol$_c$ kg^{-1} reported for zeolites as a whole. To put these zeolite *CEC* values in perspective, consider the smectite minerals, such as the montmorillonites, and the vermiculite minerals. These minerals are secondary phyllosilicates and are responsible for most of the reactivity of and solute retention in soil. The *CEC* values of the montmorillonites range between 70 and 120 cmol$_c$ kg^{-1}, while the *CEC* values for the vermiculites range between 130 and 210 cmol$_c$ kg^{-1}. By comparison, zeolite *CEC* values are approximately two to seven times greater than those of the montmorillonites, and up to four times greater than the *CEC* values of the vermiculites.

Another exploitable characteristic of the zeolites is their innate porosity and structural resistance to dehydration. Each zeolite is unique relative to structure, porosity, and channel size. The porosities of the zeolites that are identified in sedimentary environments (Table 2.5) range between 18% (analcite) and 47% (chabazite). The low porosity and pore channel size of analcite are expected, as the cross-sectional area of the 0.16-nm × 0.42-nm channel formed by an 8-tetrahedra ring (Figure 2.17) and the ~0.26 nm diameter closed-hexagonal structure are similar (Figure 2.13). The heulandite structure has a porosity of 39% and contains three different channels (Figure 2.18). Two of the channels run parallel to each other, the 10-tetrahedra 0.75 nm × 0.31 nm and the 8-tetrahedra 0.46 nm × 0.36 nm channels; while the third 0.47 nm × 0.28 nm 8-tetrahedra channel runs perpendicular to the other two. The mordenite structure (Figure 2.19) has a porosity of 28% (10% less than that of heulandite), but contains relatively large diameter

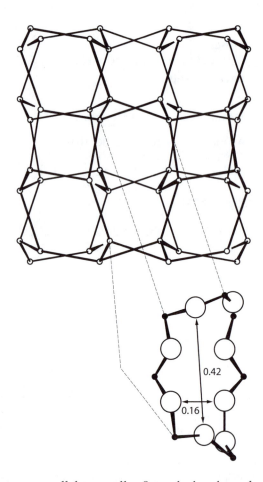

FIGURE 2.17 A ball and stick representation of the zeolite mineral analcite $[Na_{16}(Al_{16}Si_{32})O_{96} \cdot 16H_2O]$. The open circles in the main structure represent Si and Al atoms; while oxygen atoms are located at the midpoint of the connecting lines. The blown-up structure illustrates that dimensions of the channels in analcite (expressed in nanometers), and more clearly indicates the location of the oxygen atoms (open circles) in relation to the structural cations (closed circles).

12-tetrahedra channels (0.70 nm × 0.65 nm) that run parallel to smaller 8-tetrahedra channels (0.57 nm × 0.26 nm).

Because of their high exchange capacities and porosities, and rigid and uniform channel sizes, the zeolite minerals (and synthetic analogs) have many useful applications. In aquaculture and home aquariums, zeolites are used as filtration materials to remove ammonia and other toxins. In agriculture, zeolites are used in odor control in livestock operations (gas absorption) and as livestock feed additives to absorb toxins produced by molds and parasites. Zeolites are employed in slow-release nutrient delivery systems in the horticultural industry. Household products that are used for odor control contain zeolites. In water softeners, calcium and magnesium ions (water hardness) are removed by exchange with sodium ions in sodium-saturated zeolites. Zeolites are also employed in the production of gasoline and in the separation and purification of N_2 and O_2 from air, and other gas separation processes. Because zeolites can be dehydrated without collapsing their structure, they are used as absorbents for oil and other chemical spills. Zeolites are used to contain radioactive waste, in contaminated site remediation, in the removal of heavy metals, ammonia, and other contaminants in water and waste water treatment facilities.

2.5 CLAY MINERALOGY

Secondary minerals are those that are formed in a weathering environment, either through the alteration of a parent (primary) mineral or through the precipitation of soluble species. Both these processes are responsible for the formation of secondary silicates. The secondary silicates of primary interest in soils are the phyllosilicate, or aluminosilicate, minerals. The phyllosilicate minerals are included

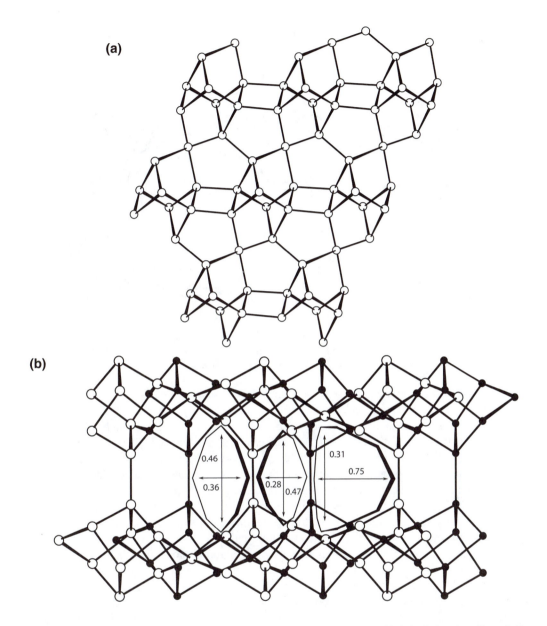

FIGURE 2.18 The heulandite [$Ca_4(Al_8Si_{28})O_{72} \cdot 24H_2O$] class of zeolites, which includes the clinoptilolite variety, are sheet zeolites. (a) Ball and stick representation of the "sheet" structure (top view, and the locations of the Si and Al atoms are represented by open circles). (b) Side view clearly showing the sheet structure, as well as the dimensions (in nm) and locations of the three channel types.

in the group of minerals defined as clay minerals, or clay materials. In the purview of soil mineralogy and chemistry, clay materials are the phyllosilicates and the soil components that are intimately associated with the phyllosilicates: accessory minerals (metal oxides, hydroxides, and oxyhydroxides) and soil organic matter. An alternate definition of clay is based on particle size and is any material that consists of particles <2 μm in size (equivalent spherical diameter). Minerals such as quartz, that are not clay minerals, can be found in the clay-sized fraction of soils, and clay minerals such as kaolinite can be found in the silt-sized fraction. However, clay materials tend to concentrate in the clay-sized fraction of soils, a property that bridges the two distinct definitions.

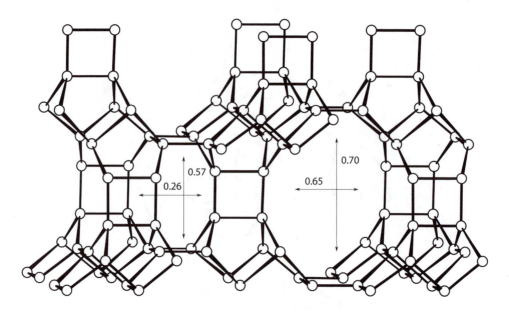

FIGURE 2.19 A ball and stick representation of the zeolite mordenite [$Na_8(Al_8Si_{40})O_{96}\cdot24H_2O$]. The dimensions of the two types of channels in this mineral are in units of nm. Structural Si and Al atoms are represented by open circles, while oxygen atoms are located at the midpoint of the connecting lines.

As with the silicates as a whole, the mineral classification scheme provides a mechanism by which to examine the clay minerals and understand their structure, chemistry, and reactivity. The phyllosilicate classification system used in clay mineralogy departs from that used in Dana's mineral classification scheme. In Dana's system, the phyllosilicates are separated into the hierarchical system of mineral *Types*, *Groups* (and *Subgroups*), and *Species* (Table 2.6). There are three phyllosilicate types, based on the ratio of tetrahedral to octahedral sheets in the mineral, which contain the clay minerals of primary environmental interest. The type categories are 1:1; 2:1; and interlayered 1:1, 2:1, and octahedra.

Within the 1:1 type category, the distinguishing characteristic is the octahedral occupation: dioctahedral (kaolinite group) or trioctahedral (serpentine group). The kaolinite group minerals are divided directly into mineral species, such as kaolinite and halloysite. The serpentine group minerals are divided into subgroups that are differentiated by crystal morphology and structure. For example, the antigorite structure consists of a modulating wave-like structure (alternating waves; Figure 2.20a); whereas, the lizardite and amesite subgroups display planar structure (unmodulated; Figure 2.20b) and the chrysolite subgroup displays cylindrical rolls (Figure 2.20c).

The 2:1 type phyllosilicate minerals are distinguished at the group level by layer charge. The 2:1 minerals that contain little to no isomorphic substitution are found in the pyrophyllite-talc group. The mineral pyrophyllite is dioctahedral and the mineral talc is trioctahedral. There are no subgroup designations to segregate the di- and trioctahedral pyrophyllite-talc minerals. The mica group minerals bear a layer charge that ranges between ~0.6 and 2, and is a result of isomorphic substitution. The muscovite subgroup (layer charge ~1) consists of the dioctahedral mineral species muscovite and paragonite. The biotite subgroup (layer charge ~1) contains the trioctahedral mineral species phlogopite, biotite, and annite. The 2:1 phyllosilicates that bear a layer charge of ~2 are found in the margarite subgroup. This subgroup contains minerals that are both dioctahedral (margarite) and trioctahedral (clintonite). In Dana's classification scheme the hydromica subgroup contains 2:1 minerals that bear a layer charge that can range between ~0.6 and ~0.9. Minerals in this subgroup include hydrobiotite (trioctahedral), illite (dioctahedral), and vermiculite (di- and trioctahedral). The smectite group minerals bear a layer charge of up to ~0.6. The smectite minerals are divided into subgroups based upon the octahedral occupation. The dioctahedral smectites contain the montmorillonite, beidellite, and

TABLE 2.6
The Classification of the Phyllosilicate Minerals According to Dana's System of Mineral Classification

Type	Group	Subgroup (Examples)	Species (Examples)	Structural Formula
1:1	Kaolinite		Dickite	$Al_2Si_2O_5(OH)_4$
			Halloysite	$Al_2Si_2O_5(OH)_4 \cdot 2H_2O$
			Kaolinite	$Al_2Si_2O_5(OH)_4$
			Nacrite	$Al_2Si_2O_5(OH)_4$
	Serpentine		Antigorite	$(Mg,Fe^{II})_3Si_2O_5(OH)_4$
		Lizardite	Lizardite	$Mg_3Si_2O_5(OH)_4$
			Greenalite	$(Fe^{II},Fe^{III})_{2-3}Si_2O_5(OH)_4$
		Amesite	Amesite	$[(Mg,Fe^{II})_2Al](SiAl)O_5(OH)_4$
		Chrysolite	Clinochrysolite	$Mg_3Si_2O_5(OH)_4$
			Orthochrysolite	$Mg_3Si_2O_5(OH)_4$
			Parachrysolite	$Mg_3Si_2O_5(OH)_4$
	Allophane		Allophane	$Al_2O_3 \cdot (SiO_2)_{1.3-2} \cdot 2.5-3H_2O$
			Imogolite	$Al_2SiO_3(OH)_4$
2:1	Pyrophyllite-talc		Pyrophyllite	$Al_2Si_4O_{10}(OH)_2$
			Talc	$Mg_3Si_4O_{10}(OH)_2$
	Mica	Muscovite	Muscovite	$KAl_2(Si_3Al)O_{10}(OH,F)_2$
			Paragonite	$NaAl_2(Si_3Al)O_{10}(OH)_2$
			Glauconite	$(Na,K)(Fe^{III},Al,Mg)_2(Si,Al)_4O_{10}(OH)_2$
		Biotite	Phlogopite	$KMg_3(Si_3Al)O_{10}(OH)_2$
			Biotite	$K(Mg,Fe)_3(Si_3Al)O_{10}(OH,F)_2$
			Annite	$KFe_3^{II}(Si_3Al)O_{10}(OH,F)_2$
			Lepidolite	$K(Li,Al)_3(Si,Al)_4O_{10}(F,OH)_2$
		Margarite	Margarite	$CaAl_2(Si_2Al_2)O_{10}(OH)_2$
			Clintonite	$Ca(Mg,Al)_3(Si_3Al)O_{10}(OH)_2$
		Hydromica	Hydrobiotite	$K(Mg,Fe)_3(Al,Fe)Si_3O_{10}(OH,F)_2 \cdot$ $(Mg,Fe^{II},Al)_3(Si,Al)_4O_{10}(OH)_2 \cdot 4H_2O$
			Illite	$(K,H_3O^+)(Al,Mg,Fe^{II})_2(Si,Al)_4O_{10}[(OH)_2,H_2O]$
			Vermiculite	$(Mg,Fe^{II},Al)_3(Si,Al)_4O_{10}(OH)_2 \cdot 4H_2O$
	Smectite	Dioctahedral smectites	Beidellite	$Na_{0.5}Al_2(Si_{3.5}Al_{0.5})O_{10}(OH)_2 \cdot nH_2O$
			Montmorillonite	$(Na,Ca)_{0.3}(Al,Mg)_2Si_4O_{10}(OH)_2 \cdot nH_2O$
			Nontronite	$Na_{0.3}Fe_2^{III}(Si,Al)_4O_{10}(OH)_2 \cdot nH_2O$
		Trioctahedral smectites	Saponite	$Ca_{0.15}Na_{0.3}(Mg,Fe^{II})_3(Si,Al)_4O_{10}(OH)_2 \cdot 4H_2O$
			Sauconite	$Na_{0.3}Zn_3(Si,Al)_4O_{10}(OH)_2 \cdot 4H_2O$
			Hectorite	$Na_{0.3}(Mg,Li)_3Si_4O_{10}(OH)_2$
Interlayered 1:1, 2:1, and octahedra	Chlorite		Donbassite	$Al_2[Al_{2.33}][Si_3AlO_{10}](OH)_8$
			Cookeite	$LiAl_4(Si_3Al)O_{10}(OH)_8$
			Sudoite	$Mg_2(Al,Fe^{III})_3Si_3AlO_{10}(OH)_8$
			Clinochlore	$(Mg,Fe^{II})_5Al(Si_3Al)O_{10}(OH)_8$
			Chamosite	$(Fe^{III},Mg,Fe^{II})_5Al(Si_3Al)O_{10}(OH,O)_8$
	Regular interstratified		Tosudite	$Na_{0.5}(Al,Mg)_6(Si,Al)_8O_{18}(OH)_{12} \cdot 5H_2O$
			Corrensite	$(Ca,Na,K)(Mg,Fe,Al)_9(Si,Al)_8O_{20}(OH)_{10} \cdot nH_2O$
			Rectorite	$(Ca,Na)Al_4(Si,Al)_8O_{20}(OH)_4 \cdot 2H_2O$

From Gaines, R.V., H.C. Skinner, E.E. Foord, B. Mason, A. Rosenzweig, et al. *Danas New Mineralogy.* J. Wiley & Sons, New York, 1997. With permission.

(a)

(b)

(c)

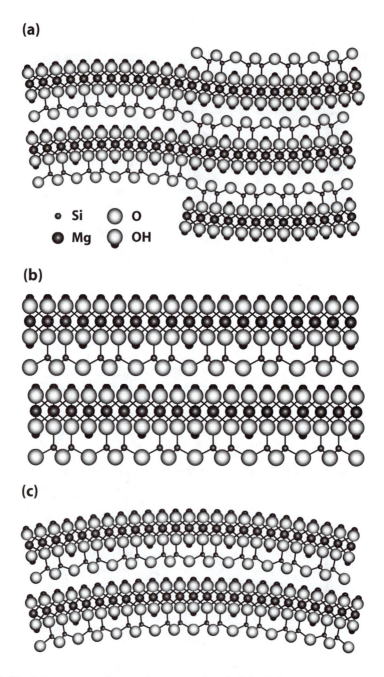

FIGURE 2.20 The 1:1 structures of serpentine group minerals [$Mg_3Si_2O_5(OH)_4$]. (a) The modulating wave-like structure of antigorite; (b) the planar structures of lizardite and amesite; (c) the cylindrical roll structure of chrysolite.

nontronite species, while the trioctahedral smectites contain the saponite and hectorite species. The interlayered 1:1, 2:1, and octahedral-type minerals contain the chlorite group (interlayered octahedral) and the regular interstratified (interlayered 1:1 and 2:1) group minerals. Neither group is differentiated by octahedral occupation, the occupation of the interlayer, or the type of interstratifications.

    Although perhaps difficult to see at this point, Dana's classification scheme is somewhat convoluted with respect to phyllosilicate classification. A utilitarian approach to clay mineral classification, and that approved by the Association Internationale pour l'Etude des Argiles (AIPEA) and the International Mineralogical Association (IMA) Committee on New Minerals and Mineral Names is presented in Table 2.7. In the AIPEA system, the phyllosilicates are separated into the hierarchical system of mineral *Divisions* (or *Types*), *Groups*, *Subgroups*, and *Species*. As in Dana's system, the phyllosilicate minerals of environmental interest are categorized according to layer type (1:1 or 2:1) at the division level. With the exception of kaolinite, illite, and the chlorite group minerals, minerals are partitioned into groups according to layer charge. For the kaolinite minerals, the distinguishing characteristic is the octahedral occupation (used at the subgroup level). Illite minerals bear a large layer charge (0.6 to 0.9per ½ unit cell), but the characteristic that distinguishes the illites from the vermiculites is nonexpansibility. Layer charge is also large in the chlorite group (generally ~1 per ½ unit cell); however, the layer charge is satisfied by an additional and charged octahedral layer rather than by an interlayer cation (archaic classification systems have labeled chlorites as 2:2 or 2:1:1). The degree of octahedral occupation is the distinguishing characteristic at the subgroup level, with the exception of the illite group which is dioctahedral by definition. There are a number of characteristics employed to define a species, including layer stacking, origin of layer charge (the layer—tetrahedral or octahedral—that contains the greatest amount of layer charge), and the type of substituting cation.

2.5.1 DIVISION 1:1 PHYLLOSILICATE MINERALS

The 1:1 phyllosilicate minerals are relatively pure in nature, having very little isomorphic substitution, or if substitution does occur, the layer charge that is generated is neutralized by additional isomorphic substitution elsewhere in the structure. These minerals are identified at the group level as the *kaolinites* (or the *serpentine-kaolin* group). The lack of isomorphic substitution in many of the kaolinite structures imparts distinct physical and chemical characteristics to these minerals. The kaolinites exhibit only a minimal amount of layer charge, as the isomorphic substitution of Si^{4+} by Al^{3+} in the tetrahedral layer, or Al^{3+} by Mg^{2+} or *vice versa* in the octahedral layer, is minor. The radius of a cation, relative to the radius of the oxygen anion, dictates the cation coordination number (Pauling rule 1); however, the substitution of the principal resident cation by a different cation invariably causes tension in the crystal structure, creating weak spots and limiting crystal size. For example, the isomorphic substitution of the larger Al^{3+} ion (0.053 nm) for Si^{4+} (0.042 nm) increases the size of the tetrahedral structure. Similarly, the substitution of relatively large ions, such as Mg^{2+} (0.086 nm), Fe^{2+} (0.075 nm), and Mn^{2+} (0.081 nm), for Al^{3+} increases the size of the octahedral structure. In the absence of isomorphic substitution, the tetrahedral and dioctahedral or trioctahedral layers are mismatched. The tetrahedral layer has larger lateral dimensions than the Al^{3+} dioctahedral layer, but smaller lateral dimensions than the Mg^{2+} trioctahedral layer. In the case of a dioctahedral mineral and in order for the sheets to fit together without excessive structural strain, adjacent silica tetrahedra rotate in opposite directions so that apical oxygen atoms align with the octahedral configuration of the adjacent layer (Figure 2.21). In the case of a trioctahedral mineral, tilting of individual tetrahedra appears to be the process by which the tetrahedral sheet and the larger trioctahedral sheet are brought into alignment. This results in tubular or rolled structures. Further, the curled morphology is generally confined to the pure trioctahedral minerals, as isomorphic substitution increases the lateral extent of the tetrahedral layer, resulting in a platy (planar) morphology (see amesite in Table 2.7 and Figure 2.20b).

The 1:1 dioctahedral phyllosilicates, where Al^{3+} occupies the octahedral layer, are designated at the subgroup as the *kaolinites*, or *kaolins*. The most common species of the kaolinites that is observed in the soil environment is the mineral *kaolinite*. Indeed, kaolinite is probably the most ubiquitous aluminosilicate mineral in soil. The ½-unit cell chemical formula for kaolinite is $Al_2Si_2O_5(OH)_4$. The kaolinite structure is illustrated in Figure 2.22. The 1:1 layers of kaolinite are held to adjacent 1:1 layers by hydrogen bonding. The interlayer hydrogen bond imparts stability to the kaolinite structure, so much so, that the 1:1 layers are not easily separated. The distance between the basal plane of oxygen atoms of one 1:1 layer and that of an adjacent layer is termed

TABLE 2.7
The Clay Mineral Classification Scheme

Division (Layer Type)	Group (x = Layer Charge Per $\frac{1}{2}$ Unit Cell)	Subgroup (Octahedral Occupation)	Species	Ideal Structural Formula
1:1	Kaolinite	Serpentines (trioctahedral)	Antigorite	$Mg_{2.8235}Si_2O_5(OH)_{3.647}$
			Lizardite	$Mg_3Si_2O_5(OH)_4$
			Greenalite	$(Fe^{II},Mg,Mn)_3Si_2O_5(OH)_4$
			Amesite	$[(Mg,Fe^{II})_2Al_1](Si_1Al_1)O_5(OH)_4$
			Clinochrysolite	$Mg_3Si_2O_5(OH)_4$
			Orthochrysolite	$Mg_3Si_2O_5(OH)_4$
			Parachrysolite	$Mg_3Si_2O_5(OH)_4$
		Kaolinites (dioctahedral)	Dickite	$Al_2Si_2O_5(OH)_4$
			Halloysite	$Al_2Si_2O_5(OH)_4 \cdot 2H_2O$
			Kaolinite	$Al_2Si_2O_5(OH)_4$
			Nacrite	$Al_2Si_2O_5(OH)_4$
2:1	Pyrophyllite ($x \sim 0$)	Talcs (trioctahedral)	Talc	$Mg_3Si_4O_{10}(OH)_2$
		Pyrophyllites (dioctahedral)	Pyrophyllite	$Al_2Si_4O_{10}(OH)_2$
	Smectite ($0.2 < x < 0.6$)	Trioctahedral smectites	Saponite	$Na_{0.4}Mg_3(Si_{3.6}Al_{0.4})O_{10}(OH)_2$
			Sauconite	$Na_{0.4}Zn_3(Si_{3.6}Al_{0.4})O_{10}(OH)_2$
			Hectorite	$Na_{0.4}(Mg_{2.6}Li_{0.4})Si_4O_{10}(OH)_2$
		Dioctahedral smectites	Beidellite	$Na_{0.4}Al_2(Si_{3.6}Al_{0.4})O_{10}(OH)_2$
			Montmorillonite	$Na_{0.4}(Al_{1.6}Mg_{0.4})Si_4O_{10}(OH)_2$
			Nontronite	$Na_{0.4}Fe_2^{II}(Si_{3.6}Al_{0.4})O_{10}(OH)_2$
	Vermiculite ($0.6 < x < 0.9$)	Trioctahedral vermiculites	Trioctahedral vermiculite	$K_{0.8}(Mg_{2.5}Fe_{0.5}^{III})(Si_{2.7}Al_{1.3})O_{10}(OH)_2$
		Dioctahedral vermiculites	Dioctahedral vermiculite	$K_{0.8}Al_2(Si_{3.2}Al_{0.8})O_{10}(OH)_2$
	Illite ($0.6 < x < 0.9$)	Illites (dioctahedral and non-expansive)	Illite	$K_{0.8}(Al_{1.8}Mg_{0.2})(Si_{3.4}Al_{0.6})O_{10}(OH)_2$
			Glauconite	$K_{0.8}(Fe_{1.0}^{III}Al_{0.4}Mg_{0.6})(Si_{3.8}Al_{0.2})O_{10}(OH)_2$
	Mica ($x \sim 1$)	Trioctahedral micas	Phlogopite	$KMg_3(Si_3Al)O_{10}(OH)_2$
			Biotite	$K(Fe_{1.5}^{II}Mg_{1.5})(Si_3Al)O_{10}(OH)_2$
			Annite	$K(Fe_{2.5}^{II}Mg_{0.5})(Si_3Al)O_{10}(OH)_2$
			Lepidolite	$K(Li_2Al)Si_4O_{10}(OH)_2$
		Dioctahedral micas	Muscovite	$KAl_2(Si_3Al)O_{10}(OH)_2$
			Paragonite	$NaAl_2(Si_3Al)O_{10}(OH)_2$
	Brittle Mica ($x \sim 2$)	Trioctahedral brittle micas	Clintonite	$Ca(Mg_2Al)(SiAl_3)O_{10}(OH)_2$
		Dioctahedral brittle micas	Margarite	$CaAl_2(Si_2Al_2)O_{10}(OH)_2$
	Chlorite (x is variable)	Trioctahedral chlorites (tri, trioctahedral chlorites)	Clinochlore	$(Mg_2Al)(OH)_6 \cdot Mg_3(Si_3Al)O_{10}(OH)_2$
			Ripidolite	$(Mg_2Al_{0.3}Fe_{0.7}^{III})(OH)_6 \cdot (Mg_{0.2}Fe_{2.8}^{2+})(Si_3Al)O_{10}(OH)_2$
		Di, trioctahedral chlorites	Cookeite	$(LiAl_2)(OH)_6 \cdot Al_2(Si_3Al)O_{10}(OH)_2$
		Dioctahedral chlorites (di, dioctahedral chlorites)	Donbassite	$Al_{2.27}(OH)_6 \cdot Al_2(Si_{3.2}Al_{0.8})O_{10}(OH)_2$

From Bailey, S.W. Structures of layer silicates. In *Crystal Structures of Clay Minerals and Their X-ray Identification.* G.W. Brindly and G. Brown (Eds.) Monogr. No. 5, Mineralogical Society, London, 1980, pp. 1–124. With permission.

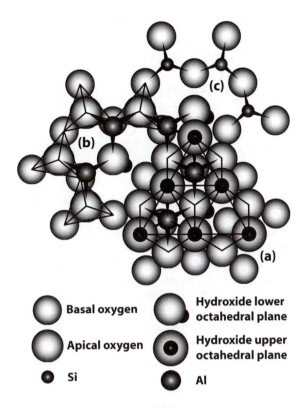

FIGURE 2.21 Representation of the spatial relationship between the Al octahedral layer (upper layer) and the distorted (ditrigonal) Si tetrahedral layer (lower layer) that occurs in the phyllosilicate minerals. (a) All oxygen atoms displayed; (b) upper hydroxyl sheet removed; (c) all octahedrally coordinated anions removed.

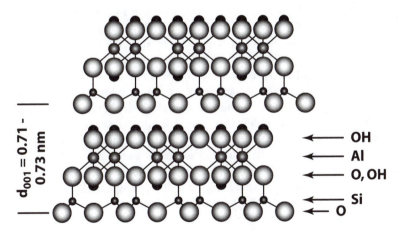

FIGURE 2.22 The 1:1 structure of kaolinite [$Al_2Si_2O_5(OH)_4$].

the d-spacing (also the *c*-dimension of the kaolinite unit cell). The d-spacing of kaolinite is 0.71 to 0.73 nm (7.1 to 7.3 Å) and does not vary, irrespective of environmental conditions (e.g., soil moisture, type and concentration of ions present in the soil solution). Because adjacent 1:1 kaolinite layers do not separate, the specific surface of kaolinite is the external extent of the particle, and ranges between 10 and 20 m^2 g^{-1}. Kaolinite also displays relatively low surface reactivity, as measured by the *CEC*. Although relatively pure, there is sufficient isomorphic substitution in kaolinite to generate a very small layer charge. Further, the mineral edges can bear charge through protonation and deprotonation reactions (discussed in Chapter 7). The two sources of charge result in a kaolinite *CEC* that ranges between 1 and 10 $cmol_c$ kg^{-1}. Deviations from the ideal structural composition of kaolinite that occur in natural samples are illustrated in the reference kaolinite (KGa-1). The natural kaolinite, from the Source Clay Repository of the Clay Mineral Society, has the ½-unit cell formula

$$(Mg_{0.01}Ca_{.0025}Na_{0.005}K_{0.005})(Al_{1.93}Fe^{3+}_{0.01}Ti_{0.055})(Si_{1.915}Al_{0.085})O_5(OH)_4$$

There is an octahedral charge in KGa-1 of +0.055 from the isomorphic substitution of Ti^{4+} for Al^{3+}, and a tetrahedral charge of −0.085 from the isomorphic substitution of Al^{3+} for Si^{4+}. The total layer charge of this mineral is −0.03, which is satisfied by base cations. The measured *CEC* of the KGa-1 kaolinite is 2.0 $cmol_c$ kg^{-1}, and the external specific surface is 10.05 m^2 g^{-1}. Morphologically, kaolinite is also an anomaly relative to other 1:1 minerals. Kaolinite generally lacks the weak spots created through isomorphic substitution and can be observed in relatively large platy crystals (rather than tubular) in the coarse clay and silt size fraction of soils.

Other minerals of the kaolinite subgroup differ from the mineral kaolinite in the manner in which the layers are stacked. If the structure depicted in Figure 2.22 is rotated 90° to obtain a side view, layer stacking differences can be seen. For example, the two structures depicted in Figure 2.23 are polytypes representing two different mineral species of the kaolinite subgroup. Polytypes are special types of polymorphs that differ in the stacking arrangement of layered structures. In Figure 2.23a the adjacent octahedral layers have the same slant (to the left). In this structure, the repeat distance of the layer structure is approximately 0.7 nm (one 1:1 layer). Kaolinite is an example of this type of structure. In the Figure 2.23b, the slant of the adjacent octahedral layers alternates, and the repeat distance of the layer structure is approximately 1.4 nm (two 1:1 layer structures). The polytypes of kaolinite, *dickite* and *nacrite* are examples of double 1:1 layer structures.

Although less common in soils relative to kaolinite, the mineral *halloysite* is also a 1:1, dioctahedral phyllosilicate. Halloysite differs from kaolinite in that the structure contains H_2O molecules in the interlayer (between the 1:1 layers, Figure 2.24). The water molecules hold the 1:1 layers to adjacent 1:1 layers by hydrogen bonding. The d-spacing of the hydrated halloysite is 1.025 nm, owing to the added dimension of the water molecules. The progressive removal of interlayer water molecules will decrease the d-spacing. Water molecules can be completely removed by heat (~60°C) or partially removed by desiccation, collapsing the halloysite structure to that of kaolinite (d-spacing of approximately 0.7 nm). The ½-unit cell chemical formula for halloysite is $Al_2Si_2O_5(OH)_4 \cdot 2H_2O$. Halloysite morphology is varied and quite distinctive. Unlike kaolinite, which is commonly planar in appearance, halloysite is observed in tubular or rolled structures. The tubular morphology is a mechanism to relieve structural strains that arise as a result of a misfit between octahedral and tetrahedral sheets. The morphological appearance of halloysite is somewhat similar to that observed for trioctahedral chrysolite (Figure 2.20c), except the halloysite structure curls in the opposite direction of chrysolite, owing to the smaller dimensions of the Al^{3+} octahedron (relative to both the Mg^{2+} octahedron and the Si^{4+} tetrahedron).

The trioctahedral kaolinites are the *serpentines*. In this subgroup, all three octahedral positions are occupied by the divalent Mg^{2+} ion (Figure 2.20). The ideal chemical formula (½-unit cell) is $Mg_3Si_2O_5(OH)_4$. Polytypes of the serpentines include the minerals *clinochrysolite* (asbestos), *lizardite*, and *antigorite*. The serpentines are unstable in a weathering environment and do not form in soil. As a result, they are rarely observed in the soil. Structurally, the serpentines are similar to the dioctahedral kaolinites, including an approximate 0.71- to 0.73-nm d-spacing and hydrogen bonding in the interlayers. The morphological characteristics of the serpentines are distinctly different from that of the kaolinite subgroup minerals (with the exception of halloysite), as curled morphologies are quite common for this subgroup (Figure 2.20c).

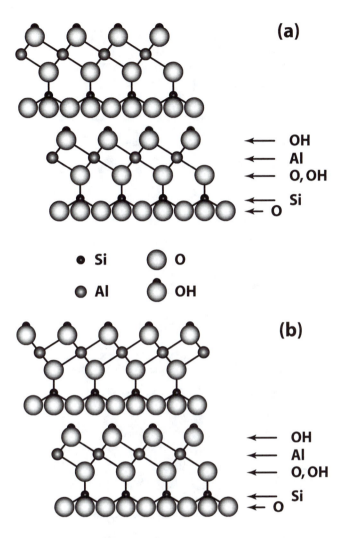

FIGURE 2.23 Layer stacking in 1:1 layer silicates. (a) Adjacent octahedral layers having the same slant; (b) alternating slant of adjacent octahedral layers.

2.5.2 DIVISION 2:1 PHYLLOSILICATE MINERALS

Unlike the kaolinites, the division 2:1 later silicates display considerable variability in the extent of layer charge created by isomorphic substitution. The 2:1 minerals are differentiated at the group-level by the amount layer charge present in the structure, although there are exceptions.

2.5.2.1 PYROPHYLLITE GROUP

The 2:1 phyllosilicates that possess negligible layer charge (layer charge ~ 0) reside in the *pyrophyllite* group (or the *pyrophyllite-talc* group). The ideal pyrophyllite mineral contains no layer charge and there are no interlayer cations present in the structure. The neutral 2:1 layers are held to adjacent 2:1 layers by van der Waals bonding, which is supplemented to a small degree by ionic interactions between the layers. In nature, the pyrophyllite minerals exhibit a small amount of substitution in both the tetrahedral and octahedral layers, providing additional attractive forces between adjacent layers. The d-spacing of the pyrophyllites ranges between 0.91 and 0.94 nm. The dioctahedral pyrophyllites are given the subgroup designation *pyrophyllite*. This subgroup contains the *pyrophyllite* species. The ideal

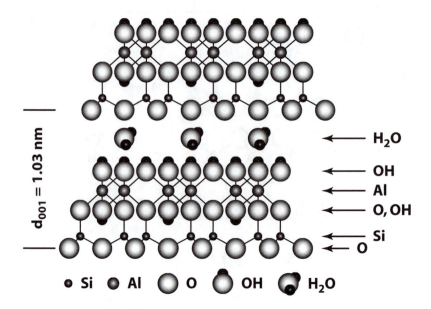

FIGURE 2.24 The structure of halloysite [$Al_2Si_2O_5(OH)_4 \cdot 2H_2O$].

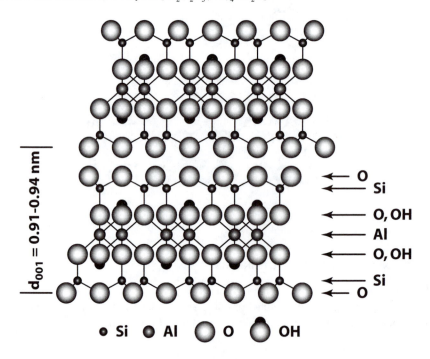

FIGURE 2.25 The structure of pyrophyllite [$Al_2Si_4O_{10}(OH)_2$].

½-unit cell formula for the mineral pyrophyllite is $Al_2Si_4O_{10}(OH)_2$, and the structure is represented in Figure 2.25. As was noted for the 1:1 structures, the octahedral layers of 2:1 structures also stack in differing fashions, yielding different polytypes. The structure diagrammed in Figure 2.26 can be observed if the structure in Figure 2.25 is rotated about the *c*-axis by 90°. Note that in the rotated structure, adjacent octahedral layers have the same slant (to the left). The trioctahedral pyrophyllites are the *talc* minerals. This subgroup contains the *talc* species. The ideal ½-unit cell formula for the mineral talc is $Mg_3Si_4O_{10}(OH)_2$, and the structure is represented in Figure 2.27.

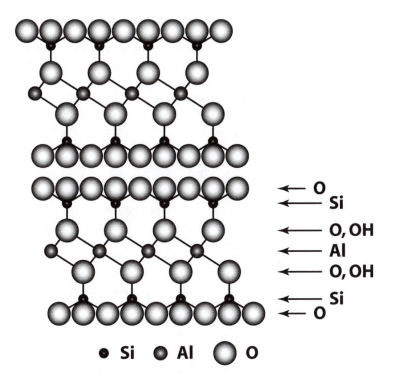

FIGURE 2.26 Layer stacking in a pyrophyllite mineral with adjacent octahedral layers having the same slant.

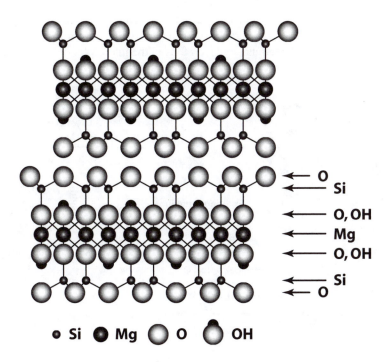

FIGURE 2.27 The structure of talc [$Mg_3Si_4O_{10}(OH)_2$].

2.5.2.2 SMECTITE GROUP

The *smectite* group contains the 2:1 phyllosilicates that have a layer charge between 0.2 and 0.6 per ½-unit cell. The layer charge in the smectites generally originates from both the tetrahedral and the octahedral layers, with one layer providing the bulk of the layer charge. The 2:1 layers are held to adjacent 2:1 layers by ionic interactions between the negatively charged layers and interlayer cations (which reside between the 2:1 layers). Because the layer charge of the smectites is relatively low, the internal surfaces (the region between the 2:1 layers) are accessible to water molecules, and to ions and molecules present in the aqueous phase. The charge-satisfying, interlayer cations remain hydrated, are not tightly held in the smectite structure, and can be displaced; they are exchangeable. The d-spacing of the smectite minerals is a function of the interlayer cation type: 1.2 nm when the mineral interlayers are K^+ or Na^+ saturated; 1.4 to 1.6 nm when Mg^{2+} or Ca^{2+} saturated; and 1.7 to 1.8 nm when Mg^{2+}-glycol or Mg^{2+}-glycerol saturated. The smectite d-spacing can be collapsed to 1.0 nm when K^+ saturated and dehydrated. Conversely, the structure will expand to very large d-spacings (>1.8 nm), the value dependent upon the size of the intercalating substance.

At the subgroup level, the dioctahedral smectites are termed *dioctahedral smectites*. This subgroup contains numerous species. The principal location of the layer charge and the chemical characteristics of the tetrahedral and octahedral layers are the primary characteristics used to classify the smectites at the species level. *Montmorillonite* is a dioctahedral smectite in which the layer charge deficit occurs principally in the octahedral layer (from the isomorphic substitution of Mg^{2+} for Al^{3+}). An ideal ½-unit cell formula for the mineral montmorillonite is $Na_{0.4}(Al_{1.6}Mg_{0.4})Si_4O_{10}(OH)_2$. Note that the substitution of 0.4 Mg^{2+} atoms for Al^{3+} atoms in the octahedral layer results in a layer charge of 0.4 (which is in the 0.2 to 0.6 range). This resulting layer charge is neutralized by 0.4 Na^+ cations that reside in the interlayer. The dioctahedral nature of this mineral is also evident: the sum of Al^{3+} plus Mg^{2+} atoms in the octahedral layer is 2. A representative structure of montmorillonite is illustrated in Figure 2.28.

Beidellite is a dioctahedral smectite in which the layer charge deficit occurs principally in the tetrahedral layer (Figure 2.28). An ideal ½-unit cell formula for the mineral beidellite is $Na_{0.4}Al_2(Si_{3.6}Al_{0.4})O_{10}(OH)_2$. In this mineral, a 0.4 layer charge is created through the isomorphic substitution of Al^{3+} for Si^{4+} in the tetrahedral layer. Ions in the interlayers of the beidellite structure are held more tightly than those in montmorillonite because the location of the negative charge in

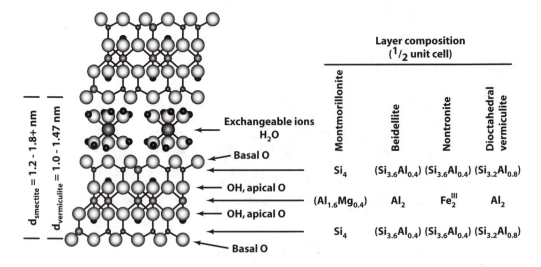

FIGURE 2.28 The structural and chemical characteristics of the expandable 2:1 dioctahedral phyllosilicates montmorillonite, beidellite, nontronite, and dioctahedral vermiculite.

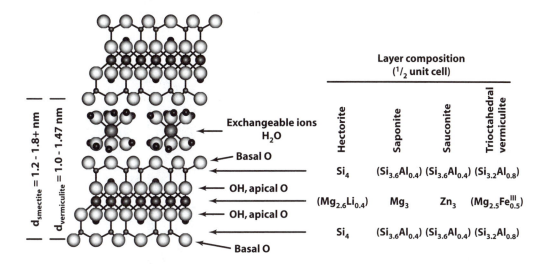

FIGURE 2.29 The structural and chemical characteristics of the expandable 2:1 trioctahedral phyllosilicates hectorite, saponite, sauconite, and trioctahedral vermiculite.

beidellite is closer to the exchangeable cation. Another dioctahedral smectite that is structurally similar to beidellite is *nontronite*. The charge deficit in nontronite occurs principally in the tetrahedral layer, and Fe^{3+} almost completely substitutes for Al^{3+} in the octahedral layer (Figure 2.28). An ideal $\frac{1}{2}$-unit cell formula for the mineral nontronite is $Na_{0.4}Fe_2^{III}(Si_{3.6}Al_{0.4})O_{10}(OH)_2$.

At the subgroup level, the trioctahedral smectites are termed *trioctahedral smectites*. The trioctahedral smectites are much less common in the soil environment than their dioctahedral counterparts, due to the instability of the trioctahedral layer, relative to the dioctahedral layer, in a weathering environment. The mineral *hectorite* is a trioctahedral smectite with a layer charge deficit of 0.2 to 0.6 per $\frac{1}{2}$-unit cell occurring in the octahedral layer (analogous to montmorillonite). The layer charge deficit in hectorite is created through the isomorphic substitution of the Li^+ ion for the Mg^{2+} ion in the octahedral layer (Figure 2.29). The ideal $\frac{1}{2}$-unit cell formula for hectorite is $Na_{0.4}(Mg_{2.6}Li_{0.4})Si_4O_{10}(OH)_2$. Note that the substitution of 0.4 Li atoms for Mg^{2+} in the octahedral layer results in a layer charge of 0.4 (which is in the 0.2 to 0.6 range). This resulting layer charge is neutralized by 0.4 Na^+ cations that reside in the interlayer. The trioctahedral nature of this mineral is also evident: the sum of Mg^{2+} plus Li^+ atoms in the octahedral layer is 3. *Saponite* is a trioctahedral smectite in which the layer charge deficit occurs principally in the tetrahedral layer, analogous to beidellite (Figure 2.29). An ideal $\frac{1}{2}$-unit cell formula for the mineral saponite is $Na_{0.4}Mg_3(Si_{3.6}Al_{0.4})O_{10}(OH)_2$. In this mineral, a 0.4-layer charge is created through the isomorphic substitution of Al^{3+} for Si^{4+} in the tetrahedral layer.

Unlike the 1:1 phyllosilicates, which are relatively pure, the smectites are chemically complex. The ideal chemical formulae introduced above for the smectite minerals oversimplify the chemical composition of the smectites as they occur in the environment. The chemical characteristics of selected reference smectite minerals obtained from natural deposits are illustrated in Table 2.8. Isomorphic substitution in both the tetrahedral and octahedral layers results in somewhat complex structural formulae. Further, incomplete occupation of the tetrahedral (less than 4 atoms per $\frac{1}{2}$-unit cell) and octahedral layers (less than 2 or 3 atoms per $\frac{1}{2}$-unit cell) may also occur. The interlayer positions also contain several different cations. Consider the STx-1 montmorillonite. The octahedral layer is composed of $(Al_{1.205}Mg_{0.355}Fe_{0.045}^{III}Ti_{0.015})$. The total number of cations that occupy this dioctahedral layer is $(1.205 + 0.355 + 0.045 + 0.015)$, or 1.62, a value that is less than the ideal occupation of 2 atoms per $\frac{1}{2}$-unit cell. In addition, the layer charge in the octahedral layer

TABLE 2.8
The Structural and Chemical Characteristics of Naturally Occurring, Reference Smectite Minerals[a]

Mineral	½-Unit Cell Formula	S	CEC
		$m^2\,g^{-1}$	$cmol_c\,kg^{-1}$
SAz-1, 2 (Cheto) montmorillonite	$(Ca_{0.195}Na_{0.18}K_{0.01})(Al_{1.355}Mg_{0.555}Fe^{III}_{0.06}Mn_{0.005}Ti_{0.015})(Si_4)O_{10}(OH)_2$	97.42	123
STx-1 (Texas) montmorillonite	$(Ca_{0.135}Na_{0.02}K_{0.005})(Al_{1.205}Mg_{0.355}Fe^{III}_{0.045}Ti_{0.015})(Si_4)O_{10}(OH)_2$	83.79	89
SWy-1, 2 (Wyoming) montmorillonite	$(Ca_{0.06}Na_{0.16}K_{0.025})(Al_{1.505}Mg_{0.27}Fe^{III}_{0.205}Mn_{0.005}Ti_{0.01})(Si_{3.99}Al_{0.01})O_{10}(OH)_2$	31.82	85
SBCa-1 beidellite	$(Mg_{0.15}Ca_{0.08}K_{0.075})(Al_{1.91}Fe^{III}_{0.09}Ti_{0.03})(Si_{3.4}Al_{0.6})O_{10}(OH)_2$	—	—
NG-1 nontronite	$(Mg_{0.075}Ca_{0.16}Na_{0.005}K_{0.015})(Fe^{III}_{1.875}Al_{0.085}Mg_{0.04}Ti_{0.005})(Si_{3.45}Al_{0.46})O_{10}(OH)_2$	—	—
SHCa-1 hectorite	$(Mg_{0.28}Na_{0.21}K_{0.025})(Mg_{0.23}Li_{0.695}Ti_{0.005})(Si_{3.875}Al_{0.085}Fe^{III}_{0.025})O_{10}(OH)_2$	63.19	66
SapCa-2 saponite	$(Ca_{0.57}Na_{0.395}K_{0.035})(Mg_{2.99}Mn_{0.005})(Si_{3.595}Al_{0.37}Fe^{III}_{0.035})O_{10}(OH)_2$	—	—

[a] *Source:* Clay Repository of the Clay Mineral Society; *S* is specific surface area by BET N_2 adsorption method; *CEC* is cation exchange capacity (Van Olphen and Fripiat, 1979; Borden and Giese, 2001). In the early 1970s the Source Clay Program was initiated to provide investigators with a source of clay minerals such that each unit of clay would be nearly identical to another. There are eight source clays which include SAz-1, 2, STx-1, SWy-1, 2, and SHCa-1. Each source clay was collected to provide one metric ton of dried and milled material. The remaining clay minerals represented in this table (SBCa-1, NG-1, and SapCa-2) are called special clays. These materials are available from the Source Clay Repository, but they have not been pretreated or homogenized.

is (0.015 − 0.355), or −0.34 (the 0.355 Mg^{2+} atoms generate −0.355 charge deficit, while the 0.015 Ti^{4+} atoms generate +0.015 charge excess). The interlayer is composed of $(Ca_{0.135}Na_{0.02}K_{0.005})$ and the positive charge generated by the interlayer cations is (2 × 0.135 + 0.02 + 0.005), or +0.295. A comparison of this value with the −0.34-layer charge value indicates that there is an unbalanced charge of approximately −0.04 per ½-unit cell. Although STx-1 is a reference clay material, it is by no means devoid of impurities. Opal, quartz, feldspar, kaolinite, and possibly talc are known to be present in this reference material, contributing Si, Al, and base cations that are not inherent to the clay, but that are reported in the chemical analysis. Impurities, slight errors in the quantitative processing of the material in preparation for analysis and in the chemical analysis, can generate errors in a measured structural formula.

The determination of a structural formula for a layer silicate requires a chemical analysis of the material and knowledge of coordination chemistry (e.g., Pauling Rule 1). X-ray fluorescence spectrometry and neutron activation analysis are two methods that can provide a direct measure of the elemental composition of a solid material. However, a more common technique involves the acid dissolution of a solid, followed by the chemical analysis of the resulting solution for the analytes of interest. One such technique is known as Bernas' method, and involves placing a known mass of clay (usually 200 mg) in contact with *aqua regia* (a mixture of concentrated nitric and hydrochloric acids) and hydrofluoric acid, in a Teflon container. The container with the mixture is placed in a pressure-bomb (a sealed stainless steel vessel), which in turn is placed in a 150°C oven. After an 8- to 12-h, high temperature and pressure reaction period, the vessel is removed from the oven and allowed to cool. Excess (unreacted) hydrofluoric acid is then neutralized with boric acid, and the resulting solution placed in a volumetric flask (e.g., 50-mL volume), and brought to volume with metal-free (Type I) water. Chemical analysis is then performed using an appropriate technique such as atomic absorption spectrophotometry or inductively coupled argon plasma (see Chapter 5). Back-calculations convert the volume-based elemental concentrations to mass-based, clay mineral composition.

The next step in the procedure requires the conversion of elemental composition data into a ½-unit cell or unit cell structural formula. The composition data for a sample of a SWy-1 reference montmorillonite is presented in Table 2.9. The composition of the silicate is presented on an oxide basis, which is a common mechanism of relating data of this type. The first step is to convert the oxide composition data into the number of moles of cation charge generated by each metal. This is done by dividing the oxide composition by the molecular weight of the oxide, and multiplying this value by the number of metal atoms that appear in the oxide formula and the valence of the cation. For SiO_2, the moles of cation charge generated by Si are:

$$\frac{629 \text{ g kg}^{-1}}{60.08 \text{ g mol}^{-1}} \times 1 \times 4 \text{ mol}_c \text{ mol}^{-1} = 41.88 \text{ mol}_c \text{kg}^{-1} \qquad (2.2)$$

For Fe_2O_3, the moles of cation charge generated by Fe^{3+} are:

$$\frac{33.5 \text{ g kg}^{-1}}{159.69 \text{ g mol}^{-1}} \times 2 \times 3 \text{ mol}_c \text{ mol}^{-1} = 1.26 \text{ mol}_c \text{kg}^{-1} \qquad (2.3)$$

The next step is to normalize the cation charge to the anion charge in the structural unit. The ½-unit cell formula contains 12 oxygen atoms. Since each oxygen atom carries a charge of 2−, the anion charge in the ½-unit cell 2:1 formula is −24. Summation of the cation charge (i.e., 41.88 + 11.53 + ⋯ + 6.14) yields a value of 63.654 mol_c kg^{-1}. The conversion factor is 24/63.654, which when multiplied by the moles cation charge per kg for each cation, yields the normalized cation charge in the structure. Finally, the normalized charge (per 12 oxygen atoms) divided by the valence

TABLE 2.9
Structural Analysis of a Wyoming Bentonite (SWy-1), a 2:1 Phyllosilicate

Oxide	g kg^{-1}	Molecular Weight, g mol^{-1}	Moles Cation Charge kg^{-1} [a]	Normalized Cation Charge[b]	Structural Formula, ½-Unit Cell[c]
SiO$_2$	629	60.08	41.88	15.79	3.95
Al$_2$O$_3$	196	101.96	11.53	4.35	1.45
TiO$_2$	0.9	79.90	0.045	0.017	0.004
Fe$_2$O$_3$	33.5	159.69	1.26	0.475	0.16
FeO	3.2	71.85	0.089	0.034	0.017
MgO	30.5	40.30	1.51	0.569	0.28
CaO	16.8	56.08	0.60	0.226	0.11
Na$_2$O	15.3	61.98	0.49	0.185	0.19
K$_2$O	5.3	94.20	0.11	0.041	0.04
H$_2$O	55.3	18.02	6.14	2.32	2.32

[a] Mole cation charge = [(g oxide kg^{-1})/(g oxide mol^{-1})] × (number of cations in oxide formula) × (cation valence).

[b] Normalized moles cation charge = (moles cation charge) × (24)/[Σ(moles cation charge)], where 24 is the total moles of negative charge in the ½-unit cell structural formula and Σ(moles cation charge) = 63.654 (the summation of the values in the "Moles cation charge" column.

[c] Structural formula = normalized moles cation charge/(cation valence).

From Van Olphen, H. and J.J. Fripiat. *Data Handbook for Clay Minerals and Other Non-metallic Minerals.* Pergamon Press, Oxford, England, 1979. With permission.

of the cation yields the number of cations in the ½-unit cell structural formula. Therefore, for example, the number of Si^{4+} atoms per ½-unit cell formula (X_{Si}) is computed as:

$$X_{Si}(\text{Si unit}^{-1}) = \frac{41.88 \text{ mol}_c \text{ kg}^{-1}}{4 \text{ charge Si}^{-1}} \times \frac{24 \text{ charge unit}^{-1}}{63.654 \text{ mol}_c \text{ kg}^{-1}} = 3.95 \text{ Si unit}^{-1} \qquad (2.3b)$$

where "unit" refers to the ½-unit cell.

The structural formula is developed with knowledge of the coordination of each element. The protons reside in the hydroxyls, resulting in an anion composition of $O_{10}(OH_{1.16})_2$. All of the Si must reside in tetrahedral location (see Table 2.3). Since there are four tetrahedral locations per ½-unit cell formula, any deficit must be overcome by the addition of another element that can reside in this coordination. There are 3.95 Si atoms in the structure, and the deficit is 0.05 atoms. The only other atom that can reside in this coordination is Al. Therefore, the tetrahedral composition is $(Si_{3.95}Al_{0.05})$. The remainder of the Al atoms (1.40) resides in the octahedral layer. Other atoms that can reside in octahedral coordination are Ti^{4+}, Fe^{3+}, Fe^{2+}, and Mg (Table 2.3). The octahedral composition is $(Al_{1.40}Fe^{III}_{0.16}Fe^{II}_{0.017}Ti_{0.004}Mg_{0.28})$. Note that the sum of octahedral cations is 1.86, a value that is less than the theoretical occupation of 2. The Ca cation can also reside in octahedral coordination; however, it does not typically do so in phyllosilicate structures. If the sum of these octahedral cations exceeded 2, then the excess Mg would be placed in the interlayer. The remaining cations, Ca, Na, and K are placed in the interlayer, which has the composition $(Ca_{0.11}Na_{0.19}K_{0.04})$. Therefore, the structural formula for the dioctahedral 2:1 phyllosilicate whose chemical characteristics are presented in Table 2.9 is:

$$(Ca_{0.11}Na_{0.19}K_{0.04}(Al_{1.40}Fe^{III}_{0.16}Fe^{II}_{0.017}Ti_{0.004}Mg_{0.28})(Si_{3.95}Al_{0.05})O_{10}(OH_{1.16})_2$$

Isomorphic substitution in the octahedral layer results in a layer charge of 0.29 per ½-unit cell (discounting the unsatisfied negative charge that results from the incomplete occupation of the octahedral layer, which is satisfied by the addition structural protons). The interlayer cations generate a positive charge of 0.45 per ½-unit cell, which exceeds the sum of the charges from the tetrahedral layer (−0.05) and the octahedral layer (−0.29). Given that the layer charge is 0.34, the cation exchange capacity of the mineral can be computed. On a unit cell basis, the layer charge is 0.68 per unit cell, and the molecular weight of the layer silicate is 746.21 g mol⁻¹. There is also 6.022×10^{23} unit cells mol⁻¹. The cation exchange capacity is computed as:

$$0.68 \frac{\text{units of charge}}{\text{unit cell}} \times \left(6.022 \times 10^{23} \frac{\text{unit cells}}{\text{mol}} \right) \div 746.21 \frac{\text{g}}{\text{mol}} = 5.488 \times 10^{20} \frac{\text{units of charge}}{\text{g}}$$

$$5.488 \times 10^{20} \frac{\text{units of charge}}{\text{g}} \times \left(1000 \frac{\text{g}}{\text{kg}} \right) = 5.488 \times 10^{23} \frac{\text{units of charge}}{\text{kg}}$$

$$5.488 \times 10^{23} \frac{\text{units of charge}}{\text{kg}} \div \left(6.022 \times 10^{23} \frac{\text{unit cells}}{\text{mol}_c} \right) = 0.911 \frac{\text{mol}_c}{\text{kg}}$$

$$0.911 \frac{\text{mol}_c}{\text{kg}} \times \left(100 \frac{\text{cmol}_c}{\text{mol}_c} \right) = 91.1 \frac{\text{cmol}_c}{\text{kg}} = CEC \tag{2.4}$$

Note that this computed *CEC* value is comparable to that measured for the reference SWy-1,2 material (85 cmol$_c$ kg⁻¹, Table 2.8).

Another characteristic of the phyllosilicates that can be computed and compared to measured values is the specific surface. Assume that the relevant unit cell dimensions of the SWy-1 montmorillonite are $a = 0.519$ nm and $b = 0.900$ nm. Neglecting edge surfaces, the surface area per unit cell is $2ab = 2 \times 0.519$ nm $\times 0.900$ nm $= 0.9342$ nm^2 per unit cell. Again, using the molecular weight of 746.21 g mol⁻¹ and Avogadro's number (6.022×10^{23} unit cells mol⁻¹), the specific surface is computed:

$$0.9342 \frac{\text{nm}^2}{\text{unit cell}} \times \left(6.022 \times 10^{23} \frac{\text{unit cells}}{\text{mol}} \right) \div 746.21 \frac{\text{g}}{\text{mol}} = 7.539 \times 10^{20} \frac{\text{nm}^2}{\text{g}}$$

$$7.539 \times 10^{20} \frac{\text{nm}^2}{\text{g}} \times 10^{-18} \frac{\text{m}^2}{\text{nm}^2} = 753.9 \frac{\text{m}^2}{\text{g}} = S \tag{2.5}$$

If we compare this computed value with the measured specific surface values presented in Table 2.8, we immediately notice a very large discrepancy. Indeed, the measured values range from approximately 4 to 13% of the computed value for SWy-1. Fortunately, neither the computation nor the measure-specific surface values are in error. The specific surface measurements reported in Table 2.8 were obtained using an N$_2$ gas adsorption technique. Because of the large size and weak adsorption characteristics, N$_2$ does not enter clay interlayers; thus, the surface area values presented in Table 2.8 represent the external surfaces of the clays. Using the data for the reference clays and the computed specific surface data, 87 to 96% of the total specific surface of the reference SWy-1 montmorillonite is internal.

Summarizing, the smectite minerals have a layer charge that ranges between 0.2 and 0.6 per ½-unit cell. This layer charge is neutralized by cations that reside in the clay interlayers and hold the structure together through electrostatic interactions. Because the charge density in the smectites is relatively low, the electrostatic interactions between the interlayer cations and the mineral structure are weak, resulting in an expandable structure and internal surface area. The specific surface of the smectites can range between 600 and 800 m^2 g⁻¹, where greater than approximately

80% of the surface is internal. The d-spacing of the smectites varies depending on the occupation of the interlayer, and ranges from approximately 1.2 nm to upwards of 1.8 nm. These minerals are quite responsive to the chemical characteristics of the environment, expanding (swelling) and contracting (shrinking) as soil moisture and the type and concentration of electrolytes in the soil solutions dictate. Since the layer charge is created during mineral formation, the cation exchange capacity is independent of the environment in which the mineral resides; it is permanent and only changes over time due to weathering. In general the *CEC* of the smectites ranges between 60 and 150 $cmol_c$ kg^{-1}. Finally, smectites in the soil environment meet both definitions of clay. They are clay minerals, in that they are phyllosilicates, and isomorphic substitution and the destabilizing influence of the common base cations that typically reside in the interlayers lead to particle sizes that range between 0.01 and 1 μm.

2.5.2.3 Vermiculite Group

The *vermiculite* group contains the 2:1 phyllosilicates that have a layer charge between approximately 0.6 and 0.9 per ½-unit cell. The layer charge in the vermiculites principally originates in the tetrahedral layer, a property that is inherited from their parent minerals, the micas. The 2:1 layers are held to adjacent 2:1 layers by ionic interactions between the negatively charged layers and interlayer cations. Although the layer charge is relatively large compared to the smectites, the vermiculites are expandable and the internal surfaces are accessible to water molecules and other ions and molecules present in the bathing solution. The charge-satisfying, interlayer cations are exchangeable. However, both K^+ and NH_4^+ ions can cause the vermiculite structure to collapse, resulting in potassium- or ammonium-fixation and a vermiculite structure that resembles that of illite. The d-spacing of the vermiculite minerals is a function of the interlayer cation: 1.0 nm when the mineral interlayers are K^+ saturated, and 1.4 to 1.5 nm when Mg^{2+} or Ca^{2+} saturated. Magnesium saturation followed by glycolation of the vermiculite structure will result in expansion of the layers to 1.45 nm, but not beyond, as only one layer of glycol will be accepted. The specific surface of the vermiculite minerals is similar to that of the smectites, 600 to 800 m^2 g^{-1}. However, due to their greater layer charge, relative to the smectites, the *CEC* of the vermiculites is greater, ranging between 100 and 200 $cmol_c$ kg^{-1}.

At the subgroup level, the dioctahedral vermiculites are termed *dioctahedral vermiculites*. This subgroup contains the species *dioctahedral vermiculite*. In this mineral, a 0.7-layer charge is created through the isomorphic substitution of Al^{3+} for Si^{4+} in the tetrahedral layer (Figure 2.28). An ideal ½-unit cell formula for the mineral dioctahedral vermiculite is:

$$K_{0.7}Al_2(Si_{3.3}Al_{0.7})O_{10}(OH)_2$$

The *trioctahedral vermiculite* subgroup contains the *trioctahedral vermiculite* species. In soils, the trioctahedral vermiculites are derived from biotite (high-iron mica) or trioctahedral chlorites. Generally, the trioctahedral vermiculites are rich in Fe^{3+}, as well as Mg (Figure 2.29). As with the dioctahedral vermiculites, the layer charge originates principally in the tetrahedral layer through the isomorphic substitution of Al^{3+} for Si^{4+}. An ideal ½-unit cell chemical formula for trioctahedral vermiculite is:

$$K_{0.7}(Mg,Fe^{3+})_3(Si_{3.3}Al_{0.7})O_{10}(OH)_2$$

The structural formula for a Llano, TX trioctahedral vermiculite is:

$$(Mg_{0.48}K_{0.01})(Mg_{2.83}Fe^{III}_{0.01}Al_{0.15})(Si_{2.86}Al_{1.14})O_{10}(OH)_2$$

In this mineral, the tetrahedral layer carries a negative charge of 1.14, which is partially neutralized by a positive layer charge of 0.16 in the octahedral layer. The net layer charge for this mineral is 0.98, which is neutralized by the interlayer cation charge of 0.97. Although this mineral has a net layer charge that exceeds the defined upper limit of 0.9 per $\frac{1}{2}$-unit cell, the mineral is expandable, a characteristic that distinguishes the vermiculites from the mica minerals.

2.5.2.4 Mica and Illite Groups

The mica and illite groups contain primary and secondary minerals, respectively. The primary micas bear a layer charge of approximately 1 per $\frac{1}{2}$-unit cell. The layer charge of the secondary illites ranges between 0.6 and 0.8 per $\frac{1}{2}$-unit cell. The charge deficit for all mica and illite minerals occurs principally in the tetrahedral layer through the isomorphic substitution of Al^{3+} for Si^{4+}. The mica layer charge is neutralized by K^+ ions (most commonly) that reside in the interlayer and bridge adjacent 2:1 layers through the formation of ionic bonds. Similarly, layer charge in the illites is neutralized by K^+ ions, but the interlayer composition also includes water molecules. Like the primary micas, the secondary illite minerals are nonexpansive (nonswelling), a characteristic that distinguishes this group of minerals from the vermiculites. Indeed, unlike the obvious structural composition differences between the smectites and the vermiculites, structural composition cannot be employed to differentiate between illite and vermiculite. The d-spacing of the mica and illite minerals is 1.0 nm, and is unaffected by environmental conditions. Because these minerals do not expand, the specific surface ranges between 70 and 120 $m^2\ g^{-1}$ and arises from the external surfaces only. Despite extensive isomorphic substitution and layer charge, the mica and illite minerals display cation exchange capacities that range between 10 and 40 $cmol_c\ kg^{-1}$ (similar to those of the kaolinites). Again, this low *CEC* is a direct result of the inaccessibility of exchangeable cations to the interlayer exchange locations.

The *dioctahedral mica* subgroup contains the primary mica *muscovite*. As previously discussed, the ideal $\frac{1}{2}$-unit cell structural formula for muscovite is $KAl_2(Si_3Al)O_{10}(OH)_2$ (Figure 2.14). An example of a structural formula for a naturally occurring muscovite is:

$$(K_{0.85}Na_{0.09})(Al_{1.81}Fe^{II}_{0.14}Mg_{0.12})(Si_{3.09}Al_{0.91})O_{9.81}(OH)_2F_{0.19}$$

Note that the layer charge arises predominantly from the tetrahedral layer (0.91), with a lesser amount occurring in the octahedral layer (0.26). The net layer charge is -0.98 (-1.17 from isomorphic substitution plus 0.19 from the substitution of F^- for O^{2-}), which is balanced with $+0.94$ from K^+ and Na^+ in the interlayer. The sodium analogue to muscovite is the mineral *paragonite*:

$$(Na_{0.85}K_{0.15})Al_2(Si_3Al)O_{10}(OH)_2$$

Paragonite is uncommon in weathering environments, due to the structural instability of having Na^+ in the interlayer.

Illite and *glauconite* are nonexpandable dioctahedral 2:1 phyllosilicates of the illite group with a layer charge that ranges between approximately 0.6 and 0.8 per $\frac{1}{2}$-unit cell. The key characteristic that differentiates these minerals from dioctahedral vermiculite is that they are nonexpandable. In addition to the nonexpandability criteria, the octahedral layer charge in illite ranges from 0.2 to 0.3, and there are 0.2 to 0.4 additional Si atoms in the tetrahedral layer of illite beyond the 3:1 Si-to-Al of muscovite. The source clay reference illite (IMt-1, 2) has the structural formula:

$$(K_{0.69}Ca_{0.03}Mg_{0.01})(Al_{1.34}Fe^{III}_{0.38}Mg_{0.25}Ti_{0.03})(Si_{3.38}Al_{0.62})O_{10}(OH)_2$$

In this example, the tetrahedral layer charge is -0.62 and there are an additional 0.38 Si atoms beyond the 3:1 Si to Al of muscovite. The octahedral layer charge is -0.25, which is within the defined range for an illite. The net layer charge is −0.87, which is satisfied by a 0.77 cation charge in the interlayer. Glauconite is an Fe-rich illite with tetrahedral Al^{3+} (or Fe^{3+}) usually greater than 0.2 and octahedral Fe^{3+} (and Al^{3+}) correspondingly greater than 1.2. In the octahedral layer, $Fe^{3+} \gg Al^{3+}$ and $Mg^{2+} \gg Fe^{2+}$. An example of the structural formula of a glauconite is:

$$(K_{0.81}Ca_{0.02})(Al_{0.36}Fe^{III}_{0.98}Fe^{II}_{0.24}Mg_{0.45})(Si_{3.75}Al_{0.25})O_{10}(OH)_2$$

Phlogopite, *biotite*, and *annite* are mica minerals in the *trioctahedral mica* subgroup. The structure of these minerals is illustrated in Figure 2.15. Phlogopite and annite are the Mg^{2+}- and Fe^{2+}-rich end members of a solid solution having the general composition:

$$K(Mg_{3-x}Fe^{II}_x)(Si_3Al)O_{10}(OH)_2$$

where $x < 3$. Phlogopite is defined as having greater than 2.1 Mg atoms per ½-unit cell in the octahedral layer; whereas, annite is defined as having greater than 2.2 Fe^{2+} atoms in octahedral coordination. The most common trioctahedral mica in soil is biotite. The Mg composition of the octahedral layer of biotite minerals can range between 0.6 and 1.8 per ½-unit cell. Example structural formula for the trioctahedral micas are:

Phlogopite:

$$(K_{0.93}Na_{0.04}Ca_{0.03})(Mg_{2.77}Fe^{II}_{0.10}Ti_{0.11})(Si_{2.88}Al_{1.12})O_{10}(OH)_{1.49}F_{0.51}$$

Biotite:

$$(K_{0.78}Na_{0.16}Ba_{0.02})(Mg_{1.68}Fe^{II}_{0.71}Fe^{III}_{0.19}Ti_{0.34}Al_{0.19}Mn_{0.01})(Si_{2.86}Al_{1.14})O_{11.12}(OH)_{0.71}F_{0.17}$$

Annite:

$$(K_{0.88}Na_{0.07}Ca_{0.03})(Fe^{II}_{2.22}Mg_{0.12}Fe^{III}_{0.19}Ti_{0.22}Al_{0.09}Mn_{0.05})(Si_{2.81}Al_{1.19})O_{10.35}(OH)_{1.38}F_{0.22}Cl_{0.05}$$

2.5.2.5 Chlorite Group

The *chlorite* group of 2:1 phyllosilicates contains both primary and secondary minerals. The magnitude and location of the layer charge in the chlorite minerals are similar to that of the mica minerals (approximately 3:1 Si to Al in the tetrahedral layer resulting in an approximate layer charge of 1). The chlorite minerals are distinguished from the other 2:1 layer silicates by the presence of an interlayer hydroxide sheet. The interlayer sheet can be dioctahedral and dominated by Al^{3+} [$Al_2(OH)_6$] or trioctahedral and dominated by Mg^{2+} and Fe^{2+} [$(Mg,Fe^{II})_3(OH)_6$]. In either case, the interlayer hydroxide sheet bears charge and acts as a layer charge-neutralizing cation. The high charge density results in a strong electrostatic attraction between 2:1 layers and the hydroxide interlayer. In addition, hydrogen bonding between the basal oxygen atoms of the Si tetrahedral layer and the hydroxides of the interlayer sheet further stabilizes the structure. The layers do not expand or collapse, and depending on chemical composition, the d-spacing ranges between 1.40 and 1.44 nm. The specific surface and the cation exchange capacity of the chlorites are similar to the values presented for the mica minerals. The specific surface ranges between 70 and 150 $m^2 g^{-1}$ (all external) and the cation exchange capacity ranges between 10 and 40 $cmol_c kg^{-1}$.

The chlorite group is divided into three subgroups depending on the occupation of the 2:1 octahedral layer and the occupation of the hydroxide interlayer. The *dioctahedral chlorite* or *di-, dioctahedral chlorite* subgroup contains minerals that are dioctahedral in both the 2:1 layer and the interlayer. *Donbassite* is an example of a dioctahedral chlorite, with a ½-unit cell structural formula:

$$Al_{2.27}(OH)_6 \cdot Al_2(Si_{3.2}Al_{0.8})O_{10}(OH)_2$$

This chemical formula illustrates the occupation of both the dioctahedral layers. However, in practice it is difficult, from a chemical analysis, to partition octahedrally coordinated elements into the 2:1 layer and the interlayer. Instead, all octahedral elements are grouped, as are all the hydroxides. The structural formula for donbassite is more commonly written:

$$Al_{4.27}(Si_{3.2}Al_{0.8})O_{10}(OH)_8$$

The *di, trioctahedral chlorite* subgroup contains the mineral *cookeite*. An ideal ½-unit cell structural formula for cookeite is:

$$(LiAl_2)(OH)_6 \cdot Al_2(Si_3Al)O_{10}(OH)_2, \text{ or } (LiAl_4)(Si_3Al)O_{10}(OH)_8$$

In this mineral, the 2:1 layer contains two Al^{3+} atoms and is dioctahedral, while the interlayer contains a total of three atoms (one Li^+ and two Al^{3+}) and is trioctahedral. The -1 layer charge generated in the tetrahedral layer is neutralized by the $+1$ interlayer charge. The *trioctahedral chlorite* or *tri-, trioctahedral chlorite* minerals are common in soils. *Clinochlore* is a Mg-rich chlorite that has the ideal ½-unit cell chemical formula:

$$(Mg_2Al)(OH)_6 \cdot Mg_3(Si_3Al)O_{10}(OH)_2, \text{ or } (AlMg_5)(Si_3Al)O_{10}(OH)_8$$

The structure of clinochlore is illustrated in Figure 2.30. In this mineral, the 2:1 layer charge is neutralized by the positive interlayer charge that is generated through the substitution of Al^{3+} for

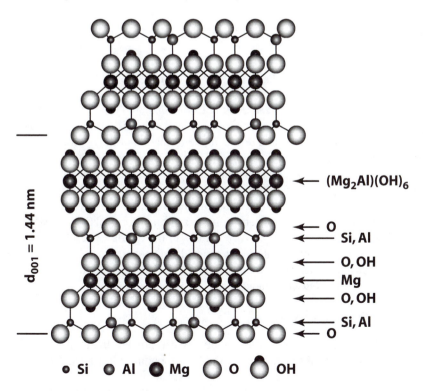

FIGURE 2.30 The ideal structure of the trioctahedral chlorite clinochlore [$(AlMg_5)(Si_3Al)O_{10}(OH)_8$].

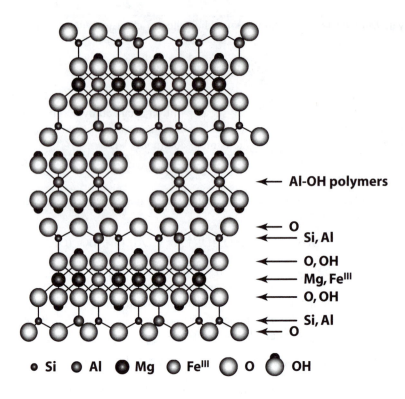

FIGURE 2.31 The structure of a hydroxy-interlayered vermiculite (HIV).

Mg^{2+}. *Ripidolite* is similar to clinochlore, but with a higher iron content. The chemical composition of the reference CCa-2 source ripidolite clay is:

$$Ca_{0.05}(Mg_{2.22}Fe^{II}_{1.51}Fe^{III}_{1.74}Al_{0.29}Mn_{0.01}Ti_{0.03})(Si_{2.25}Al_{1.75})O_{10}(OH)_8$$

In a soil environment, there are weathering reactions that are specific to smectites and vermiculites that may result in the formation of chlorite-like minerals, and ultimately *pedogenic chlorite*. This process is called chloritization, and results in the formation of the hydroxy-interlayered clays: *hydroxy-interlayered vermiculite* (*HIV*) and *hydroxy-interlayered smectite* (*HIS*). The chloritization process is common to acidic environments; thus, HIV and HIS are common in acidic soils. As we will see in Chapter 3, Al^{3+} in tetrahedral coordination is unstable, relative to Al^{3+} in octahedral coordination, and under low temperatures Al^{3+} migrates out of the 2:1 structure and into the interlayer where it undergoes hydrolysis and polymerization to form islands of hydroxy Al species (Figure 2.31). The formation of hydroxy Al polymers in the interlayer results in a vermiculite or smectite that behaves similarly to chlorite (restricted expandability, less surface area, and less surface reactivity relative to vermiculite and smectite). Unlike chlorite, however, the hydroxy Al interlayer in HIV and HIS is discontinuous.

2.5.2.6 Interstratified Layer Silicates

As illustrated in the previous sections, the analytically determined structural formulae for phyllosilicate minerals can differ substantially from the corresponding ideal formulae. The innate variability in the chemical nature of the phyllosilicates is evident at the group level of the classification scheme, for example, allowing the layer charge of montmorillonite to range between 0.2 and 0.6. Chemical variability is also expected given the heterogeneous nature of the chemical environments in which these minerals form, coupled with Pauling Rule 1, which allows several ions to occupy a given polyhedron type. It is not uncommon to see several elements occupying the octahedral layer of the 2:1 structure (e.g., Al^{3+}, Fe^{3+}, Fe^{2+}, Li^+, Mn^{2+}, Mg^{2+}, and Ti^{4+}).

Another commonly encountered deviation from the ideal is the *interstratified* or *mixed-layered* phyllosilicates (also termed *intergrades*). In the previous sections, several interlayer configurations were discussed. The mica minerals are distinguished by a 2:1 layer type that is collapsed about potassium, is nonexpandable, and has a d-spacing of 1.0 nm. The chlorite minerals are similarly described; collapsed about an interlayer hydroxy sheet, with a d-spacing of 1.4 nm. Both the smectite and vermiculite minerals have expandable 2:1 layers. The expansion of the vermiculite layers is limited to 1.4 nm, but the expansion of the smectite is a function of the size of the intercalating substance. An interstratified layer silicate contains two or more of the above interlayer configurations in the crystal. An interstatified mineral may also be defined as a layer silicate that contains two or more silicate minerals stacked along the *c*-axis of the unit cell. For example, in a dioctahedral mica crystal that has been subjected to weathering, 70% of the layers may be collapsed and remain K^+-saturated with a d-spacing of 1.0 nm. In the remaining 30% of the layers, the following condition could exist: the layer charge proximate to the interlayers has been reduced (by any number of mechanisms) and interlayer K^+ has been displaced by water and hydrated Ca^{2+} and Mg^{2+}, expanding the layers from 1.0 nm to 1.4 nm. The mineral is no longer mica, but an interstatified mica-vermiculite (if the expanded layers can not be expanded beyond 1.4 nm) or an interstatified mica-smectite (if the expanded layers can be expanded beyond 1.4 nm by intercalation with an organic compound).

There are two types of interstratification: regular and random. In regular interstratification, the layer stacking follows a periodic sequence, forming long-range mixed-layer structure. A regular interstratified structure that is composed of 50% collapsed, as in mica (designated by C), and 50% expanded, as in smectite (designated by X), layers would have the periodic stacking sequence: [...CXCXCX...] (Figure 2.32). A regular interstratified structure that is composed of 67% collapsed and 33% expanded layers would have the periodic stacking sequence [...CCXCCXCCX...]. Interstratified structures that are regular and well ordered, and contain the stacking sequence [...CXCXCX...] (50% collapsed and 50% expanded) can be given unique species names. Examples of regularly interstratified minerals that are named species are given in Table 2.10.

In random interstratification, there is no long-range order in the stacking of the layer structure. Randomly interstratified mineral assemblages are far more common in the soil environment than

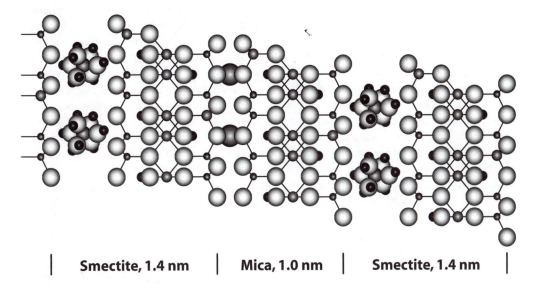

| Smectite, 1.4 nm | Mica, 1.0 nm | Smectite, 1.4 nm |

FIGURE 2.32 The stacking sequence of interstratified dioctahedral mica-dioctahedral smectite.

TABLE 2.10
Mineral Species Composed of Regularly Interstratified 2:1
Layer Structures (Ratio of Layer A to Layer B is 1:1)

Mineral	Layer A	Layer B
Aliettite	Talc	Trioctahedral smectite
Corrensite, low-charge	Trioctahedral chlorite	Trioctahedral smectite
Corrensite, high-charge	Trioctahedral chlorite	Trioctahedral vermiculite
Hydrobiotite	biotite	Trioctahedral vermiculite
Kulkeite	Talc	Trioctahedral chlorite
Rectorite	Dioctahedral mica	Dioctahedral smectite
Tosudite	Dioctahedral chlorite	Dioctahedral smectite

regularly interstratified assemblages. Indeed, it has been argued that dioctahedral mica (e.g., muscovite) does not occur as a distinct phase in a weathering environment. Instead, soil mica contains a predominance of collapsed layers, as well as a small amount of expanded layers, a consequence of the weathering process. The uniqueness of the mineral illite has been a subject of debate since Grim, Bray, and Bradley introduced the term in 1937 to describe mica minerals in argillaceous sediments. Although defined as a distinct mineral phase, many believe illite to be a mica-smectite (and other mica weathering products) interstratification. Randomly interstratified minerals are described in terms of the types and abundance of the component layers in the structure. A structure that contains 70% dioctahedral mica that is randomly interstratified with 30% dioctahedral smectite would be referred to as a dioctahedral mica-smectite interstratification, or a 70:30 dioctahedral mica-smectite interstratification. Randomly interstratified minerals that are common to the soil environment include trioctahedral mica-vermiculite and dioctahedral mica (or illite)-smectite. Again, the formation of interstratified layer silicates is a direct result of the mica weathering mechanism (discussed in Chapter 3).

2.6 HYDROUS METAL OXIDES

Hydrous metal oxides, also called accessory minerals, are weathering products that are hydrous and anhydrous oxides, hydroxides, and oxyhydroxides of metals, such as Fe^{III}, Al, Mn, and Si (Table 2.11). As their name implies, accessory minerals are accessory to and often intimately associated with the primary and secondary aluminosilicates, from which they are principally derived. Because of this intimate association, accessory minerals can mask the surface properties of layer silicates by blocking the access of solutes to interlayer adsorption sites. However, many hydrous metal oxides, particularly those that are poorly crystalline or amorphous, bear highly reactive surfaces, forming strong surface complexes with a myriad of metal, ligand, and molecular species (both inorganic and organic). Further, and unlike the layer silicates that exhibit principally negative surface charge as a result of isomorphic substitution, the hydrous metal oxides are generally amphoteric, and can develop either negative or positive surface charge depending on the chemical characteristics of the soil solution. Therefore, these minerals may exert a pronounced influence over the chemical characteristics of a soil solution and significantly impact the fate and behavior of substances in the soil environment.

2.6.1 HYDROUS ALUMINUM OXIDES

Aluminum occurs in oxides $[Al_2O_3]$, hydroxides $[Al(OH)_3]$, and oxyhydroxides $[AlOOH]$. All of these minerals, with the exception of gibbsite $[Al(OH)_3]$, are principally found in ore deposits or metamorphic rocks, and with few exceptions do not form in soil. Gibbsite is a ubiquitous metal hydroxide,

TABLE 2.11
Hydrous Metal Oxides that Commonly Occur in Soils

Mineral Name	Chemical Formula
Gibbsite	γ-Al(OH)$_3$
Boehmite	γ-AlOOH
Hematite	α-Fe$_2$O$_3$
Goethite	α-FeOOH
Maghemite	γ-Fe$_2$O$_3$
Ferrihydrite	~Fe$_5$HO$_8 \cdot$4H$_2$O
Birnessite	(Na,Ca,Mn^{II})(Mn^{III},Mn^{IV})$_7$O$_{14} \cdot$2.8H$_2$O
Lithiophorite	(Al$_2$Li)(OH)$_6$ Mn$_2^{IV}$)Mn^{III}O$_6$
Todorokite	(Mg$_{0.77}$Na$_{0.03}$)(Mg$_{0.18}$ Mn$_{0.60}^{II}$Mn$_{5.2}^{IV}$)O$_{12} \cdot$3.1H$_2$O
Vernadite	δ-MnO$_2$
Allophane	Al$_2$O$_3 \cdot$(SiO$_2$)$_{1-2} \cdot$ 2.5-3H$_2$O
Imogolite	Al$_2$SiO$_3$(OH)$_4$

particularly in highly weathered soils (e.g., Oxisols), acidic to moderately acidic soils (e.g., Ultisols), and in weathered volcanic ash (e.g., Andisols). Polymorphs of gibbsite that also occur in soils include the minerals bayerite and nordstrandite; however, they are exceedingly rare. The structure of gibbsite consists of stacked dioctahedral sheets of Al(OH)$_3$ that are held together by hydrogen bonds (Figure 2.33a). The stacking order of the hydroxide layers is the fundamental difference between gibbsite and the other Al(OH)$_3$ polymorphs. In gibbsite, the hydroxides of one octahedral sheet reside directly on top of the hydroxides of the adjacent sheets, resulting in the …ABBAAB… stacking sequence of the hydroxide layers. In bayerite, the stacking sequence is …ABABAB… and each hydroxide layer is seated in an adjacent layer (Figure 2.33b). The stacking sequence in nordstrandite is …ABABBABAAB…., which is a pattern of alternating gibbsite and bayerite units.

Boehmite and its polymorph diaspore are aluminum oxyhydroxides (AlOOH) that are rare in soils relative to gibbsite. The occurrence of these minerals is primarily restricted to bauxite ore bodies and metamorphic rocks. Although rare, boehmite and a microcrystalline, highly hydrated form of boehmite, often referred to as pseudoboehmite, have been identified in highly weathered soils. The structure of macro- and microcrystalline boehmite consists of stacked double-sheet structures that are held together by hydrogen bonds (Figure 2.33c). Each Al octahedron within a single sheet shares two edges and two corners with adjacent octahedra. Additionally, two octahedron edges are shared with octahedra in the adjacent sheet to generate the double-sheet structure that has a corrugated or zigzag appearance.

The surface area of natural gibbsite is quite variable and dependent on the inherent nature of the mineral as it exists in the environment (i.e., crystallinity). The surface area of poorly crystalline Al(OH)$_3$ may reportedly exceed 600 m^2 g^{-1}, as inferred from surface area values of laboratory prepared materials. However, the surface area of well-crystallized gibbsite generally ranges between 20 and 50 m^2 g^{-1}. Since the isomorphic substitution of divalent cations for Al^{3+} in gibbsite is negligible, structural charge deficit is insignificant. Further, all surface hydroxyls on the planar surfaces and one half of the hydroxyls on the edge surfaces are doubly-coordinated by two Al^{3+} atoms, resulting in charge-satisfied and relatively inert surface hydroxyls. However, singly-coordinated hydroxyl groups at the edges of the gibbsite sheet structures, which account for the other one-half of the edge hydroxyls, are undercoordinated and highly reactive. When solution pH values are greater than approximately 9, the gibbsite surface bears a net negative charge due to the predominance of undercoordinated $\equiv$Al—OH$^{-0.5}$ hydroxyls. As pH decreases, the proportion of $\equiv$Al—OH$^{-0.5}$ hydroxyls decreases with a concomitant increase in the proportion of protonated hydroxyls, $\equiv$Al—OH$_2^{+0.5}$. Irrespective of solution pH, a singly-coordinated hydroxyl will bear charge, and

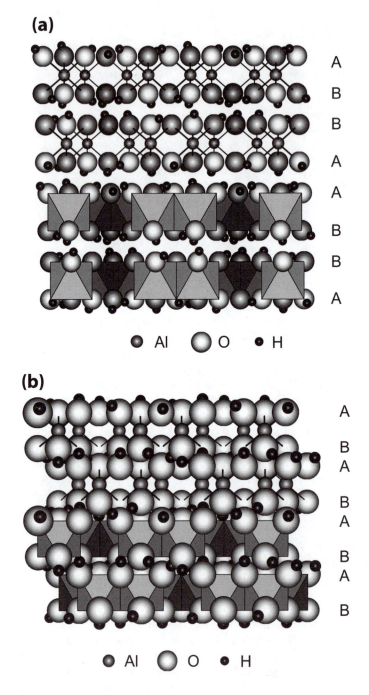

FIGURE 2.33 Combined polyhedral and ball-and-stick structural models of (a) gibbsite [Al(OH)$_3$], (b) bayerite [Al(OH)$_3$], and (c) boehmite [AlOOH]. Lighter or larger structures are superior, while darker or smaller structures are inferior. The stacking sequences of the A and B type oxygen layers in gibbsite and bayerite are also illustrated.

(c)

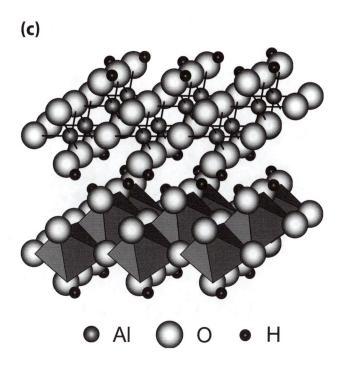

● Al   ◯ O   • H

FIGURE 2.33 (*continued*).

depending on pH, lend the surface net cation or anion exchange capacity (*CEC* or *AEC*). The maximum *CEC* of well-crystallized gibbsite in soil is approximately 1 cmol kg^{-1}, while that for poorly crystalline $Al(OH)_3$ may be 10 to 100 cmol kg^{-1}. The maximum *AEC* of gibbsite is approximately 3 cmol kg^{-1}, while that for poorly crystalline $Al(OH)_3$ may be 30 to 300 cmol kg^{-1}.

2.6.2 HYDROUS IRON OXIDES

Oxides and oxyhydroxides of Fe are also ubiquitous in soils, with goethite [FeOOH] the most common hydrous iron oxide. Goethite exists in almost every soil type and all climate regimes, although greater goethite concentrations are generally observed in cool and wet climates, where it is reportedly the only pedogenic Fe oxide in soils. A polymorph of goethite, lepidocrocite, is also common in soil. However, lepidocrocite is generally found in association with goethite and in soils that are seasonally anaerobic (redoximorphic), requiring Fe^{2+} in solution and noncalcareous conditions (low CO_2 partial pressures) for formation. The goethite structure consists of double chains of edge-sharing Fe^{III} octahedra that are bound to adjacent chains by sharing corners and by hydrogen bonds (Figure 2.34a). Lepidocrocite is isostructurally similar to (has the same structure as) boehmite (Figure 2.33c).

Closely associated with goethite in almost all soils is the Fe^{III} oxide hematite [Fe_2O_3]. Hematite formation appears to be favored in warm climates, and is common to the tropics and subtropics, arid and semiarid regions, and Mediterranean climates, where it may dominate over goethite. The hematite structure consists of stacked layers of dioctahedral sheet structures (Figure 2.34b). Within a layer, each octahedron shares edges with three adjacent octahedral units. Each octahedron also shares a face with an octahedron in an adjacent sheet, forming a common plane of oxygen atoms between neighboring layers. Magnetite [Fe_3O_4] is principally lithogenic in soils (it is inherited from parent rock) rather than pedogenic. Magnetite contains both Fe^{II} and Fe^{III}. One third of the Fe in magnetite is Fe^{II} in octahedral coordination, one third is Fe^{III} in octahedral coordination, and one third is Fe^{III} in tetrahedral coordination, yielding the unit cell formula:

$$^{oct}(Fe_8^{II}Fe_8^{III})^{tet}(Fe_8^{III})O_{32}$$

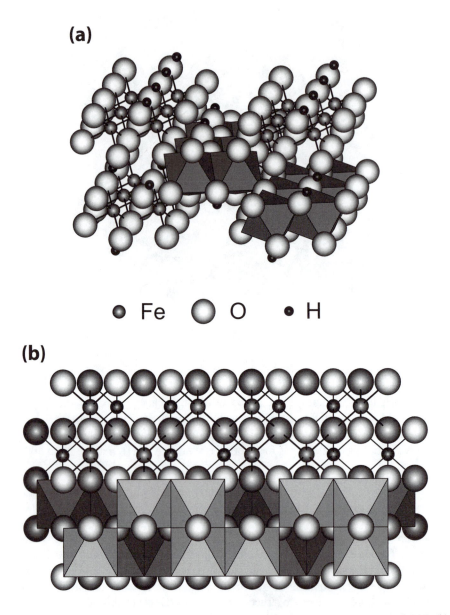

FIGURE 2.34 Combined polyhedral and ball-and-stick structural models of (a) goethite [FeOOH], (b) hematite [Fe_2O_3], and (c) magnetite [Fe_3O_4] and maghemite [Fe_2O_3]. Lighter or larger structures are superior, while darker or smaller structures are inferior.

The magnetite structure consists of sheets of tetrahedral and octahedral structures that form a common plane of oxygen atoms with adjacent sheets (Figure 2.34c). The partial or complete oxidation of Fe^{II} in the magnetite structure leads to the hematite polymorph, maghemite [Fe_2O_3], which is isostructural with magnetite. Maghemite is primarily encountered in highly weathered soils of the tropics and subtropics, and is commonly associated with goethite and hematite. Maghemite also forms as an alteration product of other pedogenic Fe oxides (e.g., goethite and hematite) in surface soils that are subjected to heat (300 to 425°C) in the presence of organic matter, a condition that can be achieved in a soil during forest or brush fires.

Ferrihydrite [$Fe_5HO_8 \cdot 4H_2O$] is a microcrystalline precipitate that forms when Fe^{II}-rich waters are rapidly oxidized in the presence of dissolved substances, such as organic acids, that inhibit

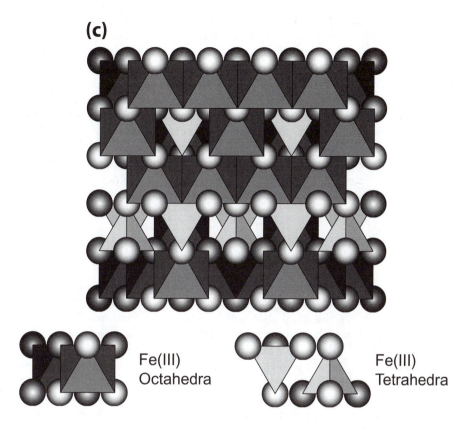

(c)

Fe(III) Octahedra

Fe(III) Tetrahedra

FIGURE 2.34 (*continued*).

crystal growth. The structure of ferrihydrite is not well established. However, it may resemble hematite with a number of structural defects, including some vacant Fe positions and the replacement of some oxygen atoms by water molecules. Although ferrihydrite forms rapidly, it is not a stable mineral phase. In warm and dry environments, ferrihydrite dehydration leads to the formation of hematite. In cool and moist environments, ferrihydrite dissolves and the solubilized Fe reprecipitates to form goethite.

The specific surface area of macrocrystalline hydrous Fe oxides varies widely and is inversely related to particle size. The specific surface areas of synthetic hematite and goethite preparations range between 6 and 115 $m^2 g^{-1}$, while that of soil goethite is reported to range between 60 and 200 $m^2 g^{-1}$. The specific surface area of poorly ordered and microcrystalline ferrihydrite ranges between 100 and 700 $m^2 g^{-1}$. The isomorphic substitution of Al^{3+} for Fe^{3+} is commonly observed in naturally occurring goethite, hematite, and maghemite. Several other cations (most notably Mn^{3+} and Cr^{3+}) may substitute for Fe^{3+} in the hydrous Fe oxide structures (as indicated by Pauling's First Rule). However, two important factors that appear to favor the isomorphic substitution of Al^{3+} for Fe^{3+}, relative to other possible substitutions, are (1) the abundance of soluble Al^{3+} in weathering environments and (2) the structural similarities between hydrous Fe and Al oxides (e.g., goethite and diaspore, hematite and corundum [Al_2O_3], and lepidocrocite and boehmite have similar structures—they are isostructural). For example, the goethite structure can reportedly accommodate up to approximately 30 mol% Al^{3+} [e.g., $(Fe_{0.7}Al_{0.3})OOH$], while hematite and maghemite may accommodate up to 15 to 18 mol% Al^{3+} [e.g., $(Fe_{0.85}Al_{0.15})_2O_3$].

Despite the potential significance of isomorphic substitution, the hydrous Fe oxides bear little if any structural charge. Even when substitution is significant, as in the substitution of Al^{3+} for Fe^{3+}, structural charge is not generated. Thus, the reactivity of the hydrous oxide surfaces is determined by the specific surface area, the type and number of proton-reactive sites, and the pH of a bathing solution. The complexity of the hydrous Fe oxide surfaces is epitomized by the goethite surface, which bears three types of surface functional groups: $\equiv Fe—OH^{-0.5}$, $\equiv Fe_2—OH^0$, and $\equiv Fe_3—O^{-0.5}$. The doubly coordinated $\equiv Fe_2—OH^0$ group is predicted to be relatively inert because the coordination environment is complete. However, singly [$\equiv Fe—OH^{-0.5}$] and triply coordinated [$\equiv Fe_3—O^{-0.5}$] hydroxyl and oxide groups are undercoordinated, and in the case of the singly-coordinated group, highly reactive. When solution pH values are greater than approximately 8.8, the goethite surface bears a net negative charge due to the predominance of undercoordinated $\equiv Fe—OH^{-0.5}$ hydroxyls (and $\equiv Fe_3—O^{-0.5}$ oxides). As pH decreases, the proportion of $\equiv Fe—OH^{-0.5}$ hydroxyls (and $\equiv Fe_3—O^{-0.5}$ oxides) decreases with a concomitant increase in the proportion of protonated hydroxyls, $\equiv Fe—OH_2^{+0.5}$ (and $\equiv Fe_3—OH^{+0.5}$ oxides). Thus, irrespective of solution pH, singly and triply coordinated hydroxyls and oxides will bear charge and, depending on pH, result in surfaces with net cation or anion exchange capacity.

2.6.3 HYDROUS MANGANESE OXIDES

Hydrous oxides of manganese (Mn) are common in soils, particularly in environments with alternating wetting (reducing) and drying (oxidizing) (redoximorphic) conditions and restricted drainage. Hydrous Mn oxides are also more abundant in soils formed from basic (mafic) rocks than from siliceous rocks. Hydrous Mn oxides precipitate during the chemical or biochemical oxidation of Mn^{2+}, which is solubilized during the weathering of primary minerals. The precipitates that form are generally microcrystalline, poorly ordered, and impure, occurring as coatings on ped faces and pore surfaces, and as concretions and nodules. The small crystal size, the poorly ordered nature, and the relatively low concentrations of the pedogenic hydrous Mn oxides have made direct characterization a challenge. However, the selective characterization of concretions and nodules, as well as the characterization of bulk soil size, density, and magnetic separates, has led to the identification of various pedogenic hydrous Mn oxides.

Manganese in pedogenic precipitates is mostly found in the Mn^{IV} oxidation state and in octahedral coordination with oxygen atoms. The minor occurrence of Mn^{III} in pedogenic hydrous Mn oxides has been mostly assumed, although a limited number of spectroscopic studies have substantiated this assumption. The hydrous Mn oxides that are commonly encountered in surface environments belong to either the phyllomanganate or the tectomanganate groups. The phyllomanganate minerals consist of layer structures of edge-linked Mn octahedra that are connected to adjacent layers by exchangeable cations and water molecules, or by an interlayer hydroxide sheet of octahedrally coordinated cations. The layer charge in the phyllomanganates arises from vacancies in the Mn octahedral layers, or from the substitution of Mn^{III} for Mn^{IV}. Birnessite [$(Na, Ca, Mn^{II})(Mn^{III}, Mn^{IV})_7O_{14} \cdot 2.8H_2O$] is a phyllomanganate that occurs as a very poorly crystalline precipitate in relatively young soils (Figure 2.35a). Vacancies in the layer structure generate a permanent structural charge that is satisfied by exchangeable interlayer cations. Lithiophorite [$(Al_2Li)(OH)_6Mn_2^{IV}Mn^{III}O_6$], also a phyllomanganate, has been identified in acid soils (Ultisols) and highly weathered acid soils (Oxisols). In these soils, lithiophorite occurs in nodules as well-developed, platy crystals. The permanent structural charge that arises from the presence of Mn^{III} in one third of the octahedral positions is satisfied by a Li^+-substituted Al^{3+} hydroxy-interlayer (Figure 2.35b). Todorokite [$(Mg_{0.77}Na_{0.03})(Mg_{0.18}Mn_{0.60}^{II}Mn_{5.22}^{IV})O_{12} \cdot 3.1H_2O$] is a tectomanganate that has been observed in soils of intermediate age (e.g., Vertisols). Tectomanganates are described as tunnel structures, because edge-sharing chains of Mn octahedra are linked by corner sharing with other chains in such a manner as to form a framework of enclosed tunnels (Figure 2.35c). The tunnels are occupied by cations and water molecules. Todorokite consists of tunnels that are three octahedrons wide by three octahedrons tall. Vernadite [MnO_2], along with birnessite, is considered to be one of the most common hydrous Mn oxides to occur in soils. Vernadite is a poorly crystalline

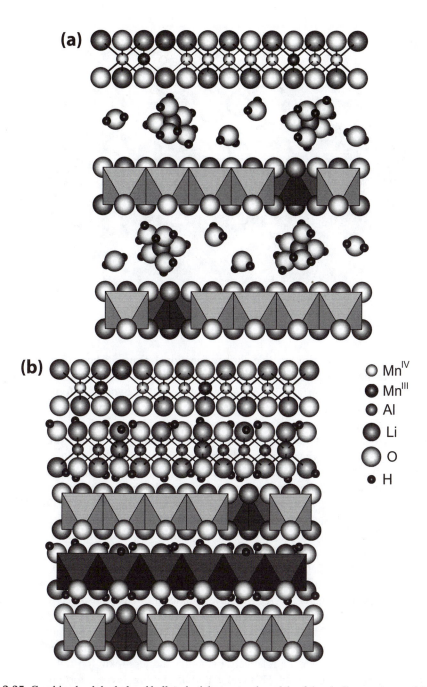

FIGURE 2.35 Combined polyhedral and ball-and-stick structural models of the phyllomanganates: (a) birnessite [(Na,Ca,Mn^{II})Mn$_7$O$_{14}$•2.8H$_2$O] and (b) lithiophorite [(Al,Li)MnO$_2$(OH)$_2$]; and the tectomanganate, (c) todokorite [(Na,Ca,K)$_{0.3-0.5}$(Mn^{IV},Mn^{III})$_6$O$_{12}$•3.5H$_2$O]. Unless noted otherwise, lighter structures and superior and darker structures are inferior.

material that has characteristics of both a phyllomanganate (a random stacked or disordered birnessite) and a tectomanganate (corner-sharing octahedra).

Despite their relatively low concentrations in surface environments, the hydrous Mn oxides have a profound impact on the fate and behavior of metals and ligands. Their small crystal size and numerous imperfections, as well as permanent structural charge and the presence of a redox-sensitive

(c)

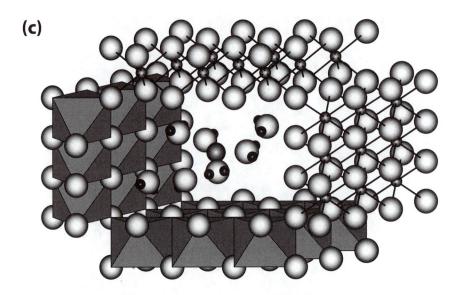

FIGURE 2.35 (*continued*).

element, lead to an abundance of reactive surface functional groups. The surface area of hydrous Mn oxides ranges between 5 and 360 m^2 g^{-1}. In addition to the permanent structural charge, numerous undercoordinated surface oxygen atoms may exist on hydrous Mn oxide surfaces. The predominant surface oxygen types are the singly and doubly coordinated $\equiv$Mn^{IV}—OH$^{-0.33}$ and $\equiv$Mn$_2^{IV}$—O$^{-0.33}$. Lesser concentrations of the singly and doubly coordinated $\equiv$Mn^{III}—OH$^{-0.5}$ and $\equiv$Mn$_2^{III}$—OH^0 may also exist, their significance depending on the structural Mn^{III} concentrations. Of these differing surface oxygen atoms, only the doubly coordinated $\equiv$Mn$_2^{III}$—OH^0 oxygen is fully coordinated and predicted to be relatively inert. However, the remaining surface oxygen atoms do not appear to be particularly proton selective, as indicated by the very high proton concentrations required to generate positively charged surface oxygen atoms (e.g., $\equiv$Mn^{IV}—OH$_2^{+0.67}$). The predominance of undercoordinated surface oxygen atoms, in conjunction with the permanent negative structural charge, results in hydrous Mn oxide surfaces that are predominantly negatively charged when solution pH values are greater than approximately 2.

In addition to the pH dependency of the surface charge on the hydrous Mn oxides, structural Mn^{IV} and Mn^{III} are oxidants that influence the redox status of various redox-sensitive metals. The adsorption of As^{III} species (e.g., H$_3$AsO$_3^0$), a relatively toxic and weakly adsorbed form of arsenic, by Mn^{IV} oxides results in Mn^{II} and the oxidative formation of As^V species (e.g., H$_2$AsO$_4^-$), which is less toxic and strongly adsorbed. The adsorption of Co^{2+} results in the oxidative formation of Co^{III}, which replaces Mn^{II} in the mineral structure and reduces the phytoavailability of this essential element. The surface-mediated oxidation of Cr^{III}, a relatively immobile and less toxic form of chromium, results in the production of Cr^{VI} species (e.g., HCrO$_4^-$), a mobile and highly toxic form of chromium.

2.6.4 Allophane and Imogolite

Allophane and imogolite are hydrated and poorly ordered aluminosilicate clay materials. Of these two aluminosilicates, allophane displays the least structure and the greater chemical variability. Allophane is composed of Al octahedrons and Si tetrahedrons that develop short-range to medium-range ordered structures. The ratio of Si to Al in allophane ranges from approximately 1:2 to 1:1, yielding the general chemical formula Al$_2$O$_3$•(SiO$_2$)$_{1-2}$•2.5–3H$_2$O. Typically, allophane occurs as a coating on soil particles, or as coagulated masses of spherical particles that range in diameter from 3.5 to 5.5 nm. Imogolite differs from allophane in chemical formula, crystallinity, and structure.

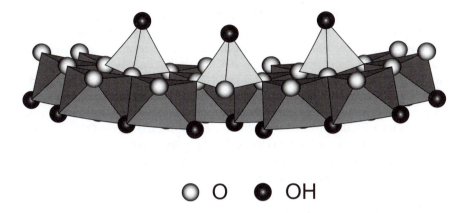

O O ● OH

FIGURE 2.36 The polyhedral structural model depicts a portion of the tubular structure of imogolite. The tube consists of three concentric anion layers: (1) the apical hydroxyls of the Si^{4+} tetrahedrons pointing into the tube; (2) the basal Si^{4+} tetrahedral oxygen atoms that are also common to the Al^{3+} octahedrons (and reside in the interior of the tube wall); and (3) the exterior layer of Al^{3+} octahedral hydroxyls.

Imogolite is reported to have a nearly fixed composition with a Si to Al ratio that ranges between 1.05:2 and 1.15:2, and an idealized structural formula of $Al_2SiO_3(OH)_4$. Imogolite may appear as 10- to 30-nm-diameter bundles of tubular crystal structures that are approximately 2.0 to 2.3 nm in outer diameter, each tube varying in length from tens to hundreds of nanometers (short-range to long-range ordered structures). The bundles may reach lengths of several micrometers. Structurally, an imogolite tube consists of a curled, edge-sharing dioctahedral Al oxyhydroxide sheet, with hydroxyls exterior to the tube and the interior oxygen atoms comprising the basal plane of insular Si tetrahedrons (the Si tetrahedral units do not polymerize) (Figure 2.36). In addition to the three basal oxygen atoms, each Si is coordinated by an apical hydroxyl which points into the tube, such that each tetrahedron has the composition $SiO_3(OH)$.

A key requirement for the development of allophane and imogolite in soils is high concentrations of dissolved Al and Si. For this reason, allophane and imogolite are most commonly associated with soils of volcanic origin (Andisols), although not strictly limited to these soils, due to the rapid release of Si and Al during the weathering of volcanic glass. Allophane and imogolite have also been identified in soils developed from a variety of other parent materials, including basalt, granite, gneiss, and sandstone, particularly in (but not limited to) acidic environments that receive high amounts of rainfall. Of the two minerals, allophane is the more common. Low solution pH values in combination with a relatively low dissolved Si concentrations ($<10^{-3.45}$ M) generally results in Al-rich allophanes with lesser amounts of imogolite. Highly soluble Si concentrations ($>10^{-3.45}$ M) favor the formation of halloysite or Si-rich allophanes.

The measured specific surface area of allophanic materials is a function of the methodology employed. When determined by polar liquid adsorption (ethylene glycol monoethyl ether or water), the specific surface area of allophanic clays are found to range between 247 and 549m^2 g^{-1}. Surface areas determined by nonpolar gas adsorption (N_2) yield values that range between 145 and 300 m^2 g^{-1}. The reactivity of allophane and imogolite surfaces is dictated by the protonation and deprotonation of surface hydroxyls (pH dependent), although the isomorphic substitution of Al^{3+} for tetrahedral Si^{4+} in allophanes has been observed. There are essentially three types of exposed surface hydroxyls: $\equiv Si-OH^0$, $\equiv Al-OH^{-0.5}$, and $\equiv Al_2-OH^0$. Both the doubly coordinated $\equiv Al_2-OH^0$ and the singly coordinated $\equiv Si-OH^0$ surface hydroxyls are fully coordinated and predicted to be relatively inert. However, this is only the case for $\equiv Al_2-OH^0$. Although the $\equiv Si-OH^0$ group accepts a proton to form $\equiv Si-OH_2^+$, the group contributes to the anion exchange characteristics of allophanic clays only when solution pH values are much less than those observed in normal soil systems (pH < 4). However, the dissociated $\equiv$Si—O$^-$ group predominates relative to $\equiv$Si—OH^0 when solution pH values are greater than approximately 7.7. The combined influence of the protonated and deprotonated $\equiv$Si—OH^0 groups on charging characteristics leads a net negative surface charge when solution pH values are greater than approximately 3.5.

As discussed for gibbsite, the singly-coordinated $\equiv$Al—OH$^{-0.5}$ hydroxyl is undercoordinated and highly reactive. When solution pH values are greater than approximately 9, this group bears a net negative charge. However, as pH decreases, the proportion of $\equiv$Al—OH$^{-0.5}$ hydroxyls decreases with a concomitant increase in the proportion of protonated hydroxyls, $\equiv$Al—OH$_2^{+0.5}$. Irrespective of solution pH, the singly coordinated $\equiv$Si—OH^0 and $\equiv$Al—OH$^{-0.5}$ hydroxyls will give the allophane and impogolite surfaces charge, and depending on pH, lend the surfaces net cation or anion exchange capacity (*CEC* or AEC). In general, the *CEC* and AEC characteristics of allophane and imogolite are a function of the Al to Si ratio and the degree of isomorphic substitution of Al^{3+} for Si^{4+} in the tetrahedral structures (which can be significant). Imogolite and Al-rich allophane (2:1 Al to Si ratio) display net *CEC*, relative to AEC, when solution pH values are greater than approximately 8, while Si-rich allophanes have net *CEC* when solution pH is greater than approximately 6 to 7 for a 1.6:1 Al to Si allophane, or pH 4 to 5.5 for a 1.2:1 Al to Si allophane. The reported *CEC* range for allophanic clays is between 10 and 40 cmol kg^{-1} at pH 7, with an AEC range of 5 to 30 cmol kg^{-1} at pH 4.

2.7 X-RAY DIFFRACTION ANALYSIS

X-ray diffraction is the premier tool for the direct identification of crystalline soil minerals. The properties of crystals that allow x-ray diffraction to provide definitive information are: (1) the locations of the atoms in the structure are fixed in three-dimensional space; (2) the arrangement of the atoms is repetitive (recall the concept of the unit cell); and (3) every mineral is structurally unique. The first two crystal characteristics lead to the viability of x-ray diffraction as a tool, while the third characteristic leads to the unambiguous identification of crystalline minerals. Just as every mineral is unique, every mineral will produce a unique diffraction pattern (akin to a fingerprint).

2.7.1 PRINCIPLES

X-radiation for diffraction is produced by bombarding a metal foil (most commonly Cu foil, but other metals are employed) with electrons. The bombarding electrons collide with electrons that reside in the electronic orbitals of the foil metal, displacing the orbital electrons and leaving vacancies in the orbitals. The resulting unstable electronic configuration is immediately stabilized by outer shell electrons that drop down (in energy and location) to fill the vacancies. The energy lost by the replacing electrons is released in the form of x-radiation. Specifically, the change in the energy of an electron, ΔE is inversely related to the wavelength of the x-ray that is produced: $\Delta E = hc/\gamma$, where h is Planck's constant, c is the velocity of the x-ray, and γ is the wavelength. The x-rays produced by this process are termed characteristic x-rays, meaning that the wavelength (or energy) of the x-rays is characteristic of the element being bombarded.

Characteristic x-radiation needed for x-ray diffraction results from very specific electron transitions within an atom. Specifically, the x-radiation produced when electrons in the orbital closest to the nucleus are displaced is required (Figure 2.37). These x-rays, termed K-series x-rays, result when a K-shell electron is displaced and electrons from the L-, M-, and N-shells drop down to the K-shell. As a result, Kα and Kβ radiation is produced. In addition to the characteristic radiation, another type of x-ray produced during electron bombardment is termed continuous or white radiation. White radiation is created as the bombarding electrons lose energy in multiple steps (through the numerous collisions that occur). Each transition (loss of energy) results in an x-ray whose energy (and wavelength) is determined by the energy lost by the electron during a collision. Since a multitude of transitions occur, a continuous range of x-ray wavelengths is produced.

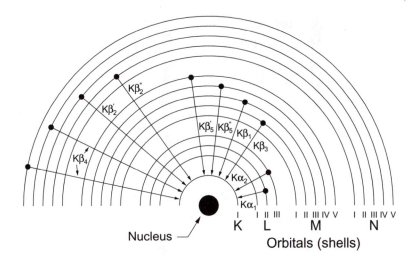

FIGURE 2.37 The electronic transitions that occur when a K-shell electron is displaced, resulting in Kα and Kβ x-radiation.

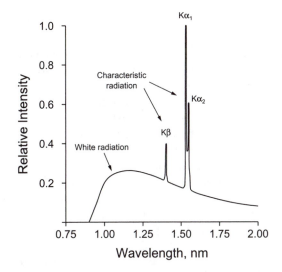

FIGURE 2.38 The K-series x-ray spectrum for copper.

The x-ray spectrum that results from the bombardment of a metal, such as Cu, consists of heterochromatic radiation, a combination of both white and characteristic radiation (Figure 2.38). The Kβ peak represents the electron transitions from primarily the M-shell (a combination of $K\beta_5$ at 0.138109 nm and $K\beta_3$ at 0.1392218 nm) (Figure 2.37). The wavelength of the $K\alpha_1$ x-ray is 0.1540562 nm, while that of $K\alpha_2$ is 0.1544390 nm. The intensity ratio $K\alpha_2/K\alpha_1$ is 0.509. Further, the ratio $K\beta/K\alpha_{1,2}$ is 0.136 (where $K\alpha_{1,2}$ is the combined intensity of $K\alpha_1$ and $K\alpha_2$). After the x-rays are produced by the electron-bombarded Cu foil, the x-rays are passed through a second metal foil that selectively absorbs much of the white and Kβ radiation. The type of metal filter foil used depends on the type of metal used to produce the x-rays. If Cu is the x-ray source metal, Ni is the filter metal. The Ni absorption edge is 0.148804 nm; x-rays with wavelengths below this value are absorbed, while those with wavelengths above pass through the Ni foil essentially

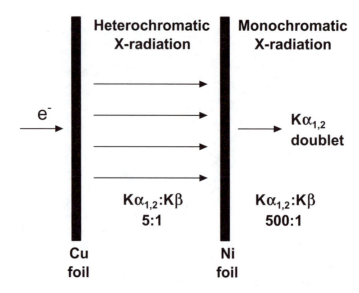

FIGURE 2.39 Schematic representation of the mechanism used to produce monochromatic Cu $K\alpha_{1,2}$ x-radiation.

unabated. After filtering, the ratio $K\alpha_{1,2}$:$K\beta$ is 500 to 1 and the radiation is essentially monochromatic (note that $K\alpha_1$ and $K\alpha_2$ are not separated, as their wavelengths are too close together and the ratio after filtering is still 2:1). The wavelength (λ) of the CuK$\alpha_{1,2}$ doublet is 0.15418 nm. The overall process of monochromatic x-ray production for x-ray diffractometry is diagrammed in Figure 2.39.

   Once produced, the monochromatic x-rays are focused on a sample. Some of the x-rays pass through the sample, while some are diffracted or scattered by the atoms in the various minerals. Although we tend to associate diffraction with reflection, the processes are not synonymous. When an x-ray impacts an atom and encounters an electron, the electron will vibrate with the same frequency as the incident x-ray. The vibrating electron emits electromagnetic radiation, or reradiates the energy (minus a small amount that is absorbed by the electron) in all directions and at the same wavelength as the incident beam. The diffraction of x-rays by atoms in different planes of equal electronic/ionic density is illustrated in Figure 2.40. In order for diffracted x-ray beams 1′ and 2′ to arrive at the detector reinforcing one another (in-phase), the additional distance traveled by beam 2′ must be equal to some whole number of the wavelength $n\lambda$, where λ is the wavelength of the monochromatic CuKα radiation. When x-rays reinforce, the distance between the planes of equal electronic/ionic density in a crystal lattice can be computed. The additional distance traveled by beam 2′ is $\overline{BC} + \overline{BD}$ (which are sides of the right triangles ABC and ABD, respectively). The side $\overline{BC}$ is numerically equal to $\overline{AB}\sin\theta$. Similarly, side $\overline{BD}$ is equal to $\overline{AB}\sin\theta$, where side $\overline{AB}$ is the distance, d, between the lattice planes (the d-spacing) and θ is the angle of incidence. The additional distance traveled ($\overline{BC} + \overline{BD}$) is $d\sin\theta + d\sin\theta$, or $2d\sin\theta$. Recall that for reinforcement the additional distance traveled must also be equal to $n\lambda$. Therefore, $n\lambda = 2d\sin\theta$. This equation is Bragg's Law, and it is the most important relationship for the use and understanding of x-ray diffraction. The CuKα wavelength is a known value ($\lambda = 0.15418$ nm) and θ is a measured property. Therefore, the d-spacing can be computed by rearranging the Bragg equation:

$$d = \frac{n\lambda}{2\sin\theta} \qquad (2.6)$$

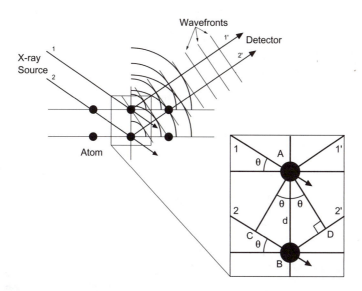

FIGURE 2.40 The diffraction of x-rays by a point lattice. Incident beams 1 and 2 impact electrons that are within atoms that lie in adjacent planes of equal electronic density. The electrons radiate (diffract) x-rays 1′ and 2′. In order for 1′ and 2′ to arrive at the detector in-phase, the extra distance traveled by beam 2′ ($\overline{CB}+\overline{BD}$) must equal a multiple of the wavelength ($n\lambda$). Both triangles ABC and ABD are right triangles; therefore, $\overline{CB}=\overline{AB}\sin\theta$ and $\overline{BD}=\overline{AB}\sin\theta$. It follows that ($\overline{CB}+\overline{BD}$) = $2\,\overline{AB}\sin\theta$. Since ($\overline{CB}+\overline{BD}$) = $n\lambda$ and $\overline{AB}$ = d (the distance between the planes of equal electronic density), $n\lambda=2\mathrm{d}\sin\theta$.

2.7.2 Clay Mineral Characterization

Phyllosilicate minerals in the soil environment do not exist as discrete particles; they are intimately associated with organic matter and accessory mineral coatings that aggregate the clays and mask their chemical and structural characteristics. These coatings also impede one's ability to characterize the clay mineralogy of a soil. There are four relevant steps in the preparation of soil samples for clay mineral characterization by x-ray diffraction: (1) removal of aggregating materials, (2) particle size separation, (3) saturation and heat treatment, and (4) preparation of an oriented mount for presentation to x-rays.

In a soil system, clay minerals are aggregated by a myriad of cementing agents. In order to facilitate clay size segregation and presentation to the x-ray beam, the cementing agents must be removed. Commonly, soils are treated in a sequential fashion to remove (1) carbonates with pH 5 sodium acetate, (2) organic matter and Mn oxides with hydrogen peroxide, and (3) iron oxides with a sodium dithionite-citrate-bicarbonate treatment. Following the sequential removal of cements, the soil sample (clay minerals) is sodium saturated (Na^+ occupies the exchange complex and interlayer positions) to attain a dispersed state, which facilitates particle size fractionation. The less than 2-μm fraction is then collected using standard Stoke's law sedimentation techniques.

As indicated previously, the d-spacings of the various phyllosilicate minerals are characteristic and can be manipulated by cation intercalation. For example, the d-spacing of montmorillonite is 1.2 nm when K^+ (or Na^+) saturated, 1.4 nm when Mg^{2+} saturated, or 1.8 nm when Mg^{2+} and glycol saturated. This characteristic of the phyllosilicate minerals is used to facilitate their identification by x-ray diffraction analysis. There are five cation saturation and temperature treatments that are commonly performed on subsamples of soil clay-sized separates prior to x-ray diffraction analysis: K^+ saturation, K^+ saturation and heated to 300°C, K^+ saturation and heated to 550°C, Mg^{2+} saturation, and Mg^{2+} and glycol saturation. The response of the phyllosilicate minerals to the

TABLE 2.12
The d-spacings for the Common Layer Silicates as a Function of Saturating Cation and Heat Treatment

Treatment	Clay Mineral d-Spacing, nm				
	Kaolinite	Mica	Vermiculite	Smectite	Chlorite
K^+-25°C	0.7	1.0	1.0	1.2	1.4
K^+-300°C	0.7	1.0	1.0	1.0	1.4
K^+-550°C	absent	1.0	1.0	1.0	1.4
Mg^{2+}	0.7	1.0	1.4	1.4	1.4
Mg^{2+}-glycol	0.7	1.0	1.4	1.8	1.4

FIGURE 2.41 The two ideal orientations of clay particles for x-ray diffraction analysis: oriented, where the clay platelets lie parallel to the support structure such that the *c* axis is preferentially presented to the x-ray beam; or random, where all planes of equal electronic density within a mineral have an equal opportunity to register a detector response (neither orientation can be achieved completely, only approached).

different treatments is shown in Table 2.12. Kaolinite neither expands nor collapses, and the d-spacing remains at 0.7 nm irrespective of cation saturation. However, kaolinite is destroyed by the 550°C treatment, which causes a phase change to an amorphous aluminosilicate. Unaffected by treatment are mica and chlorite which, like kaolinite, neither expand nor collapse, and the d-spacing remains 1.0 nm and 1.4 nm, respectively. Vermiculite interlayers have a mica configuration and the structure has a d-spacing of 1.0 nm when K^+ saturated. Because vermiculite is an expansive layer silicate, the structure will allow a larger hydrated cation into the interlayers. The d-spacing of vermiculite is 1.4 nm when Mg^{2+} saturated, a spacing that cannot be expanded further by glycolation. Finally, the smectites are differentiated from the other layer silicates by their ability to expand beyond 1.4 nm upon glycolation. The treatment of soil clay minerals and the subsequent characterization of the d-spacings will not result in a species-level qualification of the clay mineralogy (with the exception of kaolinite). As indicated in Table 2.12, only the characterization of clay mineralogy at the group level can be achieved. Thus, an identification of the actual species present in a soil environment, or even the di- or trioctahedral nature of the clay minerals, requires the application of more elaborate techniques, in addition to the detailed interpretation of x-ray diffraction data.

Following cation saturation, the clay subsamples are individually mounted for x-ray diffraction analysis. Since we are most interested in measuring the d-spacing of the clay minerals, the minerals must be mounted such that the d-spacing is preferentially measured by x-ray diffraction. This is termed an oriented mount (vs. a random mount or powder pack). There are a number of techniques used to obtain such a mount, but the desired result is to have the clay platelets lay flat on a supporting structure (Figure 2.41).

X-ray diffraction results are presented in the form of an x-ray diffractogram. A diffractogram is a plot of the intensity of the diffracted x-rays as a function of °2θ (degrees 2-theta, or two times the angle theta, where θ is the angle of incidence/diffraction as defined in Figure 2.39). In addition to the primary diffraction lines ($n = 1$ in Equation 2.6), the x-ray diffractograms also display a number of secondary and tertiary lines ($n = 2$ and $n = 3$), which may also be employed to substantiate the occurrence of a particular mineral. X-ray diffractograms for the Mg^{2+}- and Mg^{2+}-glycol–saturated subsamples of the clay fraction of mine spoil material from an abandoned surface coal mine near

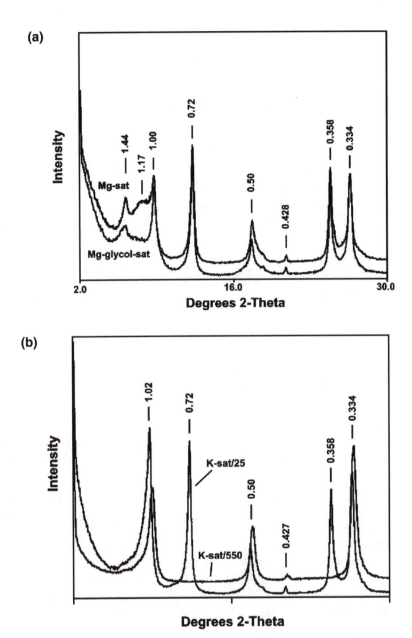

FIGURE 2.42 X-ray diffractograms of clay-sized materials separated from an abandoned mine spoil material: (a) Mg^{2+}- and Mg^{2+} glycol-saturated and (b) K^+-saturated (25 and 550°C). Peaks are identified by d-spacing (in nm).

Caryville, TN are shown in Figure 2.42a. The Mg^{2+}-saturated subsample pattern shows peaks at 1.44 nm (vermiculite and/or chlorite), which does not expand upon glycolation (indicating the absence of smectite), a broad 1.17-nm peak (possible interstratification), a 1.0-nm peak (mica), a 0.72-nm peak (kaolinite), a 0.50-nm peak (second-order mica peak), a 0.428-nm peak (quartz), a 0.358-nm peak (second order kaolinite peak), and a 0.334-nm peak (quartz and third-order mica). Figure 2.42b shows the x-ray diffractograms for K^+ saturated with no heat treatment and K^+ saturated and heated to 550°C. The disappearance of the 1.44-nm peak upon K^+ saturation indicates that vermiculite is present in the sample (collapsed to 1.0 nm), and that chlorite is absent. Disappearance

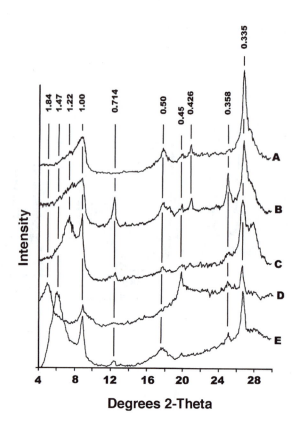

FIGURE 2.43 X-ray diffractograms of the clay fraction of a San Emigdio soil. (a) K^+- saturated, 550°C; (b) K^+- saturated, 300°C; (c) K^+- saturated, 25°C; (d) Mg^{2+}- and glycerol-saturated; (e) Mg^{2+}-saturated. Peaks are identified by d-spacing (in nm).

of the 0.72 nm and 0.428 nm peaks upon K^+ saturation and heating (550°C) indicates the presence of kaolinite in the original sample. Based on the data presented, the clay mineralogy of the mine spoil material is kaolinite, vermiculite, and mica.

The clay fraction of San Emigdio soil (coarse-loamy, mixed, superactive, calcareous, thermic Typic Xerofluvent), collected near Irvine, CA, clearly contains smectite (Figure 2.43). The 1.47 nm peak in the Mg^{2+}-saturated subsample is displaced to 1.84 nm upon intercalation with glycerol. In addition, smectite is indicated by the presence of a 1.22 nm peak in the K^+-saturated, 25°C subsample. The diffractogram of the Mg^{2+}-saturated subsample indicates the presence of mica (1.00 nm peak) and possibly kaolinite (0.714 nm peak). The occurrence of kaolinite is substantiated by loss of the 0.714 nm peak (and the 0.358 nm peak) upon heating to 550°C. Quartz is also found in the clay-sized fraction of this soil, as indicated by the presence of the 0.426 nm and 0.335 nm peaks. Based on the data in Figure 2.43, the clay minerals present in the San Emigdio soil are smectite, mica, and kaolinite.

Before urban sprawl consumed agricultural land in the valleys east of Los Angeles, CA, grapes for the production of table wines were grown in areas with Delhi soil (mixed, thermic Typic Xeropsamment). The clay fraction of this soil contains smectite, mica, and kaolinite (Figure 2.44). There is also a 1.45-nm peak in the K^+-saturated, 600°C subsample, suggesting the occurrence of chlorite.

The Loring soil (fine-silty, mixed, active, thermic Oxyaquic Fragiudalf) is typical of the loess-derived soils that are present in west Tennessee. X-ray diffraction of the Mg^{2+} and Mg^{2+}-glycol-saturated clay separates indicates the presence of mica and the absence of smectite, and suggests the presence of kaolinite and vermiculite (or chlorite) (Figure 2.45a). The presence of

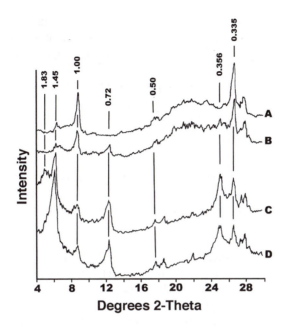

FIGURE 2.44 X-ray diffractograms of the clay fraction of a Delhi soil. (a) K^+- saturated, 600°C; (b) K^+-saturated, 25°C; (c) Mg^{2+}- and glycerol-saturated; (d) Mg^{2+}-saturated. Peaks are identified by d-spacing (in nm).

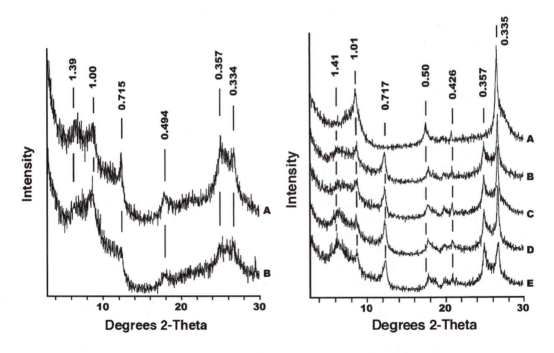

FIGURE 2.45 X-ray diffractograms of the clay fraction of a Loring soil. (a) Mg^{2+}- and glycol–saturated (A) and Mg^{2+}-saturated (B); (b) K^+- saturated, 550°C (A), K^+- and glycerol-saturated (B), K^+- saturated, 300°C (C), K^+saturated, 105°C (D), and K^+- saturated, 25°C (E). Peaks are identified by d-spacing (in nm).

kaolinite is substantiated by the absence of the 0.717-nm peak in the x-ray diffractogram of the K^+-saturated, 550°C subsample (Figure 2.45b). The absence of chlorite is also substantiated by the absence of a 1.4-nm peak in the K^+-saturated, 550°C subsample. The gradual decrease in the d-spacing of the vermiculite peak as temperature treatment increases from 25 to 550°C indicates that the vermiculite contains hydroxyl interlayers. The clay mineralogy of the Loring soil is HIV, mica, and kaolinite.

2.8   EXERCISES

1. Describe the four bonding mechanisms that are important in soil mineralogy. Indicate which mechanisms impart stability to soil minerals in a weathering environment.
2. Pauling's Fifth Rule, the Principle of Parsimony, is implied from the first four rules (i.e., it is not an explicitly stated rule in Pauling, 1960). Using the first four Pauling rules, discuss why the Principle of Parsimony must be valid.
3. Employ Pauling's electrostatic valence principle and show that the residual charge on O^{2-} bound to Si^{4+} in tetrahedral coordination can be satisfied by two Al^{3+} atoms in octahedral coordination or by three Mg^{2+} atoms in octahedral coordination.
4. The Pauling rules treat ions as hard spheres with fixed ionic radii (termed the Pauling radii). In reality, the ionic radius of a coordinated metal cation is a function of the coordination number. Verify that Al^{3+} satisfies Pauling Rule 1 in both CN 4 and CN 6 configurations, assuming the radius of O^{2-} is 0.140 nm, the radius of Al^{3+} in CN 4 is 0.053 nm, and the radius of Al^{3+} in CN 6 is 0.0675 nm.
5. Complexity in the Si tetrahedral arrangement is the criteria used in silicate classification at the class level. On the basis of this criterion, determine the class to which each of the following minerals belongs.
   a. $Be_3Al_2(Si_6O_{18})$
   b. $(Fe_{0.6}Mg_{0.4})_7Si_8O_{22}(OH)_2$
   c. $Mg_3Si_4O_{10}(OH)_2$
   d. $CaMgSiO_4$
   e. $Ca_6Al_{12}Si_{24}O_{72} \cdot 40H_2O$
   f. $CaMgSi_2O_6$
   g. $Ca_2Mg(Si_2O_7)$
6. Answer the following:
   a. What factors affect the isomorphic substitution of one element for another in a crystal structure?
   b. Why is it more difficult to remove potassium from the interlayer spaces of beidellite than from montmorillonite?
   c. How do chlorite and hydroxy-interlayered vermiculite differ?
   d. How do illite and trioctahedral vermiculite differ?
7. The Association Internationale pour l'Etude des Argiles and the International Mineralogical Association Committee on New Minerals and Mineral Names employ a hierarchical system of mineral Divisions (or Types), Groups, Subgroups, and Species to classify the phyllosilicate minerals (Table 2.7). Develop a schematic diagram that can be employed to "key out" a clay mineral to the species level when given a chemical formula. Use your classification key to provide a species name for each of the following minerals:
   a. $Ca_{0.05}(Mg_{2.22}Fe^{II}_{1.51}Fe^{III}_{1.74}Al_{0.29}Mn_{0.01}Ti_{0.03})(Si_{2.25}Al_{1.75})O_{10}(OH)_8$
   b. $Ca_{1.1}(Mg_{2.18}Al_{0.72})(Si_{1.05}Al_{2.95})O_{10}(OH)_2$
   c. $Na_{0.46}(Al_{1.98}Fe^{III}_{0.02}Mg_{0.01})(Si_{3.48}Al_{0.52})O_{9.98}(OH)_2$
   d. $K_{0.8}(Mg_{0.34}Fe^{II}_{0.04}F^{III}_{0.17}Al_{1.5})(Si_{3.43}Al_{0.57})O_{10}(OH)_2$

e. $Al_{1.98}Mg_{0.3}Si_2O_5(OH)_4$

f. $Ca_{0.81}Na_{0.19}K_{0.01}(Al_{1.99}Fe_{0.01}Mg_{0.03})(Si_{2.11}Al_{1.89})O_{10}(OH)_2$

g. $Na_{0.35}(Mg_{2.68}Li_{0.3})Si_4O_{10}F_{1.25}(OH)_{0.75}$

h. $K_{0.78}(Mg_{2.71}Fe_{0.46}^{III}Fe_{0.02}^{II}Ca_{0.06}K_{0.1})(Si_{2.91}Al_{1.14})O_{10}(OH)_2$

i. $Ca_{0.1}Na_{0.27}K_{0.02}(Al_{1.52}Fe_{0.19}^{III}Mg_{0.22})(Si_{3.94}Al_{0.06})O_{10}(OH)_2$

j. $K_{0.93}Na_{0.04}Ca_{0.03}(Mg_{2.77}Fe_{0.10}Ti_{0.11})(Si_{2.88}Al_{1.12})O_{10}(OH)_2$

k. $K_{0.88}Na_{0.07}Ca_{0.03}(Fe_{2.22}^{II}Mg_{0.12}Al_{0.09}Ti_{0.22}Mn_{0.05}Fe_{0.19}^{III})(Si_{2.81}Al_{1.19})O_{10.35}(OH)_{1.38}F_{0.22}Cl_{0.05}$

l. $K_{0.64}(Al_{1.54}Fe_{0.29}^{III}Mg_{0.19})(Si_{3.51}Al_{0.49})O_{10}(OH)_2$

m. $K_{0.94}Na_{0.06}(Al_{1.83}Fe_{0.12}^{III}Mg_{0.06})(Si_{3.11}Al_{0.89})O_{10}(OH)_2$

n. $Ca_{0.22}(Mg_{2.50}Fe_{0.26}^{III}Fe_{0.24}^{II})(Si_{3.30}Al_{0.68}Fe_{0.02}^{III})O_{10}(OH)_2$

8. The chemical composition of a clay mineral from the Montmorillone region of France is given in the table below.
   a. Calculate the unit cell structural formula for this clay mineral.
   b. Compute a value for the cation exchange capacity of the mineral.
   c. Given that the unit cell parameters are $a = 0.517$ nm and $b = 0.894$ nm, compute a value for the internal specific surface area of the mineral.
   d. Estimate a value for the total surface area of the mineral (state any assumptions).

Component	%	Component	%
SiO_2	51.14	CaO	1.62
Al_2O_3	19.76	K_2O	0.11
Fe_2O_3	0.83	Na_2O	0.04
MgO	3.22		

9. X-ray diffraction is one of the principal techniques used in mineralogical studies. The Bragg equation is used to explain how x-ray diffraction works.
   a. Derive the Bragg equation.
   b. Compute the smallest d-spacing that can be measured by x-ray diffraction using CuKα radiation ($\lambda = 0.1540598$ nm).
   c. Using CuKα radiation, calculate the incidence angles (°2θ) for diffractions that would indicate d-spacings of 0.426 nm and 0.334 nm.
   d. Using CuKα radiation, create a table similar to Table 2.12, displaying the characteristic °2θ values, instead of the characteristic d-spacings, for the phyllosilicates.

10. A 200-mg mass of a clay mineral with a structural formula of $K_{0.94}Na_{0.06}(Al_{1.83}Fe_{0.12}^{III}Mg_{0.06})(Si_{3.11}Al_{0.89})O_{10}(OH)_2$ is dissolved using the HNO_3-HCl-HF method of Bernas. After neutralizing excess HF with boric acid the solution is placed in a 50-mL volumetric flask and brought to volume with metal-free (Type I) water. Chemical analysis is then performed using inductively coupled argon plasma. What are the concentrations of K, Na, Al, Fe, Mg, and Si in this solution?

11. The smallest sphere that can be placed in CN = 6 configuration (octahedron) has a radius ratio of 0.414. Prove this to be the case, irrespective of anion size, by computing the minimum radius ratio, using geometric principles, for an octahedron using both O^{2-} ($r = 0.140$ nm) and F^- ($r = 0.133$ nm) as coordinating anions.

12. You are given a soil sample and asked to determine the clay mineralogy. Beginning with the bulk soil sample, describe a methodology that could be employed to determine the clay mineralogy of the soil by x-ray diffraction. Include a discussion of any pretreatments, clay isolation techniques, and clay saturation treatments. Also, indicate how the various

clay minerals respond to the imposed saturation treatments (it will be necessary to consult a reference that describes Stoke's Law sedimentation).

13. Determine the clay mineralogy of the clay-sized fraction of an Alfisol using the x-ray diffraction data presented below.

Treatment	Peak location, °2θ
K^+saturated:	
Room	6.90, 8.73, 12.28, 17.72, 20.81, 24.85, 26.58
400°C	8.77, 12.30, 17.74, 20.80, 24.80, 26.60
550°C	8.70, 17.63, 20.69, 26.50
Mg^{2+}-glycol saturated	6.08, 8.75, 12.27, 17.75, 20.82, 24.82, 26.57

14. Describe how the surface charge characteristics of phyllosilicates and hydrous metal oxides differ.

15. For each of the following minerals: (1) describe (diagram) the structure and provide a structural formula; (2) indicate the locations on the mineral surface where ions may be adsorbed; and (3) identify the characteristic of the mineral that results in surface charge (how is surface charge developed and where is it located).
    a. Vermiculite
    b. Montmorillonite
    c. Gibbsite
    d. Goethite
    e. Clinoptilolite
    f. Kaolinite
    g. Birnessite
    h. Todokorite
    i. Allophane

REFERENCES

Baerlocher, Ch., W.M. Meier, and D.H. Olsen. *Atlas of Zeolite Framework Type* (5th ed.). Elsevier, Amsterdam, 2001.

Bailey, S.W. Structures of layer silicates. In *Crystal Structures of Clay Minerals and Their X-ray Identification.* G.W. Brindly and G. Brown (Eds.) Monogr. No. 5, Mineralogical Society, London, 1980, pp. 1–124.

Borden, D. and R.F. Giese. Baseline studies of the clay minerals society source clays: cation exchange capacity measurements by the ammonium-electrode method. *Clays Clay Min.* 49:444–445, 2001.

Gaines, R.V., H.C. Skinner, E.E. Foord, B. Mason, and A. Rosenzweig. *Danas New Mineralogy.* J. Wiley & Sons, New York, 1997.

Grim, R.E., R.H. Bray, and W.F. Bradley. The mica in argillaceous sediments. *Am. Mineralogist.* 22:813–829, 1937.

Ming, D.W. and F.A. Mumpton. Zeolites in soils. In *Minerals in Soil Environments* (2nd ed.). J.B. Dixon and S.B. Weed (Eds.) SSSA Book Series, no. 1. SSSA, Madison, WI, 1989.

Pauling, L. *The Nature of the Chemical Bond.* Cornell Press, Ithaca, NY, 1960.

Van Olphen, H. and J.J. Fripiat. *Data Handbook for Clay Minerals and Other Non-Metallic Minerals.* Pergamon Press, Oxford, England, 1979.

3 Chemical Weathering

Chemical weathering is a natural soil process that occurs under the prevailing environmental conditions and results in the transfer of matter from unstable mineral phases to more stable mineral phases or soluble species. Weathering is not limited to indigenous soil minerals. Indeed, any material that is generated under a set of conditions that differs from those of the soil environment will undergo transformation if introduced into the soil. These transformations include the dissolution or alteration of the introduced or indigenous materials and the precipitation of relatively more stable mineral phases. For example, metallic Pb from spent ammunition rounds or lead pipe will oxidize in soil to form litharge (PbO), which is transformed further to cerussite ($PbCO_3$) in alkaline soil. The metallic Pb is unstable in the presence of exceeding low levels of $O_2(g)$, and is oxidized to the lead oxide:

$$Pb^0(s) + \tfrac{1}{2}O_2(g) \rightarrow PbO(s) \tag{3.1}$$

In this reaction, metallic lead (Pb) is the primary phase of lead, while litharge is a secondary phase. However, litharge is also unstable relative to a host of other lead-bearing minerals, including lead silicates, phosphates, sulfates, and carbonates. In an alkaline soil environment, the most stable lead-bearing mineral is the carbonate, $PbCO_3$, which is also a secondary phase:

$$PbO(s) + CO_2(g) \rightarrow PbCO_3(s) \tag{3.2}$$

Because cerussite is the stable mineral phase for chemical conditions in an alkaline soil environment, additional weathering can only result in the dissolution of cerussite and the release of Pb^{2+} to be mobilized or to participate in other retention processes (e.g., adsorption, plant uptake):

$$PbCO_3(s) + 2H^+(aq) \rightarrow Pb^{2+}(aq) + CO_2(g) + H_2O(l) \tag{3.3}$$

The transfer of lead from the unstable metallic Pb phase, through the progressively more stable litharge and cerussite phases, to the soluble divalent cation, Pb^{2+}, is an example of a weathering sequence. Typically, however, weathering is a term and phenomenon that is specific to the transformation of naturally occurring soil minerals, particularly the silicates.

3.1 HYDROLYSIS AND OXIDATION

Understanding the transformations that occur when anthropogenic materials are deposited in soil is required for predicting substance fate and behavior and for the development, assessment, and implementation of remediation options. However, indigenous soil minerals are also in a constant state of transformation; transformations that occur because many soil minerals are unstable. A soil is a natural and dynamic entity that forms because of chemical weathering processes. Chemical weathering is defined as the chemical action of substances in the atmosphere (air), hydrosphere (water), and biosphere (biota) on soil minerals that redistributes elements at the Earth's surface. Typically, the chemical weathering processes of most interest in soils are those involving the primary

TABLE 3.1
Common Primary Silicate Minerals in Soils

Mineral	Formula[a]
Nesosilicates	
Olivines	$(Mg,Fe)_2SiO_4$
Garnets	$(Ca,Mg,Fe^{2+},Mn)(Al,Fe^{3+},Cr^{3+})(SiO_4)_3$
Sphene	$CaTiO(SiO_4)$
Zircon	$(Zr,Hf)SiO_4$
Sorosilicates	
Epidote	$Ca_2(Al,Fe)Al_2O(SiO_4)(Si_2O_7)(OH)$
Cyclosilicates	
Tourmaline	$(Na,Ca)(Li,Mg,Al)(Al,Fe,Mn)_6(BO_3)_3(Si_6O_{18})(OH)_4$
Inosilicates	
Augite	$Ca(Mg,Fe)Si_2O_6$
Hornblende	$NaCa_2Mg_5Fe_2(AlSi_7)O_{22}(OH)]$
Phyllosilicates	
Mica	$KAl_2(Si_3Al)O_{10}(OH)_2$
Biotite	$K(FeMg_2)(Si_3Al)O_{10}(OH)_2$
Chlorite	$(AlMg_5)(Si_3Al)O_{10}(OH)_8$
Tectosilicates	
Quartz	SiO_2
Microcline (K-feldspar)	$KAlSi_3O_8$
Orthoclase (K-feldspar)	$KAlSi_3O_8$
Albite (Na-plagioclase)	$NaAlSi_3O_8$
Anorthite (Ca-plagioclase)	$CaAl_2Si_2O_8$

[a] Unless noted otherwise, the oxidation state of Fe and Mn is II.

silicate minerals and their transformation products, the secondary silicates. The primary silicates are the most abundant group of minerals in soils and, as indicated in Table 3.1, all of the silicate classes are represented. These primary minerals are reactants, and when acted upon by weathering processes, are altered to secondary minerals, such as the clay and accessory minerals. However, just as litharge (PbO, a weathering product of metallic Pb oxidation) weathers to cerussite ($PbCO_3$) in the above example, secondary silicates can also weather to other secondary phases. Thus, mineral alteration by chemical weathering in the soil is not limited to primary silicates.

In any soil, the minerals are inherited from underlying parent rock, transported and deposited by wind and water, or formed in place. Primary or secondary minerals that are derived from underlying rock are termed authigenic; they have formed from sources originally found within the depositional basin. Minerals that are transported from a rock source outside the depositional basin, and the weathering products of the transported minerals, are termed detrital. The fact that the observed mineral assemblages in a soil can be inherited from multiple sources is particularly troublesome when attempting to establish weathering sequences for the silicates (unlike the metallic Pb example where the Pb form and source are known).

The driving force for all chemical weathering mechanisms is water; for without water, chemical weathering would not occur and soils would not form. Water can mediate chemical reactions, such as the simple dissolution and solvation (solutioning) of a salt ($CaSO_4 \cdot 2H_2O(s)$ [gypsum] $\rightarrow Ca^{2+}(aq) + SO_4^{2-}(aq) + 2H_2O(l)$); carry chemical agents, such as root exudates and protons that enhance mineral dissolution by chelation and acidification ($FeOOH(s) + citrate^{3-}(aq) + 3H^+ \rightarrow Fe\text{-}citrate^0(aq) + 2H_2O(l)$); or directly participate in the weathering process by a combined hydrolysis and ion exchange reaction, as shown in the following example:

$$KAl_2(Si_3Al)O_{10}(OH)_2(s) \text{ [muscovite]} + 0.4Mg(H_2O)_6^{2+} (aq) + 0.2H_4SiO_4^0 + 0.2H^+(aq)$$

$$\rightarrow [Mg(H_2O)_6^{2+}]_{0.4} Al_2(Si_{3.2}Al_{0.8})O_{10}(OH)_2(s) \text{ [dioctahedral vermiculite]} + K^+(aq)$$

$$+ 0.2Al(OH)_3(s) \text{ [gibbsite]} + 0.2H_2O(l) \tag{3.4}$$

In addition to its function as a solvent and a source of protons, water is a transport medium. In a leaching environment, elements that are solubilized and solvated during the weathering process are removed from the reaction zone and ultimately leached through the soil profile. Because weathering is a chemical reaction and obeys the Law of Mass Action, the removal of reaction products (the solubilized elements) forces the reaction to proceed and enhances the potential for additional weathering. For example, the dissolution of the olivine forsterite ($Mg_2SiO_4(s) + 4H^+(aq) \rightarrow 2Mg^{2+}(aq) + H_4SiO_4^0(aq)$) will be facilitated by the removal of Mg^{2+} or $H_4SiO_4^0$ from the reaction zone. Thus, soils that are leached are said to be older than soils that receive limited moisture, irrespective of actual age.

There are several chemical weathering processes, including solutioning (simple dissolution) and chelation. However, the two predominant chemical weathering mechanisms responsible for the alteration of the silicate minerals are hydrolysis and oxidation. Hydrolysis involves the action of water and protons resulting in a change in the chemical form of a substance. In natural systems, protons are derived from water (through the hydrolysis of Al^{3+} and Fe^{3+} that are produced during weathering) or from naturally occurring acids, such as carbonic acid and a multitude of organic acids. When Al^{3+} and Fe^{3+} are released to the soil solution during mineral decomposition, they are enveloped by numerous water molecules in a process termed hydration (or solvation). Both Al^{3+} and Fe^{3+} have high charge densities, which strongly attract the electrons in the oxygen atoms of the waters of hydration. As a result, the O—H bond of a water of hydration lengthens, such that the proton can dissociate, depending on the pH of the soil solution. This process is termed metal hydrolysis. The reaction between a water of hydration and an Al^{3+} or Fe^{3+} ion results in a change in the chemical form of the aluminum or iron in solution and the release of a proton:

$$Al^{3+}(aq) + H_2O(l) \rightarrow AlOH^{2+}(aq) + H^+(aq) \tag{3.5a}$$

$$Fe^{3+}(aq) + H_2O(l) \rightarrow FeOH^{2+}(aq) + H^+(aq) \tag{3.5b}$$

In soil solutions, the process generally does not end with the production of just one proton, but instead continues. For example, the stepwise hydrolysis of Al^{3+} and Fe^{3+} to form gibbsite or goethite would produce two additional protons that could eventually be consumed in mineral weathering processes:

$$AlOH^{2+}(aq) + H_2O(l) \rightarrow Al(OH)_2^+ + H^+ \tag{3.5c}$$

$$Al(OH)_2^+ + H_2O(l) \rightarrow Al(OH)_3(s) + H^+ \tag{3.5d}$$

$$FeOH^{2+}(aq) + H_2O(l) \rightarrow Fe(OH)_2^+ + H^+ \tag{3.5e}$$

$$Fe(OH)_2^+ \rightarrow FeOOH(s) + H^+ \tag{3.5f}$$

Carbonic acid ($H_2CO_3^0(aq)$) is in equilibrium with $CO_2(g)$ in the soil atmosphere. Soil microbial respiration elevates $CO_2(g)$ partial pressures to between 10 and 1000 times the partial pressure in air, which in turn increases the total acidity of a soil solution. Soil microorganisms also produce organic acids, many of which have metal complexation and chelation capabilities. Similarly, plant roots respire CO_2, release protons, and also exude organic acids (and chelates), all of which effectively reduce the pH of the rhizosphere (the volume of soil that is in intimate contact with plant roots) 0.1 to 0.2 pH units below that of the bulk soil solution.

Hydrolysis is arguably the most important process in the chemical weathering of the silicate minerals. In general, characteristic hydrolysis reactions are a function of the complexity of the Si-tetrahedral linkages. For instance, the hydrolysis of a nesosilicate is generalized in the reaction:

$$\text{Silicate} + H_2O + H_2CO_3^0 \rightarrow \text{base cation} + HCO_3^- + H_4SiO_4^0 + \text{accessory mineral} \qquad (3.6)$$

In this reaction, a base cation would commonly represent Mg^{2+} or Ca^{2+}, $H_2CO_3^0$ is a proton source, HCO_3^- is bicarbonate, $H_4SiO_4^0$ is silicic acid, and gibbsite ($Al(OH)_3$) is a representative accessory mineral. Hydrolysis reactions of the more complex silicates, such as the inosilicates, phyllosilicates, and tectosilicates are similar to Equation 3.6 in form:

$$\text{Aluminosilicate} + H_2O + H_2CO_3^0$$

$$\rightarrow \text{clay mineral} + \text{base cations} + HCO_3^- + H_4SiO_4^0 + \text{accessory mineral} \qquad (3.7)$$

However, the greater complexity of the Si-tetrahedral linkages in an aluminosilicate allows for the generation of a clay mineral (secondary phyllosilicate) in addition to the components generated in Equation 3.6. Comparison of Equations 3.6 and 3.7 indicates that there are common characteristics associated with the silicate mineral hydrolysis reactions: consumption of H_2O and H^+, the release of base cations and silicic acid, and the formation of accessory minerals. An actual mineral hydrolysis reaction may include any number of these common characteristics.

Oxidation, which generally occurs in combination with a hydrolysis reaction, is a chemical weathering mechanism in which an element in a crystal structure loses an electron. The most common element to be oxidized during the weathering of primary soil minerals is Fe^{II} (as is Mn^{II}, but to a lesser extent), which is always found in the reduced state in primary minerals. There are two mechanisms by which oxidation proceeds: (1) oxidation of structural Fe^{II} as a process that initiates mineral breakdown and (2) oxidation of Fe^{2+} after Fe^{II} is released to the soil solution via hydrolysis. The oxidation of Fe^{II} to Fe^{III} in the mineral structure disrupts the electrostatic neutrality of the mineral. As a result, cations (other than Fe^{III}) leave the crystal so that electroneutrality is maintained. This leaves vacancies (holes) in the mineral structure. The mineral may then collapse, or may become more susceptible to other chemical weathering processes (e.g., hydrolysis, chelation, solutioning). The oxidation of Fe^{2+} after it is released by hydrolysis is a two-step process, as illustrated by the hydrolysis and oxidation of the olivine fayalite:

Step 1. Fayalite hydrolysis:

$$Fe_2SiO_4(s) + 4H^+(aq) \rightarrow 2Fe^{2+}(aq) + H_4SiO_4^0(aq) \qquad (3.8a)$$

Step 2. Fe^{2+} oxidation and the precipitation of $Fe^{III}OOH$ [goethite]:

$$2Fe^{2+}(aq) + 0.5O_2(g) + 3H_2O(l) \rightarrow 2FeOOH(s) + 4H^+(aq) \qquad (3.8b)$$

The summation of Equations 3.8a and 3.8b leads to the overall hydrolysis and oxidation reaction:

$$Fe_2SiO_4(s) + 0.5O_2(g) + 3H_2O(l) \rightarrow 2FeOOH(s) + H_4SiO_4^0(aq) \qquad (3.8c)$$

A more general olivine hydrolysis and oxidation reaction is:

$$MgFeSiO_4(s) + 0.25O_2(g) + 1.5H_2O(l) + 2H^+(aq) \rightarrow FeOOH(s) + Mg^{2+}(aq) + H_4SiO_4^0(aq) \quad (3.9)$$

Again, note the characteristics of this reaction: consumption of H_2O and H^+, release of silicic acid ($H_4SiO_4^0$) and base cation (Mg^{2+}), and the formation of the accessory mineral goethite (FeOOH).

3.2 BALANCING CHEMICAL REACTIONS

Reactions that occur in any environment must obey the Law of Mass Action: what goes in (reactant) must come out (product). A balanced chemical reaction has the following characteristics: mass (or mole) balance, charge balance, and reactant and product species that are appropriate for the chemical conditions of the environment in which the reaction occurs. In order to successfully create a balanced chemical weathering reaction, there are four sequential steps that must be followed: (1) determine the chemical formula of the minerals involved; (2) choose appropriate soluble species and balance all reactant and product components except hydrogen and oxygen; (3) balance the oxygen atoms using water; and (4) balance the hydrogen atoms using the hydrogen ion (proton). Consider the following three examples:

Example 1: The Congruent Dissolution of Diopside

The dissolution of a mineral to produce only soluble species, such that the molar stoichiometry of components in the reactant mineral and in an equilibrating solution phases are identical (congruent), is called a congruent dissolution reaction. In this type of reaction, the reaction products are all soluble species and the formation of one or more solids does not occur.

Step 1: Determine the chemical formula of the minerals involved. In this example, the chemical formula for diopside can be obtained from any number of reference sources: $CaMgSi_2O_6$. Because this is a congruent dissolution reaction, the products are all soluble species.

Step 2: Choose appropriate soluble species and balance all components except hydrogen and oxygen. In soil solutions, both Ca and Mg exist exclusively as divalent cations, Ca^{2+} and Mg^{2+}. The speciation of Si in soil solutions is principally a function of pH, as it exists as the polyprotic silicic acid, $H_4SiO_4^0$, which can dissociate to form $H_3SiO_4^-$, $H_2SiO_4^{2-}$, $HSiO_4^{3-}$, or SiO_4^{4-}, depending on solution pH. In typical soil solutions (pH 4 to 9), Si predominantly exists as the undissociated species $H_4SiO_4^0$. Thus, this species should be included as a soluble product in the dissolution reaction. Our Step 2 result is:

$$CaMgSi_2O_6(s) \rightarrow Ca^{2+}(aq) + Mg^{2+}(aq) + 2H_4SiO_4^0(aq) \qquad (3.10a)$$

Note that there is 1 mol of Ca in diopside balanced by 1 mol of Ca^{2+} in solution, 1 mol of Mg in diopside balanced by 1 mol of Mg^{2+} in solution, and 2 mol of Si in diopside balanced by 2 mol of $H_4SiO_4^0$ (2 mol of Si^{4+}) in solution (again, a requirement of congruency).

Step 3: Balance the oxygen atoms using water. In Equation 3.10a, there are 6 mol of oxygen in diopside on the reactant side of the expression, and 8 mol in the 2 mol of silicic acid on the product side. Therefore, an additional 2 mol of oxygen are required and must be added as a reactant, which is accomplished through the inclusion of 2 mol of water:

$$CaMgSi_2O_6(s) + 2H_2O(l) \rightarrow Ca^{2+}(aq) + Mg^{2+}(aq) + 2H_4SiO_4^0(aq) \qquad (3.10b)$$

Step 4: Balance the hydrogen atoms using the hydrogen ion (proton). In Equation 3.10b there are 4 mol of hydrogen in the two water molecules on the reactant side of the expression, and 8 mol in the 2 mol of silicic acid on the product side. Therefore, an additional 4 mol of hydrogen are required as a reactant:

$$CaMgSi_2O_6(s) + 2H_2O(l) + 4H^+(aq) \rightarrow Ca^{2+}(aq) + Mg^{2+}(aq) + 2H_4SiO_4^0(aq) \quad (3.10c)$$

This is the balanced chemical reaction for the hydrolysis and congruent dissolution of diopside. Compare the characteristics of the diopside hydrolysis reaction to the generalized hydrolysis reaction in Equation 3.6: water and protons are consumed and base cations and soluble silica are released. Also note that the source of the proton in Equation 3.10c is not included, such as $H_2CO_3^0$ in Equation 3.6. Equation 3.10c can also be written:

$$CaMgSi_2O_6(s) + 2H_2O(l) + 4H_2CO_3^0(aq)$$

$$\rightarrow Ca^{2+}(aq) + Mg^{2+}(aq) + 2H_4SiO_4^0(aq) + 4HCO_3^-(aq) \qquad (3.10d)$$

However, since there are numerous sources of protons in a soil solution in addition to carbonic acid, Equation 3.10c is preferred.

Example 2: The Incongruent Dissolution of Kaolinite to Form Gibbsite

The dissolution of a mineral to produce one or more mineral species is called an incongruent dissolution reaction. In this type of reaction, the molar stoichiometry of components in the reactant mineral and in an equilibrating solution phase is different (incongruent), and the formation of one or more product minerals occurs.

Step 1: Determine the chemical formula of the minerals involved. The chemical formula for kaolinite is $Al_2Si_2O_5(OH)_4$ and that for gibbsite is $Al(OH)_3$. This is termed an incongruent dissolution reaction because a solid is one of the products.

Step 2: Choose appropriate soluble species and balance all components except hydrogen and oxygen. For this example, 2 mol of gibbsite will balance the 2 mol of Al^{3+} in kaolinite, and 2 mol of silicic acid will balance the two moles of Si^{4+} in kaolinite:

$$Al_2Si_2O_5(OH)_4(s) \rightarrow 2Al(OH)_3(s) + 2H_4SiO_4^0(aq) \qquad (3.11a)$$

The reaction will be incongruent because the hydrolysis of kaolinite does not produce 2 mol of $H_4SiO_4^0$ and 2 mol of a soluble Al^{3+} species. Instead, the Al^{3+} is consumed by gibbsite.

Step 3: Balance the oxygen atoms using water. In Equation 3.11a there are 9 mol of oxygen in kaolinite on the reactant side of the expression, and 14 mol in 2 mol of gibbsite and 2 mol of silicic acid on the product side. Therefore, an additional 5 mol of oxygen is required as reactant:

$$Al_2Si_2O_5(OH)_4(s) + 5H_2O(l) \rightarrow 2Al(OH)_3(s) + 2H_4SiO_4^0(aq) \qquad (3.11b)$$

Step 4: Balance the hydrogen atoms using protons. In Equation 3.11b there are 14 mol of hydrogen in kaolinite and 5 mol of water on the reactant side of the expression, and 14 mol in 2 mol of gibbsite and 2 mol of silicic acid on the product side. Since the moles of proton in the products and reactants are equal, Equation 3.11b is balanced.

Example 3: The Weathering of Biotite to Form Trioctahedral Vermiculite and Accessory Goethite

This example of an incongruent dissolution reaction is slightly more complicated than the two previous examples. However, the steps required to generate a balance weathering reaction remain unchanged.

Step 1: The chemical formulas of the minerals involved are $K(Mg_2Fe^{II})(Si_3Al)O_{10}(OH)_2$ (biotite), $K_{0.7}(Mg_{2.9}Fe^{III}_{0.1})(Si_{3.2}Al_{0.8})O_{10}(OH)_2$ (trioctahedral vermiculite), and the accessory mineral $FeOOH$ (goethite).

Step 2: Choose appropriate soluble species and balance all components except hydrogen and oxygen. For this example, the following intermediate is obtained:

$$0.8[K(Mg_2Fe^{II})(Si_3Al)O_{10}(OH)_2(s)] + 0.8H_4SiO_4^0(aq) + 1.65Mg^{2+}(aq)$$

$$\rightarrow Mg_{0.35}(Mg_{2.9}Fe^{III}_{0.1})(Si_{3.2}Al_{0.8})O_{10}(OH)_2(s) + 0.7FeOOH(s) + 0.8K^+(aq) \quad (3.12a)$$

This expression is written such that the Al^{3+} in biotite is conserved in vermiculite. In addition to the mole masses of the component ions, there is one other component that must be considered and balanced in this reaction. In this weathering reaction, Fe^{II} in biotite is oxidized to Fe^{III} in vermiculite and goethite. For every mole of Fe^{II} oxidized to Fe^{III}, one mole of electrons is transferred; or the oxidation of 0.8 mol of Fe^{II} yields 0.8 mol of electron:

$$0.8Fe^{II} \rightarrow 0.8Fe^{III} + 0.8e^- \quad (3.12b)$$

Thus, Equation 3.12a can be balanced with respect to electrons using Equation 3.12b:

$$0.8K(Mg_2Fe^{II})(Si_3Al)O_{10}(OH)_2(s) + 0.8H_4SiO_4^0(aq) + 1.65Mg^{2+}(aq)$$

$$\rightarrow Mg_{0.35}(Mg_{2.9}Fe^{III}_{0.1})(Si_{3.2}Al_{0.8})O_{10}(OH)_2(s) + 0.7FeOOH(s) + 0.8K^+(aq) + 0.8e^-$$
$$(3.12c)$$

Unlike protons (H^+ ions), electrons are not true aqueous species; they are transferred from one electroreactive species to another. Therefore, an electroreactive species is needed to accept the electrons and complete the oxidation-reduction coupling. For this example, $O_2(g)$ will be used to accept the electrons from Fe^{II}, resulting in the production of O^{2-}. A further complicating factor in the $O_2(g)$ reduction process is that O^{2-} is not a species that exists in aqueous solutions. However, the oxidation state of oxygen in water is $-II$, which leads to the reduction half reaction:

$$0.2O_2(g) + 0.8e^- + 0.8H^+ \rightarrow 0.4H_2O(l) \quad (3.12d)$$

Combining Equations 3.12c and 3.12d, such that electrons are conserved yields:

$$0.8K(Mg_2Fe^{II})(Si_3Al)O_{10}(OH)_2(s) + 0.8H_4SiO_4^0(aq) + 1.65Mg^{2+}(aq)$$

$$+ 0.2O_2(g) + 0.8H^+(aq)$$

$$\rightarrow Mg_{0.35}(Mg_{2.9}Fe^{III}_{0.1})(Si_{3.2}Al_{0.8})O_{10}(OH)_2(s) + 0.7FeOOH(s)$$

$$+ 0.8K^+(aq) + 0.4H_2O(l) \quad (3.12e)$$

Step 3: Balance the oxygen atoms using water. In Equation 3.12e there are 13.2 mol of oxygen on the reactant side of the expression and 13.8 mol on the product side. Therefore, an additional 0.6 mol of oxygen (as water) is required as reactant:

$$0.8K(Mg_2Fe^{II})(Si_3Al)O_{10}(OH)_2(s) + 0.8H_4SiO_4^0(aq) + 1.65Mg^{2+}(aq)$$

$$+ 0.2O_2(g) + 0.8H^+(aq) + 0.6H_2O(l)$$

$$\rightarrow Mg_{0.35}(Mg_{2.9}Fe_{0.1}^{III})(Si_{3.2}Al_{0.8})O_{10}(OH)_2(s) + 0.7FeOOH(s)$$

$$+ 0.8K^+(aq) + 0.4H_2O(l) \tag{3.12f}$$

The 0.6 reactant H_2O minus the 0.4 product H_2O yields 0.2 reactant H_2O, simplifying Equation 3.12f to:

$$0.8K(Mg_2Fe^{II})(Si_3Al)O_{10}(OH)_2(s) + 0.8H_4SiO_4^0(aq) + 1.65Mg^{2+}(aq)$$

$$+ 0.2O_2(g) + 0.8H^+(aq) + 0.2H_2O(l)$$

$$\rightarrow Mg_{0.35}(Mg_{2.9}Fe_{0.1}^{III})(Si_{3.2}Al_{0.8})O_{10}(OH)_2(s) + 0.7FeOOH(s) + 0.8K^+(aq) \tag{3.12g}$$

Step 4: Balance the hydrogen atoms using protons. In Equation 3.12g there are 6 mol of hydrogen on the reactant side of the expression and 2.7 mol on the product side. Therefore, the reaction in Equation 3.12g requires an additional 3.3 mol of H^+ on the product side:

$$0.8K(Mg_2Fe^{II})(Si_3Al)O_{10}(OH)_2(s) + 0.8H_4SiO_4^0(aq) + 1.65Mg^{2+}(aq)$$

$$+ 0.2O_2(g) + 0.8H^+(aq) + 0.2H_2O(l)$$

$$\rightarrow Mg_{0.35}(Mg_{2.9}Fe_{0.1}^{III})(Si_{3.2}Al_{0.8})O_{10}(OH)_2(s) + 0.7FeOOH(s)$$

$$+ 0.8K^+(aq) + 3.3H^+(aq) \tag{3.12h}$$

The 3.3 product H^+ minus the 0.8 reactant H^+ yields 2.5 product H^+, simplifying Equation 3.12h to:

$$0.8K(Mg_2Fe^{II})(Si_3Al)O_{10}(OH)_2(s) + 0.8H_4SiO_4^0(aq) + 1.35Mg^{2+}(aq)$$

$$+ 0.2O_2(g) + 0.2H_2O(l)$$

$$\rightarrow Mg_{0.35}(Mg_{2.9}Fe_{0.1}^{III})(Si_{3.2}Al_{0.8})O_{10}(OH)_2(s) + 0.7FeOOH(s)$$

$$+ 0.8K^+(aq) + 2.5H^+(aq) \tag{3.12i}$$

This is a balanced weathering reaction for the hydrolysis and oxidation of the primary phyllosilicate biotite to form the clay mineral trioctahedral vermiculite and the accessory mineral goethite.

3.3 MINERAL STABILITY: PRIMARY SILICATES IN THE SAND- AND SILT-SIZED FRACTIONS

The susceptibility of the primary silicate minerals to weathering is a function of the environment, particle size, and crystal chemistry. Goldich (1938) examined the mineral assemblages present in soil under a variety of environmental conditions. From his observations he developed a stability series for sand- and silt-sized particles that illustrates the relative stability (weatherability) of the more common primary silicate minerals (Figure 3.1). Minerals that reside at the highest positions of the stability diagram, such as olivine and Ca-plagioclase (anorthite), are the most easily weathered and are the first to disappear from the sand- and silt-sized fraction of a soil as a result of chemical weathering. Minerals at the bottom of the diagram, such as quartz and muscovite, are the most resistant to weathering, and are expected to remain in soils that have been extensively weathered.

Goldich (1938) noted that the order of decreasing mineral stability was identical to the order of silicate crystallization from a cooling magma, formally described as the Bowen reaction series (Bowen, 1922). Minerals that are the most stable at high magma temperatures, such as the olivines, Ca-plagioclase, and the pyroxenes, are the first to crystallize from a cooling magma and correspondingly have high base and divalent cation contents; whereas, minerals that are most stable at low magma temperatures, such as K-feldspar and quartz, crystallize last from a cooling magma and are principally composed of Si and Al (with a low base cation content, with the exception of K^+). The observation that relative mineral stability (Goldich stability series) closely follows the order of crystallization (Bowen reaction series) has been explained by the degree of disequilibrium between the high temperature magma environment in which a mineral forms and the low temperature

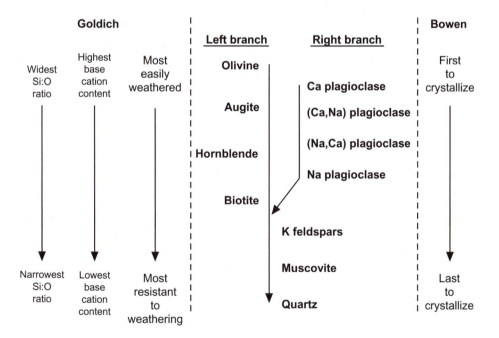

FIGURE 3.1 The Goldich stability series for sand- and silt-sized particles illustrates the relative weatherability of the primary silicate minerals in soil and sediment. Increasing resistance to alteration is a function of base cation (Ca^{2+}, Mg^{2+}, and Na^+) content and the degree of covalent character in the chemical bonds, as indicated by the ratio of Si:O in the mineral. The Bowen reaction series, which illustrates the order of mineral crystallization from a cooling magma, is identical to the Goldich stability series, indicating that minerals formed at low magma temperatures are more stable in the terrestrial environment than minerals formed at high magma temperatures.

terrestrial environment in which a mineral weathers. A mineral that forms first from a cooling magma, such as an olivine, is forming at a temperature that is vastly different from that of a weathering environment. A mineral that forms at a lower temperature, such as a K-feldspar, is forming at a temperature much closer to that of a weathering environment. Because the temperature of formation of the K-feldspar is closer to soil temperatures, it is relatively more stable in the environment than an olivine, and thus would be more resistant to weathering.

The degree of disequilibrium between the temperature of mineral formation and that of the soil environment qualitatively explains the relative stabilities of the primary silicates (and was the explanation offered by Goldich). However, there are a number of mineral-specific factors that serve to explain the relative stabilities of the sand- and silt-sized primary silicates. The relative susceptibility of the primary silicates to alteration is due to the number of tetrahedral linkages, the presence of cations other than Al^{3+} and Si^{4+} (such as the base cations), the relative number of Al^{3+} to Si^{4+} tetrahedra, and other structural factors (presence of Fe^{II} and other oxidizable metals, empty coordinated positions, and packing density of the oxygen atoms). The Goldich stability series (Figure 3.1) is essentially a road map that illustrates the influence of these factors on mineral weathering. The left branch of the stability series represents the number of tetrahedral linkages and the base and divalent cation content of the minerals, which are inversely related properties. Olivines, which are nesosilicates, are composed of individual Si-tetrahedra linked through divalent cations. This class of silicates contain the smallest Si to O ratio (0.25) and the smallest amount of covalent character in the crystal-chemical bonds. Recall that the Si—O bond is approximately 50% covalent, while bonds involving divalent cations and oxygen are predominantly ionic (see Chapter 2). Augite, which is a pyroxene, is relatively more stable than an olivine. The pyroxene class of silicate minerals is composed of single chains of Si tetrahedra linked through basal oxygen atoms, with the individual chains linked through divalent cations (Ca^{2+}, Mg^{2+}, and Fe^{II} in the case of augite). Each Si tetrahedral unit is linked to two other units of a chain, such that the ratio of Si to O in the pyroxenes (0.33) is greater than that of the nesosilicates, indicating a greater degree of covalent character, and thus greater stability. The trend in increasing Si to O ratio and increasing stability continues through the amphiboles (0.364), the micas (0.4), to the tectosilicates (0.5). Quartz, a very persistent mineral in the sand- and silt-sized fraction of soil, is composed of a three-dimensional network of Si tetrahedra in which each oxygen atom is shared by neighboring tetrahedra. This maximum sharing of oxygen atoms imparts greater structural stability as a result of the high percentage of covalent character in the Si—O bond.

The trend in increasing mineral stability with increasing Si to O ratio is related to the trend in increasing mineral stability with decreasing base cation content. The degree of stability of a mineral is determined by the weakest bond in the structure, as these bonds are the first to be broken in a weathering environment. In all the silicate minerals, with the exception of quartz, the weakest bond is that which holds the tetrahedral structural units together or balances the charge of the aluminum tetrahedron. Many of these bond types are predominantly ionic in nature, such as the alkali and alkaline earth element bonds with oxygen (the Mg^{2+}, Ca^{2+}, Na^+, and K^+ bonds with oxygen are 74%, 79%, 82%, and 84% ionic). However, even bonds that approach the Si—O bond in covalent character, such as the Fe^{II}—O (~50% ionic) and Mn^{II}—O (~60% ionic) bonds, are relatively weak bonds due to the low valency of the cations (typically +2) and the high coordination they occupy (typically octahedral or CN = 6). For example, both the Fe^{II}—O and Mn^{II}—O bonds in CN 6 coordination have a strength of +0.33, compared to the bond strengths of +1 for Si in CN 4, +0.75 for Al in CN 4, and +0.5 for Al in CN 6. In the nesosilicates, the weakest bond is that which links the individual tetrahedra, including the Mn^{II}—O, Fe^{II}—O, Mg—O, and Ca—O linkages. These are also the weakest bond types in the sorosilicates and cyclosilicates, linking the double tetrahedral and cyclic units. The weakest bonds in the inosilicates involve the divalent base cations and other cations that link the single and double tetrahedral chains. The weakest bond in the primary phyllosilicates (the mica minerals) involves the cation which binds the basal oxygen atoms of one tetrahedral sheet to the basal oxygen atoms of an adjacent tetrahedral layer; typically this cation is K^+.

Although mineral stability is related to the number of bonds that bridge the structural tetrahedral units (base cation content), there are exceptions to the rule. For example, zircon ($ZrSiO_4$) is a nesosilicate that is very resistant to weathering. The Zr—O bond is approximately 67% ionic, and Zr^{4+} is coordinated by eight oxygen atoms (bond strength is +0.5). None of these characteristics would suggest a structure that is particularly resistant to weathering. The difference between olivine, one of the least resistant minerals, and zircon is the packing density of the oxygen atoms. The volume of the olivine unit cell is 0.309 nm^3, while the volume of the zircon unit cell, containing the same number of atoms as the olivine unit cell, is 0.261 nm^3, a characteristic that is related to the high charge density of Zr^{4+}. The greater packing density in zircon imparts stability to the mineral.

The isomorphic substitution of Al^{3+} for Si^{4+} in tetrahedral coordination tends to reduce mineral stability. The impact of such a substitution on mineral weathering can be seen in the plagioclase series (the right branch of the stability series). The stability of the plagioclase minerals increases with decreasing Ca^{2+} content and increasing Na^+ content. Anorthite, the calcium end-member of the plagioclase series ($CaAl_2Si_2O_8$), has the greatest amount of isomorphic substitution of Al^{3+} for Si^{4+}. Conversely, albite, the sodium end-member of the plagioclase series ($NaAlSi_3O_8$), has the smallest amount of isomorphic substitution. There are two principal factors that explain the greater stability of albite relative to anorthite. First, the Al—O bond has greater ionic character (~60%) relative to the Si—O bond (~50%). Anorthite, with a greater amount Al^{3+} substitution, has less covalent character in the crystal-chemical bonds than does albite, and thus is relatively less stable. Second, the radius ratio of Al^{3+} with oxygen as the anion is 0.427 (see Pauling's rules in Chapter 2), which is greater than the limiting radius ratio for sixfold (octahedral) coordination (0.414). Consequently, the Al-tetrahedron is distorted to accommodate the greater radius of Al^{3+} relative to Si^{4+}, which distorts and destabilizes the mineral structure. The degree of distortion and thus mineral instability is directly related to the degree of isomorphic substitution of Al^{3+} for Si^{4+}.

Iron is a common element in the primary silicate minerals, particularly in those minerals with high base cation content and high susceptibility to weathering. Almost all iron in the primary silicates occurs in the Fe^{II} oxidation state. In the presence of O_2, Fe^{II} and the less common Mn^{II} are oxidized to Fe^{III} and Mn^{VI} (as well as Mn^{III} and Mn^{IV}), which creates a charge imbalance in the mineral structure. To compensate for this electrostatic imbalance, cations other than Fe^{III} and Mn^{VI} must leave the structure. Departing cations leave empty positions, causing the structure to collapse or increasing the susceptibility of the mineral to other weathering mechanisms (such as hydrolysis, acidification, and solutioning). In addition, the loss of an electron reduces the radius of an element. The radius of Fe^{II} in octahedral coordination is 0.075 nm, while that of Fe^{III} is 0.069 nm. The radius of Mn^{II} in octahedral coordination is 0.081 nm. Manganese (VI), which has a radius of 0.040 nm, is unstable in octahedral coordination, existing instead in tetrahedral coordination. The reduction in cation size and the potential change in the stable coordination of an element places additional stress on mineral structures and increases mineral susceptibility to decomposition.

3.4 MINERAL STABILITY: CLAY-SIZED FRACTION

The mineral fraction of soil consists of sand-, silt-, and clay-sized materials. The sand- and silt-sized particles are largely composed of primary minerals that are inherited from parent rock. The relative susceptibility of sand- and silt-sized primary silicate minerals to weathering in soil and sediment is established in the Goldich stability series. The series remains a useful tool for establishing the relative importance of the various mineral-specific factors that impact stability. However, the weathering series does not correspond to a weathering sequence. For example, olivine is an unstable mineral relative to an amphibole in a weathering environment; however, olivine does not weather to amphibole, nor does anorthite weather to albite.

The clay-sized fraction contains minerals that represent the weathering stage of a soil. In a general sense, the occurrence of primary silicates in the clay fraction is indicative of a soil that has not been subjected to the environmental forces that promote weathering. The predominance of

TABLE 3.2
The Jackson Weathering Stages of Clay-Sized Minerals[a]

Index	Typical Minerals	Mineral Characteristics
	Early Stages (young soils)	
1	Gypsum, halite, sulfides, soluble salts	Soluble evaporites
2	Calcite, dolomite, apatite	Sparingly soluble carbonates and phosphates
3	Olivine, amphiboles, pyroxenes	Primary silicates with high base cation content
4	Biotite, glauconite, mafic chlorite, nontronite	Primary phyllosilicates with high Fe^{II} and Mg content (trioctahedral)
5	Feldspars (albite and K-feldspars)	Tectosilicates with isomorphic substitution
6	Quartz	Tectosilicates without isomorphic substitution
	Intermediate Stages	
7	Muscovite, illite	Primary phyllosilicates with low Fe^{II} and Mg content and their initial alteration products
8	Vermiculite, interstratified 2:1 layer silicates	Mica alteration products with high Fe^{III} and Mg
9	Smectites, HIV/HIS, Al-chlorite	Low Fe^{III} and Mg and Al hydroxyl-interlayered phyllosilicates (pedogenic chlorite)
	Advanced Stages (highly weathered soils)	
10	Kaolinite, halloysite	Base cation-poor 1:1 phyllosilicates
11	Gibbsite, allophane	Al-bearing accessory minerals
12	Iron oxides (goethite, hematite)	Fe^{III}-bearing accessory minerals
13	Titanium oxides (anatase, rutile, ilmenite), zircon, corundum	Remnant minerals

[a] Jackson et al. (1948), Jackson and Sherman (1953), and Jackson (1964).

secondary phyllosilicate minerals in the clay fraction is a clear indication that weathering has occurred, the degree of weathering depending on the type of clay present. Finally, the predominance of accessory minerals in the clay fraction, or primary minerals that are highly resistant to weathering, indicates that the soil has been intensively weathered.

The relationship between weathering intensity and the mineral assemblages present in the clay fraction of a soil is described in the weathering sequence of Jackson et al. (1948) and Jackson and Sherman (1953) (Table 3.2). Jackson reasoned that the mineral composition of the clay-sized fraction could be employed to establish the weathering stage of a soil. Common minerals that are found in the clay fraction of a soil are identified by an index number: low numbers represent minerals that are easily weathered and most abundant in young soils; whereas, high numbers represent minerals that are relatively resistant to weathering and abundant in old soils. The clay fraction of a soil is typically composed of three to five dominant minerals. A weighted average of the index numbers for these minerals establishes the stage of weathering of the soil. In the context of mineral weathering, soil age is not related to temporal age, but to the amount of water that has leached through the soil profile. The rate of mineral alteration and the associated index number are not linearly related. Minerals in unweathered and slightly weathered soils (up to and including index number 5), including evaporites (e.g., gypsum and calcite) and the primary silicates (except of quartz and muscovite), may rapidly weather given favorable conditions (water). Minerals described in indices 6 through 10, primarily the secondary phyllosilicates, weather slowly, while minerals in indices 10 through 13, kaolinite and the accessory minerals, are virtually resistant to weathering (as many formed in a weathering environment).

The Goldich stability series and the Jackson weathering sequence provide similar information on the relative stability of common primary soil minerals. However, the Jackson weathering stages are only applicable to clay-sized minerals, and in addition to primary minerals, include evaporites, secondary phyllosilicates, and accessory minerals. In both schemes, increasing stability is associated with decreasing degree of tetrahedral linkages, decreasing base cation content, decreasing Fe^{II} content, and decreasing Al content. It is notable that the Goldich series mirrors the Jackson weathering sequence for the primary minerals, with the exception of quartz and muscovite which are inverted in the sequence; however, these similarities end with index numbers greater than 6.

3.5 WEATHERING AND FORMATION CHARACTERISTICS OF THE PHYLLOSILICATES

The general weathering sequence offered by Jackson et al. (1948) and Jackson and Sherman (1953) describes the binary or incongruent weathering and the sequence of formation of the phyllosilicates. The Jackson weathering sequence remains a valid tool for describing the general transitions that occur in a weathering environment. However, the specific weathering and alteration reactions of a phyllosilicate mineral in soil are a function of the physical–chemical characteristics of the mineral and the environmental conditions, specifically the chemical composition of the soil solution. The following sections discuss the mineral characteristics and soil solution properties that impact the alteration of primary phyllosilicates and the alteration and formation of secondary phyllosilicates.

3.5.1 MICA

The general weathering characteristics of the primary phyllosilicate minerals (the mica minerals) are unique in comparison to other silicate minerals. While minerals in many silicate classes dissolve and the components leach or precipitate as accessory minerals (complex transformation), the mica minerals are altered via solid–solid transformations to form secondary phyllosilicates, or clay minerals (simple transformation). Consider the simple and incongruent weathering of muscovite and its transformation into dioctahedral vermiculite:

$$KAl_2(Si_3Al)O_{10}(OH)_2(s) + 0.3H_4SiO_4^0(aq) + 0.3H^+(aq) + 0.35Mg^{2+}(aq)$$

$$\rightarrow Mg_{0.35}Al_2(Si_{3.3}Al_{0.7})O_{10}(OH)_2(s) + 0.3Al(OH)_3(s) + K^+(aq) + 0.3H_2O(l) \tag{3.13}$$

This idealized reaction contains numerous salient features and indicates the environmental conditions necessary for mica weathering and vermiculite stability. First, examine the essential differences between the parent mica mineral and the daughter vermiculite. One of the main differences is the reduced layer charge of vermiculite relative to muscovite. This is a defining characteristic of phyllosilicate minerals at the group level of the classification scheme. The reduction in layer charge (from -1 to -0.7) results from the loss of Al^{3+} from the tetrahedral layer (from 1 to 0.7 atoms per $\frac{1}{2}$-unit cell). The reduction in tetrahedral Al^{3+} is also accompanied by a widening of the Si to Al tetrahedral ratio from 3 to 1 in mica to 3.3:0.7 in the vermiculite. The presence of Al^{3+} in tetrahedral coordination is a high temperature phenomenon (recall that the radius ratio for Al^{3+} is not consistent with tetrahedral coordination). Under terrestrial and acidic conditions, noting the proton consumption in Equation 3.13, Al is stable in octahedral coordination, as found in the accessory mineral gibbsite which is formed during the weathering process. The widening Si to Al ratio also requires the incorporation of Si^{4+} into the vacated tetrahedral positions, or the exchange of Si^{4+} for Al^{3+}. According to Equation 3.13, this is accomplished through the consumption of $H_4SiO_4^0$. Another salient feature of mica weathering is the displacement of interlayer K^+ by another (hydrated) base cation, represented by Mg^{2+}. Thus, K^+ is a product and Mg^{2+} is a reactant. Finally, vermiculite inherits the octahedral occupation of the mica; the structural framework is maintained. In simple transformation processes, a dioctahedral mineral can only weather to another dioctahedral mineral and a trioctahedral can only weather to another trioctahedral mineral. Based on the reaction in Equation 3.13, the environmental conditions that should hasten the weathering of muscovite are: high soluble silica and base cation (Mg^{2+}, Ca^{2+}, or Na^+) concentrations, low K^+ concentrations, and acidic conditions.

Another example of a simple and incongruent weathering process is the transformation of biotite into trioctahedral vermiculite:

$$K(Mg_2Fe^{II})(Si_3Al)O_{10}(OH)_2(s) + 0.25O_2(g) + 0.2H_4SiO_4^0(aq)$$

$$+ 1.1H_2O(l) + 1.25Mg^{2+}(aq)$$

$$\rightarrow Mg_{0.35}(Mg_{2.9}Fe_{0.1}^{III})(Si_{3.2}Al_{0.8})O_{10}(OH)_2(s) + 0.2Al(OH)_3(s)$$

$$+ 0.9FeOOH(s) + K^+(aq) + 1.5H^+(aq) \tag{3.14}$$

Again, the reduction in layer charge (from -1 to -0.7) confirms the transformation of biotite into the vermiculite. However, unlike the alteration of muscovite into dioctahedral vermiculite, the reduction in layer charge results from the loss of tetrahedral Al^{3+}, as well as the oxidation and expulsion of Fe from the mineral structure. The reduction in tetrahedral Al^{3+} is accompanied by a widening of the Si to Al tetrahedral ratio from 3:1 in biotite to 3.2:0.8 in the vermiculite. This alteration is facilitated by the exchange of Si^{4+} for Al^{3+} and the precipitation of gibbsite. Most of the Fe^{II} in the octahedral layer is oxidized to Fe^{III} which is expelled (possibly replaced by Mg^{2+}) to form goethite. In this example (Equation 3.14), the Fe^{III} that remains in the octahedral layer contributes positive charge to the mineral structure and partially satisfies the tetrahedral layer charge, yielding a net layer charge of -0.7. The layer charge is satisfied by Mg^{2+}, which displaces interlayer K^+. Again, the vermiculite inherits the octahedral occupation of the biotite, and the structural framework is maintained. Based on the reaction in Equation 3.14, the environmental conditions that should hasten the weathering of biotite are: high concentrations of soluble silica and base cations (other than K^+, such as Mg^{2+}, Ca^{2+}, or Na^+), low K^+ concentrations, the presence of dissolved oxygen (or some other suitable oxidant), and basic conditions.

Many of the characteristics of the above mica weathering reactions are discussed in detail below. However, one can compile a list of characteristics that are generally applicable to primary phyllosilicate hydrolysis and oxidation weathering reactions:

- Reduction in the amount of Al^{3+} in tetrahedral coordination and a widening of the Si to Al ratio
- Oxidation of Fe^{II} and expulsion of Fe^{III} from the crystal lattice
- Reduction in layer charge
- Formation of accessory minerals
- Protons consumed or released depending on extent of Fe^{II} oxidation
- Displacement of K^+ by exchange with water and hydrated base cations
- Consumption of soluble silica and base cations (other than K^+)

In general, the resistance of the micas to weathering is a function of the temperature under which they were formed: the higher the formation temperature, the lower the resistance to weathering. Biotite and other trioctahedral micas are rich in Mg and Fe^{II} and crystallize at a higher temperature than muscovite and other dioctahedral and base-poor micas. As a result, biotite is less resistant to weathering. The weathering of K^+-bearing micas to form expandable 2:1 minerals, such as vermiculite and illite, occurs through simple transformations (skeletal structure maintained) through the reduction in layer charge and the replacement of interlayer K^+ by hydrated cations (e.g., Mg^{2+}, Ca^{2+}, and Na^+) and water. The reduction in layer charge reduces the force of attraction between the interlayer K^+ and the mica layers, allowing for the hydration of adsorbed K^+ and its subsequent replacement by cations that bear greater charge and that are more common: Ca^{2+} and Mg^{2+} (electrostatic and mass action principles). The net result of this process is an expansive interlayer. There are two general mechanisms by which micas weather depending on particle size: layer weathering and edge weathering. In layer weathering, some layers are opened up all the way through the mica particle. This type of weathering is common to clay-sized mica particles and typically leads to interstratified mica-vermiculite and mica-smectite minerals (Figure 3.2). Edge weathering,

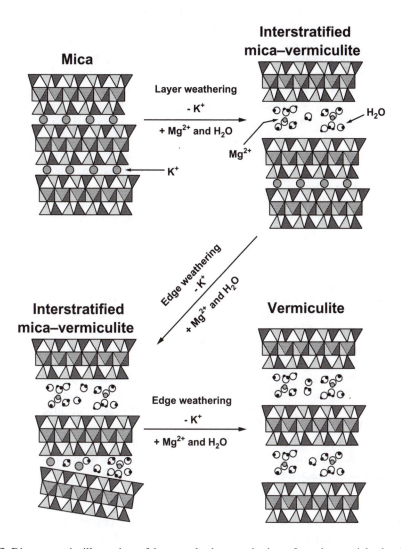

FIGURE 3.2 Diagrammatic illustration of layer and edge weathering of a mica particle that leads to the formation of vermiculite. Clay-sized mica particle can develop preferential weathering planes (layer weathering). Coupled with the reduction in layer charge, the K^+ in the interlayer is replaced by hydrated cations, resulting in an interstratified mica-vermiculite particle. The process is facilitated by elevated solution Mg^{2+} concentrations and depressed solution K^+ concentrations, and stabilized through the realignment of structural protons in the octahedral layer of the mica. The weathering of mica particles also proceeds through the progressive wedging action of hydrated cations at the particle edge that displace K^+. Ultimately, an expansive 2:1 layer silicate (e.g., vermiculite) is formed; the interlayers containing hydrated cations.

commonly associated with sand- and silt-sized particles, involves the simultaneous expansion of many layers along edges of and fractures in mica particles.

Under natural conditions, the weathering of mica to expandable 2:1 minerals is accompanied by a reduction in layer charge. As described above, the principle mechanisms thought to be responsible for the reduction in layer charge are the loss of Al^{3+} from the tetrahedral layer (accompanied by the incorporation of Si^{4+}) and the oxidation of structural Fe^{II} and expulsion of Fe^{III}. These two mechanisms are supported by observations that the simple weathering products of the mica minerals, the vermiculites, contain a wider ratio of Si^{4+} to Al^{3+} in the tetrahedral layer (between 3.1:0.9 and 3.4:0.6 compared to 3:1), as well as a lower Fe content compared to the micas. In addition to the mechanisms described above, several other possible mechanisms have been proposed to account for layer charge reduction. The oxidation of structural Fe^{II} to Fe^{III} that occurs during or following the opening of the biotite interlayers is often cited as an important mechanism for layer charge reduction and is thought to enhance the release of K^+; both are required for the formation of expandable 2:1 minerals. However, the positive charge generated through the oxidation of Fe^{II} may be neutralized by mechanisms other than the release of interlayer K^+. The oxidation of octahedral Fe^{II} can be accompanied by loss of protons from structural octahedral hydroxyl groups. Not only does the release of the proton conserve the layer charge, but there is a stabilization of interlayer K^+, retarding its release. The oxidation of structural Fe^{II} can be accompanied by the expulsion of cations instead of or in addition to Fe^{III}. The expulsion of cations, such as octahedrally coordinated Mg^{2+} ions, may also conserve layer charge. It has also been suggested that the incorporation of protons into the mica structure by combination with apical oxygen atoms of the Si tetrahedron reduce layer charge. Other changes in the mica structure that occur under acidic conditions and following K^+ release are the dissolution of the octahedral sheet, particularly in trioctahedral minerals, and the production of Al hydroxy-interlayers, resulting in the conversion of mica to pedogenic chlorite.

The factors that impact the rate and extent of mica weathering by simple transformation are the nature of the mineral, particle size, and the environment. In addition to the destabilizing effects of Fe^{II} and octahedral Mg^{2+} on trioctahedral micas, the orientation of the proton on structural octahedral hydroxyls also serves to destabilize the structure. In the 2:1 layer silicates, structural OH is a component of the octahedral layer and bonded to two Al^{3+} ions in a dioctahedral mineral, or three Mg^{2+} ions in a trioctahedral mineral. In the dioctahedral structure, the proton is shifted toward the open position in the octahedral layer (Figure 3.3). In the trioctahedral structure, the proton points directly toward the interlayer K^+ ion (Figure 3.4). Thus, the proton is closer to the interlayer K^+ in trioctahedral micas, resulting in greater electrostatic instability. Because of this difference in proton orientation, the dioctahedral micas hold K^+ more tightly than trioctahedral micas. The orientation of the structural hydroxyl proton is also a factor in the stabilization of alternating mica layers that occurs during layer weathering. With layer weathering, O—H bonds shift toward expanded interlayers so that the K^+ is held more tightly in adjacent interlayers (Figure 3.5).

The soil solution chemical composition is the principal environmental factor that influences mica alteration. Low K^+ concentration in solution promotes the release of K^+ from mica interlayers. The weathering reactions of muscovite and biotite (Equations 3.13 and 3.14) illustrate the release of K^+ during simple transformation to vermiculites. Since these reactions are subject to the laws of mass action, high soluble K^+ levels coupled with low Mg^{2+} levels will retard vermiculite formation. Conversely, low soluble K^+ levels coupled with high Mg^{2+} levels will promote vermiculite formation. The actual concentration of soluble K^+ in a soil solution required to stabilize a mica structure is an innate property of the particular mineral. One way to assess the impact of soluble K^+ on mica stability is through the application of equilibrium solubility techniques (see Chapter 6). For simplicity, consider the incongruent dissolution of phlogopite to form gibbsite and quartz (instead of vermiculite):

$$KMg_3(Si_3Al)O_{10}(OH)_2(s) + 7H^+(aq)$$

$$\rightarrow Al(OH)_3(s) + K^+(aq) + 3Mg^{2+}(aq) + 3SiO_2(s) + 3H_2O(l) \qquad (3.15a)$$

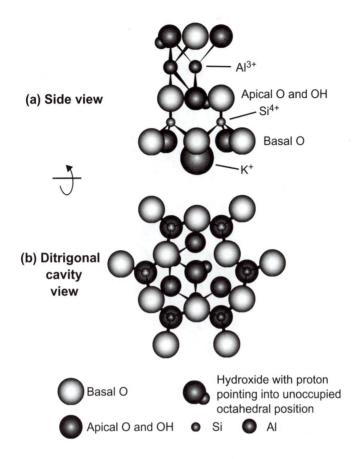

(a) Side view

Al³⁺

Apical O and OH
Si⁴⁺

Basal O

K⁺

(b) Ditrigonal cavity view

Basal O

Apical O and OH ● Si ● Al

Hydroxide with proton pointing into unoccupied octahedral position

FIGURE 3.3 In dioctahedral minerals, the proton on a structural OH is shifted to the empty octahedral location, pointing away from the interlayer K^+. The relatively low electrostatic repulsion between structural protons and interlayer K^+ ions is thought to make dioctahedral 2:1 layer silicates more stable than their trioctahedral counterparts.

The equilibrium constant and relationship that describes this reaction is:

$$K_{\text{Phlogopite}} = \frac{(K^+)(Mg^{2+})^3}{(H^+)^7} = 10^{31.06} \tag{3.15b}$$

where $K_{\text{Phlogopite}}$ is the equilibrium constant and the parentheses denote the concentration variable of the enclosed species (in a dilute soil solution, molar concentration units are applicable). For a pH 7 soil solution that contains $10^{-5}\ M\ Mg^{2+}$, the molar concentration of K^+ in equilibrium with phlogopite is computed using Equation 3.15b to be $1.15 \times 10^{-3}\ M$, or 28 mg L⁻¹ K^+. Therefore, phlogopite is predicted to support a K^+ concentration of 28 mg L⁻¹. Potassium ion concentrations that are greater than this value will stabilize phlogopite relative to gibbsite. However, solution K^+ concentrations less than the predicted value will result in the release of K^+ and the weathering of phlogopite. The identical analysis can be performed for muscovite. The incongruent dissolution of muscovite to form gibbsite and quartz is given by:

$$KAl_2(Si_3Al)O_{10}(OH)_2(s) + 3H_2O(l) + H^+(aq) \rightarrow 3Al(OH)_3(s) + K^+(aq) + 3SiO_2(s) \tag{3.15c}$$

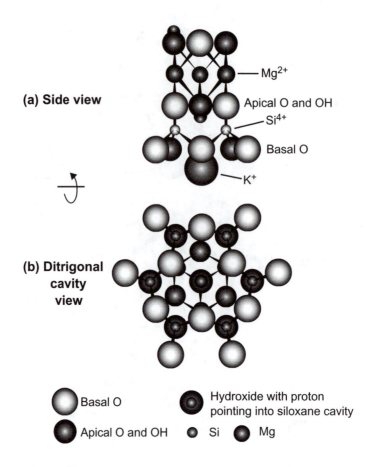

FIGURE 3.4 In trioctahedral minerals, the proton on a structural OH points directly toward the interlayer K^+ ion, as all three octahedral positions are occupied by a divalent cation. This destabilizes the 2:1 trioctahedral structure relative to dioctahedral minerals, resulting in the greater weatherability of trioctahedral minerals.

The equilibrium constant for this reaction is:

$$K_{\text{Muscovite}} = \frac{(K^+)}{(H^+)} = 10^{1.59} \tag{3.15d}$$

Again, assuming a soil pH of 7, the predicted concentration of K^+ that is supported by muscovite in an equilibrated solution is 3.94×10^{-6} M, or 0.15 mg L^{-1}. Solution K^+ concentrations less than this value will result in K^+ release from muscovite; whereas, greater K^+ concentrations will stabilize muscovite. Comparison of the K^+ concentrations supported by phlogopite (28 mg L^{-1}) and muscovite (0.15 mg L^{-1}) indicates the relative stabilities of the two minerals. The trioctahedral phlogopite requires high concentrations of solution K^+ to remain a stable phase. Solution K^+ concentrations below 28 mg L^{-1} will result in phlogopite weathering, while dioctahedral muscovite remains stable.

Another solution component important in the weathering of the mica minerals is dissolved organic carbon. The presence of soluble organic compounds bearing carboxylic and amine functional groups, or other electron donor groups, can promote the rapid dissolution of the octahedral layer of the mica minerals. These organic moieties have a high affinity for Fe^{3+} and Al^{3+}, forming aqueous complexes (e.g., chelated metal species), shifting the chemical equilibrium toward mineral dissolution. This effect is enhanced under acidic soil conditions, as Fe^{3+} and Al^{3+} display greater inherent solubility as soil solution pH decreases. The trioctahedral micas, particularly biotite, are

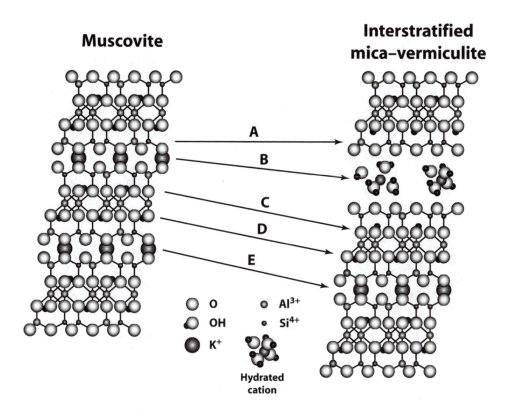

FIGURE 3.5 This schematic depiction illustrates the layer weathering of mica minerals to form interstratified mica-vermiculite phases. (a) Muscovite weathering is initiated through the exchange of tetrahedrally coordinated Al^{3+} by Si^{4+} (A), leading to a reduction in layer charge. As a result, interlayer K^+ is replaced by hydrated cations (B). The loss of K^+ and the expansion of the layer allow the structural protons to shift toward the interlayer (C). Additionally, protons in the adjacent structural layer (D) shift even further toward the open octahedral position, minimizing the repulsive electrostatic interaction with interlayer K^+. This shift in proton orientation, and the resulting reduction in electrostatic interactions, stabilizes the collapsed layer (E), leading to the formation and stability of the interlayered mica-vermiculite. (b) The layer weathering of biotite to form an interstratified mica-vermiculite is initiated by the oxidation of octahedral Fe^{II} and the expulsion of an Mg^{2+} ion from the octahedral layer (A). The exchange of Si^{4+} for tetrahedral Al^{3+} leads to reduction of layer charge (B). As a result, interlayer K^+ is displaced by hydrated cations (C). Loss of octahedral cations allows octahedral protons to shift toward the open positions (D and E), reducing the electrostatic repulsion between structural protons and K^+ ions in the interlayer adjacent to the expansive layer (F); thus, stabilizing the interstratified mineral.

most susceptible to this weathering mechanism. Indeed, in the presence of organic chelates the biotite structure can disintegrate at a rate that exceeds the rate of K^+ exchange from the interlayers. Further, these conditions can also result in complex transformations: the complete dissolution of the mica and the precipitation of kaolinite and accessory minerals (i.e., structure is not inherited). The formation of kaolinite and gibbsite from mica is common in intensely weathered soils, such as soils found in tropical climates.

Soil conditions that promote mica weathering—elevated concentrations of organic acids, soil acidity, and reduced K^+ concentrations—are typical of rhizosphere soil and surface soil horizons containing an active biological component. Barring the complete dissolution of a mica mineral, the mica alteration products formed during weathering are determined by the octahedral occupation of the primary mineral, particle size, soil moisture regime, and the intensity of weathering. As described above, trioctahedral mica minerals are particularly susceptible to weathering, due to the presence of Fe^{II}, the repulsive electrostatic interactions between structural hydroxyl protons and

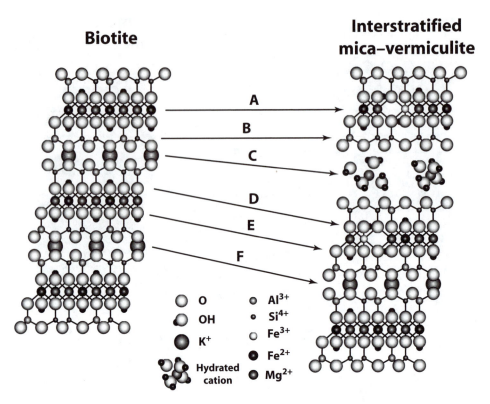

FIGURE 3.5 (*continued*).

interlayer K^+ ions, and a high base cation (Mg^{2+}) content. The initial product of biotite weathering is the interstratified layer silicate hydrobiotite (biotite-trioctahedral vermiculite) (Figure 3.2). Additional transformation of the biotite layers leads to trioctahedral vermiculite, then dioctahedral vermiculite through the loss of divalent octahedral ions and their replacement by Al^{3+} from the tetrahedral sheet. In addition to vermiculite, kaolinite, gibbsite, and Fe-bearing accessory minerals (goethite, hematite, and ferrihydrite) can be found on weathering biotite surfaces.

Muscovite, or dioctahedral mica, is more resistant to weathering than biotite. Low structural Fe^{II} content, an orientation of structural hydroxyl protons that minimize repulsive electrostatic interactions with interlayer K^+ ions, and low Mg^{2+} content (Al^{3+}-dominated octahedral layer) account for the greater structural stability. Muscovite weathers in a manner similar to that described for biotite. Layer weathering, through K^+ displacement by hydrated cations and layer expansion, stabilizes adjacent layers to a greater degree than in the parent mica (Figure 3.2), producing mica-vermiculite interstratification. However, the commonly encountered product of muscovite weathering is interstratified mica-smectite, indicating the high degree of stabilization offered by shifting hydroxyl protons. In moist acidic environments, the vermiculite component of the interstratified mica alters to beidellite, eventually resulting in discrete beidellite crystals. Kaolinite and gibbsite are also formed during muscovite weathering.

3.5.2 CHLORITE

The chlorite minerals in soil are both primary and secondary. The primary chlorites are commonly trioctahedral (clinochlore and chamosite), with the octahedral layer and polymerized hydroxyl interlayer dominated by Mg^{2+} and Fe^{2+} and other divalent metal cations (e.g., Mn^{2+}, Ni^{2+}, and Zn^{2+}). Trioctahedral chlorites are unstable in soil environments and are typically found in soils where only slight weathering has occurred (they occupy the same Jackson weathering stage as biotite, Table 3.2).

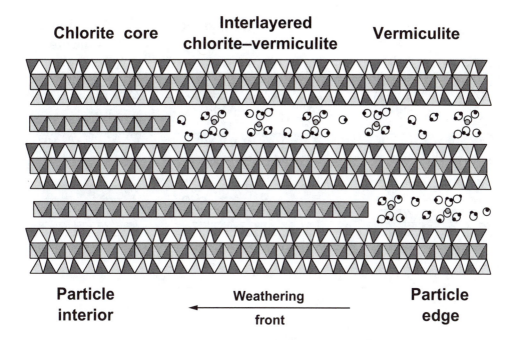

FIGURE 3.6 The alteration of chlorite occurs through a combined layer and edge weathering mechanism. Beginning at the particle edge, acidic conditions promote the dissolution of the hydroxy interlayer and the intrusion of layer charge-satisfying hydrated cations. The loss of octahedral cations in response to Fe^{II} oxidation and the reduction in layer charge by Si^{4+} substitution for tetrahedral Al^{3+} stabilizes alternating interlayers and results in an interstratified chlorite-vermiculite. As the layer weathering front proceeds into the particle interior, adjacent hydroxy-interlayers are dissolved at the particle edge and replaced by hydrated cations. The process continues into the chlorite core, until the hydroxyl-interlayers are dissolved and the transformation to the expansive vermiculite is complete.

The Mg-bearing chlorites are unstable relative to chlorites composed primarily of Fe^{II}, due to the greater stability of the Fe^{II}—O bond. However, the oxidation of Fe^{II} to Fe^{III} destabilizes the hydroxyl interlayer, and plays a significant role in the weathering of chlorite to secondary phyllosilicates, with the concomitant formation of Fe^{III} oxyhydroxide accessory minerals. Both Mg- and Fe^{II}-dominated interlayers are susceptible to acid attack.

The chlorite minerals weather in a manner similar to that described for the mica minerals, and edge weathering appears to be the predominant weathering mechanism (Figure 3.6). The initial weathering product is an interstratified chlorite-vermiculite (or chlorite-smectite) phase that is created through the preferential dissolution of the Mg^{2+} and Fe^{II} (and Fe^{III}) hydroxide interlayer sheets. With additional weathering, the interstratified phase is altered to trioctahedral vermiculite, which in turn weathers to a smectite (e.g., nontronite) or kaolinite (or halloysite). The decomposition of the trioctahedral vermiculite to form accessory minerals and soluble products also occurs. Interstratified illite-smectite and illite-vermiculite are chlorite alteration products when mica is present in the weathering environment. Apparently, mica is a K^+ source for illite formation and stability.

Secondary chlorite is an alteration product of primary chlorite, vermiculite, or smectite. Moderately acidic environments (pH 5 to 6) promote Al^{3+} expulsion from the tetrahedral layer of the chlorite, vermiculite, and smectite structures. During the weathering of primary chlorite, Al^{3+} expelled from the tetrahedral layer displaces divalent cations (e.g., Mg^{2+}) in the hydroxyl interlayer sheet. The Al-chlorite or pedogenic chlorite is an intermediate in the weathering process that leads to hydroxy-interlayer vermiculites (HIV), then vermiculite or kaolinite (or halloysite). Pedogenic chlorite, created through the process of chloritization, is a product of vermiculite or smectite weathering. Early in the chloritization process, Al^{3+} hydrolysis and polymerization results in the

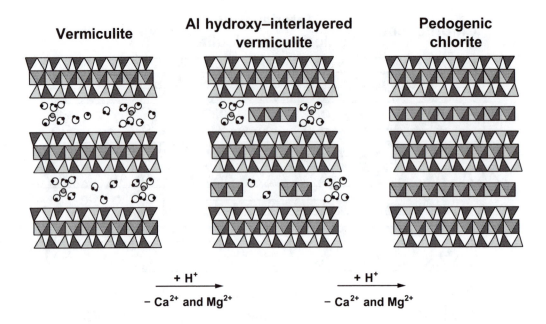

FIGURE 3.7 In slightly acid soil conditions, the Al^{3+} released from tetrahedral coordination hydrolyzes and polymerizes to partially occupy the vermiculite interlayer (at the expense of hydrated base cations). As the process proceeds, the Al hydroxy-interlayers become continuous, and the transformation of vermiculite to pedogenic chlorite is complete.

formation of islands of dioctahedral hydroxide sheet structures within the expansive 2:1 interlayers of vermiculite or smectite, resulting in HIV or hydroxy-interlayered smectite (HIS) (Figure 3.7). Low organic matter content also facilitates the chloritization process, as complexation or chelation of the released Al^{3+} would inhibit hydrolysis and hydroxy-interlayer formation. Continued loss of tetrahedral Al^{3+} eventually results in pedogenic chlorite, which contains a continuous Al hydroxy-interlayer. Oxidizing conditions, and frequent wetting and drying cycles, are also conducive to the formation and stability of pedogenic chlorite. Ultimately, pedogenic chlorite is altered to kaolinite through a structural rearrangement process (e.g., inversion of the Si tetrahedral sheet), and chlorite dissolution-kaolinite precipitation.

3.5.3 VERMICULITE

In soil, vermiculites are most often associated with or are alteration products of muscovite, biotite, or chlorite. Trioctahedral vermiculite forms from biotite, phlogopite, and primary chlorite, and can accumulate under moderate weathering conditions. Trioctahedral vermiculite is also macrocrystalline, existing as discrete particles in the sand, silt, and clay fractions of the soil. Dioctahedral vermiculite forms from muscovite, possibly through illite, and is more common under severe weathering conditions. Dioctahedral vermiculite may also form from biotite, through trioctahedral vermiculite. Expulsion of octahedral cations as a result of Fe^{II} oxidation, and the replacement of Fe^{III} by Al^{3+} from the tetrahedral layer, is required for the trioctahedral to dioctahedral transition. Dioctahedral vermiculite is seldom seen as discrete crystals in silt and sand fractions.

In general, the factors that influence the weathering of the micas also influence the formation of vermiculite. The replacement of K^+ by Mg^{2+} (and other hydrated cations) is prevented by even small concentrations of soluble K^+. This is particularly the case for the alteration of muscovite to dioctahedral vermiculite, which requires lower soluble K^+ levels than the alteration of biotite to trioctahedral vermiculite. The exchange of Si^{4+} by Al^{3+} in the tetrahedral layers is also required, as the ratio of Si to Al increases from the 3:1 of mica to the ~3.1:0.9 to 3.4:0.6 of the vermiculites.

Thus, the transformation of mica to vermiculite requires effective removal of soluble K^+, while hydrated cation (Mg^{2+}, Ca^{2+}, and Na^+) and soluble silica concentrations are maintained. Iron in biotite is principally in the +II oxidation state; whereas, Fe in vermiculite is principally in the +III oxidation state. Therefore, the oxidation of Fe^{II}, which creates an electrostatic imbalance in the 2:1 structure and the expulsion of cations, is required for the biotite to trioctahedral vermiculite transformation. The orientation of the proton on the structural octahedral hydroxyl determines the rate of mica transformation. Muscovite and other dioctahedral micas will slowly alter to dioctahedral vermiculite because the repulsive force between the hydroxyl proton and interlayer K^+ is minimized. However, trioctahedral micas alter rather rapidly to trioctahedral vermiculite because the repulsive forces are larger.

Trioctahedral vermiculite that is derived from trioctahedral mica is considered to be a fast-forming, unstable intermediate in the soil environment, which rapidly weathers to kaolinite and smectite. In general, all trioctahedral 2:1 layer silicates are short-term residents of soils. This is because the trioctahedral Mg^{2+} layer forms at a higher temperature than the dioctahedral Al^{3+} layer, and thus is relatively unstable in the low-temperature soil system. However, the formation and presence of Al hydroxy-interlayers stabilize both the dioctahedral and trioctahedral vermiculites. Indeed, dioctahedral vermiculite with an Al hydroxy-interlayer is more stable in soil than kaolinite.

3.5.4 SMECTITE

Smectite minerals exist in soil environments that contain high soluble silica ($H_4SiO_4^0$) and base cations (Ca^{2+}, Mg^{2+}, and Na^+). Typically, these are environments in which leaching is at a minimum, such as with poorly drained soils. Parent materials that contain rocks that are high in Mg and Fe (e.g., mafic or ferromagnesium minerals) and low in Al weather to Fe^{III}-rich dioctahedral smectite, rather than saponite, a Mg-rich trioctahedral smectite. This supports the general rule that dioctahedral smectites (and dioctahedral phyllosilicates in general) form under low temperature (surficial) conditions. Beidellite, a dioctahedral smectite with layer charge principally in the tetrahedral layer (Al^{3+} for Si^{4+}) weathers from rock containing micas and chlorites (substitution already present). Without this framework, beidellite formation is not favorable, as Al^{3+} prefers octahedral coordination at surficial temperatures.

Montmorillonite is expected to form as a result of complex weathering reactions (pedogenically) in alkaline soils (pH > 7.5) high in soluble silica, Al, and Mg. This process is termed neoformation. The olivines, inosilicates, and feldspars contribute the raw materials (Mg^{2+}, Ca^{2+}, Al^{3+}, and $H_4SiO_4^0$), as well as structural components (linked Si tetrahedra), for the neoformation of smectites. The type of smectite formed is dependent on the availability of the various different raw materials. For example, the presence of Mg^{2+}-rich primary silicates may lead to the neoformation of both trioctahedral and dioctahedral smectite, while Mg^{2+}-poor K-feldspars can alter to beidellite. In well-drained environments, smectites are transitory, as the leaching of base cations and soluble silica are not conducive to smectite preservation; they are present in parent materials or they are detrital.

The essential changes required for the transformation of mica to smectites are: loss of interlayer K^+ (depotassification), loss of tetrahedral Al^{3+} (dealumination), and incorporation of Si^{4+} into the tetrahedral sheet (silication). Mica depotassification generally results in the formation of beidellite (tetrahedral substitution of Al^{3+} for Si^{4+} essentially maintained). Trioctahedral micas may weather to form trioctahedral smectites. However, these minerals are unstable, tending to lose Mg and Fe from the octahedral sheet with additional weathering. Similarly, mafic trioctahedral chlorites that are subjected to oxidizing and acid conditions (pH < 6) in a leaching environment tend to lose their Mg and Fe hydroxyl interlayer to form trioctahedral smectite, although nontronite (Fe^{III}-rich dioctahedral smectite) is a more common product of iron-rich chlorite alteration. Again, the trioctahedral minerals are unstable and short-lived in soil. The environmental conditions required for smectite formation and stability are surficial temperature to destabilize tetrahedral Al^{3+}, low K^+ concentration to promote depotassification of mica (and illite), high Ca^{2+} and Mg^{2+} activities to force the displacement of K^+ from the interlayer, low Al^{3+} and high $H_4SiO_4^0$ activities, and alkaline conditions (pH > 7.5) (when pH < 6 mica weathers to vermiculite then to kaolinite or HIV).

Smectite becomes unstable when leaching becomes pronounced. It is common to see kaolinite as the dominant clay mineral in soils on well-drained hilltop soils and smectite as the dominant clay in the poorly drained bottoms. Indeed, any improvement in drainage will result in smectite transformation to kaolinite. Under weak weathering conditions, hydroxy-interlayered smectite will form. Weak weathering conditions are described as those that result in slightly acidic soil pH values (destabilizing tetrahedral Al^{3+}), but leaching is not sufficient to reduce soluble $H_4SiO_4^0$ and Mg^{2+} activities. Ultimately, this process can result in the formation of pedogenic chlorite. As the weathering environment becomes moderate, the leaching of base cations is sufficient to result in the formation of interstratified kaolinite-smectite. Under strong weathering conditions, the continued loss of base cations and soluble silica results in kaolinite, crystalline and noncrystalline hydrated oxyhydroxides materials, or the complete dissolution of the smectite.

3.6 GENERAL WEATHERING SCHEME FOR THE PHYLLOSILICATES

The mineral transformations that occur in the soil environment are controlled by numerous physical, chemical, and biological factors (as described above). Detailed discussions concerning the weathering and formation of phyllosilicates in specific environments can be found in Dixon and Weed (1989), Churchman (1999), and Dixon and Schulze (2002). However, the two most important factors controlling mineral weathering are the proton concentration and the extent of leaching, which are inexorably linked. While every environment differs with regard to the factors controlling mineral alteration and the mineral assemblages present, there is a high degree of commonality in phyllosilicate weathering products. These generalities are described in the weathering scheme illustrated in Figure 3.8.

Simple silicates, such as the olivines, dissolve to release their base cations (Ca and Mg), and Ti, Fe, Mn, Al, and Si. With the exception of the base cations, which remain in solution or are consumed, for example, during the formation of authogenic smectites, metals released through olivine hydrolysis and oxidation rapidly precipitate to form amorphous hydrous oxides. With time, these materials crystallize to form accessory minerals (e.g., anatase, goethite, hematite, and gibbsite). The weathering products of the more complex primary silicates (inosilicates) can polymerize, forming Mg^{2+}-rich trioctahedral or dioctahedral smectites. The K-feldspars may also weather to an Al-rich dioctahedral smectite, with the required Mg obtained from olivine and inosilicate dissolution.

All primary phyllosilicates initially weather to interstratified minerals. Biotite will undergo simple transformations to form mica-trioctahedral vermiculite, while muscovite will initially weather to mica-dioctahedral vermiculite, and may further weather, conditions permitting, to illite or interstratified mica-smectite. The mica transformations are hastened by the exchange of hydrated base cations for interlayer K^+. The primary chlorites take a similar weathering path; the preferential dissolution of Mg hydroxide interlayer sheets results in the formation of chlorite-trioctahedral vermiculite. Eventually, the interstratified phases are completely altered to vermiculite with the octahedral occupation inherited from the parent mineral. Conversion of trioctahedral chlorite and biotite to trioctahedral vermiculite occurs rapidly in the soil environment due to the inherent instability of the primary minerals. Under basic conditions, trioctahedral vermiculite may undergo further transformation to form a trioctahedral smectite. However, the more probable weathering pathways involve the migration of tetrahedral Al to octahedral positions and the formation of dioctahedral vermiculite or smectite (as trioctahedral smectites are uncommon in soils). Alternatively, trioctahedral vermiculite may dissolve if leaching is sufficient to remove soluble Mg^{2+},

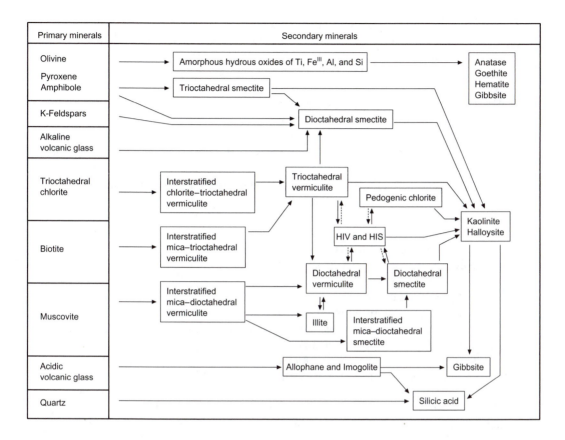

FIGURE 3.8 This diagram illustrates the weathering and formation of primary and secondary phyllosilicate minerals.

forming kaolinite, accessory minerals, and soluble silica and base cations. Alteration of any of the mafic minerals will be enhanced in soil environments having an active microbial community, or in direct contact with plant roots. The relative abundance of organic moieties and chelates in these environments facilitates the solubilization of Fe^{III} and hastens weathering.

In basic environments, those high in soluble silica and base cations, neoformation of dioctahedral smectite, principally montmorillonite, is favored. Dioctahedral smectite will weather to kaolinite in well-drained soils as soluble silica and base cations are leached. Kaolinite will weather further with the additional leaching of soluble silica to form gibbsite. In moderately acidic soils, the dealumination of the tetrahedral layer leads to the formation of Al HIV and HIS. Pedogenic chlorite is formed when the Al hydroxyl-interlayer becomes continuous. Both the Al hydroxyl-interlayered and pedogenic chlorite phases ultimately weather to kaolinite.

3.7 EXERCISES

1. The layer charge of vermiculite minerals is less than the layer charge of the minerals from which they are derived.
   a. What are the minerals from which trioctahedral vermiculites are derived and what are possible mechanisms of layer charge reduction in these minerals?
   b. What are the minerals from which dioctahedral vermiculites are derived and what are possible mechanisms of layer charge reduction in these minerals?

2. The Goldich stability series states that the weatherability of the tectosilicates decreases from anorthite $\rightarrow$ albite $\rightarrow$ orthoclase $\rightarrow$ quartz. For each mineral describe the crystal-chemical characteristics that account for their placement in the series.

3. The weathering sequence, biotite $\rightarrow$ trioctahedral illite $\rightarrow$ saponite $\rightarrow$ kaolinite, cannot occur. Discuss why this is the case.

4. The Goldich stability series and the Bowen reaction series are very similar in their presentation. Discuss why this should be the case.

5. For each of the following weathering sequences: (1) indicate if the sequence can occur in the soil environment; (2) if the sequence is implausible indicate why; or (3) if the sequence is plausible, describe the structural transformations that must occur and the soil environmental conditions that must be present for each step in the sequence:
   a. Chlorite $\rightarrow$ HIV $\rightarrow$ montmorillonite $\rightarrow$ kaolinite
   b. Biotite $\rightarrow$ trioctahedral vermiculite $\rightarrow$ pedogenic chlorite
   c. Biotite $\rightarrow$ talc $\rightarrow$ saponite
   d. Muscovite $\rightarrow$ dioctahedral vermiculite $\rightarrow$ illite
   e. Muscovite $\rightarrow$ illite $\rightarrow$ dioctahedral vermiculite $\rightarrow$ kaolinite
   f. Biotite $\rightarrow$ trioctahedral vermiculite $\rightarrow$ soluble species $\rightarrow$ montmorillonite $\rightarrow$ kaolinite

6. Why are trioctahedral minerals uncommon in soils relative to dioctahedral minerals?

7. List the general characteristics of primary aluminosilicate chemical weathering reactions.

8. Develop a balanced chemical reaction for the weathering of biotite $[K(Mg_2Fe)(Si_3Al)O_{10}(OH)_2]$ to kaolinite, gibbsite, and goethite.

9. Develop a balanced chemical reaction for the weathering of biotite $[K_{0.98}(Fe^{III}_{2.31}Mg_{0.43}Al_{0.18})(Si_{2.72}Al_{1.28})O_{10}(OH)_2]$ to trioctahedral vermiculite $[K_{0.8}(Mg_{2.36}Fe^{III}_{0.48}Al_{0.16})(Si_{2.72}Al_{1.28})O_{10}(OH)_2]$ and goethite.

10. Develop a balanced chemical reaction for the weathering of Camp Berteau montmorillonite $[Na_{0.67}(Al_{2.92}Fe^{III}_{0.41}Fe^{II}_{0.03}Mg_{0.64})Si_8O_{20}(OH)_4]$ to kaolinite, goethite, and silicic acid $[am\text{-}Si(OH)_4]$.

11. Write balanced chemical reactions for the following (*Note:* only the mineral reactants and products are listed; you must choose the appropriate aqueous reactants and products and obtain the correct chemical formulas for the minerals indicated; you may use idealized formula for the phyllosilicates):
    a. The oxidation of biotite to form hematite and vermiculite
    b. The formation of vermiculite from clinochlore
    c. The congruent dissolution of diopside
    d. The weathering of andesine to form kaolinite
    e. The weathering of oligoclase to form kaolinite
    f. The formation of vermiculite and gibbsite from muscovite
    g. The neoformation of montmorillonite from soluble species
    h. The weathering of donbassite to form vermiculite
    i. The congruent dissolution of hornblende by hydrolysis
    j. The weathering of microcline to kaolinite and gibbsite
    k. The oxidation of annite to form vermiculite and lepidocrocite

12. In the Jackson weathering stages a "young" soil does not connote the temporal age of a soil. What does "young" imply?

13. The weathering of the primary phyllosilicate minerals (chlorites and micas) is a relative complex process. Describe how the process occurs, as well as the crystal-chemical alterations in the minerals' structure that lead to a relatively more stable weathering product for each of the following transitions:
    a. Trioctahedral chlorite to vermiculite
    b. Biotite to vermiculite
    c. Muscovite to vermiculite

14. Figure 3.8 illustrates the weathering and formation of primary and secondary silicates and other clay materials. For all transition that can be traced back to a primary phyllosilicate, describe the chemical characteristics of the soil solution that would facilitate the weathering process.

REFERENCES

Bowen, N.L. The reaction principle in petrogenesis. *J. Geol.* 30:177–198, 1922.

Churchman, G.J. The alteration and formation of soil minerals by weathering. In M.E. Sumner (ed.) *Handbook of Soil Science*. CRC Press, Boca Raton, FL, 1999, pp. F3–F76.

Dixon, J.B. and D.G. Schulze (Eds.) *Soil Mineralogy with Environmental Applications*. SSSA Book Ser. 7. Soil Science Society of America, Madison, WI, 2002.

Dixon, J.B. and S.B. Weed (Eds.) *Minerals in Soil Environments*, 2nd ed. SSSA Book Ser. 1. Soil Science Society of America, Madison, WI, 1989.

Goldich, G.G. A study in rock weathering. *J. Geol.* 46:17–58, 1938.

Jackson, M.L. Chemical composition of soils. In *Chemistry of the Soil*. F.E. Bear (Ed.) Reinhold Publishing Corp., New York, 1964, pp. 71–141.

Jackson, M.L., S.A. Tyler, A.L. Willis, G.A. Bourbeau, and R.P. Pennington. Weathering sequence of clay-size minerals in soils and sediments. I: Fundamental generalizations. *J. Phys. Coll. Chem.* 52:1237–1260, 1948.

Jackson, M.L. and G.D. Sherman. Chemical weathering of minerals in soils. *Adv. Agron.* 5:219–318, 1953.

4 Organic Matter in Soil

Soil contains a virtual cornucopia of living organisms that obtain energy, structure, and function from organic compounds. Living organisms in soil and their associated organic compounds are collectively called the soil biomass. The living organisms consist of macrofauna (e.g., earthworms and insects), mesofauna (e.g., nematodes), microfauna (e.g., protozoa and archezoa), microorganisms (e.g., algae, fungi, actinomycetes, and bacteria), burrowing animals, and plant roots. These organisms produce a diverse abundance of biochemical compounds required for cellular function that collectively constitute the fraction of organic carbon in soil called cellular or biomass carbon (Figure 4.1). Living organisms also produce and excrete extracellular substances, such as enzymes and other organic compounds that facilitate nutrient uptake and microbial mobility (e.g., microbial mucilages), as well as metabolic byproducts (wastes). Biochemical compounds are also released to the extracellular environment through the death and decay of organisms (cell lysis), and material derived from higher plant residues (leaf litter), sloughing cells from plant roots, and animal excrement. All of these mechanisms add to the innumerable array of biochemical compounds in the soil environment that make up the soil biomass carbon pool.

Undecayed or partially decomposed organic materials that enter the soil system such as leaf litter, fallen trees, animal carcasses, and other detritus are called organic residues. Essentially, the source of organic matter in this pool is still recognizable. Another pool of organic matter in soil is called humus, or soil organic matter (SOM). Soil organic matter consists of two groups of compounds, nonhumic and humic substances. Organic substances that belong to chemically recognizable classes in biochemistry are the nonhumic substances. These substances include, for example, low-molecular-mass organic acids; simple carbohydrates and polysaccharides; amino sugars; amino acids, peptides, and proteins; lipids and phospholipids; nucleic acids; and lignin. In general, nonhumic substances are transitory in soil, as they are utilized as substrates by soil microorganisms, and are ultimately decomposed to base inorganic components, principally CO_2 and H_2O. However, on the degradative journey from a recognizable biochemical compound to CO_2, organic carbon may become a component of a group of natural organic substances that are refractory (recalcitrant) in soil, sediment, and natural waters, and do not fall into any discrete category of biochemistry. These compounds are collectively known as humic substances.

Both nonhumic and humic substances play important roles in the environment, affecting the biochemical, physical, and chemical properties of soil. For example, SOM is a reservoir of metabolic energy for soil microbes, and when mineralized by microbes, a source of macronutrients for plants (NH_4^+, NO_3^-, $H_2PO_4^-$, HPO_4^{2-}, and SO_4^{2-}). Soil organic matter is intimately associated with soil minerals and is important for the maintenance of aggregate stability. Organic matter in soils also increases the water retention capacity and impacts the thermal properties of soil. Soil organic matter plays a significant role in buffering soil solution pH, is responsible for a sizeable fraction of the cation exchange capacity of soil, enhances the dissolution of soil minerals, complexes and chelates metal cations, and retains nonionic organic compounds (altering their efficacy or potential toxicity). Despite the substantial influence exerted by SOM on soil properties, the concentrations of soil organic carbon (SOC) in mineral surface soils, on average, do not exceed 60 g kg^{-1} (6%) (where SOM $\approx 1.72 \times$ SOC). The organic carbon content of surface soils is highly variable and influenced by the soil-forming factors of climate, biota (vegetation and soil organisms), topography, and parent material operating over time. Since these factors vary for different soils, the SOM content can also

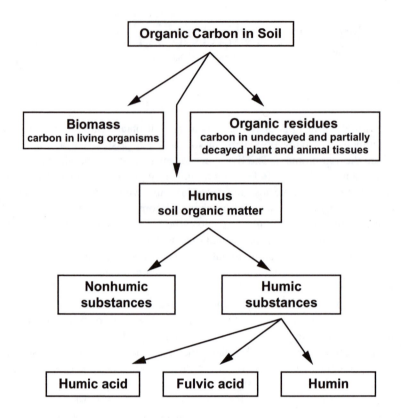

FIGURE 4.1 The pools of organic carbon in soil.

be described as a function of soil type. This is illustrated by the descriptive statistics for organic carbon content of surface soils given in Table 4.1. The elemental content of Histosols, which are organic soils, is dominated by organic carbon, containing on average 419 g C kg^{-1} and up to 100% SOM. Andisols and Spodosols have greater median concentrations of organic carbon (38.3 g C kg^{-1} and 27.9 g C kg^{-1}) than Alfisols, Ultisols, and Aridisols (3.8 g C kg^{-1}, 4.6 g C kg^{-1}, and 5.4 g C kg^{-1}), due principally to differences in climatic conditions (temperature, precipitation, and humidity) and vegetation. The data in Table 4.1 also illustrate a broad range of organic carbon concentrations within each soil order. In general, the maximum organic carbon concentrations in Table 4.1 are associated with the uppermost surface horizons (A horizons) of mineral soils, while the minimum values are associated with the lower surface horizons (B horizons). This trend is displayed in Figure 4.2 for the mineral surface horizons of the Clarion (Typic Hapludoll), Unsel (Durinodic Haplargid), Kirvin (Typic Hapludult), and Herman (Typic Haplorthod) soils.

Despite the relatively low concentrations of SOC in mineral soils compared to inorganic components (see Chapter 1), this pool of carbon is an important component of the global carbon budget and cycle. Eswaran et al. (1993) estimate that world soils contain 1576 Pg (1 Pg = 10^{15} g) of SOC to a depth of 1 m, while soils contain an estimated 1738 Pg of inorganic carbon, mainly as carbonates (Swift, 2001). The soil orders that primarily contribute to the global SOC budget are Histosols (390 Pg) and Inceptisols (352 Pg) (Table 4.1). Although these numbers are quite large, their true significance lies in the fact that the estimated size of the global SOC pool is more than twice the estimated size of the atmospheric carbon pool (750 Pg), and approximately three times the organic carbon estimated to occur in vegetation biomass (550 Pg). The ocean and geologic carbon reservoirs contain the greatest quantities of carbon (38,000 and 65.5 × 10^6 Pg carbon), but the latter is not in dynamic equilibrium with the ocean and terrestrial reservoirs.

TABLE 4.1
Organic Carbon Content and Global Organic Carbon Mass of Surface Soils (A and B Horizons)[a]

Soil Order	Mean ± Standard Error[b]	Median	Range	Global Mass
		g C kg^{-1}		(Pg)
Alfisols (292)[c]	5.80 ± 0.39	3.8	0.2–50.0	127
Andisols (36)	57.1 ± 10.5	38.3	0.9–308	78
Aridisols (36)	6.67 ± 0.92	5.4	1.6–33.1	110
Entisols (53)	14.4 ± 2.38	6.6	0.3–94.2	148
Histosols (310)[d]	419 ± 7.7	—	306–724	357
Inceptisols (362)	18.8 ± 1.09	10.5	0.3–113.7	352
Mollisols (239)	12.2 ± 0.71	8.4	0.4–54.5	72
Oxisols (231)	13.4 ± 1.03	7.8	0.6–117.3	119
Spodosols (24)	57.6 ± 15.0	27.9	0.6–331.4	71
Ultisols (218)	8.64 ± 0.75	4.6	0.3–72.0	105
Vertisols (81)	12.4 ± 1.04	8.4	1.5–46.7	19

[a] Data were obtained from the International Soil Reference and Information Centre, Wageningen, The Netherlands (Batjes, 1995) and Eswaran et al. (1993).

[b] The standard error is the standard deviation divided by the square-root of the number of observations included in the mean.

[c] The value in parentheses represents the number of soil samples (observations) included in the mean, median, and range.

[d] Data, in part, from Batjes and Dijkshoon (1999) and van Noordwijk et al. (1997).

4.1 DETERMINATION OF SOIL ORGANIC CARBON CONCENTRATIONS

Numerous methods have been developed and employed to facilitate the extraction and isolation of both specific organic compounds and specific classes of organic compounds from soils (see Senesi and Loffredo [1999] and Baldock and Nelson [2000] for reviews of these methods). Briefly, the method-specific variables employed in soil extraction techniques include: extractant type (aqueous-polar or nonaqueous-nonpolar) and composition (acidic or basic, oxidant or reductant), solid-to-solution ratio, length of extraction period, temperature and pressure of extraction, and order in which the various solvents are employed (if the procedure is sequential). Once obtained, the soil extracts may be subjected to any number of compound class-specific isolation techniques, including solvent extraction, dialysis or ultrafiltration, or fractionation by employing ion exchange or adsorption resins. Additional isolation may be performed using chromatographic techniques (gas chromatography and high and low pressure liquid chromatography) which are commonly coupled to analytical detectors (UV absorbance, fluorescence, or mass spectrometry). Structural characteristics of organic substances in the isolates are elucidated through the application of various spectrometric techniques. These include UV-visible, fluorescence, infrared, nuclear magnetic resonance, and electron spin resonance spectrometry. The development and use of the various extraction, isolation, quantization, and structural characterization techniques have provided a wealth of information, and have led to significant advances in our understanding of the organic chemistry of soil. However, the ability to perform these critical investigations, either with respect to facilities or instrumentation, is not common.

One measurement that is routinely performed as part of a standard array of soil physical-chemical characterizations is the determination of total SOC content. Several methods are available for the determination of SOC concentrations (Nelson and Sommers, 1996). These methods can be grouped

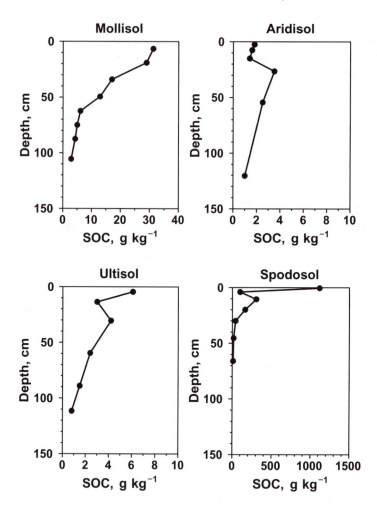

FIGURE 4.2 The distribution of soil organic carbon as a function of depth in a Mollisol (Clarion soil; Typic Hapludoll), an Aridisol (Unsel soil; Haplic Haplargid), an Ultisol (Kirvin soil; Typic Hapludult), and a Spodosol (Jesso soil: Typic Fragiorthod). (Data from Batjes, 1995.)

into two categories: dry combustion or wet digestion. Dry combustion has become the most widely employed method for determining the total carbon (TC) content of a soil, as the technique can be implemented with many commercially available instruments. The method does not provide a direct measure of SOC if the soil contains inorganic carbon (SIC) forms, such as carbonates (calcite and dolomite) and elemental carbon (charcoal, graphite, or coal). However, in noncalcareous and acidic soils, dry combustion may offer a direct measure of SOC, because the SIC concentrations are often negligible. Generally, TC is determined by first mixing a soil sample of known mass with a catalyst or an accelerant (e.g., CuO, Sn metal, or Fe metal) to facilitate the complete oxidation of organic compounds. The sample is then placed in a furnace that may achieve temperatures from 1000 to >1500°C (depending on instrument manufacturer) for organic carbon combustion and the conversion of carbonates to $CO_2(g)$. The released $CO_2(g)$ is then trapped (e.g., by an alkaline solution or a mixture of solid hydroxides, e.g., Ascarite) and analyzed by titrimetric, gravimetric, or conductimetric techniques. Organic carbon is then determined by difference: SOC = TC − SIC, where SIC concentrations are determined by a separate analysis, which typically involves the determination of the acid consumed by reaction with soil carbonates.

The direct determination of SOC may be accomplished by the wet digestion of a soil sample in a $K_2Cr_2O_7$-H_2SO_4 solution. This technique, known as the Walkley-Black method, was once the standard for SOC determinations; however, it is labor intensive, requires a great deal of analytical skill, employs strong oxidants and acids that must be heated, and generates a hazardous waste (and has been replaced by the dry combustion methods described above). In this method, a mass of soil is placed in a glass digestion tube, to which is added a standardized $K_2Cr_2O_7$ solution and concentrated H_2SO_4. The tube is then placed in a digestion block and heated to 150°C for 30 minutes. During this period, the following oxidation-reduction reaction occurs:

$$1.5CH_2O(s) + Cr_2O_7^{2-}(aq) + 8H^+(aq) \rightarrow 1.5CO_2(g) + 2Cr^{3+}(aq) + 5.5H_2O(l) \qquad (4.1)$$

where $CH_2O(s)$ represents an average chemical formula for SOC. In this reaction, the carbon in SOC, assumed to have an average oxidation state of zero (C^0), is oxidized to $CO_2(g)$ (C^{IV}) by the $Cr_2O_7^{2-}(aq)$ (Cr^{VI}) in an excess of acid (provided by H_2SO_4). Following digestion, the soil suspension is cooled and titrated to the *N*-phenylanthranilic acid end point with a solution containing standard $Fe(NH_4)_2(SO_4)_2$. During the titration, the $Cr_2O_7^{2-}(aq)$ that was not consumed by SOC is reduced by Fe^{2+} according to the reaction:

$$Cr_2O_7^{2-}(aq) + 6Fe^{2+}(aq) + 14H^+(aq) \rightarrow 2Cr^{3+}(aq) + 6Fe^{3+}(aq) + 7H_2O(l) \qquad (4.2)$$

The difference between the moles of dichromate initially reacted with the soil and the moles of dichromate that remain after soil digestion is proportional to the SOC content of the soil sample.

For example, a 0.50-g overdried surface soil sample is digested with 5 mL of a standard 0.167 M $K_2Cr_2O_7$ solution, with 7.5 mL of concentrated H_2SO_4, according to the aforementioned procedure. After cooling, the suspension is transferred using deonized-distilled water to an Erlenmeyer flask and indicator solution is added. The suspension is then titrated to the *N*-phenylanthranilic acid end point, which is attained after the addition of 3.90 mL of standard 0.2 M $Fe(NH_4)_2(SO_4)_2$ solution. The moles of Fe^{2+} consumed by the unreacted dichromate are computed as the volume of titrant added (in mL, V) multiplied by the concentration of Fe^{2+} in the titrant (in mmol mL^{-1}, M) (assuming degradative loss of dichromate is minimal): mmoles = V•M. Therefore, the consumed quantity of Fe^{2+} is 3.9 mL multiplied by 0.2 mmol mL^{-1}, or 0.78 mmol Fe^{2+}. According to Equation 4.2, the reduction of 1 mol of $Cr_2O_7^{2-}$ requires 6 mol of Fe^{2+}. Thus, the amount of $Cr_2O_7^{2-}$ remaining after reaction with the soil is 0.13 mmol $Cr_2O_7^{2-}$ (0.78 ÷ 6). The quantity of $Cr_2O_7^{2-}$ consumed by the soil is the difference between the moles initially added and the moles left unreacted. The quantity of $Cr_2O_7^{2-}$ initially added is 5 mL multiplied by 0.167 mmol mL^{-1}, or 0.84 mmol $Cr_2O_7^{2-}$. The quantity of $Cr_2O_7^{2-}$ consumed by SOC is 0.84 mmol − 0.13 mmol, or 0.71 mmol of $Cr_2O_7^{2-}$. Equation 4.1 indicates that 1 mol of $Cr_2O_7^{2-}$ consumes 1.5 mol of SOC; therefore, the quantity of SOC present in the 0.5-g soil sample is 1.065 mmol C (0.71 × 1.5). The SOC concentration in the soil is:

$$SOC = \frac{1.065 \text{ mmol C}}{0.50 \text{ g}} \times \frac{12.011 \text{ mg C}}{\text{mmol C}} = 25.6 \text{ mg g}^{-1} = 25.6 \text{ g kg}^{-1} \qquad (4.3)$$

Again, implicit to the wet oxidation method is the assumption that the average valency of carbon in SOC is zero.

4.2 ORGANIC FUNCTIONAL GROUPS: A REVIEW

The chemical reactivity of SOM is directly related to the quantities and types of organic functional groups and structural components that are present. The ability of SOM to form bonds with the reactive surface functional groups of soil minerals, to complex inorganic and organic solutes, to partition (absorb) hydrophobic organic solutes, and to structurally rearrange (as dictated by the

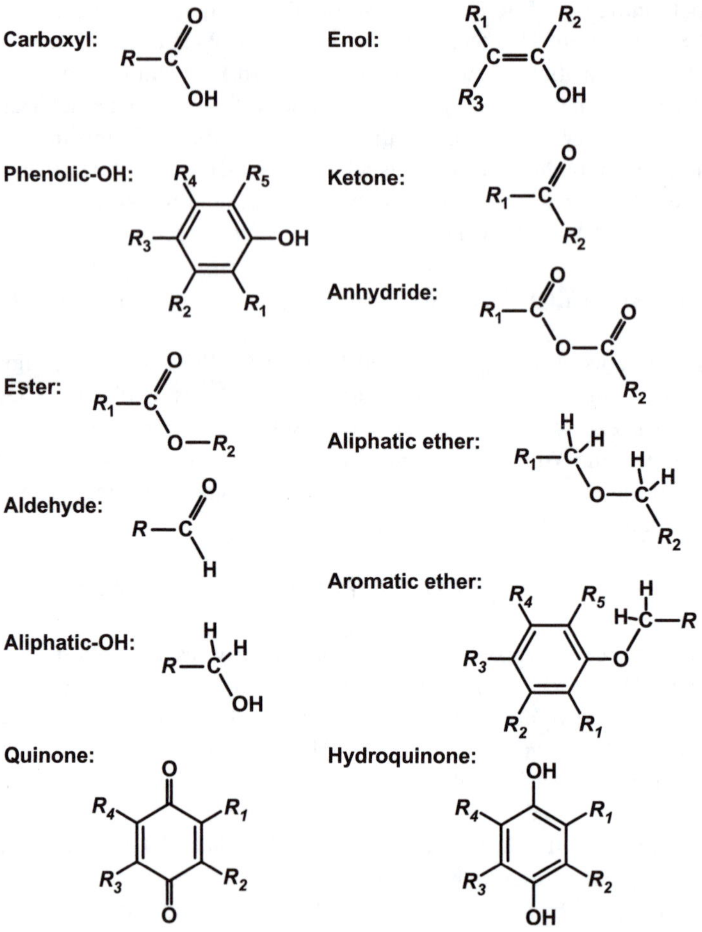

FIGURE 4.3 Organic functional groups and classes of compounds with similarities in structure that are commonly observed in soil organic matter.

chemical conditions of the soil environment), are controlled by the reactivity of organic functional groups, as well as by the hydrophilic and hydrophobic character of various organic structural components. Humic and nonhumic substances are polyfunctional in that they contain a variety of organic functional groups on a carbon backbone structure. The vast majority of SOM functional groups contain oxygen (Figure 4.3), and these are either neutral (do not ionize, but could be polar) or acidic (develop negative charge through proton dissociation). The acidic functional groups include the carboxyl (where R represents either an aliphatic or an aromatic carbon structure):

$$R{-}COOH^0 \rightarrow R{-}COO^- + H^+ \qquad (4.4a)$$

the phenolic-OH (where Ar is an aromatic structure):

$$Ar{-}OH^0 \rightarrow Ar{-}O^- + H^+ \qquad (4.4b)$$

and the enol:

$$R{-}CH{=}CROH^0 \rightarrow R{-}CH{=}CRO^- + H^+ \qquad (4.4c)$$

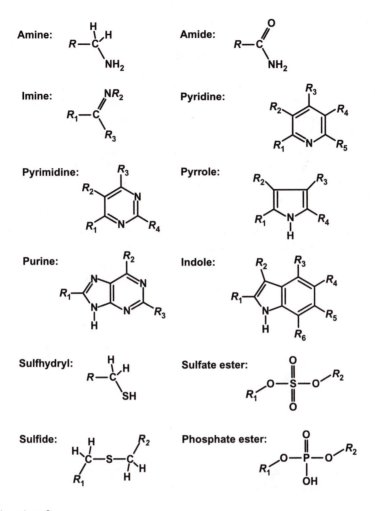

FIGURE 4.3 *(continued)*.

Neutral SOM functional groups and structures that contain oxygen are the carbonyl (found in esters, ketones, aldehydes, and anhydrides), the aliphatic-OH (alcoholic-OH), the aliphatic and aromatic ether, and the quinone (aromatic carbonyl). The organic functional groups and compounds that are bases (develop positive charge through proton complexation) are principally those that contain nitrogen. These include the amino group:

$$R-NH_2^0 + H^+ \rightarrow R-NH_3^+ \tag{4.5a}$$

the amide:

$$R-CO-NH_2^0 + H^+ \rightarrow R-CO-NH_3^+ \tag{4.5b}$$

the imine:

$$R-CH=NH^0 + H^+ \rightarrow R-CH=NH_2^+ \tag{4.5c}$$

and the aromatic ring nitrogen, such as found in pyridine:

$$(ArN)^0 + H^+ \rightarrow (ArN)—H^+ \tag{4.5d}$$

Although much less common in SOM than the oxygen- and nitrogen-containing functional groups, functional groups that contain sulfur and phosphorus also contribute to the structural stability and reactivity of SOM. The principal sulfur-containing groups and compounds include the sulfhydryl (thiol), the sulfide, and the sulfate ester, while the principal phosphorus group is the phosphate ester.

An important characteristic of organic functional groups is the ability to ionize and form anionic or cationic species. The reactions in Equations 4.4 and 4.5 illustrate the ionization of organic functional groups and indicate that ionization occurs through proton dissociation or association reactions. The ability of a functional group to donate protons to a solution is a characteristic of Lowry-Brønsted acidity. Conversely, Lowry-Brønsted basicity is the ability of a functional group to accept a proton from the solution, or to donate a hydroxide. Lowry-Brønsted acids and bases are described by their strength and a parameter called the acid dissociation constant, K_a. All reactions that involve the acceptance or the donation of a proton are uniquely characterized by a K_a value. The dissociation of a carboxyl group, as might be found in SOM, is illustrated for methoxyacetic acid:

$$CH_3OCH_2COOH^0 \rightarrow CH_3OCH_2COO^- + H^+ \tag{4.6a}$$

At equilibrium, the distribution of the carboxyl functional group between the protonated and deprotonated forms is described by the acid dissociation constant:

$$K_a = \frac{(CH_3OCH_2COO^-)(H^+)}{(CH_3OCH_2COOH^0)} \tag{4.6b}$$

where the parentheses ( ) denote the activity of the enclosed species; however, in a dilute solution, molar concentration units are applicable. Rearranging Equation 4.6b:

$$\frac{K_a}{(H^+)} = \frac{(CH_3OCH_2COO^-)}{(CH_3OCH_2COOH^0)} \tag{4.6c}$$

This expression indicates that the concentrations of the protonated and deprotonated forms of the organic acid will be equal [$(CH_3OCH_2COOH^0) = (CH_3OCH_2COO^-)$] when the hydrogen ion concentration of the solution is equal to the acid dissociation constant [$K_a = (H^+)$]. Alternatively, $(CH_3OCH_2COOH^0) = (CH_3OCH_2COO^-)$ when $pK_a = pH$, where "p" denotes the negative common logarithm of K_a or H^+ concentration. For the dissociation reaction in Equation 4.6a, $K_a = 2.9 \times 10^{-4}$, and $pK_a = 3.54$. Therefore, when the pH of a dilute solution containing the methoxyacetic acid is 3.54, the concentrations of $CH_3OCH_2COOH^0$ and $CH_3OCH_2COO^-$ will be equal. When the solution pH is less than the pK_a, the ratio $K_a/(H^+)$ on the left side of the equal sign in Equation 4.6c will be less than 1 (note that as the negative logarithm of a parameter decreases the actual value of the parameter increases; thus, a solution with a pH of 2 has ten times the proton concentration of a solution with a pH of 3: 0.01 M vs. 0.001 M). Therefore, the concentration of the undissociated acid ($CH_3OCH_2COOH^0$) will be greater than that of the dissociated acid ($CH_3OCH_2COO^-$). Conversely, when the solution pH is greater than the pK_a, the ratio $K_a/(H^+)$ will be greater than 1, and the concentration of the dissociated acid will be greater than that of the undissociated acid. Therefore, as the pH of a solution varies, relative to the pK_a for a particular acid, so do the concentrations of the charged and neutral species (Figure 4.4).

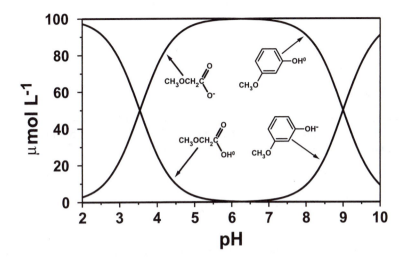

FIGURE 4.4 The distributions of methoxyacetic acid (pK_a = 3.54) and methoxyphenol (pK_a = 9.65) as a function of solution pH. The total concentration of each analyte is 10^{-4} M.

All acidic organic functional groups are characterized as weak Lowry-Brønsted acids. This means that the pK_a values of organic acids lie between 0 and 14, and their degree of dissociation is determined by the pH of the solution. Strong Lowry-Brønsted acids have pK_a values that are less than 0. These types of acids, which are represented by inorganic acids (e.g., nitric acid and hydrochloric acid), are completely dissociated in aqueous solutions. The pK_a values associated with weak acids are indicators of relative acid strength. The pK_a for m-methoxyphenol ($CH_3OC_6H_4OH$) is 9.65, which is greater than the pK_a of 3.54 for methoxyacetic acid. Therefore, methoxyacetic acid is a stronger acid than the phenol (or the phenolic-OH is a weaker acid than methoxyacetic acid). As shown in Figure 4.4, relatively strong weak acids ionize at lower pH values than do weak acids that are relatively weak as well.

The distribution of a weak acid functional group between the acid species (protonated species) and the conjugate base (deprotonated species) may be determined given the solution pH. Consider the methoxyacetic acid distribution described in Figure 4.4. The total concentration of the acid compound in solution is given by:

$$[CH_3OCH_2COOH]_T = [CH_3OCH_2COOH^0] + [CH_3OCH_2COO^-] \tag{4.7a}$$

where the brackets [ ] represent molar concentrations. Equation 4.6b can be rearranged to yield an expression that describes the concentration of the protonated acid group as a function of (H^+) and K_a:

$$[CH_3OCH_2COOH^0] = \frac{[CH_3OCH_2COO^-][H^+]}{K_a} \tag{4.7b}$$

Substituting Equation 4.7b into 4.7a for [$CH_3OCH_2COOH^0$] yields:

$$[CH_3OCH_2COOH]_T = \frac{[CH_3OCH_2COO^-][H^+]}{K_a} + [CH_3OCH_2COO^-] \tag{4.7c}$$

Additional rearrangement leads to:

$$[CH_3OCH_2COOH]_T = [CH_3OCH_2COO^-]\left(\frac{[H^+]}{K_a} + 1\right) \tag{4.7d}$$

or

$$[CH_3OCH_2COOH]_T \left(\frac{[H^+]}{K_a} + 1 \right)^{-1} = [CH_3OCH_2COO^-] \qquad (4.7e)$$

Using the parameters used in constructing Figure 4.4 ($[CH_3OCH_2COOH]_T = 10^{-4}\ M$ and $K_a = 10^{-3.54}$), the concentration of the conjugate base can be computed for any value of solution pH. For example, in a pH 4.5 solution the concentration of $[CH_3OCH_2COO^-]$ is:

$$[CH_3OCH_2COO^-] = (10^{-4}) \left(\frac{10^{-4.5}}{10^{-3.54}} + 1 \right)^{-1} = 9.012 \times 10^{-5}\ M \qquad (4.7f)$$

The concentration of the undissociated acid ($9.881 \times 10^{-6}\ M$) is then computed using Equation 4.7a (where $[CH_3OCH_2COOH^0] = 10^{-4}\ M - 9.012 \times 10^{-5}\ M$).

The basicity of organic functional groups may also be described by pK_a values, in a manner similar to that for the acidic organic functional groups. For example, the pyridinium ion dissociates according to the reaction:

$$C_6H_5NH^+ \rightarrow C_6H_5N^0 + H^+ \qquad (4.8a)$$

with a pK_a of 5.2, while pK_a equals 9.78 for the dissociation of the ammonium ion of the amino acid glycine:

$$^-OOCCH_2NH_3^+ \rightarrow {}^-OOCCH_2NH_2^0 + H^+ \qquad (4.8b)$$

The basicity of the N moiety on these two compounds is not readily apparent from the reactions in Equations 4.8a and 4.8b. Indeed, it would appear as though both the protonated pyridine N and amino groups are acidic, and the former is a stronger acid than the latter. The basicity of these functional groups, however, is illustrated in the following reactions:

$$C_6H_5N^0 + H_2O \rightarrow C_6H_5NH^+ + OH^- \qquad (4.9a)$$

and

$$^-OOCCH_2NH_2^0 + H_2O \rightarrow {}^-OOCCH_2NH_3^+ + OH^- \qquad (4.9b)$$

At equilibrium, the distribution of the pyridine and amino functional group between the protonated and deprotonated forms is described by a basicity constant, K_b. Using the pyridine reaction (Equation 4.9a) as an example:

$$K_b = \frac{(C_6H_5NH^+)(OH^-)}{(C_6H_5N^0)} \qquad (4.9c)$$

The pK_b value provides a measure of base strength, just as the pK_a provides a measure of acid strength. Basic compounds that have pK_b values that lie between 0 and 14 are called weak bases.

Further, the closer the pK_b value is to 0, the stronger is the base strength. The pK_a and pK_b values for any given compound are related; one can be computed if the other is known. The K_a for the pyridinium dissociation reaction (Equation 4.8a) is:

$$K_a = \frac{(C_6H_5N^0)(H^+)}{(C_6H_5NH^+)} \tag{4.10a}$$

The concentrations of the proton and the hydroxide ion are related through the dissociation of water:

$$H_2O \rightarrow H^+ + OH^- \tag{4.10b}$$

At equilibrium:

$$K_w = (H^+)(OH^-) \tag{4.10c}$$

and

$$(H^+) = \frac{K_w}{(OH^-)} \tag{4.10d}$$

where K_w equals 10^{-14} ($pK_w = 14$) at standard temperature (25°C) and pressure (1 atm). Substituting Equation 4.10d into Equation 4.10a for (H^+) yields:

$$K_a = \frac{(C_6H_5NH^+)K_w}{(C_6H_5N^0)(OH^-)} \tag{4.10e}$$

Rearranging:

$$\frac{K_w}{K_a} = \frac{(C_6H_5N^0)(OH^-)}{(C_6H_5NH^+)} \tag{4.10f}$$

The right side of Equation 4.10f is equal to K_b, as described by Equation 4.9c. Equation 4.10f can then be written:

$$\frac{K_w}{K_a} = K_b \tag{4.10g}$$

or

$$pK_w - pK_a = pK_b \tag{4.10h}$$

The pK_b value for pyridinium is $14 - 5.2 = 8.8$. Similarly, pK_b for the glycine amino group is $14 - 9.78 = 4.22$. Therefore, glycine is a stronger base and protonates at a higher pH (lower pOH) than does pyridine (Figure 4.5a and b).

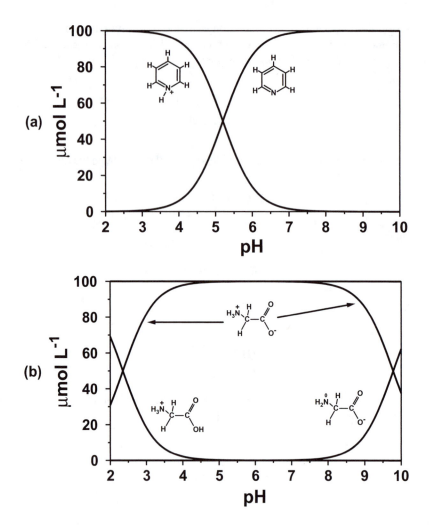

FIGURE 4.5 The distributions of (a) pyridine (pK_a = 5.2; pK_b = 8.8) and (b) glycine (amino group pK_a = 9.78; pK_b = 4.22) as a function of solution pH. The total concentration of each analyte is 10^{-4} M.

4.3 NONHUMIC SUBSTANCES

Simply stated, nonhumic substances are the parent material of humic substances. Nonhumic substances are the principal metabolic energy source of soil microbes, and the degradation products of these substances are the building blocks of the humic substances. As a group, nonhumic substances consist of compounds that belong to the known classes of biochemicals; they are the biomolecules required for function, structure, and reproduction of living organisms. These compounds may be broadly grouped into the following classes: carbohydrates (monosaccharides, oligosaccharides, and polysaccharides), nitrogen compounds (amino acids, proteins, nucleic acids, amino sugars, and teichoic acids), lipids (fats, waxes, resins, sterols, and terpenes), and lignin.

4.3.1 CARBOHYDRATES

The term carbohydrates was first used to express the belief that these compounds represent hydrates of carbon, as illustrated in the chemical formula for glucose: $C_6H_{12}O_6$ or $(C{\bullet}H_2O)_6$. Although this view is erroneous, the terminology has survived to describe a group of natural compounds that are comprised of simple sugars. Carbohydrates account for the largest component of nonhumic substances (50 to 250 g kg^{-1} of SOC), and the second largest component of SOC behind the humic substances. The carbohydrates range in character from the monosaccharides (simple sugars), to the oligosaccharides (compound sugars that yield from two to six monosaccharides upon hydrolysis), to the polysaccharides (high-molecular-mass biopolymers consisting of numerous—eight or more—monosaccharides, but generally consisting of 100 to 3000 monosaccharide units). The majority of the carbohydrates found in soils is derived from plant tissue and consists of cellulose and hemicellulose. Plants also contribute simple sugars and polysaccharides to the soil environment as root exudates and mucilaginous materials. Structural polysaccharides (e.g., chitin), extracellular mucilages, and intracellular polysaccharides are also contributed to the soil environment by microorganisms.

The basic structural unit of the oligosaccharides and the polysaccharides are the monosaccharides, or simple sugars. The concentrations of free simple sugars in soils are exceedingly low. They are readily available to microorganisms and are converted to CO_2 or other carbohydrates, or incorporated into other microbial products, such as amino acids or lipids, at a far faster rate than the polysaccharides. Monosaccharides are essentially consumed about as fast as they are produced. The naturally occurring monosaccharides generally result from the hydrolysis of plant and microbial polysaccharides and include neutral sugars, amino sugars, acidic sugars, methylated sugars, and sugar alcohols.

The neutral sugars do not contain a carboxyl or an amino functional group and may be divided into three groups: hexoses, pentoses, and deoxyhexoses. The hexoses contain six carbon atoms and are derived from both plant and microbial sources. This group of sugars is dominated by glucose, with lesser amounts of galactose and mannose (Figure 4.6). Data compiled by Stevenson (1994)

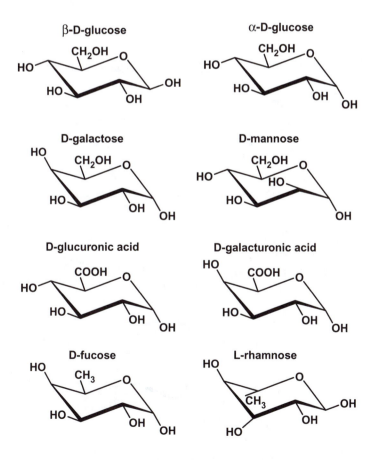

FIGURE 4.6 Hexoses that are commonly observed in soil.

and Senesi and Loffredo (1999) indicate that glucose accounts for 208 to 540 g kg^{-1} (with a mean value of 346 g kg^{-1}) of the neutral sugars extracted by hydrolysis (H$_2$SO$_4$ extraction) from soil (a process that releases the monomeric units of the polysaccharides), while galactose and mannose concentrations range from 119 to 231 g kg^{-1} (mean of 163 g kg^{-1}) and from 104 to 219 g kg^{-1} (mean of 160 g kg^{-1}).

The pentoses contain five carbon atoms and are predominately derived from plant polysaccharides. The common pentoses in soils are arabinose and xylose (ribose), accounting for 50 to 180 g kg^{-1} (120 g kg^{-1} mean) and 40 to 236 g kg^{-1} (102 g kg^{-1} mean) of the neutral sugars extracted from soil (Figure 4.7). The deoxyhexoses, rhamnose and fucose, which are of microbial origin, account for 35 to 188 g kg^{-1} (78 g kg^{-1} mean) and 11 to 41 g kg^{-1} (28 g kg^{-1} mean) of neutral sugars (Figure 4.6). The amino sugars (hexosamines) account for 20 to 60 g kg^{-1} of SOC and are principally of microbial origin. This group of sugars is dominated by glucosamine (Figure 4.8), with lesser amounts of galactosamine, muramic acid, mannosamine, and fucosamine. The uronic or acidic sugars are estimated to be approximately 10 to 50 g kg^{-1} of SOC. This group of sugars

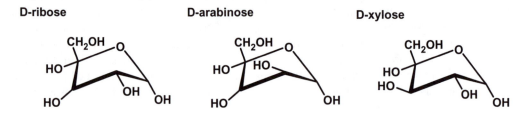

FIGURE 4.7 Pentoses that are commonly observed in soil.

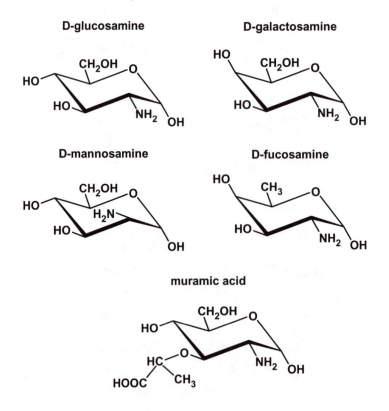

FIGURE 4.8 Amino sugars that are commonly observed in soil.

is dominated by glucuronic acid and galacturonic acid (Figure 4.6). Also included in the carbohydrate pool of substances are the water-soluble low-molecular-mass organic acids, such as oxalic, formic, citric, acetic, malic, and succinic acids (described in Chapter 5).

Glucose, the most common monosaccharide extracted from soil, is the monomer of which the polysaccharides cellulose (β-D-glucose monomers) and amylose (α-D-glucose monomers) are composed. Cellulose is the principal structural component of plant cells and comprises 10 to 20% of plant leaves (dry weight basis) and approximately 50% of tree wood and bark. The individual β-D-glucose units in cellulose are connected by a β-glucoside linkage in which the number 1 carbon of one unit is linked to the hydroxyl of the number 4 carbon of the next unit (β-*O*-4 linkage, Figure 4.9). Amylose (a starch) consists of α-D-glucose units linked through an α-*O*-4 linkage. Hemicelluloses are the second most common group of structural polysaccharides in plants. They consist of complex, branched structures that contain a variety of monosaccharides. Principal among these monosaccharides are the pentoses D-xylose, of which xylan is composed (occurs in association with cellulose), and L-arabanose (Figure 4.9). Pectin, a polymer of D-galacturonic acid, is also a hemicellulose (Figure 4.9). Microbial polysaccharides are generally lacking in the pentoses and primarily consist of (in addition to glucose) galactose, mannose, rhamnose, fucose, glucosamine, and galactosamine monomers. Chitin, the structural component of fungal cell walls, structural membranes, and mycelia, is a polymer of 1,4-β-*N*-acetylglucosamine (Figure 4.9). *N*-Acetylglucosamine and *N*-acetylmuramic acid are components of peptidoglycan, which is found in bacterial cell walls.

4.3.2 NITROGEN, SULFUR, AND PHOSPHORUS COMPOUNDS

Nitrogen, sulfur, and phosphorus are relatively minor components of SOM; however, SOM is the principal repository for these elements in the soil environment. Total soil N ranges from 200 to 5000 mg kg^{-1}, with a median value of 2000 mg kg^{-1}. Total soil S ranges from 30 to 1600 mg kg^{-1}, with a median value of 700 mg kg^{-1}. Total soil P concentrations range between 35 and 5300 mg kg^{-1}, with a median value of 800 mg kg^{-1}. Soil organic P ranges from 19 to 70% of the total P. However, the organic forms of N and S account for approximately 90% of the total concentrations of these elements in the surface layers of most soils. Approximately 10 to 15% of total N and 1 to 3% of total S in surface soils is estimated to be associated with the soil biomass. Therefore, 75 to 80% of the total soil N, and 87 to 89% of the total soil S is associated with SOM.

Soil N occurs in a great variety of N-containing organic compounds that are excreted by living organisms or released after death. Principal among these are the amino acids, peptides, proteins, nucleic acids, and phospholipids. Other organic N compounds in SOM include the amino sugars (described in the previous section), which occur in chitin (a major structural component of fungi and arthropods) and the teichoic acids (a major structural component of bacteria); vitamins; and antibiotics excreted by fungi and actinomycetes. The organic S forms are primarily associated with amino acids and proteins, coenzymes and vitamins, and sulfolipids.

The organic forms of soil N are often extracted from soil by refluxing soil samples in hot 6 *M* HCl for 6 to 12 h. This acid hydrolysis procedure generates an extract that is analyzed for total N, ammonia N, amino acid N, and amino sugar N. As a result, there are five operationally defined forms of soil N (Senesi and Loffredo, 1999):

NH$_3$-N: NH$_4^+$ extracted from clay minerals, amino sugars, amino acid amides (asparagines and glutamine), and other amino acids (tryptophan, serine, and threonine)

Amino acid-N: amino acid monomers hydrolyzed from peptides and proteins

Amino sugar N: Amino sugar monomers hydrolyzed from the exudates and structural polysaccharides of fungi, bacteria, and other microorganisms

Hydrolyzable unidentifiable N: Total hydrolyzable N minus the sum of hydrolyzable NH$_3$, amino acid, and amino sugar N; believed to be derived from non-α-amino acid N

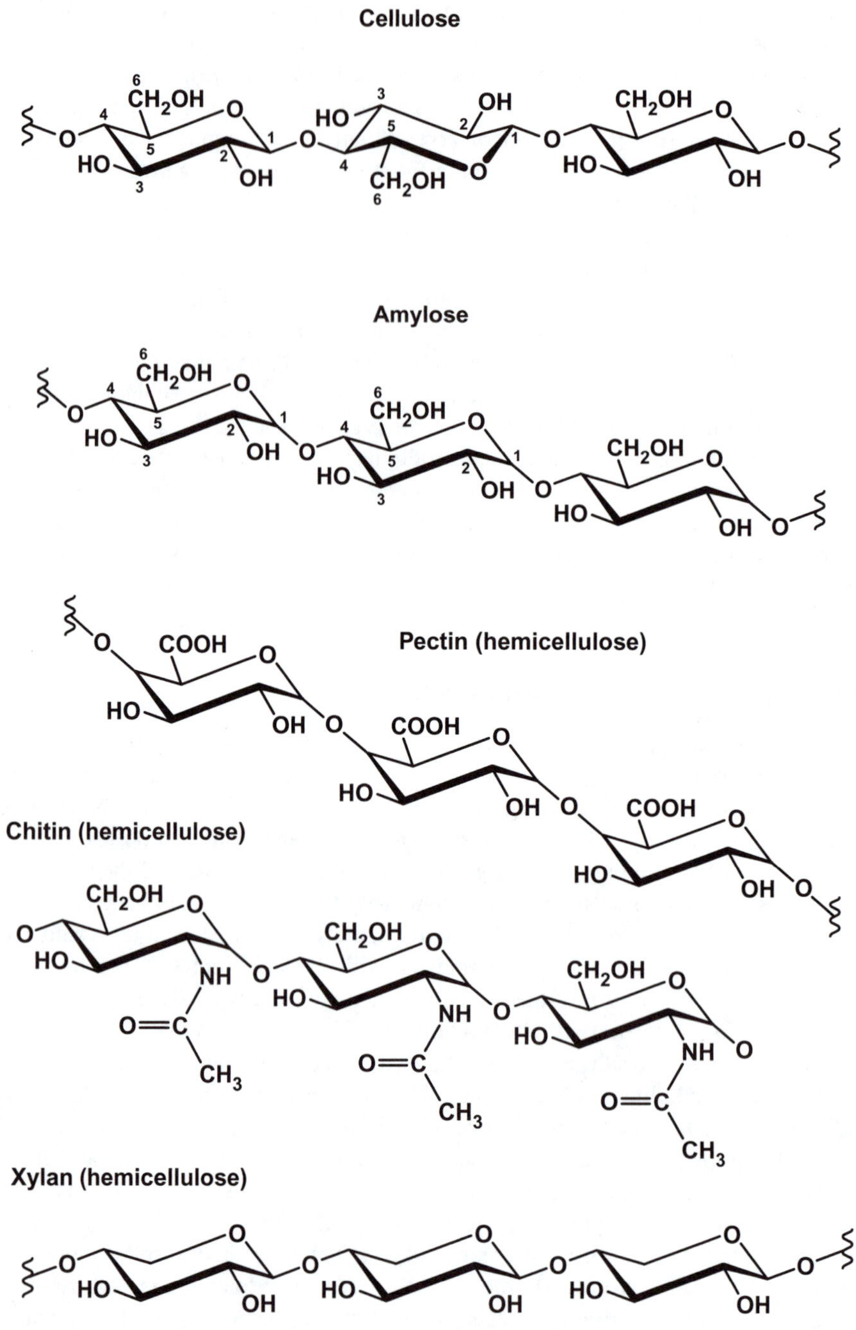

Cellulose

Amylose

Pectin (hemicellulose)

Chitin (hemicellulose)

Xylan (hemicellulose)

FIGURE 4.9 Examples of common polysaccharides in soil are cellulose (polymer of β-D-glucose); the starch, amylose (polymer of α-D-glucose); and the hemicelluloses, pectin (polymer of D-galacturonic acid), chitin (polymer of 1,4-β-N-acetylglucosamine), and xylan (polymer of D-xylose). The individual monosaccharide units are linked through a β-glucoside linkage (β-O-4 linkage), as in cellulose, or an α-glucoside linkage (α-O-4 linkage), as in amylose.

(β- or γ-amino acids including, arginine, tryptophan, lysine, histidine, and proline) with a small percentage arising from purines and pyrimidines

Acid insoluble N: Total soil N minus total hydrolyzable N; a portion of this fraction of soil N is believed to be associated with humic substances

In general, the distribution of total soil N found in each pool is variable for soils, perhaps affected by the wide variation in soil type and total soil N content. The proportions of each form of total soil N in mineral surface soils are (from the compilation of Stevenson [1994], mean and standard deviation are given in parentheses where appropriate):

- NH_3-N, 9 to 37% (25.4% ± 8.2)
- Amino acid N, 13 to 50% (37.9% ± 9.1)
- Amino sugar N, 1 to 14% (6.0% ± 1.8)
- Hydrolyzable unidentifiable N, 4 to 40% (17.3%)
- Acid-insoluble N, 4 to 44% (13.6% ± 5.5)

The amino acids comprise one of the largest classes of N-bearing substances in soils and account for approximately 30 to 45% of the total soil N. Amino acids are compounds that contain an amino group ($R-NH_2$) and a carboxyl group ($R-COOH$), and therefore exhibit both acidic and basic properties. Amino acids are amphoteric; they can exist as either anions, when the carboxyl group is dissociated ($R-COO^-$) and the amino group is neutral, or as cations, when the amino group is protonated ($R-NH_3^+$) and the carboxyl group is neutral. It is very common for an amino acid to exist as a zwitterion (an inner salt), where both the amino and carboxyl functional groups are ionized, lending the molecule both positive and negative charge concurrently. For example, the distribution of the protonated and deprotonated forms of glycine (Figure 4.5) illustrates that this amino acid is principally anionic when solution pH is above 9.78, and principally cationic below pH 2.35. When the solution pH is between 2.35 and 9.78, the zwitterion form of glycine predominates.

The most common type of amino acid is the α-type. In α-amino acids, the amino group is located on the carbon adjacent to the carboxyl carbon. In contrast, the amino group in β-amino acids is located on the carbon that is two carbons away from the carboxyl carbon (Figure 4.10). Several amino acids have been identified in soil hydrolysates (described above); many of these are also shown in Figure 4.10. The amino acids are classified in one of two ways: (1) according to the nature of the carbon backbone structure (aliphatic, aromatic, or heterocyclic), or (2) according to the relative abundance or ratio of amino to carboxyl groups in the molecule (neutral, $R-COOH = R-NH_2$; acidic, $R-COOH > R-NH_2$; and basic, $R-COOH < R-NH_2$). The amino acids illustrated in Figure 4.11 are identified using both classifications. Amino acids are the monomeric structural units of proteins and peptides. A peptide is a compound that consists of two or more amino acid units, with each unit linked to adjacent units by an amide bond, formally called a peptide bond (Figure 4.11). Peptides that contain greater than 50 amino acids are known as proteins.

Similar to free sugars, free amino acids are readily metabolized by soil microbes. Thus, the bulk soil concentrations of free amino acids are generally quite low. However, free amino acids are excreted by plant roots and found in relative abundance in rhizosphere soil because they are excreted faster than microbes can metabolize them. They are also intermediates in the microbial conversion of protein- and peptide-N to ammonia (NH_4^+). Although soil amino acids predominately occur in peptides and proteins, many also occur in complex polymeric structures that are thought to be of microbial origin (waste products of microbial metabolism). Indeed, some amino acids identified in soils are not normal constituents of proteins. These include α, ε-diaminopimelic acid, ornithine, β-alanine, α-aminobutyric acid, and γ-aminobutyric acid (Figures 4.10 and 4.12).

In addition to the major N pools in soils (amino acids, amino sugars, and their associated polymeric forms), there are a number of nitrogenous compounds that occur in very low concentrations. These compounds, which include the nucleic acids and their derivatives, porphyrins

Neutral and Aliphatic Amino Acids

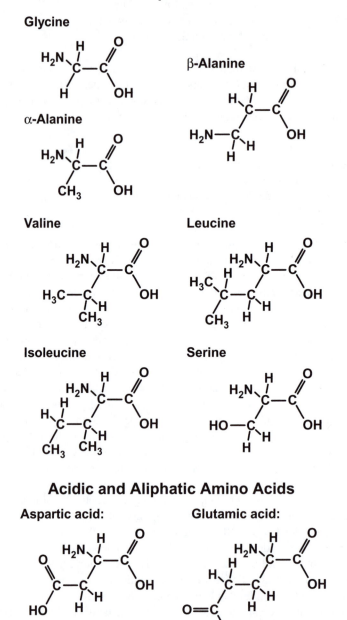

FIGURE 4.10 Common amino acids found in soils.

(chlorophyll) and derivatives, vitamins, antibiotics, and phospholipids, are present in soils because they are exuded by living organisms or because they occur in cells of living organisms and are released upon their death. Of this minor pool, the nucleic acids are perhaps the most common, accounting for up to 1% of the total soil N, although higher values have been reported. The nucleic acids include the ribonucleic acids (RNA) and deoxyribonucleic acids (DNA) and contain a pentose (ribose for DNA and deoxyribose for RNA), a phosphate, and an N heterocycle in the form of a

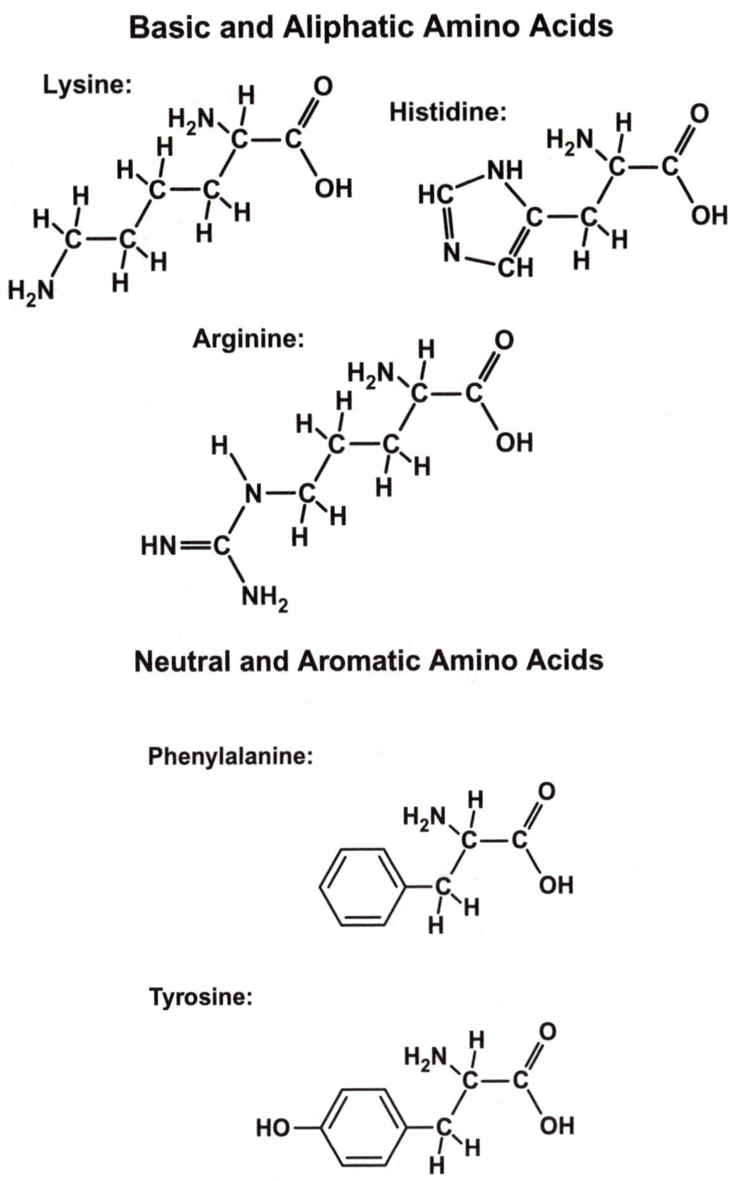

Basic and Aliphatic Amino Acids

Lysine:

Histidine:

Arginine:

Neutral and Aromatic Amino Acids

Phenylalanine:

Tyrosine:

FIGURE 4.10 (*continued*).

purine (adenine and guanine) or a pyrimidine base (uracil, cytosine, and thymine) (purine and pyrimidine ring structures are illustrated in Figure 4.3).

Similar to N, S is an important component of several groups of biomolecules, including amino acids and proteins, coenzymes and vitamins, biotin, thiamine, and sulfolipids. Further, from 95 to 99% of the total S in soil is found in organic forms. The distribution of soil S into inorganic S (SO_4) and organic S pools and the characterization of organic soil S is commonly determined using a selective sequential extraction procedure. The fractionation procedure (described by Tabatabai [1996]) first removes sulfate-S, and then partitions the organic S forms into ester sulfate-S and carbon bonded-S forms (Figure 4.3). The latter category is subjected to further fractionation to provide information on the amount of S in the form of amino acids (cysteine and methionine). Data compiled by Senesi and Loffredo (1999) revealed that, on average, 51% of the total S in surface soils collected from around the world is associated with the ester sulfate-S fraction ($R—O—S$, which also

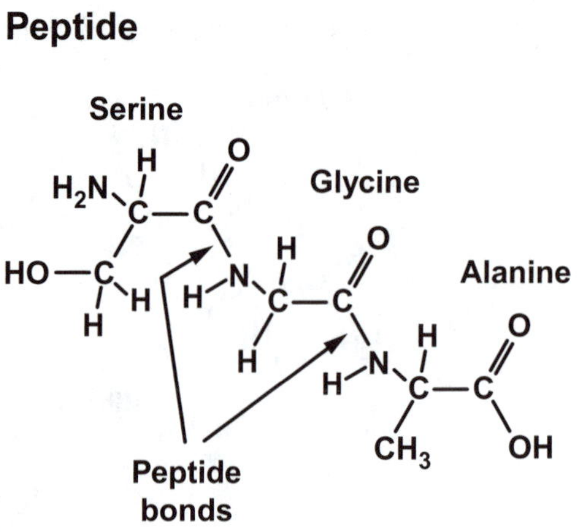

Peptide

FIGURE 4.11 A peptide composed of the amino acids serine, glycine, and alanine. The individual amino acid monomers in peptides and proteins are linked through peptide bonds.

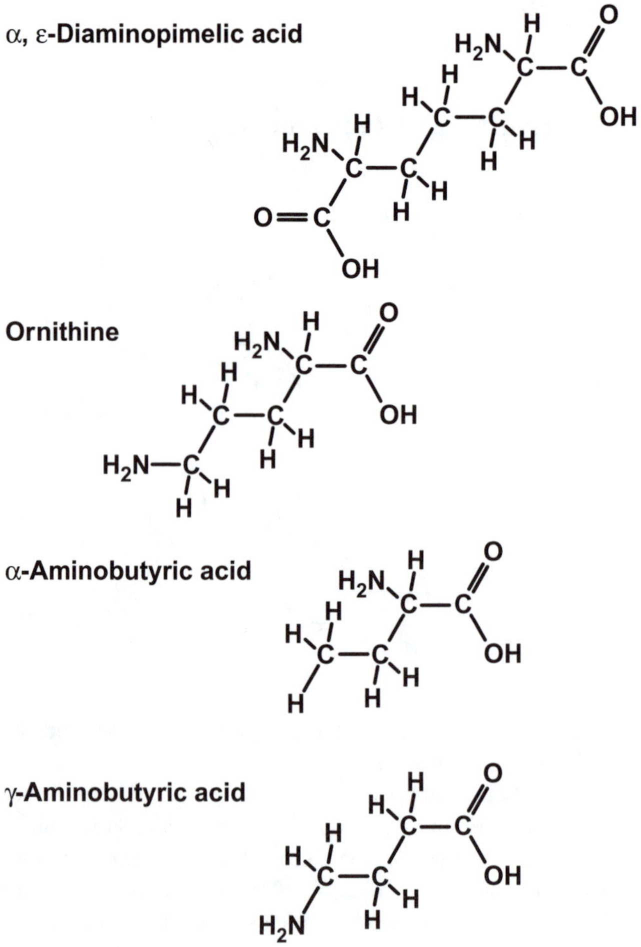

FIGURE 4.12 Amino acids that are not normally found in proteins (in addition to β-alanine, Figure 4.10). These amino acids are generated by soil microorganisms and occur, for example, in antibiotics or as structural components of bacteria.

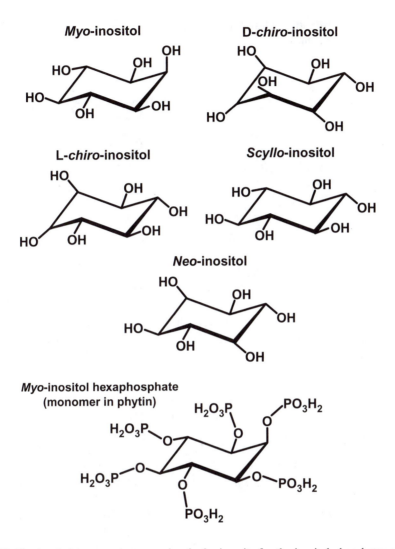

FIGURE 4.13 The inositol isomers that comprise the basic units for the inositol phosphates, such as *myo*-inositol hexaphosphate, the monomer found in phytin.

includes nonester S in R—N—S—R' and R—S—S—R' linkages), 18% occurs in Raney-Ni–reducible S forms (principally R—SH, R—SO—CH_3, R—SO—OH, and Ar—SO_2—OH), and 29% is in nonreducible S forms (R—SO_2—R' and R—SO_2—OH).

Organic P accounts for a very small percentage (1% to approximately 3%) of SOM. The most abundant forms of organic soil P reside in a yet to be identified pool. Of the forms that have been identified, the most common and the most resistant to decomposition are the inositol phosphates, which are reported to range from <1 to 60% of the soil organic P, and up to 40% of the total soil P. The inositol phosphates are phosphate esters of hexahydroxycyclohexane (inositol, Figure 4.13). Several isomers of inositol occur in nature, the most common of which, *myo*-inositol, is found in higher plants. Indeed, *myo*-inositol hexaphosphate (phytic acid) is the monomer of which phytin is composed (a substance found in cereal grains). In addition to the hexaphosphate, the mono-, di-, and triphosphates of inositol also occur in plant tissues. In soils, the *myo*-, *scyllo*-, *neo*-, and D- and L-*chiro*-inositol hexaphosphates have been identified, as well as the *myo*-, *scyllo*-, and *chiro*-inositol pentaphosphates. The isomer *scyllo*-inositol hexaphosphate has been found to account for

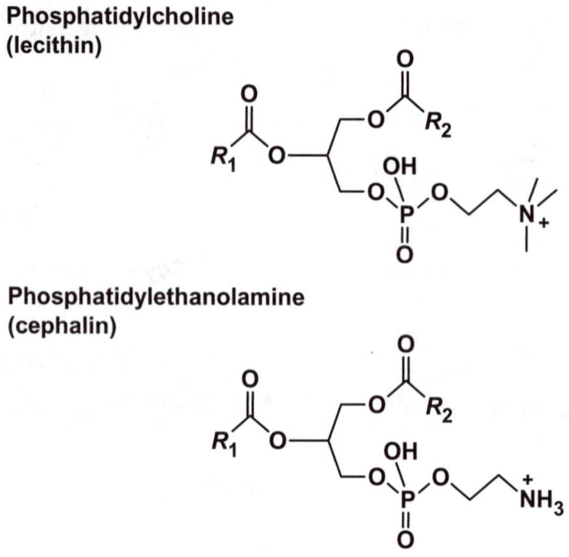

**Phosphatidylcholine
(lecithin)**

**Phosphatidylethanolamine
(cephalin)**

FIGURE 4.14 The two common phospholipids in soil are phosphatidylcholine and phosphatidylethanolamine. These compounds are esters of glycerol (1,2,3-propantriol) and long-chain alkanoic acids (fatty acids) and phosphate (R_1 and R_2 represent the alkyl chains).

approximately 46% of the inositol phosphates extracted from soil. Since this particular isomer is only found in microorganisms, a large percentage of the soil inositol phosphates are thought to be synthesized *in situ* by microorganisms.

The phospholipids account for 0.5 to 7% of organic soil P. Nucleic acids also contain a small percentage of soil organic P (0.2 to 2.4%). Like other lipids (described below), the phospholipids consist of compounds that are sparingly soluble in water and soluble in organic (fat) solvents (ether, benzene, chloroform). The commonly identified phospholipids in soils are the glycerophosphatides, phosphatidylcholine (lecithin), and phosphatidylethanolamine (cephalin) (Figure 4.14). These compounds are esters of glycerol (1,2,3-propanetriol) and long-chain alkanoic acids (fatty acids) and phosphate. The soil phospholipids are also thought to be of microbial origin. For example, the teichoic acids are a component of microbial cell walls and cell membranes and are composed of alanine, ribose, and polyglycerol phosphate esters (Figure 4.15).

4.3.3 LIPIDS

Lipids are a class of compounds that are operationally defined by their sparing solubility in water and their solubility in organic (fat) solvents, such as ether, chloroform, or benzene. Included in this category are fatty acids, fats, waxes, resins, sterols, terpenes, porphyrins, and polynuclear aromatic hydrocarbons. The bulk of the lipid content of soils is derived from partially decomposed and undecomposed plant and animal resides; microbial lipids are present in only small quantities. Lipids are generally considered to be short-lived in aerobic soils, but some fats and waxes are found to accumulate in highly acidic (pH < 4) or anaerobic soils. The exceptions to this rule are the waxes, sterols, and the long-chain fatty acids, alcohols, and alkanes, which tend to resist microbial decomposition. In general, fats, waxes, and resins usually account for 1 to 6% of SOM. Some of the other lipids, such as sterols, terpenes, and porphyrins occur naturally, but are present in exceedingly low concentrations in soils. Polynuclear aromatic hydrocarbons have also been detected in soils, but these are primarily present as a result of human activities.

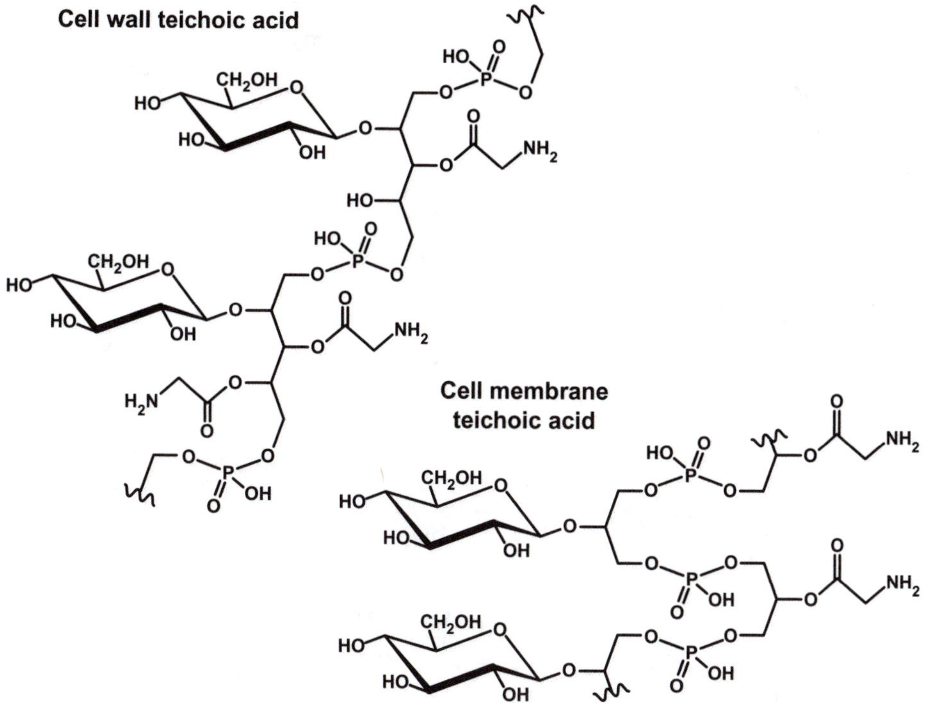

Cell wall teichoic acid

Cell membrane teichoic acid

FIGURE 4.15 Teichoic acids are components of microbial cell walls and cell membranes and are composed of alanine, ribose, and polyglycerol phosphate esters.

Fats and waxes comprise a group of compounds that are esters of long-chained fatty acids and aliphatic and aromatic alcohols (Figure 4.16). The fatty acids that are components of fats and waxes are typically unbranched (*n*-fatty acids), contain from 12 to 34 carbon atoms (even numbers predominate), and are saturated (no C—C double bonds). Examples of aliphatic *n*-fatty acids are palmitic acid [$CH_3(CH_2)_{14}COOH$], and stearic acid [$CH_3(CH_2)_{16}COOH$]. An example of an unsaturated fatty acid is oleic acid [$CH_3(CH_2)_7CH=CH(CH_2)_7COOH$]. The most common alcohols in fats and waxes are also aliphatic. The alcohol present in fats is glycerol, a propanetriol; whereas, waxes contain long-chained (greater than 16 carbons) *n*-alcohols (Figure 4.16). Schnitzer et al. (1986), using a supercritical *n*-pentane extraction of Aquoll, Boroll, and Aquod soils, found that *n*- and branched-alkanes accounted for 0.15 to 0.45% of SOC, with the *n*-alkanes accounting for 87 to 95% of the total alkanes extracted. The number of carbons in the extracted alkanes varied from 12 to 32, with the C_{24} and C_{26} alkanes being found in the greatest abundance. The total quantities of fatty acids extracted accounted for 4.2 to 6.2% of SOC. The average distribution of the fatty acids was (as a percentage of the total extracted):

- Free *n*-fatty acids (C_7 to C_{29}), 53%
- Bound *n*-fatty acids (C_7 to C_{29}), 19%
- Unsaturated fatty acids (C_{16} and C_{18} with a single C=C double bond, and C_{18} with two double bonds), 6%
- Branched fatty acids (C_{12} to C_{19}), 19%
- Hydroxy fatty acids (C_{12} to C_{16}), 2%
- α-ω-Diacids (C_{15} to C_{25}), 5%

These authors also concluded that the *n*-fatty acids were primarily derived from microbial sources.

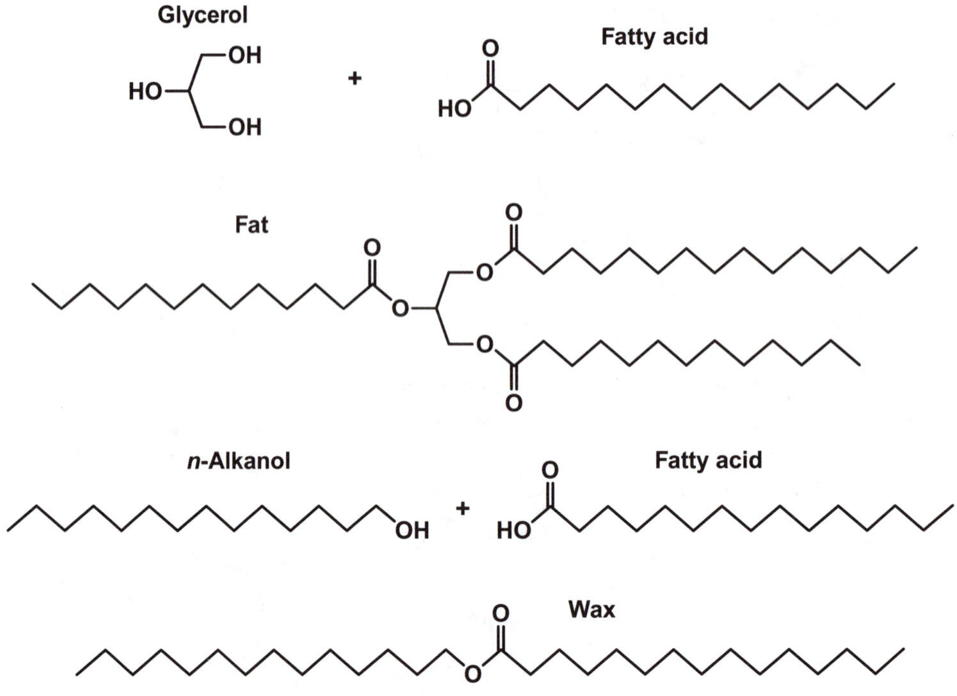

FIGURE 4.16 Fats are esters of glycerol and fatty acids; whereas, waxes are esters of an *n*-alkanol and a fatty acid. The fatty acids in this figure are saturated, because all bonds in the alkyl carbon chain are single bonds.

4.3.4 LIGNINS

Lignins are perhaps some of the most refractory nonhumic substances present in soils in significant quantities, and are perceived to be some of the more important precursors to the humic substances. Other highly aromatic and refractory biopolymers are also present in soils, but their true significance relative to SOM content and to the formation of humic substances has not been resolved. One group of refractory biopolymers is the sporopollenins. These are compounds that comprise the outer walls of spores and pollen grains and constitute a group of substances that play a key role in their protection and longevity. The sporopollenins are similar to lignins in that they contain phenylpropanoid units (described below). The refractory nature of the sporopollenins is evidenced by the observation that they enjoy wide distribution and occurrence in plant fossils and can constitute up to 50% of the mass of some coals.

Another group of refractory biopolymers that are highly aromatic are the tannins. Tannins are a significant constituent of terrestrial plants, exceeded in abundance only by cellulose, hemicelluloses, and lignins. There are two groups of recalcitrant tannins: the proanthocyanidin polymers and the phlorotannins. The proanthocyanidins are composed of flavonoid monomers, while the phlorotannins contain phloroglucinal units (Figure 4.17). Refractory aliphatic substances are also produced by higher plants. These include the cutins and suberins, which are extracellular compounds that coat the epidermal cells of leaves and are also found in bark and root tissues. These compounds are mainly polyesters composed of hydroxyl and epoxy C_{16} and C_{18} fatty acids. The refractory nature of these substances is suggested by their presence in the fossil record; however, their longevity and significant presence in soils has yet to be validated.

The lignins are components of and are only produced by vascular plants, accounting for 5 to 10% of the dry weight of leaves and up to 30% of the dry weight of wood. Thus, in soils

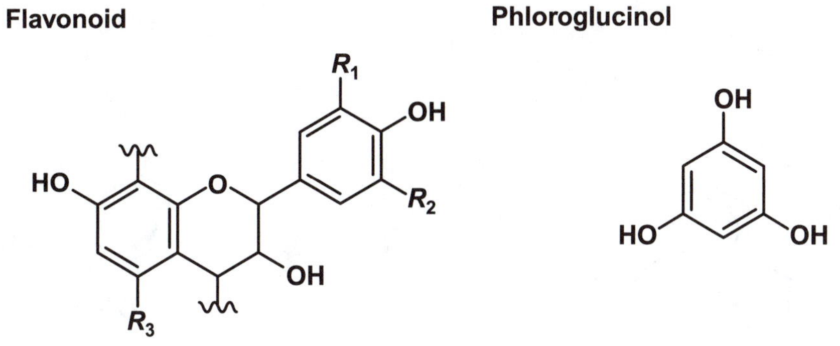

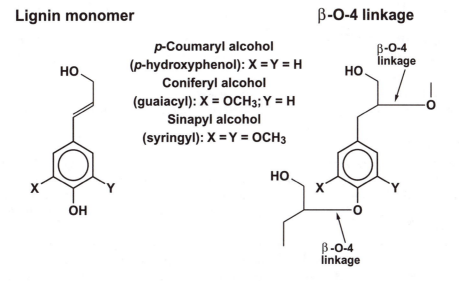

FIGURE 4.17 The refractory tannins consist of the proanthocyanidins, composed of flavonoid units, and the phlorotannins, which contain phloroglucinal units.

FIGURE 4.18 There are three basic types of lignin monomers. Coniferyl alcohol, or the guaiacyl unit, is a monomethoxyphenol common to lignin in the softwood gymnosperms (conifers). Sinapyl alcohol, or the syringyl unit, is a dimethoxyphenol found in the lignin of dicotyledonous angiosperms (hardwoods). The lignin in monocotyledonous angiosperms (grasses) is composed of p-hydroxyphenol units (polymers of p-coumaryl alcohol, also known as p-hydroxycinnamyl alcohol). The lignin monomer couple with other monomers via the β-O-4 linkage.

they are only subject to degradation, principally by white rot and brown rot fungi (they are not synthesized by microorganisms). Lignins are best described as a group of high-molecular-mass polymers of phenolpropanoid units, of which there are three basic types (Figure 4.18). In the conifers, the softwood gymnosperms, the lignin structure is based upon the monomethoxyphenol (coniferyl alcohol, guaiacyl) unit. The conifers do not contain the dimethoxyphenol unit (sinapyl alcohol, syringyl). Lignin in hardwoods (dicotyledonous angiosperms) contains both the coniferyl and sinapyl alcohol units. In grasses (monocotyledonous angiosperms), lignin is composed of p-coumaryl alcohol (p-hydroxycinnamyl alcohol, p-hydroxyphenol) units. Therefore, the relative abundances of the various types of lignin present in the soil are dictated to a large

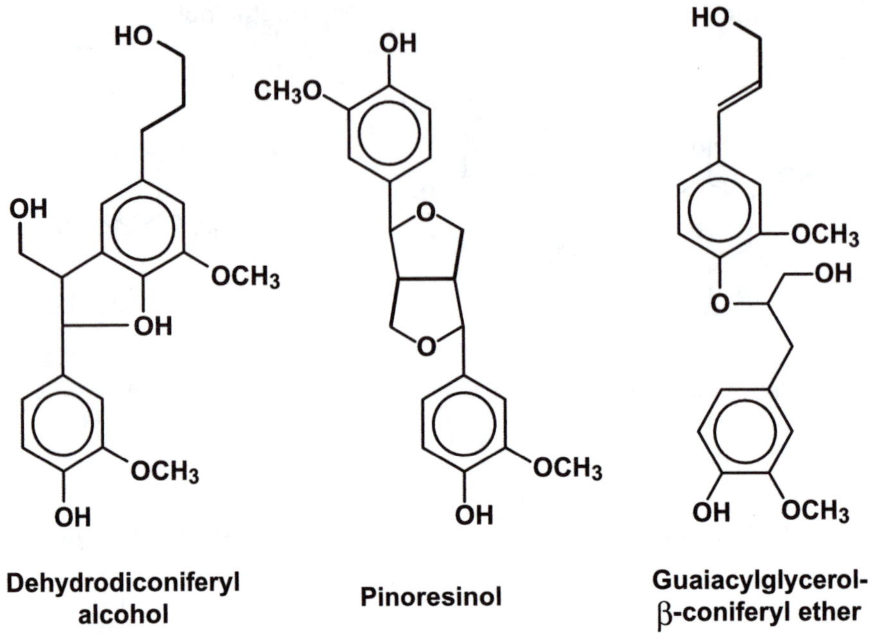

| Dehydrodiconiferyl alcohol | Pinoresinol | Guaiacylglycerol-β-coniferyl ether |

FIGURE 4.19 Lignin monomers dimerize to form dilignols. The three common dilignols, using coniferyl alcohol as the monomer, are guaiacylglycerol-β-coniferyl ether, pinoresinal, and dehydrodiconiferyl alcohol.

degree by the indigenous vegetation. The lignin monomers are linked through β-O-4 linkages to form dilignol units. For example, the common dilignols formed by coniferyl alcohol are guaiacylglycerol-β-coniferyl ether, pinoresinol, and dehydrodiconiferyl alcohol (Figure 4.19). Additional polymerization results in a highly complex and branched structure similar to that illustrated in Figure 4.20.

Soil lignin characterizations are performed by first cleaving the β-O-4 linkages in the lignin macromolecular structure, followed by the quantization of the *p*-hydroxyphenol, guaiacyl, and syringyl units. The most common techniques for generating the lignin monomers are CuO oxidation and tetramethylammonium hydroxide (TMAH) thermochemolysis. Once produced, the concentrations of the individual lignin monomers, which occur as aldehyde, carboxylic acid, and methyl ketone derivatives, are readily determined using high pressure liquid chromatography or gas chromatography-mass spectrometry.

The longevity of lignin in soils is due to its aromatic nature and the diversity of its bonds (cross-linkages). Lignin is virtually indestructible in anaerobic environments, accumulating, for example, in peat bogs. However, in aerobic soils lignin decomposition does occur and is performed by a group of filamentous basidiomycetes. Of these, the white-rot fungi (*Coriolus* and *Phanaerochaete*) are the most active lignin decomposers. These organisms secrete extracellular enzymes that principally cleave the β-O-4 bonds and oxidize the side chains (Figures 4.18 and 4.19). A second group of basidiomycetes, the brown rot fungi (*Poria* and *Gloephyllum*) are also involved in lignin decomposition. The brown rot fungi are principally responsible for the demethylation of the methyl groups on the syringyl and guaiacyl units yielding hydroquinone structures (also through the excretion of extracellular enzymes). Neither the white-rot nor the brown-rot fungi derive metabolic energy from the decomposition of lignin, as lignin degradation is a secondary metabolism. Lignin-degrading organisms obtain energy from the polysaccharides (cellulose and hemicelluloses) that are intimately associated with the lignin.

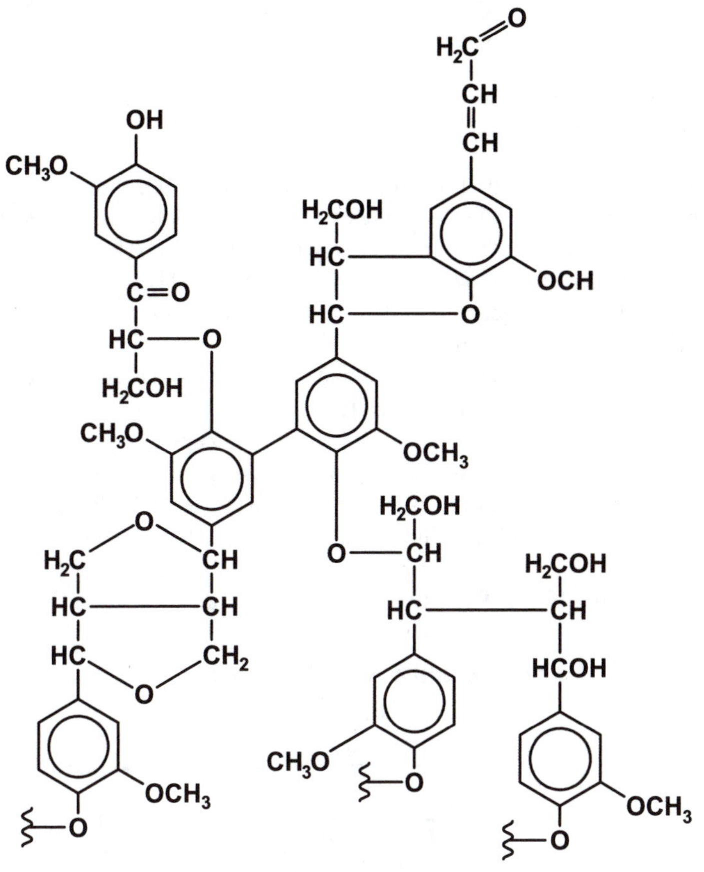

FIGURE 4.20 A possible structural configuration of lignin.

4.4 HUMIC SUBSTANCES

As classically defined, humic substances are (Stevenson, 1994):

> A series of relatively high-molecular-weight, brown- to black-colored substances formed by secondary synthesis reactions; the term is used as a generic name to describe the colored material or its fractions obtained on the basis of solubility characteristics; these materials are distinctive to the soil (or sediment) environment in that they are dissimilar to the biopolymers of microorganisms and higher plants (including lignin).

Although this definition is commonly cited, it provides very little information about the chemical nature, properties, and environmental role of humic substances. Indeed, the definition is vague relative to the genesis and function of humic substances in soil, instead focusing on color characteristics and the solubility characteristics of its various fractions. However, it clearly sets forth the long-held view that humic substances are macromolecular (high-molecular-mass), formed naturally (and indeed a substance that is unique to soil), and structurally similar to microbial and plant biopolymers (like lignin and other phenylpropanoids), but derived by the re-polymerization of the byproducts of biopolymer decomposition (secondary synthesis).

A more recent definition of humic substances (after MacCarthy, 2001) offers insight into their chemical nature and origin:

Humic substances comprise an extraordinarily complex, amorphous mixture of highly heterogeneous, chemically reactive yet refractory molecules that serve a key role in the Earth's ecological system, produced during early diagenesis in the decay of biomatter, and formed ubiquitously in the environment via processes involving chemical reaction of species randomly chosen from a pool of diverse molecules and through random chemical alteration of precursor molecules.

The characteristics of humic substances that are described in this definition are discussed below. It cannot be overstated that humic substances are exceedingly complex, heterogeneous, and reactive components of the soil environment that are relatively resistant to microbial decomposition. Further, and seemingly contradictory to the definition of Stevenson (1994), the definition of MacCarthy (2001) makes no mention of the perception that humic substances are macromolecular. This view has been hotly debated by humic scientists and there is no consensus as to whether humic substances are (1) large macromolecules formed by biotic and abiotic synthesis reactions that repolymerize the products of biomolecule decomposition, or (2) supramolecular associations of heterogeneous and relatively small molecules that are derived from the decomposition of biomolecules.

Humic substances are partitioned into three main fractions: humic acid, fulvic acid, and humin (Figure 4.1). These classes of humic substances are operationally defined by their differential aqueous solubilities in acidic and alkaline solutions, not by their innate structural or chemical characteristics. One method for isolating the fractions of humic substances from soil is illustrated in Figure 4.21 (after Swift, 1996). The extraction of soil humus or acid-washed soil with a 0.5 *M*

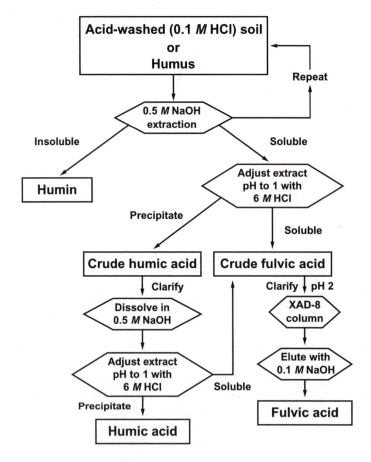

FIGURE 4.21 A fractionation scheme commonly employed to isolate the humic substances (after Swift, 1996). Note that the three separates (humin, humic acid, and fulvic acid) are strictly defined by their differential solubilities in acidic and alkaline solutions.

NaOH solution solubilizes the humic and fulvic acid fractions of soil humus. The organic component that is not solubilized after repeated alkali extraction of soil humus is defined as the humin fraction. The alkaline extract is then treated with concentrated (6 M) HCl in order to adjust the pH to approximately 1. The precipitate that forms as a result of the pH adjustment is the crude humic acid fraction, while the organic material remaining in solution is the crude fulvic acid fraction. Both these fractions may be clarified to (1) separate nonhumic substances and fulvic acid from the crude humic acid precipitate, and to (2) separate nonhumic substances from the soluble crude fulvic acid fraction. In the case of the former, the crude humic acid fraction is first redissolved in 0.5 M NaOH, and then reprecipitated by acidification with 6 M HCl. The resulting precipitate is called generic humic acid. The acid solution is combined with the solution containing the crude fulvic acid fraction. The separation of nonhumic substances from fulvic acid is performed by passing the acidic crude fulvic acid solution through an XAD-8 resin column (or equivalent). The XAD-8 resin is a nonionic, macroporous (25-μm pore size), methyl methacrylate ester polymer. In acidic eluents, the acidic functional groups of fulvic acid are protonated and adsorbed by the resin, while inorganic solutes and nonhumic substances (such as polysaccharides and low-molecular-mass organic acids) pass through the column. The fulvic acid fraction is then removed from the resin by eluting with an alkaline solution, yielding generic fulvic acid.

4.4.1 GENESIS OF HUMIC SUBSTANCES

Several mechanisms, or pathways, have been proposed to describe the genesis of humic substances. The five principal and credible mechanisms are illustrated in Figure 4.22. There are essentially two categories of pathways: (1) those that are purely biological and involve the enzymatic decomposition of biopolymers and the enzymatic recombination of the microbial byproducts (pathways 2, 3, and 4),

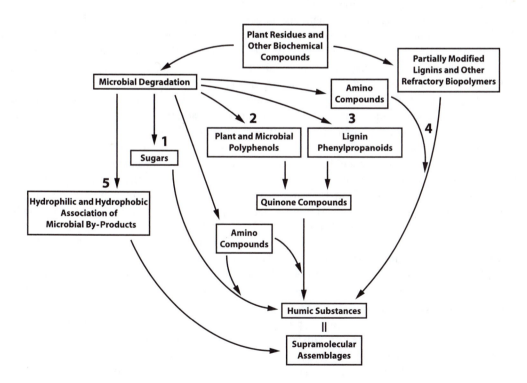

FIGURE 4.22 Numerous pathways have been proposed to describe the genesis of humic substances. The plausible pathways, either producing humic substances alone or in combination, are (1) sugar-amine theory; (2) polyphenol theory; (3) and (4) lignin-protein theory; and (5) self-aggregation theory.

and (2) those that involve the biotic decomposition of biopolymers and the abiotic assemblage of macromolecular structures (pathway 1) or aggregates (pathway 5). One of the oldest concepts of humic substance formation is the sugar-amine theory (pathway 1). In this pathway, monosaccharides and amino compounds produced during microbial metabolism of nonhumic substances recombine via purely abiotic condensation reactions (water is released) and result in the formation of multiple and principally covalent bonds. The products of these reactions are brown nitrogenous polymers, or melanoidins, that contain furan, pyrrole, pyridine, and other heterocyclic aromatic residues. A process that is commonly cited to explain abiotic condensation in soils is the Maillard reaction: a random condensation of amino acids and monosaccharides that results in the formation of complex macromolecular structures of varying size and solubility. Laboratory-formed melanoidins are similar to natural humic substances, particularly with respect to their high degree of cross-linkage, insolubility in water, and nonhydrolyzability. The Maillard reaction is highly sensitive to reaction conditions, being favored in alkaline systems that contain large amounts of monosaccharides, proteins, peptides, and amino acids. However, as discussed by Burdon (2001), alkaline soils do not contain greater quantities of humic substances than acidic soils, and the concentrations of free sugars and amino acids in soils are exceedingly low. Further, the products of the Maillard reaction contain primarily heterocyclic N, rather than the amide N common to humic substances.

Two of the pathways illustrated in Figure 4.22 (3 and 4) are actually subpathways described by the lignin-protein theory, or simply the lignin theory. Both of these pathways view lignin as the source of humic substances (even though humic substances form in soils that lack lignin input from higher plants). In pathway 4, plant lignins are only partially (superficially) modified by soil microbes, such that the polymeric structure of the macromolecules essentially remains intact. Pathway 3 builds on pathway 4 by allowing for the enzymatic cleavage of the β-O-4 bonds in the lignin macromolecule and the generation of lignin and dilignol polypropanoid units (Figure 4.19). Irrespective of the degree of lignin decomposition, the modification that is required for the repolymerization of the lignin monomers and larger lignin structures is the demethylation of the phenolic ester leading to a catechol (p-benzenediol, Figure 4.23). In addition, the oxidation of the propanol side-chains results in the formation of carboxyl groups. Carboxyl groups may also result from the cleavage of the aromatic rings in lignin. The catechol is enzymatically oxidized further to the o-quinone (by polyphenoloxidases), which undergoes condensation with amino compounds and other quinones. The amine-quinone condensation reactions illustrated in Figure 4.23 result in the production of an aromatic amine, which will undergo additional condensation with other quinones to form a polymeric humic substance. The macromolecules that initially form in pathways 3 and 4 tend to be relatively oxygen-poor, and relatively insoluble in the alkaline solutions employed to operationally describe humic and fulvic acids, thus, these substances would be defined as humin. As the oxidation of these macromolecules continues, they become more enriched in oxygen-bearing functional groups (carboxyl, carbonyl, phenolic-OH). As a result, they display greater solubility in alkaline media and insoluble in acidic solutions; thus, they are operationally defined as humic acids. With continued oxidation, these substances would ultimately develop into both acid- and base-soluble compounds—the fulvic acids. Therefore, lignin theory describes the least humified substances as humins, and the most humified substances as fulvic acids.

Perhaps the most widely accepted theory of humic substance formation is the polyphenol theory (pathway 2). Unlike the lignin theory, the polyphenol theory states that the quinones originate from both plant and microbial sources. This theory also considers all biopolymers to be decomposed to their monomeric units before enzymatic repolymerization occurs. By this mechanism, the complexity of humic substances increases with age, with the formation of fulvic acids occurring initially. The continued incorporation of monomeric decay products into humic macromolecules ultimately leads to the most humified substance, humin. This sequence of humification is counter to that proposed in the lignin theory.

The polyphenol theory is diagrammed in Figure 4.24. Lignin is still considered to be the principal source for phenolic substances, although these substances may also be derived from

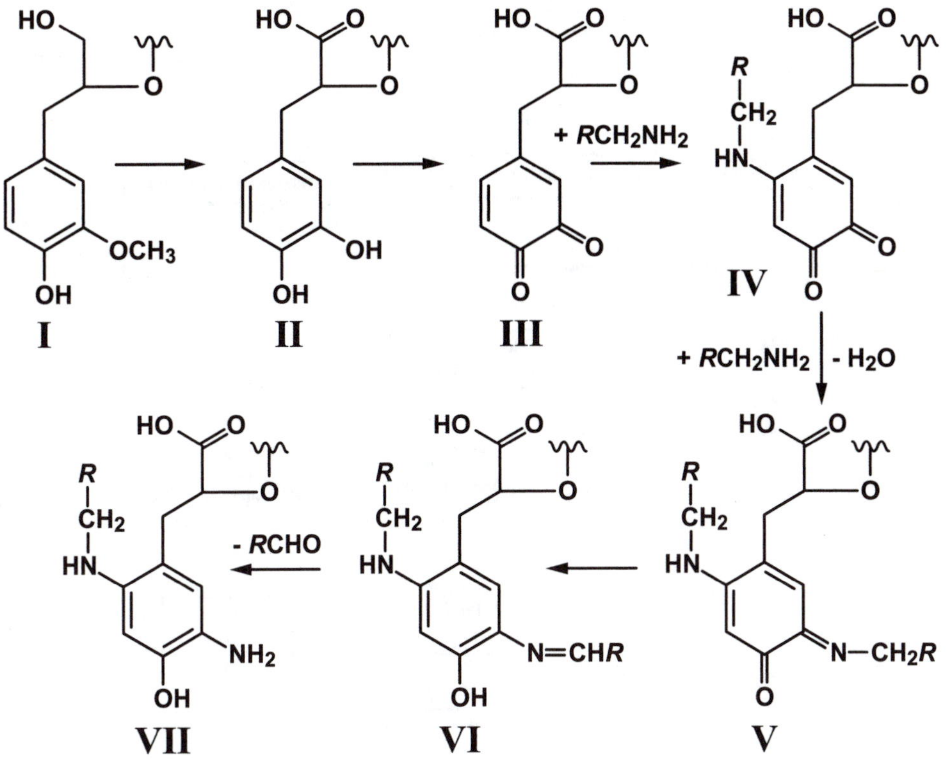

FIGURE 4.23 The polymerization of phenylpropanoid units is theorized to occur through a series of microbially mediated processes as illustrated for the guaiacyl unit (I). Demethoxylation and oxidation of the propanoid result in a catechol (II), which is then oxidized to the *o*-quinone by polyphenoloxidases (III). Amino acids and other amino compounds (amino sugars) react with the quinone to produce the substituted quinone (IV). Additional combination of the quinone with amino compounds (V), reorganization (VI) and removal of an aldehyde results in an aminophenol (VII). Additional combination of the aminophenol (VII) with the quinone (III) and the substituted quinone (IV) leads to a polymeric humic substance.

glycosides that contain aromatic structures (e.g., antibiotics), tannins (which contain flavonoids and phloroglucinol, Figure 4.17), aromatic amino acids (e.g., phenylalanine and tyrosine, Figure 4.10), and microbial synthates. Once produced, the phenolic substances are oxidized to quinones by polyphenoloxidases. Self-polymerization of quinones may occur in soils and has been shown to occur in laboratory cultures. However, the condensation of quinones is greatly enhanced in the presence of amino compounds (Figure 4.23), such as amino sugars and amino acids, and results in a product that has many of the characteristics of humic substances.

In the pathways described above (1 through 4), humic substances are presumed to be comprised of macromolecular, polymeric structures. These substances result from condensation reactions that form numerous cross-linkages that involve strongly covalent bonds, such as the C—C and C—N bonds. However, there is an alternate paradigm to the traditional view of humic substances as macromolecules. This view is the basis for pathway 5 (self-aggregation theory), which considers humic substances to be supramolecular associations of relatively small molecules that have self-organized into relatively large molecular entities. The small molecules in these suprastructures consist of plant and microbial residues and their microbial degradation products. During the enzymatic and oxidative depolymerization of plant biopolymers, such as lignin (shown in Figure 4.20), tannins, and cutins, carboxyl groups are formed. Some portions of these monomeric or larger units may remain relatively unaltered and relatively hydrophobic. The unaltered portion of the unit

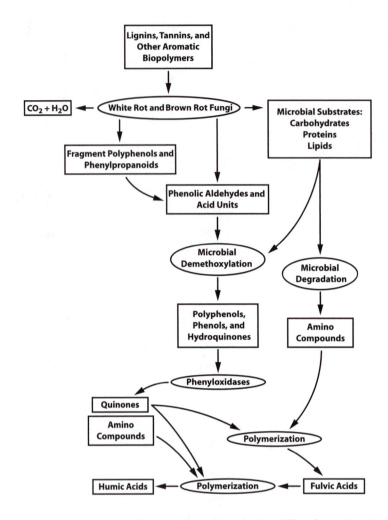

FIGURE 4.24 The polyphenol theory of humic substance production differs from other theories in that the phenolic building blocks may be derived from biopolymers other than lignin. The individual phenolic units are then polymerized according to the scheme shown in Figure 4.23.

will contain aromatic and aliphatic structures and will be relatively nonpolar or hydrophobic; whereas, that portion of the molecule containing the carboxyl group will be polar or hydrophilic, if not ionic (Figure 4.25a). In essence, the depolymerization of biomolecules may result in substances that are amphophilic and surfactant-like. These amphophiles may aggregate on mineral surfaces or in solution. At low concentrations in soil solutions, the amphophilic units remain relatively dispersed. However, as solution concentrations increase above some critical concentration, the units aggregate to form a micelle. The critical concentration is called the critical micelle concentration. A micelle is a globular unit composed of several amphophiles arranged such that the hydrophobic portions of the molecules are in the interior and the hydrophilic portions are on the exterior (Figure 4.25b). These micelles are stabilized by π-π bonds and weak forces that do not contain covalent bonds. These weak forces include attractive hydrophobic interactions, such as van der Waals forces, and hydrogen bonds (described in Chapter 2). The solubility of the micelle units is dictated by the number and types of acidic functional groups that make up the exterior of the micelles. If the units contain a large number of acidic functional groups that are deprotonated, the units will remain soluble as part of the fulvic acid fraction.

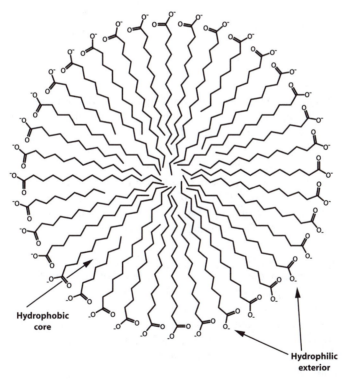

FIGURE 4.25 The self-aggregation theory of humic substance production is based on the premise that biopolymer degradation products are amphophilic, consisting of hydrophilic (polar or ionic) and hydrophobic molecular segments (a). In solution, the degradation byproducts form micelle-like structures (b), composed of a hydrophilic outer layer and a hydrophobic core. Accumulation of humic substances on soil surfaces (c) is initiated by the reaction of hydrophilic moieties with surface oxygens (electrostatic interactions) or surface metal cations (direct ionic/covalent bonds). The hydrophobic portions of these surface molecules are shielded from the polar aqueous phase by a second layer of amphophiles to form a bilayer.

c) Bilayer

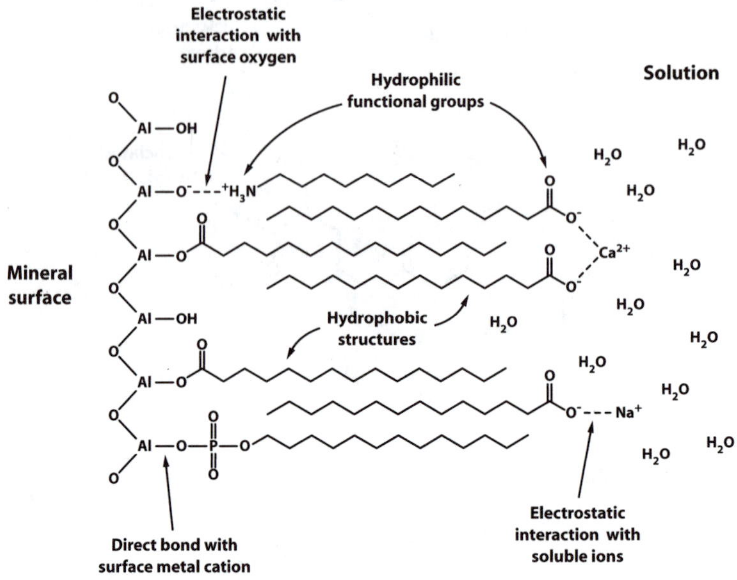

FIGURE 4.25 (*continued*).

Two mechanisms have been proposed to explain supramolecule formation on soil surfaces. One theory suggests that polar portions of the amphophilic molecules, the dissociated carboxyl group, reacts with inorganic, constant potential surface functional groups via a ligand exchange mechanism (described in Chapter 7) to form direct and relatively covalent bonds with surface metal cations (Figure 4.25c). Adsorption may also occur if the surface and the organic moiety have opposing charges, as illustrated by the electrostatic retention of a protonated amino group by a negatively charged surface site. The hydrophobic portions of these surface-bound molecules are then stabilized by the additional accumulation of amphophiles at the surface, arranged such that the polar portions point toward the aqueous phase. The resulting organic agglomeration at the soil surface is called a bilayer.

A second theory that describes the accumulation of solid humic substances assigns relatively unique hydrophilic-hydrophobic character to the different humus fractions. Fulvic acids are described as associations of small hydrophobic molecules that contain a sufficient number of acidic functional groups to maintain supramolecule solubility at any solution pH (again, the operational definition of fulvic acid). Humic acids are described as supramolecular associations composed of predominantly hydrophobic units that are stabilized by weak attractive forces. As solution pH decreases, intermolecular hydrogen bonding stabilizes the units further such that precipitation occurs (acid insoluble). These hydrophobic units may also bond to clay minerals that have a small amount of isomorphic substitution and/or an interlayer with hydrophobic character, such as the smectites (see Chapter 7).

Arguments for and against the view that humic substances are supramolecular entities are quite persuasive (for examples, see Swift (1999) and Burdon (2001)). However, it is important to recognize that humic substances are not generated by any one single pathway, but by several pathways with one generally dominating. For example, in poorly drained soils or anaerobic systems lignin decomposition is severely restricted. In these environments, the humic substances may arise principally via pathways 3 and 4 (Figure 4.22). However, in well-aerated soils almost all of the lignin and other biopolymers may be completely degraded to their base monomers with only a small amount of larger biopolymer fragments repolymerizing with the monomers to form humic substances (pathway 2). Further, it is entirely conceivable that monomers and larger fragments may self-associate or form associations with secondary synthesis products through weak van der Waals and electrostatic forces (pathway 5), irrespective of the environmental conditions.

4.4.2 CHEMICAL AND STRUCTURAL CHARACTERISTICS OF HUMIC SUBSTANCES

As discussed in the previous section, humic substances are either exceedingly complex macromolecules formed by secondary abiotic or biotic synthesis reactions that repolymerize the polymeric fragments or monomeric units of biopolymers; or, they are associates, aggregates, or micelles of relatively small molecules held together by weak van der Waals forces and hydrogen bonding; or, they are all of the above. Irrespective of their mechanisms of formation or their structural character, humic substances are most often studied after they have been isolated and purified according (typically) to the extraction procedure detailed in Figure 4.21. Within this framework, a considerable amount of chemical information, such as elemental and functional group composition, has been accumulated. However, the macrostructural characteristics of humic substances remain an enigma.

4.4.2.1 Elemental Content

One of the most common means of characterizing humic substances is to determine the total elemental content. Typically, this involves the determination of C, H, O, N, and S (and sometimes P) concentrations in the isolates. The separates that comprise the humic substances tend to differ in their elemental character (Table 4.2). Relative to fulvic acids, humic acids generally have greater C and lower O content. The disparity between C and O concentrations also results in differences in the O/C mole ratios between the two fractions, with fulvic acids having a greater O/C ratio than the humic acids. The greater oxygen content and O/C ratio of the fulvic acids is related to the

TABLE 4.2
Mean Elemental Content (in g kg^{-1}), O/C and H/C Mole Ratios (in mol mol^{-1}) of Soil Humic and Fulvic Acids Collected from Around the World

Element	Humic Acids		Fulvic Acids	
	Mean	Range	Mean	Range
C	562	536–58 7	457	407–50 6
H	47	32–62	54	38–70
O	355	328–38 3	448	397–49 8
N	32	8–43	21	9–33
S	8	1–15	19	1–36
O/C[a]	0.51		0.74	
H/C[a]	1.00		1.42	

[a] The mole ratio values were computed using the mean elemental contents of C, O, and H.

From Steelink, C. Elemental characteristics of humic substances. *In* G.R. Aiken et al. (ed.) *Humic Substances in Soil, Sediment and Water.* John Wiley & Sons, New York, 1985, pp. 457–476. With permission.

higher concentrations of oxygen containing functional groups, such as the carboxyl group (R—COOH), and to the higher concentrations of carbohydrates in the fulvic acid materials. The higher concentrations of oxygen-bearing moieties in the fulvic acids is also expected and related to their operational definition. A relatively large number of acidic groups are required to maintain fulvic acid solubility when the initially alkaline soil extract is acidified (Figure 4.21). Another distinguishing characteristic of the fulvic and humic acids is the H/C mole ratio. The relative magnitude of the H/C mole ratio is used to indicate the degree of aromaticity and unsaturation of carbon chains. Relatively small H/C mole ratios suggest that the humic substances have a high degree of aromaticity; whereas, larger values indicate a greater abundance of aliphatic structures. Typically, H/C mole ratios for the fulvic acids are greater than those of the humic acids, indicating that the fulvic acids have greater aliphatic character than the humic acids (or that the humic acids have greater aromatic character than the fulvic acids).

Rice and MacCarthy (1991) surveyed the elemental contents of humins, humic acids, and fulvic acids extracted from soil, freshwater, marine, and peat sources from around the world (Table 4.3). One of the more interesting aspects of the data compiled by these researchers is the distribution of the concentration and mole ratio data about their respective mean values. As illustrated in Table 4.3, the observed ranges of elemental composition data for the three humic substances are quite broad. For example, the C content of one humic acid sample contained 372 g C kg^{-1}, while another contained 758 g C kg^{-1}. Similarly, the fulvic acids contained from 351 to 754 g C kg^{-1}, and the humins from 483 to 616 g C kg^{-1}. This can be explained given that these humic substances are products of numerous and variable extraction procedures (but based on the common theme shown in Figure 4.21) performed by many different individuals, and that the isolates are derived from a variety of sources (soils, freshwater, marine, and peat) from different geographical locations and parent materials (vegetation). It would appear from these data that the different humic substances (humic acids, fulvic acids, and humins) are chemically diverse in nearly all elemental respects, and on average chemically indistinguishable from one another, because their elemental concentration ranges overlap (as illustrated for carbon above). However, it must be recognized that these data values are the extremes and do not represent the respective populations of the humic substances; they are the outliers.

Figure 4.26 illustrates the normal distribution of the elemental contents for the humins, humic acids, and fulvic acids computed using the mean and standard deviation data reported by Rice and MacCarthy (1991). Also illustrated are the 95% confidence intervals of the mean values. The x-axis represents the elemental concentration (or mole ratio), and the y-axis represents the number of samples (or frequency of occurrence) having elemental contents that are a particular concentration (the curves model the histograms of the actual data). Instead of a broad and diffuse distribution that might be expected given that the data are obtained from diverse sources by diverse means, it is observed that the elemental contents of the substances are normally distributed about mean values that have relatively small standard deviations (with the exception of the S data). For example, the mean C content of the humic substances is 551 g kg^{-1} with a standard deviation of 50 g kg^{-1}. Thus, 68% of the 410 humic acid samples examined have C concentrations in the 501 to 601 g kg^{-1} range (±1 standard deviation); 96% of the 410 samples have C concentrations within the 451 to 651 g kg^{-1} range (±2 standard deviation).

Humic substances are compositionally unique relative to the environmental conditions in which they form. However, based on the statistical evaluation of Rice and MacCarthy (1991), the mean chemical compositions (concentrations of C, H, O, N, and S) and the mole ratios (O/C and H/C) of humic acids and humins are statistically similar (the 95% confidence intervals for the mean values overlap, as shown in Figure 4.26). The fulvic acids differed from the humic acids with respect to their mean C, N, and O concentrations and O/C and H/C ratios. On average, the fulvic acids contain less C than humic acids (462 g kg^{-1} vs. 551 g kg^{-1}), more O (456 g kg^{-1} vs. 356 g kg^{-1}), and less N (25 g kg^{-1} vs. 35 g kg^{-1}) (Table 4.3). The O/C mole ratio of the fulvic acids is also greater than that of the humic acids (0.76 mol mol^{-1} vs. 0.50 mol mol^{-1}). The mean concentrations of C, O, and

TABLE 4.3

Mean Elemental Content (in g kg^{-1}) and O/C and H/C Mole Ratio Values (in mol mol^{-1}) of Humic Acids, Fulvic Acids, and Humin Obtained from a Geographically Diverse Range of Soil, Freshwater, Marine, and Peat Sources[a]

	C	H	O	N	S	O/C	H/C
			Humic Acids				
All sources	551 ± 50	50 ± 11	356 ± 58	35 ± 15	18 ± 16	0.50 ± 0.13	1.10 ± 0.25
(n = 410)[b]	(372–758)	(16.4–117)	(79.3–566)	(5.0–105)	(1–83)	(0.08–1.20)	(0.08–1.85)
Soil	554 ± 38	48 ± 10	360 ± 37	36 ± 13	8 ± 6	0.50 ± 0.09	1.04 ± 0.25
(n = 215)	(372–641)	(16.4–80)	(271–520)	(5.0–70.0)	(1–48.8)	(0.33–0.97)	(0.08–1.77)
Freshwater	512 ± 30	47 ± 6	404 ± 38	26 ± 16	19 ± 14	0.60 ± 0.08	1.12 ± 0.17
(n = 56)	(438–560)	(35–65.4)	(309–482)	(6.3–79.7)	(3.5–43.1)	(0.42–0.80)	(0.79–1.69)
Marine	563 ± 66	58 ± 14	317 ± 78	38 ± 15	31 ± 14	0.45 ± 0.18	1.23 ± 0.23
(n = 95)	(375–758)	(37.6–117)	(79.3–566)	(9.7–105)	(12–83)	(0.08–1.20)	(0.67–1.85)
Peat	571 ± 25	50 ± 8	352 ± 27	28 ± 10	4 ± 2	0.47 ± 0.06	1.04 ± 0.17
(n = 23)	(505–628)	(36–65.7)	(307–432)	(6.0–39)	(1–7)	(0.37–0.64)	(0.73–1.35)
			Fulvic Acids				
All sources	462 ± 54	49 ± 10	456 ± 55	25 ± 16	12 ± 12	0.76 ± 0.16	1.28 ± 0.31
(n = 214)	(351–754)	(4.3–72)	(169–558)	(4.5–81.6)	(1.0–36)	(0.17–1.19)	(0.77–2.13)
Soil	453 ± 54	50 ± 10	462 ± 52	26 ± 13	13 ± 11	0.78 ± 0.16	1.35 ± 0.34
(n = 127)	(351–754)	(32–70.0)	(169–559)	(4.5–58.7)	(1–36)	(0.17–1.19)	(0.77–2.13)
Freshwater	467 ± 43	42 ± 7	459 ± 51	23 ± 21	12 ± 9	0.75 ± 0.14	1.10 ± 0.13
(n = 63)	(392–563)	(4.3–59)	(347–558)	(4.7–81.6)	(1.6–30.5)	(0.49–1.07)	(0.81–1.53)
Marine	450 ± 40	59 ± 9	451 ± 60	41 ± 23	—	0.77 ± 0.17	1.56 ± 0.13
(n = 12)	(384–500)	(43–68.0)	(369–545)	(10–68.3)		(0.55–1.07)	(1.31–1.73)
Peat	542 ± 43	53 ± 11	378 ± 37	20 ± 5	8 ± 6	0.53 ± 0.094	1.20 ± 0.33
(n = 12)	(469–608)	(42–72)	(311–443)	(12–26)	(12–26)	(0.38–0.71)	(0.85–1.84)
			Humins				
All sources	561 ± 26	55 ± 10	347 ± 34	37 ± 13A	4 ± 3	0.46 ± 0.06	1.17 ± 0.24
(n = 26)	(483–616)	(42–72.8)	(288–451)	(9.0 – 60.0)	(1–9)	(0.37–0.61)	(0.82–1.72)

[a] Mean ± standard deviation (range is shown in parentheses).

[b] Within each separate, the data are either aggregated (as in the "All sources" row) or separated by sample type (soil, freshwater, marine, or peat). The n values represent the number of samples included in the mean, standard deviation, and range. However, the n values differ for S in all categories and for N in some categories. For the humic acids, the n values for S are: all sources, n = 160; soil, n = 67; freshwater, n = 13; marine, n = 66; peat, n = 12. For N in peat humic acids, n = 21. For the fulvic acids, the n values for S are: all sources, n = 71; soil, n = 45; freshwater, n = 14; marine, n = 12; peat, n = 11. For the humins, n = 24 for N data and n = 16 for S data.

From Rice, J.A. and P. MacCarthy. Statistical evaluation of the elemental composition of humic substances. *Org. Geochem.* 17:635–648, 1991. With permission.

N, and the O/C ratio of fulvic acids also differ from those of the humins, with the fulvic acids containing less C, more O, less N, and having a higher O/C. Finally, soil fulvic acids have lower C concentrations than soil humic acids (453 g C kg^{-1} vs. 554 g C kg^{-1}), lower N concentrations (26 g N kg^{-1} vs. 36 g N kg^{-1}), higher O concentrations (462 g O kg^{-1} vs. 360 g O kg^{-1}), higher O/C mole ratios (0.78 vs. 0.50), and higher H/C mole ratios (1.35 vs. 1.04) (Table 4.3).

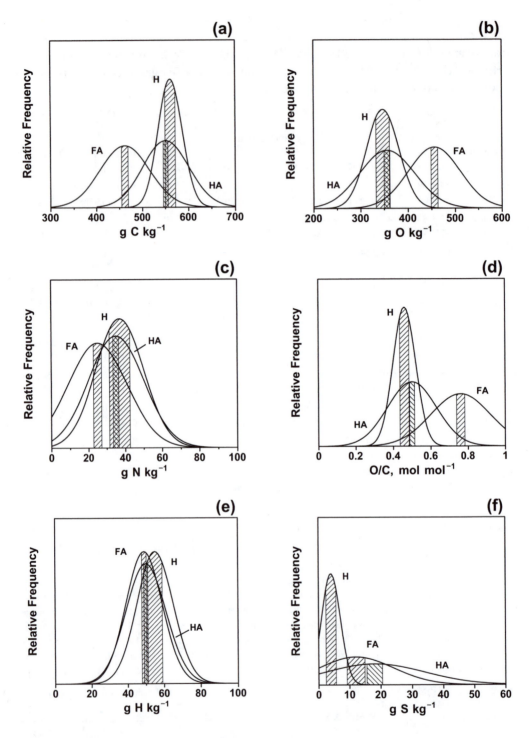

FIGURE 4.26 Frequency diagrams illustrating the normal distribution of elemental composition data for the humic substances. The solid lines represent the normal distribution curves produced using the statistical data of Rice and MacCarthy (1991). The hatched areas represent the 95% confidence intervals about the mean values. The concentrations are presented on a g kg^{-1} of humic substance basis: (a) organic carbon; (b) oxygen; (c) nitrogen; (d) mole ratio of oxygen to carbon; (e) hydrogen; (f) sulfur; and (g) mole ratio of hydrogen to carbon.

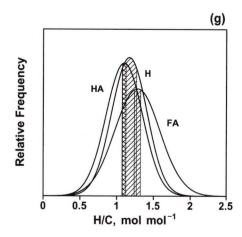

FIGURE 4.26 (*continued*).

The segregation of the humic and fulvic acids by source tends to reduce the variability associated with the mean composition data (standard deviations are decreased). Further, compositional differences within the humic and fulvic acids as a function of source may also be illustrated. Humic acids obtained from different environments differ in their elemental contents. On average, soil humic acids have higher C (554 g kg^{-1} vs. 512 g kg^{-1}) and N (36 g kg^{-1} vs. 26 g kg^{-1}) contents than freshwater humic acids, and lower O (360 g kg^{-1} vs. 404 g kg^{-1}) contents and O/C (0.50 vs. 0.60) ratios (Table 4.3). Degree of aliphaticity of humic acids also decreases from freshwater (H/C = 1.12) to soil sources (H/C = 1.04). Within the fulvic acids, soil sources are more aliphatic than freshwater sources (H/C is 1.35 for soil and 1.10 for freshwater).

4.4.2.2 Functional Groups and Structural Components

The major functional groups in humic substances are oxygen containing and include carboxyls, alcoholic and phenolic hydroxyls, carbonyls, and methoxyls. These groups are illustrated in Figure 4.3. Techniques for determining the types and concentrations of the various functional groups in humic substances can be grouped into two categories: wet-chemical methods and spectroscopic procedures. Typically, wet-chemical methods exploit the acidic properties of the various functional groups, or their derivatives. The more commonly used wet-chemical methods are described in Table 4.4. Many of the methods suffer from an inability to react specifically with the target functional groups on the humic macromolecular structure. For example, a common method for determining the carboxyl content of humic substances is the calcium acetate procedure. In this method, a 10-mL volume of a standard 0.5 M calcium acetate solution is reacted with a 50- to 100-mg mass of humic substance. After a 24-h reaction period, the humic residues are quantitatively separated from the solution by filtration and the solution titrated with standard 0.1 M NaOH. The excess volume of NaOH required to attain a pH of 9.8 in the humic filtrate, relative to the volume needed to attain pH 9.8 in a 0.5 M calcium acetate blank, is directly related to the concentration of acid functional groups. Although this is a very straightforward analytical procedure, there are a number of confounding factors that may lead to questionable analytical results. A major interference may arise from the dissociation of acidic hydroxyl groups. Normally, hydroxyl groups are only weakly acidic. For example, phenol and most substituted phenols have pK_a values that range between 9 and 10. However, hydroxyl groups in humic substances are attached to a diverse variety of aliphatic and aromatic structures. These different structures influence the Lowry-Brønsted acidity of the hydroxyls such that the groups may display pK_a values that are less than the pH of the calcium acetate solution (approximately 9).

TABLE 4.4
Summary of the Wet-Chemical Methods Commonly Employed to Characterize Various Functional Groups in Humic Substances

Functional Group	Method	Reaction	Analysis	Main Interferences
Total acidity	Barium hydroxide	$2(R\text{—COOH} + R\text{—OH} + Ar\text{—OH}) + Ba(OH)_2 \rightarrow (R\text{—COO} + R\text{—O} + Ar\text{—O})_2Ba(s) + 2H_2O$	Titration of unused $Ba(OH)_2$ with standard acid	1) Not all humic material may be precipitated 2) Excess mineral acids from humic substance isolation may be present
Carboxyl	Calcium acetate	$2R\text{—COOH} + Ca(CH_3COO)_2 \rightarrow (R\text{—COO})_2Ca + 2CH_3COOH$	Titration of liberated acetic acid with standard base	Reaction with acidic OH groups occurs
	Direct titration	$R\text{—COOH} + NaOH \rightarrow R\text{—COONa} + H_2O$	NaOH consumed to reach pH 8 is the carboxyl content	Assumes that only carboxyl groups have pK_a values less than 7
Total-OH	Acetylation	$(R\text{—OH} + Ar\text{—OH}) + (CH_3CO)_2O \rightarrow (R\text{—O—C}(=O)\text{—CH}_3 + Ar\text{—O—C}(=O)\text{—CH}_3) + CH_3COOH$	Titration of liberated acetic acid with standard base	Significant acetylation of carboxyl groups occurs requiring a correction
Phenolic-OH	Difference	$[\text{Phenolic-OH}] = [\text{total acidity}] - [R\text{—COOH}]$		Compounded errors
	Ubaldini method	1) $Ar\text{—OH} + KOH \rightarrow Ar\text{—O—K} + H_2O$ 2) $2Ar\text{—O—K} + CO_2 + H_2O \rightarrow 2Ar\text{—OH} + K_2CO_3$	Titration of K_2CO_3 with standard acid	1) Proton-K exchange reaction is nonspecific 2) Hydrolysis by KOH
Alcoholic-OH	Difference	$[\text{Alcoholic-OH}] = [\text{total-OH}] - [\text{phenolic-OH}]$		Compounded errors
Total carbonyl	Derivitization to oxime using hydroxylamine	$RC=O + NH_2OH \rightarrow RC=NOH + H_2O$	Titration of unused hydroxylamine using standard perchloric acid	Reagent may react with groups other than $RC=O$
Quinones	Ferrous iron reduction in alkaline triethanolamine	$Ar=O + Fe^{2+} + H^+ \rightarrow Ar\text{—OH} + Fe^{3+}$	Titration of unused ferrous Fe with standard chromate	
Ether linkages	Zeisel method	1) $(R\text{—OCH}_3 + Ar\text{—OCH}_3) + HI \rightarrow (R\text{—OH} + Ar\text{—OH}) + CH_3I$ 2) $CH_3I + 6Br_2 + 5H_2O \rightarrow HIO_3 + 12HBr + CO_2$ 3) $2HIO_3 + 10KI + 5H_2SO_4 \rightarrow 6I_2 + 6H_2O + 5K_2SO_4$	Titration of liberated I_2 with standard sodium thiosulfate $(Na_2S_2O_3)$ with starch as an indicator	

TABLE 4.5
Mean Total Acidity and Functional Group Content (in mol kg^{-1})
of Soil Humic and Fulvic Acids Collected from around the World

Functional Group	Humic Acids		Fulvic Acids	
	Mean	Range	Mean	Range
Total acidity	67	56–89	103	64–142
Carboxyl	36	15–57	82	52–112
Phenolic-OH	39	21–57	30	3–57
Alcoholic-OH	26	2–49	61	26–95
Carbonyl	29	1–56	27	12–42
Methoxyl	6	3–8	8	3–12

From Stevenson, F.J. *Humus Chemistry: Genesis, Composition, Reactions.* John Wiley & Sons, New York, 1994. With permission.

Thus, these hydroxyls are ionized (as are the carboxyl groups) and they dissociate in the alkaline calcium acetate solution. An additional confounding factor is that the Ca^{2+} is capable of displacing protons from a small number of protonated functional groups that would otherwise not deprotonate in the alkaline solution if, for example, Na^+ were the principal cation. Both interferences tend to overestimate the carboxyl content of humic substances.

Despite the questionable selectivity of the various wet-chemical methods, they do provide information on the nature and trends of the distribution of functional groups in the various humic substances. The humic and fulvic acids differ relative to their compositions of oxygen functional groups (Table 4.5). On average, humic substances contain approximately equal contents of carboxyl (36.0 mol kg^{-1}) and phenolic-OH (39.0 mol kg^{-1}) groups. Although the carboxyl and phenolic-OH groups appear to dominate, the abundances of alcoholic-OH and carbonyl groups are also highly significant (26.0 and 29.0 mol kg^{-1}). Fulvic acids are dominated by carboxyl groups (82.0 mol kg^{-1}), followed by alcoholic-OH (61.0 mol kg^{-1}). As a result of the greater carboxyl content, the fulvic acids have greater acidity (103.0 mol kg^{-1}) than the humic acids (67.0 mol kg^{-1}). The concentrations of the phenolic-OH and carbonyl groups are approximately equal in the fulvic acids (30.0 and 27.0 mol kg^{-1}), but are substantially lower than the carboxyl and alcoholic-OH groups. Methoxyl groups (OCH$_3$) are minor components of both the fulvic and humic acids.

Studies of the structural components of the humic substances, as well as their functional group contents, are most commonly performed using nondestructive spectroscopic techniques. Chemical degradation (destructive) methods, which employ pyrolysis (heat), acids, oxidants, or reductants (or some combination) to cleave humic structures into smaller units for analysis, have also been extensively employed to provide information on the structural character of humic substances. Once produced, the relatively simple compounds created by the degradation process are related back to the original macromolecular structure (either directly or through deduction based on potential degradation mechanisms). Although some feel that the degradation methods provide the best descriptions of humic structures, their popularity has been supplanted by nondestructive techniques.

Of the spectroscopic techniques, which include the ultraviolet-visible, fluorescence, and infrared spectroscopy, nuclear magnetic resonance (NMR) spectroscopy provides the most definitive information on the structural components and functional groups in humic substances. Further, NMR provides semiquantitative and quantitative information and is applicable to the analysis of both liquid- and solid-state samples. Nuclear magnetic resonance exploits the property of nuclear spin. A magnetic moment is associated with nuclear spin, such that the nuclei behave like tiny magnets. In the absence of a magnetic field, the poles of these nuclear magnets are randomly aligned.

TABLE 4.6

Chemical Shift Ranges and Assignments Associated with Cross-Polarization, Magic Angle Spinning ^{13}C Nuclear Magnetic Resonance (CPMAS ^{13}C NMR) Using a Tetramethylsilane Standard (the Regions Are Illustrated in Figure 4.27)

Chemical Shift, ppm	Assignments
0–25	Primary alkyl ($-CH_3$)
25–35	Secondary alkyl ($-CH_2-$)
35–50	Complex aliphatic (CH_2, CH, C)
50–60	Methoxyl, methyne, tertiary and quarternary alkyls ($-OCH_3$, $CH-NH$, CH, C)
60–96	Saccharide, alcohol, ether ($CHOH$, CH_2OH, CH_2-O-)
96–108	Anomeric ($O-C-O$)
108–120	Aromatic ($=CH=$)
120–145	Aromatic ($=CH=$, $=C-$)
145–162	Phenolic ($C-O-$, $C-OH$)
162–190	Carboxyl, ester, quinone ($COOH$, $COO-$, $C=O$)
190–220	Ketone, aldehyde, quinone ($C=O$, $HC=O$)

However, when a magnetic field is imposed, certain atomic nuclei align with the field in only one of two ways: with the field (low energy, $+\frac{1}{2}$ spin) or against the field (high energy, $-\frac{1}{2}$ spin). The nuclei that have spin $\pm\frac{1}{2}$, and that are most important to the characterization of humic substances, are ^1H, ^{13}C, ^{15}N, and ^{31}P. Nuclei that are placed in a magnetic field will predominantly exist in the low energy, $+\frac{1}{2}$ spin state. However, transition of the nuclei to the high energy spin state (the resonance condition) can be achieved by imposing an oscillating magnetic field of electromagnetic radiation that is perpendicular to a steady magnetic field, and that has a frequency that corresponds to the energy separation of the $+\frac{1}{2}$ and $-\frac{1}{2}$ spin states. The frequency required to achieve the resonance is a function of the chemical environment in which a nucleus resides, as the atoms that surround a nucleus provide shielding from the magnetic field. Thus, there is a shift in the resonance frequency relative to a standard. For ^{13}C NMR the standard compound is commonly tetramethylsilane (TMS, $(CH_3)_4Si$). The degree of shift is indicative of the types and arrangement of the surrounding atoms. The chemical shift between the resonance frequency of the standard and that of a moiety in the compound of interest is denoted by the symbol δ and is expressed in units of parts per million (ppm).

Perhaps the most common NMR technique employed to characterize humic substances is a solid-state ^{13}C NMR method called cross-polarization, magic angle spinning ^{13}C NMR (CPMAS ^{13}C NMR). The chemical shift assignments, relative to TMS, for CPMAS ^{13}C NMR are given in Table 4.6 and the example spectra in Figure 4.27. Quantitative analysis is performed by determining the area under the spectra that is defined by a given chemical shift region. This area is then divided by the total peak area of the spectra to yield the fractional amount of C associated with the particular chemical shift region. Mahieu et al. (1999) surveyed the published CPMAS ^{13}C NMR data for soil humic substances obtained from different environments (Table 4.7). On average, the alkyl and O-alkyl content of humic and fulvic acids are similar. However, the humic acids have a greater abundance of aromatic C than the fulvic acids (25.4 mol kg^{-1} vs. 19.1 mol kg^{-1} of organic C), while the fulvic acids have a greater abundance of carbonyl C (20.6 mol kg^{-1} vs. 13.7 mol kg^{-1} of organic C). These findings are generally consistent with the wet-chemical data presented in Table 4.5. However, the greater abundance of alcoholic-OH functional groups in the fulvic acids, determined by wet-chemical methods, is not confirmed by the CPMAS ^{13}C NMR data; that is, a greater abundance of O-alkyl C in fulvic acids relative to humic acids is not seen by CPMAS ^{13}C NMR. Ussiri and Johnson (2003) performed a direct comparison of CPMAS ^{13}C NMR spectra for humin,

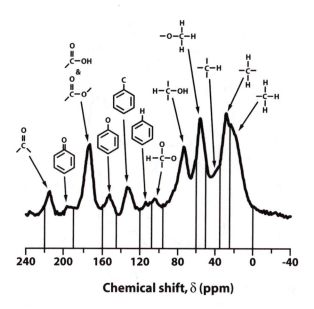

FIGURE 4.27 Example of a cross-polarization, magic angle spinning ^{13}C nuclear magnetic resonance (CPMAS ^{13}C NMR) spectra of humic acid from Conte et al. (1997). The vertical lines delineate the typical chemical shift assignments (from Table 4.6).

TABLE 4.7
Structural Composition (Mean ± Standard Deviation in mol kg^{-1} of Organic Carbon) of Soil Humic and Fulvic Acids in Soils Collected from a Wide Variety of Climates Using Cross-Polarization, Magic Angle Spinning (CPMAS) ^{13}C NMR

Functional Groups[a]	Humic Acids (n = 208)[b]		Fulvic Acids (n = 66)	
	Mean	Range	Mean	Range
Alkyl	22.1 ± 9.16	8.33–58.4	22.1 ± 8.33	6.99–39.6
O-Alkyl	22.1 ± 6.58	0–40.8	21.6 ± 9.91	7.24–50.8
Aromatic	25.4 ± 7.41	7.41–57.4	19.1 ± 5.58	5.83–30.7
Carbonyl	13.7 ± 3.75	3.33–25.0	20.6 ± 3.66	1.25–28.1

[a] Alkyls, between 0 and 50 ppm; O-alkyls, between 50 and 110 ppm; aromatics, between 110 and 160 ppm; carbonyls, between 160 and 220 ppm. See Table 4.6 for the specific carbon types included in each chemical shift region.
[b] n denotes the number of samples included in the statistical analysis.

From Mahieu, N., D.S. Powlson, and E.W. Randall. Statistical analysis of published carbon-13 MNR spectra of soil organic matter. *Soil Sci. Soc. Am. J.* 63:307–319, 1999. With permission.

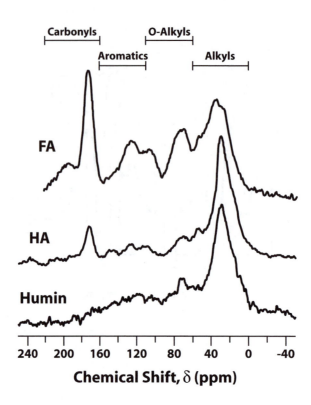

FIGURE 4.28 A comparison of the chemical characteristics of carbon in humic substance isolates from the Bh horizon of a Typic Haplorthod (65.3 g kg^{-1} SOC) determined by cross-polarization, magic angle spinning ^{13}C nuclear magnetic resonance (CPMAS ^{13}C NMR) (modified from Ussiri and Johnson, 2003). The four general chemical shift regions are alkyl C (0–50 ppm), O-alkyl C (50–110 ppm), aromatic C (110–160 ppm), and carbonyl C (160–220 ppm). See Table 4.6 for specific carbon types included in each shift region.

humic acid, and fulvic acid obtained from the Bh horizon of a Typic Haplorthod. The varied chemical nature of the three separates is illustrated in Figure 4.28, and in the discussion that follows. Carbon in the humin fraction (residuum remaining after base extraction of SOM) of the Bh horizon is principally found in alkyl structures, such as n-alkanes, fatty acids, and waxes (440 g kg^{-1} of C), followed by O-alkyls (297 g kg^{-1} of C), aromatics (213 g kg^{-1} of C), and carbonyls (57 g kg^{-1} of C). In the humic acid fraction, alkyl structures contain 533 g kg^{-1} of C, followed by O-alkyls (207 g kg^{-1} of C), aromatics (130 g kg^{-1} of C), and carbonyls (133 g kg^{-1} of C). Both the humin and humic acids substances are dominated by alkyl carbon; however, the humic acid fraction contains less carbon in O-alkyl and aromatic structures, and more C in carbonyls. Finally, the fulvic acid fraction of the Bh horizon contains the most C in alkyl structures (343 g kg^{-1} of C), followed by O-alkyls (279 g kg^{-1} of C), carbonyls (228 g kg^{-1} of C), and aromatics (161 g kg^{-1} of C). Again, the fulvic acid fraction contains a greater abundance of oxygen-bearing functional groups (507 g kg^{-1} of C; sum of O-alkyl and carbonyl concentrations) than either the humic acid (340 g kg^{-1} of C) or the humin fractions (354 g kg^{-1} of C).

Nuclear magnetic resonance has also been employed to characterize the forms of N and P in humic substances. Solid-state ^{15}N NMR spectra of humic substances are consistently similar, showing a major peak for amide N (–230 to 285 ppm relative to the nitromethane standard) which indicates the presence of proteinaceous materials (Figure 4.29). Other forms of organic N-containing structures are detected in humic substances, including indoles, pyrroles, and imidazoles (–140 to –250 ppm),

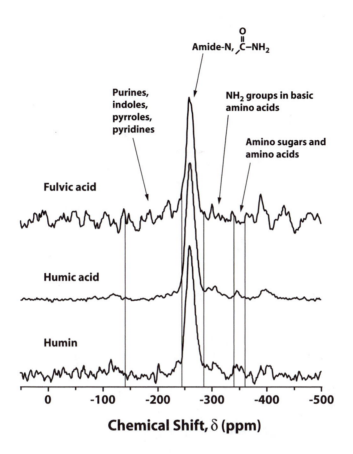

FIGURE 4.29 Example of a cross-polarization, magic angle spinning ^{15}N solid-state nuclear magnetic resonance (CPMAS ^{15}N NMR) spectra of the humic fractions extracted from plant residues (modified from Knicker, 2002). The general chemical shift regions and chemical assignments are also illustrated.

and amino groups (−285 to −320 ppm); however, these groups only occur in minor abundance. Liquid-state ^{31}P NMR (Figure 4.30) indicates the presence of both inorganic and organic P forms in humic substances. The most commonly identified forms of P are inorganic orthophosphate (6.0 to 7.5 ppm, referenced to an external 85% phosphoric acid solution), monoester organic P (found primarily in inositol phosphates) (4.2 to 6.0 ppm), diester organic P (−0.1 to −1.1 ppm), and organic pyrophosphates (−2.5 to −4.3 ppm). Other organic P forms that are detectable using ^{31}P NMR are the organic polyphosphates (−19 to −20 ppm) and the phosphonates (20 to 21 ppm). The liquid-state ^{31}P NMR spectra of a soil NaOH extract in Figure 4.29 (insert) indicates that the monoester P forms account for 79.4% of the total organic P (794 g kg^{-1} of P), while diester P and pyrophosphate P account for 4.5 and 5.1% of the total organic P. In general, liquid-state ^{31}P NMR studies indicate that monoester organic P accounts for approximately 75 to 85% of the total organic P extracted from soil with alkaline extractants.

4.4.2.3 Molecular Mass and Configuration

Humic substances are formed through the random polymerization or aggregation of a diverse array of compounds (monomers and fragments) from a pool comprised of the microbial degradates of biopolymers. The probability of finding two humic molecules that are exactly alike is exceedingly

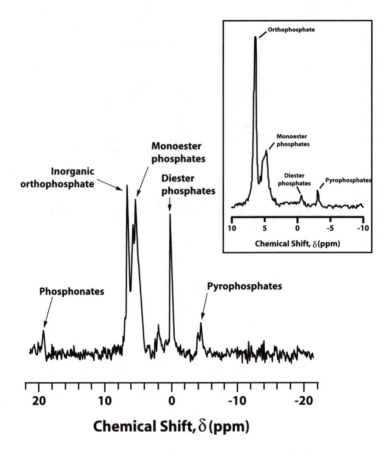

FIGURE 4.30 Liquid-state ^{31}P nuclear magnetic resonance spectra of NaOH extracts of an Alfisol and an Inceptisol from Hawkes et al. (1984) and Zhang et al. (1999) (insert).

small, particularly as molecules increase in size. In addition and despite their refractive nature, humic substances evolve and degrade, further enhancing their random character. Despite the randomness associated with their formation and degradation, there is relative uniformity among the average chemical properties of the humins and the humic and fulvic acids, such as elemental and functional group content (discussed in the previous sections). Indeed, these findings have been used to suggest that there exists an optimal chemical composition for humic substances in nature. However, the molecular masses of the humic substances do not appear to be constrained to the relatively narrow distributions exhibited by the chemical characteristics.

A number of methods have been employed to determine the molecular masses of humic substances. Ultrafiltration, small-angle x-ray scattering, and gel chromatography represent just a few of the techniques that have been employed (see Senesi and Loffredo [1999] for a review of these and other techniques). In general, soil humic acid extracts are composed of compounds that display a continuum of molecular masses ranging from approximately 1000 Da (daltons, a non-SI mass unit equal to one twelfth the atomic mass of ^{12}C; a dalton is equivalent to g mol^{-1}) to 500,000 Da, with the majority of substances having masses that center around 50,000 Da. Cameron et al. (1972) performed a pivotal study, the results of which have influenced the concepts of humic substance size and shape. They employed gel chromatography and pressure filtration through graded porosity membranes to fractionate the humic acid isolate from a Sapric Histosol. The molecular masses of the humic substances in each fraction were then determined by ultracentrifugation.

The molecular mass values for the humic acids ranged from approximately 2000 to 1,300,000 Da. Approximately 75% of the humic acids had molecular mass values <100,000 Da, and 25% had values <10,000 Da. A majority of the humic acids had molecular masses in the 20,000- to 50,000-Da range. Less than 20% of the substances were >100,000 Da in mass. Using their ultracentrifugation data, Cameron et al. (1972) convincingly proposed that humic acid macromolecules structurally consist of flexible, expanding, and branched random coils. Fulvic substances are quite small in comparison to the humic substances. Their masses tend to fall in a very limited range; from approximately 300 to 2000 Da.

Perhaps the most nebulous of characteristics of the humic substances is their molecular structure. It has been stated that any representation of the molecular configuration of a humic substance is essentially a cartoon. Humic scientists recognize that it is not possible to write a molecule structure or set of structures that truly defines the configuration of a humic substance. Structural representations are meant to convey information about the structural moieties that are thought to exist in humic substances, rather than represent the precise molecular structure. Such structures are called psuedostructures (after MacCarthy [2001]). A humic substance psuedostructure is a hypothetical molecular construct having elemental, structural, and functional group features that are consistent with some or all of the observed composition and mass properties. A pseudostructure of humic acid, developed by Schulten and Schnitzer (1997), is shown in Figure 4.31. Suwannee River fulvic acid pseudostructures have been proposed by Leenheer et al. (1998) (Figure 4.32a and b) and Kubicki and Apitz (1999) (Figure 4.32c).

The pseudostructure illustrated in Figure 4.31 was developed using the available chemical and structural information from the literature on humic acids. The pseudostructure contains numerous

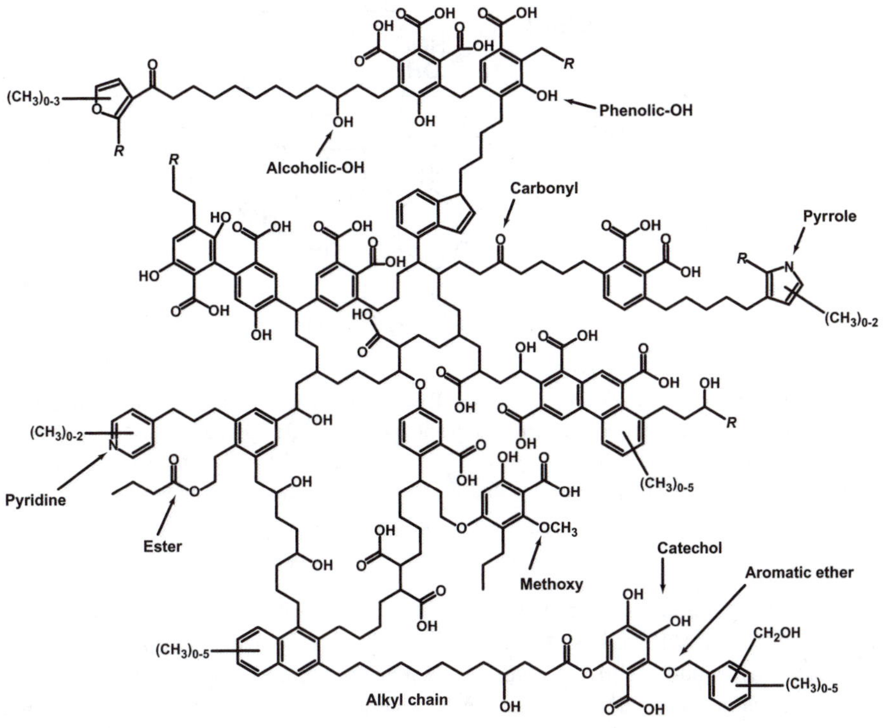

FIGURE 4.31 Partial pseudostructure of humic acid. (Modified from Schulten, H.R. and M. Schnitzer. Chemical model structures for soil organic matter and soils. *Soil Sci.* 162:115–130, 1997. With permission.)

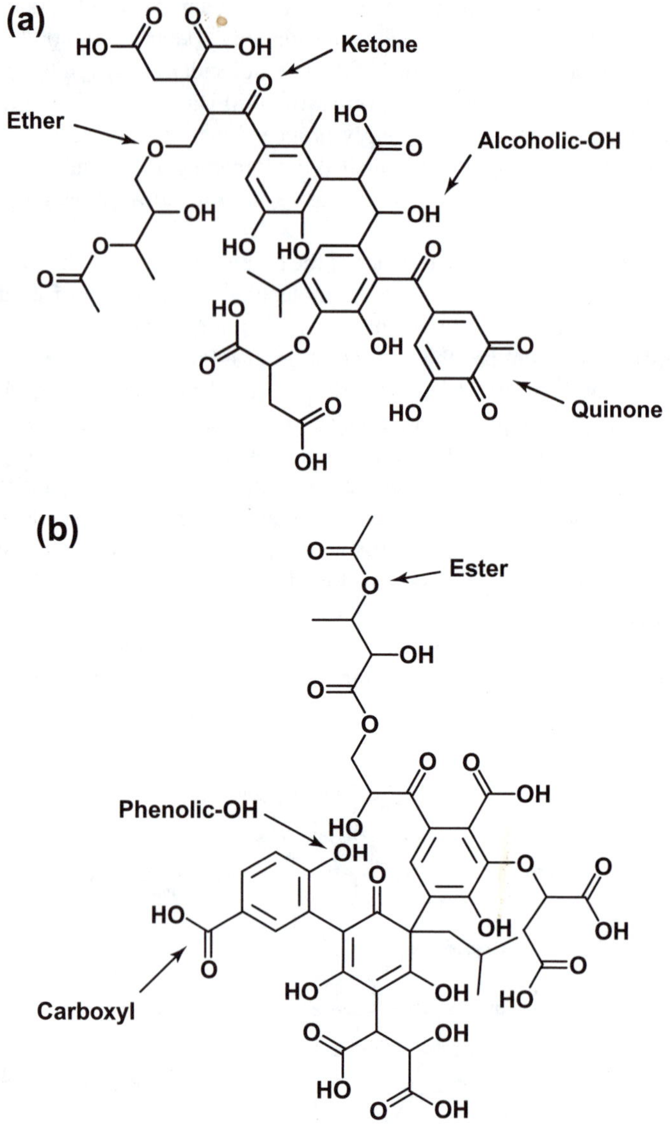

FIGURE 4.32 Pseudostructural models of Suwannee River fulvic acid. Fulvic acids in (a) and (b) are from Leenheer et al. (1998) and are based on different plant precursors. The model in (a) is derived from the degradation of proanthocyanidine and phloroglucinal tannins (Figure 4.17). The model in (b) is based upon a cutin-lignin-tannin complex. The pseudostructure in (c) is a three-dimensional representation of fulvic acid (Kubicki and Apitz, 1999).

oxygen-bearing structures and functional groups (carboxyls, phenolic and alcoholic hydroxyls, ketones, esters, and ethers), as well as a small number of N-bearing structures (pyrrole and pyridine). These groups are hydrophilic and are readily solvated (encircled by water molecules). Many of these groups, particularly the carboxyls and the N moieties, ionize and interact with inorganic and other organic ions or molecules from the aqueous phase. This type of interaction is responsible for the cation exchange capacity of SOM, which can range from 60 to 300 cmol kg^{-1} at pH 7, and account for 25 to 90% of the cation exchange capacity of mineral soils.

(C)

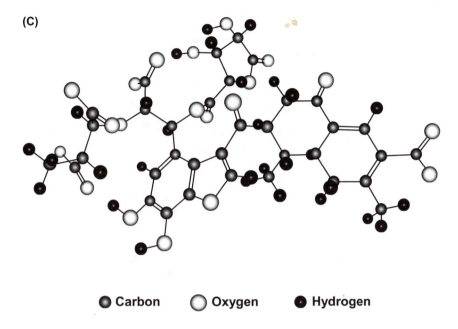

● Carbon   ○ Oxygen   ● Hydrogen

FIGURE 4.32 (*continued*).

The Suwannee River fulvic acid pseudostructures illustrated in Figure 4.32 were derived using measured chemical properties of the fulvic acid isolate. The fulvic acid pseudostructures are noticeably smaller than the humic acid structure (Figure 4.31). Indeed, the molecular mass of these fulvic acids is approximately 950 Da, while that of the humic acid described above is approximately 6650 Da. The fulvic acid pseudostructures characteristically have a large number of carboxyl groups, as well as alcoholic and phenolic hydroxyls, ester and ether linkages, and ketone groups. Relative to the humic acid structure, the fulvic acids are lacking in long-chain alkyl carbons, and the hydrophobic portions of the molecules (aromatic rings) are highly oxygenated (each is a substituted phenol). The fulvic acid molecules are also structurally labile, as described above for the humic acids, and the abundance of polar and ionizable functional moieties account for their solubility in both acidic and alkaline solutions.

The humic acid structural moieties will also interact intramolecularly through hydrogen bonding, metal complexation, and other electrostatic interactions (discussed in Chapter 5). Combined, these inter- and intramolecular interactions stabilize (minimize) the electrostatic energy of the humic acid molecule. However, this structure is not static; indeed, the molecular configuration of a humic acid molecule is flexible (labile) and a function of the salt concentration and pH of the aqueous environment. High aqueous salt concentrations or low pH values collapse the flexible, random coils into globular aggregates or ring-like structures (Figure 4.33); whereas, low aqueous salt concentrations and high pH conditions expand the flexible coils to form thread- or net-like structures. The humic acid pseudostructure in Figure 4.30 also contains numerous nonpolar aromatic units and alkyl carbon chains, illustrating the amphophilic character of the substance. Because these structures are hydrophobic, they tend to aggregate in the interior of the humic acid molecule and weakly interact with one another via van der Waals forces. These nonpolar structural components are shielded from the aqueous solution by the polar and ionic portions of the molecule, which tend to form an external shell (akin to a micelle structure).

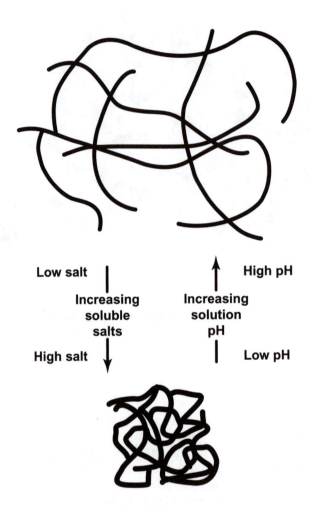

Low salt | **High pH**

Increasing   **Increasing**
soluble   **solution**
salts   **pH**

High salt ↓   **Low pH**

FIGURE 4.33 The molecular configuration of the flexible, random coils of the humic acid molecule as a function of the salt concentration and pH of the aqueous environment.

4.5 EXERCISES

1. *Polyfunctionality* and *structurally labile* are terms often used to describe the structural characteristics of humic substances. Define these terms.
2. Describe how the structural characteristics of polyfunctionality and structural lability combine to impart *hydrophilicity* to humic substances in aqueous environments.
3. In your own words, develop definitions for:
   a. Humic substances
   b. Humic acids
   c. Fulvic acids
   d. Humin
4. Compare and contrast the pathways that have been proposed to describe the formation of humic substances (see Figures 4.22 and 4.24).
5. A 1.0-g surface soil sample is digested with 5 mL of a standard 0.167 M K$_2$Cr$_2$O$_7$ solution and 7.5 mL of concentrated H$_2$SO$_4$ according to the Walkley-Black procedure. Titration of the digestate to the N-phenylanthranilic acid end point requires 3.11 mL of a standard 0.2 M Fe(NH$_4$)$_2$(SO$_4$)$_2$ solution. Compute the SOC content of the soil sample.

6. A soil humic acid is found to have O/C and H/C of 0.50 and 1.00. Answer the following:
   a. What is the average chemical formula of the humic acid?
   b. Write the reaction that describes the complete oxidation of this humic acid by $K_2Cr_2O_7$ to form Cr^{3+} and CO_2 (as done for CH_2O in Equation 4.1).
   c. Equation 1.4 indicates that 1 mol of $Cr_2O_7^{2-}$ consumes 1.5 mol of SOC. Is this assumption valid for the reaction generated in part 6(b) above?

7. Lignin theory describes the least humified humic substances as humins and the most humified substances as fulvic acids. The polyphenol theory, however, states that fulvic acids are the initial products of humification and humin is the most humified substance. Which of these two pathways is supported by the elemental content data presented in Table 4.3 (use the "All sources" data for each humic substance in your evaluation)?

8. Comment on the statement: "humic acids have lower total acidity and carboxyl contents and greater aromaticity than fulvic acids. The molecular masses of fulvic acids are also much less than those of humic acids. Therefore, humic acids may be isolated by acidifying an alkaline soil extract."

9. The elemental composition of the humic and fulvic acid fractions of an Elliott soil (an Aquic Argiudoll) is given in the table below. Answer the following:
   a. Calculate the molar O/C and H/C of the humic and fulvic acids and compare your results to the O/C and H/C of acetic acid (a simple aliphatic acid) and benzoic acid (a simple aromatic acid). Discuss your findings.
   b. Calculate the maximum potential negative charge that can be created on the fulvic and humic acids, assuming that the oxygen content is assigned to carboxyl groups. Express your results in $cmol_c$ kg^{-1}. Compare your results to the cation exchange capacity values attributed to smectites and vermiculites.
   c. Calculate the maximum potential positive charge that can be created on the fulvic and humic acids, assuming that the nitrogen content is assigned to amino groups. Express your results in $cmol_c$ kg^{-1}.

Elemental Content (in g kg^{-1}) of Humic and Fulvic Acids Extracted from an Elliott Soil

Compound	C	H	O	N	S
Humic acid	581	36.8	341	41.4	4.4
Fulvic acid	506	37.7	437	27.2	5.6

Data from the International Humic Substances Society.

10. An antiquated method for determining the carbonate and bicarbonate concentrations in soil solutions involves titration with standardized sulfuric acid, first to the phenolphthalein end point, then to the methyl orange endpoint. It is commonly observed during the titration of soil extracts that a brown precipitate forms in the titrand as the titration approaches the methyl orange endpoint. Describe the chemical process occurring in the titrand.

11. A pH 4.8 soil solution contains 0.05 mmol L^{-1} of succinic acid, a dicarboxylic acid that dissociates according to the following reactions:

$$HOOCC_2H_4COOH \rightarrow HOOCC_2H_4COO^- + H^+ \ (pK_{a1} = 4.21)$$

$$HOOCC_2H_4COO^- \rightarrow {}^-OOCC_2H_4COO^- + H^+ \ (pK_{a2} = 5.64)$$

What are the concentrations of $HOOCC_2H_4COOH$, $HOOCC_2H_4COO^-$, and $^-OOCC_2H_4COO^-$ in the soil solution?

12. A pH 5.2 soil solution contains 0.5 mmol L^{-1} of acetic acid, a monocarboxylic acid that dissociates according to the following reactions:

$$CH_3COOH \rightarrow CH_3COO^- + H^+ \ (pK_{a1} = 4.76)$$

What are the concentrations of CH_3COOH and CH_3COO^- in the soil solution?

13. Comment on this statement: "molecular models that illustrate the structure of humic or fulvic acids are cartoons."

REFERENCES

Baldock, J.A. and P.N. Nelson. Soil organic matter. In *Handbook of Soil Science*. M.E. Sumner (Ed.) CRC Press, Boca Raton, FL, 2000, pp. B25–B84.

Batjes, N.H. (Ed.) A homogenized soil data file for global environmental research: a subset of FAO, ISRIC and NRCS profiles (Version 1.0). Working Paper and Peprint 95/10b, International Soil Reference and Information Centre, Wageningen, The Netherlands, 1995.

Batjes, N.H. and J.A. Dijkshoon. Carbon and nitrogen stocks in soils of the Amazon Region. *Geoderma* 89:273–286, 1999.

Burdon, J. Are the traditional concepts of the structures of humic substances realistic? *Soil Sci.* 166:752–769, 2001.

Cameron, R.S., B.K. Thornton, R.S. Swift, and A.M. Posner. Molecular weight and shape of humic acid from sedimentation and diffusion measurements on fractionated extracts. *J. Soil Sci.* 23:394–408, 1972.

Conte, P., A. Piccolo, B. van Lagen, B. Buurman, and P.A. de Jager. Quantitative differences in evaluating soil humic substances by liquid- and solid-state ^{13}C-NMR spectroscopy. *Geoderma* 80:339–352, 1997.

Eswaran, H., E. van den Berg, and P. Reich. Organic carbon in soils of the world. *Soil Sci. Soc. Am. J.* 57:192–194, 1993.

Hawkes, G.E., D.S. Powlson, E.W. Randall, and K.R. Tate. A ^{31}P nuclear magnetic resonance study of the phosphorus species in alkali extracts of soils from long-term field experiments. *J. Soil Sci.* 35:35–45, 1984.

Knicker, H. The feasibility of using DCPMAS ^{15}N ^{13}C NMR spectroscopy for a better characterization of immobilized ^{15}N during incubation of ^{13}C- and ^{15}N-enriched plant material. *Org. Geochem.* 33:237–246, 2002.

Kubicki, J.D. and S.E. Apitz. Models of natural organic matter and interactions with organic contaminants. *Org. Geochem.* 30:911–927, 1999.

Leenheer, J.A., G.K. Brown, P. MacCarthy, and S.E. Cabaniss. Models of metal binding structures in fulvic acid from the Suwannee River, Georgia. *Environ. Sci. Technol.* 32:2410–2418, 1998.

MacCarthy, P. The principles of humic substances. *Soil Sci.* 166:738–751, 2001.

Mahieu, N., D.S. Powlson, and E.W. Randall. Statistical analysis of published carbon-13 MNR spectra of soil organic matter. *Soil Sci. Soc. Am. J.* 63:307–319, 1999.

Nelsom, D.W. and L.E. Sommers. Total carbon, organic carbon, and organic matter In *Methods of Soil Analysis*. Part 3. D.L. Sparks et al. (Eds.) Soil Sci. Soc. Am. Book Series Number 5. SSSA and ASA, Madison, WI, 1996, pp. 961–1010.

Rice, J.A. and P. MacCarthy. Statistical evaluation of the elemental composition of humic substances. *Org. Geochem.* 17:635–648, 1991.

Schnitzer, M., C.A. Hindle, and M. Meglic. Supercritical gas extraction of alkanes and alkanoic acids from soils and humic materials. *Soil Sci. Soc. Am. J.* 50:913–919, 1986.

Schulten, H.R. and M. Schnitzer. Chemical model structures for soil organic matter and soils. *Soil Sci.* 162:115–130, 1997.

Senesi, N. and E. Loffredo. The chemistry of soil organic matter. In *Soil Physical Chemistry*. 2nd ed. D.L. Sparks (Ed.) CRC Press, Boca Raton, FL, 1999, pp. 239–370.

Steelink, C. Elemental characteristics of humic substances. In *Humic Substances in Soil, Sediment and Water*. G.R. Aiken et al. (Eds.) John Wiley & Sons, New York, 1985, pp. 457–476.

Stevenson, F.J. *Humus Chemistry: Genesis, Composition, Reactions*. John Wiley & Sons, New York, 1994.

Swift, R.S. Organic matter characterization. In *Methods of Soil Analysis*. Part 3. D.L. Sparks et al. (Eds.) Soil Sci. Soc. Am. Book Series Number 5. SSSA and ASA, Madison, WI, 1996, pp. 1011–1069.

Swift, R.S. Macromolecular properties of soil humic substances: fact, fiction, and opinion. *Soil Sci.* 164:790–802, 1999.

Swift, R.S. Sequestration of carbon by soil. *Soil Sci.* 166:858–871, 2001.

Tabatabai, M.A. Sulfur. In *Methods of Soil Analysis*. Part 3. D.L. Sparks et al. (Eds.) Soil Sci. Soc. Am. Book Series Number 5. SSSA and ASA, Madison, WI, 1996, pp. 921–960.

Ussiri, D.A.N. and C.E. Johnson. Characterization of organic matter in a northern hardwood forest soil by ^{13}C NMR spectroscopy and chemical methods. *Geoderma* 111:123–149, 2003.

Van Noordwijk, M., C. Cerri, P.L. Woomer, K. Nugroho, and M. Bernoux. Soil carbon dynamics in the humid tropical forest zone. *Geoderma* 79:187–225, 1997.

Zhang, T.Q., A.F. Mackenzie, and F. Sauriol. Nature of soil organic phosphorus as affected by long-term fertilization under continuous corn (*Zea Mays* L.): a ^{31}P NMR study. *Soil Sci.* 164:662–670, 1999.

5 Soil Water Chemistry

Soil water, or the aqueous phase, is arguably the most important phase in the soil. Almost all chemical reactions in soil are mediated by or occur in the soil solution. Some of the more important types of chemical reactions that occur in soil water, affecting the fate and behavior of substances in the environment, are: hydration-hydrolysis, acid-base, oxidation-reduction, and complexation. The soil solution mediates many of the reactions that control the retention of substances by soil solids, such as precipitation-dissolution, adsorption-desorption, and ion exchange. The extent to which these retention-controlling reactions occur is dictated by substance behavior in soil water. The soil solution is also the principal phase in which substances move in and through the soil. It is for these reasons that considerable emphasis is placed on understanding the processes that occur in the soil solution, as well as the ability to predict how these processes will impact substance fate and behavior.

5.1 NATURE OF WATER

Water is a highly reactive substance and an exceedingly effective solvent. It is a compound that has a high dielectric constant, which is a measure of a solvent's ability to overcome the attraction between a dissolved cation and an anion. The dielectric constant may also be defined as the ability to oppose the electrical attraction between ions of opposite charge. This definition is illustrated mathematically in the expression:

$$F = \left(\frac{Z_- Z_+}{r^2}\right)\left(\frac{1}{\varepsilon}\right) \tag{5.1}$$

where F is the force of attraction between ions of opposite charge of magnitude Z_- and Z_+ that are separated by a radius of r in a solution having a dielectric constant of ε. The force of attraction between two oppositely charged ions will be less in solvents that have high dielectric constants, relative to solvents that have low dielectric constants.

A relatively high dielectric constant is a rather unique property of water that is a result of a nonlinear molecular configuration (Figure 5.1). In isolation (gaseous state), the angle between the two protons in a water molecule is 104.45°, and the O—H bond length is 95.84 pm. In liquid, the angle between the two protons in a water molecule reportedly ranges from 104.52 to 109.5°, and the O—H bond length ranges from 95.7 to 100 pm. These deviations in the molecular configuration (in liquid relative to in isolation) are attributed to the adhesive character of water molecules. Within the water molecule, the H—O bond is approximately 61% covalent (see Chapter 2 for a discussion of chemical bond character). Oxygen is highly electronegative relative to the proton and attracts electrons from the hydrogen atoms. The unequal sharing of electrons between O and H results in a partial negative charge on the oxygen and a partial positive charge on the proton. Moreover, oxygen has a greater supply of electrons than hydrogen, further enhancing the partial charges on oxygen and hydrogen. Since H_2O is a neutral molecule, the partial charge on oxygen must exactly balance that on the two hydrogen atoms. If δ is used to designate partial charge, then the δ+ charge on each hydrogen atom must be balanced by the 2δ– charge on the oxygen atom. For the water molecule, δ reportedly ranges from 0.241 to 0.520, depending on the molecular model employed to describe water. Commonly employed molecular models predict δ to range from 0.41 to 0.4238.

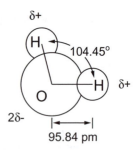

Isolated Water Molecule

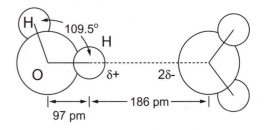

Water Molecules in Liquid

FIGURE 5.1 Physical characteristics of an isolated water molecule and two water molecules associated through a hydrogen bond. $\delta+$ and $2\delta-$ represent the partial positive and negative charges on the proton and oxygen, $\delta \approx 0.42$ (Reference; Mahoney and Jorgensen, 2000). The bond energy of the covalent $O-H$ bond in the water molecule is 470 kJ mol^{-1}; while the bond energy of the hydrogen bond between water molecules is approximately 23.3 kJ mol^{-1} (Suresh and Naik, 2000).

The polarity (separation of charge) of water molecules makes them mutually attractive (adhesive), as well as attractive to ionic and other polar substances (cohesive), a characteristic that makes water an excellent solvent. The mutual attractiveness of water molecules results in a secondary linkage between the partially positive charged H portion of one molecule and the partially negative charged O portion of another (Figure 5.1). This secondary linkage is weak and undirected (does not depend on the geometric configuration of electronic bonding orbitals) and is a special case of a dipole–dipole interaction that is termed a hydrogen bond. The bond energy of the hydrogen bond between water molecules is approximately 23.3 kJ mol^{-1}, compared to 470 kJ mol^{-1} for the 61% covalent $O-H$ bond in the water molecule. It is typical to diagram the hydrogen bond in such a way as to show the bridging proton as "belonging" to and in close association with a single water molecule, while only weakly electrostatically attracted to and distanced from a second water molecule (Figure 5.1). However, in reality the bridging proton oscillates between the two hydrogen-bonded molecules, and in effect, lengthens and weakens the $O-H$ covalent bond distance relative to that in an isolated molecule. Indeed, the hydrogen bond is approximately 90% ionic and 10% covalent, with the bridging proton distributed into three configurations: covalent, $HO-H - - -OH_2$; ionic, $HO^{\delta-}-H^{\delta+} - - - O^{\delta}H_2$; and covalent, $HO^- - - - H-O^+H_2$ (the latter is the least evident type of covalent bond). Further, liquid water has a character similar to that of slush, but on a molecular level. Through hydrogen bonding, the water molecules form molecular ice structures. These structures are transitory, yet individual water molecules deform slightly to accommodate a tetrahedral configuration, which has a 109.5° $H-O-H$ angle, thus the increased angle between the two protons, relative to that in isolation (Figure 5.1).

A comparison of the properties of water to those of its sister compounds, H_2S, H_2Se, and H_2Te, and to other solvents that are liquid at room temperature illustrates the unique character of water (Table 5.1). The data clearly indicate that H_2O has strong self-attraction, relative to the other solvents that are liquid at room temperature. While weak in comparison to the covalent $O-H$ bond, the $H-$bond is of sufficient strength to provide liquid water with internal structure, and the divergent properties are shown in Table 5.1. Indeed, a linear water molecule might have properties similar to those of H_2S, which is a linear molecule.

TABLE 5.1
Physical-Chemical Characteristics of Water, Solvents That Are Liquid at Room Temperature, and Other Dihydrogen Compounds of the Periodic Group VIB Elements[a]

Solvent	ΔH_v kJ mol^{-1}	ΔH_f kJ mol^{-1}	ε (25°C)	Melting Point °C	Boiling Point °C
H_2O	40.70	6.009	78.5	0	100
Methanol	35.21	2.196	33	–97	64.6
Ethanol	38.56	4.973	24	–117.3	78.5
Acetone	29.1	4.770	21	–95.4	56.2
Benzene	30.72	9.300	2.3	5.5	80.1
H_2S	—	16.8	9.1 (–78.5°C)	–85.5	–60.7
H_2Se	—	—		–60.4	–41.5
H_2Te	—	12.9		–48.9	–2.2

[a] ΔH_v is the heat of vaporization (latent heat) and is energy required to convert a liquid to a gas at constant temperature. ΔH_f is the heat of fusion and is the energy required to convert a solid to a liquid at constant temperature.

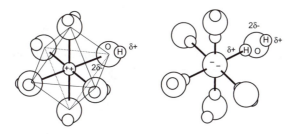

Hydrated Cation   **Hydrated Anion**

FIGURE 5.2 Illustrations of the primary hydration sphere of a cation and an anion. Both ions are in octahedral coordination, surrounded by six waters of hydration (δ is the partial charge arising from the polarization of water and the bond lengths are exaggerated to illustrate structure and alignment of water molecules).

5.2 ION HYDRATION

When an electrolyte ion (charged ion) is introduced into pure liquid water, the structure of bulk water (molecular slush) is perturbed. Water molecules in close proximity to the charged ion will become ordered and form a shell around the ion. If the ion is a cation, the partially negative portion of the water molecule (oxygen portion) will be attracted to and pointing toward the cation; while the protons will be repelled and pointing outward (Figure 5.2). Conversely, water molecules in close proximity to an anion will be oriented with the protons pointing toward the anion. Water molecules that reside in the region closest to an ion occupy the primary hydration sphere of the ion, also termed the coordination sphere of the ion. In this sphere there is strong attraction between the water molecules and the ion. With few exceptions (e.g., large monovalent ions), the influence of the ion on bulk water extends beyond the primary hydration sphere. Water molecules that are outside the primary sphere, but influenced by the ion, reside in the secondary hydration sphere. Water molecules in this sphere are poorly structured and considerably more mobile than coordinated water. The effect of the primary and secondary hydration spheres is to insulate the ion, such that the ion's charge is dissipated and shielded from other ions in the solution. Thus, electrolyte ions in water will interact less electrostatically and will behave more ideally (no interactions) than when in other solvents.

The degree to which the structure of bulk water is disrupted, and to which water molecules are ordered in ion hydration spheres, is quantified by the enthalpy change (ΔH_h) that occurs when an ion goes from the gaseous to the aqueous phase:

$$M^{m+}(g) + nH_2O(l) \rightarrow [M(H_2O)_n]^{m+}(aq) \qquad (5.2)$$

where M is a metal ion with a charge of $m+$, and n is the number of waters of hydration. The ΔH_h ($= H_{aq} - H_g$) is the enthalpy of hydration and relates the energy release associated with the disruption of bulk water structure and the formation of a hydration sphere (free state to hydrated state releases energy). The enthalpy of the metal in the gaseous state is greater than that in the aqueous state because the formation of the hydration sphere constricts the ion, reducing its chemical potential. Thus, the more negative the ΔH_h, the greater the disruption to bulk water and the greater the degree of ion hydration.

The impact of ion size and valence on ΔH_h is illustrated in Table 5.2. For example, within the periodic group IIIA, trivalent ion size increases: Sc^{3+} (88 pm) < Y^{3+} (104 pm) < La^{3+} (117 pm);

TABLE 5.2
Enthalpy (ΔH_h) and Entropy (ΔS_h) of Hydration for Metal Cations and Select Anions as a Function of Ion Size and Charge

	±1			±2			±3	
Ion	r, pm[a]	$\Delta H_h^b (\Delta S_h^b)$	Ion	r, pm	$\Delta H_h (\Delta S_h)$	Ion	r, pm	$\Delta H_h (\Delta S_h)$
H^+	—	−1091	Be^{2+}	59	−2487	Sc^{3+}	88	−3960
Li^+	90	−515	Mg^{2+}	86	−1922 (−183)	Y^{3+}	104	−3620
Na^+	116	−405 (59)	Ca^{2+}	114	−1592 (−53)	La^{3+}	117	−3283
K^+	152	−321 (101)	Sr^{2+}	132	−1445			
Rb^+	166	−296	Ba^{2+}	149	−1304	Al^{3+}	67	−4660 (−322)
Cs^+	181	−263	Ra^{2+}	162	−1259	Cr^{3+}	75	−4402
						Fe^{3+}	78	−4376 (−300)
Cu^+	91	−594	Zn^{2+}	88	−2044 (−110)	Ga^{3+}	76	−4685
Ag^+	129	−475	Cu^{2+}	91	−2100	In^{3+}	94	−4109
Tl^+	164	−326	Ni^{2+}	83	−2106	Tl^{3+}	102	−4184
			Co^{2+}	88	−2054			
F^-	119	−505 (−10)	Fe^{2+}	92	−1920			
Cl^-	167	−363 (55)	Mn^{2+}	97	−1845			
Br^-	182	−336	Cr^{2+}	94	−1850			
I^-	206	−295	Cd^{2+}	109	−1806 (−76)			
NO_3^-	264	−328 (125)	Pb^{2+}	133	−1480			
			SO_4^{2-}	290	−1145 (17)			

[a] Radii pertain to sixfold coordination, with the exception of Ra^{2+} (CN = 8).

[b] ΔH_h has units of kJ mol^{-1}, ΔS_h has units of J K^{-1} mol^{-1}, Burgess, J. 1978. *Metal ions in solutions.* Ellis Horwood, Chichester, England, and Wulfsberg, G. 1987. *Principles of descriptive chemistry.* Brooks/Cole Publishing, Monterey, CA.

---

and ΔH_h values increase (grow less negative): Sc^{3+} (–3960 kJ mol^{-1}) $<$ Y^{3+} (–3620 kJ mol^{-1}) $<$ La^{3+} (–3283 kJ mol^{-1}), indicating that the smaller ions disrupt the structure of bulk water to a greater degree than do larger ions. The influence of ion charge on the extent of hydration sphere formation can be seen by comparing the ΔH_h values of Li^+ (–515 kJ mol^{-1}), Mg^{2+} (–1922 kJ mol^{-1}), and Sc^{3+} (–3960 kJ mol^{-1}), which are all approximately the same size (90 pm, 86 pm, and 88 pm).

Entropies of hydration (ΔS_h) indicate the degree of disorder that occurs in a solution when an ion goes from the gaseous to the aqueous phase (Equation 5.2). The ΔS_h values reveal additional information regarding the impact of ions on the nature of water (Table 5.2). Positive ΔS_h values, such as those for SO_4^{2-} and large monovalent ions (Na^+, K^+, Cl^-, and NO_3^-), indicate that the net effect of hydration is the disruption of the structure of bulk water (less order). Negative ΔS_h values, like those for di- and trivalent cations, indicate that the net effect of hydration is the introduction of order, through the formation of hydration spheres. The more negative the ΔS_h value, the more ordered the waters of hydration.

The nature of hydration spheres (e.g., number of water molecules in the sphere, resistance to displacement, and residence time) is impacted by a number of factors. Electrolyte valence and ionic radius combined have the greatest impact on the nature of the hydration spheres. The more concentrated the charge of the ion, through a combination of high valence and small size, the greater the force of attraction between the ion and the waters of hydration will be (Equation 5.1). A third factor is the concentration of electrolytes in the solution. An increase in ion concentration reduces the chemical potential of the bulk water, resulting in waters of hydration being drawn back into bulk solution. For example, in a concentrated salt solution (4.49 m $CaCl_2$) the hydrated Ca species is $[Ca(H_2O)_{5.5}]^{2+}$; whereas, $[Ca(H_2O)_{10}]^{2+}$ reportedly exists in a less concentrated solution (1 m $CaCl_2$).

Additional insight into the stability of cation hydration spheres is evidenced by the mean residence time of water in the coordination sphere. The rate of substitution of water molecules from the primary coordination sphere of cations falls over a wide range of values (Figure 5.3). Mean residence times range from $\tau = 10^7$ to 10^{-10} sec (17 orders of magnitude) for the ions illustrated in Figure 5.3. Correspondingly, first-order rate constants range from $k = 10^{10}$ to 10^{-7} sec^{-1} ($k = \tau^{-1}$). Metal cations are placed into four classes based on water substitution rates. For a Class I metal,

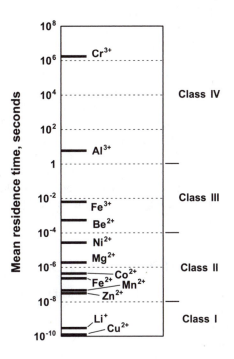

FIGURE 5.3 Mean residence times of waters of hydration in the coordination spheres of various cations (Hunt and Friedman, 1983; Cusanelli et al., 1996).

TABLE 5.3
Hydration Numbers and Hydrated Radii for Selected Cations

Ion	Probable Hydrated Species	$\dfrac{Z^2}{r}$	Hydration Number[a]	Hydrated Radius, pm[b]	Hydrated Radius, pm[c]
Li^+	$[Li(H_2O)_{4-6}]^+$	0.0111	22	340	382
Na^+	$[Na(H_2O)_4]^+$	0.0088	13	276	358
K^+	$[K(H_2O)_4]^+$	0.0066	7	232	331
Cs^+	$[Cs(H_2O)_6]^+$	0.0055	6	228	329
Mg^{2+}	$[Mg(H_2O)_6]^{2+}$	0.0465	36		428
Ca^{2+}	$[Ca(H_2O)_6]^{2+}$	0.0351	29		412
Sr^{2+}	$[Sr(H_2O)_6]^{2+}$	0.0303	29		412
Ba^{2+}	$[Ba(H_2O)_6]^{2+}$	0.0268	28		404
Ni^{2+}	$[Ni(H_2O)_6]^{2+}$	0.0482	15		404
Zn^{2+}	$[Zn(H_2O)_6]^{2+}$	0.0599	44		430
Cd^{2+}	$[Cd(H_2O)_6]^{2+}$	0.0549	39		426

[a] Rutgers and Hendrikx (1962) and Hunt and Friedman (1983).
[b] Cotton et al. (1999).
[c] Nightingale, E.R., Jr. (1959).

the exchange of water in the primary hydration sphere is very fast ($\tau = 10^{-8} - 10^{-10}$ sec) and essentially diffusion controlled. Metals in this class include the alkali metals (Li^+, Na^+, K^+, Rb^+, and Cs^+), the alkaline earth metals (Ca^{2+}, Sr^{2+}, and Ba^{2+}), and the divalent ions of Cr, Cu, Cd, and Hg. The primary coordination numbers for the alkali metals range from 4 to 6 (Table 5.3). It is also apparent from Figure 5.3 and Table 5.3 that the residence time and the total hydration number for the monovalent ions (primary plus secondary waters of hydration) increase with decreasing ion size. The divalent ions in Class I are coordinated by six water molecules with the formula $[M(H_2O)_6]^{2+}$. Residence times for the waters of hydration of Class II metals range from 10^{-4} to 10^{-8} sec. Divalent Mg^{2+}, the trivalent lanthanide ions, and the first row transition metal ions, Ni^{2+}, Co^{2+}, Fe^{2+}, Fe^{3+}, Mn^{2+}, and Zn^{2+}, are Class II metals and coordinated by six water molecules. The greater charge density of Zn^{2+}, relative to Ni^{2+}, is responsible for the greater numbers of primary and secondary waters of hydration for Zn^{2+} (Table 5.3). The mean residence times for water coordinated by Class III metals, Al^{3+}, Be^{2+}, and V^{2+}, range from 10 to 10^{-4} sec. The primary waters of hydration of Class IV metals, such as Cr^{3+}, are exceedingly difficult to displace, having residence times that range from 1 sec to approximately 10 days. Six water molecules predominantly coordinate Class III and IV metals.

The hydration of anions also occurs, but is considerably less significant than the coordinated hydration of cations. As previously discussed, the introduction of anions into a solution tends to result in a net increase in disorder. Indeed, the mean residence time for a water of hydration in the species $[Cl(H_2O)_6]^-$ is less than 10 psec (10^{-11} sec), indicating that Cl^- has virtually no impact on the motion of bulk water molecules.

5.3 ELECTROLYTE SOLUTIONS

The importance of the primary and secondary hydration spheres to the behavior of cations in soil solutions cannot be overstated. As will be seen in later chapters, the hydrated radius of an ion and the ease of replacement of waters of hydration are important characteristics that influence aqueous speciation, adsorption, and ion exchange processes. It has already been stated that hydration spheres dissipate and shield ion charge. This results in electrolyte ions interacting less electrostatically. As ions interact less, solutions behave more ideally. An ideal solution is one in which there are no interactions. More specifically, an ideal solution is one in which the concentration-dependent properties of the solution are linearly dependent upon concentration. The concentration-dependent properties of a solution are known as the colligative properties (or collective properties; if one is known, all others can be computed). The colligative properties of a solution are:

- Vapor pressure: $P = P^oX$ (Raoult's law), where P is the solution vapor pressure, P^o is the pure solvent vapor pressure, and X is the mole fraction of the solvent in the solution.
- Freezing point: $\Delta T_f = iK_f m$, where ΔT_f is the freezing point depression, i is the van't Hoff factor (number of ions a salt dissociates into; 2 for NaCl, 3 for $CaCl_2$), K_f is the freezing point depression constant, and m is molality of the solute.
- Boiling point: $\Delta T_b = iK_b m$, where ΔT_b is the boiling point elevation, i is the van't Hoff factor, K_b is the boiling point elevation constant, and m is molality of the solute.
- Osmotic pressure: $\pi = MRT$, where π is the osmotic pressure, M is molarity, R is the natural gas constant, and T is the temperature.

For all these colligative properties, the deviation from the properties of a pure solvent is a function of solute concentration variables (molality, mol kg^{-1}; molarity, mol L^{-1}; or mole fraction, moles of a substance divided by the summation of the moles of all substances present).

Although ionic interactions are dissipated due to the formation of hydration spheres, soil solutions are nonideal because the electrolytes interact. When salts are introduced into a solution, two principal types of interactions occur: (1) solute-solvent interactions—the formation of hydration spheres and the disruption to the structure of bulk water, and (2) long-range (>500 pm) electrostatic solute-solute interactions. It is because of the solute-solvent and solute-solute interactions that soil solutions are nonideal. Correspondingly, the colligative properties of a soil solution are not concentration dependent, as described in the above concentration-dependent linear expressions. However, we can account for solution nonideality by defining a concentration-dependent term; a term upon which the colligative properties are linearly dependent. This term is known as the activity.

The activity of an ion in solution is defined in the expression $a_i = \gamma_i m_i$, where γ_i is the single-ion activity coefficient for ion i with units of kg mol^{-1}, a_i is the activity of i and is unitless, and m_i is the molal concentration (mol kg^{-1}) of i. If the density of the soil solution is assumed to be unity, then molal concentrations can be replaced by molar concentrations, and $a_i = \gamma_i M_i$. As solutions become more dilute in electrolytes, the disruption to the structure of bulk water decreases and solutes interact less. Thus, solutions behave more ideally, with the limiting condition, $\lim_{c \to 0} \gamma_i = 1$, defining the activity coefficient for an infinitely dilute solution, as well as a reference state for the single-ion activity coefficient. Further, the degree of deviation of γ_i from unity indicates the degree of deviation from ideality.

Despite the apparent simplicity of the $a = \gamma M$ expression, there are a number of considerations that detract from this simplicity. First, it is impossible to directly measure the activities of individual ions in solution, as cations and anions cannot be physically isolated (although an ion-selective electrode theoretically responds to changes in the activity of a specific ion, as illustrated in Chapter 10 for the H^+ ion-selective electrode). Second, one can measure the activities of salts (neutral electrolytes) by measuring a colligative property. Finally, γ_i is not measurable. However, γ_i can be estimated using theoretical or empirical methods. Two theoretical expressions used to compute single-ion activity coefficients are the Debye-Hückel limiting law equation and the extended Debye-Hückel equation. The Debye-Hückel limiting law equation is applicable to very dilute solutions (ionic strength < 0.01 M):

$$\log \gamma_i = -AZ_i^2 \sqrt{I} \qquad (5.3)$$

where A is a temperature-dependent constant (0.5116 $L^{\frac{1}{2}}$ $mol^{-\frac{1}{2}}$ at 25°C), Z_i is the charge on species i, and I is the ionic strength of the solution. Each ion in a solution may have a different activity coefficient. However, a solution has only one ionic strength, which is computed by:

$$I = \tfrac{1}{2} \Sigma_j (Z_j^2 M_j) \tag{5.4}$$

Consider a solution that contains 0.01 M NaCl, 0.02 M CaCl$_2$, and 0.005 M MgCl$_2$. The ionic strength of the solution is:

$$I = \tfrac{1}{2}\{(+1)^2[Na^+] + (+2)^2[Ca^{2+}] + (+2)^2[Mg^{2+}] + (-1)^2[Cl^-]\} \tag{5.5a}$$

where brackets [ ] denote molar concentrations. Substituting for the concentrations of the individual ions,

$$I = \tfrac{1}{2}\{(+1)^2[0.01\ M_{Na}] + (+2)^2[0.02\ M_{Ca}] + (+2)^2[0.005\ M_{Mg}] + (-1)^2[0.06\ M_{Cl}]\} \tag{5.5b}$$

$$I = \tfrac{1}{2}\{0.01 + 0.08 + 0.02 + 0.06\} \tag{5.5c}$$

$$I = \tfrac{1}{2}\{0.17\} \tag{5.5d}$$

$$I = 0.085\ M \tag{5.5e}$$

Note that the ionic strength of this solution is too high for the application of the Debye-Hückel limiting law equation. Instead, one might employ the extended Debye-Hückel equation, which is applicable to solutions having $I < 0.1$ M:

$$\log \gamma_i = \frac{-AZ_i^2 \sqrt{I}}{(1 + B\alpha_i \sqrt{I})} \tag{5.6}$$

where A, I, and Z are as previously defined, B is a temperature-dependent constant (0.33 for aqueous solutions at 25°C), and α_i is the effective diameter of the ion in solution (representative of distance of closest approach or the hydrated diameter of the ion) expressed in angstroms (Table 5.4).

Another expression that is applicable to soil solutions is the Davies equation. The Davies equation is an empirical expression that is valid for solutions with $I < 0.5$ M:

$$\log \gamma_i = -AZ_i^2 \left(\frac{\sqrt{I}}{(1 + \sqrt{I})} - 0.3I \right) \tag{5.7}$$

where the various parameters have been previously defined. The Davies equation is widely used because of its applicability to a broader concentration range, and because it requires a smaller number of parameters than the extended Debye-Hückel equation. Consider the above NaCl-CaCl$_2$-MgCl$_2$ solution ($I = 0.085$ M). The activity coefficient for Na$^+$ computed using the Davies equation is:

$$\log \gamma_{Na^+} = -0.5116(+1)^2 \left(\frac{\sqrt{0.085}}{(1 + \sqrt{0.085})} - 0.3(0.085) \right) \tag{5.8a}$$

$$\log \gamma_{Na^+} = -0.1024 \tag{5.8b}$$

$$\gamma_{Na^+} = 0.7899\ L\ mol^{-1} \tag{5.8c}$$

TABLE 5.4
A Comparison of Single-Ion Activity Coefficients for Ions Computed Using the Extended Debye-Hückel and the Davies Equation[a]

α_i ($\times 10^{10}$ m)[a]	Ionic Strength, M						Ion
	0.001	0.005	0.01	0.05	0.10	0.25	
Monovalent Ions							
Davies	0.9649	0.9268	0.9016	0.8207	0.7806	0.7376	
9	0.9665	0.9335	0.9132	0.8536	0.8252		H^+
6	0.9656	0.9295	0.9063	0.8331	0.7953		Li^+, benzoate$^-$, salicylate$^-$
4	0.9649	0.9266	0.9012	0.8160	0.7689		Na^+, $CdCl^+$, HCO_3^-, $H_2PO_4^-$, $H_2AsO_4^-$, acetate$^-$
4.5	0.9650	0.9274	0.9025	0.8258	0.7761		
3.5	0.9647	0.9259	0.8998	0.8111	0.7612		OH^-, F^-, HS^-, ClO_4^-, formate$^-$, H_2-citrate$^-$
3	0.9645	0.9251	.08984	0.8060	0.7530		K^+, Cl^-, Br^-, I^-, CN^-, NO_2^-, NO_3^-
2.5	0.9643	0.9243	0.8969	0.8006	0.7442		Rb^+, Cs^+, NH_4^+, Tl^+, Ag^+
Divalent Ions							
Davies	0.8667	0.7378	0.6608	0.4537	0.3713	0.2960	
8	0.8715	0.7552	0.6888	0.5155	0.4439		Mg^{2+}, Be^{2+}
6	0.8692	0.7466	0.6748	0.4818	0.4000		Ca^{2+}, Cu^{2+}, Zn^{2+}, Mn^{2+}, Fe^{2+}, Ni^{2+}, Co^{2+}, glutarate^{2-},
5	0.8680	0.7420	0.6673	0.4632	0.3756		Sr^{2+}, Ba^{2+}, Ra^{2+}, Cd^{2+}, Hg^{2+}, S^{2-}, WO_4^{2-}, malonate^{2-}, succinate^{2-}, tartarate^{2-}
4.5	0.8673	0.7397	0.6635	0.4534	0.3628		Pb^{2+}, SO_4^{2-}, $S_2O_3^{2-}$, SeO_4^{2-}, CrO_4^{2-}, HPO_4^{2-}, H-citrate^{2-}, oxalate^{2-}
Trivalent Ions							
Davies	0.7248	0.5045	0.3938	0.1689	0.1076	0.0646	
9	0.7360	0.5382	0.4416	0.2406	0.1775		Al^{3+}, Fe^{3+}, Cr^{3+}, Y^{3+}, Ln^{3+}, PO_4^{3-},
5	0.7271	0.5110	0.4025	0.1770	0.1105		Citrate^{3-}

[a] α_i values and corresponding ionic species were obtained from Kielland (1937).

If we assume that all of the sodium in the solution exists as Na^+, then the activity of Na^+ in this solution is $(Na^+) = 0.7899$ L mol^{-1} $\times$ 0.01 mol L^{-1} = 0.0079 (unitless). The activity coefficient for Ca^{2+} can be determined in a similar manner. Alternatively, close inspection of Equation 5.8a leads to equalities that will compute di- and trivalent ion activity coefficients ($\gamma_{2\pm}$ and $\gamma_{3\pm}$):

$$\log \gamma_{2\pm} = 4 \times \log \gamma_\pm \tag{5.9}$$

$$\log \gamma_{3\pm} = 9 \times \log \gamma_\pm \tag{5.10}$$

Equation 5.9 states that the $\log \gamma$ value for a divalent ion is the square of the valence multiplied by the $\log \gamma$ value for a monovalent ion. Therefore, using Equation 5.9,

$$\log \gamma_{Ca^{2+}} = 4 \times (-0.1024) = -0.4098 \tag{5.11a}$$

$$\gamma_{Ca^{2+}} = 0.3893 \tag{5.11b}$$

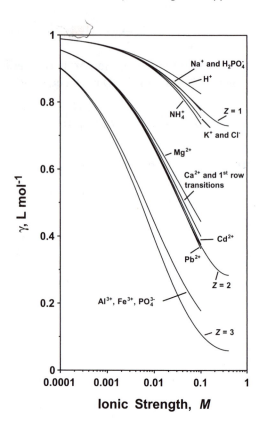

FIGURE 5.4 Single-ion activity coefficients at 25°C as a function of ionic strength as computed using the extended Debye-Hückel and Davies equations. Activity coefficients for monovalent ($Z = 1$), divalent ($Z = 2$), and trivalent ($Z = 3$) species are computed using the Davies equation. Activity coefficients for specific species are computed using the extended Debye-Hückel equation and associated α_i values (Table 5.4).

Again, assuming all of the calcium in the solution exists as Ca^{2+}, the activity of Ca^{2+} is $(Ca^{2+}) = 0.3893$ L mol^{-1} × 0.02 mol L^{-1} = 0.0078 (unitless). Also,

$$\log \gamma_{Cl^-} = \log \gamma_{Na^+} \tag{5.12a}$$

$$\log \gamma_{Mg^{2+}} = \log \gamma_{Ca^{2+}} \tag{5.12b}$$

when computed using the Davies equation.

The extended Debye-Hückel and Davies equations are valid for charged species that occur in soil solutions that meet the specified ionic strength requirements. A comparison of the $\log \gamma$ values predicted by these two expressions is shown in Figure 5.4. Tabulated values of γ, computed as a function of I and α_i, clearly illustrate the deviation of Debye-Hückel-predicted from Davies-predicted γ values as ionic strength increases (Table 5.4). Neutral species also occur in soil solutions (e.g., $H_4SiO_4^0$), and ionic strength effects can impact their activities, particularly if the neutral species exhibits dipolar character. The Setchenov equation is commonly employed to provide an empirical estimate for the $\log \gamma$ for neutral species: $\log \gamma_0 = k_s I$, where k_s is termed the salting-out coefficient when positive, or the salting-in coefficient when negative. Typically, k_s is assigned a value of 0.1 L mol^{-1} for neutral species in soil solutions.

Another colligative property of a solution is the chemical potential. In an ideal solution, the chemical potential of a substance is defined by the expression:

$$\mu_i = \mu_i^\circ + RT \ln x_i \tag{5.13}$$

where μ_i is the chemical potential of substance i, μ_i° is the standard state chemical potential of i, x_i is the mole fraction of i, R is the natural gas constant, and T is the temperature in K. In a nonideal solution, the chemical potential is similarly defined, providing an alternate definition of the activity of a substance (instead of $a = \gamma M$):

$$\mu_i = \mu_i^\circ + RT \ln a_i \tag{5.14}$$

The standard state chemical potential of i is a known quantity and a reference point. For an ionic solute, the standard state is a hypothetical state at 298.15 K (25°C), 1 atm pressure (0.101 MPa), molal concentration (m°) = 1 mol kg^{-1}, and no interactions. Expression 5.14 indicates that the activity of a species is a measure of the deviation of the chemical potential of a substance from the standard state chemical potential. Consider the gibbsite ($Al(OH)_3$) precipitation reaction:

$$Al^{3+}(aq) + 3OH^-(aq) \rightarrow Al(OH)_3(s) \tag{5.15}$$

At equilibrium, the following condition exists:

$$\mu_{Al^{3+}} + 3\mu_{OH^-} = \mu_{gibbsite} \tag{5.16}$$

This expression states that the sum of chemical potentials of the reactants (multiplied by the appropriate stoichiometric coefficients) is equal to the sum of the chemical potentials of the products (also multiplied by the appropriate stoichiometric coefficients) at equilibrium *in the environment of interest*. The following expressions are also valid [( ) denotes activity]:

$$\mu_{Al^{3+}} = \mu_{Al^{3+}}^\circ + RT \ln(Al^{3+}) \tag{5.17a}$$

$$\mu_{OH^-} = \mu_{OH^-}^\circ + RT \ln(OH^-) \tag{5.17b}$$

$$\mu_{gibbsite} = \mu_{gibbsite}^\circ + RT \ln(gibbsite) \tag{5.17c}$$

Substituting Equations 5.17a through 5.17c into the expression that defines equilibrium (Equation 5.16) yields:

$$\mu_{Al^{3+}}^\circ + RT \ln(Al^{3+}) + 3\mu_{OH^-}^\circ + 3RT \ln(OH^-) = \mu_{gibbsite}^\circ + RT \ln(gibbsite) \tag{5.18a}$$

Rearranging to collect common terms,

$$\mu_{Al^{3+}}^\circ + 3\mu_{OH^-}^\circ - \mu_{gibbsite}^\circ + RT \ln(Al^{3+}) + 3RT \ln(OH^-) - RT \ln(gibbsite) = 0 \tag{5.18b}$$

Collecting terms within the natural logarithm:

$$\mu_{Al^{3+}}^\circ + 3\mu_{OH^-}^\circ - \mu_{gibbsite}^\circ + RT \ln\left[\frac{(Al^{3+})(OH^-)^3}{(gibbsite)}\right] = 0 \tag{5.18c}$$

or

$$-RT\ln\left[\frac{(\text{gibbsite})}{(\text{Al}^{3+})(\text{OH}^-)^3}\right] = \mu^\circ_{\text{gibbsite}} - \mu^\circ_{\text{Al}^{3+}} - 3\mu^\circ_{\text{OH}^-} \qquad (5.18\text{d})$$

The standard state chemical potential is numerically identical to the Gibbs standard free energy of formation (ΔG°_f). Then, Equation 5.18d becomes:

$$-RT\ln\left[\frac{(\text{gibbsite})}{(\text{Al}^{3+})(\text{OH}^-)^3}\right] = \Delta G^\circ_{f,\text{gibbsite}} - \Delta G^\circ_{f,\text{Al}^{3+}} - 3\Delta G^\circ_{f,\text{OH}^-} \qquad (5.18\text{e})$$

Equation 5.18e can be simplified further using the following definitions:

$$K \equiv \frac{(\text{gibbsite})}{(\text{Al}^{3+})(\text{OH}^-)^3} \qquad (5.19)$$

where K is the equilibrium constant (note that the equilibrium constant is defined as a function of activities, NOT concentrations), and

$$\Delta G^\circ_r = \Delta G^\circ_{f,\text{gibbsite}} - \Delta G^\circ_{f,\text{Al}^{3+}} - 3\Delta G^\circ_{f,\text{OH}^-} \qquad (5.20)$$

where ΔG°_r is the standard free energy change for the reaction. Substituting K and ΔG°_r for the appropriate terms in Equation 5.18e yields:

$$-RT\ln K = \Delta G^\circ_r \qquad (5.21)$$

At 298.15 K (25°C), $R = 8.314$ J K^{-1} mol^{-1}, and converting the natural logarithm to $\log_{10}$ (with ΔG°_r in units of kJ mol^{-1}), the relationship between the equilibrium constant and the standard free energy change is:

$$\log K = -\frac{\Delta G^\circ_r}{5.708} \qquad (5.22)$$

Numerous reference compilations of standard free energy of formation (ΔG°_f) values for a vast array of aqueous, solid, and gaseous species are available. Some of these compilations, those that contain critically reviewed data, are employed in the following sections. Table 5.5 compares ΔG°_f values from established sources for some of the more environmentally relevant species. Returning to the gibbsite precipitation example, from Robie et al. (1978) (Table 5.5) we obtain the ΔG°_f values for Al^{3+} (–489.4 kJ mol^{-1}), OH$^-$ (–157.3 kJ mol^{-1}), and gibbsite (–1154.9 kJ mol^{-1}). The standard free energy change for the reaction is:

$$\Delta G^\circ_r = -1154.9 - (-489.4) - 3(-157.3) = -193.6 \text{ kJ mol}^{-1} \qquad (5.23)$$

Because ΔG°_r is negative, the reaction product, gibbsite, is more stable relative to the reactants, Al^{3+} and OH$^-$, *if the reaction were to occur in the standard state*. Conversely (but not the case in this example), a positive ΔG°_r indicates that the reactants are more stable than the products *if the*

TABLE 5.5
Standard Free Energies of Formation (ΔG_f°) in Units of kJ mol^{-1} for Free Aqueous Species from Three Reference Compilations (25°C and 0.101 MPa)

Species	Naumov et al. (1974)	Robie et al. (1978)	Wagman et al. (1982)	Species	Naumov et al. (1974)	Robie et al. (1978)	Wagman et al. (1982)
Al^{3+}	−492.0	−489.4	−485	Li^+	−292.6	−292.6	−293.3
AsO_4^{3-}	−652.0	—	−652.0	Mg^{2+}	−455.3	−454.8	−454.8
Ba^{2+}	−547.5	−560.7	−560.8	Mn^{2+}	−230.0	−228.0	−228.1
$B(OH)_3^0$	−968.6	—	−968.8	MoO_4^{2-}	−838.0	—	−836.3
Br^-	−104.2	−104.0	−104.0	Ni^{2+}	−45.1	−45.6	−45.6
Cd^{2+}	−77.9	−77.6	−77.6	NO_3^-	−111.4	−111.5	−108.7
Ca^{2+}	−552.7	−553.5	−553.6	NH_4^+	−78.7	−79.4	−79.3
$CO_2(g)$	−394.4	−394.4	−394.4	$NH_3(g)$	−16.5	−16.4	−16.4
CO_3^{2-}	−527.9	−527.9	−527.8	OH^-	−157.3	−157.2	−157.3
Cl^-	−131.3	−131.3	−131.2	H_2O	−237.2	−237.2	−237.1
Cr^{3+}	−203.9	—	—	PO_4^{3-}	−1018.8	−1019.0	−1018.7
CrO_4^{2-}	−720.9	—	−727.8	K^+	−282.7	−282.5	−283.3
$Cr_2O_7^{2-}$	−1287.6	—	−1301.1	SeO_3^{2-}	−363.8	—	−369.8
Cu^{2+}	65.3	65.5	65.5	SeO_4^{2-}	−441.4	—	−441.3
F^-	−280.0	−281.7	−278.8	$H_4SiO_4^0$	−1306.9	−1308.0	−1316.6
H^+	0	0	0	Na^+	−262.2	−261.9	−261.9
Fe^{2+}	−92.2	−78.9	−78.9	Sr^{2+}	−571.4	−559.4	−559.5
Fe^{3+}	−17.9	−4.6	−4.7	SO_4^{2-}	−743.8	−744.6	−744.5
Pb^{2+}	−24.4	−24.4	−24.4	Zn^{2+}	−147.2	−147.3	−147.1

reaction were to occur in the standard state. Continuing with the computation of the equilibrium constant:

$$\log K = -\frac{-193.6}{5.708} = 33.9 \tag{5.24a}$$

$$K = 10^{33.9} = \frac{(\text{gibbsite})}{(Al^{3+})(OH^-)^3} \tag{5.24b}$$

and

$$33.9 = \log(\text{gibbsite}) - \log(Al^{3+}) - 3\log(OH^-) \tag{5.24c}$$

Equation 5.24b states that the activity product $[(Al^{3+})(OH^-)^3]$ will exactly equal the product $[(\text{gibbsite})(10^{-33.9})]$ if equilibrium conditions exist *in the environment of interest.*

5.4 HYDROLYSIS OF CATIONS

As was previously indicated, all ions hydrate to some degree in water. The heat released when the gaseous form of an ion is immersed in water (ΔH_h values) indicates the strength and degree of hydration. Most metal cations form strong bonds to oxygen. In aqueous solution, the metal-oxygen interaction induces a strong polarizing effect on the waters of hydration, which can split or decompose a water molecule, to form a new ionic metal species (a hydroxide or an oxide) and protons.

TABLE 5.6
Ionic Potentials (Z/r) of Metal Cations and Their Degree of Hydration or Hydrolysis in Aqueous Solutions[a]

Element	r, pm	Z/r	Aqueous Species	Element	r, pm	Z/r	Aqueous Species
			IP < 0.03: Ions Exist Principally as Hydrated Cations				
Li^+	73	0.014	$[Li(H_2O)_{4-6}]^+$	Zn^{2+}	88	0.023	$[Zn(H_2O)_6]^{2+}$
Na^+	113	0.009	$[Na(H_2O)_4]^+$	Cu^{2+}	91	0.022	$[Cu(H_2O)_6]^{2+}$
K^+	151	0.007	$[K(H_2O)_4]^+$	Ni^{2+}	83	0.024	$[Ni(H_2O)_6]^{2+}$
Mg^{2+}	86	0.012	$[Mg(H_2O)_6]^{2+}$	Co^{2+}	88	0.023	$[Co(H_2O)_6]^{2+}$
Ca^{2+}	114	0.009	$[Ca(H_2O)_6]^{2+}$	Fe^{2+}	92	0.022	$[Fe(H_2O)_6]^{2+}$
Sr^{2+}	132	0.008	$[Sr(H_2O)_6]^{2+}$	Mn^{2+}	97	0.021	$[Mn(H_2O)_6]^{2+}$
				Cr^{2+}	94	0.021	$[Cr(H_2O)_6]^{2+}$
				Cd^{2+}	109	0.018	$[Cd(H_2O)_6]^{2+}$
				Pb^{2+}	133	0.015	$[Pb(H_2O)_6]^{2+}$
			0.03 < IP < 0.10: Hydrolysis Occurs as Well as the Formation of Sparingly Soluble Hydroxide				
Al^{3+}	67	0.045	$Al(H_2O)_{6-n}(OH)_n^{3-n}$				
Fe^{3+}	78	0.038	$Fe(H_2O)_{6-n}(OH)_n^{3-n}$				
			IP > 0.10: Formation of Oxyanion or Hydroxyanion				
B^{3+}	25	0.120	$B(OH)_3^0$	Cr^{6+}	40	0.150	CrO_4^{2-}
Si^{4+}	40	0.100	$Si(OH)_4^0$	Mn^{6+}	40	0.150	MnO_4^{2-}
C^{4+}	29	0.140	CO_3^{2-}	As^{5+}	48	0.104	AsO_4^{3-}
N^{5+}	27	0.185	NO_3^-	Se^{6+}	42	0.143	SeO_4^{2-}
S^{6+}	43	0.140	SO_4^{2-}	Mo^{6+}	55	0.109	MoO_4^{2-}
Cl^{7+}	22	0.318	ClO_4^-	P^{5+}	31	0.161	PO_4^{3-}

[a] r is the ionic radius for sixfold coordination and Z is the valence.

This process is termed hydrolysis. Hydrolysis is a process that can significantly impact the chemistry of a metal in an aqueous environment. In essence, hydrolysis changes the chemical form of a metal, impacting reactivity with other soluble species, adsorption and exchange characteristics, and mineral solubility.

Cation hydration and hydrolysis behavior in an aqueous solution can be generalized into three categories, depending on ion charge, Z, and radius, r. The ionic potential (*IP*) of an ion is defined as Z/r. Cations with small *IP* values (<0.03) tend to remain hydrated (Table 5.6) and are not hydrolyzed throughout the normal pH range of soil solutions. Ions with moderate to high *IP* values (0.03 < *IP* < 0.1) tend to strongly polarize water and promote hydrolysis: $[M(H_2O)_n]^{m+} = [MOH(H_2O)_{n-1}]^{m-1} + H^+$, where M is a metal cation with charge $m+$. This type of hydrolysis behavior is common to Al^{3+} and Fe^{3+}. Ions with high *IP* values (>0.1) promote the complete dissociation of waters of hydration, resulting in the formation of stable oxyanions: $[M(H_2O)_n]^{m+} = [MO_n]^{m-2n} + 2nH^+$. Indeed, metals that display a high *IP* do not exist as cations in aqueous solutions.

The degree of hydrolysis is also pH dependent. In acidic solutions, proton activity is high and dissociation of a proton from a water of hydration (to essentially put more protons into solution) is not favored. In alkaline solutions, proton activity is low and hydrolysis, or the dissociation of protons from waters of hydration, is favored. The degree of hydrolysis is described by an equilibrium constant. Consider the hydrolysis of $Al(H_2O)_6^{3+}$:

$$Al(H_2O)_6^{3+} \rightarrow AlOH(H_2O)_5^{2+} + H^+ \tag{5.25}$$

TABLE 5.7
Hydrolysis Reactions and Equilibrium Constants for the Formation of Mononuclear Species at 25°C[a]

Hydrolysis Reaction	$\log K$	Hydrolysis Reaction	$\log K$
$B(OH)_3^0 + H_2O \rightarrow B(OH)_4^- + H^+$	−9.236	$Fe^{3+} + H_2O \rightarrow FeOH^{2+} + H^+$	−2.19
$Al^{3+} + H_2O \rightarrow AlOH^{2+} + H^+$	−5.00	$Fe^{3+} + 2H_2O \rightarrow Fe(OH)_2^+ + 2H^+$	−5.67
$Al^{3+} + 2H_2O \rightarrow Al(OH)_2^+ + 2H^+$	−10.1	$Fe^{3+} + 3H_2O \rightarrow Fe(OH)_3^0 + 3H^+$	−12
$Al^{3+} + 3H_2O \rightarrow Al(OH)_3^0 + 3H^+$	−16.8	$Fe^{3+} + 4H_2O \rightarrow Fe(OH)_4^- + 4H^+$	−21.6
$Al^{3+} + 4H_2O \rightarrow Al(OH)_4^- + 4H^+$	−22.7	$Co^{2+} + H_2O \rightarrow CoOH^+ + H^+$	−9.65
$La^{3+} + H_2O \rightarrow LaOH^{2+} + H^+$	−8.5	$Ni^{2+} + H_2O \rightarrow NiOH^+ + H^+$	−9.86
$Ce^{3+} + H_2O \rightarrow CeOH^{2+} + H^+$	−8.3	$Cu^{2+} + H_2O \rightarrow CuOH^+ + H^+$	−7.7
$Nd^{3+} + H_2O \rightarrow NdOH^{2+} + H^+$	−8.0	$Cu^{2+} + 2H_2O \rightarrow Cu(OH)_2^0 + 2H^+$	−13.8
$VO^{2+} + H_2O \rightarrow VO(OH)^+$	−2.26	$Zn^{2+} + H_2O \rightarrow ZnOH^+ + H^+$	−8.96
$VO_2^+ + 2H_2O \rightarrow VO(OH)_3^0 + H^+$	−3.3	$Cd^{2+} + H_2O \rightarrow CdOH^+ + H^+$	−10.08
$VO_2^+ + 2H_2O \rightarrow VO_2(OH)_2^- + 2H^+$	−7.3	$Hg^{2+} + H_2O \rightarrow HgOH^+ + H^+$	−3.40
$VO_2(OH)_2^- \rightarrow VO_3(OH)^{2+} + H^+$	−8.55	$Hg^{2+} + 2H_2O \rightarrow Hg(OH)_2^0 + 2H^+$	−6.17
$Cr^{3+} + H_2O \rightarrow CrOH^{2+} + H^+$	−4.0	$Si(OH)_4^0 \rightarrow SiO(OH)_3^- + H^+$	−9.86
$Cr^{3+} + 2H_2O \rightarrow Cr(OH)_2^+ + 2H^+$	−9.7	$Si(OH)_4^0 \rightarrow SiO_2(OH)_2^{2-} + 2H^+$	−22.92
$Mn^{2+} + H_2O \rightarrow MnOH^+ + H^+$	−10.59	$Pb^{2+} + H_2O \rightarrow PbOH^+ + H^+$	−7.71
$Fe^{2+} + H_2O \rightarrow FeOH^+ + H^+$	−9.5	$Pb^{2+} + 2H_2O \rightarrow Pb(OH)_2^0 + 2H^+$	−17.12

[a] Data compiled from Baes and Mesmer (1986) and Smith and Martell (1976).

The equilibrium constant for this reaction is (25°C and 0.101 MPa; Table 5.7):

$$K_a = \frac{(AlOH(H_2O)_5^{2+})(H^+)}{(Al(H_2O)_6^{3+})} = 10^{-5.00} \tag{5.26}$$

and K_a is termed the acid dissociation constant. The negative log of the K_a ($-\log K_a$) is designated pK_a (p simply means −log). For this reaction, the pK_a is 5.00. The acid strength of all Lowry-Brønsted acids (compounds that donate protons to solutions) can be described by their associated pK_a values: the lower this value, the stronger the acid. If we rearrange Equation 5.26 and simplify the notation for the Al species by removing the waters of hydration:

$$\frac{K_a}{(H^+)} = \frac{(AlOH^{2+})}{(Al^{3+})} \tag{5.27a}$$

or

$$pK_a - pH = \log(Al^{3+}) - \log(AlOH^{2+}) \tag{5.27b}$$

The following points are noted:

- When $pH > pK_a$ [$(H^+) < K_a$], the deprotonated hydrolysis product, $AlOH^{2+}$, is the dominant species present in the solution.
- When $pH < pK_a$ [$(H^+) > K_a$], the hydrated free cation species Al^{3+} is the dominant species present in the solution.
- When $pH = pK_a$, the hydrated free cation and hydrolysis product have equal activities in solution: $(Al^{3+}) = (AlOH^{2+})$.

Metal hydrolysis is not strictly limited to the decomposition of a single water of hydration, but may involve the stepwise dissociation of the waters of hydration. The hydrolysis behavior of Al^{3+} typifies the polyprotic behavior of select metals. In addition to the formation of the $AlOH^{2+}$ species, the following Al^{3+} hydrolysis reactions occur in the normal pH range of soil:

$$AlOH^{2+} + H_2O \rightarrow Al(OH)_2^+ + H^+ \tag{5.28a}$$

$$Al(OH)_2^+ + H_2O \rightarrow Al(OH)_3^0 + H^+ \tag{5.28b}$$

$$Al(OH)_3^0 + H_2O \rightarrow Al(OH)_4^- + H^+ \tag{5.28c}$$

Using the data in Table 5.7, the pK_a values for the formation of $Al(OH)_2^+$, $Al(OH)_3^0$, and $Al(OH)_4^-$ are 5.10, 6.70, and 5.9. This information can be employed to construct a predominance diagram of Al speciation as a function of pH (Figure 5.5). As the diagram illustrates, the Al hydrolysis product that predominates in a solution (present in the greatest quantity) will be a function of solution pH. Further, the positive charge of Al and the hydrolysis products decreases from +3 in acidic solutions, through +1 in near-neutral solutions, to −1 in neutral and basic solutions. The characteristic of a substance to display both positive and negative charge, depending on solution pH, is termed amphoterism. Thus, Al is amphoteric, and can exist as a cation or an anion, depending on soil pH. The speciation of other metals that participate in hydrolysis reactions are also illustrated in Figure 5.5. Clearly, the nonhydrolyzed metal cation predominates for the vast majority of metals in the pH range of soils (pH 4 to 9). Notable exceptions include Fe^{3+}, Cr^{3+}, and Hg^{2+} (in addition to Al^{3+}).

Although useful, predominance diagrams do not provide a complete picture of the distribution of metals and associated hydrolysis products in solutions. Predominance diagrams indicate the species that is found in the greatest abundance for the given condition. However, less abundant species also occur, their existence masked by the predominant species. For example, the $Al(OH)_3^0$ species is absent from the Al predominance diagram (Figure 5.5), even though this species does occur in the pH 5 to 8 range. Another mechanism employed to graphically relate the aqueous speciation of a substance is the distribution diagram. These diagrams relate the relative abundance of all significant species that may exist for a given substance. Typically, the abundance of a species is expressed as a percentage of the total elemental concentration as a function of solution pH. The distribution diagrams for Al, Fe(III), Cr(III), Cu, Zn, and Pb illustrate important points relative to metal hydrolysis (Figure 5.6):

- For the trivalent cations, the free metal species can be of little significance in soil solutions, particularly in the pH 6 to 8 range.
- For the divalent cations, the free metal species will occur in soil solutions throughout a broad pH range. However, the concentrations of the free species will begin to diminish with increasing pH, beginning at pH values well below the pK_a.
- The free metal cation and one or more hydrolysis products can coexist in a solution, depending on pH.
- Not all metal hydrolysis products will predominate in a solution; for example, $Al(OH)_3^0$ and $CuOH^+$ form, but are minor species under all pH conditions.

5.5 LOWRY-BRØNSTED ACIDS AND BASES

Lowry-Brønsted acidity is defined as the ability to lose or donate protons. The hydrolysis reactions described above for Al^{3+} and other metal cations are examples of the Lowry-Brønsted acidity of the metal ions. Lowry-Brønsted acids and bases are characterized by their strength. Strong acids are those that have negative pK_a values and completely dissociate in water (Table 5.8). Hydrochloric acid (HCl) is a strong acid because it has a negative pK_a value and is completely dissociated in

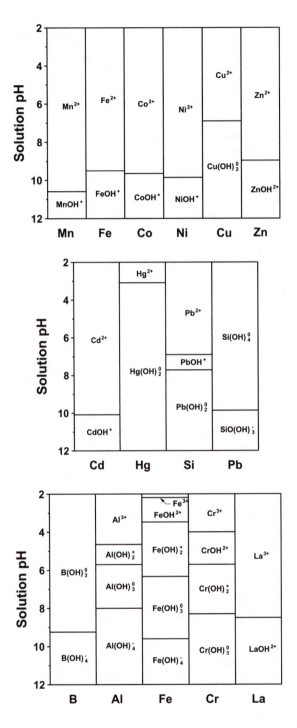

FIGURE 5.5 Predominance diagrams illustrating the hydrolytic metal species that represent greater than 50% of the total content of a solution as a function of pH at 25°C. The pH where the free metal cation transitions to the hydrolysis product is the pK_a value for the hydrolysis reaction (normalized to represent the production of a single proton). The pH where a metal hydrolysis product transitions to a higher order hydrolysis product (MOH to M(OH)$_2$) is also representatived of the pK_a value for the reaction. The pK_a values employed to construct the diagrams were derived from the data compiled in Table 5.7.

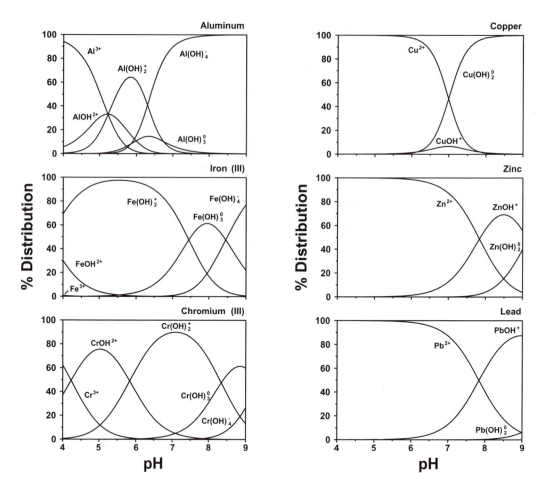

FIGURE 5.6 Distribution diagrams illustrating metal speciation as a function of solution pH at 25°C. The percent distribution is the concentration of a specified species divided by the total concentration of the element in the solution. The diagrams were constructed using a total metal concentration of 10^{-5} M in a 10^{-2} M NaClO$_4$ background electrolyte. The pK_a values employed to construct the diagrams were derived from the data compiled in Table 5.7.

aqueous solution; HCl^0 does not exist in solution, only H$^+$ and Cl$^-$. Weak acids are those that have pK_a values that lie between 0 and 14: the larger the pK_a the weaker the acid strength; the smaller the pK_a the greater the acid strength. Hydrofluoric acid (HF) is a weak acid with a pK_a of 3.45. At pH values less than 3.45, the HF^0 species predominates relative to F$^-$. Conversely, when solution pH is greater than 3.45, HF^0 dissociates and the F$^-$ species predominates. Further, HF is a stronger acid than carbonic acid (H$_2$CO$_3^*$), which has a pK_a of 6.35 (H$_2$CO$_3^*$ represents the total analytical concentration of dissolved CO$_2$; [H$_2$CO$_3^*$] = [CO$_2^0$] + [H$_2$CO$_3^0$]).

Several acids have the ability to donate more than one proton to a solution. These acids are termed polyprotic. The Al^{3+} and Fe^{3+} ions are polyprotic acids, as are many of the oxyanions represented in Table 5.8. For some oxyanion acids (e.g., H$_2$SO$_4$, H$_2$SeO$_4$, and H$_2$CrO$_4$), the initial deprotonation is a strong acid reaction. The resulting conjugate base anions (HSO$_4^-$, HSeO$_4^-$, and HCrO$_4^-$) are weak acids themselves, and deprotonate to form divalent oxyanions that predominate in soil solutions (SO$_4^{2-}$, SeO$_4^{2-}$, and CrO$_4^{2-}$ for pH > 6.5). For other polyprotic acids (e.g., H$_3$PO$_4$, H$_3$AsO$_4$, and H$_2$CO$_3^*$), all protons are weak acid protons. With the exception of molydbic acid, for which the MoO$_4^{2-}$ species predominates above pH 4.24, the fully dissociated oxyanions do not

TABLE 5.8
Selected Strong and Weak Lowry-Brønsted Acids and Acid
Dissociation Constants (pK_a Values) at 25°C[a]

Acid Reaction	pK_a	Acid Reaction	pK_a
Strong Acids			
$HClO_4^0 \rightarrow ClO_4^- + H^+$	−7	$H_2SeO_4^0 \rightarrow HSeO_4^- + H^+$	~−3
$HCl^0 \rightarrow Cl^- + H^+$	~−3	$HNO_3^0 \rightarrow NO_3^- + H^+$	−1
$H_2SO_4^0 \rightarrow HSO_4^- + H^+$	~−3	$H_2CrO_4^0 \rightarrow HCrO_4^- + H^+$	−0.2
Weak Inorganic Acids			
$HSeO_4^- \rightarrow SeO_4^{2-} + H^+$	1.70	$HF^0 \rightarrow F^- + H^+$	3.17
$HSO_4^- \rightarrow SO_4^{2-} + H^+$	1.99	$H_2MoO_4^0 \rightarrow HMoO_4^- + H^+$	4.00
$H_3PO_4^0 \rightarrow H_2PO_4^- + H^+$	2.15	$HMoO_4^- \rightarrow MoO_4^{2-} + H^+$	4.24
$H_2PO_4^- \rightarrow HPO_4^- + H^+$	7.20	$H_2CO_3^* \rightarrow HCO_3^- + H^+$	6.35
$HPO_4^{2-} \rightarrow PO_4^{3-} + H^+$	12.35	$HCO_3^- \rightarrow CO_3^{2-} + H^+$	10.33
$H_3AsO_4^0 \rightarrow H_2AsO_4^- + H^+$	2.24	$HCrO_4^- \rightarrow CrO_4^{2-} + H^+$	6.51
$H_2AsO_4^- \rightarrow HAsO_4^{2-} + H^+$	6.96	$B(OH)_3^0 \rightarrow B(OH)_4^- + H^+$	9.23
$HAsO_4^{2-} \rightarrow AsO_4^{3-} + H^+$	11.50	$HAsO_3^{2-} \rightarrow AsO_3^{3-} + H^+$	9.29
$H_2SeO_3^0 \rightarrow HSeO_3^- + H^+$	2.75	$H_4SiO_4^0 \rightarrow H_3SiO_4^- + H^+$	9.86
$HSeO_3^- \rightarrow SeO_3^{2-} + H^+$	8.50	$H_3SiO_4^- \rightarrow H_2SiO_4^{2-} + H^+$	13.1

[a] Data compiled from Baes and Mesmer (1986) and Smith and Martell (1976).

predominate in soil solutions. Instead, an intermediate protonated species will predominate, the actual species depending on solution pH (Figure 5.7). In the normal pH range of soils, the orthophosphate species $H_2PO_4^-$ and HPO_4^{2-} predominate. Similarly, $H_2AsO_4^-$ and $HAsO_4^{2-}$ predominate, and arsenate distribution is nearly identical to that of phosphate. Of course, this similarity in speciation is not coincidental, as As is in the same periodic group as P.

Weak organic acids are ubiquitous to soil solutions, ranging in concentration from less than 0.01 mmol L^{-1} to upward of 5 mmol L^{-1} (Figure 5.8). The pK_a values for selected low-molecular-mass organic acids that have been identified in soil solutions are presented in Table 5.9. Like their inorganic counterparts, organic acids have the ability to donate (or consume) one to several protons to the soil solution, depending on the compound and the solution pH. Formic and acetic acids are monoprotic, with each molecule containing one carboxyl group. Oxalic and succinic acid are examples of diprotic acids, with each molecule containing two carboxyl groups. Citric acid is an example of a triprotic acid.

The ability of an organic compound to donate or consume protons is not limited to only those that contain carboxyl groups. Phenolic-OH groups may also dissociate in the soil pH range (e.g., gallic acid). The dissociation of a protonated amine group, such as found in amino acids, can also be described by a pK_a value (Table 5.9). However, because the amine group is a base (accepts protons), it is common to describe base strength using the pK_b. The pK_b describes the dissociation of a hydroxide, or the production of hydroxide through hydrolysis:

$$NaOH^0 \rightarrow Na^+ + OH^- \tag{5.29a}$$

$$K_b = \frac{(Na^+)(OH^-)}{(NaOH^0)} \tag{5.29b}$$

$$NH_3^0 + H_2O \rightarrow NH_4^+ + OH^- \tag{5.30a}$$

$$K_b = \frac{(NH_4^+)(OH^-)}{(NH_3^0)} \tag{5.30b}$$

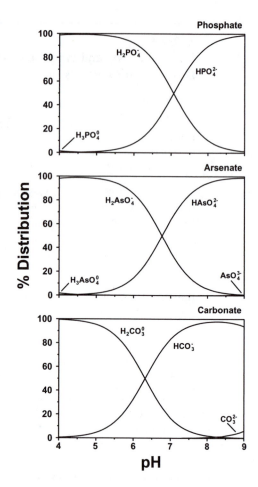

FIGURE 5.7 The distribution of weak Lowry-Brøn-sted acids as a function of solution pH at 25°C. The diagrams were constructed using total concentrations of: 10^{-5} M for PO_4, 10^{-6} M for AsO_4, 10^{-3} M for CO_3, and a background electrolyte of 10^{-2} M for $NaClO_4$. The pK_a values employed to construct the diagrams were obtained from Table 5.8.

Strong bases are characterized as having pK_b values that are less than zero. This is the case for the dissociation of $NaOH^0$ (Equations 5.29a and 5.29b) which has a pK_b value of –0.2. A weak base is characterized as having a pK_b value that lies between 0 and 14, as is the case for the reaction in Equation 5.30a (p$K_b = 4.76$) and the amine functional groups of amino acids. The acid dissociation constant (pK_a) and pK_b for a given compound are related through the ionization constant for water:

$$pK_b = pK_w - pK_a \tag{5.31}$$

Thus, the pK_a for the hydrolysis of Na^+:

$$Na^+ + H_2O \rightarrow NaOH^0 + H^+ \tag{5.32}$$

is p$K_a = 14 - (-0.2) = 14.2$. The pK_a for the dissociation of ammonium:

$$NH_4^+ \rightarrow NH_3^0 + H^+ \tag{5.33}$$

is p$K_a = 14 - 4.76 = 9.24$.

Weak acids dominate soil solution chemistry and they contribute to the buffering capacity of the soil. One of the more prevalent weak acids in the soil solution is carbonic acid. The carbonic

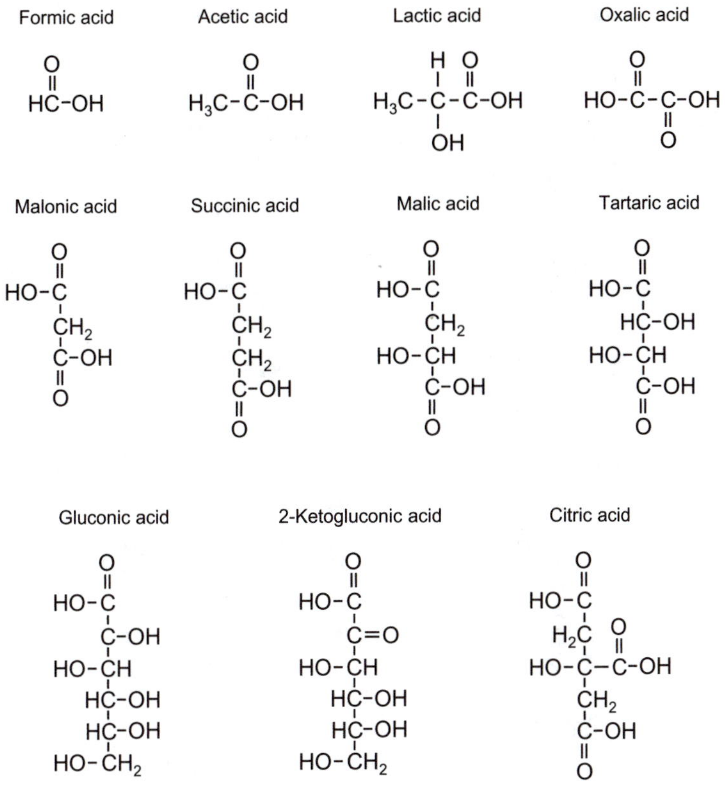

Formic acid

$$O$$
$$\|$$
$$HC-OH$$

Acetic acid

$$O$$
$$\|$$
$$H_3C-C-OH$$

Lactic acid

$$H \quad O$$
$$| \quad \|$$
$$H_3C-C-C-OH$$
$$|$$
$$OH$$

Oxalic acid

$$O$$
$$\|$$
$$HO-C-C-OH$$
$$\|$$
$$O$$

Malonic acid

$$O$$
$$\|$$
$$HO-C$$
$$|$$
$$CH_2$$
$$C-OH$$
$$\|$$
$$O$$

Succinic acid

$$O$$
$$\|$$
$$HO-C$$
$$|$$
$$CH_2$$
$$CH_2$$
$$C-OH$$
$$\|$$
$$O$$

Malic acid

$$O$$
$$\|$$
$$HO-C$$
$$|$$
$$CH_2$$
$$HO-CH$$
$$C-OH$$
$$\|$$
$$O$$

Tartaric acid

$$O$$
$$\|$$
$$HO-C$$
$$|$$
$$HC-OH$$
$$HO-CH$$
$$C-OH$$
$$\|$$
$$O$$

Gluconic acid

$$O$$
$$\|$$
$$HO-C$$
$$|$$
$$C-OH$$
$$HO-CH$$
$$HC-OH$$
$$HC-OH$$
$$HO-CH_2$$

2-Ketogluconic acid

$$O$$
$$\|$$
$$HO-C$$
$$|$$
$$C=O$$
$$HO-CH$$
$$HC-OH$$
$$HC-OH$$
$$HO-CH_2$$

Citric acid

$$O$$
$$\|$$
$$HO-C$$
$$|$$
$$H_2C \quad O$$
$$\quad \|$$
$$HO-C-C-OH$$
$$|$$
$$CH_2$$
$$C-OH$$
$$\|$$
$$O$$

FIGURE 5.8 Low-molecular-mass organic acids and amino acids that have been identified in soil solutions.

acid activity in a soil solution is controlled by the partial pressure of $CO_2(g)$ in the soil atmosphere. The pH that results from this equilibrium can be determined with knowledge of the Henry's law constant for CO_2 (K_H describes the partitioning of a volatile substance between a solution and a gaseous phase in contact with the solution), the pK_a values that describe the dissociation of carbonic acid to eventually form the carbonate ion, and the water ionization constant. The following reactions are pertinent (25°C and 0.101 MPa):

$$CO_2(g) + H_2O(l) \rightarrow H_2CO_3^* \qquad pK_H = 1.47; \text{ Henry's Law constant} \qquad (5.34a)$$

$$H_2CO_3^* \rightarrow HCO_3^- + H^+ \qquad pK_{a1} = 6.4; \text{ first dissociation constant} \qquad (5.34b)$$

$$HCO_3^- \rightarrow CO_3^{2-} + H^+ \qquad pK_{a2} = 10.3; \text{ second dissociation constant} \qquad (5.34c)$$

$$H_2O \rightarrow H^+ + OH^- \qquad pK_w = 14; \text{ ionization constant of water} \qquad (5.34d)$$

The charge balance for the solution and the master expression that will lead to a computed solution pH is:

$$[H^+] = [HCO_3^-] + 2[CO_3^{2-}] + [OH^-] \qquad (5.35)$$

Each term on the right side of Equation 5.35 must be expressed in terms of $[H^+]$ in order to compute solution pH. The following manipulations are relevant and assume a nearly ideal solution,

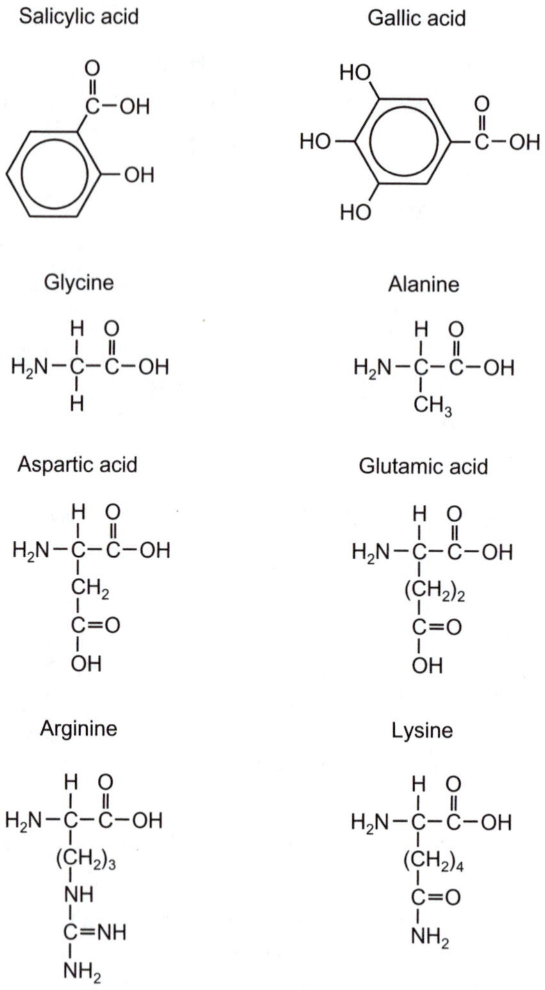

Salicylic acid

Gallic acid

Glycine

Alanine

Aspartic acid

Glutamic acid

Arginine

Lysine

FIGURE 5.8 (*continued.*)

i.e., activity equals concentration:

$$K_{\mathrm{H}} = \frac{[\mathrm{H_2CO_3^*}]}{P_{\mathrm{CO_2}}} \tag{5.36a}$$

$$[\mathrm{H_2CO_3^*}] = K_{\mathrm{H}}P_{\mathrm{CO_2}} \tag{5.36b}$$

$$K_{a1} = \frac{[\mathrm{HCO_3^-}][\mathrm{H^+}]}{[\mathrm{H_2CO_3^*}]} \tag{5.36c}$$

Using Equation 5.36b and substituting for $[\mathrm{H_2CO_3^*}]$ in Equation 5.36c:

$$K_{a1} = \frac{[\mathrm{HCO_3^-}][\mathrm{H^+}]}{K_{\mathrm{H}}P_{\mathrm{CO_2}}} \tag{5.36d}$$

$$[\mathrm{HCO_3^-}] = \frac{K_{a1}K_{\mathrm{H}}P_{\mathrm{CO_2}}}{[\mathrm{H^+}]} \tag{5.36e}$$

TABLE 5.9
Acid Dissociation Constants for Naturally Occurring Organic Acids and Amino Acids, and for the Synthetic Chelates EDTA and DTPA at 25°C[a]

| Organic Acid | $RCOOH = RCOO^- + H^{+b}$ | | | | |
	pK_{a1}	pK_{a2}	pK_{a3}	pK_{a4}	pK_{a5}
Formic	3.75				
Acetic	4.76				
Lactic	3.86				
Gluconic	3.96				
2-Ketogluconic	2.33				
Oxalic	1.25	4.27			
Succinic	4.21	5.64			
Malic	3.46	5.10			
Tartaric	3.04	4.37			
Malonic	2.85	5.70			
Salicylic	2.97	13.74			
Citric	3.13	4.76	6.40		
Gallic	4.26	8.7	11.45		
EDTA	2.29	3.23	6.94	11.28	
DTPA	2.3	2.6	4.17	8.26	9.48

| Amino acid | $RCOOH = RCOO^- + H^+$ | | $RNH_3^+ = RNH_2^0 + H^+$ | |
	pK_{a1}	pK_{a2}	pK_{a3}	K_{a4}
Glycine	2.35			9.78
Alanine	2.35			9.87
Aspartic	1.99	3.90		10.00
Glutamic	2.23	4.42		9.95
Arginine	1.82			8.99
Lysine	2.20		8.86	10.25

[a] Martell and Smith (1974, 1977, 1982), Smith and Martell (1975).
[b] pK_a values refer to stepwise deprotonation reactions.

Dissociation of bicarbonate is described by:

$$K_{a2} = \frac{[CO_3^-][H^+]}{[HCO_3^-]} \tag{5.36f}$$

Using Equation 5.36e and substituting for $[HCO_3^-]$ in Equation 5.36f:

$$K_{a2} = \frac{[CO_3^{2-}][H^+]^2}{K_{a1}K_H P_{CO_2}} \tag{5.36g}$$

$$[CO_3^{2-}] = \frac{K_{a2}K_{a1}K_H P_{CO_2}}{[H^+]^2} \tag{5.36h}$$

The ionization of water is described by:

$$K_w = [H^+][OH^-] \tag{5.36i}$$

$$[OH^-] = \frac{K_w}{[H^+]} \tag{5.36j}$$

Substituting Equations 5.36e, h, and j into Equation 5.35 yields:

$$[H^+] = \frac{K_{a1}K_H P_{CO_2}}{[H^+]} + \frac{2K_{a2}K_{a1}K_H P_{CO_2}}{[H^+]^2} + \frac{K_w}{[H^+]} \tag{5.37}$$

Substituting the numerical values for the equilibrium constants and $P_{CO_2} = 10^{-3.52}$ (atmospheric level) and rearranging:

$$[H^+] = \frac{10^{-6.4}10^{-1.47}10^{-3.52}}{[H^+]} + \frac{(2\times10^{-10.3})10^{-6.4}10^{-1.47}10^{-3.52}}{[H^+]^2} + \frac{10^{-14}}{[H]} \tag{5.38a}$$

$$[H^+] = \frac{10^{-11.39}}{[H^+]} + \frac{(2\times10^{-21.69})}{[H^+]^2} + \frac{10^{-14}}{[H^+]} \tag{5.38b}$$

$$[H^+]^3 = 10^{-11.39}[H^+] + (2 \times 10^{-21.69}) + 10^{-14}[H^+] \tag{5.38c}$$

$$[H^+]^3 = (10^{-11.39} + 10^{-14})[H^+] + (2 \times 10^{-21.69}) \tag{5.38d}$$

$$0 = (10^{-11.39} + 10^{-14})[H^+] + (2 \times 10^{-21.69}) - [H^+]^3 \tag{5.38e}$$

Solving this cubic expression yields $[H^+] = 2.02 \times 10^{-6}$, or pH = 5.695. Therefore, an infinitely dilute solution in contact with a gaseous phase containing atmospheric CO_2 levels will have a pH of 5.695 at 25°C and 0.101 MPa (assuming equilibrium exists).

In this instance, the mathematical computation of solution pH can be greatly simplified by assuming that the only source of protons in the solution will be carbonic acid ($H_2CO_3^*$); HCO_3^- does not contribute a proton. It then follows that the proton concentration in the solution will be equal to the bicarbonate concentration: $[H^+] = [HCO_3^-]$ (a requirement for mass balance, as one molecule of $H_2CO_3^*$ will dissociate into one H^+ and one HCO_3^-). Substituting this equality into Equation 5.36d yields:

$$K_{a1} = \frac{[H^+][H^+]}{K_H P_{CO_2}} \tag{5.39a}$$

Rearranging,

$$K_{a1}K_H P_{CO_2} = [H^+]^2 \tag{5.39b}$$

Substituting the numerical values for the equilibrium constants and the partial pressure of CO_2:

$$(10^{-6.4})(10^{-1.47})(10^{-3.52}) = [H^+]^2 \tag{5.39c}$$

Solving yields $[H^+] = 2.02 \times 10^{-6}$, or pH = 5.695, which is identical to the pH computed by solving 5.38e. In hindsight, it is easy to see the reason for the unique solution to the two methods for computing pH. The pK_{a2} value that describes the dissociation of HCO_3^- to form CO_3^{2-} and to produce an additional proton is 10.3. Thus, HCO_3^- doesn't contribute a proton until the solution is highly alkaline (Figure 5.7). Similarly, the hydroxide concentration will be insignificant relative to the proton concentration in the slightly acidic solution; when pH = 6, pOH = 8 ($[OH^-]$ is two orders of magnitude less than $[H^+]$). Therefore, both $[CO_3^{2-}]$ and $[OH^-]$ terms in Equation 5.35 are insignificant relative to the $[HCO_3^-]$ term.

5.6 COMPLEX IONS AND ION PAIRS

Short-range (<500 pm) interactions between ions or molecules that occur in soil solutions lead to the formation of complex ions and ion pairs. A general ion pair formation reaction is:

$$aM^{m+}(aq) + bL^{n-}(aq) \rightarrow M_aL_b^q(aq) \qquad (5.40)$$

where M^{m+} is the free metal ion with charge $m+$, L^{n-} is the free ligand ion (or molecule) with charge $n-$, and $M_aL_b^q$ ($q = am - bn$) is the soluble complex or ion pair. A ligand ion or molecule is a species that can reside in the coordination sphere of a central metal cation. A ligand may be a charged species (anion) or a neutral molecule. For instance, an undissociated (weak) acid, such as HF^0, is an example of an ion pair, where H^+ is the metal cation and F^- is the ligand. Hydrated cations are also examples of soluble complexes. For example, the metal Al^{3+} in solution is surrounded by six waters of hydration; each H_2O molecule is a ligand because it occupies a position in the coordination sphere of Al^{3+}.

There are three basic types of soluble complexes. A solvation complex is the metal ion plus its primary hydration sphere $[Al(H_2O)_6^{3+}]$ (Figure 5.2); H_2O is the ligand. A complex ion is a soluble species in which the metal ion and ligand (other than H_2O) are directly linked; there are no water molecules between the metal and ligand, and the ligand has replaced a water of hydration (Figure 5.9). This type of complex is also termed an inner-sphere complex. An ion pair is a soluble species in which the metal ion and ligand are attached outside the hydration sphere; the ligand does not

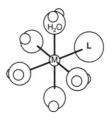

Inner-Sphere Complex

FIGURE 5.9 Diagrammatic illustrations of inner-sphere and outer-sphere complexes involving a metal cation (M) and a ligand (L). An inner-sphere complex is formed when the ligand displaces a water of hydration from the coordination sphere (primary hydration sphere) of the metal ion. An outer-sphere complex is formed when the ligand does not displace a primary water of hydration, but still results in a chemical unit whose stability depends on the intensity of electrostatic attraction.

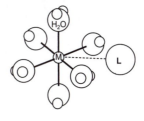

Outer-Sphere Complex

displace water from the primary hydration sphere. This type of complex is termed an outer-sphere complex.

It is very important to recognize that when ions and molecules react to form soluble complexes, they lose their separate identities. For example, Ca^{2+} and MoO_4^{2-} will react in solution to form the ion pair $CaMoO_4^0$. All three species will exist in solution, each displaying unique chemical and environmental behavior. A typical soil solution may contain significant concentrations of H, Li, NH_4, Na, Mg, Al, K, Ca, Mn, Fe, CO_3, NO_3, OH, F, SiO_4 PO_4, SO_4, and Cl. These substances may be present as the free metal or cationic species: H^+, Li^+, NH_4^+, Na^+, Mg^{2+}, Al^{3+}, K^+, Ca^{2+}, Mn^{2+}, Fe^{2+}, and Fe^{3+}; and the free ligand species (excluding organic ligands and H_2O): CO_3^{2-}, NO_3^-, OH^-, F^-, SiO_4^{4-}, PO_4^{3-}, SO_4^{2-}, and Cl^-. Assuming that each cation forms only one complex with each ligand, a typical soil solution would contain over 100 soluble complexes and free species. Further, both inner-sphere and outer-sphere complexes can exist for a given metal-ligand pairing, with one type dominating.

5.6.1 Stability of Soluble Complexes

The formation of soluble complexes, as well as their dissociation, is generally considered to be an instantaneous process in natural waters. The rate for the displacement of a water of hydration from the coordination sphere of a metal by ligands, such as SO_4^{2-}, Cl^-, F^-, and acetate (CH_3COO^-), are in good agreement with the water exchange rate (Figures 5.3 and 5.10). With the exception of the formation of $CrCl^{2+}$ and CrF^{2+} ($t_{1/2} > 10^6$ sec, formation half-life greater than approximately 12 d),

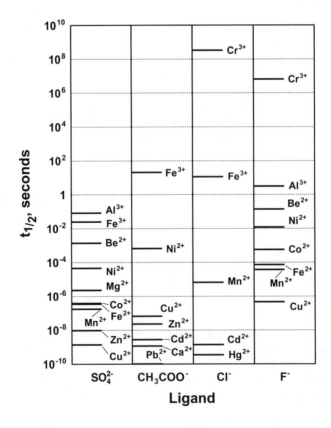

FIGURE 5.10 The rate of displacement (represented as the half-life, $t_{1/2}$ in seconds) of a water of hydration from the coordination sphere of metal ions by SO_4^{2-}, acetate (CH_3COO^-), Cl^-, and F^- (Margerum et al., 1978; Pankow and Morgan, 1981; Plankey et al., 1986).

CH_3COOFe^{2+} ($t_{1/2} \approx 20$ sec), $FeCl^{2+}$ ($t_{1/2} \approx 11$ sec), and AlF^{2+} ($t_{1/2} \approx 3$ sec), half-lives for the formation are essentially instantaneous ($t_{1/2} < 0.1$ sec). Because of the instantaneous nature of ion association reactions, ion speciation in soil solutions can be described using equilibrium concepts.

Ion speciation and the stability of the soluble species are determined using the Law of Mass Action. For the reaction in Equation 5.40, the distribution of free and complex species in a soil solution at equilibrium is given by:

$$K_f = \frac{(M_a L_b^q)}{(M^{m+})^a (L^{n-})^b} \tag{5.41}$$

where ( ) denote activities and K_f is an equilibrium constant, also termed the ion association constant or ion formation constant. In general, the magnitude of K_f is a function of the radii and valences of the component ions and the type of complex formed (inner- vs. outer-sphere).

Equation 5.1 states that the force of electrostatic attraction between two ions of opposite charge is directly related to the magnitude of the charge of the ions and inversely related to the distance of closest approach. Clearly, the force of attraction, and thus the K_f value, will be greater for inner-sphere complexes relative to outer-sphere complexes (Figure 5.9). Greater stability is also observed with increasing valence and decreasing size of the component ions. Further, the stability of complexes is a function of the degree of covalent character in the bonds formed: the greater the covalent character, the greater the stability (and the higher the K_f value). Complex stability is also a function of the polarizability and electronegativity of the component ions. Ion pairs, where a water molecule separates metal and ligand, are held together by electrostatic attraction. In complex ions, where the metal and ligand are directly attached, there is the opportunity for covalent bonding to occur. Finally, the number of metals and ligands in a complex will impact complex stability. The more species that interact to form a soluble complex, the less stable the complex.

The number of bonds a central metal cation forms with a single complexing ligand also influences the stability of an aqueous complex. Monodentate ligands, such as Cl^-, acetate (CH_3COO^-), and lactate ($CH_3CHOHCOO^-$), form a single bond with a central metal cation (Figure 5.11). Oxalate ($^-OOCCOO^-$), CO_3^{2-}, and SO_4^{2-} may complex a central metal cation at two locations, and are bidentate ligands (CO_3^{2-} and SO_4^{2-} also act as mondentate ligands). Organic ligands that form two or more bonds with a central metal cation are collectively termed chelates, or polydentate (or multidentate) ligands. The naturally occurring citric acid and the synthetic compounds EDTA and DTPA are examples of polydentate ligands that are chelates. The chelation of a metal by a ligand greatly enhances the stability of an aqueous complex, compared to the stability of the monodentate complex. This increase in stability is known as the chelate effect, and results from the formation of chelate ring structures that include the metal ion. For example, consider the following equilibrium constants for the complexation of Cu^{2+} by acetate (monodentate), oxalate and carbonate (bidentate), and EDTA (polydentate) (25°C and 0.101 MPa):

$$Cu^{2+} + CH_3COO^- \rightarrow [CH_3COO - Cu]^+ \qquad \log K_f = 2.22 \tag{5.42}$$

$$Cu^{2+} + (COO^-)_2 \rightarrow [(COO^-)_2 - Cu]^0 \qquad \log K_f = 6.23 \tag{5.43}$$

$$Cu^{2+} + CO_3^{2-} \rightarrow CuCO_3^0 \qquad \log K_f = 6.75 \tag{5.44}$$

$$Cu^{2+} + EDTA^{4-} \rightarrow [EDTA - Cu]^{2-} \qquad \log K_f = 20.9 \tag{5.45}$$

Both the oxalate and carbonate complexes, which contain a single-ring structure, are 10^4 times as stable as the acetate complex. The EDTA complex, which is a five-ring structure (Figure 5.12), is more than 10^{18} times as stable as the acetate complex, and nearly 10^{14} times as stable as the oxalate and carbonate complexes.

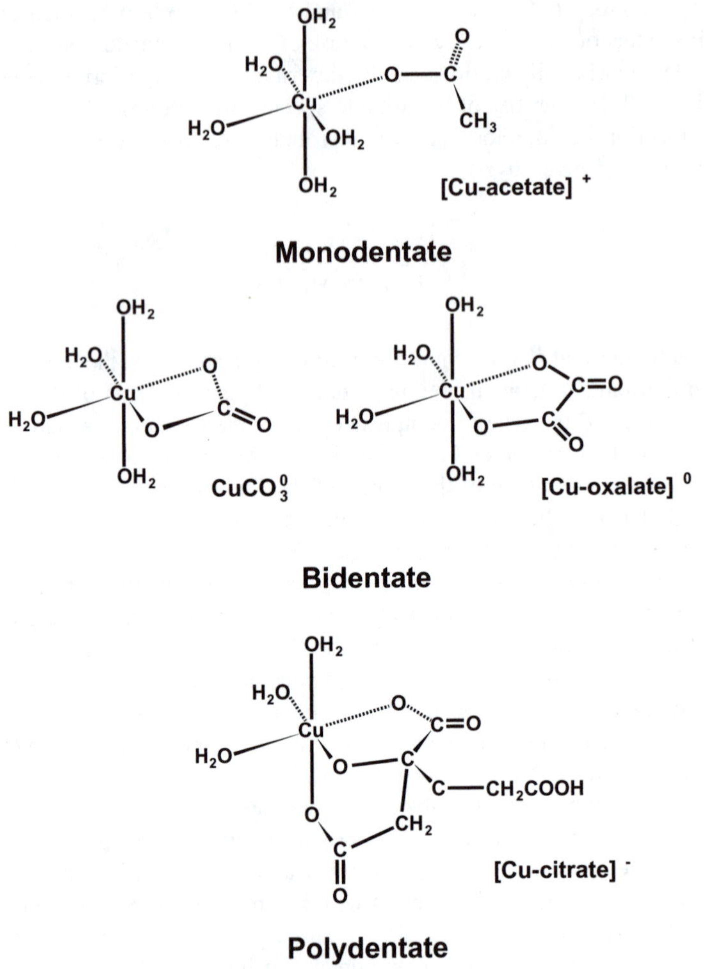

Monodentate

Bidentate

Polydentate

FIGURE 5.11 Representations of the complexation of Cu^{2+} by acetate (monodentate), carbonate and oxalate (bidentate), and citrate (tri-, poly- or multidentate). Single-ring structures involving Cu^{2+} are formed in the carbonate and oxalate complexes, and a double-ring structure is formed in the citrate complex.

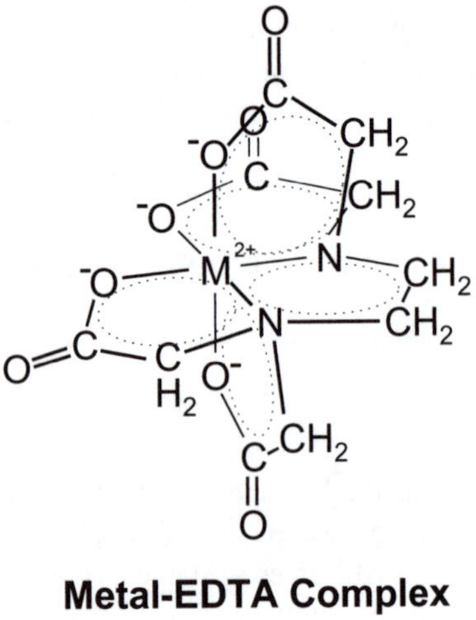

FIGURE 5.12 Representation of the complexation of a divalent metal ion by the synthetic chelate, ethyl-enediaminetetraacetate (EDTA). All six positions of the metal ion coordination sphere are occupied by functional moieties of the EDTA molecule, resulting in five ring structures that involve the metal cation. Four metal coordinate positions are occupied by carboxylic oxygen atoms, and two positions are occupied by amine nitrogen atoms.

Metal-EDTA Complex

5.6.2 THE IRVING AND WILLIAMS SERIES

Another trend in the complexation chemistry of metal ions is described in the Irving and Williams series. Irving and Williams (1953) compiled and examined the equilibrium constants for the formation of complexes involving divalent metal ions of the first row transition elements and organic ligands that contain nitrogen. Irrespective of the nature of coordinating ligand (monodentate or polydentate) or the number of ligand molecules involved in the complex, the stability of the metal complexes ($\log K_f$ values) for a single ligand nearly always increased with atomic number, with the largest values found for Cu^{2+}: $Mn^{2+} < Fe^{2+} < Co^{2+} < Ni^{2+} < Cu^{2+} > Zn^{2+}$. For example, the stability sequence ($\log K_f$ for $M^{2+} + L^{n-} \rightarrow ML^{2-n}$) of metal complexation for the following organic nitrogen-bearing ligands clearly illustrates the series ($\log K_f$ values are for 25°C and 0.101 MPa):

Ethylenediamine ($I = 0.5$ M): 2.77 (Mn^{2+}) < 4.34 (Fe^{2+}) < 5.96 (Co^{2+}) < 7.56 (Ni^{2+}) < 10.74 (Cu^{2+}) > 5.86 (Zn^{2+})

Cyclohexane-1, 2-diamine ($I = 0.1$ M): 2.94 (Mn^{2+}) < 6.37 (Co^{2+}) < 7.99 (Ni^{2+}) < 11.13 (Cu^{2+}) > 6.37 (Zn^{2+})

1,2,3,-Triaminopropane ($I = 0.1$ M): 6.80 (Co^{2+}) < 9.30 (Ni^{2+}) < 11.1 (Cu^{2+}) > 6.75 (Zn^{2+})

1,4,7-Triazaheptane ($I = 1$ M): 3.99 (Mn^{2+}) < 6.23 (Fe^{2+}) < 8.57 (Co^{2+}) < 10.96 (Ni^{2+}) < 16.34 (Cu^{2+}) > 9.22 (Zn^{2+})

1,4,7,10-Tetraazadecane ($I = 0.1$ M): 4.93 (Mn^{2+}) < 7.84 (Fe^{2+}) < 11.09 (Co^{2+}) < 14.0 (Ni^{2+}) < 20.4 (Cu^{2+}) > 12.05 (Zn^{2+})

1,4,7,10,13-Pentaazatridecane ($I = 0.1$ M): 6.60 (Mn^{2+}) < 9.96 (Fe^{2+}) < 13.5 (Co^{2+}) < 17.6 (Ni^{2+}) < 23.1 (Cu^{2+}) > 15.3 (Zn^{2+})

Pyridine ($I = 0.5$ M): 0.14 (Mn^{2+}) < 0.6 (Fe^{2+}) < 1.19 (Co^{2+}) < 1.87 (Ni^{2+}) < 2.56 (Cu^{2+}) > 0.99 (Zn^{2+})

The Irving-Williams series also holds for a number of organic acids:

Lactate: 1.43 (Mn^{2+}) < 1.90 (Co^{2+}) < 2.22 (Ni^{2+}) < 3.02 (Cu^{2+}) > 2.22 (Zn^{2+})

Oxalate: 3.95 (Mn^{2+}) < 4.72 (Co^{2+}) < 5.16 (Ni^{2+}) < 6.23 (Cu^{2+}) > 4.87 (Zn^{2+})

Malonate: 3.28 (Mn^{2+}) < 3.74 (Co^{2+}) < 4.05 (Ni^{2+}) < 5.70 (Cu^{2+}) > 3.84 (Zn^{2+})

Succinate: 2.26 (Mn^{2+}) < 2.32 (Co^{2+}) < 2.34 (Ni^{2+}) < 3.28 (Cu^{2+}) > 2.52 (Zn^{2+})

Phthalate: 2.74 (Mn^{2+}) < 2.83 (Co^{2+}) < 2.95 (Ni^{2+}) < 4.04 (Cu^{2+}) > 2.91 (Zn^{2+})

Citrate ($I = 0.1$ M): 3.70 (Mn^{2+}) < 4.4 (Fe^{2+}) < 5.00 (Co^{2+}) < 5.40 (Ni^{2+}) < 5.90 (Cu^{2+}) > 4.98 (Zn^{2+})

EDTA ($I = 0.1$ M): 13.87 (Mn^{2+}) < 14.32 (Fe^{2+}) < 16.31 (Co^{2+}) < 18.62 (Ni^{2+}) < 18.80 (Cu^{2+}) > 16.50 (Zn^{2+})

DTPA ($I = 0.1$ M): 15.60 (Mn^{2+}) < 16.5 (Fe^{2+}) < 19.27(Co^{2+}) < 20.32 (Ni^{2+}) < 21.55 (Cu^{2+}) > 18.40 (Zn^{2+})

However, for inorganic complexes the order is not as apparent. Indeed, with the exception of NH_3^0 the Irving and Williams series order is not followed:

NH_3^0: 0.95 (Mn^{2+}) < 1.34 (Fe^{2+}) < 2.03 (Co^{2+}) < 2.76 (Ni^{2+}) < 4.11 (Cu^{2+}) > 2.28 (Zn^{2+})

OH^-: 3.41 (Mn^{2+}) < 4.50 (Fe^{2+}) > 4.35 (Co^{2+}) > 4.14 (Ni^{2+}) < 6.3 (Cu^{2+}) > 5.04 (Zn^{2+})

HPO_4^{2-}: 3.66 (Mn^{2+}) > 3.60 (Fe^{2+}) > 3.03 (Co^{2+}) > 2.98 (Ni^{2+}) < 4.14 (Cu^{2+}) > 3.29 (Zn^{2+})

HCO_3^-: 1.27 (Mn^{2+}) > 1.10 (Fe^{2+}) < 2.40 (Co^{2+}) < 2.45 (Ni^{2+}) > 2.09 (Cu^{2+}) > 1.50 (Zn^{2+})

CO_3^{2-}: 4.49 (Mn^{2+}) < 5.00 (Fe^{2+}) > 4.51 (Co^{2+}) < 4.92 (Ni^{2+}) < 6.83 (Cu^{2+}) > 4.71 (Zn^{2+})

SO_4^{2-}: 2.40 (Mn^{2+}) > 2.26 (Fe^{2+}) < 2.36 (Co^{2+}) > 2.26 (Ni^{2+}) < 2.39 (Cu^{2+}) < 2.41 (Zn^{2+})

5.6.3 METAL–FULVATE INTERACTIONS

The aqueous complexation characteristics of metal cations with inorganic ligands and synthesized organic ligands (such as chelates) are generally well established. In addition, metal complexation by ligands that belong to classes of biochemistry (low-molecular-mass organic acids, sugar acids, amino acids, amines, phenols, and phenolic acids) is also well documented, even though the chemical nature of a soil solution is exceedingly complex and rarely ever characterized to the level necessary to utilize this information. Aquatic environments also contain a chemically diverse group of substances formed through secondary synthesis reactions, collectively termed humic and fulvic acids (Chapter 4). Relative to humic acid, fulvic acid is soluble and the predominant secondary synthesis compound in soil solutions and other natural waters, largely due to its lower molecular mass and greater total acidity. Indeed, fulvic acids are ubiquitous in natural waters. In addition, fulvic acids are important complexing agents of metal cations, directly impacting metal bioavailability and transport.

It has been recognized (Chapter 4) that the structures developed for fulvic (and humic) acids are merely conceptual constructs that may bear little or no resemblance to reality. However, this fact does not diminish the importance of these natural compounds relative to their impact on metal fate and behavior, nor does it translate into an inability to characterize metal-fulvate interactions. Indeed, there exists a large body of knowledge that seeks to elucidate the mechanisms of metal-fulvate interactions, the specific fulvate functional groups involved, and the metal-fulvate stability, or binding constants. Polyfunctionality is an important characteristic of humic materials. The predominant functional groups associated with fulvates are the carboxylic and the phenolic-O moieties. It should not be surprising then that the moieties on the fulvate molecular framework that are responsible for metal complexation are the O-bearing functional groups, principally carboxylic and phenolic-O. Nitrogen-bearing moieties, such as those associated with the amino acid components of the fulvate framework, may also play an important role in metal complexation, albeit minor, as the N content of fulvic acids only ranges from 1 to 3% of the total oxygen content. Reduced sulfur-bearing functional groups are theorized to potentially participating in metal complexation processes; however, evidence of metal interactions with fulvate-S has not been forthcoming.

Fulvates form both inner-sphere and outer-sphere complexes with metal cations. The interaction of metals with strong, high energy binding sites, referred to as type 1 sites, results in inner-sphere complexation. The high energy sites represent only a small fraction of the total number of binding sites on the fulvate molecule. Further, high-energy metal binding appears to be the preferred binding mechanism when fulvic acid concentrations in solution far exceed that of the metal ion (metal/fulvate mole ratio < 0.001), a condition that is likely to be present in uncontaminated soil or water systems. As the metal/fulvate mole ratio is narrowed (>0.004), either through increased metal concentrations or decreased fulvate concentrations, the strong binding sites become saturated, and weak outer-sphere metal binding mechanisms dominate. The low energy sites, referred to as type 2 sites, represent the bulk of the metal binding sites on the fulvate molecule.

The outer-sphere complexation of metal cations by fulvates involves weak bonding by water and O-bearing ligands (type 2 sites), including ketonic, phenolic, and predominantly carboxylic functional groups (Figure 5.13). In general, the outer-sphere complexes are monodentate, forming on the exterior of the fulvate molecule. Numerous complexation models have been employed to mechanistically characterize inner-sphere metal complexation by fulvates (type 1 sites). Metal-fulvate bonding mechanisms have been derived from numerous spectrometric studies, such as NMR spectroscopy, molecular fluorescence spectrometry, and x-ray absorption near-edge spectroscopy. These methods monitor the molecular environment of electron donor atoms in specific functional moieties (such as phenolic or carboxylic oxygens) by examining spectral changes that occur upon introduction of a metal ion. The degree of change is then compared to those observed when model compounds (that contain the functional moieties) react with the metal ion. Inner-sphere complexes may be monodentate or polydentate, occupying a single position or up to five positions in the

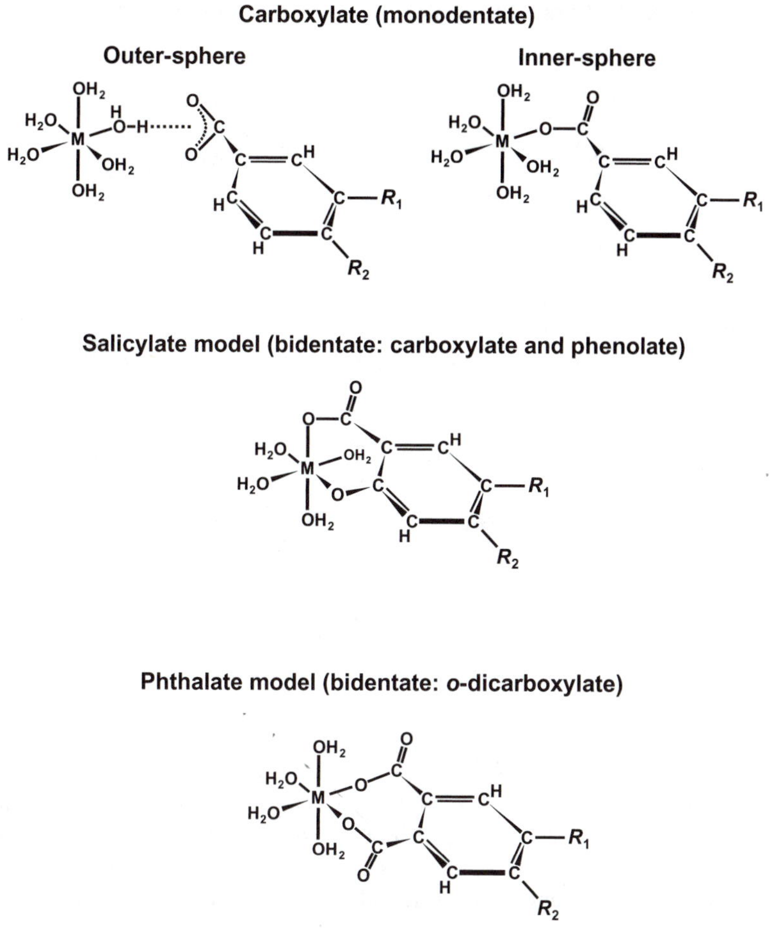

FIGURE 5.13 Molecular models used to characterize metal complexation by fulvic acid. The monodentate aromatic carboxylate models (inner- and outer-sphere) are typically employed to characterize type 2 (weak) metal binding by fulvates. The multidentate aromatic dicarboxylate (phthalate), aromatic hydroxyl and carboxylate (salicylate), aliphatic dicarboxylate (malonate), aliphatic hydroxyl and dicarboxylate (citrate), and the aliphatic ester and tetracarboxylate (oxy-succinate) models have all been employed (singly or in combination) to describe type 1 (very strong and strong) metal binding by fulvates (Otto et al., 2001b; Smith and Kramer, 2000; Sekaly et al., 1999; Leenheer et al., 1998; Li et al., 1998; Nantsis and Carper, 1998; Luster et al., 1996).

coordination sphere of a metal ion (although monodentate inner-sphere complexation is not wholly substantiated in the literature). Figure 5.13 illustrates examples of inner-sphere metal binding models that have been used to describe metal-fulvate complexation. An individual model may uniquely describe metal complexation, or the models may be combined to describe complexation. For example, O atoms from a 1,2-dihydroxyphenol functional group on the fulvate structure (salicylate model) may form a bidentate, coordinated structure with a metal cation. Alternatively, two O atoms from 1,2-dihydroxyphenol and two O atoms from o-dicarboxylate (phthalate model) may form a polydentate complex with a metal cation. In this coordinated structure, four positions in the metal ion coordination sphere are occupied by fulvate O atoms. Metal complexation by fulvates may also involve aliphatic carboxylic and hydroxyl O atoms, as illustrated by the malonate and citrate models (Figure 5.13). The oxy-succinate model, also shown in Figure 5.13, illustrates the high-energy (type 1), coordinated bonding of a metal ion by five oxygen moieties: four O atoms arise from two succinate functionaries, plus O from aliphatic ether.

Malonate model (bidentate)   **Citrate model (tridentate)**

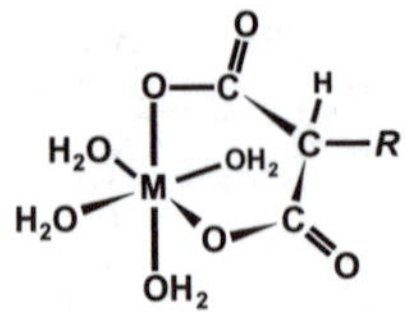

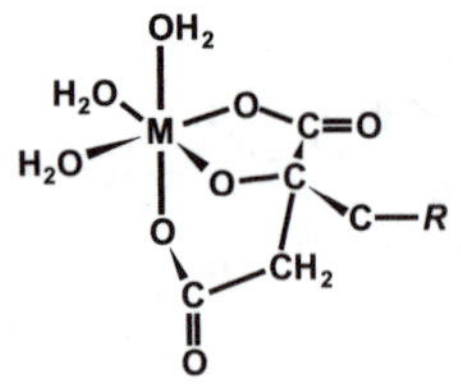

Oxy-succinate model (polydentate)

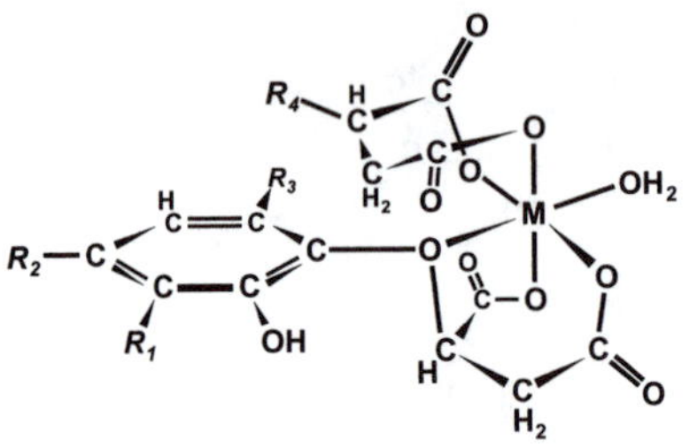

FIGURE 5.13 *(continued.)*

Metal binding by soluble humic substances is (semi-) quantitatively described by a binding constant, often termed a conditional association or stability constant (see Equation 5.48). Metal-fulvate binding constants are not true thermodynamic constants, as defined in Equation 5.41. Instead, it is recognized that the association of a metal ion and a fulvate ligand is a function of solution chemistry. Thus, a binding constant determined under a unique set of pH, ionic strength, and metal and fulvate concentrations is unique and cannot be translated to solutions whose chemical properties differ. A principal reason binding constants are not transferable is that the structure and functionality of a fulvate ligand are not static. The solution pH, composition, and ionic strength influence the intermolecular and intramolecular interactions of the metal binding functional groups of the fulvate ligand (i.e., the molecule is labile). The metal-to-fulvate ligand concentration ratio in a solution also influences the intensity of the complexation process, and thus the measured binding constant. Further, the amounts and types of different functional groups associated with fulvic acids vary significantly between substances of different origin, potentially limiting the utility of an experimentally determined binding constant.

Despite the shortcomings listed above, conditional metal-fulvate association constants for metal ions and fulvates from various sources have been determined (Table 5.10). The type of metal complexation process—very strong inner-sphere and polydentate (type 1a), strong inner-sphere and polydentate (type 1b), and weak outer-sphere and monodentate (type 2)—is evidenced by the magnitude of the conditional association constant. Using the data for Cu^{2+} as an example, very strong Cu^{2+}-fulvate complexes are described by $\log^c K$ values that range from 7.05 to 8.7 for pH 6 and $I = 0.1$ M solutions (25°C and 0.101 MPa). The $\log^c K$ values for strong Cu^{2+}-fulvate complexes range from 4.65 to 5.78. The $\log^c K$ values for weak Cu^{2+}–fulvate complexes are reported to range from 4.08 to 4.94, with an anomalous value of 0.0 reported from one source. Despite the fact that fulvic acids from numerous sources are employed to generate the data in Table 5.10 (soils, rivers, lakes, and leaf litter), there exists a surprising level of agreement among the $\log^c K$ values within the various binding intensity (type) categories. Indeed, the type 2 binding constant for the formation Cd^{2+}–fulvate ($\log^c K = 3.3$) is an average value computed from the $\log^c K$ values of 3.4, 3.3, 3.1,

TABLE 5.10
Conditional Association Constants (log^{c}K) for the Formation of Metal-Fulvate Complexes

Ion	log^{c}K		
	Type 1a	Type 1b	Type 2
Cu^{2+}		5.46[a]	
		4.65[b]	
	7.47[c]	5.82[c]	4.79 (types 1b and 2)[c]
		5.24[d]	
	8.6[e]		0.0[e]
	7.05[f]		4.94[f]
	7.34[g]	5.78[g]	4.08[g]
	8.7[h]		4.2[h]
	6.19 (pH 6.5)[i]		
Pb^{2+}		4.59[a]	
	6.6[e]		0.6[e]
	8.38[h]		3.79[h]
Cd^{2+}		4.17[a]	
			3.3 (pH 6.4)[j]
		4.42[d]	
	6.29[h]		2.3[h]
Ca^{2+}			2.8 (pH 6.4)[k]
		3.02[d]	
Ni^{2+}		3.80[d]	
Zn^{2+}		3.77[d]	
Al^{3+}	8.05[c]	5.79[c]	

Unless noted otherwise, log^{c}K values pertain to pH 6 and $I = 0.1$ M conditions at 25°C. Type 1a and 1b complexes are representative of very strong and strong inner-sphere metal binding sites. Type 2 sites are representative of weak outer-sphere metal binding.

[a] Esteves da Silva et al. (2002): complexation by a soil fulvate determined using potentiometric and acid-base conductimetric titrations. A 1:1 metal-to-fulvate mole ratio in the complex is indicated from the experimental data.

[b] Esteves da Silva et al. (2002): complexation by a soil fulvate determined using molecular fluorescence spectroscopy. log^{c}K values of 3.57, 4.07, and 4.52 were also obtained for pH 3, pH 4, and pH 5 conditions and type 1 metal binding. A 1:1 metal-to-fulvate mole ratio in the complex is indicated from the experimental data.

[c] Luster et al. (1996): complexation by fulvates extracted from a juniper leaf litter. Binding constants were determined using an ion exchange method and molecular fluorescence spectroscopy and an assumed 1:1 metal-to-fulvate mole ratio in the complex. Very strong complexes (type 1a), strong complexes (type 1b), and weak complexes (type 2) are resolved by varying the metal to fulvate ligand concentration ratio. Binding constants pertain to $I = 0.01$ M. Type 1a complexes are inner-sphere and polydentate, involving four O ligands or two O ligands and two N ligands. The O ligands are from 1,2-dihydroxyphenols and/or amino acids and the N ligands are from amino acids. Type 1b complexes are also inner-sphere and polydentate, involving four O ligands from carboxylic and/or phenolic moieties in an arrangement that resembles that in salicylic acid or aliphatic α-dicarboxylic acids. The type 2 complex is outer-sphere and the metal is weakly associated with carboxylic moieties and water molecules.

[d] Brown et al. (1999a): complexation by Suwannee River fulvic acid is characterized using an ion exchange method. A 1:1 metal-to-fulvate stoichiometry in the complex is inferred from the experimental data. For Cd^{2+} complexation, the ancillary data and inferences of Leenheer et al. (1998) were employed to suggest a polydentate binding mechanism that contains a mixture of inner- and outer-sphere character. The oxy-succinate model diagrammed in Figure 5.13 is proposed.

[e] Christl et al. (2001): complexation by a soil fulvate and 1:1 metal-to-fulvate mole stoichiometry determined using potentiometric titrations. Association constants for strong (type 1) and weak (type 2) complexes determined using data from a global analysis of titration data from pH 4, 6, and 8 systems.

(continued)

TABLE 5.10
Conditional Association Constants (log^{c}K) for the Formation of Metal-Fulvate Complexes (Continued)

[f] Brown et al. (1999b): complexation by Suwannee River fulvic acid determined using an ion exchange method. Assumes 1:1 metal-to-fulvate mole ratio in the complex and two metal binding sites per fulvate molecule: strong (type 1) and weak (type 2).

[g] See footnote for Brown et al. (1999b). Assumes 1:1 metal-to-fulvate mole ratio in the complex and three metal binding sites per fulvate molecule: very strong (type 1a), strong (type 1b), and weak (type 2).

[h] Berbel et al. (2001): complexation by a fulvic acid (source undefined) and 1:1 metal-to-fulvate mole stoichiometry determined using potentiometric and voltammetric titrations. Two metal binding sites assumed: type 1 and type 2. Data indicate that there are approximately twice as many type 2 sites as there are type 1.

[i] Carballeira et al. (2000): complexation by a soil fulvic acid and 1:1 metal-to-fulvate determined using voltammetric titrations. Data analysis assumes monodentate or monodentate plus bidentate complexation. Both models yield identical association constants. log^{c}K values as a function of I for the Cu^{2+}-fulvate complex are: 6.59 ($I = 0.001$ M), 6.48 ($I = 0.005$ M), 6.33 ($I = 0.01$ M), and 6.12 ($I = 0.05$ M). An intrinsic constant (independent of I) is computed: log$K^{int} = 4.62$. Constants represent type 1 complexation.

[j] Otto et al. (2001a and b): complexation by four fulvates: Suwannee River, Wakarusa River, Clinton Lake, and Laurentian soil fulvic acids, determined using ^{113}Cd NMR. The association constant is an average of the binding constants for the four samples: log^{c}K = 3.4 (Suwannee River), 3.3 (Wakarusa River), 3.1 (Clinton Lake), and 3.5 (Laurentian soil). A 1:1 metal-to-fulvate mole ratio in the complex is evidenced by the experimental data, as is metal complexation by a bidentate aliphatic dicarboxylate or hydroxycarboxylate. The binding constants represent the average of type 1 and type 2 sites.

[k] Otto et al. (2001a): complexation by four fulvates: Suwannee River, Wakarusa River, Clinton Lake, and Laurentian soil fulvic acids, determined using ^{113}Cd NMR and competitive interactions with Cd-fulvate. The association constant is an average of the binding constants for the four samples: log^{c}K = 2.8 (Suwannee River), 2.7 (Wakarusa River), 2.7 (Clinton Lake), and 2.9 (Laurentian soil). A 1:1 metal-to-fulvate mole ratio in the complex is evidenced by the experimental data, as is metal complexation by a monodentate aliphatic carboxylate. The binding constants represent the average of type 1 and type 2 sites.

and 3.5 for Suwannee River, Wakarusa River, Clinton Lake, and Laurentian soil–derived fulvates, respectively.

Conditional metal-fulvate association constants vary as a function of solution pH and ionic strength. For example, the binding constants (as log^{c}K values) for Cu^{2+}–fulvate in 0.1 M ionic strength solutions are 3.57, 4.07, 4.52, and 4.65 for pH 3, 4, 5, and 6 (see Table 5.10 footnotes). Metal competition from protons for fulvate binding sites increases with decreasing pH; thus, conditional binding constants decrease as proton activity increases. Solution ionic strength influences metal ion activity and the electrical potential and intermolecular configuration of the fulvate molecule. In a pH 6.0, 0.1 M ionic strength solution, the binding constants for Cu^{2+}–fulvate formation (as log^{c}K values) are 7.34, 5.78, and 4.08 for type 1a, 1b, and 2 complexes (Table 5.10). In a pH 6.0, 0.001 M ionic strength solution, the type 1a, 1b, and 2 binding constants are 8.40, 6.42, and 4.74. Thus, as ionic strength decreases, conditional metal-fulvate binding constants increase.

5.7 THE ION ASSOCIATION MODEL

In soil solutions, direct measurement of selected free metal and ligand concentrations is possible using ion selective electrodes. Ion selective electrodes are available and may be employed to determine aqueous concentrations of (for example): H$^+$, Na$^+$, K$^+$, NH$_4^+$, NO$_3^-$, F$^-$, Cl$^-$, Ca^{2+}, water hardness (Ca^{2+} + Mg^{2+}), Cu^{2+}, Cd^{2+}, and Pb^{2+}. However, for the vast majority of free species and soluble complexes, analytical methods are not available to directly measure their concentrations (only total concentrations). Thus, the distribution of an element between free and complex species must be estimated (predicted). The computation of aqueous speciation is not a minor exercise, as soil solutions contain a great many inorganic and organic, metals and ligands at significant concentrations, leading to a multitude of soluble complexes (as previously indicated). However, very good estimations of solution chemistry can be obtained given the following information:

- The concentrations of all elements in a soil solution that can influence the speciation of a target element, including organic ligands
- K_f values for all possible soluble complex formation reactions

Ion association constants are available for a large number of metal-ligand complexes. These data are compiled as $\log K_f$ values for specified complexation reactions, or as standard state chemical potential ($\mu°$) or as Gibbs free energy values ($\Delta G_f^°$), from which $\log K_f$ values can be derived (Equation 5.22). Numerous compilations of these thermodynamic parameters exist. Some of the more notable sources are: Baes and Mesmer (1986), Martell and Smith (1974–1982), Naumov et al. (1974), Robie et al. (1978), Sadiq and Lindsay (1979), Wagman et al. (1982), and Woods and Garrels (1987). Selected $\log K_f$ values for the formation of soluble divalent cation complexes are shown in Table 5.11. With this information, one can then employ a chemical model known as the ion association model to quantitatively determine ion speciation in soil solutions.

TABLE 5.11
Ion Association Constants ($\log K_f$ Values) for the Formation of Aqueous Complexes of Divalent Metal Cations at 25°C According to the Generalized Reaction:
$$M^{2+} + L^{n-} \rightarrow ML^{2-na}$$

Ligand	Metal						
	Ca^{2+}	Mg^{2+}	Ni^{2+}	Cu^{2+}	Zn^{2+}	Cd^{2+}	Pb^{2+}
SO_4^{2-}	2.32	2.23	2.34	2.36	2.34	2.46	2.75
Cl^-	−1.00	−0.03	−0.4	0.40	0.46	1.98	1.59
CO_3^{2-}	3.15	2.92	4.92	6.75	4.71	4.09	7.04
HCO_3^-	1.1	1.01	2.45	2.09	1.50	2.09	3.23
HPO_4^{2-}	2.74	2.91	2.00	3.2	3.1	3.8	3.1
$H_2PO_4^-$	1.4	1.9	0.5	1.59	1.60	3.20	1.5
SeO_4^{2-}	2.67	2.24	2.67			2.24	
SeO_3^{2-}	3.17	2.87					
MoO_4^{2-}	2.57	3.03					
Acetate[b]	1.27	1.18	1.43	2.22	1.57	1.93	2.68
Lactate[b]	1.37	1.45	2.22	3.02	2.22	1.70	2.78
Oxalate[c]	3.00	3.43	5.16	6.23	4.87	3.89	4.91
Malonate[c]	2.35	2.85	4.05	5.70	3.84	3.22	
Succinate[c]	2.00	2.05	2.34	3.28	2.52	2.72	
Malate[c]	2.7	2.3	3.8	4.0	3.5	3.0	
Citrate[d]	4.8	4.7	6.6	7.2	6.1	5.0	5.7
EDTA[e]	(10.61)	(8.83)	(18.52)	(18.70)	(16.44)	(16.36)	(17.88)
DTPA[f]	(10.75)	(9.34)	(20.17)	(21.38)	(18.29)	(19.0)	(18.66)

[a] Data obtained from Smith and Martell (1976), Martell and Smith (1982), Lindsay (1979), Quinn (1985), and Essington (1992). The structures of the organic acids are illustrated in Figure 5.8. $\log K_f$ values in parentheses refer to the 0.1 M ionic strength system.

[b] $\log K_f$ values for the reaction: $M^{2+} + L^- \rightarrow ML^+$.

[c] $\log K_f$ values for the reaction: $M^{2+} + L^{2-} \rightarrow ML^0$.

[d] $\log K_f$ values for the reaction: $M^{2+} + L^{3-} \rightarrow ML^-$.

[e] $\log K_f$ values for the reaction: $M^{2+} + EDTA^{4-} \rightarrow MEDTA^{2-}$. EDTA is ethylenediaminetetraacetate.

[f] $\log K_f$ values for the reaction: $M^{2+} + DTPA^{5-} \rightarrow MDTPA^{3-}$. DTPA is diethylenetriaminepentaacetate.

The total concentration of a dissolved component in any solution reflects the sum of the free, uncomplexed species (solvation complex) concentration and the concentrations of the various soluble complexes that may form. This "mass balance" relationship can be expressed for any metal (M) and any ligand (L):

$$M_T = [M^{m+}] + \sum_i a_i [M_{a_i} L_{b_i}^{q_i}]$$

(5.46)

$$L_T = [L^{n-}] + \sum_i b_i [M_{a_i} L_{b_i}^{q_i}]$$

(5.47)

where a and b are stoichiometric coefficients, q is the charge associated with the soluble complex ($q = am - bn$), and brackets [ ] denote concentrations. The formation of a soluble complex can be quantitatively described as a function of free ion concentrations by the conditional stability constant:

$$^c K_f = \frac{[M_a L_b^q]}{[M^{m+}]^a [L^{n-}]^b}$$

(5.48)

for the ion association reaction described in Equation 5.40. A value for $^c K_f$ can be experimentally determined (since the concentrations of species are involved) or determined from the true thermodynamic equilibrium constant, which is a function of activities (recall Equation 5.41) and the ionic strength of the solution. A relationship between $^c K_f$ and K_f can be developed by first recalling the relationship between concentration and activity:

$$[M^{m+}] = \frac{(M^{m+})}{\gamma_{M^{m+}}}$$

(5.49a)

$$[L^{n-}] = \frac{(L^{n-})}{\gamma_{L^{n-}}}$$

(5.49b)

$$[M_a L_b^q] = \frac{(M_a L_b^q)}{\gamma_{M_a L_b^q}}$$

(5.49c)

Substituting Equations 5.49a through 5.49c into Equation 5.48 yields:

$$^c K_f = \frac{\gamma_{M^{m+}}^a \gamma_{L^{n-}}^b}{\gamma_{M_a L_b^q}} \times \frac{(M_a L_b^q)}{(M^{m+})^a (L^{n-})^b}$$

(5.50a)

or

$$^c K_f = \frac{\gamma_{M^{m+}}^a \gamma_{L^{n-}}^b}{\gamma_{M_a L_b^q}} \times K_f$$

(5.50b)

which relates K_f and $^c K_f$.

Returning to Equation 5.48 and rearranging yields:

$$[M_a L_b^q] = {}^c K_f [M^{m+}]^a [L^{n-}]^b$$

(5.51)

Substitution of this expression into the mass balance expressions (Equations 5.46 and 5.47) yields:

$$M_T = [M^{m+}] + \sum_i a_i {}^c K_{f,i} [M^{m+}]^{a_i} [L^{n-}]^{b_i} \tag{5.52}$$

$$L_T = [L^{n-}] + \sum_i b_i {}^c K_f [M^{m+}]^{a_i} [L^{n-}]^{b_i} \tag{5.53}$$

Consider the speciation of dissolved Pb in pH-neutral soil leachate. The pertinent complex-forming ligands are OH^-, Cl^-, SO_4^{2-}, CO_3^{2-}, and fulvic acid (Ful^-). The complexes formed are $PbOH^+$, $PbCl^+$, $PbSO_4^0$, $PbCO_3^0$, and $PbFul^+$. The list of Pb species considered here is probably incomplete for the true soil solution, as these are not the only Pb-bearing soluble complexes, nor is the particular complex formed for each ligand the only complex possible (for instant, $PbCl_2^0$ and $Pb(SO_4)_2^{2-}$ may also form). However, for the purpose of this example, these species will suffice.

Total soluble lead (Pb_T) concentration can be determined analytically. The measured value represents the sum of the free and complex species and is described by the mass balance equation:

$$Pb_T = [Pb^{2+}] + [PbOH^+] + [PbCl^+] + [PbSO_4^0] + [PbCO_3^0] + [PbFul^+] \tag{5.54}$$

where [ ] denote species concentrations in mol L^{-1}. The formation of each complex in Equation 5.54 can be described in terms of the free species concentrations and a conditional stability constant:

$$[PbOH^+] = {}^c K_1 [Pb^{2+}][OH^-] \tag{5.55a}$$

$$[PbCl^+] = {}^c K_2 [Pb^{2+}][Cl^-] \tag{5.55b}$$

$$[PbSO_4^0] = {}^c K_3 [Pb^{2+}][SO_4^{2-}] \tag{5.55c}$$

$$[PbCO_3^0] = {}^c K_4 [Pb^{2+}][CO_3^{2-}] \tag{5.55d}$$

$$[PbFul^+] = {}^c K_5 [Pb^{2+}][Ful^-] \tag{5.55e}$$

Substituting Equations 5.55a through 5.55e into Equation 5.54 and rearranging yields:

$$Pb_T = [Pb^{2+}]\{1 + {}^c K_1[OH^-] + {}^c K_2[Cl^-] + {}^c K_3[SO_4^{2-}] + {}^c K_4[CO_3^{2-}] + {}^c K_5[Ful^-]\} \tag{5.56}$$

For convenience, the ratio of $[Pb^{2+}]$ to Pb_T is termed the distribution coefficient for Pb^{2+}:

$$\alpha_{Pb^{2+}} = \frac{[Pb^{2+}]}{Pb_T} = (1 + {}^c K_1[OH^-] + {}^c K_2 [Cl^-] + {}^c K_3[SO_4^{2-}] + {}^c K_4[CO_3^{2-}] + {}^c K_5[Ful^-])^{-1} \tag{5.57}$$

The distribution coefficient can be calculated if the free concentrations of the ligands are known. In practice, however, α values cannot be computed directly, as one must also consider ligand speciation in detail.

The mechanics of obtaining an empirical solution to an aqueous speciation problem is illustrated for a very simple system. Consider a solution with an ionic strength of 0.01 M controlled by KNO_3 at 25°C and an applied pressure of 0.101 MPa. The solution contains 10^{-3} M Ca_T and 10^{-4} M MoO_{4T}.

Assume that K^+ and NO_3^- are conservative (do not participate in ion association reactions), and that only the formation of the $CaMoO_4^0$ complex occurs:

$$Ca^{2+}(aq) + MoO_4^{2-}(aq) \rightarrow CaMoO_4^0(aq) \tag{5.58a}$$

$$\log K_f = 2.57 \tag{5.58b}$$

From Equation 5.50, the conditional equilibrium constant is computed using the expression:

$$\log^c K_f = \log K_f + (a + b)mn\, F(I) - k_s I \tag{5.58c}$$

where $F(I) = -0.5116\{ I^{1/2}/[1 + I^{1/2}] - 0.3I\}$, $k_s = 0.1$, and a, b, m, and n are defined in Equation 5.40. Substituting,

$$\log^c K_f = 2.57 + 8(-0.045) - 0.001 = 2.21 \tag{5.58d}$$

The mole balance expressions (Equations 5.52 and 5.53) that describe the system are:

$$Ca_T = [Ca^{2+}] + 10^{2.21}[Ca^{2+}][MoO_4^{2-}] \tag{5.58c}$$

$$MoO_{4T} = [MoO_4^{2-}] + 10^{2.21}[Ca^{2+}][MoO_4^{2-}] \tag{5.58d}$$

and

$$Ca_T = [Ca^{2+}]\{1 + 10^{2.21}[MoO_4^{2-}]\} \tag{5.58e}$$

$$MoO_{4T} = [MoO_4^{2-}]\{1 + 10^{2.21}[Ca^{2+}]\} \tag{5.58f}$$

The fractional amount of the total concentration that is present as the free, uncomplexed species is:

$$\alpha_{Ca^{2+}} = \frac{[Ca^{2+}]}{Ca_T} = \{1 + 10^{2.21}[MoO_4^{2-}]\}^{-1} \tag{5.58g}$$

$$\alpha_{MoO_4^{2-}} = \frac{[MoO_4^{2-}]}{MoO_{4T}} = \{1 + 10^{2.21}[Ca^{2+}]\}^{-1} \tag{5.58h}$$

Equations 5.58g and 5.58h can be rewritten strictly in terms of the distribution coefficients and the total concentrations of Ca and MoO_4 in the solution:

$$\alpha_{Ca^{2+}} = \{1 + 10^{2.21}\alpha_{MoO_4^{2-}} MoO_{4T}\}^{-1} \tag{5.58i}$$

$$\alpha_{MoO_4^{2-}} = \{1 + 10^{2.21}\alpha_{Ca^{2+}} Ca_T\}^{-1} \tag{5.58j}$$

Equations 5.58i and 5.58j illustrate the coupled nature of the mass balance expression. Further, there are two unknowns (the two α values) in these two equations, which lead to an empirical solution. Algebraic manipulation of Equations 5.58i and 5.58j yields the quadratic expression:

$$0 = 10^{2.21} Ca_T \alpha_{Ca^{2+}}^2 + (1 + 10^{2.21} MoO_{4T} - 10^{2.21} Ca_T)\alpha_{Ca^{2+}} - 1 \tag{5.58k}$$

Solving, $\alpha_{Ca^{2+}} = 0.9862$. Using Equation 5.54j, $\alpha_{MoO_4^{2-}}$ is computed to be 0.8621. Thus, 98.62% of the total Ca occurs as Ca^{2+} (9.862×10^{-4} M) and 86.21% of the total MoO_4 occurs as MoO_4^{2-} (8.621×10^{-5} M). The concentration of the soluble complex is $[CaMoO_4^0] = Ca_T - [Ca^{2+}] = 0.001$ $M -$ 9.862×10^{-4} $M = 0.0000138$ M (which also equals $MoO_{4T} - [MoO_4^{2-}]$). Using the Davies equation (Equation 7), the single ion activity coefficient for a divalent ion, when $I = 0.01$ M, $\log\gamma_{2\pm} =$ -0.1246, or $\gamma_{2\pm} = 0.7505$ M^{-1}. The activities of the free, uncomplexed ions, Ca^{2+} and MoO_4^{2-}, are readily computed:

$$(Ca^{2+}) = \gamma_{2\pm}[Ca^{2+}] = (0.7505\ M^{-1})\ (9.862 \times 10^{-4}\ M) = 7.401 \times 10^{-4} \tag{5.58l}$$

$$(MoO_4^{2-}) = \gamma_{2\pm}[MoO_4^{2-}] = (0.7505\ M^{-1})\ (8.621 \times 10^{-5}\ M) = 6.470 \times 10^{-5} \tag{5.58m}$$

The speciation of Ca and MoO_4 in the above example illustrates the mechanics of determining aqueous speciation. However, unlike the above example, the inherent complexity of soil solutions does not generally allow for an empirical solution to the aqueous speciation problem. Complexation chemistry in soil solutions is not limited to the formation of 1-to-1 complexes, as in the above illustration of $CaMoO_4^0$ formation. Thus, the mass balance expressions, although coupled, will be nonlinear. Further, for true soil solutions, the number of coupled, nonlinear mass balance expressions can be quite large. Due to the sheer number of equations, combined with the nonlinearity of the expressions, the realization of an analytical (empirical) solution is not a practical objective. Instead, numerical iterative approximation techniques are employed to solve the mass balance expressions and generate an estimate of ion speciation.

Figure 5.14 outlines a generalized procedure used by computer codes to estimate ion speciation. The total concentrations of all metals and ligands in the solution, as well as solution pH, are the data required to initiate the speciation algorithm. Clearly, the more detailed the understanding of the chemical composition of a solution, the more accurate the model predictions will be. The number of essential metals and ligands to be included in the speciation computations will be determined by the concentrations of the components of interest. As a general rule, all metals and ligands that have a concentration equal to or greater than the components of interest should be included in the model. In addition, any metal or ligand that may have a significant influence on the aqueous chemistry of any component of interest must be included, irrespective of concentration. The operation of the ion speciation model proceeds as follows (the bullet numbers for the descriptions below correspond to the numbers that identify the steps in Figure 5.14):

1. The algorithm is initiated by defining the chemical nature of the soil solution. This involves identifying to the computer code the metals and ligands present in the soil solution, as well as their total concentrations: M_T, L_T, and pH. Also input by the user is an estimate of the ionic strength of the solution and estimates of the free concentrations of the metals and ligands. Depending on the model, an estimate of the ionic strength may also be determined by generating a stoichiometric ionic strength ($I_s = \frac{1}{2}\{\sum_i m_i^2 M_{Ti} + \sum_j n_j^2 L_{Tj}\}$) by assuming that the total concentrations of the components are equivalent to the free and uncomplexed species concentrations: $[M^{m+}] = M_T$ and $[L^{n-}] = L_T$. Note that I_s is an estimate of the true ionic strength of the solution. The I_s values are used as a starting point for the mathematical algorithm and allow for an initial estimation of activity coefficients using an appropriate expression, such as the Davis equation.

2. A thermodynamic database, which contains K_f values for a multitude of soluble complexes, is accessed. The K_f values are converted to ^cK_f values by employing the estimated ionic strength and computed activity coefficients (Equation 5.48). Routinely, the K_f values are retained in a database that is an integral component of the computer code. Thus, the validity of the computer model-predicted ion speciation will be a function of the totality of the database, as well as the accuracy of the ion association constants compiled in the database.

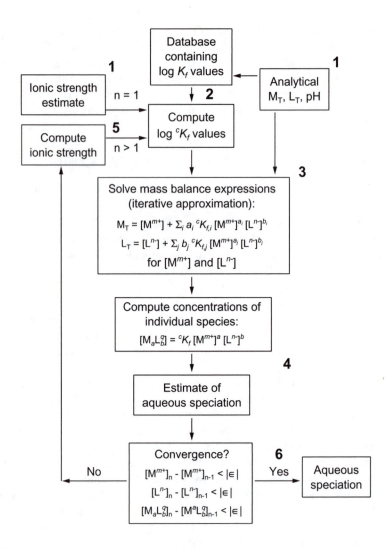

FIGURE 5.14 Schematic representation of the algorithm employed to compute ion speciation in aqueous solutions.

3. Based on the complex species retained in the thermodynamic database, the mass balance expressions are constructed as a function of free concentrations and ^cK_f values. The set of mass balance expressions are then solved for the free concentrations of all metals and ligands present in the solution. This is accomplished using an iterative approximation technique (e.g., the Newton-Raphson method), as an empirical solution is generally not feasible.

4. The resultant free species concentrations are then used to compute the concentrations of the soluble complexes, based on the ^cK_f expressions. What then results after completing the first cycle of the algorithm is an initial approximation of the aqueous speciation.

5. At this point, the algorithm is initiated a second time by computing an improved estimate of the ionic strength based on the estimated speciation (the estimated concentrations of the various species), which in turn is employed to compute improved estimates of activity coefficients. The improved estimate of the ionic strength is computed using the estimated concentrations of all the species in the soil solution. Thus, the model-computed ionic

strength is a prediction of the effective or true ionic strength (I_e). The thermodynamic data is again modified to produce a new set of ^cK_f values, which are employed to generate an improved estimate of ion speciation. With each successive loop through the algorithm, the estimation of aqueous speciation converges to a unique distribution of metals and ligands into free and complexed species.

6. Convergence occurs and the computations are complete when the predicted aqueous speciation from iteration n is essentially identical to that of iteration $n - 1$ (as determined by a user defined level of acceptable computational error).

In the speciation algorithm described above, it is assumed that the ionic strength of the solution is an unknown quantity requiring computation by the geochemical model. It is often stated that the ionic strength of a solution is the "effective salt content" of a solution. It seems natural then that the ionic strength would be related to the salinity of a solution. The salinity of a soil solution is conveniently determined by measuring the electrical conductivity (EC). Pure water is a perfect insulator and does not conduct electricity. When electrolytes are introduced, water attains the ability to carry an electric current, with the EC increasing with increasing electrolyte concentration. There are numerous factors that influence solution EC; but the concentrations and the valences of the specific ions present are among the most important. Consider two solutions, a 0.015 M Na_2SO_4 solution and a 0.015 M $CaSO_4$ solution. The stoichiometric ionic strength of the Na_2SO_4 solution is 0.045 M, and that of the $CaSO_4$ solution is 0.060 M. However, the EC of the Na_2SO_4 solution (1.64 dS m^{-1}) is greater than that of the $CaSO_4$ solution (1.27 dS m^{-1}). The reasons for this discrepancy lie in the nature of the species present in the two solutions. In the Na_2SO_4 solution, only the free ions Na^+ and SO_4^{2-} exist and the effective ionic strength is equal to the stoichiometric ionic strength. However, in the $CaSO_4$ solution, three species exist: Ca^{2+}, SO_4^{2-}, and $CaSO_4^0$. Since $CaSO_4^0$ bears no charge, it does not contribute to the effective ionic strength, or to the EC of the solution. It then follows that the EC of the $CaSO_4$ solution should be less than that of the Na_2SO_4 solution. Indeed, the effective (true) ionic strength of the 0.015 M $CaSO_4$ solution is 0.040 M, which is less than the ionic strength of the Na_2SO_4 solution (0.045 M).

The preceding example suggests that the true ionic strength of a solution should be significantly and positively correlated to the EC of the solution. This is indeed the case, as models that have been generated to describe the variance of I_e as a function of EC enjoy highly significant linear regression coefficients (r^2 values, Table 5.12). The true ionic strength of a soil solution is essentially a linear function of EC. For low salinity solutions ($EC < 1.5$), I_e (mol L^{-1}) $\approx 0.012 \times EC$ (dS m^{-1}). However, when an extended range of solution EC is examined ($EC < 32$), the slope of the regression models increase slightly, such that I_e (mol L^{-1}) $\approx 0.0135 \times EC$ (dS m^{-1}) (average slope of the Griffin and Jurinak [1973] and Marion and Babcock [1976] models). The slope of the I_e (mol L^{-1}) vs. EC (dS m^{-1}) relationship is also a function of the dominant salts present in the soil solution. The model derived by Pasricha (1987) employs a slope that is less than those used in other models. Presumably, this results from the dominance of sodium in the soil extracts, rather than calcium.

5.8 ION SPECIATION IN SOIL SOLUTIONS

The results obtained from a geochemical model are best appreciated through an actual application. To demonstrate speciation chemistry in acidic and alkaline soil solutions, the geochemical model GEOCHEM-PC is employed. The concentrations of the major components were obtained from literature sources; they are measured values from actual soil solution samples. Trace concentrations of the elements, Cd, Cu, Ni, Pb, and Zn, are not actually measured; they are imposed on the solution to allow for a prediction of the trace element speciation.

TABLE 5.12

Empirical Models That Predict the True, or Effective, Ionic Strength (I_e) of a Soil Solution Using Measured Electrical Conductivity (EC) Values

Model[a]	r^2	EC Range, dS m^{-1}	Comments and Source
$I_e = 0.0127 \times EC - 0.0003$	0.992	<32	Griffin and Jurinak (1973) analyzed 124 river waters and 27 soil extracts from the arid western U.S.
$\log I_e = -1.841 + 1.009 \times \log EC$ [$I_e = 0.01442 \times EC^{1.009}$]	0.994	0.05 to 12.9	Marion and Babcock (1976) analyzed 19 river waters, soil extracts, and soil suspensions. The log-linear regression line very closely follows the theoretical lines for $CaCl_2$ and $MgCl_2$ solutions.
$I_e = 0.012 \times EC - 0.004$	0.986	0.034 to 1.0	Gillman and Bell (1978) analyzed 18 soil solutions.
$I_e = 0.01162 \times EC - 0.00105$	0.964	1 to 30	Pasricha (1987) incubated soils with NaCl or $NaHCO_3$ prior to extraction. Resulting regression equation valid for sodic soils.
$I_e = 0.012 \times EC - 0.0002$	0.853	<0.02 to 1.4	Alva et al. (1991) analyzed 211 soil extracts.

[a] The units of I_e are mol L^{-1} and the units of EC are dS m^{-1}.

5.8.1 ACID SOIL SOLUTION

For the acidic soil solution, chemical information for eleven metals (H, Ca, Mg, K, Na, Al, Cd, Cu, Ni, Pb, and Zn) and four ligands (SO_4, Cl, NO_3, and fulvate) are input to the model (Table 5.13). The initial estimate of the ionic strength of the solution is user defined as 7.01 mM and is computed by assuming $[M^{m+}] = M_T$ and $[L^{n-}] = L_T$:

$$I_s = \tfrac{1}{2}\{(2)^2[Ca_T] + (2)^2[Mg_T] + (1)^2[K_T] + (1)^2[Na_T] + (3)^2[Al_T]$$

$$+ (-2)^2[SO_{4T}] + (-1)^2[Cl_T] + (-1)^2[NO_{3T}]\} \tag{5.59}$$

where I_s is the stoichiometric ionic strength. The Davies equation (Equation 5.7) is then used to compute single-ion activity coefficients which, along with K_f values from a thermodynamic database, are used to compute ^cK_f values (Equation 5.50). The coupled-mass balance expressions are solved to provide an initial estimate of ion speciation in the soil solution. The thermodynamic database for the model contains K_f values for 101 soluble complexes that may form from the metals and ligands present in the system. Following the initial estimation of aqueous speciation, the predicted concentrations of all the soluble species are used to compute a new estimate of the ionic strength, which is then employed to compute new ^cK_f values for the second iteration of the speciation algorithm. When convergence is achieved (after six iterations of the algorithm), the true ionic strength (I_e) of the solution is computed as 6.36 mM (compared to the initial estimate of 7.01 mM). The measured EC for this solution is 0.41 dS m^{-1}. Based on this EC and the Marion-Babcock Equation ($I_e = 0.01442 \times EC^{1.009}$), the predicted I_e is 5.86 mM (a value that is consistent with geochemical model predicted).

The aqueous chemistry of the major soluble components in the acidic soil solution is dominated by the free, uncomplexed species: Ca^{2+}, Mg^{2+}, K^+, Na^+, SO_4^{2-}, Cl^-, and NO_3^-, with approximately 100% of the total Ca, Mg, K, Na, Cl, and NO_3; and 84% of the total SO_4, occurring as the free species (Table 5.13). Although it is difficult to speak in generalities when soils are concerned, the predominance of free base cation (Ca^{2+}, Mg^{2+}, K^+, and Na^+) and common anion species in the acidic soil solution is not unusual. This is particulary the case for Na, K, Cl, and NO_3, which are conservative in their speciation chemistry. Approximately 98% of the total soluble Al exists in the

TABLE 5.13
Total Composition and Aqueous Speciation of Metals and Ligands in an Acidic Soil Solution at 25°C[a]

Component	mmol L^{-1}	Aqueous Speciation (% Distribution)
pH	5.97	
Ca	1.55	Ca^{2+} (97.5), $CaSO_4^0$ (1.1), $CaFul^+$ (1.5)
Mg	0.38	Mg^{2+} (98.7), $MgSO_4^0$ (0.9), $MgFul^+$ (0.5)
K	0.21	K^+ (99.9), KSO_4^- (0.1)
Na	0.28	Na^+ (100)
Al	0.14	Al^{3+} (1.8), $AlSO_4^+$ (0.2), $AlOH^{2+}$ (10.9), $Al(OH)_2^+$ (62.8), $Al(OH)_3^0$ (10.7), $Al(OH)_4^-$ (13.6)
SO$_4$	0.13	SO_4^{2-} (84.3), $CaSO_4^0$ (12.7), $MgSO_4^0$ (2.5), KSO_4^- (0.1), $NaSO_4^-$ (0.1), $AlSO_4^+$ (0.2)
Cl	0.13	Cl^- (99.9)
NO$_3$	3.90	NO_3^- (100)
Fulvate[b]	0.03	Ful^- (17.9), $CaFul^+$ (76.1), $MgFul^+$ (6.0)
	µmol L^{-1}	
Cd	0.01	Cd^{2+} (95.8), $CdSO_4^0$ (1.7), $CdCl^+$ (0.9), $CdNO_3^+$ (0.5), $CdFul^+$ (1.2)
Cu	0.10	Cu^{2+} (89.5), $CuSO_4^0$ (1.2), $CuNO_3^+$ (0.8), $CuFul^+$ (6.7), $CuOH^+$ (2.2)
Ni	0.10	Ni^{2+} (92.8), $NiSO_4^0$ (1.0), $NiNO_3^+$ (0.6), $NiFul^+$ (5.6)
Pb	0.10	Pb^{2+} (78.5), $PbSO_4^0$ (1.7), $PbCl^+$ (0.3), $PbNO_3^+$ (3.4), $PbFul^+$ (14.9), $PbOH^+$ (1.1)
Zn	0.10	Zn^{2+} (93.4), $ZnSO_4^0$ (1.0), $ZnNO_3^+$ (0.7), $ZnFul^+$ (3.5), $ZnOH^+$ (1.4)

[a] Data obtained from Wolt (1994). Speciation computation performed by GEOCHEM-PC.

[b] Molar fulvate concentration is based on measured dissolved organic carbon (1.00 mmol L^{-1}), and assumes 500 g kg^{-1} C in fulvic acid and a fulvate molecular weight of 800 g mol^{-1}.

hydrolysis products, consistent with the Al distribution shown in Figure 5.6, and 76% of the total fulvate is complexed by Ca. The free, divalent trace element species (Cd^{2+}, Cu^{2+}, Ni^{2+}, Pb^{2+}, and Zn^{2+}) are predicted to predominate in this acidic soil solution. The only other significant trace element-bearing species predicted to occur are the metal fulvate complexes. Indeed, an average 96% of the total soluble trace element content is predicted to occur in the free divalent metal ion and metal-fulvate complexes in the acidic soil solution (Table 5.13 and Figure 5.15).

5.8.2 ALKALINE SOIL SOLUTION

For the alkaline solution, chemical information for eleven metals (H, Ca, Mg, K, Na, Al, Cd, Cu, Ni, Pb, and Zn) and five ligands (CO_3, SO_4, Cl, NO_3, and fulvate) are input to the geochemical model (Table 5.14). There are 123 possible complexes that can form in this solution, based on the available data in the thermodynamic database of the GEOCHEM-PC code. The initial estimate of ionic strength for this soil solution is 5.76 mM. When convergence is achieved (after seven iterations of the algorithm), the true ionic strength (I_e) of the solution is computed to be 3.91 mM. The distribution of the major elements in the alkaline soil solution is very similar to that observed in the acidic solution (Tables 5.13 and 5.14). Approximately 100% of the total K, Na, Cl, and NO_3 occur as free species (again, due to the conservative aqueous speciation nature of these ions). Eighty-nine percent of the total Ca and 88% of the total SO_4 occur as free species. Bicarbonate is the principle inorganic C species, accounting for approximately 89% of the total soluble inorganic C. The presence of bicarbonate (and carbonate) in the alkaline soil solution has a minimal impact on major cation speciation. However, base cation complexation with the fulvate is enhanced, relative to that observed in the acidic soil solution. Nearly all of the soluble Al is found in the anionic species, $Al(OH)_4^-$ (a condition that is consistent with solution pH; see Figure 5.6).

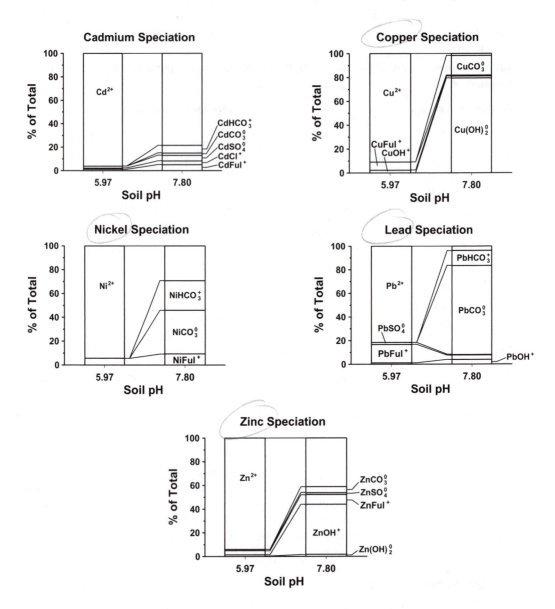

FIGURE 5.15 Metal speciation at 25°C in pH 5.97 and pH 7.80 soil solutions computed using the GEOCHEM-PC computer code. The chemical characteristics of the solutions are given in detail in Tables 5.13 and 5.14.

The predicted trace metal speciation chemisty in the alkaline soil is drastically different from that found in the acidic soil solution. There are two principal reasons for this difference: (1) higher pH results in the greater stability of metal hydrolysis products, and (2) metal complexation by carbonate and bicarbonate ions (Figure 5.15). The stability of metal hydrolysis products in the alkaline soil is consistent with their pK_a values (Table 5.7 and Figures 5.5 and 5.6). Of the trace elements examined, pH and inorganic C least affect Cd speciation, as Cd^{2+} dominates in both acidic and alkaline solutions. Copper speciation is dominated by the hydrolysis product, $Cu(OH)_2^0$ in the pH 7.8 solution, with the bulk of the remaining soluble Cu present in the $CuCO_3^0$ species. Nickel is nearly equally distributed between the Ni^{2+}, $NiHCO_3^+$, and $NiCO_3^0$ species, with the $NiFul^+$ accounting for approximately 9% of the total Ni. The insignificance of the $NiOH^+$ in this solution is due to the high pK_a value (9.86) for Ni^{2+} hydrolysis. The $PbHCO_3^+$ and $PbCO_3^0$ complexes dominate

TABLE 5.14

Total Composition and Aqueous Speciation of Metals and Ligands in an Alkaline Soil Solution at 25°C[a]

Component	mmol L^{-1}	Aqueous Speciation (% Distribution)
pH	7.8	
Ca	1.23	Ca^{2+} (88.5), $CaHCO_3^+$ (0.9), $CaCO_3^0$ (0.2), $CaSO_4^0$ (3.3), $CaFul^+$ (7.1)
Mg	0.07	Mg^{2+} (93.7), $MgHCO_3^+$ (0.8), $MgCO_3^0$ (0.3), $MgSO_4^0$ (2.8), $MgFul^+$ (2.4)
K	0.04	K^+ (99.8), KSO_4^- (0.2)
Na	0.03	Na^+ (99.7), $NaHCO_3^0$ (0.2), $NaSO_4^-$ (0.1)
Al	0.004	$Al(OH)_2^+$ (0.1), $Al(OH)_3^0$ (1.2), $Al(OH)_4^-$ (98.8)
CO_3	1.00	CO_3^{2-} (0.4), HCO_3^- (88.9), $H_2CO_3^0$ (3.3), $CaHCO_3^+$ (1.1), $CaCO_3^0$ (0.2),
SO_4	0.37	SO_4^{2-} (88.4), $CaSO_4^0$ (11.1), $MgSO_4^0$ (0.5)
Cl	0.54	Cl^- (100)
NO_3	0.19	NO_3^- (100)
Fulvate[b]	0.115	Ful^- (23.0), $CaFul^+$ (75.5), $MgFul^+$ (1.4)
	μmol L^{-1}	
Cd	0.01	Cd^{2+} (78.3), $CdHCO_3^+$ (6.6), $CdCO_3^0$ (2.0), $CdSO_4^0$ (4.7), $CdCl^+$ (3.2), $CdFul^+$ (5.0), $CdOH^+$ (0.3)
Cu	0.10	Cu^{2+} (1.7), $CuHCO_3^+$ (0.1), $CuCO_3^0$ (16.4), $CuSO_4^0$ (0.1), $CuFul^+$ (0.7), $CuOH^+$ (1.7), $Cu(OH)_2^0$ (79.3)
Ni	0.10	Ni^{2+} (28.9), $NiHCO_3^+$ (24.4), $NiCO_3^0$ (36.2), $NiSO_4^0$ (1.1), $NiFul^+$ (9.2), $NiOH^+$ (0.2)
Pb	0.10	Pb^{2+} (3.8), $PbHCO_3^+$ (12.7), $PbCO_3^0$ (74.9), $PbSO_4^0$ (0.3), $PbFul^+$ (3.8), $PbOH^+$ (3.9)
Zn	0.10	Zn^{2+} (41.0), $ZnCO_3^0$ (5.1), $ZnSO_4^0$ (1.5), $ZnFul^+$ (8.2), $ZnOH^+$ (42.0), $Zn(OH)_2^0$ (1.8)

[a] Data obtained from Wolt (1994). Speciation computation performed by GEOCHEM-PC.

[b] Molar fulvate concentration is based on measured dissolved organic carbon (3.83 mmol L^{-1}), and assumes 500 g kg^{-1} C in fulvic acid and a fulvate molecular weight of 800 g mol^{-1}.

Pb speciation, while the influence of the carbonate on Zn speciation is minor. Interestingly, the pK_a value for the formation of $ZnOH^+$ (8.96) is higher than the pK_a value for $PbOH^+$ formation (7.7); yet, the percentage of total Zn occurring as $ZnOH^+$ (42.0%) is greater than the percentage of total Pb occurring as $PbOH^+$ (3.9%). This observation illustrates an important point: *the aqueous speciation of a substance in a soil solution cannot be determined by examining a single speciation reaction in isolation, but instead requires holistic understanding of solution chemistry.*

It is evident from the aqueous speciation results described above that element distribution in a soil solution is element and environment specific. It is also apparent that the total soluble concentration of an element may not correspond to the free concentration of the element. This is particularly the case for the trace elements. Therefore, predictions of metal fate and behavior in a soil environment that are based on the errant assumption that, for example, $Pb_T = Pb^{2+}$, would indeed be unreliable.

5.9 QUALITATIVE ASPECTS OF ION SPECIATION

The computations involved in predicting ion speciation in a soil solution can be quite involved. Even though computer programs facilitate the determination of aqueous speciation, there is an investment of time involved in learning to use the codes and in inputting and formatting the required data. There are tools available that (1) provide qualitative information on the types of complexes to expect in soil solutions, (2) indicate how changing the concentration of a ligand (or metal) will influence metal (or ligand) speciation, and (3) identify the K_f values that may be needed to quantify aqueous speciation.

Gilbert Newton Lewis defined a base as an atom, molecule, or ion that has at least one pair of valence electrons that are not already being shared in a covalent bond. A Lewis base is a ligand, and can be neutral or possess negative charge. Similarly, an acid is a unit in which at least one atom has a vacant orbital in which a pair of electrons can be accommodated. A Lewis acid is a metal and is positively charged. Lewis bases generally fall into two categories: those that bind strongly to protons (weak Lowry-Brønsted acid anions) and those that form weak proton bonds (strong Lowry-Brønsted acid anion). Consistent with this categorization is the relative polarizability of the conjugate base species. A strong Lowry-Brønsted acid anion is polarizable, or soft in character, while a weak Lowry-Brønsted acid anion is nonpolarizable, or hard. The base species that contain a coordinating atom that is in the first row of the Periodic groups V, VI, and VII (i.e., N, O, and F) are the hardest in each group, and possess weak Lowry-Brønsted acid anion character. As the coordinating atoms in each group increase in atomic weight (i.e., from N to P to As to Sb), they become progressively softer and their ability to bind protons decreases (stronger Lowry-Brönsted acid anion character).

Ahrland et al. (1958) noted that Lewis acid metal ions could be classified according to the stability of the complexes formed with Lewis bases. Specifically, metal ions that form their most stable complexes with the ligand atoms in the first row of Periodic groups V, VI, and VII (N, O, and F) are termed *class a*. In general, the class a metal preference for ligands is:

$$N > P; \ N > As; \ F > O > N = Cl > Br > I > S$$
$$O > S \approx Se; \ OH^- > R{-}O^- > RCOO^-$$
$$F > Cl; \ CO_3^{2-} \gg NO_3^-; \ PO_4^{3-} \gg SO_4^{2-} \gg ClO_4^-$$

Metals that form their most stable complexes with ligand atoms in the second or subsequent row (group V: P, As, and Sb; group VI: S, Se, and Te; group VII: Cl, Br, I) are termed *class b*. The general class b metal preference for ligands is:

$$P > N; \ As > N$$
$$Se \approx S > O; \ S > I > Br > Cl = N > O > F$$
$$Cl > F$$

Pearson (1963) noted that the salient characteristics that promote the class a or class b behavior of a Lewis acid are size and oxidation state. Metals bearing small size and high positive oxidation state (low polarizability) feature class a behavior; while those that are large in size with low or zero oxidation state (high polarizability) are class b. Other Lewis acid and base characteristics that are associated with class a (hard) and class b (soft) behavior are listed in Table 5.15. As a matter of convenience, Pearson termed the class a Lewis acids *hard* acids and class b Lewis acids *soft* acids (Table 5.16).

Pearson (1963) noted the following generalization: hard Lewis acids prefer to associate with hard Lewis bases, and soft Lewis acids prefer soft Lewis bases. This generalization is known as the hard and soft acids and bases principle (HSAB principle). The HSAB principle was further clarified by Pearson (1968) through the inclusion of acid or base strength concepts: hard Lewis acids prefer to complex hard Lewis bases and soft Lewis acids prefer to complex soft Lewis bases *under conditions of comparable acid–base strength*. Thus, two properties of a Lewis acid and base are needed to utilize the HSAB principle and predict the stability of a complex: strength (intrinsic strength) and softness. Concerning strength, it is expected that OH^- will always form stronger complexes and be a stronger base than H_2O, even though both are hard bases (OH^- bears charge, H_2O does not). Similarly, Mg^{2+} will be a stronger acid than Na^+, even though both are hard acids (Mg^{2+} is divalent, Na^+ is monovalent). Acid or base strength (a charge characteristic) can mask the softness (or hardness) character of a species, making it difficult to determine, for example, whether H_2O is a harder base species than OH^-.

TABLE 5.15
Characteristics Associated with Hard and Soft Lewis Acid and Base Species

Hard	Soft
Low polarizability	High polarizability
Bases have high pK_a values	Bases have low pK_a values
Acids with high oxidation states	Acids with low oxidation states
Bases are hard to oxidize, have high redox potentials (E°_{red} values), and large ionization potentials	Bases are easy to oxidize with low E°_{red} values, and low ionization potentials
Acids are hard to reduce (low E°_{red} values) and are high on the electronegative series	Acids are easy to reduce (high E°_{red} values) and are low on the electronegative series
Small size	Large size
Acids have high positive charge density	Acids have low positive charge density
High negative charge density on donor atom (bases)	Low negative charge density on donor atom (bases)

TABLE 5.16
Classification of Metals and Ligands According to the Principle and Hard and Soft Acids and Bases

Lewis Acids

Hard Acids
H^+, Li^+, Na^+, K^+, Rb^+, Cs^+, Be^{2+}, Mg^{2+}, Ca^{2+}, Sr^{2+}, Ba^{2+}, Sc^{2+}, La^{3+} to Lu^{3+}, Ac^{4+}, Ti^{4+}, Zr^{4+}, Hf^{4+}, Cr^{3+}, Mn^{2+}, Mn^{7+}, MoO^{3+}, Fe^{3+}, As^{3+}, Co^{3+}, Al^{3+}, Si^{4+}, UO^{2+}, VO^{2+}

Borderline Acids
V^{2+}, Fe^{2+}, Co^{2+}, Ni^{2+}, Cu^{2+}, Zn^{2+}, Rh^{3+}, Ir^{3+}, Ru^{3+}, Os^{2+}, Sb^{3+}, Bi^{3+}, Pb^{2+}, Pb^{4+}

Soft Acids
Cu^+, Ag^+, Cd^{2+}, Au^+, Hg^{2+}, Tl^+, Pt^{2+}, CH_3Hg^+, RSe^+, RTe^+, Ga^+, Sn^{2+}, Tl^{3+}, Au^{3+}, In^{3+}

Lewis Bases

Hard Bases
NH_3, RNH_2, H_2O, OH^-, O^{2-}, ROH, CH_3COO^-, CO_3^{2-}, NO_3^-, PO_4^{3-}, SO_4^{2-}, F^-

Borderline Bases
$C_6N_5NH_2$, C_5H_5N, N_2, NO_2^-, SO_3^{2-}, Br^-, Cl^-

Soft Bases
C_2H_4, C_6H_6, R_3P, $(RO)_3P$, R_3As, R_2S, RSH, $S_2O_3^{2-}$, S^{2-}, I^-

Within a given category (e.g., hard) there exists a spectrum of hardness/softness character: metals (and ligands) vary in the degree of hardness or softness. There is also no clear dividing line between hard and soft categories, thus the existence of the borderline catetory. A borderline metal is able to form stable complexes with all ligands and will show some degree of preference for ligands in either hard or soft categories, reflecting the degree of hard or soft character. For instance, a metal may display hard character in preferring to form complexes with N donor atoms, and soft character in preferring to form complexes with S and I donor atoms. In essence, borderline species are ambivalent in their complexation characteristics.

Several investigators have attempted to quantify the HSAB principle by defining absolute hardness or absolute softness parameters (Table 5.17). Parr and Pearson (1983) used the ionization

TABLE 5.17
Absolute Hardness and Softness Parameters for Lewis Acids[a]

				Valence							
+1				+2				+3			
Metal[b]	$\eta_{S,NM}$	$\eta_{S,M}$	η_H	Metal	η_S	$\eta_{S,M}$	η_H	Metal	η_S	$\eta_{S,M}$	η_H
Li(H)	1.39	0.36	35.1	Ba(H)	1.74	2.62	12.8	La(H)	2.42	2.45	15.4
K(H)	1.46	0.92	—	Sr(H)	1.78	2.08	—	Ce(H)	2.45	—	—
Na(H)	1.55	0.93	21.1	Ca(H)	1.84	1.62	19.7	Nd(H)	2.51	—	—
Rb(H)	1.56	2.27	11.7	Mg(H)	2.58	0.87	32.5	Al(H)	3.50	0.70	45.8
Cs(H)	1.57	2.73	—	Be(H)	2.85	—	—	In(S)	5.26	—	—
Tl(S)	5.90	—	7.2	Mn(H)	3.96	3.03	9.3	Fe(H)	5.72	2.37	13.1
Cu(S)	6.53	3.45	6.3	Zn(B)	4.54	2.34	10.8	Cr(H)	—	2.70	—
Ag(S)	7.85	3.99	6.9	V(B)	4.60	—	—	Tl(S)	7.49	—	10.5
Au(S)	14.32	—	5.7	Fe(B)	5.39	3.09	7.3				
				Cd(S)	5.51	3.04	10.3				
				Co(B)	5.61	2.96	—				
				Ni(B)	5.73	2.82	8.5				
				Cu(B)	6.41	2.89	8.3				
				Pb(B)	7.18	3.58	8.5				
				Hg(S)	7.80	4.25	7.7				
				Pt(S)	9.41	—	—				

[a] $\eta_{S,NM}$ is the softness parameter of Nieboer and McBryde (1973); $\eta_{S,M}$ the softness parameter of Misono et al. (1967); η_H is the hardness parameter of Parr and Pearson (1983).

[b] S, B, and H: soft, borderline, and hard, respectively.

potential (I_A) and the electron affinity (A_A) of species to calculate an absolute hardness parameter: $\eta_H = \frac{1}{2}(I_A - A_A)$. Nieboer and McBryde (1973) employed metal ion electronegativities (X_M) and crystal radii (r_M) to compute class b metal character [$\eta_{S,NM} = (X_M)^2(r_M + 0.85)$]; whereas, Misono et al. (1967) employed X_M values and ionization energies (I_n) [$\eta_{S,M} = (X_M^2 + \Sigma I_n + 2 X_M \Sigma I_n^{1/2})/10$]. It is readily apparent from the data in Table 5.17 that absolute hardness and softness parameters lay on a continuum. Further, there are varying degrees of hardness or softness. For example, both Li^+ and Na^+ are hard Lewis acids; but Na^+ is a softer species than Li^+ (or Li^+ is a harder species than Na^+). It is also apparent that the softness parameters are inversely related to the hardness parameter; high η_H values generally correspond to low $\eta_{S,NM}$ and $\eta_{S,M}$ values, and *vice versa*.

The HSAB principle is an exceedingly useful tool and the basis for many generalities associated with the complexation behavior of metals and ligands in soil solutions and other natural waters. Soil solutions are dominated by hard species: Ca^{2+}, Mg^{2+}, K^+, Na^+, Al^{3+} and hydrolysis products, the hydrolysis products of Fe^{3+}, H_2O (one of the hardest species), OH^-, and the oxyanions CO_3^{2-}, HCO_3^-, NO_3^-, HPO_4^{2-}, $H_2PO_4^-$, and SO_4^-. In general, these hard species form outer-sphere complexes, and the bonding mechanism is principally ionic (electrostatic). Thus, complex stability is proportional to the ionic potentials (Z/r values) of the ions involved. Soft species have a greater innate capability to form inner-sphere complexes (relative to hard species), due in part to their greater polarizability. Thus, the stability of complexes involving soft species will be greater than that predicted from ionic potentials. The HSAB principle also predicts that a change in the concentration of a hard base will primarily influence hard acid complexation chemistry. Similarly, a change in the concentration of a soft base will primarily influence soft acid complexation chemistry. If a base is not of sufficient strength to displace H_2O

TABLE 5.18

Metal Ion Toxicity Sequences for a Variety of Organisms[a]

Organism	Toxicity Sequence[b]
Algae	$Hg^{2+}(S)>Cu^{2+}(B)>Cd^{2+}(S)>Fe^{2+}(B)>Cr^{3+}(H)>Zn^{2+}(B)>Ni^{2+}(B)>$ $Co^{2+}(B)>Mn^{2+}(H)$
Fungi	$Ag^{+}(S)>Hg^{2+}(S)>Cu^{2+}(B)>Cd^{2+}(S)>Cr^{3+}(H)>Ni^{2+}(B)>Pb^{2+}(B)>$ $Co^{2+}(B)>Zn^{2+}(B)>Fe^{2+}(B)>Ca^{2+}(H)$
Barley	$Hg^{2+}(S)>Pb^{2+}(B)>Cu^{2+}(B)>Cd^{2+}(S)>Cr^{3+}(H)>Ni^{2+}(B)>Zn^{2+}(B)$
Protozoa	$Hg^{2+}(S){\sim}Pb^{2+}(B)>Ag^{+}(S)>Cu^{2+}(B){\sim}Cd^{2+}(S)>Ni^{2+}(B){\sim}Co^{2+}(B)>$ $Mn^{2+}(H)>Zn^{2+}(B)$
Platyhelminthe	$Hg^{2+}(S)>Ag^{+}(S)>Au^{+}(S)>Cu^{2+}(B)>Cd^{2+}(S)>Zn^{2+}(B)>H^{+}(H)>$ $Ni^{2+}(B)>Co^{2+}(B)>Cr^{3+}(H)>Pb^{2+}(B)>Al^{3+}(H)>K^{+}(H)>Mn^{2+}(H)>$ $Mg^{2+}(H)>Ca^{2+}(H)>Sr^{2+}(H)>Na^{+}(H)$
Annelida	$Hg^{2+}(S)>Cu^{2+}(B)>Zn^{2+}(B)>Pb^{2+}(B)>Cd^{2+}(S)$
Vertebrata	$Ag^{+}(S)>Hg^{2+}(S)>Cu^{2+}(B)>Pb^{2+}(B)>Cd^{2+}(S)>Au^{+}(S)>Al^{3+}(H)>$ $Zn^{2+}(B)>H^{+}(H)>Ni^{2+}(B)>Cr^{3+}(H)>Co^{2+}(B)>Mn^{2+}(H)>K^{+}(H)>$ $Ba^{2+}(H)>Mg^{2+}(H)>Sr^{2+}(H)>Ca^{2+}(H)>Na^{+}(H)$
Mammalia	$Ag^{+}(S){\sim}Hg^{2+}(S){\sim}Tl^{+}(S){\sim}Cd^{2+}(S)>Cu^{2+}(B){\sim}Pb^{2+}(B){\sim}Co^{2+}(B){\sim}$ $Sn^{2+}(S){\sim}Be^{2+}(H)>In^{3+}(S){\sim}Ba^{2+}(H)>Mn^{2+}(H){\sim}Zn^{2+}(B){\sim}Ni^{2+}(B){\sim}$ $Fe^{2+}(B){\sim}Cr^{3+}(H)>Y^{3+}(H){\sim}La^{3+}(H)>Sr^{2+}(H){\sim}Sc^{3+}(H)>Cs^{+}(H){\sim}$ $Li^{+}(H){\sim}Al^{3+}(H)$

[a] Compilation by Nieboer and Richardson (1980).

[b] S, B, and H: soft, borderline, and hard, respectively.

from the hydration sphere, an outer-sphere complex will develop. For example, SO_4^{2-} is marginally capable of displacing waters of hydration from the coordination sphere of Al^{3+}, resulting in both outer- and inner-sphere complexes. The measured $\log K_f = 3.5$ for $AlSO_4^{+}$ formation is a composite value representing the formation of both inner-and outer-sphere complexes. However, F^- clearly displaces waters of hydration, resulting in an inner-sphere (and quite stable) complex ($\log K_f = 7.0$ for AlF^{2+} formation). The species $CaCO_3^{0}$ and $MgCO_3^{0}$ are outer-sphere complexes with $\log K_f = 3.15$ and 2.12, respectively. The $CuCO_3^{0}$ complex is inner-sphere (with bidentate character) with $\log K_f = 6.75$.

The classification of metals and ligands on the basis of Lewis acidity and basicity, hardness or softness, allows for a qualitative determination of the potential for aqueous complexes to form. In addition, the hardness or softness of a metal or ligand can be used as an indicator of biological toxicity. The mechanisms for toxicity in biological systems tend to fall into three major categories: (1) blocking of an essential functional group in a biomolecule, (2) displacement of an essential metal from a biomolecule, and (3) modification of the configuration of an active biomolecule. It has been noted that soft species can participate in all three toxicity mechanisms, due to their ability to deform (polarize) and take on the configuration required to interact with a biomolecule. Borderline metals can also participate in all three mechanisms, but not with the same tenacity as the soft species. The metal ion toxicity sequences for a number of organisms as a function of Lewis acid character indicate that, in general, soft metals are more toxic than borderline metals, which in turn are more toxic than hard metals (Table 5.18). Hard species can be highly toxic to organisms. Examples include beryllium when inhaled and fluoride when adsorbed through the skin. However, there is no finesse involved in the mechanism of toxicity; their high ionic potential allows them to interrupt biomolecule configuration through brute force interference.

5.10 SOIL WATER SAMPLING METHODOLOGIES

In this chapter, considerable emphasis is placed on the acid-base and the complexation chemistry of metals and ligands in soil water. It has been noted that water is a highly reactive substance and the soil solution is the location of reactions that control the fate and behavior of substances in the soil. Yet, for all the emphasis placed on attaining a detailed understanding of the chemical reactions that occur in the soil solution, there are severe limitations in the ability to isolate and sample the true soil water. This difficulty arises from the fact that soil water exists under a myriad of tensions (tension free or gravitational to hygroscopic water, depending on soil moisture) and in a variety of locations within the soil matrix (micropore vs. macropore water; interstatial and extrastatial water). Further, soil water flow paths differ as a function of soil moisture, and may even change from one precipitation event to the next (as will water flow rates and contact times with the solid phase). It is not surprising that the chemical composition of soil water is naturally dynamic and spatially heterogeneous. It should also be evident that the extraction of water from the soil matrix, for the express purpose of obtaining a true soil water sample, is not a feasible objective. However, numerous mechanisms are available and employed to obtain soil water and soil solution samples that estimate the true character of the soil water, providing a realistic measure of the processes operating in a soil environment. Wolt (1994) presents a thorough discussion of the many soil water–sampling methodologies; some of the more common are discussed below.

In a broad sense, there are two types of soil water samples: *in situ* and *ex situ*. Soil water collected from the field environment reflects *in situ* soil water. *In situ* water samples are normally obtained from undisturbed soils. Drainage water from a soil horizon, a whole soil profile, or a whole field can be obtained and reflect *in situ* soil water. Similarly, water extracted by vacuum through porous ceramic cups that have been emplaced in a soil profile represents an *in situ* soil water sample. Water collected from soil samples that have been removed from the environment and that have been disturbed by mixing or through the homogenization of composite samples is *ex situ* soil water. Typically, *ex situ* soil solutions are not true soil water samples, as water is often added and equilibrated with soil in the laboratory so that a volume of solution suitable for analysis can be extracted.

An integral component of any soil water sampling device is the barrier material that allows for the transfer and isolation of water from a soil matrix. Barrier materials that have been employed in soil water samplers are numerous and include: ceramics, silica, stainless steel, Teflon, fritted glass, and an array of plastics (polycarbonate, polysulfone, polyvinylalcohol). All barrier (or filter) materials, and materials employed in the construction of the sampling device, have the potential to impact the integrity of the soil water samples. Reactions between dissolved substances in the soil water and the construction and filtration materials employed in the water collection device can alter the chemical characteristics of the soil water, an occurrence that is of particular concern when sampling for trace constituents. Such interactions can have both negative interferences (removal of substances) and positive interferences (addition of substances) on the chemical characteristics of the collected soil water. Positive interferences can be minimized by rinsing the filter material with distilled/deionized water, or a weak acid solution (such as 0.1 M HCl) followed by distilled/deionized water, prior to use or installation. The removal of trace substances (negative interferences) by filter material remains problematic, as prewashing procedures will do nothing to the inherent retention properties of the filter material. However, filter material can be selected to minimize negative interferences for the substances of interest.

5.10.1 *IN SITU* SOIL WATER SAMPLING

Tension-free water from a soil horizon or profile may be collected using pan lysimeters (Figure 5.16). Water collected by such a device is assumed to be reflective of an undisturbed soil, even though considerable disruption of soil structure may occur during installation (Figure 5.17). Typically, pan

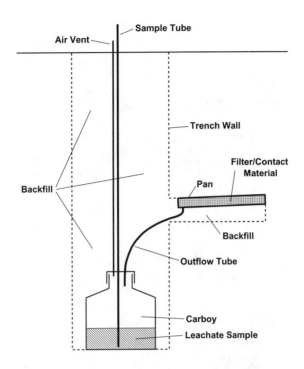

FIGURE 5.16 Schematic diagram of a tension-free pan lysimeter. The pan collects gravitational water that flows through soil macropores.

lysimeter installation entails (1) the excavation of a pit; (2) the lateral excavation of a volume of soil sufficient for pan installation at a desired depth and under undisturbed soil; (3) the construction of a pan using non-reactive material and filled with a nonreactive porous media; (4) emplacement of the pan in the lateral excavation such that the fill material in the pan is in intimate contact with the overlying soil; (5) backfilling around the pan; (6) connection of the pan drain to a carboy to collect and hold the leachate; (7) attachment of vent and sampling lines that run from the carboy to the surface; and (8) backfilling the soil pit. Clearly, the installation of a pan lysimeter is a cumbersome process that disturbs the soil structure in the vicinity of the pan and may indeed introduce confounding soil water flow patterns. It is also assumed that the volume of soil from which the soil water sample is obtained is localized in a region above and proximate to the pan (an assumption of questionable validity). On a larger scale, soil water from a large spatial region of a field can be obtained from the discharge of tile drains or other subsoil devices used to maintain the flow of water through the soil profile.

   For soil water to enter a pan or drainage device, the contacting soil must be saturated. Thus, soil water collected by tension-free pan lysimeters, or by sampling drainage water, generally represents water that has moved through the macropores (large pores). Therefore, it is water that has bypassed the water that is in contact with the reactive surfaces of the soil, also termed micropore water. The chemical composition of macropore water is generally not reflective of water that is intimately associated with and under the influence of soil solids. An additional confounding factor associated with macropore water is that the flow path soil water takes before it is intercepted by the drainage collection device (drainage pattern), and thus the source of the collected water, is largely unknown.

   Water collection by replicate pan lysimeters is very sporadic and inconsistent. The variable water collection characteristics of pan lysimeters is demonstrated for three pans emplaced at a 90-cm depth in a 1-acre experimental plot consisting of loess-derived soil and located in west Tennessee (Figure 5.18). The volume of water that moves through a soil profile during a precipitation event

FIGURE 5.17 Photograph showing an installed tension-free pan lysimeter. Directly above the right corner of the pan is a suction cup lysimeter (diagrammed in Figure 5.20). The berm contains the excavated soil that will be returned to the soil pit after pan installation. (Photograph courtesy of Dr. Donald Tyler, West Tennessee Experiment Station, The University of Tennessee.)

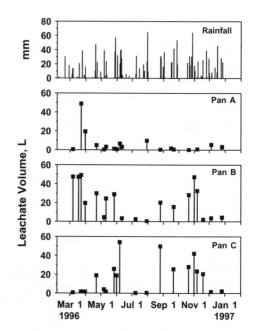

FIGURE 5.18 The variability in leachate collection by three tension-free pan lysimeters. The pans are located within a 1-acre production cotton field at Ames Plantation (near La Grange, TN) and emplaced at a depth of 90 cm. Rainfall data are collected on-site.

is a function of several factors, including the intensity and duration of an event, soil topography, presence of surface residues, and antecedent soil water. In general, these variables are similar, if not identical, for the three pans. During the March, 1996 through December, 1996 period, 1486 mm of precipitation, or 668.6 L of precipitation over the 4500-cm^2 area of a pan, was received. During the same period, pan A collected 114 L of leachate, or 17% of the total volume of precipitation received at the soil surface. Pan B collected 413 L (62% of total precipitation), and pan C collected 322 L (48% of total precipitation). The pans also differed in their leachate collection characteristics during specific rainfall events. During the rainfall event of March 22 to 23, 1996, pans A and B each collected 49 L of leachate, while pan C collected 1.6 L. However, during the rainfall event of June 7, 1996, pan A collected 10 L and pan B collected 3.6 L, while pan C collected 54 L. These data clearly indicate that there is no consistency among replicated pan lysimeters, and that the natural heterogeneity of the soil plays the definitive role in water flow characteristics (even though the surficial parameters that influence leachate volume are not spatially variable, or minimally so). It is also evident that there may be no ability to truly obtain replicate information from subsurface flow studies.

In addition to the leachate volume, the chemical characteristics of leachates collected by replicated pan lysimeters can be highly variable. The concentrations of the herbicide fluometuron in leachates, collected by the field replicated pan lysimeters described above, illustrate leachate chemical variability (Figure 5.19). Peak fluometuron concentrations in leachates collected at the 90-cm depth following a May 7, 1996 rainfall event were 256 µg L^{-1} (pan B), 395 µg L^{-1} (pan C), and 870 µg L^{-1} (pan A). The mass of fluometuron received by the lysimeters during the May 7 rainfall event was 539 µg (pan B), 745 µg (pan C), and 102 µg (pan A). Despite the potential shortcomings indicated above, the collection and analysis of macropore water (drainage water) can provide valuable information concerning the movement of water and solutes through the soil profile. Studies that have involved the use of tension-free pan lysimeters illustrate the rapid movement of substances through the soil profile, movement that is inconsistent with the soil retention characteristics for the substances (O'Dell et al., 1992; Essington et al., 1995).

The withdrawal of soil water through a porous material (porous ceramic cup, fritted glass, polymer hollow fiber) by the application of a vacuum is a commonly employed *in situ* sampling technique. Vacuum extraction has the advantage of being convenient, particularly with respect to ease of installation of the sampling devices, and in the ability to sample soil water from a number of soil depths with minimal soil disturbance. Soil solution samplers may be constructed using a

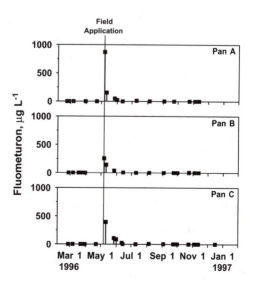

FIGURE 5.19 The concentration of the herbicide fluometuron collected by three tension-free pan lysimeters. The pans are located within a 1-acre production cotton field at Ames Plantation (near La Grange, TN) and emplaced at a depth of 90 cm. Field application of the herbicide occurred on May 6, 1996; whereupon rapid leaching of the herbicide is observed during a rainfall event.

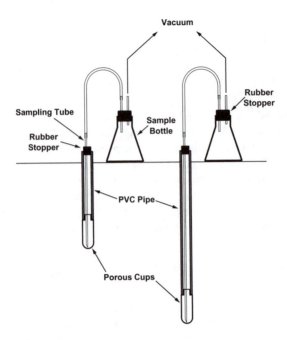

FIGURE 5.20 Schematic diagram of suction-cup lysimeters. The soil solution sampling devices collect gravitational water and water under tension (macropore and micropore water).

porous ceramic cup cemented to a length of PVC pipe (Figure 5.20). A soil auger, whose diameter matches that of the PVC pipe, is used to bore a hole into the soil to the desired depth of cup emplacement. The solution sampler is inserted into the hole and seated such that the ceramic cup is in intimate contact with the surrounding soil. A rigid plastic tube is then threaded through a one-hole rubber stopper, such that the end of the sampling tube resides inside the ceramic cup. The stopper seals the PVC pipe, and additional tubing is used to attach the solution sampler tubing to a sample collection bottle (which is also a vacuum reservoir).

Soil water collected using vacuum extraction represents gravitational water and water that is held under tension (from tension free down to that equal to the applied vacuum). Thus, water from both macropores and micropores is extracted, and soil water saturation is not required to obtain a soil water sample. However, the flow path that water takes to reach the soil solution sampler is largely unknown (a shortcoming noted for pan lysimeters). Because vacuum extractors access a different pool of soil water than do pan lysimeters, the chemical characters of the waters are expected to differ. Indeed this is the case, as evidenced by the fluometuron leachate data in Figure 5.21. Both sampling devices indicate the rapid leaching of the herbicide through the soil profile to a depth of at least 120 cm. However, during the period immediately following herbicide application, herbicide concentrations are greater in the vacuum extracted samples relative to the lysimeter samples. During the fall of 1996, the concentrations of fluometuron in the lysimeter leachates are below detectable levels (<1 μg L^{-1}), while concentrations in the vacuum-extracted samples range up to 100 μg L^{-1} fluometuron.

5.10.2 *Ex Situ* Soil Water Sampling

Vacuum extraction and pressure filtration (compression) techniques are employed to obtain *ex situ* soil solution from disturbed soil samples that have been combined with water in the laboratory. Vacuum extraction involves the withdrawal of water through a filter (filter paper and membrane filters composed of Teflon, nylon, polycarbonate, and polysulfone) by the application of a vacuum.

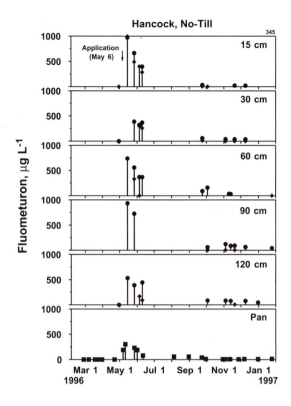

FIGURE 5.21 The concentration of the herbicide fluometuron collected by a tension-free pan lysimeter installed at a depth of 90 cm, and suction-cup lysimeters installed at various depths. The suction-cup lysimeters are installed within a 1 meter radius of the pan.

Pressure filtration uses pressure to force the soil solution out of a soil sample through a filter. Both techniques are convenient, and while the composition of the solution may be quite different from that of a true soil solution, the techniques do provide a soil solution sample whose composition closely reflects something of the true reactions between the soil solution and the soil solids.

The use of centrifugal force is commonly employed to separate the solid and solution phases of disturbed soil samples. The method is convenient and also provides an aqueous solution whose composition reflects something of the true reactions between the soil solution and solid constituents. However, water saturation (or oversaturation) is required, and the composition of the extract may be quite different from that of a true soil solution. An additional disadvantage to the centrifugal separation technique is that an additional filtration step may also be required, as smaller particles may remain suspended; vortices formed in the centrifuge tubes during deceleration can resuspend smaller particles.

Immiscible displacement is a centrifugal separation method that relies upon a heavy density, nonreactive organic liquid of low water solubility (carbon tetrachloride, ethyl benzoylacetate) to displace soil water from a disturbed soil sample. The application of this method does not require water saturation (or over saturation) of a soil sample, but the method is specific to disturbed and moist soil samples and requires specialized equipment (high-speed centrifuge, centrifuge tubes that are resistant to attack by the organic liquid). If inorganic elements are of primary interest and if the soil organic matter content is low, the contamination potential is negligible. While these organic liquids are perceived to be nonreactive, they do have some water solubility, which may confound organic carbon determinations. Further, soil organic matter may partition into the organic liquid, carrying associated inorganic elements and confounding both the organic and the inorganic characteristics of the soil solution.

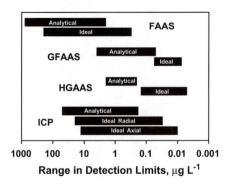

FIGURE 5.22 The range of detection limits observed for atomic spectrometry techniques (FAAS, flame atomic absorption spectrophotometry; GFAAS, graphite furnace AAS; HGAAS, hydride generation AAS; ICP, inductively coupled argon plasma spectrometry with the torch viewed from the side (radial) or end-on (axial); analytical ICP range is for radial orientation).

5.11 METHODS OF CHEMICAL ANALYSIS: ELEMENTAL ANALYSIS

A variety of analytical techniques are available to establish the elemental composition of soil solutions. The common techniques are: flame atomic absorption spectrophotometry (flame AAS), atomic emission spectrometry, graphite-furnace AAS, hydride generation AAS (and cold-vapor AAS), and inductively coupled argon plasma spectrometry (ICP). The factors that dictate the most suitable technique to employ include: instrument availability, ease of use, and cost per sample or element; the element or elements of interest; the level of detection required; and the range of elemental concentrations expected. With the exception of ICP, which is a multielemental technique, the atomic spectroscopy techniques are single-element. Further, the atomic spectroscopy techniques can all be performed by a single instrument (with the appropriate attachments).

Probably the most significant criterion in establishing the appropriate analytical technique (aside from access and cost) is the required level of detection. Detection limits are technique (instrument) and element specific (Figure 5.22). In general, the lowest elemental detection limits are obtained with graphite-furnace AAS. For elements that form hydrides and mercury, the hydride generation AAS and cold vapor AAS techniques offer the lowest detection limits. An additional criterion for selecting the appropriate instrumental technique is the analytical working range (also known as the linear range) of the instrument. The analytical range of an instrument is the concentration range of over which instrument response is linearly related to element concentration. The advantage of a broad analytical working range is that sample handling prior to analysis is minimized (such as performing dilutions which are a potential source of error). The instrument with the greatest analytical range is the ICP, with a dynamic range of six orders of magnitude. Flame AAS has an analytical range of three orders of magnitude, while the most sensitive techniques (graphite furnace, hydride generation, and cold-vapor AAS) have the most limiting analytical range (two orders of magnitude).

5.11.1 ATOMIC SPECTROMETRY

Atomic spectrometric techniques are the standard for the elemental analysis of solutions. There are two principal mechanisms of operation: atomic emission and atomic absorbance. Both techniques rely on thermal energy to either (1) promote atoms to a high-energy and unstable excited state, whereupon they decay to the ground state and produce light energy, or (2) to position an element in an atomic vapor, allowing for the absorption of light energy which promotes the element from the ground state to the excited state. As described below, the light energy emitted or absorbed is the basis for elemental quantitation.

5.11.1.1 Flame Atomic Absorption Spectrophotometry

Flame AAS is a method of determining the total concentrations of elements in aqueous solutions by measuring the light energy absorbed by an atom during excitation. Every element has a specific number of electrons associated with its nucleus. In the ground state, which is the normal configuration of

Atomic Absorption

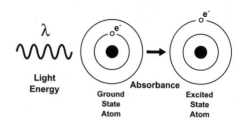

Atomic Emission

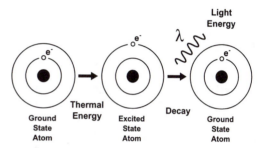

FIGURE 5.23 Schematic representation of the electronic transition that occurs in an atom as electrons moves between the ground and excited states, resulting in the absorption or the emission of characteristic radiation.

an element, electrons reside in their most stable (lowest energy) electronic orbitals. When an element is placed in an atomic vapor, light energy can be absorbed. The absorption process results in the excitation of outer shell electrons to higher energy levels (Figure 5.23). Each electronic transition requires a specific quantity of light energy. The specific energy requirement for an electronic transition is both element and transition dependent, and directly related to the wavelength of light that is absorbed by the atom: $E = h\nu$, where E is energy, h is Planck's constant, and ν is the velocity of light in a vacuum divided by the wavelength (c/λ). Since there are a number of electronic transitions that can potentially occur in an atom and each element has specific energy requirements to affect the transitions, every element is characterized as having a unique set of excitation energies. Therefore, when a specific wavelength of light that passes through an atomic vapor exactly matches an excitation energy of an element in the vapor, some of the light energy will be absorbed. The decrease in the intensity of the light is proportional to the number of atoms in the atomic vapor that are absorbing the light energy. Thus, by measuring the decrease in light intensity (or the amount of light absorbed), a quantitative measure of element concentration can be made.

For every element there are multiple exicitation energies that result in the promotion of electrons to higher energy levels, as there are numerous electronic transitions possible. Figure 5.24 illustrates the transitions that can occur in atomic Na. The spectrum of radiation that results from the multiple excitations is termed resonance line radiation. Thus, the number of atoms (element concentration) in an atomic vapor can be determined by determining the absorption of any one of the resonance lines. However, some of the resonance lines are absorbed to a greater degree than others, due to the fact that some electronic transitions are favored more than others. For example, the resonance spectrum of Cu contains several lines that can be employed to determine Cu concentrations in solutions. The 324.8 nm line is the most sensitive, with a detection limit of 0.02 mg L^{-1} and a linear range up to 5 mg L^{-1} (Table 5.19). The 324.8 line is approximately 75 times more sensitive than the 249.2 nm line, which has a detection limit of approximately 1.5 mg L^{-1} and a linear range of 100 mg L^{-1}. The different sensitivities of the resonance lines for a given element can be exploited. A solution that contains a concentration of Cu of 50 mg L^{-1} would require at least a tenfold dilution to be brought into the linear range of the 324.8 nm line. However, analysis of Cu in the solution would be possible without dilution by examining the absorbance of the 249.2 nm line.

TABLE 5.19
Characteristic Absorbance Lines Commonly Used in Flame Atomic Absorption Spectrophotometry (the Method Detection Limit and the Dynamic [Linear] Analytical Range Are Also Shown)

Element	λ nm	Sensitivity	Linear Range mg L^{-1}	Element	λ nm	Sensitivity	Linear Range mg L^{-1}
Al	309.3	1.1	100	K	766.5	0.043	2.0
Ba	553.6	0.46	20		769.9	0.083	20.0
Ca	422.7	0.092	5.0	Li	670.8	0.035	3.0
	239.9	13.0	800	Mg	285.2	0.0078	0.50
Cd	228.8	0.028	2.0		202.6	0.19	10.0
Co	247.7	0.12	3.5	Mn	279.5	0.052	2.0
	252.1	0.28	7.0		280.1	0.11	5.0
Cr	357.9	0.078	5.0	Na	589.0	0.012	1.0
	359.4	0.10	7.0	Ni	232.0	0.14	2.0
Cu	324.8	0.077	5.0		352.5	0.39	20.0
	249.2	5.8	100	Pb	283.3	0.45	20.0
Fe	248.3	0.11	6.0	Sr	460.7	0.11	5.0
	302.1	0.4	10.0		407.8	2.0	20.0
	246.3	1.1	20.0	Zn	213.9	0.018	1.0

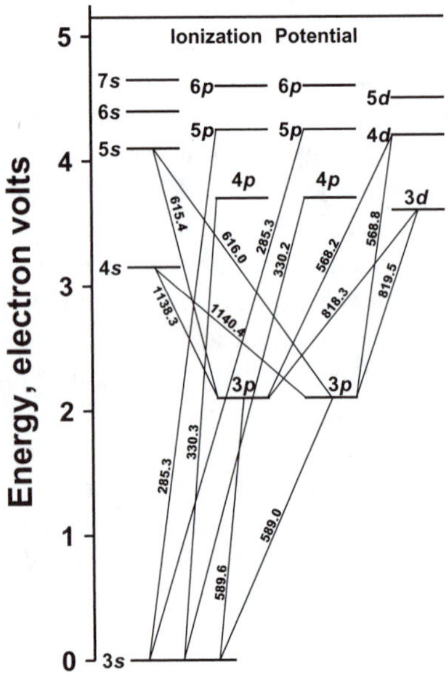

FIGURE 5.24 Schematic representation of the electronic transitions that occur in a sodium atom (Na^0). The energy associated with each transition is identified by a characteristic wavelength, expressed in nm.

Elemental detection limits reported for FAAS cover a broad range and depend on the element of interest, the compositional properties of the sample solution, and instrument set-up variables (Figure 5.22 and Table 5.20). If it is assumed that the instrument is set up to provide maximum sensitivity, then elemental detection limits can be divided into two classes: method detection limits (MDL values) and ideal detection limits (IDL values). Method detection limits represent the minimum quantitation levels obtained for real-world samples, such as soil solutions, using optimal instrument conditions. Ideal detection limits represent the minimum quantitation levels obtained using high-purity standards and optimal instrument conditions. The MDL values are greater than

TABLE 5.20
Typical Method Detection Limits (MDL Values) and Instrument Detection Limits (IDL Values) for Spectrometric Elemental Analysis Techniques[a]

Element	FAAS MDL[b]	FAAS IDL[b]	GFAAS MDL	GFAAS IDL	HGAAS	ICP-AES MDL	ICP-AES Radial	ICP-AES Axial
				$\mu g\ L^{-1}$				
Ag	10	2	—	0.05		5	2	0.5
Al	100	30	3	0.25		25	6	1.5
As	—	300	1	0.33	2	50	12	2
B	—	20	—	43		6	0.5	0.2
Ba	100	20	2	0.4		1	0.2	0.04
Be	5	1	0.2	0.025		0.2	0.2	0.06
Ca	10	1	—	0.04		10	0.03	0.03
Cd	5	1.5	0.1	0.02		2	1	0.01
Co	50	5	1	0.5		3	2	0.5
Cr	50	6	1	0.025		5	2	0.4
Cs	20	4	—	0.3		—	3200	—
Cu	20	3	1	0.07		2	2	0.3
Fe	30	6	1	0.06		5	1	0.3
Hg	—	145	—	18	0.2	30	9	1.2
K	10	2	—	0.02		400	6.5	0.5
Mg	10	0.3	0.2	0.01		15	0.1	0.03
Mn	10	2	0.2	0.03		1	0.3	0.05
Mo	100	20	1	0.14		5	4	0.5
Na	2	0.3	—	0.05		10	1	0.2
Ni	40	10	—	0.24		10	6	0.4
P	—	4000	—	100		60	18	13
Pb	100	10	1	0.04		25	14	1
S	—		—			80	20	28
Se	—	500	2	0.65	2	50	20	5
Si	300	200	—	0.8		20	5	2
Sn	800	95	—	0.6		17	0.1	0.01
Sr	30	2	—	0.1		1	0.1	0.01
Ti	300	70	—	1.6		2	0.6	0.09
Tl	100	20	1	0.75		30	16	3
V	200	50	4	0.7		5	2	0.5
Zn	5	1	0.05	0.0075		4	1	0.06
Zr	—	1500	—	—		3	0.8	—

[a] FAAS, flame atomic absorption spectrophotometry and emission spectrometry; GFAAS, graphite-furnace AAS; HGAAS, hydride generation AAS or cold-vapor AAS for Hg; ICP-AES, inductively-coupled argon plasma-atomic emission spectrometry. Radial views the plasm from the side, axial views the plasma end-on. ICP MDL values are for radial view. Detection limit values for both radial and axial are for ideal condition. Data are derived from instrument manufacturers, U.S. Environmenatal Protection Agency (2002), American Public Health Association (1985), Wright and Stuczynski (1996), and Soltanpour et al. (1996).

[b] MDL values represent the practical level of quantification that would be obtained through the analysis of actual samples. IDL values are those reported by various instrument manufacturers and expected under ideal conditions.

TABLE 5.21
Interferences Associated with Atomic Absorption Spectrophotometry, Atomic Emission Spectrometry, and Inductively Coupled Plasma Spectrometry

Interference	Description
Chemical	Compounds of the analyte are thermally stable in the environment of the flame and do not completely decompose to the atomic components, reducing the number of atoms capable of absorbing light energy
Ionization	Thermal energy of the flame is sufficient to expel an electron from an atom, creating an ion and depleting the number of ground state atoms
Matrix	The physical properties of the sample and standard solution are drastically different, resulting in differing flow rates to the flame or burning characteristics in the flame
Emission	At high analyte concentrations, the emission characteristics of an element can mask the absorption signal
Spectral	A spectral line of another element in the sample are identical, or nearly so, to that of the analyte
Background absorption	Light from the lamp is scattered by particles in the flame or molecules in the flame, other than the analyte, absorb the light energy

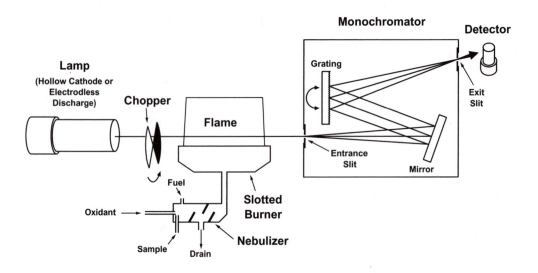

FIGURE 5.25 Schematic representation of an atomic absorption spectrophotometer.

the corresponding IDL values because there will be a greater number and degree of interferences in a chemically complex solution, relative to those in a standard solution, encountered during absorption analysis. Typical categories of interferences include: chemical interferences, ionization interferences, matrix interferences, emission interferences, spectral interferences, and background absorption (Table 5.21).

All flame atomic absorption spectrophotometers have a number of features in common (Figure 5.25). An atomic vapor is generated by aspirating a liquid sample into a flame. The solution is drawn into a nebulizer by the venturi effect (the vacuum created by the fuel flow to the flame) where the solution is converted into an aerosol. The aerosol flows into the flame and the elements in the sample enter the light path of resonance line radiation. There is no universally applicable mixture of gases (fuel-oxidant) for flame AAS. The more common fuel-oxidant gases are air-acetylene, with a flame tempertature of 2100 to 2400°C, and nitrous oxide-acetylene, with a temperature of 2600 to 2800°C. Even within a fuel-oxidant combination, gas proportions for peak sensitivity are element specific. For example, the optimal condition for Cu determination requires an air-acetylene flame that is lean on fuel (oxidizing flame). Conversely, Cr requires an air-acetylene flame that is rich on fuel (reducing flame) and Al requires a nitrous oxide-acetylene flame that is rich on fuel (reducing flame). Element-specific fuel combinations and proportions are required because each element has a specific thermal energy requirement for atomization. The resonance line radiation is generated by a primary light source, either a hollow cathode lamp or an electrodeless discharge lamp, which emits the resonance line spectrum of the element of interest. Before entering the flame, the resonance line spectrum passes through a chopper. The light source is chopped to differentiate between the light energy generated by the lamp and the light energy emitted by the flame (see next section). As the resonance line radiation passes through the flame, the element of interest adsorbs the light energy in direct proportion to concentration. Upon exiting the flame, the light enters a monochrometer which isolates the specific absorbed resonance line used for quantitation. Resonance line intensity is measured by a detector, such as a photomultiplier tube, that translates light intensity into electrical current.

5.11.1.2 Atomic Emission Spectrometry

In addition to creating an atomic vapor, the thermal energy of the flame environment in atomic absorption spectroscopy also produces excited atoms. The excited atoms are created through collision processes that raise outer-shell electrons to an excited state. Atoms remain in the excited state for only a short time, approximately 10^{-8} sec, before they decay back to the ground state (Figure 5.23). Each electronic transition generates a specific quantity of light energy ($=hc/\lambda$). The specific energy produced during an electronic transition is both element and transition dependent, as was the wavelength of light required to excite an atom during atomic absorption. Since there are a number of electronic transitions that can potentially occur in an atom, every element is characterized as having a unique set of emission lines, which reflect the energy held by the excited atom. Indeed, the spectrum of light energies emitted during the decay of an excited atom is exactly the same as that required to excite an atom. The intensity of the emission lines will increase as the number of excited atoms increases, thus allowing for the quantitative analysis of an element.

Atomic emission spectroscopy can be, and generally is, conducted using the atomic absorption spectrophotometer (Figure 5.25). However, a primary light source is not needed. Emission spectroscopy is only practical for a small number of elements, as detection limits are almost universally lower for atomic absorption. In the soil sciences, the two elements that are commonly analyzed using atomic emission are Na and K. The detection limit for Na by emission is similar to that for absorption by measuring the intensity of the 589.0 nm line (Table 5.20). However, emission holds a distinct advantage over absorption for the determination of K. The K MDL is approximately four times lower for emission, relative to absorption, using the 766.5 line.

5.11.1.3 Graphite Furnace Atomic Absorption Spectrophotometry

Flame AAS has considerable utility as a method for determining the total concentrations of elements in a solution. However, the nebulizer and flame system of producing an aerosol and creating an atomic vapor is relatively inefficient. A large percentage of the sample drains from the nebulizer during atomization (90 to 95%), and the residence time of atoms in the light path of the flame is very short. In graphite furnace AAS, the nebulizer and flame system of a flame spectrophotometer is replaced by a small graphite tube, or cuvette, which sits inside an electric furnace and is placed horizontally to the light path (Figure 5.26). A micropipette delivers a small volume of sample (20 to 50 μL)

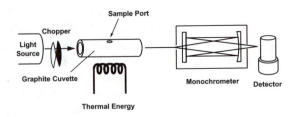

Graphite Furnace AAS

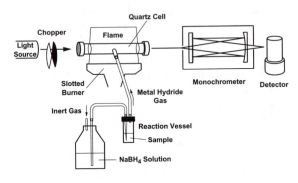

Hydride Generation AAS
Cold Vapor AAS

FIGURE 5.26 Schematic representations of graphite furnace and hydride generation (and cold-vapor) atomic absorption spectrophotometers.

through a small orifice in the top of the cuvette. The cuvette is then heated through a series of steps: drying, charring, and atomization. In the drying step, the furnace achieves a temperature that is slightly greater than the boiling point of the solvent (100 to 110°C for dilute aqueous solutions), and the temperature is held until the solvent is driven off (a function of the sample volume). The charring step is included to remove matrix components that might cause absorbance interferences. The charring step also results in the volatilization of high-boiling point matrix components and the pyrolysis matrix materials, such as fats and oils, which will crack and carbonize. Charring temperatures range from 250 to 1800°C held from 20 to 90 sec, depending on the analyte of interest. In the atomization step the analyte is placed in an atomic vapor where absorbance can occur. In general, atomization temperatures range between 2300 and 2800°C for approximately 10 sec.

The sensitivity of graphite furnace AAS is approximately three orders of magnitude greater than that of flame AAS (Table 5.20). There are two principal reasons for the increased sensitivity (reduced detection limits). First, the entire volume of the sample introduced into the graphite cuvette is atomized (there is no wasted sample), resulting in a dense atomic cloud. Second, the atomized sample is contained inside the cuvette and in the light path for an extended period of time. While there are definite advantages to graphite furnace AAS, with respect to detection limits and sensitivity, instrument response is a function of numerous variables, such as sample volume and characteristics, analyte of interest, temperatures of each stage, time at temperature during each stage, and elemental interferences (of which background absorbance and matrix effects are particularly troublesome; Table 5.21). Example instrument conditions for the analysis of select elements in water samples are shown in Table 5.22. Note that the absorbance lines for the elements are identical to those for flame AAS (Table 5.19), and that the drying temperature and time are uniform (as these values are

TABLE 5.22
Standard Conditions for Elemental Analysis Using Graphite Furnace Atomic Absorption Spectrophotometry

Element	λ, nm	Dry		Char		Atomization	
		T, °C	Time, s	T, °C	Time, s	T, °C	Time, s
Al	309.3	105	50	1200	30	2700	10
As	193.7	105	50	1100	30	2700	10
Cd	228.8	105	50	300	90	2000	10
Cr	357.9	105	50	1350	20	2700	10
Cu	324.8	105	50	1000	20	2700	10
Fe	248.3	105	50	1100	20	2700	10
Mn	279.5	105	50	1100	20	2700	10
Ni	232.0	105	50	1000	20	2700	10
Pb	283.3	105	50	500	90	2700	10
Zn	213.8	105	50	300	90	2100	10

solvent specific). Charring and atomization temperatures and times do differ, indicating the inherent volatility of each element.

5.11.1.4 Hydride Generation Atomic Absorption Spectrophotometry

Several elements, including As and Se, are able to form volatile hydrides ($AsH_3(g)$ and $SeH_2(g)$), that are readily decomposed to elemental forms (As^0 and Se^0) in the flame of an atomic absorption spectrophotometer. Metal hydrides are produced by reaction of an acidified sample solution with sodium borohydride ($NaBH_4$). The acid condition of the sample hastens the liberation of hydrogen gas from the borohydride, which in turn reduces the metal to form the gaseous hydride. The reactions for the formation of AsH_3 and SeH_2 are:

$$6BH_4^- \rightarrow 3B_2H_6 + 3H_2(g) + 6e^- \tag{5.60a}$$

$$AsO_3^{3-}(aq) + 3H^+ + 3H_2(g) + 6e^- \rightarrow AsH_3(g) + 3H_2O \tag{5.60b}$$

$$SeO_3^{2-}(aq) + 3H^+ + 3H_2(g) + 6e^- \rightarrow SeH_2(g) + 3H_2O \tag{5.60c}$$

The summation of Equations 5.60a and 5.60b (or Equations 5.60a and 5.60c) results in the redox reaction:

$$AsO_3^{3-}(aq) + 3H^+ + 6BH_4^- \rightarrow AsH_3(g) + 3H_2O + 3B_2H_6 \tag{5.60d}$$

The above reactions are typically performed in a hydride generator device that consists of a reaction flask that contains the acidified sample and a reservoir for the borohydride solution. Elemental analysis is initiated by despensing a small volume of the reductant solution into the reaction vessel. As described above, the ionic forms of the susceptible elements are converted to their respective metal hydrides. The metal hydrides are then carried by an inert gas stream to a quartz cell that is mounted in the flame of an AAS unit (Figure 5.26). The metal hydrides are decomposed in the heat of the flame to their elemental forms. The elemental forms are then able to absorb the primary light energy (193.7 nm for As and 196.0 nm for Se) that is passing through the cell. Like graphite furnace AAS, all hydride-forming elements in a sample are introduced into the quartz cell, resulting in a dense atomic cloud. Further, the atomized sample is contained inside the cell and in the light path for an extended period of time. Thus, elemental detection limits for the technique are very low and comparable to those obtained by graphite furnace (Table 5.20).

All forms of an element in an aqueous solution are not equally susceptible to reduction and hydride formation. Thus, hydride generation AAS does not typically result in a total element analysis, unless samples are prepared accordingly. This is particularly the case for As and Se. The reduction of arsenite ($As^{III}O_3^{3-}$) by $H_2(g)$ is approximately 20% more efficient than the reduction of arsenate ($As^VO_4^{3-}$). Further, $SeH_2(g)$ is only generated from selenite ($Se^{IV}O_3^{2-}$). Thus, solutions that contain the oxidized forms of As and Se, or mixtures of oxidized and reduced forms, require pretreatment prior to hydride generation AAS analysis, if total concentrations are required. Typically, reduction is performed by acidification and heating (for Se), or by acidification and the addition of a reductant (for As).

The hydride generation system described above is also suitable for the determination of Hg. The purpose of generating metal hydrides is to create a gaseous form of an element that can be easily oxidized to the elemental form by thermal energy. Instead of a hydride, the reduction of ionic mercury (Hg^{2+}) by $H_2(g)$ in a sample solution results in the direct production of volatile elemental Hg^0:

$$2BH_4^- \rightarrow B_2H_6 + H_2(g) + 2e^- \tag{5.61a}$$

$$Hg^{2+}(g) + H_2(g) + 2e^- \rightarrow Hg^0(g) + 2H^+ \tag{5.61b}$$

The summation of Equations 5.61a and 5.61b yields:

$$Hg^{2+}(g) + 2BH_4^- \rightarrow Hg^0(g) + B_2H_6 + 2H^+ \tag{5.61c}$$

The Hg^0 is transported to the quartz cell of the AAS unit using inert gas (Figure 5.26), where it is positioned in the path of primary light (253.6 nm). Since the mercury is already in the elemental form, thermal energy from a flame is not normally required for Hg determinations. This technique, termed cold-vapor AAS, is very sensitive, with a detection limit of 0.2 µg L^{-1} for Hg (Table 5.20). The method does require mercury to be present in the divalent ion form; thus, pretreatment to decompose organic Hg and to oxidize other reduced inorganic Hg species may be required.

5.11.2 Inductively Coupled (Argon) Plasma Spectrometry

Inductively coupled argon plasma–atomic emission spectroscopy (ICP) works under the same atomic excitation principles as atomic emission spectroscopy. A solution is aspirated into a thermal environment, creating a cloud of completely atomized elements that are excited through collision processes, which raise electrons to an excited state. As atoms decay back to the ground state, characteristic radiation is released in the form of light energy. The characteristic resonance lines are then detected by a spectrometer (Figure 5.27). However, the differences between flame atomic emission and ICP are substantial. First, as indicated in the names, flame atomic emission creates the thermal environment for atomic excitation through the burning of a fuel, such as acetylene. The flame achieves temperatures that range between 2100 and 2800°C, depending on the gas combinations and proportions. This temperature range lacks sufficient thermal energy to efficiently atomize and excite all but a few elements. The ICP creates a thermal environment through the interaction of ionized argon gas and a radio frequency field to form argon plasma. The temperature of the plasma ranges between 4000 and 8000°C, and is sufficient to atomize almost all elements in a sample. Further, the argon plasma is close to an ideal thermal source, providing a high degree of excitation and a multiplicity of resonance lines from which to quantify elemental concentrations.

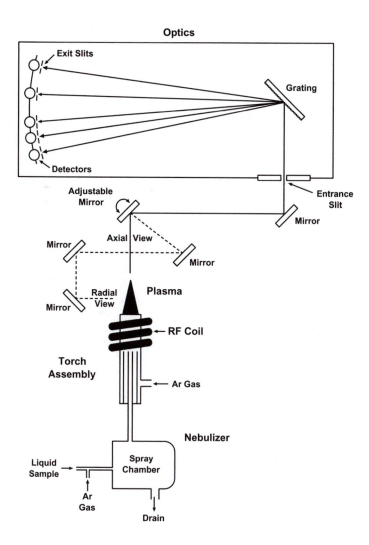

FIGURE 5.27 Schematic representation of an inductively coupled argon plasma-atomic emission spectrometer.

Inductively coupled argon plasma spectrometry also differs from atomic emission and atomic absorption techniques in that ICP is a truly simultaneous, multielemental analysis technique. Sequential ICP instruments, designed to quantify one element at a time can be obtained; however, they do not fully utilize the capabilities of the technique. Simultaneous ICP instruments employ a polychrometer to disperse the light energy from the plasma, focusing each resonance line to be measured on its own dedicated photomultiplier tube (Figure 5.27). Photodiode array detectors and charge coupled devices may also be employed, offering greater flexibility in the number of resonance lines that can be simultaneously detected for a single element.

Light energy from the plasma can be obtained by viewing the plasma in one of two orientations: radial or axial. The radial configuration is the traditional orientation, offering analytical detection limits that range from 1 to 80 μg L^{-1} (Table 5.20), and an analytical range of up to five orders of magnitude (Table 5.23). The axial configuration is a relatively recent innovation, yielding a five- to tenfold increase in sensitivity for most elements relative to the radial configuration. The overall sensitivity, multielemental capabilities, and broad analytical range of the ICP, relative to flame AAS, have made the technique a preferred analytical tool in soil and environmental research and soil testing labs.

TABLE 5.23
Emission Lines and the Dynamic Analytical Range for Elemental Analysis by Inductively Coupled Argon Plasma Spectrometry

Element	λ nm	Analytical Range Low mg L^{-1}	Analytical Range High mg L^{-1}	Element	λ nm	Analytical Range Low mg L^{-1}	Analytical Range High mg L^{-1}
Al	308.215	0.025	500	Mn	257.610	0.001	100
As	193.696	0.050	500	Mo	202.030	0.005	500
B	249.678	0.006	200	Na	588.995	0.010	500
Ba	493.409	0.001	100	Ni	231.604	0.010	500
Ca	317.933	0.01	1000	P	214.914	0.060	1000
Cd	228.802	0.002	200	Pb	220.353	0.025	1000
Co	228.616	0.003	300	S	182.040	0.080	1000
Cr	267.716	0.005	500	Se	196.026	0.050	1000
Cu	324.754	0.002	200	Si	288.158	0.02	500
Fe	259.940	0.005	500	Sr	421.552	0.001	100
Hg	184.950	0.03	500	Ti	334.941	0.002	200
K	766.491	0.40	1000	Zn	213.856	0.002	200
Mg	279.079	0.015	1000	Zr	339.198	0.003	200

EXERCISES

1. In soil solutions and other natural waters, the $PbCl^+$ complex forms according to the reaction: $Pb^{2+}(aq) + Cl^-(aq) \rightarrow PbCl^+(aq)$. Given the standard free energy of formation (ΔG_f°) values below, compute the formation constant $(\log K_f)$ for this ion association reaction (the correct answer is $\log K_f = 1.63$).

Species	ΔG_f°, kJ mol^{-1}
Pb^{2+}	−24.4
Cl^-	−131.3
$PbCl^+$	−164.8

- Calculate the value of $\log {}^cK_f$ in solutions of ionic strength 0.01 M and 0.1 M. Does increasing the ionic strength of a solution increase or decrease the concentration of $PbCl^+$ formed?
- If the ionic strength of a soil solution is 0.01 M, the Pb^{2+} concentration is 10^{-7} M, and the Cl^- concentration is 10^{-3} M, what is the total concentration of Pb in the soil solution (assuming only the Pb^{2+} and $PbCl^+$ species occur)?

2. Soil solutions are nonideal. Define nonideality and indicate the interactions in soil solutions that are responsible for nonideality. Identify the characteristics of solutes and the solvent that affect solution nonideality.

3. On the basis of the data in the table below, what can be said about the relative Lewis base character of ligands A$^-$ and B$^-$ (indicate which base is relatively soft, A$^-$ and B$^-$)? Discuss how you obtained your answer.

Complex	$\log K_f$	Complex	$\log K_f$
CuA^+	-0.03	CuB^+	0.40
FeA^{2+}	0.60	FeB^{2+}	1.48
PbA^+	1.77	PbB^+	1.59
TlA^0	0.91	TlB^0	0.49
TlA^{2+}	9.70	TlB^{2+}	7.72
AgA^0	4.68	AgB^0	3.31
CdA^+	2.14	CdB^+	1.98

4. The toxic metal cadmium (Cd^{2+}) has a tendency to complex with as many as four Cl^- ions. The complexation reactions can be written as:

$$Cd^{2+} + Cl^- \rightarrow CdCl^+ \qquad \log K_{f,1} = 1.98$$
$$Cd^{2+} + 2Cl^- \rightarrow CdCl_2^0 \qquad \log K_{f,2} = 2.60$$
$$Cd^{2+} + 3Cl^- \rightarrow CdCl_3^- \qquad \log K_{f,3} = 2.40$$
$$Cd^{2+} + 4Cl^- \rightarrow CdCl_4^{2-} \qquad \log K_{f,4} = 2.50$$

Compute the percentage of total Cd in a solution that remains uncomplexed (free Cd^{2+}) if the Cl^- concentration is 0.005 M (ignore activity corrections – activities equal concentrations).

5. The effective ionic strength of a solution (I_e) is closely related to the electrical conductivity (EC) of a solution. This relationship is quantified by the Marion-Babcock equation: $\log I_e = 1.159 + \log EC$ (where I_e is in units of mol m^{-3} and EC is in dS m^{-1}). The value of $\log K_f$ for the formation of $CuSO_4^0$ from Cu^{2+} and SO_4^{2-} is 2.36. The value of $\log K_f$ for the formation of $CuH_2PO_4^+$ from Cu^{2+} and $H_2PO_4^-$ is 1.59. Calculate $\log{}^c K_f$ for both the $CuSO_4^0$ and $CuH_2PO_4^+$ formation reactions in a soil solution that has an EC of 1.2 dS m^{-1}. Does increasing EC enhance or diminish formation of the soluble complexes (the answer may differ depending on the complex formed)?

6. In a pH 7 soil solution Pb is distributed into Pb^{2+}, $PbCO_3^0$, and $PbHCO_3^+$. Develop an expression for α_{Pb} (the distribution coefficient). Compute the value of α_{Pb} if $[CO_3^{2-}] = 10^{-4}$ M and $[HCO_3^-] = 10^{-3}$ M. Use the $\log K_f$ values presented in Table 5.11 for the formation of $PbCO_3^0$ and $PbHCO_3^+$, and assume an ionic strength of 0.01 M. What percentage of the total Pb is present as the free cation?

7. A soil solution is found to contain the following Cu and Cd species (where $-\log M$ is the negative common logarithm of the molar concentration):

Species	$-\log M$
Cu^{2+}	8.24
$CuHCO_3^+$	8.72
Cu–org^0	5.72
Cd^{2+}	6.06
$CdHCO_3^+$	7.62
$CdSO_4^0$	7.16
$CdCl^+$	6.61
Cd–org^0	6.76

- What are the total concentrations of Cu and Cd in this soil solution, as might be measured by an applicable analytical technique? Express your answer in units of $-\log M$.
- Using Davies equation for charged species and the Setchenov equation for neutral species, calculate the single-ion activity coefficient for each species ($I = 0.01$ M).
- Compute the activity of each species in the soil solution.
- What percentage of the total Cu and Cd present in the soil solution exists in each species? What are the essential differences between the aqueous speciation of Cu and Cd?
- Based on the results of your computations, which element would be predicted to be more mobile in the soil from which this solution was obtained and why?

8. Humic substances (e.g., fulvic acids) and nonhumic substances (low-molecular-mass organic acids) are ubiquitous in soil solutions. Discuss how these substances impact the fate and behavior of metal cations in soils.

9. Increased Fe uptake by plants tends to reduce Mn uptake. Based on this information, and the relatively low solubility of Mn in soils, explain how the application of a Mn-chelate fertilizer to a marginally Mn-deficient soil may actually induce Mn deficiency symptoms.

10. Using the ion association constants presented in Table 5.11 and the standard free energy of formation (ΔG_f°) values in Table 5.5, compute the ΔG_f° values for $CaMoO_4^0$ and $MgMoO_4^0$.

11. It is stated in the text that the extraction of soil water from the soil matrix, for the express purpose of obtaining a true soil water sample, is not a feasible objective. Comment on this statement and discuss the reasons why soil solution sampling techniques cannot generate a sample of the true soil solution.

12. ICP and atomic absorption spectrophotometry (AAS) are two common mechanisms for determining the total concentrations of elements in aqueous solutions. Compare and contrast these two analytical techniques.

13. Table 5.18 lists several toxicity sequences for metals in a variety of organisms. The biochemical mechanisms of metal toxicity tend to fall into three classes: (1) blocking of an essential functional group of a biomolecule; (2) displacement of an essential metal in a biomolecule; and (3) modification of the configuration of a biomolecule. Give a rationale for the toxicity sequences in the table in terms of the HSAB principle. What is the apparent Lewis base character (hard, soft) of biomolecule complexation sites?

14. TM Industries produces an organic pesticide that they wish to have labeled for application to cropland to control broadleaf weeds. They fund a researcher at the local experiment station to examine the behavior of this compound in the environment. If the researcher can show that this compound is immobile and rapidly degraded, the compound will be labeled for use. The organic compound is applied to a soil surface to examine fate and behavior. Shortly after application, a rainfall of sufficient duration and intensity to generate drainage occurs. When sufficiently dry, the study area is entered and soil samples are obtained as a function of depth down to 1 m and subjected to analysis for the organic compound. The results indicate that the compound is concentrated in the surface soil where the organic matter content is highest. Further, they do not detect any compound below a depth of 30 cm. However, a mass balance indicates that greater than 50% of the applied compound cannot be accounted for. The researchers conclude that the absence of compound in the soil profile, coupled with the finding that no detectable levels of the compound are found below a depth of 30 cm, is direct evidence of compound immobility and rapid degradation. These results and interpretations are then reported to TM Industries. Are the interpretations of the researcher correct? (Defend your response.) What additional information might be useful to elucidate this compound's environmental fate and behavior?

REFERENCES

Ahrland, S., J. Chatt, and N.R. Davies. The relative affinities of ligand atoms for acceptor molecules and ions. *Q. Rev. Chem. Soc.* 12:265–276, 1958.

Alva, A.K., M.E. Sumner, and W.P. Miller. Relationship between ionic strength and electrical conductivity of soil solutions. *Soil Sci.* 152:239–242, 1991.

American Public Health Association. *Standard Methods for the Examination of Water and Wastewater.* APHA, Washington, D.C., 1985.

Baes, C.F. and R.E. Mesmer. *The Hydrolysis of Cations.* R.E. Krieger, Malabar, FL, 1986.

Berbel, F., J.M. Diaz-Cruz, C. Ariño, M. Esteban, F. Mas, J.L. Garcés, and J. Puy. Voltammetric analysis of heterogeneity in metal ion binding by humics. *Environ. Sci. Technol.* 35:1097–1102, 2001.

Brown, G.K., P. MacCarthy, and J.A. Leenheer. Simultaneous determination of Ca, Cu, Ni, Zn and Cd binding strengths with fulvic acid fractions by Schubert's method. *Anal. Chim. Acta* 402:169–181, 1999a.

Brown, G.K., S.E. Cabaniss, P. MacCarthy, and J.A. Leenheer. Cu(II) binding by a pH-fractionated fulvic acid. *Anal. Chim. Acta* 402:183–193; Voltammetric analysis of heterogeneity in metal binding by humics. *Environ. Sci. Technol.* 35:1097–1102, 1999b.

Burgess, J. *Metal Ions in Solutions.* Ellis Horwood, Chichester, England, 1978.

Carballeira, J.L., J.M. Antelo, and F. Arce. Analysis of the Cu^{2+}-soil fulvic acid complexation by anodic stripping voltammetry using an electrostatic model. *Environ. Sci. Technol.* 34:4969–4973, 2000.

Christl, I., C.J. Milne, D.G. Kinniburgh, and R. Kretzschmar. Relating ion binding by fulvic and humic acids to chemical composition and molecular size. II. Metal binding. *Environ. Sci. Technol.* 35:2512–2517, 2001.

Cotton, F.A., G. Wilkerson, C.A. Murillo, and M. Bochmann. *Advanced Inorganic Chemistry.* J. Wiley & Sons, New York, 1999.

Cusanelli, A., U. Frey, D.T. Richens, and A.E. Merbach. The slowest water exchange at a homoleptic mononuclear metal center: variable-temperature and variable-pressure ^{17}O NMR study on $[Ir(H_2O)_6]^{3+}$. *J. Am. Chem. Soc.* 118:5265–5271, 1996.

Essington, M.E. Formation of calcium and magnesium molybdate complexes in dilute aqueous solutions. *Soil Sci. Soc. Am. J.* 56:1124–1127, 1992.

Essington, M.E., D.D. Tyler, and G.V. Wilson. Fluometuron behavior in long-term tillage plots. *Soil Sci.* 160:405–414, 1995.

Esteves da Silva, J.C.G. and C.J.S. Oliveira. Metal ion complexation properties of fulvic acids extracted from composted sewage sludge as compared to a soil fulvic acid. *Water Res.* 2002.

Gillman, G.P., and L.C. Bell. Soil solution studies on weathered soils from tropical Northern Queensland. *Aust. J. Soil Res.* 16:67–77, 1978.

Griffin, R.A., and J.J. Jurinak. Estimation of activity coefficients from the electrical conductivity of natural aquatic systems and soil extracts. *Soil Sci.* 116:26–30, 1973.

Hunt, J.P. and H.L. Friedman. Aquo complexes of metal ions. *Progress Inorg. Chem.* 30:359–387, 1983.

Irving, H. and R.J.P. Williams. The stability of transition-metal complexes. *J. Am. Chem. Soc.* 75:3192–3210, 1953.

Kielland, J. Individual activity coefficients of ions in aqueous solutions. *J. Am. Chem. Soc.* 59:1675–1678, 1937.

Leenheer, J.A., G.K. Brown, P. MacCarthy, and S.E. Cabaniss. Models of metal binding structures in fulvic acid from the Suwannee River, Georgia. *Environ. Sci. Technol.* 32:2410–2416, 1998.

Li, J., E.M. Perdue, and L.T. Gelbaum. Using Cadmium-113 NMR spectrometry to study metal complexation by natural organic matter. *Environ. Sci. Technol.* 32:483–487, 1998.

Lindsay, W.L. *Chemical Equilibria in Soils.* John Wiley & Sons, New York, 1979.

Luster, J., T. Lloyd, G. Sposito, and I.V. Fry. Multi-wavelength molecular fluorescence spectrometry for quantitative characterization of copper(II) and aluminum(III) complexation by dissolved organic matter. *Environ. Sci. Technol.* 30:1565–1574, 1996.

Mahoney, M.W. and W.L. Jorgensen. A five-point model for liquid water and the reproduction of the density anomaly by rigid, nonpolarizable potential functions. *J. Chem. Phys.* 112:8910–8922, 2000.

Margerum, D.W., G.R. Cayley, D.C. Weatherburn, and G.K. Pagenkopf. Kinetics and mechanisms of complex formation and ligand exchange. In *Coordination Chemistry.* Vol. 2. A.E. Martell (Ed.) ACS Monograph 174, ACS, Washington, D.C., 1978, pp. 1–220.

Marion, G.M. and G.L. Babcock. Predicting specific conductance and salt concentration in dilute aqueous solutions. *Soil Sci.* 122:181–187, 1976.

Martell, A.E. and R.M. Smith. *Critical Stability Constants. Vol. 1: Amino Acids*. Plenum Press, New York, 1974.

Martell, A.E. and R.M. Smith. *Critical Stability Constants. Vol. 3: Other Organic Acids*. Plenum Press, New York, 1977.

Martell, A.E. and R.M. Smith. *Critical Stability Constants. Vol. 5: First Supplement*. Plenum Press, New York, 1982.

Misono, M., E. Ochiai, Y. Saito, and Y. Yoneda. A new dual parameter scale for the strength of Lewis acids and bases with the evaluation of their softness. *J. Inorg. Nucl. Chem.* 29:2685–2691, 1967.

Nantsis, E.A. and W.R. Carper. Molecular structure of divalent metal ion-fulvic acid complexes. *J. Molecular Structure (Theochem.)* 423:203–212, 1998.

Naumov, G.B., R.N. Ryzhenko, and I.L. Khodakovsky. *Handbook of Thermodynamic Data*. Translated by G.J. Soleimani. PB-226 722. National Technology Information Service, Springfield, VA, 1974.

Nieboer, E. and W.A.E. McBryde. Free-energy relationships in coordination chemistry. III. A comprehensive index to complex stability. *Can. J. Chem.* 51:2512–2524, 1973.

Nieboer, E. and D.H.S. Richardson. The replacement of the nondescript term 'heavy metals' by a biologically significant classification of metal ions. *Environ. Pollut. (Series B)* 1:3–26, 1980.

Nightingale, E.R., Jr. Phenomenological theory of ion salvation. Effective radii of hydrated ions. *J. Phys. Chem.* 63:1381–1387, 1959.

O'Dell, J.D., J.D. Wolt, and P.M. Jardine. Transport of imazethapyr in undisturbed soil columns. *Soil Sci. Soc. Am. J.* 56:1711–1715, 1992.

Otto, W.H., W.R. Carper, and C.K. Larive. Measurement of cadmium(II) and calcium(II) complexation by fulvic acids using [113]Cd NMR. *Environ. Sci. Technol.* 35:1463–1468, 2001a.

Otto, W.H., S.D. Burton, W.R. Carper, and C.K. Larive. Examination of cadmium(II) complexation by the Suwannee River fulvic acid using [113]Cd NMR relaxation measurements. *Environ. Sci. Technol.* 35:4900–4904, 2001b.

Pankow, J.F. and J.J. Morgan. Kinetics for the aquatic environment. *Environ. Sci. Technol.* 15:1155–1164, 1981.

Parr, R.G. and R.G. Pearson. Absolute hardness: companion parameter to absolute electronegativity. *J. Am. Chem. Soc.* 105:7512–7516, 1983.

Pasricha, N.S. Predicting ionic strength from specific conductance in aqueous soil solutions. *Soil Sci.* 143:92–96, 1987.

Pearson, R.G. Hard and soft acids and bases. *J. Am. Chem. Soc.* 85:3533–3539, 1963.

Pearson, R.G. Hard and soft acids and bases, HSAB. I. Fundamental principles. *J. Chem. Educ.* 45:581–587, 1968.

Plankey, B.J., H.H. Patterson, and C.S. Cronan. Kinetics of fluoride complexation in acidic waters. *Environ. Sci. Technol.* 20:160–165, 1986.

Quinn, T.R. Chemical Speciation of Selenium in Fly Ash. Ph.D. dissertation. University of California, Riverside, 1985.

Robie, R.A., B.S. Hemingway, and J.R. Fisher. Thermodynamic properties of minerals and related substances at 298.15 K and 1 bar (10^5 pascals) pressure and at higher temperature. U.S. Geological Survey Bull. 1452. U.S. Government Printing Office, Washington, D.C., 1978.

Rutgers, A.T. and Y. Hendrikx. *Trans. Faraday Soc.* 58:2184, 1962.

Sadiq, M. and W.L. Lindsay. Selection of standard free energies of formation for use in soil chemistry, Colorado State Univ. Tech. Bull. 134 and Supplements. Colorado State University Experimental Station, Fort Collins, CO, 1979.

Sekaly, A.L.R., R. Mandal, N.M. Hassan, J. Murimboh, C.L. Chakrabarti, M.H. Back, D.C. Grégoire, and W.H. Schroeder. Effect of metal/fulvic acid mole ratios on the binding of Ni(II), Pb(II), Cu(II), Cd(II), and Al(III) by well-characterized fulvic acids in aqueous model solutions. *Anal. Chim. Acta* 402:211–221, 1999.

Smith, D.S. and J.R. Kramer. Multisite metal binding to fulvic acid determined using multiresponse fluorescence. *Anal. Chim. Acta* 416:211–220, 2000.

Smith, R.M. and A.E. Martell. *Critical Stability Constants. Vol. 2: Amines*. Plenum Press, New York, 1975.

Smith, R.M. and A.E. Martell. *Critical Stability Constants. Vol. 4: Inorganic Complexes*. Plenum Press, New York, 1976.

Soltanpour, P.N., G.W. Johnson, S.M. Workman, J.B. Jones, Jr., and R.O. Miller. Inductively coupled plasma emission spectrometry and inductively coupled plasma-mass spectrometry. In *Methods of Soil Analysis*. Part 3. Chemical methods. D.L. Sparks et al. (Ed.) SSSA Book Series 5, SSSA and ASA, Madison, WI, 1996, pp. 91–139.

Suresh, S.J. and V.M. Naik. Hydrogen bond thermodynamic properties of water from dielectric constant data. *J. Chem. Phys.* 113:9727–9732, 2000.

U.S. Environmental Protection Agency. Test methods for evaluating solid waste. Physical/chemical methods. U.S. EPA SW 846. 2002. URL: http://www.epa.gov/epaoswer/hazwaste/test/txmain.htm.

Wagman, D.D., W.H. Evans, V.B. Parker, R.H. Schumm, I. Harlow, S.M. Bailey, K.L. Churney, and R.L. Nutall. Selected values for inorganic and C_1 and C_2 organic substances in SI units. *J. Phys. Chem. Ref. Data* 11, Supplement No. 2, 1982.

Wolt, J.D. *Soil Solution Chemistry. Applications to Environmental Science and Agriculture.* John Wiley & Sons, New York, 1994.

Woods, T.L. and R.M. Garrels. *Thermodynamic Values at Low Temperature for Natural Inorganic Materials: An Uncritical Summary.* Oxford University Press, New York, 1987.

Wright, R.J. and T. Stuczynski. Atomic absorption and flame emission spectrometry. In *Methods of Soil Analysis. Part 3. Chemical Methods.* D.L. Sparks et al. (Eds.) SSSA Book Series 5, SSSA and ASA, Madison, WI, 1996, pp. 65–90.

Wulfsberg, G. *Principles of Descriptive Chemistry.* Brooks and Cole Publishing, Monterey, CA, 1987.

6 Mineral Solubility

Mineral dissolution and precipitation processes influence the distribution of inorganic substances between the soil solid and solution phases, and therefore have a pronounced influence on the chemical composition of soil solutions. The extent to which either process occurs is determined by the chemical properties of the soil solution (specifically, ion activities) and the intrinsic stability of the minerals involved. While the misconception that insoluble minerals do indeed exist in the terrestrial environment persists, it is a fact that no mineral is insoluble in water, and minerals that appear to be insoluble are in reality sparingly soluble. All minerals, indeed all solids, are soluble to some degree in water. Mineral solubility principles are used to elucidate pedogenic processes, to predict the concentrations of elements in soil solutions, to predict trace element-bearing minerals formed in natural and waste-affected environments, and to verify the effectiveness of processes designed to enhance the *in situ* stabilization of trace elements that are potentially harmful to human health and the environment.

6.1 MINERAL SOLUBILITY: BASIC PRINCIPLES

The dissolution of a soil mineral can be described by the following generalized reaction:

$$M_aL_b(OH)_c(s) + cH^+(aq) \rightarrow aM^{m+}(aq) + bL^{n-}(aq) + cH_2O(l) \tag{6.1}$$

where M^{m+} is a metal cation and L^{n-} is a ligand anion. The true dissolution equilibrium constant for this reaction is:

$$K_{dis} = \frac{(M^{m+})^a(L^{n-})^b(H_2O)^c}{(M_aL_b(OH)_c)(H^+)^c} \tag{6.2}$$

where the parentheses denote activities, and $am + c - bn = 0$ is a condition of electroneutrality. For example, consider the dissolution of the copper(II) hydroxycarbonate, malachite:

$$Cu_2(OH)_2CO_3(s) + 2H^+(aq) \rightarrow 2Cu^{2+}(aq) + CO_3^{2-}(aq) + 2H_2O(l) \tag{6.3}$$

The equilibrium constant for the malachite dissolution reaction is:

$$K_{dis} = \frac{(Cu^{2+})^2(CO_3^{2-})(H_2O)^2}{(Cu_2(OH)_2CO_3)(H^+)^2} \tag{6.4}$$

Recall from Chapter 5 that the equilibrium constant for any reaction can be determined if the standard state chemical potentials ($\mu°$ values) or standard Gibbs free energies of formation ($\Delta G_f°$ values) of the reactants and products are known. For the malachite dissolution reaction (Equation 6.3), the $\Delta G_f°$ values for Cu^{2+} (65.52 kJ mol^{-1}), CO_3^{2-} (−527.9 kJ mol^{-1}), and H_2O (−237.141 kJ mol^{-1})

are obtained from Table 5.5. The ΔG_f° value for malachite (-903.7 kJ mol^{-1}) is obtained from Symes and Kester (1984). The standard free energy change for the reaction is:

$$\Delta G_r^\circ = 2\Delta G_{f,Cu^{2+}}^\circ + \Delta G_{f,CO_3^{2-}}^\circ + 2\Delta G_{f,H_2O}^\circ - \Delta G_{f,malachite}^\circ \tag{6.5}$$

Substituting the numeric ΔG_f° values into Equation 6.5 yields:

$$\Delta G_r^\circ = 2(65.52) + (-527.9) + 2(-237.141) - (-903.7) = 32.56 \text{ kJ mol}^{-1} \tag{6.6}$$

Because ΔG_r° is positive, the reaction products (Cu^{2+}, CO_3^{2-}, and liquid water) are less stable than the reactants (malachite and H^+) if the reaction were to occur in the standard state. Using Equation 5.22, the equilibrium constant for malachite dissolution is computed:

$$\log K_{dis} = -\frac{32.56}{5.708} = -5.70 \tag{6.7}$$

Equation 6.4 becomes:

$$10^{-5.70} = \frac{(Cu^{2+})^2(CO_3^{2-})(H_2O)^2}{(Cu_2(OH)_2CO_3)(H^+)^2} \tag{6.8}$$

Equations 6.2 and 6.8 can be simplified by assuming the activities of the mineral and of liquid water are unity $[(M_aL_b(OH)_c) = (H_2O) = 1.0]$. By definition, a well-crystallized and uncontaminated mineral phase existing in an environment with a temperature of 25°C and an applied pressure of 0.101 MPa (1 atm) is in the standard state. This means that the chemical potential of a mineral in such an environment will be equal to the standard state chemical potential: $\mu_{mineral} = \mu_{mineral}^\circ$. Recalling Equation 5.14 from Chapter 5, the chemical potential of a mineral in any environment is a function of the standard-state chemical potential and the activity of the mineral:

$$\mu_{mineral} = \mu_{mineral}^\circ + RT \ln a_{mineral} \tag{6.9}$$

Since $\mu_{mineral} = \mu_{mineral}^\circ$, the term $RT \ln a_{mineral}$ must equal 0. The natural gas constant (R) and the standard state temperature (T) are nonzero values; therefore, the term $RT \ln a_{mineral}$ can only equal 0 when $a_{mineral} = 1.0$. If it is assumed that the malachite in Equation 6.3 is well crystallized, and that it does not contain chemical impurities (through isomorphic substitution), then $(Cu_2(OH)_2CO_3) = 1.0$.

The activity of liquid water is defined as the partial pressure of water vapor in contact with the liquid divided by the standard state partial pressure: P/P° (assuming pressure and fugacity are equivalent). The chemical potential of water deviates from the standard state chemical potential according to:

$$\mu_w = \mu_w^\circ + \frac{RT}{M_w} \ln\left(\frac{P_w}{P_w^\circ}\right) \tag{6.10}$$

where M_w is the molecular weight of water. By convention, $\psi_w \equiv \mu_w - \mu_w^\circ$ is the soil water potential expressed in J kg^{-1}, or equivalent units. Equation 6.10 becomes:

$$\Psi_w = \left[\frac{RT}{M_w}\right] \ln\left(\frac{P_w}{P_w^\circ}\right) \tag{6.11}$$

At 25°C (298.15 K) and substituting numerical values for R and M_w, Equation 6.11 becomes,

$$\Psi_w = (137,646) \times \ln\left(\frac{P_w}{P_w^{\circ}}\right) \tag{6.12}$$

or

$$\frac{\Psi_w}{137,646} = \ln\left(\frac{P_w}{P_w^{\circ}}\right) \tag{6.13}$$

and

$$\exp\left[\frac{\Psi_w}{137,646}\right] = \left(\frac{P_w}{P_w^{\circ}}\right) \tag{6.14}$$

In water-saturated soils and under conditions required for normal plant growth, Ψ_w can range from 0 J kg^{-1} (tension-free water) to -1500 J kg^{-1} (permanent wilting point). Using Equation 6.14, it is apparent that the activity of water will only range between 1.0 and 0.989 under normal plant growth conditions. In addition, the activity of water in sea water (containing 19.4 g Cl kg^{-1}) is 0.981. However, in unsaturated (moist to dry) and saline soils, water activity can range down to 0.2.

Applying the above assumptions (unit activity of liquid water and the solid), Equation 6.2 for the generalized dissolution reaction becomes:

$$K_{sp} = \frac{(M^{m+})^a (L^{n-})^b}{(H^+)^c} \tag{6.15}$$

where K_{sp} is the solubility product constant. The solubility product constant is a true constant. For example, the K_{sp} for the malachite dissolution reaction (Equation 6.3) is:

$$K_{sp,malachite} = 10^{-5.70} = \frac{(Cu^{2+})^2 (CO_3^{2-})}{(H^+)^2} \tag{6.16}$$

6.1.1 MINERALOGICAL CONTROLS ON ION ACTIVITIES IN SOIL SOLUTIONS

The solubility product concept is a powerful tool for predicting the chemical composition of a soil solution that is in contact with a mineral phase at equilibrium. Consider an alkaline soil environment that receives anthropogenic Cu from a local smelting operation. If ancillary evidence suggests that malachite is present in the affected soil, the information contained in Equation 6.16 can be used to predict the activity of Cu^{2+} in the soil solution. However, in order to predict Cu^{2+} activity, two of the reaction components, the activities of CO_3^{2-} and H^+, must be established. There are three ways to obtain values for the activities of reaction components: (1) analytically determine the total concentrations, use the ion speciation model to compute the concentrations of the reaction components, then multiply the species concentration by the single-ion activity coefficient to generate the species activity (or in the case of the proton, directly measure solution pH); (2) poise or fix (or control) the activity of a component using a second mineral solubility reaction or another controlling process that is known or predicted to occur; or (3) assign a value for the activity of a component based on known values obtained through the characterization of similar environments.

The first step in determining the activity of Cu^{2+} controlled by malachite in an alkaline soil is to establish a value for the activity of CO_3^{2-}. There are two approaches that can be used, both leading to the same result. The soil atmosphere is a reservoir for $CO_2(g)$, which is in equilibrium with CO_3^{2-} in the soil solution according to the reaction:

$$CO_2(g) + H_2O(l) \rightarrow CO_3^{2-}(aq) + 2H^+(aq) \tag{6.17}$$

where $\log K = -18.17$. At equilibrium, the following equality describes the distribution of inorganic C between the gaseous CO_2 and the carbonate ion:

$$K = 10^{-18.17} = \frac{(CO_3^{2-})(H^+)^2}{P_{CO_2}} \tag{6.18}$$

Rearranging, the activity of CO_3^{2-} can be described as a function of CO_2 partial pressure in the soil atmosphere and the solution proton activity:

$$(CO_3^{2-}) = \frac{10^{-18.17} P_{CO_2}}{(H^+)^2} \tag{6.19}$$

For convenience, taking the common logarithm of both sides of Equation 6.19 leads to:

$$\log(CO_3^{2-}) = -18.17 + \log P_{CO_2} + 2pH \tag{6.20}$$

Thus, the activity of CO_3^{2-} in the soil solution is determined by the partial pressure of CO_2 in the soil atmosphere and solution pH. The atmospheric partial pressure of CO_2 is $10^{-3.52}$. Typically, however, the partial pressure of CO_2 in the soil atmosphere can be 10 to 100 times greater than that in the above-ground atmosphere (principally a result of microbial respiration). Therefore, the selection of a P_{CO_2} value of $10^{-2.5}$ is appropriate. Since the soil is alkaline, the pH will be greater than 7. For this example, the soil is assumed to have a solution pH of 7.5. Substituting the numerical values for CO_2 partial pressure and pH into Equation 6.20 yields:

$$\log(CO_3^{2-}) = -18.17 + (-2.5) + 2(7.5) = -5.67 \tag{6.21}$$

Rearranging Equation 6.16 to obtain Cu^{2+} activity as a function of CO_3^{2-} and proton activities:

$$(Cu^{2+})^2 = \frac{10^{-5.70}(H^+)^2}{(CO_3^{2-})} \tag{6.22}$$

or

$$(Cu^{2+}) = \frac{10^{-2.85}(H^+)}{(CO_3^{2-})^{0.5}} \tag{6.23}$$

Taking the common logarithm of Equation 6.23 yields:

$$\log(Cu^{2+}) = -2.85 - \tfrac{1}{2}\log(CO_3^{2-}) - pH \tag{6.24}$$

Substituting the computed $\log(CO_3^{2-}) = -5.67$ and pH $= 7.5$ into Equation 6.24 yields:

$$\log(Cu^{2+}) = -2.85 - \tfrac{1}{2}(-5.67) - 7.5 = -7.52 \tag{6.25}$$

The activity of Cu^{2+} in a pH 7.5 soil solution in equilibrium with the mineral malachite and a soil atmosphere having a CO_2 partial pressure of $10^{-2.5}$ is predicted to be $10^{-7.52}$. Further, the computed Cu^{2+} activity is only valid for a soil at 25°C, under an applied pressure of 0.101 MPa, and assuming unit activity of malachite and soil water. If the ionic strength of the soil solution is known, the concentration of the free Cu^{2+} species can be determined using the equality: $(Cu^{2+})/\gamma_{2+} = [Cu^{2+}]$, or $\log(Cu^{2+}) - \log\gamma_{2+} = \log[Cu^{2+}]$. For example, the single-ion activity coefficient for a divalent ion in a 0.01 M ionic strength solution is $\gamma_{2+} = 0.6608$ L mol^{-1} ($\log\gamma_{2+} = -0.18$) using the Davies equation. It follows that $\log[Cu^{2+}] = -7.52 - (-0.18) = -7.34$, or $[Cu^{2+}] = 10^{-7.34}$ M.

A less cumbersome method of predicting Cu^{2+} activity in the alkaline soil solution requires that the malachite dissolution reaction be written such that CO_2 is generated, rather than CO_3^{2-}. Chemical reactions can be summed, just as mathematical expresses are summed. Further, the $\log K$ value for the resulting reaction is the sum of the $\log K$ values from the summed reactions:

$$Cu_2(OH)_2CO_3\ (s) + 2H^+(aq) \rightarrow 2Cu^{2+}(aq) + CO_3^{2-}(aq) + 2H_2O(l) \qquad \log K = -5.70$$

$$CO_3^{2-}(aq) + 2H^+(aq) \rightarrow CO_2(g) + H_2O(l) \qquad \log K = 18.17$$

$$\overline{Cu_2(OH)_2CO_3\ (s) + 4H^+(aq) \rightarrow 2Cu^{2+}(aq) + CO_2(g) + 3H_2O(l) \qquad \log K = 12.47}$$

The solubility product constant for the malachite dissolution reaction that results in the production of CO_2 is:

$$K_{sp} = 10^{12.47} = \frac{(Cu^{2+})^2\, P_{CO_2}}{(H^+)^4} \tag{6.26}$$

The common logarithm of Equation 6.26 is:

$$12.47 = 2\log(Cu^{2+}) + \log P_{CO_2} + 4pH \tag{6.27}$$

Rearranging to obtain (Cu^{2+}) as a function of CO_2 partial pressure and pH:

$$\log(Cu^{2+}) = 6.51 - \tfrac{1}{2}\log P_{CO_2} - 2pH \tag{6.28}$$

Again, applying the stipulated conditions $\log P_{CO_2} = -2.5$ and pH $= 7.5$,

$$\log(Cu^{2+}) = 6.24 - \tfrac{1}{2}(-2.5) - 2(7.5) = -7.52 \tag{6.29}$$

which is identical to the $\log(Cu^{2+})$ value computed in Equation 6.25.

In the preceding example, the activity of Cu^{2+} (a dependent variable) is uniquely determined (controlled) by the dissolution of malachite once the two independent variables (CO_2 partial pressure and pH) are established. Altering either one of the independent variables, or both, would result in a different value for the activity of Cu^{2+} controlled by malachite. The result of modifying either CO_2 partial pressure or pH on the Cu^{2+} activity controlled by malachite can be deduced by examining the solubility relationship (Equation 6.28). A unit increase in $\log P_{CO_2}$ (from -2.5 to -1.5, or a tenfold increase in P_{CO_2}) will decrease $\log(Cu^{2+})$ by 0.5 unit from -7.52 to -8.02 (or the activity of Cu^{2+} will decrease by approximately one third), if the pH is held constant. However, a one unit increase in solution pH (from 7.5 to 8.5) will decrease $\log(Cu^{2+})$ by two units, from -7.52 to -9.52 (or the activity of Cu^{2+} will decrease 100-fold) if P_{CO_2} is held constant.

It is important to recognize that the solubilities of trace element-bearing minerals in the soil environment do not control the soil solution composition of all the components contained in the mineral. For example, the dissolution of malachite will not control the soil solution composition of Cu^{2+}, CO_3^{2-}, and H^+ as might be inferred from Equation 6.16. Instead, the activity of Cu^{2+} will be the mineral-controlled property of an equilibrated solution (through the K_{sp} of malachite), as determined by the system properties: pH and CO_2 partial pressure (or CO_3^{2-} activity). Solution pH is controlled by the buffering capacity of the soil. The dissolution of a trace mineral, such as malachite, will not impact soil solution pH. Also, CO_3^{2-} activity in the soil solution is controlled by the reservoir of atmospheric CO_2. The release of CO_3^{2-} into the soil solution as a result of malachite dissolution will not alter the content of the CO_2 reservoir. This evaluation is not limited to trace element-bearing carbonate minerals. Corollary evaluations can also be performed for trace element-bearing minerals composed of ligands other than carbonate that are major soil constituents, such as metal sulfates, phosphates, and silicates.

6.1.2 THE ION ACTIVITY PRODUCT AND RELATIVE SATURATION

In the previous section, the activity of a component in a soil solution is predicted by assuming the presence of a controlling mineral phase and equilibrium conditions. The procedure works equally well in reverse and can be used to predict the presence of a controlling mineral phase in a soil. Even in soils contaminated by trace elements, the concentrations of trace element-bearing minerals will be below the levels required for direct characterization by x-ray diffraction. Thus, indirect methods of characterization are employed. One such technique, termed the equilibrium solubility method, is commonly employed to either establish or discount the presence of trace mineral phases in soil.

The solubility product constant (K_{sp}) describes the activity ratio of products to reactants that will exist in a soil solution at equilibrium with a mineral. The value of K_{sp} is specific to the mineral of interest and to the specific mineral dissolution reaction. Therefore, *if* the solution activities of the ions involved in a mineral dissolution reaction are determined and substituted accordingly into the K_{sp} expression for the mineral (see Equation 6.15 for the generalized expression), the expression will yield a numerical value that is equal to the K_{sp}, if the mineral is present and in equilibrium with the soil components. However, if the mineral is not controlling the activities of the soluble constituents, the numerical result will not equal the K_{sp}. Irrespective of the value obtained, the numerical result to the computation is defined as the ion activity product (*IAP*). Both the K_{sp} and the *IAP* are mineral-specific and have the same mathematical form. The K_{sp} unequivocally states the product or ratio of ion activities that *will be* present in a solution at equilibrium. The *IAP* is the actual product or ratio of ion activities that is present in a soil solution, irrespective of the equilibrium status of the mineral and solution components.

The *IAP* is a property of a soil solution, just as ion activities are compositional properties of a soil solution. However, the mechanism for accurately determining ion activities is fraught with numerous pitfalls that can lead to inaccurate predictions of ion activities. The quality and the completeness of the thermodynamic data employed to describe ion speciation, as well as the quality and completeness of the soil solution composition data, will dictate the validity of the computed ion activities. A detailed description of the ion association model is given in Chapter 5. Briefly, the chemical composition of a soil solution must be characterized in detail, such that all substances that may impact the chemistry of an element of interest are quantified. Typically, substances whose concentrations in the soil solution are equal to or greater than that of the element of interest must be known (including organic ligands). The ion associate model is then employed to compute ion speciation and the activities of individual species. The computation process requires ion formation constants (K_f values that are contained in a thermodynamic data file) to establish the mass balance expressions and to describe the partitioning of a substance into component species. Once known, ion activities can be used to determine the status of a solution with respect to the potential presence of trace minerals.

Consider a lead-contaminated and slightly acidic soil where a lead phosphate mineral is thought to control aqueous Pb^{2+} activity. Independent analysis of the soil solids (e.g., scanning electron microscopy–energy dispersive x-ray analysis) indicates a lead–phosphorus association in the solid phase, suggesting the occurrence of lead in a phosphate mineral. There are three lead-bearing phosphates that may potentially control Pb^{2+} in this soil, and for which solubility product constants are available: lead phosphate [$Pb(PO_4)_{0.67}(s)$], hydroxypyromorphite [$Pb(PO_4)_{0.6}(OH)_{0.2}(s)$], and plumbogummite [$PbAl_3(PO_4)_2(OH)_5(s)$]. Lead phosphate dissolves according to the following dissolution reaction:

$$Pb(PO_4)_{0.67} \rightarrow Pb^{2+} + 0.67PO_4^{3-}$$

$$K_{sp,Pb(PO_4)_{0.67}} = (Pb^{2+})(PO_4^{3-}) = 10^{-14.79} \tag{6.30}$$

Hydroxypyromorphite (HP) dissolves according to the reaction:

$$Pb(PO_4)_{0.6}(OH)_{0.2} + 0.2H^+ \rightarrow Pb^{2+} + 0.6PO_4^{3-} + 0.2H_2O$$

$$K_{sp,HP} = \frac{(Pb^{2+})(PO_4^{3-})^{0.6}}{(H^+)^{0.2}} = 10^{-12.56} \tag{6.31}$$

Plumbogummite (PG) dissolves according to the reaction:

$$PbAl_3(PO_4)_2(OH)_5 \cdot H_2O + 5H^+ \rightarrow Pb^{2+} + 3Al^{3+} + 2PO_4^{3-} + 6H_2O$$

$$K_{sp,PG} = \frac{(Pb^{2+})(Al^{3+})^3(PO_4^{3-})^2}{(H^+)^5} = 10^{-32.79} \tag{6.32}$$

From the solubility reactions, it is evident that the status of the soil solution with respect to the three minerals can be determined once solution pH and the activities of Pb^{2+}, PO_4^{3-}, and Al^{3+} are known. Detailed chemical analysis of the acidic (pH 5.97) soil solution, followed by the application of an ion speciation model (e.g., GEOCHEM-PC), yields a computed Pb^{2+} activity of $10^{-7.25}$, a PO_4^{3-} activity of $10^{-13.13}$, and an Al^{3+} activity of $10^{-9.86}$ (Table 6.1). The IAP of the soil solution with respect to each of the three solids is:

$$IAP_{Pb(PO_4)_{0.67}} = (Pb^{2+})(PO_4^{3-})^{0.67} = (10^{-7.25})(10^{-13.13})^{0.67} = 10^{-16.00} \tag{6.33}$$

$$IAP_{HP} = \frac{(Pb^{2+})(PO_4^{3-})^{0.6}}{(H^+)^{0.2}} = \frac{(10^{-7.25})(10^{-13.13})^{0.6}}{(10^{-5.97})^{0.2}} = 10^{-13.93} \tag{6.34}$$

$$IAP_{PG} = \frac{(Pb^{2+})(Al^{3+})^3(PO_4^{3-})^2}{(H^+)^5} = \frac{(10^{-7.25})(10^{-9.86})^3(10^{-13.13})^2}{(10^{-5.97})^5} = 10^{-33.24} \tag{6.35}$$

TABLE 6.1
Solubility Product Constants ($\log K_{sp}$) for Lead Phosphate Minerals and Corresponding Ion Activity Products ($\log IAP$) and Saturation Indices (SI Values) Computed for an Acidic (pH 5.97) Soil Solution with $\log(Pb^{2+}) = -7.25$, $\log(PO_4^{3-}) = -13.13$, and $\log(Al^{3+}) = -9.86$[a]

Mineral Dissolution Reaction and $\log IAP$ Function	$\log K_{sp}$[b]	$\log IAP$	SI
Lead phosphate			
$Pb(PO_4)_{0.67} \rightarrow Pb^{2+} + 0.67\,PO_4^{3-}$ $\log IAP = \log(Pb^{2+}) + 0.67\log(PO_4^{3-})$	−14.79	−16.00	−1.21
Hydroxypyromorphite			
$Pb(PO_4)_{0.6}(OH)_{0.2} + 0.2H^+ \rightarrow Pb^{2+} + 0.6PO_4^{3-} + 0.2H_2O$ $\log IAP = \log(Pb^{2+}) + 0.6\log(PO_4^{3-}) + 0.2pH$	−12.56	−13.93	−1.37
Plumbogummite			
$PbAl_3(PO_4)_2(OH)_5 \cdot H_2O + 5H^+ \rightarrow Pb^{2+} + 3Al^{3+} + 2PO_4^{3-} + 6H_2O$ $\log IAP = \log(Pb^{2+}) + 3\log(Al^{3+}) + 2\log(PO_4^{3-}) + 5pH$	−32.79	−33.24	−0.45

[a] Ion activities were computed using GEOCHEM-PC (Parker et al. 1995).
[b] Values obtained from Nriagu (1974) and Lindsay (1979) representing 25°C and 0.101 MPa conditions.

The corresponding log IAP values are shown in Table 6.1. By comparing the ion activity products, which are solution properties, to the solubility product constants, which are unique to each mineral, one can establish the potential presence of the mineral in the soil and the status of a solution with respect to the mineral. Parameters that have been defined to facilitate this comparison are the relative saturation, Ω:

$$\Omega = \frac{IAP}{K_{sp}} \tag{6.36}$$

and the saturation index, SI:

$$SI = \log IAP - \log K_{sp} \tag{6.37}$$

The relative saturation of the acidic soil solution with respect to each of the three minerals is:

$$\Omega_{Pb(PO_4)_{0.67}} = \frac{10^{-16.00}}{10^{-14.79}} = 10^{-1.21}(=6.17 \times 10^{-2}) \tag{6.38}$$

$$\Omega_{HP} = \frac{10^{-13.93}}{10^{-12.56}} = 10^{-1.37}(=4.27 \times 10^{-2}) \tag{6.39}$$

$$\Omega_{PG} = \frac{10^{-33.24}}{10^{-32.79}} = 10^{-0.45}(=0.35) \tag{6.40}$$

The saturation indices are SI = −1.21, −1.37, and −0.45 (note that $\Omega = 10^{SI}$, or SI = $\log \Omega$) for $Pb(PO_4)_{0.67}$, HP, and PG, respectively (Table 6.1).

The following general conclusions can be made based on a comparison of IAP and K_{sp} values (or the computed Ω or SI values):

- If *IAP* lies between K_{sp} and $0.1\ K_{sp}$ ($0.1 < \Omega < 1$ and $-1 < SI < 0$) for a specified mineral phase, the solution is said to be *saturated with respect to the solid.* It may then be concluded that the solid is present and controlling the activities of the soluble species, if equilibrium exists.
- If *IAP* $< 0.1K_{sp}$ ($\Omega < 0.1$ and $SI < -1$) for a specific mineral phase, the solution is said to be *undersaturated with respect to the solid.* This condition indicates that the solid is not stable in the environment, and if present would dissolve.
- If *IAP* $> K_{sp}$ ($\Omega > 1$ and $SI > 0$) for a specific mineral phase, the solution is said to be *supersaturated with respect to the solid.* This condition indicates that there is the potential for the solid to precipitate in the environment.

A comparison of the *IAP* values for $Pb(PO_4)_{0.67}$ and HP to the respective K_{sp} values indicates that the acidic soil solution is undersaturated with respect to both minerals, as the $\log K_{sp}$ values are greater than the log *IAP* values. If equilibrium exists, the $Pb(PO_4)_{0.67}$ and HP phases cannot be present. However, the soil solution is saturated with respect to PG, $\Omega_{PG} = 0.35$ and $SI_{PG} = -0.45$; values that lie within the range of Ω and SI values that indicate saturation. Therefore, the equilibrium solubility method indicates the presence of plumbogummite in the soil assuming the following: (1) equilibrium conditions exist, (2) the activities of the minerals are unity, (3) the activity of the soil water is unity, (4) the soil system exists at standard temperature (25°C) and pressure (0.101 MPa), (5) the concentrations of all significant components in the soil solution have been determined, and (6) all significant aqueous species were taken into account during the ion speciation modeling.

Despite that apparent theoretical soundness of the equilibrium solubility method, an observation that the *IAP* and the K_{sp} do not match for a specified mineral phase is not, in itself, sufficient evidence to discount the occurrence of the mineral. Similarly, an observation that the *IAP* and the K_{sp} are equivalent for a specified mineral phase is not unambiguous evidence for the existence of the mineral. When the *IAP* and K_{sp} are not consistent for a specified mineral, there are a number of possible conclusions:

- Equilibrium conditions do not exist.
- The specified mineral is not controlling ion activities in the soil solution.
- The specified mineral is controlling ion activities in the soil solution, but the mineral is not in the standard state.
- The specified mineral is controlling ion activities in the soil solution, but the ion activities are incorrectly determined.

6.2 APPLICATION OF MINERAL SOLUBILITY
PRINCIPLES: IMPEDIMENTS

It is often stated that the soil is in a state of disequilibrium and a state of constant flux. This statement is indisputable, but is often employed to downplay or discount the utility of the equilibrium solubility method. The reality, however, is that solubility equilibrium is relative to one's view of the soil system and the time frame during which the view is applicable. As a whole, soil is unstable, and as long as water carries dissolved substances through the soil profile, minerals will weather. Equilibrium can also be defined relative to specific mineral dissolution reactions rather than to the soil as a whole. If soil water movement is slow relative to the kinetics of dissolution or precipitation (for a specified mineral), the attainment of solubility equilibrium in a soil is feasible. The true difficulty arises when dissolution and precipitation kinetics are slow, relative to the duration of the soil water contact time.

Unlike the formation of aqueous species, which is generally considered to be an instantaneous process, the equilibration of a mineral and aqueous phase can be kinetically restricted. Several factors

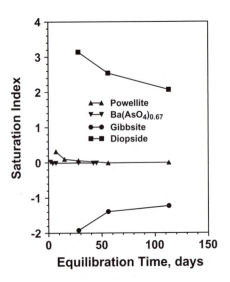

FIGURE 6.1 Saturation indices ($SI = \log IAP - \log K_{sp}$) for gibbsite [Al(OH)$_3$] and diopside [CaMg(SiO$_3$)$_2$] in a spent oil shale leachate tend toward zero with equilibration time, but solution IAP values are one to two orders of magnitude removed from the respective K_{sp} values after 130 days (Essington and Spackman, 1988). Powellite [CaMoO$_4$] and Ba(AsO$_4$)$_{0.67}$(c) saturation indices rapidly approach zero in constant ionic strength background electrolyte solutions (Essington, 1988a; 1990).

influence the time required for a mineral to achieve equilibrium with a solution. One factor influencing equilibration times is the mineral itself (Figure 6.1). In alkaline (pH = 10.03) oil shale combustion byproduct solutions, the SI values clearly indicate solution supersaturation with respect to diopside [CaMg(SiO$_3$)$_2$] and undersaturation with respect to gibbsite [Al(OH)$_3$] during a 113-day equilibration period. However, there is a tendency for the solutions to shift toward saturation (SI = 0) with respect to both minerals as the equilibration period increases. The SI for diopside decreases, perhaps indicating the precipitation of the mineral, a response that is expected based on the supersaturated condition of the solution. In contrast, the SI for gibbsite increases, perhaps indicating the dissolution of the mineral. Again, if gibbsite is initially present in the solid phase the observed response is expected. Also illustrated in Figure 6.1 are the SI values for two solids that rapidly equilibrate with a solution phase. The Ba(AsO$_4$)$_{0.67}$ phase is essentially equilibrated with the bathing solution following a 2-day equilibration; whereas, the powellite phase [CaMoO$_4$] requires approximately 28 days to equilibrate with the solution phase.

In the lead phosphate illustration above, the selection of minerals to evaluate was aided by ancillary information (direct evidence that lead and phosphorus were associated in soil particles). Such information is not always forthcoming; indeed, ancillary information is typically not available, due to the inability to detect trace minerals by standard direct identification techniques (e.g., x-ray diffraction), or the need to perform special preconcentration techniques (e.g., density and magnetic-susceptibility fractionation) to enhance the utility of direct methods of analysis. Thus, in order to establish trace mineralogy, one must evaluate the soil solution chemistry with respect to a vast array of potential phases. For example, if the lead–phosphorus association were not known, the list of minerals for which IAP values could be computed might include: cerussite [PbCO$_3$], hydrocerussite [Pb$_3$(CO$_3$)$_2$(OH)$_2$], leadhillite [Pb$_4$SO$_4$(CO$_3$)$_2$(OH)$_2$], plumbogummite [PbAl$_3$(PO$_4$)$_2$(OH)$_5$•H$_2$O], lead phosphate [Pb$_3$(PO$_4$)$_2$], hydroxypyromorphite [Pb$_5$(PO$_4$)$_3$OH], chloropyromorphite [Pb$_5$(PO$_4$)$_3$Cl], fluoropyromorphite [Pb$_5$(PO$_4$)$_3$F], angelsite [PbSO$_4$], hinsdalite [PbAl$_3$PO$_4$SO$_4$(OH)$_6$], lead silicate [Pb$_2$SiO$_4$], and alamosite [PbSiO$_3$]. Further, a solubility product constant for each phase would be required, values that for a large number of minerals are simply not available.

6.2.1 The Estimation of Standard Free Energies of Formation

As indicated above, one limitation to the equilibrium solubility characterization of trace elements in soils and waste-affected environments is the lack of standard free energy of formation values (ΔG_f°) for trace element-bearing solids. To circumvent this problem, several methods have been developed to estimate ΔG_f° values for minerals for which measured values are unavailable. For the most part, methods of estimating the ΔG_f° values of minerals are purely empirical; they rely on measured ΔG_f° values to establish a trend relating mineral ΔG_f° values to those of aqueous ions, metal oxides or hydroxides, or less complex mineral phases. The majority of the estimation techniques were developed for estimating the ΔG_f° values of silicate minerals. The Chen extrapolation method (Chen, 1975) is one such technique that can be employed to elucidate the ΔG_f° of trace element-bearing minerals having complex chemical composition. The application of the Chen extrapolation method is illustrated for the lead mineral hinsdalite [$PbAl_3PO_4SO_4(OH)_6$]. The formation of this mineral can be expressed in terms of component oxides. For example, hinsdalite may be formed using the following reactants, each with a known ΔG_f° value:

$$PbO(c) + \tfrac{3}{2}Al_2O_3(c) + \tfrac{1}{2}P_2O_5(c) + SO_3(c) + 3H_2O(l) \rightarrow PbAl_3PO_4SO_4(OH)_6(s) \quad (6.41)$$

Summation of the reactant ΔG_f° values is performed according to the expression:

$$\sum_i n_i \Delta G_{f,i}^\circ = \Delta G_{f,PbO}^\circ + \tfrac{3}{2}\Delta G_{f,Al_2O_3}^\circ + \tfrac{1}{2}\Delta G_{f,P_2O_5}^\circ + \Delta G_{f,SO_3}^\circ + 3\Delta G_{f,H_2O}^\circ \quad (6.42)$$

Using ΔG_f° values from the compilation of Naumov et al. (1974), $\sum n_i \Delta G_{fi}^\circ$ in Equation 6.42 is -4319.6 kJ mol^{-1}. If $PbO(c)$ is replaced by hydroxypyromorphite as the Pb source, and $Al_2O_3(c)$ replaced by diaspore as the Al source in Equation 6.41, a second hinsdalite formation reaction is generated:

$$\tfrac{1}{5}Pb_5(PO)_3OH(c) + \tfrac{1}{5}P_2O_5(c) + 3AlOOH(c) + SO_3(c) + \tfrac{7}{5}H_2O(l) = PbAl_3PO_4SO_4(OH)_6(s) \quad (6.43)$$

The computed $\sum_i n_i \Delta G_{fi}^\circ$ for this reaction is -4482.2 kJ mol^{-1}, which is more negative than the $\sum_i n_i \Delta G_{fi}^\circ$ computed for Equation 6.41. The process of forming hinsdalite from component solids can be continued to produce a series of reactions, each successive reaction generating a more negative value for $\sum_i n_i \Delta G_{fi}^\circ$ and using a lesser number of reactants. The series of reactions used to form hinsdalite and their associated $\sum_i n_i \Delta G_{fi}^\circ$ values are shown in Table 6.2. Each reaction is then

---

TABLE 6.2
Reactions that Result in the Production of Hinsdalite
[$PbAl_3PO_4SO_4(OH)_6$] for Use in Chen's Extrapolation Method
(Chen, 1975)[a]

Reactants	$\sum_i n_i \Delta G_{f,i}^\circ$	x
$PbO(c) + \tfrac{3}{2}Al_2O_3(c) + \tfrac{1}{2}P_2O_5(c) + SO_3(c) + 3H_2O(l)$	-4319.6[b]	0
$\tfrac{1}{5}Pb_5(PO)_3OH(c) + \tfrac{1}{5}P_2O_5(c) + 3AlOOH(c) + SO_3(c) + \tfrac{7}{5}H_2O(l)$	-4482.2	1
$\tfrac{1}{5}Pb_5(PO)_3OH(c) + \tfrac{1}{3}Al_2(SO_4)_3(c) + \tfrac{2}{3}Al_2PO_4(OH)_3(c) + \tfrac{23}{15}Al(OH)_3(c)$	-4658.7	2
$\tfrac{1}{3}Pb_3(PO)_2(c) + \tfrac{1}{3}Al_2(SO_4)_3(c) + \tfrac{1}{3}AlPO_4(c) + 2Al(OH)_3(c)$	-4697.3	3
$PbSO_4(c) + Al_2O_3(c) + AlPO_4(c) + 3H_2O(c)$	-4724.4	4
$PbSO_4(c) + Al_2PO_4(OH)_3(c) + Al(OH)_3(c)$	-4734.6	5

[a] $\sum_i n_i \Delta G_{f,i}^\circ$ is the summation of the reactant ΔG_f° values, where n_i is a reactant stoichiometric coefficient, and x is an integer counter.

[b] ΔG_f° values for the reactants were obtained from Naumov et al. (1974).

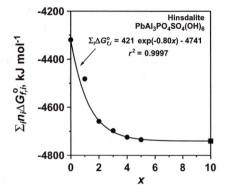

FIGURE 6.2 A plot illustrating the application of Chen's extrapolation method for determining the Gibb's free energy of formation (ΔG_f°) for hinsdalite. The closed circles correspond to the $\sum_i n_i \Delta G_{fi}^\circ$ vs. x data pairs in Table 6.2. The closed square is the predicted ΔG_f° for hinsdalite.

assigned an integer value (denoted as x) with the reaction with the least negative (or greatest) value for $\sum_i n_i \Delta G_{fi}^\circ$ assigned $x = 0$. Thereafter, x is assigned in order of decreasing $\sum_i n_i \Delta G_{fi}^\circ$. A plot of $\sum_i n_i \Delta G_{fi}^\circ$ against x indicates that as x increases, $\sum_i n_i \Delta G_{fi}^\circ$ asymptotically approaches a minimum value as $x \to \infty$, a value that is defined as the estimated ΔG_{fi}° value for the solid (Figure 6.2). The estimated ΔG_f° is computed by nonlinear regression analysis using the exponential expression:

$$\sum_i n_i \Delta G_{f,i}^\circ = a\exp(-bx) + \Delta G_f^\circ \tag{6.44}$$

where a, b, and ΔG_f° are regression parameters. Using the data in Table 6.2, the exponential expression for hinsdalite is:

$$\sum_i n_i \Delta G_{f,i}^\circ = 421\exp(-0.80x) - 4741 \tag{6.45}$$

where -4741 kJ mol^{-1} is the predicted ΔG_f° and r^2 is 0.999. According to Naumov et al. (1974) the ΔG_f° for hinsdalite is -4700.7 kJ mol^{-1}. The error associated with the estimated ΔG_f° is 0.85% of the measured value, a value that is consistent with the error associated with the prediction of aluminosilicate ΔG_f° values using the Chen extrapolation method.

The Chen extrapolation technique is restricted to predicting the free energies of formation of complex minerals, as one could not amass a series of formation reactions sufficient to establish a trend in $\sum_i n_i \Delta G_{fi}^\circ$ for simple minerals, such as $PbSO_4(s)$. Tardy and others (reviewed by Essington, 1988) established a correlation technique that has been employed to estimate the ΔG_f° values for numerous classes of compounds, including hydroxides, silicates, sulfates, nitrates, carbonates, phosphates, arsenates, selenates, and selenites. The Tardy correlation technique defines the parameter $\Delta_G O^{2-} M$ for a given cation M as:

$$\Delta_G O^{2-} M = \tfrac{1}{x}\{\Delta G_{f,MO_x}^\circ - \Delta G_{f,M^{2x+}}^\circ\} \tag{6.46}$$

where the free energy of formation of the metal oxide is $\Delta G_{f,MO_x}^\circ$, the free energy of formation of the metal cation is $\Delta G_{f,M^{2x+}}^\circ$, and x is the cation valence divided by 2. Further, the parameter $\Delta_G[M_z LO_v(c)]$ is defined as:

$$\Delta_G[M_z LO_v(c)] = \Delta G_{f,M_z LO_v}^\circ - z\Delta G_{f,MO_x}^\circ - \tfrac{1}{y}\Delta G_{f,L_y O_w}^\circ \tag{6.47}$$

Rearranging,

$$\Delta G_{f,M_z LO_v}^\circ = \Delta_G[M_z LO_v(c)] + z\Delta G_{f,MO_x}^\circ + \tfrac{1}{y}\Delta G_{f,L_y O_w}^\circ \tag{6.48}$$

TABLE 6.3
Linear Free Energy Correlations Used in the Tardy Correlation Method to Predict ΔG_f° Values for Various Compound Types[a]

Compound Type	Tardy Correlation	r^2	Source
Hydroxide	$\Delta_G[MOH_{2x}(c)] = -0.210\Delta_G O^{2-}M - 46.57$	0.974	Tardy and Garrels (1976)
Orthosilicate	$\Delta_G[M_z SiO_4(c)] = -1.01\Delta_G O^{2-}M - 189.12$	0.970	Tardy and Garrels (1977)
Metasilicate	$\Delta_G[M_z SiO_3(c)] = -0.673\Delta_G O^{2-}M - 126.90$	0.996	Tardy and Garrels (1977)
Phosphate	$\Delta_G[M_z PO_4(c)] = -1.285\Delta_G O^{2-}M - 419.24$	0.962	Tardy and Vieillard (1977)
Sulfate	$\Delta_G[M_z SO_4(c)] = -0.999\Delta_G O^{2-}M - 385.31$	0.964	Tardy and Gartner (1977)
Nitrate	$\Delta_G[M_z NO_3(c)] = -0.541\Delta_G O^{2-}M - 159.79$	0.968	Tardy and Gartner (1977)
Carbonate	$\Delta_G[M_z CO_3(c)] = -0.771\Delta_G O^{2-}M - 199.41$	0.974	Tardy and Gartner (1977)
Arsenate	$\Delta_G[M_z AsO_4(c)] = -1.25\Delta_G O^{2-}M - 315.28$	0.98	Essington (1988b)
Selenate	$\Delta_G[M_z SeO_4(c)] = -0.91\Delta_G O^{2-}M - 380.00$	0.99	Essington (1988b)
Selenite	$\Delta_G[M_z SeO_3(c)] = -0.88\Delta_G O^{2-}M - 236.01$	0.96	Essington (1988b)

[a] The $\Delta_G[M_z LO_v(c)]$ and $\Delta_G O^{2-}M$ parameters are defined in the text.

where $M_z LO_v(c)$ is the compound of interest, z is the number of cations (M) for every ligand unit (LO_v) in the compound, $L_y O_w$ is the crystalline oxide of the ligand LO_v, and y and w are subject to the electroneutrality condition $2y = lw$ (l is the oxidation state of L). For the mineral classes identified above, the parameter $\Delta_G[M_z LO_v(c)]$ is linearly correlated to $\Delta_G O^{2-}M$:

$$\Delta_G[M_z LO_v(c)] = \alpha\Delta_G O^{2-}M + \beta \tag{6.49}$$

where α and β are the regression constants specific to the ligand group. For example, the linear expression that relates $\Delta_G[M_z AsO_4(c)]$ and $\Delta_G O^{2-}M$ is:

$$\Delta_G[M_z AsO_4(c)] = -1.25\Delta_G O^{2-}M - 315.28 \tag{6.50}$$

with an r^2 value of 0.98 (Essington, 1988). The α and β values for other compound types are shown in Table 6.3. Using known ΔG_f° values for any metal $M^{2x+}(aq)$ and the corresponding metal oxide $MO_x(c)$, $\Delta_G O^{2-}M$ could be computed using Equation 6.46. Equation 6.49 could then be employed to generate a corresponding $\Delta_G[M_z LO_v(c)]$ value for the solid of interest. Finally, Equation 6.48 is used to compute a predicted ΔG_f° value for the mineral $M_z LO_v(c)$.

The application of the Tardy correlation technique is illustrated for predicting the ΔG_f° value of $Ba_3(AsO_4)_2(c)$. The $\Delta_G O^{2-}Ba$ parameter is computed using Equation 6.46 and the ΔG_f° values for $BaO(c)$ (−520.39kJ mol^{-1}) and $Ba^{2+}(aq)$ (−560.74 kJ mol^{-1}):

$$\Delta_G O^{2-}Ba = \{-520.39 - (-560.74)\} = 40.35 \text{ kJ mol}^{-1} \tag{6.51}$$

Employing Equation 6.50, the computed $\Delta_G[Ba_{1.5}AsO_4(c)]$ is −365.72 kJ mol^{-1}. Using Equation 6.48 and the free energy values for $BaO(c)$ and $As_2O_5(c)$ (−782.53 kJ mol^{-1}), the computed free energy of formation for $Ba_{1.5}AsO_4(c)$ is:

$$\Delta G_{f,Ba_{1.5}AsO_4}^O = -365.72 + 1.5(-520.39) + \tfrac{1}{2}(-782.53) = -1537.6 \text{ kJ mol}^{-1} \tag{6.52}$$

Thus, the ΔG_f° for $Ba_3(AsO_4)_2(c)$ is -3075 kJ mol^{-1} (two times the ΔG_f° for $Ba_{1.5}AsO_4(c)$). For this example, and relative to the measured ΔG_f° for $Ba_3(AsO_4)_2(c)$ of -3101.2 kJ mol^{-1}, the error of prediction is 26 kJ mol^{-1}, a value that is approximately equal to the method standard error of 27 kJ mol^{-1} for the metal arsenates.

Although methods for estimating mineral ΔG_f° values have utility and result in a thermodynamic quantity that can be employed to examine the solubility of a potentially important mineral phase, the predicted values cannot supplant measured values. Consider the dissolution of $Ba_3(AsO_4)_2(c)$:

$$Ba_3(AsO_4)_2(c) \rightarrow 3Ba^{2+} + 2AsO_4^{3-} \tag{6.53}$$

The solubility product constant for the reaction is:

$$K_{sp} = (Ba^{2+})^3(AsO_4^{3-})^2 = 10^{-21.7} \tag{6.54}$$

Using the ΔG_f° value predicted for $Ba_3(AsO_4)_2(c)$ in Equation 6.52 would lead to an estimated K_{sp} of $10^{-17.2}$, a value that is more than four orders of magnitude less than the actual K_{sp}. This error would result in a predicted AsO_4^{3-} activity in a solution that is more than two orders of magnitude greater than should actually occur if the solid were present and in equilibrium with the solution. At best, ΔG_f° estimates fill a gap by allowing for inclusion of potentially important mineral phases in a system analysis, and by identifying the minerals for which measured thermodynamic data are required.

6.2.2 METASTABILITY AND THE STEP RULE

An additional complicating factor associated with the application of the equilibrium solubility method is that the most stable mineral for a particular soil solution condition may not be the mineral present in the soil and controlling ion activities in the solution. The model soil system that epitomizes this concept was first investigated in the late 1950s by Lindsay and Stephenson (1959a and b). They observed when soluble phosphate fertilizer (as monocalcium phosphate monohydrate, $Ca(H_2PO_4)_2 \cdot H_2O$) is applied to calcareous soil, the phosphate minerals that initially form are not the most stable for the soil conditions. Indeed, of the calcium phosphate minerals that could possibly form (in order of increasing stability: brushite $[CaHPO_4 \cdot 2H_2O]$ < monetite $[CaHPO_4]$ < octacalcium phosphate $[Ca_8H_2(PO_4)_6 \cdot 5H_2O]$ < apatite $[Ca_5(PO_4)_3(OH,F)]$, the minerals that initially form are the least stable of the group (brushite and monetite), rather than the stable phase (apatite). A mineral that forms rapidly in a soil system, yet is not the stable phase for an element, is termed a metastable fast-former. A metastable phase will control soil solution composition until complete dissolution of the phase occurs.

Calcium phosphate minerals are very stable in neutral to alkaline soil environments. When a highly soluble phosphate fertilizer is applied to an alkaline soil, thermodynamics would predict the precipitation of hydroxyapatite, the most stable calcium phosphate. However, the first minerals to form are observed to be the metastable fast-formers brushite and monetite. In the case of calcium phosphate precipitation, mineral precipitation from a supersaturated alkaline soil solution follows the general order:

$$\text{Soluble P} \xrightarrow{k_1} CaHPO_4 \cdot 2H_2O \text{ and } CaHPO_4 \xrightarrow{k_2} Ca_8H_2(PO_4)_6 \cdot 5H_2O \xrightarrow{k_3} Ca_5(PO_4)_3OH \tag{6.55}$$

where the rate constants decrease in the order, $k_1 > k_2 \gg k_3$. The activity of HPO_4^{2-} in equilibrium with brushite is $(HPO_4^{2-}) = 10^{-3.32}$ (assuming soil solution pH = 7.5, CO_2 partial pressure = 10^{-2}, and Ca^{2+} activity controlled by calcite dissolution). In an alkaline soil that has received soluble phosphate, the mineral phase that forms after brushite (and monetite) is octacalcium phosphate, which controls a soil solution HPO_4^{2-} activity of $10^{-3.93}$. The time required for the complete conversion of brushite to octacalcium phosphate can range from a few weeks to several months, depending on solution pH, the presence of calcite, temperature, and the concentration of soluble organic carbon. Theoretically, apatite would then supplant octacalcium phosphate and support an alkaline soil solution HPO_4^{2-} activity of $10^{-6.95}$. However, direct evidence of apatite formation is either not available or is inconclusive.

The precipitation of metastable minerals, and the transition to relatively more stable minerals, is described in the *step rule* (also known as the Gay-Lussac-Ostwald step rule). For a specific ion, the mineral that precipitates first from a soil solution that is supersaturated with respect to several minerals (that contain the ion) will be the mineral whose K_{sp} is closest to and below the *IAP* of the solution. Thereafter, minerals that contain the ion of interest will form in order of increasing stability and the rate of formation will decrease with increasing mineral stability. When a metastable fast-former precipitates, the mineral will control the composition of the soil solution with respect to the specified ion. However, the solution is still supersaturated with respect to other minerals that may potentially form. The next mineral to form will be the phase whose K_{sp} is closest to and below the *IAP* controlled by the initial metastable precipitate. The precipitation of the second and more stable mineral drains the soil solution of the component ion, forcing the metastable phase to dissolve in an attempt to maintain solution *IAP*. Ultimately, the initial precipitate will completely dissolve and solution *IAP* will be consistent with the K_{sp} of the second and more stable mineral. This process, the dissolution of an unstable phase and the precipitation of a more stable phase, will continue until eventually the most stable mineral is the only phase present. Based on the step rule and the fact that metastable phases may remain in soil for extended time periods, the application of the equilibrium solubility method requires not only the consideration of stable minerals, but metastable minerals as well, in order to indirectly identify trace mineralogy.

6.3 THE DEVIATION OF K_{SP} FROM K_{DIS}

The simplification of the K_{dis} expression (Equation 6.2) by assuming unit activity of the solid and of liquid water results in an expression that defines K_{sp} (Equation 6.15). Only in desiccated soils does the activity of water significantly vary from unity, and under these conditions water activity can be determined by measuring the partial pressure of water (P_w/P_w°). The activity of water can be incorporated into the equilibrium solubility method as the need arises. The assumption that soil minerals can be described as existing in the standard state has considerably less validity than the assumption of unit water activity. There are two principle reasons why a soil mineral may not reside in the standard state: (1) microcrystallinity or poor crystallinity, and (2) impurities.

6.3.1 POORLY CRYSTALLINE AND MICROCRYSTALLINE SOLIDS

A solid in the standard state is pure and well crystalline (macrocrystalline with negligible specific surface). However, precipitates that form in the soil environment rarely meet the standard-state criteria. The equilibrium solubility ($\log K_{sp}$) of poorly crystalline minerals is greater than that of their well-crystalline counterparts, a characteristic that is associated with increased specific surface. Indeed, the positive deviation of a poorly crystalline (or microcrystalline) mineral $\log K_{sp}$ value from that of the well-crystalline counterpart increases with specific surface area. For example, gibbsite dissolution can be described by:

$$Al(OH)_3(s) + 3H^+(aq) \rightarrow Al^{3+}(aq) + 3H_2O(l)$$

$$K_{dis} = \frac{(Al^{3+})}{(Al(OH)_3)(H^+)^3} = \frac{K_{sp}}{(Al(OH)_3)} \tag{6.56}$$

TABLE 6.4
The pH, Al^{3+} Activity, and the Ion Activity Product (IAP) of Solutions in Equilibrium with Gibbsite [$Al(OH)_3$] in the Presence and Absence of 2-Ketogluconate, a Low-Molecular-Mass Organic Acid Anion

Without 2-ketogluconate			With 2-ketogluconate		
pH	$\log(Al^{3+})$	$\log IAP$	pH	$\log(Al^{3+})$	$\log IAP$
5.11	−6.60	8.73	6.35	−7.95	11.10
3.82	−3.63	7.83	6.52	−8.37	11.19
4.88	−6.56	8.08	5.30	−5.78	10.12
3.90	−3.78	7.92	5.41	−5.63	10.60
4.25	−4.11	8.64	5.65	−5.67	11.28
4.35	−5.38	7.67	6.28	−8.04	10.80
4.49	−5.38	8.09	6.28	−7.75	11.09
4.37	−5.68	7.43	6.36	−7.60	11.48
4.11	−3.98	8.35	6.12	−7.30	11.06
4.39	−5.20	7.97	5.77	−6.52	10.79
	Mean $\log IAP$:	8.07 ± 0.17	5.18	−4.67	10.87
			6.43	−7.19	12.10
			5.54	−4.74	11.88
			5.59	−4.77	12.00
			6.17	−6.07	12.44
			7.05	−9.18	11.97
				Mean $\log IAP$:	11.30 ± 0.40

Based on the findings of Holden (1996) and the gibbsite solubility studies of Mr. Jeffery Scott, University of Tennessee, Knoxville.

The true equilibrium constant for the reaction is $\log K_{dis} = 8.05$. Assuming a well-crystalline solid with negligible surface area (unit activity) and a unit water activity, the solubility product constant ($\log K_{sp}$) is also 8.05. Correspondingly, the $\log IAP$ for a solution in equilibrium with gibbsite will be 8.05. However, for microcrystalline gibbsite, $\log K_{sp}$ (or $\log IAP$) is 9.35 (Sposito, 1989; 1994). The activity of the microcrystalline gibbsite is computed as:

$$(Al(OH)_3) = \frac{K_{sp}}{K_{dis}} = \frac{10^{9.35}}{10^{8.05}} = 19.95 \tag{6.57}$$

Poorly crystalline gibbsite displays an even greater solubility ($\log IAP = 10.78$; Sposito, 1989) and a resulting $(Al(OH)_3)$ value of 537.

The presence of low-molecular-mass organic acids, such as citric acid and fulvic acid, are known to inhibit the growth of gibbsite crystals (as well as other soil minerals), resulting in the formation of poorly crystalline gibbsite. The data presented in Table 6.4 illustrate the influence of 2-ketogluconate, a low-molecular-mass organic acid anion, on gibbsite $\log IAP$ values. In the absence of the organic acid, the mean $\log IAP$ value of 8.07 ± 0.17 is essentially $\log K_{dis}$ (8.05). However, in the presence of 2-ketogluconate, the gibbsite $\log IAP$, or $\log K_{sp}$, is 11.30 ± 0.40. Employing Equation 6.57, the activity of the poorly crystalline gibbsite is 1778. Similar observations have been made for kaolinite and calcite:

$$Al_2Si_2O_5(OH)_4(s) + 6H^+(aq) \rightarrow 2Al^{3+}(aq) + 2H_4SiO_4^0(aq) + H_2O(l) \tag{6.58}$$

$$CaCO_3(s) \rightarrow Ca^{2+}(aq) + CO_3^{2-}(aq) \tag{6.59}$$

The corresponding $\log K_{dis}$ values are 5.72 for kaolinite and -8.48 for calcite. The $\log K_{sp}$ (or $\log IAP$) values for a poorly crystalline kaolinite and a microcrystalline calcite are 10.5 and -7.96, respectively (Sposito, 1981; 1994). Using Equation 6.57, the activity of the poorly crystalline kaolinite is $10^{4.78}$, and the activity of the microcrystalline calcite is 3.3.

While the influence of poor crystallinity and microcrystallinite on mineral solubility is known, quantization of the effect and application to predicting the composition of the soil aqueous phase has not been realized for a great many mineral phases. At best, the knowledge that microcrystalline (specific surface) and poor crystalline mineral phases occur in natural systems can be employed to develop limits of expected solution composition values based on the deviation of $\log K_{sp}$ from $\log K_{dis}$. For example, a pH 7.5 soil solution in equilibrium with calcite and a $P_{CO_2} = 10^{-2}$ soil atmosphere would be expected to support a Ca^{2+} activity between $10^{-2.54}$ (2.88×10^{-3}) and $10^{-3.33}$ (4.67×10^{-4}). Deviation outside this range would suggest other controls of Ca^{2+} activity.

Poorly crystalline solids, particularly amorphous metal (hydr-)oxides, are generally dealt with by defining the amorphous character of the solid as the standard-state condition. Thus, the activity of the amorphous solid is unity and the observed IAP is the K_{sp}. The IAP of $10^{10.78}$ cited by Sposito (1989) for poorly crystalline gibbsite or the IAP of $10^{11.30}$ for the poorly crystalline gibbsite described in Table 6.4 would then describe different measurements of the K_{sp} for $Al(OH)_3(am)$. In essence, there is no impact on the predicted Al^{3+} activity in an equilibrated solution, whether one chooses to assign unit activity to an amorphous $Al(OH)_3$ phase or to impose a gibbsite activity at a value other than unity.

6.3.2 Solid Solutions

The poor crystallinity of secondary precipitates in a soil can also be associated with the inclusion of impurities in the crystal structure, impurities that further impact the activity of a mineral. The assumption of purity in natural minerals is generally not appropriate. A commonly encountered characteristic of soil minerals is the isomorphic substitution of cations and anions that are unique to the mineral by ions of similar size during mineral formation. The result of this process is embodied in the phyllosilicates (e.g., montmorillonite) where Si^{4+} is replaced by Al^{3+} or Fe^{3+} in the tetrahedral layer and Al^{3+} by several ions, including Ti^{4+}, Fe^{3+}, Mg^{2+}, Mn^{2+}, Co^{2+}, Ni^{2+}, Cu^{2+}, Zn^{2+}, or Pb^{2+} in the octahedral layer. Numerous other examples also occur in soil: the replacement of Mn^{IV} by a wide variety of metals, including Fe^{3+}, Fe^{2+}, Co^{2+}, Ni^{2+}, Zn^{2+}, or Pb^{2+} in manganese oxides; the Ca^{2+} ion in the calcite $[CaCO_3]$ structure can be replaced by Mg^{2+}, Sr^{2+}, Ba^{2+}, Fe^{2+}, Mn^{2+}, Zn^{2+}, or Cd^{2+}; Zn^{2+} in smithsonite $[ZnCO_3]$ may be replaced by Cd^{2+}; during goethite formation, Fe^{3+} may be replaced by Al^{3+}, Mn^{2+}, Ni^{2+}, Cu^{2+}, or Zn^{2+}; Ba^{2+} in barite $[BaSO_4]$ may be replaced by Sr^{2+}; and in hydroxyapatite $[Ca_5(PO_4)_3OH]$, Ca^{2+} may be replaced by Sr^{2+} or Pb^{2+}, OH^- by F^- or Cl^-, and PO_4^{3-} by AsO_4^{3-} or CO_3^{2-}. The phenomenon described above is an example of a process known as coprecipitation. Specifically, the examples above illustrate the substitution of the major constituent (e.g., Zn^{2+} in smithsonite) by another ion (e.g., Cd^{2+}) to a minor degree. When a substitution occurs such that the minor component is distributed uniformly throughout the mineral, the resulting phase is termed a solid solution.

Consider the coprecipitation of Cd^{2+} and Ca^{2+} with the ligand CO_3^{2-} to form the mineral calcite $[(Ca,Cd)CO_3]$. The end members of the solid solution are calcite and otavite $[CdCO_3]$. For each end member in the solid solution, the following expressions are valid:

$$(Ca^{2+})(CO_3^{2-}) = K_{dis,calcite}(CaCO_3) \tag{6.60}$$

$$(Cd^{2+})(CO_3^{2-}) = K_{dis,otavite}(CdCO_3) \tag{6.61}$$

where $K_{dis,calcite}$ is $10^{-8.48}$ and $K_{dis,otavite}$ is $10^{-12.1}$. The replacement of Ca^{2+} by Cd^{2+} in the calcite structure can be described by the reaction:

$$CaCO_3(s) + Cd^{2+} \rightarrow CdCO_3(s) + Ca^{2+} \tag{6.62}$$

The equilibrium constant for this reaction is commonly termed the distribution coefficient:

$$D_{eq} = \frac{(Ca^{2+})(CdCO_3)}{(Cd^{2+})(CaCO_3)} \tag{6.63}$$

The activity of a solid, such as $CaCO_3(s)$ or $CdCO_3(s)$, in a solid solution is defined in a manner similar to the activity of an ion in an aqueous solution by the expression:

$$(CaCO_3) = f_{calcite}N_{calcite} \tag{6.64}$$

$$(CdCO_3) = f_{otavite}N_{otavite} \tag{6.65}$$

where the f values are rational activity coefficients and the N values are the mole fractions of each solid end member in the solid solution. An ideal solid solution is described when f values for both solids equal 1.0 (or nearly so) over the entire range of isomorphic substitution. An ideal solid solution might be expected to occur when select properties of the major (e.g., Ca^{2+}) and minor (e.g., Cd^{2+}) component in the mixture are similar. Properties that are important in determining the ideality of a solid solution are the ionic radii and electronegativities of the major and minor component ions, and the crystal classes in which the two end member minerals tend to form. The more similar these properties are, the more ideal the mixture. In the case of Ca^{2+} and Cd^{2+}, the crystallographic radii are 114 pm for Ca^{2+} and 109pm for Cd^{2+} (octahedral coordination); the electronegativities are 1.0 for Ca^{2+} and 1.7 for Cd^{2+}; and both calcite and otavite occur in a rhombohedral structure.

Equations 6.64 and 6.65 may be substituted into Equation 6.63 to arrive at an expression that relates the distribution coefficient to the aqueous activities of the substituting ions, and the rational activity coefficients and mole fractions of the end member components in the solid solution:

$$D_{eq} = \frac{(Ca^{2+})f_{otavite}N_{otavite}}{(Cd^{2+})f_{calcite}N_{calcite}} \tag{6.66}$$

Alternatively, Equations 6.60 and 6.61 may be substituted into Equation 6.63 to obtain an expression for the distribution coefficient that is only a function of the dissolution equilibrium constants of the end member components:

$$D_{eq} = \frac{K_{dis,calcite}}{K_{dis,otavite}} \tag{6.67}$$

If ideality is assumed, Equation 6.67 may be written:

$$D_{ideal} = \frac{K_{sp,calcite}}{K_{sp,otavite}} \tag{6.68}$$

For an ideal mixture of calcite and otavite, the equilibrium constant (D_{ideal}) for the replacement reaction (Equation 6.62) is computed using Equation 6.68:

$$D_{ideal} = \frac{10^{-8.48}}{10^{-12.1}} = 10^{3.62} \tag{6.69}$$

Tesoriero and Pankow (1996) experimentally determined that the distribution coefficient for a Cd^{2+}-substituted calcite has the formula $(Ca_{0.937}Cd_{0.063})CO_3$ (which may also be written $[0.937CaCO_3 \cdot 0.063CdCO_3]$). The average D_{eq} value determined from solubility experiments was $10^{3.09}$. Equation 6.69 can be written to indicate the expected ratio of Ca^{2+} to Cd^{2+} assuming an ideal solid solution:

$$D_{ideal} = \frac{(Ca^{2+})(CO_3^{2-})}{(Cd^{2+})(CO_3^{2-})} = \frac{(Ca^{2+})}{(Cd^{2+})} \tag{6.70}$$

Substituting Equation 6.70 into Equation 6.66 yields an expression that relates the ideal distribution coefficient to the measured value:

$$D_{eq} = D_{ideal} \frac{f_{otavite} N_{otavite}}{f_{calcite} N_{calcite}} \tag{6.71}$$

This expression can be rewritten such that the ratio of rational activity coefficients can be computed:

$$\frac{f_{otavite}}{f_{calcite}} = \frac{D_{eq} N_{calcite}}{D_{ideal} N_{otavite}} = \frac{10^{3.09} \times 0.937}{10^{3.62} \times 0.063} = 4.39 \tag{6.72}$$

Since $N_{calcite}$ is approximately equal to 1, it can be assumed that $f_{calcite}$ is also approximately 1. Thus, the ratio of rational activity coefficients generated in Equation 6.72 is equivalent to $f_{otavite}$, or $f_{otavite} = 4.39$.

The solubility of a minor or trace component in a stable solid solution will always be depressed relative to that of the pure end member precipitate; otherwise, the solid solution would not be stable and the pure end member phase would result. This is easily seen through the evaluation of Equation 6.61 and the assumption of an ideal solid solution mixture, $(Cd^{2+})(CO_3^{2-}) = K_{dis,otavite}(CdCO_3)$. Since the activity of otavite in the solid solution is given by the mole fraction of Cd^{2+} in the coprecipitate ($N_{otavite} = 0.063$), the ion activity product for a solution in equilibrium with the ideal solid solution would be $IAP = 10^{-13.3}$. Thus, the solubility of otavite in the calcite-otavite solid solution is only 6.3% of the solubility of the pure otavite phase. If the same analysis is performed for a nonideal solid solution, the ion activity product for a solution in equilibrium with the solid solution would be $IAP = 10^{-12.7}$. While solid solution nonideality results in greater solubility relative to the ideal solid solution, the solubility of otavite in the nonideal calcite-otavite solid solution is only 27.7% of the solubility of the pure otavite phase. Thus, otavite $[CdCO_3]$ is unstable relative to the calcite-otavite coprecipitate $[(Ca_{0.937}Cd_{0.063})CO_3]$.

As described above, suitable mechanisms are available to predict the influence of coprecipitation (solid solution type) on trace element solubility in soil solutions. However, the use of these mechanisms requires three principle items of information: (1) the variation in end member rational activity coefficients throughout the range of isomorphic substitution; (2) the variation of the thermodynamic distribution coefficient (D_{eq}) throughout the range of isomorphic substitution; and (3) the degree of isomorphic substitution in the coprecipitate in the soil environment of interest. The first two items can be derived through experimentation; however, the third item is not easily

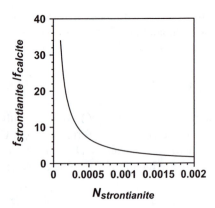

FIGURE 6.3 Variation in the ratio of rational activity coefficients as a function of the mole fraction of strontianite [$SrCO_3$] in a strontianite–calcite [$CaCO_3$] solid solution.

substantiated. Again, as suggested for poorly crystalline minerals, the solubility of a trace element in a soil environment can be described as falling within a solubility range that is defined by the anticipated limits of isomorphic substitution. For example, D_{eq} is a constant 0.021 for the isomorphic substitution of Sr^{2+} in the calcite structure for $N_{strontianite}$ values up to 0.002 (the miscibility limit is $N_{strontianite} = 0.0035$). Since D_{eq} is constant, the rational activity coefficient for strontianite in the solid solution will vary as a function of the mole fraction ratio (Equation 6.71 and Figure 6.3). However, because D_{eq} is constant, the IAP is also constant (for $N_{strontianite}$ values up to 0.002). In essence, changes in $f_{strontianite}$ are corrected by changes in $N_{strontianite}$ such that the IAP is a constant value equal to $10^{-11.74}$ for the solid solution $(Ca_{(1-x)}Sr_x)CO_3$, $x < 0.002$. Therefore, the occurrence of strontianite in an alkaline soil environment could not be discounted if the strontianite IAP value ranged between $10^{-11.74}$ and $10^{-9.27}$ (the K_{dis} for strontianite).

6.4 MINERAL SOLUBILITY AND SOLUTION COMPOSITION

The activities of ions in a solution in equilibrium with a solid will be strictly determined by the solubility product constant of the solid. However, the soluble concentrations of the ions will vary depending on the chemical properties of the solution. Three main factors will impact ion concentrations under the condition of solid-solution equilibrium: (1) ionic strength effect, (2) ion complexation or ion pairing, and (3) common ion effect. These factors are illustrated in the findings of Longnecker and Lyerly (1959), which examine the influence of various salts, or background electrolytes, and their concentrations on gypsum [$CaSO_4 \cdot 2H_2O$] solubility (Table 6.5). Note that solubility, as used here, is defined as the concentration of the gypsum dissolved in the equilibrating solution, expressed as mmol L^{-1} of gypsum dissolved. This definition should not be confused with gypsum stability or solubility product, which are thermodynamic characteristics of the mineral and not influenced by solution composition. The dissolution of gypsum and the associated K_{sp} value are described by:

$$CaSO_4 \cdot 2H_2O(s) \rightarrow Ca^{2+}(aq) + SO_4^{2-}(aq) + 2H_2O(l)$$

$$K_{sp} = (Ca^{2+})(SO_4^{2-}) = 10^{-4.62} \qquad (6.73)$$

At equilibrium, the IAP [$= (Ca^{2+})(SO_4^{2-})$] of a solution in contact with gypsum will *always* be $10^{-4.62}$.

Table 6.5 shows that gypsum solubility (the amount of gypsum dissolved) decreases with increasing $CaCl_2$ concentration. Consider the gypsum dissolution reaction above (Equation 6.73). The reaction is controlled by the solubility product constant $K_{sp} = (Ca^{2+})(SO_4^{2-})$, which states that at equilibrium the product of calcium ion activity and sulfate ion activity is a constant value (or

TABLE 6.5
Gypsum Solubility in CaCl$_2$, NaCl, and MgCl$_2$ Solutions[a]

Salt Concentration mmol L^{-1}	EC dS m^{-1}	Gypsum Solubility mmol L^{-1}	% of Solubility in H$_2$O
		CaCl$_2$	
0	Pure H$_2$O	15.01	100.0
0.5	0.21	14.51	96.6
2.5	0.64	14.12	94.0
5	1.20	13.69	91.2
10	2.25	12.77	85.1
20	4.35	10.85	72.3
		NaCl	
0	Pure H$_2$O	15.01	100.0
1	0.20	15.03	100.1
5	0.65	15.74	104.9
10	1.21	16.58	110.5
20	2.40	17.44	116.2
40	4.60	18.36	122.3
		MgCl$_2$	
0	Pure H$_2$O	15.01	100.0
0.5	0.21	15.35	102.2
2.5	0.63	16.04	106.8
5	1.19	17.23	114.8
10	2.30	18.74	124.9
20	4.30	20.47	136.3

[a] Data were obtained from Longnecker and Lyerly (1959). EC is the solution electrical conductivity. Gypsum solubility is the concentration of soluble CaSO$_4$.

$IAP = K_{sp} = 10^{-4.62}$). If one adds Ca to the solution, as in this case by the addition of CaCl$_2$, one is adding a reaction product to the system (increasing the activity of Ca^{2+}). This forces the chemical reaction to move to the left and results in gypsum precipitation (or less gypsum dissolution). This is called the common ion effect, where calcium is a common ion in the gypsum dissolution reaction. A similar result is observed when Na$_2$SO$_4$ is added to solutions in equilibrium with gypsum. Indeed, increasing the concentration of a common ion decreases the CaSO$_4$ concentration in the equilibrium solution, but has no impact on IAP_{gypsum} (Figure 6.4a).

In an NaCl system, gypsum solubility increases with increasing salt concentration (Table 6.5). This result is explained by the ionic strength effect. As the salt content of the equilibrating solution increases, so does the ionic strength of the solution. As ionic strength increases, activity coefficients decrease, as do the activities of Ca^{2+} and SO$_4^{2-}$. Since the solubility product constant (K_{sp}) for gypsum is constant and is numerically equal to the product of Ca^{2+} and SO$_4^{2-}$ activities: (Ca^{2+})(SO$_4^{2-}$), a decrease in Ca^{2+} and SO$_4^{2-}$ activities would result in more gypsum dissolution to return the solution to the equilibrium state [$K_{sp} = $ (Ca^{2+})(SO$_4^{2-}$)] (Figure 6.4b).

Increasing the concentration of MgCl$_2$ results in an increase in gypsum solubility, similar to that observed in the NaCl system (Table 6.5). Clearly, the ionic strength effect has a significant influence on gypsum solubility in the MgCl$_2$ system. However, more gypsum dissolves in the MgCl$_2$ systems than in comparable ionic strength NaCl systems. A 20-mmol L^{-1} solution of MgCl$_2$ results in a 36.3% increase in gypsum solubility compared with a 22.3% increase in the 40 mmol L^{-1}

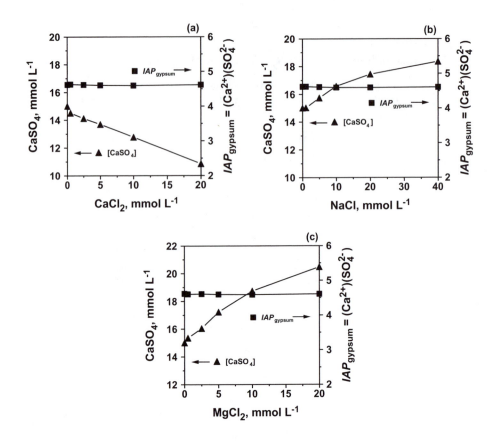

FIGURE 6.4 Gypsum [$CaSO_4 \cdot 2H_2O$] solubility, expressed in mmol L^{-1} dissolved $CaSO_4$ (left axis; closed triangles), and gypsum ion activity product ($IAP_{gypsum} = (Ca^{2+})(SO_4^{2-})$; right axis; closed squares) as a function of (a) $CaCl_2$, (b) $NaCl$, and (c) $MgCl_2$ concentrations.

NaCl system. We have established the fact that the ionic strength effect is active in both the $MgCl_2$ and NaCl systems. In the NaCl systems, the only significant aqueous complexation is the formation of the $CaSO_4^0$ species. Neither Na nor Cl substantially contributes to ion pair formation in these systems (Table 6.6). However, the presence of Mg results in the formation of an additional ion pair: $MgSO_4^0$. By virtue of this reaction, some of the SO_4^{2-} is removed from solution (into the soluble $MgSO_4^0$ species), forcing more gypsum to dissolve to maintain the K_{sp} (Figure 6.4c). This ion pairing or complexation effect has the opposite impact of the common ion effect on mineral solubility.

6.5 STABILITY DIAGRAMS

The comparison of the *IAP* value of a solution to the K_{sp} of a particular mineral is a rather inefficient and tedious mechanism to investigate the status of a soil solution relative to the minerals present, particularly if many minerals may potentially control an element's solution chemistry. More efficient mechanisms are available, two of which are the activity diagram (and the activity ratio diagram) and the predominance diagram. The activity diagrams are also used to evaluate the relative stability of minerals as a function of solution chemistry. Predominance diagrams also provide this information, and in addition identify stability fields, or the ranges of solution variables over which a particular mineral is stable.

TABLE 6.6

Ion Speciation in $CaCl_2$, NaCl, and $MgCl_2$ Solutions in Equilibrium with Gypsum Expressed as a Percentage of Total Soluble Calcium (Ca_T) or Sulfate (SO_{4T})[a]

Salt Concentration mmol L^{-1}	% of Ca_T		% of SO_{4T}		
	Ca^{2+}	$CaSO_4^0$	SO_4^{2-}	$CaSO_4^0$	$NaSO_4^-$ or $MgSO_4^0$
$CaCl_2$					
0	68.16	31.84	68.16	31.84	—
0.5	68.93	31.07	67.91	32.09	—
2.5	71.08	28.91	65.96	34.04	—
5	73.68	26.29	64.05	35.96	—
10	77.95	21.98	60.83	39.17	—
20	84.55	15.29	56.52	43.48	—
NaCl					
0	68.16	31.84	68.16	31.84	—
1	68.53	31.46	68.38	31.46	0.16
5	69.05	30.94	68.30	30.94	0.76
10	69.83	30.14	68.41	30.14	1.45
20	71.13	28.80	68.44	28.80	2.76
40	73.65	26.22	68.78	26.22	5.00
$MgCl_2$					
0	68.16	31.84	68.16	31.84	—
0.5	68.37	31.63	67.49	31.63	0.88
2.5	69.89	30.10	65.93	30.10	3.97
5	70.96	29.01	63.90	29.01	7.10
10	73.25	26.68	61.42	26.68	11.90
20	76.56	23.31	57.87	23.31	18.82

[a] Computed using GEOCHEM-PC (Parker) and the data of Longnecker and Lyerly (1959).

6.5.1 ACTIVITY DIAGRAMS

The activity diagram is a two- or three-dimensional plot. The axes variables are the activities of ions in solution. The stability of a mineral is represented on a two-dimensional diagram as a line and on a three-dimensional diagram as a surface. When the stability lines (or surfaces) of two or more minerals are included in the diagram, the relative stability of the minerals is indicated. Further, by plotting the measured activities of ions, an identification of the status of a solution with respect to the various solids can be determined.

All activity diagrams are constructed by employing a common procedure. First, a set of minerals that contain the element of interest, and that may control the free ion activity of the element in the environment of interest, is identified and dissolution reactions are written. For, example, the mineralogical characterization of a sewage sludge and a sewage sludge-amended soil has shown that Pb in the mineral phase is associated with either silica or phosphate (found in both the sewage sludge and the amended soil) (Essington and Mattigod, 1991). Two plausible Pb-bearing minerals are hydroxypyromorphite [$Pb_5(PO_4)_3OH$] and alamosite [$PbSiO_3$]. An activity diagram can be employed to determine the relative stability of the two Pb-bearing phases, and to indicate the phase that will ultimately and uniquely exist in the amended soil. The dissolution reactions for these two minerals are:

Hydroxypyromorphite

$$Pb_5(PO_4)_3OH(s) + 7H^+(aq) \rightarrow 5Pb^{2+}(aq) + 3H_2PO_4^-(aq) + H_2O(l) \qquad (6.74)$$

Alamosite

$$PbSiO_3(s) + 2H^+(aq) + H_2O(l) \rightarrow Pb^{2+}(aq) + H_4SiO_4^0(aq) \qquad (6.75)$$

The next step is to obtain the solubility product constant for each mineral dissolution reaction and express K_{sp} as a function of the ion activities of the reactants and products. For hydroxypyromorphite the solubility product constant is:

$$K_{sp,\text{HPM}} = \frac{(Pb^{2+})^5(H_2PO_4^-)^3}{(H^+)^7} = 10^{-4.14} \qquad (6.76)$$

A linear expression is then obtained by performing a logarithmic transformation:

$$5\log(Pb^{2+}) + 3\log(H_2PO_4^-) + 7pH = -4.14 \qquad (6.77)$$

Similarly, the K_{sp} for alamosite is:

$$K_{sp,\text{ALA}} = \frac{(Pb^{2+})(H_4SiO_4^0)}{(H^+)^2} = 10^{5.94} \qquad (6.78)$$

$$\log(Pb^{2+}) + \log(H_4SiO_4^0) + 2pH = 5.94 \qquad (6.79)$$

In order to plot the stability lines that represent the two minerals the independent (x-axis) and dependent (y-axis) axis variables must be selected. For the two Pb-bearing minerals, there are four ion activity variables from which to select the axis variables: $\log(Pb^{2+})$, $\log(H_2PO_4^-)$, $\log(H_4SiO_4^0)$, and pH. The natural choice for the dependent variable (y-axis) is $\log(Pb^{2+})$, since the objective of the analysis is to determine the Pb^{2+} activity levels controlled by the Pb-bearing minerals. The dependent variable (x-axis) will be $\log(H_2PO_4^-)$. The sewage sludge contains an abundance of phosphate in the form of brushite [$CaHPO_4 \cdot 2H_2O$], as well as soluble organic P forms that are readily mineralized to phosphate in the soil environment. Brushite is an unstable mineral phosphate form that will weather to more stable P-bearing minerals, such as octacalcium phosphate [$Ca_8H_2(PO_4)_6 \cdot 5H_2O$]. It is expected that soluble phosphate levels in the sewage sludge–amended soil environment will vary, perhaps over several orders of magnitude, decreasing with time. Conversely, pH and the activity of silicic acid are expected to be relatively stable, and representative values for these variables can be selected to suitably describe their status in the soil solution.

Equations 6.77 and 6.78 are rewritten to yield linear equations with $\log(Pb^{2+})$ as the dependent variable and $\log(H_2PO_4^-)$ as the independent variable. For hydroxypyromorphite, Equation 6.77 becomes:

$$5\log(Pb^{2+}) = -3\log(H_2PO_4^-) - 7pH - 4.14 \qquad (6.80)$$

or

$$\log(Pb^{2+}) = -\tfrac{3}{5}\log(H_2PO_4^-) - \tfrac{7}{5}pH - 0.83 \qquad (6.81)$$

The expression representing alamosite stability (Equation 6.79) becomes:

$$\log(\text{Pb}^{2+}) = -\log(\text{H}_4\text{SiO}_4^0) - 2\text{pH} + 5.94 \tag{6.82}$$

The equation that describes the stability of hydroxypyromorphite (Equation 6.81) has three variables: $\log(\text{Pb}^{2+})$ and $\log(\text{H}_2\text{PO}_4^-)$, which are the axis variables, and pH. In order to plot the hydroxypyromporphite stability line on a two-dimensional plot, the pH must be fixed, i.e., the pH of the environment to which this evaluation is pertinent must be known or estimated. The pH of this sewage sludge–amended soil is approximately 7. Therefore, a pH of 7 is selected as the fixed pH. The hydroxypyromorphite stability line (Equation 6.81) becomes:

$$\log(\text{Pb}^{2+}) = -\tfrac{3}{5}\log(\text{H}_2\text{PO}_4^-) - 10.63 \tag{6.83}$$

The equation that describes the stability of alamosite (Equation 6.82) also has three variables: $\log(\text{Pb}^{2+})$ (an axis variable), $\log(\text{H}_4\text{SiO}_4^0)$, and pH. The fixed pH of the soil is 7. It remains for us to select a $\log(\text{H}_4\text{SiO}_4^0)$ value, for which there are two alternatives. One alternative is to simply select a value for $\log(\text{H}_4\text{SiO}_4^0)$ that is consistent with values observed in pH 7 sewage sludge–amended soil environments. This is the mechanism we employed in selecting a pH value. A second mechanism is to allow $\log(\text{H}_4\text{SiO}_4^0)$ to be controlled by the dissolution of a soil mineral phase. In many soil systems, $(\text{H}_4\text{SiO}_4^0)$ is controlled by the dissolution of amorphous silica [$\text{H}_4\text{SiO}_4(am)$], a metastable phase. The dissolution reaction is $\text{H}_4\text{SiO}_4(am) = \text{H}_4\text{SiO}_4^0(aq)$, with a $\log K_{sp} = -2.71$. Therefore, $\log(\text{H}_4\text{SiO}_4^0) = -2.71$. Substituting for pH and $\log(\text{H}_4\text{SiO}_4^0)$ in the alamosite stability equation (Equation 6.82) yields:

$$\log(\text{Pb}^{2+}) = -5.35 \tag{6.84}$$

which is the equation of a line parallel to the x-axis (no slope).

Equations 6.83 and 6.84 represent the stability of hydroxypyromorphite and alamosite in two-dimensional space defined by $\log(\text{Pb}^{2+})$ and $\log(\text{H}_2\text{PO}_4^-)$ (Figure 6.5). The activity diagram is valid assuming the following: unity activities of the solids and of liquid water, 25°C and 0.101 MPa, pH = 7, and silicic acid activity controlled by $\text{H}_4\text{SiO}_4(am)$. Changes in any one of the above assumptions would produce a different stability diagram and alter one's interpretation of mineral stability and lead activity in the environment. The rule that is necessary for the interpretation of stability diagrams is: the mineral that controls the activity of a species in the soil solution is the one that produces the smallest activity of the species. Throughout a broad range of H_2PO_4^- activity ($10^{-8.8}$ to 10^{-2}),

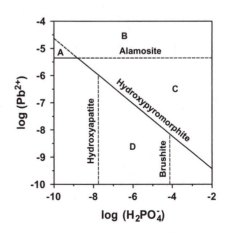

FIGURE 6.5 An activity diagram illustrating the stability of hydroxypyromorphite [$\text{Pb}_5(\text{PO}_4)_3\text{OH}$], alamosite [$\text{PbSiO}_3$], hydroxyapatite [$\text{Ca}_5(\text{PO}_4)_3\text{OH}$], and brushite [$\text{CaHPO}_4{\cdot}2\text{H}_2\text{O}$]. The diagram is valid assuming unity activities of the solids and of liquid water, 25°C and 0.101 MPa, pH 7, H_4SiO_4^0 activity controlled by $\text{H}_4\text{SiO}_4(am)$, CO_2 partial pressure is 10^{-2}, and Ca^{2+} activity controlled by calcite [CaCO_3].

Pb^{2+} activity is controlled by hydroxypyromorphite. Only at very low phosphate activities [$(H_2PO_4^-)$ < $10^{-8.8}$)] is Pb^{2+} activity controlled by the lead silicate phase (alamosite). If the solubilities of hydroxyapatite and brushite are employed to bracket the expected $H_2PO_4^-$ activity in the soil solution (assuming CO_2 partial pressure is 10^{-2} and calcite controls Ca^{2+} activity), $H_2PO_4^-$ activity will range from $10^{-7.75}$ (hydroxyapatite) to $10^{-4.11}$ (brushite). It is apparent that alamosite will always be unstable in the sewage sludge–amended soil, under the stated conditions (Figure 6.5). Therefore, occurrence of alamosite in the amended soil is temporary—a relic of the sewage sludge amendment, while the Pb-phosphate phase is pedogenic.

An activity diagram can also be employed to establish the status of an equilibrated solution relative to several minerals (instead of performing one-to-one comparisons of K_{sp} and IAP values). If a soil solution exists under the conditions employed to construct the activity diagram in Figure 6.5 and the activities of Pb^{2+} and $H_2PO_4^-$ in the solution are determined, when plotted the [$\log(Pb^{2+})$, $\log(H_2PO_4^-)$] data pair will fall into one of four general regions, labeled A through D. If the [$\log(Pb^{2+})$, $\log(H_2PO_4^-)$] data pair plots in region A, the soil solution is supersaturated with respect to alamosite ($IAP_{ALA} > K_{sp,ALA}$) and undersaturated with respect to hydroxypyromorphite ($IAP_{HPM} < K_{sp,HPM}$). One might then conclude that alamosite will eventually precipitate and that hydroxypyromorphite is not present or, if present, will eventually dissolve. In region B, the soil solution is supersaturated with respect to both minerals ($IAP > K_{sp}$). This would suggest that either mineral has a potential to precipitate. In region C, the soil solution is undersaturated with respect to alamosite ($IAP_{ALA} < K_{sp,ALA}$) and supersaturated with respect to hydroxypyromorphite ($IAP_{HPM} > K_{sp,HPM}$). One might then conclude that hydroxypyromorphite will eventually precipitate and that alamosite is not present, or if present, will dissolve. Finally, a soil solution with a composition consistent with that of region D is undersaturated with respect to both alamosite and hydroxypyromorphite ($IAP < K_{sp}$), and both phases are unstable and will dissolve if present. Of course, if the composition of the soil solution plots on either the alamosite or the hydroxypyromorphite stability line, it may be concluded that the mineral is present and controlling the activity of Pb^{2+}.

6.5.2 ACTIVITY RATIO DIAGRAMS

Another type of stability diagram is the activity ratio diagram. The activity ratio diagram is a two- or three-dimensional plot. The axes variables are functions of the activities of ions in solution. Specifically, the axes variables are of the form:

$$[pH - \tfrac{1}{m} pM^{m+}] \text{ and } [pH + \tfrac{1}{n} pL^{n-}] \tag{6.85}$$

where pM^{m+} and pL^{n-} are the negative logarithm values of metal and ligand activities. The advantage an activity ratio diagram offers over an activity diagram is that three activity variables are expressed on a two-dimensional diagram: pH, (M^{m+}), and (L^{n-}). Activity ratio diagrams are constructed in a manner identical to that of an activity diagram. Consider the above example that describes the speciation of lead in a sewage sludge–amended soil. The hydroxypyromorphite dissolution reaction can be written as:

$$Pb_5(PO_4)_3OH(s) + 10H^+(aq) \rightarrow 5Pb^{2+}(aq) + 3H_2PO_4^-(aq) + H_2O(l) + 3H^+(aq) \tag{6.86}$$

This reaction differs from Equation 6.74 in that three protons are added to both sides of the dissolution reaction. Since both the reactant and product sides of the hydroxypyromorphite reaction are equally treated to proton additions, there is no net effect, and the $\log K_{sp}$ value of -4.14 for Equation 6.86 is identical to that of Equation 6.74. The $\log K_{sp}$ for the reaction in Equation 6.86 is expressed as:

$$\log K_{sp} = 5\log(Pb^{2+}) + 10pH + 3\log(H_2PO_4^-) - 3pH \tag{6.87}$$

Converting to "p" notation (where p = −log):

$$\log K_{sp} = 10\text{pH} - 5\text{pPb}^{2+} - 3\text{pH} - 3\text{pH}_2\text{PO}_4^- \qquad (6.88)$$

$$\log K_{sp} = 10[\text{pH} - \tfrac{1}{2}\text{pPb}^{2+}] - 3[\text{pH} + \text{pH}_2\text{PO}_4^-] \qquad (6.89)$$

Rearranging such that $[\text{pH} - \tfrac{1}{2}\text{pPb}^{2+}]$ is the dependent variable and the $[\text{pH} + \text{pH}_2\text{PO}_4^-]$ is the dependent variable:

$$[\text{pH} - \tfrac{1}{2}\text{pPb}^{2+}] = \tfrac{3}{10}[\text{pH} + \text{pH}_2\text{PO}_4^-] + \tfrac{1}{10}\log K_{sp} \qquad (6.90)$$

Substituting −4.14 for $\log K_{sp}$ yields the straight-line stability expression for hydroxypyromorphite:

$$[\text{pH} - \tfrac{1}{2}\text{pPb}^{2+}] = \tfrac{3}{10}[\text{pH} + \text{pH}_2\text{PO}_4^-] - 0.414 \qquad (6.91)$$

The stability line for alamosite is similarly obtained:

$$[\text{pH} - \tfrac{1}{2}\text{p}(\text{Pb}^{2+})] = \tfrac{1}{2}\text{p}(\text{H}_4\text{SiO}_4^0) + 2.97 \qquad (6.92)$$

If it is again assumed that silicic acid activity is controlled by $\text{H}_4\text{SiO}_4(am)$ $[\text{pH}_4\text{SiO}_4^0 = 2.71]$, Equation 6.92 becomes:

$$[\text{pH} - \tfrac{1}{2}\text{pPb}^{2+}] = 4.33 \qquad (6.93)$$

Equations 6.91 and 6.93 are plotted, along with the $[\text{pH} + \text{pH}_2\text{PO}_4^-]$ values controlled by brushite and hydroxyapatite dissolution, and assuming a CO_2 partial pressure of 10^{-2} and $[\text{pH} - \tfrac{1}{2}\text{pCa}^{2+}] = 5.87$ (controlled by calcite) (Figure 6.6). The activity diagram (Figure 6.5) and the activity ratio diagram (Figure 6.6) are very similar in appearance, and provide comparable information about the relative stability of hydroxypyromorphite and alamosite. However, in order to investigate relative mineral stability at a pH other than 7 (Figure 6.5), a separate activity diagram must be constructed. Using the activity ratio diagram (Figure 6.6), one can easily examine the influence of solution pH on mineral stability.

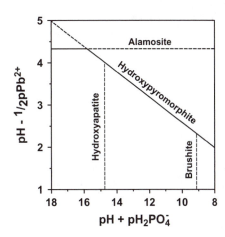

FIGURE 6.6 An activity ratio diagram illustrating the stability of hydroxypyromorphite $[\text{Pb}_5(\text{PO}_4)_3\text{OH}]$, alamosite $[\text{PbSiO}_3]$, hydroxyapatite $[\text{Ca}_5(\text{PO}_4)_3\text{OH}]$, and brushite $[\text{CaHPO}_4 \cdot 2\text{H}_2\text{O}]$. The diagram is valid assuming unity activities of the solids and of liquid water, 25°C and 0.101 MPa, H_4SiO_4^0 activity controlled by $\text{H}_4\text{SiO}_4(am)$, CO_2 partial pressure is 10^{-2}, and Ca^{2+} activity controlled by calcite $[\text{CaCO}_3]$.

It was established in Section 6.3 that poor crystallinity (and microcrystallinity) will enhance mineral solubility relative to the well-crystallized, standard-state phase. It was also established that the solubility of a minor constituent in a solid solution will be reduced relative to the pure end member phase. Both phenomena minimize the utility of solubility equilibria evaluations for predicting the activities of elements in the soil solution, and for indirectly characterizing the mineral phases that may be present in the soil environment. That minerals in a soil environment may not be well crystallized or pure can be considered during the development of activity or activity ratio diagrams, as well as in the assessment of environmental fate and behavior. Indeed, the formation of metastable, amorphous phases is favored in terrestrial environments, as the most thermodynamically stable minerals will not readily precipitate (step rule). For example, the activity diagram in Figure 6.7 illustrates the stability of poorly crystalline gibbsite [gibbsite activity, $(Al(OH)_3) = 1778$] relative to that of standard-state gibbsite [$(Al(OH)_3) = 1.0$]. The poorly crystalline gibbsite is a metastable fast-former (also known as amorphous $Al(OH)_3$), and may be expected to control aqueous Al^{3+} activity upon initial formation. As the $Al(OH)_3(am)$ ages, conversion to the thermodynamically stable gibbsite phase will occur, resulting in a solution Al^{3+} activity that is consistent with the presence of gibbsite. Therefore, a range of Al^{3+} activities, denoted by the hatched region in Figure 6.7, can be controlled by gibbsite dissolution.

A diagram similar to Figure 6.7 can be constructed to illustrate the Cd^{2+} activity levels controlled by the calcite-otavite solid solution and the pure otavite end member (Figure 6.8). The diagram establishes a range of Cd^{2+} activity values that might be expected in a soil solution in equilibrium with otavite, as a function of the carbonate ion activity. The diagram also establishes the greater stability of the solid solution relative to the pure otavite end member. The fact that trace elements can exist in solid solutions in the soil environment, coupled with the fact that solid solutions control lower activities of trace elements in soil solutions relative to the pure end member phases, have not been lost on the environmental scientist. Specifically, it is recognized that predicted trace element

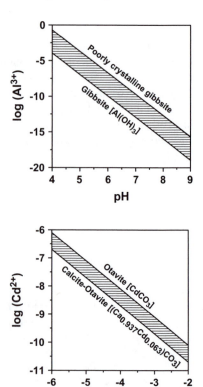

FIGURE 6.7 An activity diagram illustrating the stability of gibbsite [$(Al(OH)_3) = 1.0$] and poorly crystalline (amorphous) gibbsite [$(Al(OH)_3) = 1778$] (activity of liquid water is unity, 25°C and 0.101 MPa). Solution pH and corresponding $\log(Al^{3+})$ values that fall in the hatched region may indicate the presence of a gibbsite phase if equilibrium exists.

FIGURE 6.8 An activity diagram illustrating the stability of pure otavite [$(CdCO_3) = 1.0$] and otavite in a calcite–otavite solid solution [$(Ca_{0.937}Cd_{0.063})CO_3$] ($f_{otavite} = 4.39$, $N_{otavite} = 0.063$, activity of liquid water is unity, 25°C and 0.101 MPa). Solution $\log(CO_3^{2-})$ and corresponding $\log(Cd^{2+})$ values that fall in the hatched region may indicate the presence of an otavite phase if equilibrium exists.

solubility and transport based on pure mineral dissolution may overestimate potential mobility and off-site contamination. Unfortunately, there exists a general lack of information necessary to quantify coprecipitation processes and more accurately predict environmental behavior.

6.5.3 THE TEMPERATURE DEPENDENCE OF THE EQUILIBRIUM CONSTANT

The standard free energy of reaction (ΔS_r°) and the equilibrium constant for the reaction are related through the expression:

$$\Delta S_r^\circ = - RT \ln K \tag{6.94}$$

If ΔS_r° is expressed in units of kJ mol^{-1} and $T = 298.15$ K:

$$\Delta S_r^\circ = - 5.708 \log K \tag{6.95}$$

This expression has great utility in the environmental sciences. However, chemical reactions that occur in the environment do not often occur at 298.15 K (25°C). For a reaction in standard state, the standard free energy change for the reaction is:

$$\Delta S_r^\circ = \Delta H_r^\circ - T\Delta S_r^\circ \tag{6.96}$$

where ΔH_r° is the standard enthalpy change for the reaction ($\Delta H_r^\circ = \sum_i v_i h_{f,i}^\circ$ and $h_{f,i}^\circ$ is the standard enthalpy of formation and v_i is the stiochiometry of substance i) and ΔS_r° is the standard entropy change for the reaction ($\Delta S_r^\circ = \sum_i v_i s_{f,i}^\circ$ and $s_{f,i}^\circ$ is the partial molal entropy). Combining Equations 6.94 and 6.96:

$$\ln K = -\left(\frac{\Delta H_r^\circ}{RT}\right) + \left(\frac{\Delta S_r^\circ}{R}\right) \tag{6.97}$$

The temperature dependence of the equilibrium constant can be determined by differentiating Equation 6.97 with respect to T:

$$\frac{\partial(\ln K)}{\partial T} = -\left(\frac{1}{RT}\right)\left(\frac{\partial \Delta H_r^\circ}{\partial T}\right) + \left(\frac{\Delta H_r^\circ}{RT^2}\right) + \left(\frac{1}{R}\right)\left(\frac{\partial \Delta S_r^\circ}{\partial T}\right) \tag{6.98}$$

For a relatively narrow temperature range, the first and third terms on the right side of Equation 6.98 are negligible. The resulting expression is van't Hoff's equation:

$$\frac{\partial(\ln K)}{\partial T} = \frac{\Delta H_r^\circ}{RT^2} \tag{6.99}$$

Over small ranges of temperature it may be assumed that ΔH_r° is approximately independent of temperature. Transforming to the common logarithm and integrating Equation 6.99 leads to an expression that computes $\log K$ as a function of T:

$$\int_{K^\circ}^{K_T} d\log K = \left(\frac{\Delta H_r^\circ}{2.303R}\right)\int_{T^\circ}^{T}\left(\frac{1}{T^2}\right)dT \tag{6.100}$$

$$\log K_T - \log K^\circ = -\left(\frac{\Delta H_r^\circ}{2.303R}\right)\left(\frac{1}{T} - \frac{1}{T^\circ}\right) \tag{6.101}$$

where K is the equilibrium constant at any temperature T, and ΔH_r° and K° are the enthalpy and equilibrium constant values for the reaction in a reference system at T° (usually the standard state at 298.15 K).

For wide temperature ranges (ΔH_r° is dependent on temperature), additional evaluation of Equation 6.99 is required. By definition:

$$\frac{\partial \Delta H_r^\circ}{\partial T} = \Delta C_P^\circ \tag{6.102}$$

$$\frac{\partial \Delta S_r^\circ}{\partial T} = \frac{\Delta C_P^\circ}{T} \tag{6.103}$$

where ΔC_P° is the heat capacity of a reaction at fixed pressure. Substituting Equations 6.102 and 6.103 into 6.98 yields:

$$\frac{\partial (\ln K_T)}{\partial T} = -\left(\frac{1}{RT}\right)\Delta C_P^\circ + \frac{\Delta H_r^\circ}{RT^2} + \left(\frac{1}{R}\right)\left(\frac{\Delta C_P^\circ}{T}\right) \tag{6.104}$$

Rearranging,

$$\partial(\ln K_T) = -\left(\frac{1}{RT}\right)(\Delta C_P^\circ)\partial T + \left(\frac{\Delta H_r^\circ}{RT^2}\right)\partial T + \left(\frac{1}{R}\right)\left(\frac{\Delta C_P^\circ}{T}\right)\partial T \tag{6.105}$$

Integrating Equation 6.105,

$$\int_{K^\circ}^{K_T} \partial(\ln K_T) = -\left(\frac{1}{RT}\right)\int_{T^\circ}^{T}(\Delta C_P^\circ)\partial T - \frac{\Delta H_{r,T}^\circ}{RT} + \frac{\Delta H_{r,T^\circ}^\circ}{RT^\circ} + \left(\frac{1}{RT}\right)\int_{T^\circ}^{T}\partial \Delta H_r^\circ + I + \left(\frac{1}{R}\right)\int_{T^\circ}^{T}\left(\frac{\Delta C_P^\circ}{T}\right)\partial T \tag{6.106}$$

where I is a constant of integration that equals 0 when $T = T^\circ$. Continuing,

$$\ln K_T - \ln K^\circ = -\left(\frac{1}{RT}\right)\int_{T^\circ}^{T}(\Delta C_P^\circ)\partial T + \left(\frac{\Delta H_{r,T}^\circ}{R}\right)\left[\left(\frac{1}{T^\circ}\right) - \left(\frac{1}{T}\right)\right] + \left(\frac{1}{R}\right)\int_{T^\circ}^{T}\left(\frac{\Delta C_P^\circ}{T}\right)\partial T \tag{6.107}$$

Utilizing Equation 6.97 and rearranging, with standard state as the reference temperature:

$$\ln K_T = \left(\frac{T^\circ}{T}\right)\ln K^\circ + \left[\left(\frac{\Delta S_r^\circ}{RT}\right)(T - T^\circ)\right] - \left(\frac{1}{RT}\right)\int_{T^\circ}^{T}(\Delta C_P^\circ)\partial T + \left(\frac{1}{R}\right)\int_{T^\circ}^{T}\left(\frac{\Delta C_P^\circ}{T}\right)\partial T \tag{6.108}$$

This expression may be employed to determine the equilibrium constant of any reaction at any temperature. The solution to Equation 6.108 requires values for the equilibrium constant and entropy of reaction at a reference temperature (and pressure), and the heat capacity of the reaction as a function of temperature. Typically, the reference system is standard state, as compilations of thermodynamic data are quite extensive for this condition. The heat capacity of a solid at fixed pressure and as a function of temperature is typically expressed as

$$C_P = a + bT + \frac{c}{T^2} \tag{6.109}$$

For aqueous species:

$$C_P = bT \tag{6.110}$$

The coefficients, a, b, and c are constant for any given species and can also be found in thermodynamic data compilations.

The solution to Equation 6.108 is essentially an algebraic exercise that is most appropriately solved using common data manipulation software. The Van't Hoff equation is a convenient mechanism to investigate the influence of temperature on the relative stability of trace mineral phases in a soil environment. As indicated in Equation 6.101, a $\log K_{sp}$ value for any temperature T can be approximated, given the standard enthalpy change for the mineral dissolution reaction (ΔH_r°) and $\log K_{sp}$ at a reference T, typically 298.15 K (denoted as $\log K_{sp}^\circ$ and T° in Equation 6.101). Consider the relative stability of various Zn-bearing minerals. In keeping with the procedure outlined above for the construction of an activity diagram, the Zn-bearing minerals that may potentially control aqueous Zn^{2+} activities in terrestrial environments are: zincite [ZnO], smithsonite [$ZnCO_3$], hydrozincite [$Zn_5(OH)_6(CO_3)_2$], hopeite [$Zn_3(PO_4)_2 \cdot 4H_2O$], willemite [Zn_2SiO_4], and franklinite [$ZnFe_2O_4$]. The dissolution reactions for these minerals are shown in Table 6.7, as are the $\log K_{sp}$ values for the standard-state temperature of 25°C. The $\log K_{sp}$ value for mineral dissolution at any temperature other than the reference temperature requires a ΔH_r° value for the dissolution reaction. The ΔH_r° for any reaction can be computed using the h_f° values for the reactants and the products. For example, hopeite dissolution can be represented by the reaction:

$$Zn_3(PO_4)_2 \cdot 4H_2O(s) + 4H^+(aq) \rightarrow 3Zn^{2+}(aq) + 2H_2PO_4^-(aq) + 4H_2O(l) \tag{6.111}$$

The ΔH_r° for the reaction is given by:

$$\Delta H_r^\circ = 3h_{f,Zn^{2+}}^\circ + 2h_{f,H_2PO_4^-}^\circ + 4h_{f,H_2O}^\circ - h_{f,\text{hopeite}}^\circ \tag{6.112}$$

Using the enthalpy of formation values in Table 6.8, the standard enthalpy change for the hopeite dissolution reaction is:

$$\Delta H_r^\circ = 3(-153.4) + 2(-1296.3) + 4(-285.8) - (-4093.5) = -102.5 \text{ kJ mol}^{-1} \tag{6.113}$$

Rearranging Equation 6.101 and substituting the numerical values for the natural gas constant (R) and the standard-state temperature (T°) yields an expression that can be used to compute the $\log K_{sp}$ for mineral dissolution at any temperature:

$$\log K_T = -\left(\frac{\Delta H_r^\circ}{0.01915}\right)\left(\frac{298.15 - T}{298.15 \times T}\right) + \log K^\circ \tag{6.114}$$

The $\log K_{sp}$ values for the Zn-bearing minerals for $T = 283.15$ K (10°C) and $T = 313.15$ K (40°C) are presented in Table 6.7 and computed using the standard state ΔH_r° and $\log K_{sp}$ values.

TABLE 6.7
Enthalpies of Reaction (ΔH_r° at 25°C and 0.101 MPa; Computed Using the Enthalpy of Formation Values in Table 6.9), $\log K_{sp}$ Values at 25°C, and $\log K_{sp}$ Values at 10°C and 40°C (Computed Using the van't Hoff Equation (Equation 6.101)) for the Mineral Dissolution Reactions Used to Construct the Activity Diagrams Displayed in Figures 6.9a through 6.9c[a]

Reaction	ΔH_r°, kJ mol^{-1}	$\log K_{sp}$ 10°C	$\log K_{sp}$ 25°C	$\log K_{sp}$ 40°C
Zincite $ZnO(s) + 2H^+(aq) \rightarrow Zn^{2+}(aq) + H_2O(l)$	−88.5	10.52	11.34	12.08
Smithsonite $ZnCO_3(s) + 2H^+(aq) \rightarrow Zn^{2+}(aq) + CO_2(g) + H_2O(l)$	−19.9	7.97	8.15	8.32
Hydrozincite $Zn_5(OH)_6(CO_3)_2(s) + 10H^+(aq) \rightarrow 5Zn^{2+}(aq) + 2CO_2(g) + 8H_2O(l)$	−375	45.52	49.00	52.15
Hopeite $Zn_3(PO_4)_2 \cdot 4H_2O(s) + 4H^+(aq) \rightarrow 3Zn^{2+}(aq) + 2H_2PO_4^-(aq) + 4H_2O(l)$	−102.5	2.85	3.80	4.66
Franklinite $ZnFe_2O_4(s) + 8H^+(aq) \rightarrow Zn^{2+}(aq) + 2Fe^{3+}(aq) + 4H_2O(l)$	−221.66	4.81	6.87	8.73
Willemite $Zn_2SiO_4(s) + 4H^+(aq) \rightarrow 2Zn^{2+}(aq) + H_4SiO_4^0(aq)$	79.5	16.07	15.33	14.66
Goethite $FeOOH(s) + 3H^+(aq) \rightarrow Fe^{3+}(aq) + 2H_2O(l)$	−60.8	−2.21	−1.65	−1.14
Amorphous silica $SiO_2(am) + 2H_2O(l) \rightarrow H_4SiO_4^0(aq)$	−39.5	−3.03	−2.66	−2.33
Calcite $CaCO_3(s) + 2H^+(aq) \rightarrow Ca^{2+}(aq) + CO_2(g) + H_2O(l)$	−14.7	9.60	9.74	9.86
Hydroxyapatite $Ca_5(PO_4)_3OH(s) + 7H^+(aq) \rightarrow 5Ca^{2+}(aq) + 3H_2PO_4^-(aq) + H_2O(l)$	−167.1	12.91	14.46	15.86

[a] $\log K_{sp}$ values obtained from Robie et al. (1978), Naumov et al. (1974), Smith and Martell (1976), and Lindsay (1979).

The activity diagrams will illustrate the relative stability of the Zn-bearing minerals as a function of temperature and pH. Thus, the axis variables will be the dependent variable $\log(Zn^{2+})$ and the independent variable pH. In order for all of the Zn-bearing minerals to appear on a single $\log(Zn^{2+})$ vs. pH activity diagram representing any one of the temperature regimes, the activities of a number of components must be controlled or fixed. Specifically, the activities of CO_3^{2-}, $H_2PO_4^-$, $H_4SiO_4^0$, and Fe^{3+} are required to establish the stability lines for smithsonite, hydrozincite, hopeite, willemite, and franklinite. The activity of CO_3^{2-} will be controlled by a CO_2 partial pressure of 10^{-2}. Silicic acid activity will be controlled by the dissolution of $H_4SiO_4(am)$, and Fe^{3+} activity will be controlled by goethite dissolution. Phosphate (as $H_2PO_4^-$) activity will be controlled by the dissolution of hydroxyapatite and calcite in equilibrium with a CO_2 partial pressure of 10^{-2}. The stability of each Zn-bearing mineral on a $\log(Zn^{2+})$ vs. pH activity diagram is then characterized by a linear expression that predicts the variation in $\log(Zn^{2+})$ as a function pH for each temperature condition (Table 6.9).

The Zn-bearing mineral that is predicted to control Zn^{2+} activities in the soil environment is the phase that generates the lowest Zn^{2+} activity for the specified condition. In a 25°C environment, the zinc silicate mineral (willemite) is the predicted stable zinc phase under the conditions employed to construct the activity diagram (Figure 6.9a). Throughout the pH 6 to 9 range, willemite is

TABLE 6.8
Enthalpy of Formation Values (h_f° at 25°C and 0.101 MPa) Used to Determine the Temperature-Dependence of Zinc- and Lead-Bearing Mineral Dissolution Reactions[a]

Species	h_f°, kJ mol^{-1}	Species	h_f°, kJ mol^{-1}
$Ca^{2+}(aq)$	−542.8	$CaHPO_4 \cdot 2H_2O(s)$	−1207.4
$CO_2(g)$	−393.5	$ZnO(s)$	−350.7
$Cl^-(aq)$	−167.1	$ZnCO_3(s)$	−812.8
$Fe^{3+}(aq)$	−48.5	$Zn_5(OH)_6(CO_3)_2(s)$	−3465.4[b]
$H_2O(l)$	−285.8	$Zn_3(PO_4)_2 \cdot 4H_2O(s)$	−4093.5[c]
$H_2PO_4^-(aq)$	−1296.3[e]	$ZnFe_2O_4(s)$	−1171.94[d]
$H_4SiO_4^0(aq)$	−1460.0	$Zn_2SiO_4(s)$	−1636.5
$Pb^{2+}(aq)$	−1.7	$PbO(s)$	−219.4
$Zn^{2+}(aq)$	−153.4	$PbCO_3(s)$	−699.2
$Al(OH)_3(s)$	−1293.1	$Pb_3(CO_3)_2(OH)_2(s)$	−1914.2[b]
$FeOOH(s)$	−559.3	$Pb_5(PO_4)_3OH(s)$	(−3786)[f]
$SiO_2(am)$	−848.9	$Pb_5(PO_4)_3Cl(s)$	(−4191)[f]
$CaCO_3(s)$	−1207.4	$PbAl_3(PO_4)_2(OH)_5 \cdot H_2O(s)$	(−5863)[f]
$Ca_5(PO_4)_3OH(s)$	−6721.6		

[a] Values were obtained from Robie et al. (1978) unless noted otherwise.

[b] Sangameshwar and Barnes (1983).

[c] Almaydama et al. (1992).

[d] Naumov et al. (1974).

[e] Wagman et al. (1982).

[f] Estimated using the entropies of formation (s_f°) values for hydroxyapatite and chloroapatite and Equation 6.96.

TABLE 6.9
Zinc Mineral Stability Lines Plotted in Figures 6.9a through 6.9c

Mineral	Stability Functions		
	10°C	25°C	40°C
Zincite	$\log(Zn^{2+}) = 10.52 - 2pH$	$\log(Zn^{2+}) = 11.34 - 2pH$	$\log(Zn^{2+}) = 12.08 - 2pH$
Smithsonite[a]	$\log(Zn^{2+}) = 9.97 - 2pH$	$\log(Zn^{2+}) = 10.15 - 2pH$	$\log(Zn^{2+}) = 10.32 - 2pH$
Hydrozincite[a]	$\log(Zn^{2+}) = 9.90 - 2pH$	$\log(Zn^{2+}) = 10.60 - 2pH$	$\log(Zn^{2+}) = 11.28 - 2pH$
Hopeite[b]	$\log(Zn^{2+}) = 10.97 - 2pH$	$\log(Zn^{2+}) = 11.10 - 2pH$	$\log(Zn^{2+}) = 11.21 - 2pH$
Franklinite[c]	$\log(Zn^{2+}) = 9.23 - 2pH$	$\log(Zn^{2+}) = 10.16 - 2pH$	$\log(Zn^{2+}) = 11.01 - 2pH$
Willemite[d]	$\log(Zn^{2+}) = 9.55 - 2pH$	$\log(Zn^{2+}) = 9.00 - 2pH$	$\log(Zn^{2+}) = 8.50 - 2pH$

[a] $P_{CO_2} = 10^{-2}$.

[b] ($H_2PO_4^-$) controlled by hydroxyapatite, calcite, and $P_{CO_2} = 10^{-2}$.

[c] (Fe^{3+}) controlled by goethite.

[d] ($H_4SiO_4^0$) controlled by $H_4SiO_4(am)$.

predicted to support a Zn^{2+} activity that is approximately one order of magnitude less than that supported by the metastable phases, zinc ferrite (franklinite) and zinc carbonate (smithsonite). The zinc hydroxycarbonate (hydrozincite) is unstable relative to smithsonite and the zinc silicate and ferrite phases. Zinc phosphate (hopeite) is slightly more stable than the oxide (zincite), which is the least stable zinc-bearing mineral. The predicted order of increasing mineral stability or

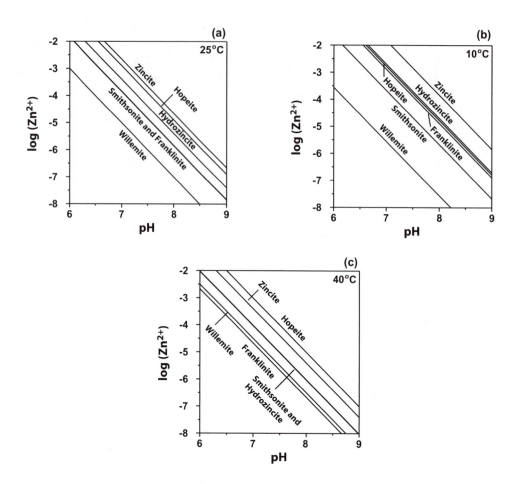

FIGURE 6.9 Stability of the Zn-bearing minerals, zincite [ZnO], smithsonite [$ZnCO_3$], hydrozincite [$Zn_5(OH)_6(CO_3)_2$], hopeite [$Zn_3(PO_4)_2 \cdot 4H_2O$], franklinite [$ZnFe_2O_4$], and willemite [Zn_2SiO_4] as a function of solution pH and temperature: (a) 25°C, (b) 10°C, and (c) 40°C. The following conditions apply: 0.101 MPa pressure, solid and water activities are unity, CO_2 partial pressure is 10^{-2}, solution activity of $H_2PO_4^-$ is controlled by hydroxyapatite and calcite, $H_4SiO_4^0$ activity is controlled by $H_4SiO_4(am)$, and Fe^{3+} is controlled by goethite [FeOOH].

decreasing Zn^{2+} activities supported by the mineral is: zincite ($10^{-3.66}$) > hopeite ($10^{-3.90}$) > hydrozincite ($10^{-4.40}$) > franklinite ($10^{-4.84}$) ≈ smithsonite ($10^{-4.85}$) > willemite ($10^{-6.00}$) (values in parentheses are the Zn^{2+} activities supported by each mineral at pH 7.5).

The relative stability of the zinc minerals is clearly illustrated in the activity diagram. In addition, the stability lines are parallel to one another, with a slope of −2 (see also Table 6.9), indicating that each mineral controls a specific activity ratio, (pH − ½pZn^{2+}). For every unit decrease in soil solution pH, a twofold increase in logZn^{2+} activity is predicted, irrespective of zinc speciation in the solid phase. Also notable is the relative instability of the zinc phosphate mineral hopeite. Metal phosphates are often a desired result of *in situ* soil stabilization technologies for their perceived stability in a soil environment and in the gastrointestinal tract of humans and animals. Clearly, the relative instability of hopeite suggests that phosphate-induced stabilization of a zinc-contaminated soil would not provide long-term remediation. However, zinc phosphate minerals do exist for which thermodynamic data are unavailable, which precludes any assessment of their relative environmental stability.

In a cooler environment (10°C, Figure 6.9b), willemite is predicted to control Zn^{2+} in the aqueous phase. The silicate mineral is predicted to support Zn^{2+} activities that are approximately two orders of magnitude less than that support by smithsonite. The relative stability of the two zinc carbonates differs by approximately one order of magnitude, with smithsonite predicted to be more stable than hydrozincite. The stability of hopeite is similar to that of hydrozincite. The oxide (zincite) is the least stable mineral under the controlling conditions. The predicted order of increasing mineral stability or decreasing Zn^{2+} activities supported by the mineral at 10°C is: zincite ($10^{-2.84}$) > hydrozincite ($10^{-3.70}$) > hopeite ($10^{-3.78}$) > franklinite ($10^{-3.89}$) > smithsonite ($10^{-4.67}$) > willemite ($10^{-6.56}$) (values in parentheses are the Zn^{2+} activities supported by each mineral at pH 7.5).

In the 40°C environment (Figure 6.9c), franklinite is predicted to control Zn^{2+} activity in the pH 6 to 9 range. Willemite supports greater Zn^{2+} activities than in the 25°C system, and is metastable relative to franklinite in the 40°C system. Both zinc carbonates support lower Zn^{2+} activities with increasing temperature, but are still unstable relative to the silicate and ferrite phases. Hopeite is the least stable zinc mineral in the 40°C environment; however, the increase in temperature has little impact on the absolute stability of the mineral (the Zn^{2+} activity supported was unchanged). Finally, zincite solubility decreased with increasing temperature. The predicted order of increasing mineral stability or decreasing Zn^{2+} activities supported by the mineral at 40°C is: hopeite ($10^{-4.01}$) > zincite ($10^{-4.40}$) > smithsonite ($10^{-5.02}$) > hydrozincite ($10^{-5.03}$) > willemite ($10^{-5.50}$) > franklinite ($10^{-5.67}$) (values in parentheses are the Zn^{2+} activities supported by each mineral at pH 7.5).

The predictions offered by the activity diagrams in Figures 6.9a through 6.9c indicate that temperature can influence the relative stability of minerals. Under all temperature conditions, zinc silicate is predicted to be the most stable Zn-bearing mineral, or nearly so, given that solid and water activities are unity, the partial pressure of CO_2 is 10^{-2}, the solution activities of $H_2PO_4^-$, $H_4SiO_4^0$, Fe^{3+}, and Ca^{2+} are controlled by the specified minerals, and that equilibrium conditions exist. When the stated conditions are met, the predictions offered by the activity diagrams should be valid. For example, detailed mineralogical analysis of zinc smelter–contaminated sediment dredged from a ship canal in France and subsequently deposited on an agricultural soil substantiates the presence of sphalerite [ZnS], zincite, and willemite (Isaure et al., 2002). Sphalerite in the sediment is anthropogenic, arising from the smelting of zinc sulfide ores. The anoxic nature of the sediment environment of the canal is conducive to the stability of sphalerite, and its presence in the dredge material is directly related to the reducing conditions. Both zincite and willemite may also originate from the smelting process, as both are high temperature minerals. According to the zinc activity diagrams, zincite is an unstable zinc-bearing mineral and it is expected to dissolve in favor of a more stable zinc phase as the dredge material weathers. Furthermore, the zinc activity diagrams support the occurrence and stability of willemite, even though the mineral is reportedly a high temperature phase.

If the assumptions do not reflect reality, the predictions offered in the activity diagrams would be of minimal value. For example, in alkaline soil systems that are zinc contaminated through medieval mining activities (implies an environment at equilibrium), neither the zinc silicate nor the zinc ferrite phase have been observed. Instead, the zinc carbonate mineral, smithsonite, has been directly identified (Mattigod et al., 1986). The direct observation of the carbonate phase in the contaminated soil suggests that (1) the assumptions used to create the activity diagrams (Figures 6.9a through 6.9c) are apparently inappropriate for the environmental conditions, particularly with respect to soluble $H_4SiO_4^0$ activity which establishes the stability of willemite or with respect to the partial pressure of CO_2 which establishes the stability of smithsonite, (2) true thermodynamic equilibrium is not achieved with respect to the zinc silicate phase resulting in the formation and persistence of the metastable zinc carbonate, or (3) the thermodynamic data that describe the stability of zinc silicate (willemite) is in error. Since smithsonite is predicted to be more stable than franklinite in the 25 and 40°C systems, it may also be concluded that these soil temperature conditions predominate in the contaminated environment (assuming the formation of the ferrite is not kinetically restricted as well).

6.5.4 *In Situ* Stabilization: Observed and Predicted Transformations of Metallic Lead in Alkaline Soil

Organic contaminants in the soil environment can be detoxified or decomposed to carbon dioxide and water *in situ* through the proper management of indigenous biota, or through the introduction and management of genetically modified organisms. In essence, organic contaminants can be made to simply disappear. The *in situ* amelioration of metals-contaminated soil is not so easily addressed, as inorganic compounds cannot be made to simply disappear. Rather, a change must occur in the chemical form that results in a metal-bearing phase that is (1) stable in the soil environment (sparingly soluble) and resistant to the actions of acidic root exudates (low plant availability), (2) stable against the solvating action of regulatory extractions (e.g., TCLP), and (3) stable against solubilization in the gastrointestinal tract of animals and humans.

There are essentially two mechanisms available for the *in situ* stabilization of metals-contaminated soil. The first, but often overlooked mechanism is to do nothing. Research has clearly illustrated that the solid-phase speciation of metals in waste materials that are exposed to the environment or incorporated into the soil will be altered. Solids that are stable under the conditions in which a waste is produced are generally unstable in a soil environment; thus, a phase change occurs resulting in minerals that are stable under the prevailing environmental conditions. For example, the lead silicate phase found in sewage sludge will dissolve in amended soil with the concomitant formation of a lead phosphate phase, a phase that is stable under the prevailing soil conditions. The problem with the "do nothing" scenario is that the rate of stabilization may be exceedingly slow relative to the movement of water through the soil. However, an important reality is indicated in the "do nothing" scenario: the degree and mechanism of stabilization must be possible for the given environmental conditions, particularly if there is the expectation of long-term stabilization.

The second *in situ* treatment option is to enhance the natural process by increasing the rate at which the final, stable state of the metal is achieved. It is a requirement for the successful *in situ* stabilization of metals that the ultimate solid phase formed must be the most stable in the soil under the prevailing natural conditions. The most commonly proposed mechanism of *in situ* stabilization of metal cations (such as Pb, Cd, Cu, Ni, Zn), particularly Pb, is metal phosphate precipitation. Metal phosphates are generally orders of magnitude more stable than metal sulfides, sulfates, carbonates, ferrites, and (hydr-)oxides over a range of environmental conditions. The application of apatite [$Ca_5(PO_4)_3(OH, Cl, F)$] as a phosphate source to Pb-affected soils generally results in the formation of a pyromorphite [$Pb_5(PO_4)_3(OH, Cl, F)$]. Alkaline environments present a particularly challenging problem to the formation and long-term stability of pyromorphite, as a lead carbonate (such as cerussite, $PbCO_3$) may be thermodynamically stable relative to the lead phosphate. Although stable in the soil environment, lead carbonates may be easily solubilized through rhizosphere acidity, easily extracted by the solvating action of acidic regulatory extractants, and easily dissolved in the gastrointestinal tract of animals and humans, relative to a lead phosphate.

Lead in surface soil at military, police, and civilian firing ranges occurs in the form of thermodynamically unstable, but nonlabile (kinetically restricted dissolution) metallic Pb. Closure of these sites requires the remediation of the surface soils with respect to the metallic Pb, as well as other trace elements found in depleted ammunitions. One remediation option is to remove and dispose of the affected surface soil. This option seems relatively straightforward; however, the high cost associated with excavation, transportation, and land filling, coupled with the desire of government agencies to be better stewards of monetary resources, have provided the impetus to explore *in situ* treatment options. Laboratory evaluations have demonstrated the utility of mixing apatite with soil treated or contaminated with water-soluble forms (labile forms) of lead to hasten the precipitation of lead phosphate. Since apatite is the stable phosphate mineral in calcareous soils, it is expected that this phosphate source would provide long-term lead stabilization potential.

The potential for lead phosphate to form in an alkaline soil, as well as the long-term stability of lead phosphate in a stabilized soil, can be investigated using activity diagrams. Further, the resulting predictions can be validated through the direct observation of lead solid phase speciation in lead-affected soils. A model system that reflects the long-term impact of metallic Pb on the surrounding soil and provides an assessment of the redistribution of lead is the archaeological excavation at Horace's Villa. The Villa is located in the Sabine (Alpine) Hills in south central Italy approximately 50 km northeast of Rome. The Villa is situated in the deeply dissected limestone region of Italy with topographic changes from 360 to 410 m in the valleys to over 980 to1059 m on the high peaks. The elevation of the Villa is approximately 420 to 430 m. The geologic materials at Horace's Villa are mainly sedimentary rocks dominated by limestone and shale. The soil in the vicinity of the Villa is alluvial, strongly calcareous, and contains small limestone chips.

As part of the archaeological investigation of the Villa, a study was conducted to determine the general stratigraphy of the site and characterize the soils of the area. During the site characterization process, a number of lead pipes were located in the Villa gardens and excavated. Inscriptions on the pipes indicate they were manufactured by the same individual over 2,000 years BP. A mineralogical study was conducted on the pipe crust material and surrounding soils to determine the extent of lead migration and to identify Pb-bearing mineral phases.

Soil samples obtained at incremental distances from a lead pipe, as well as from within the pipe, indicate lead migration. The concentration of lead in the soil samples is a function of the distance from the pipe. Soil inside the ~12 cm inner diameter pipe contains 25.4 g kg^{-1} Pb (25,400 mg kg^{-1}). Lead concentrations in the soil surrounding the pipe are 30,800 mg kg^{-1} (0 to 1 cm from pipe), 4,020 mg kg^{-1} (1 to 3 cm), 3,340 mg kg^{-1} (3 to 6 cm), and 1,400 mg kg^{-1} (6 to 9 cm). The lead content of a soil sample collected 0.5 m from the pipe (at the same elevation) is 157 mg kg^{-1}. The lead content of a background soil sample collected from a buried A-horizon of a profile in the vicinity of the Villa, but from an uninhabited area, is 50 mg kg^{-1}. Higher lead concentrations in the Villa control sample, relative to that of the off-site background sample, is a finding consistent with those from other Roman archaeological sites; most notably, soil lead levels are elevated relative to adjacent and uninhabited areas. Thus, it is unlikely that the source of the elevated lead levels in the control sample was the lead pipe. The phosphorus content of the Villa soils is also elevated relative to uninhabited areas. Phosphorus concentrations in the soil surrounding the pipe are 885 mg kg^{-1} (0 to 1 cm from pipe), 653 mg kg^{-1} (1 to 3 cm), 983 mg kg^{-1} (3 to 6 cm), and 1,102 mg kg^{-1} (6 to 9 cm). Within the lead pipe, the phosphorus concentration is 2,920 mg kg^{-1}. The phosphorus content of a soil sample collected 0.5 m from the pipe is 671 mg kg^{-1}. The P content of a background soil sample collected from a buried A-horizon of a profile in the vicinity of the Villa, but historically an uninhabited area, is <100 mg kg^{-1}. Phosphorus is a minor element relative to lead in the soil inside the pipe and immediately surrounding the pipe and may not have an impact on the solid phase speciation of lead.

The crust on the lead pipe is primarily composed of litharge [PbO], with smaller amounts of cerussite [PbCO$_3$], quartz, and metallic Pb (Figure 6.10a). Scanning electron microscopy and energy-dispersive x-ray analysis (SEM-EDXA) of the crust clearly illustrates the presence of lead-bearing particles that are fibrous (needlelike) and platy (Figure 6.11), reflecting the crystal habits and chemistry of cerussite and litharge. In the 0 to 1 cm soil sample, calcite is the dominant mineral, accompanied by small amounts of quartz and cerussite (Figure 6.10b). Similarly, the soil within the pipe is predominantly composed of cerussite and calcite, with a small amount of quartz (Figure 6.10c). X-ray diffraction analysis of the particle-sized separates of the soil inside the lead pipe suggests that cerussite is concentrated in the silt- and clay-sized separates. The lead-affected soil clay fraction is composed of calcite, smectite, mica, cerussite, and quartz. This mineral assemblage indicates that the soil is young in relative age, having received very little precipitation to hasten weathering processes (mean annual precipitation is 72 cm).

The x-ray diffraction results substantiate the presence of litharge and cerussite in association with the lead pipe, and cerussite in the soil surrounding the pipe. The activity ratio diagrams in

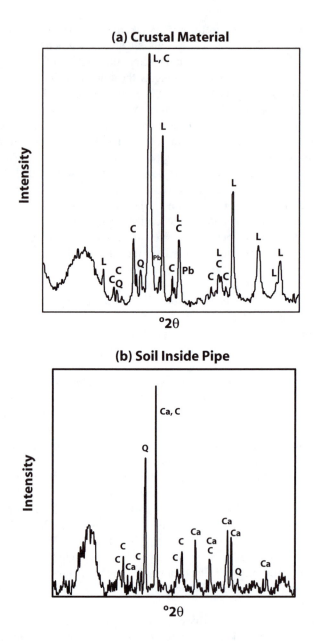

FIGURE 6.10 X-ray diffraction profiles of soil surrounding a lead pipe that has been buried for more then 2000 years at Horace's Villa near Rome, Italy: (a) crustal material at the pipe–soil interface, (b) soil from the interior of the pipe, and (c) soil in the 0- to 1-cm increment exterior to the pipe (C, cerussite; Ca, calcite; L, litharge; Q, quartz; Pb, metallic lead).

Figure 6.12 (25°C) examine the relative stability of litharge; hydroxyapatite; brushite (DCPD); the lead carbonates: cerussite and hydrocerussite; and the lead phosphates, hydroxypyromorphite, chloropyromorphite, and plumbogummite as a function of solution pH, Pb^{2+} activity, $H_2PO_4^-$ activity, and CO_2 partial pressure. It is a characteristic of activity ratio diagrams that three components are incorporated into the two axis variables: ($pH - \frac{1}{2}pPb^{2+}$) and ($pH + pH_2PO_4^-$). For the development of Figure 6.12, additional components—CO_2 partial pressure, Cl^- activity, Al^{3+} activity, and Ca^{2+} activity—are controlled by either a reservoir or a controlling mineral phase. The partial

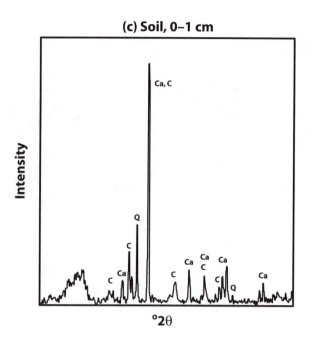

FIGURE 6.10 (*continued*).

pressure of CO_2 is fixed at two levels: 0.01 or 0.1, resulting in two diagrams (one for each CO_2 partial pressure). The activity of Cl^- is fixed by assuming $(pH + pCl^-) = 11.3$ (the activity of Cl^- is 5×10^{-4} at pH 8). The activity of Ca^{2+}, required to establish the hydroxyapatite and brushite stability lines, is controlled by calcite. The activity of Al^{3+}, required to establish the plumbogummite stability line, is controlled by gibbsite.

At 25°C and under a CO_2 partial pressure of 0.01, the minerals predicted to control the solution activities of Pb^{2+} are cerussite, chloropyromorphite, and plumbogummite (Figure 6.12). At equilibrium, litharge, hydroxypyromorphite, and hydrocerussite will not control Pb^{2+} activity under the constraints used to construct to diagram. The specific Pb^{2+}-controlling mineral is determined by the activity of $H_2PO_4^-$ in the soil solution. Cerussite is the controlling phase when $(pH + pH_2PO_4^-)$ is greater than 16.28 (or when the activity of $H_2PO_4^-$ is less than $10^{-8.28}$ in a pH 8 environment). When $(pH + pH_2PO_4^-)$ is less than 16.28, chloropyromorphite is the Pb^{2+}-controlling mineral. Plumbogummite is predicted to control Pb^{2+} activities when $(pH + pH_2PO_4^-)$ is less than 10.78, a condition that is not likely to occur in a stable environment due to the high soluble phosphate levels required (note that plumbogummite stability requires phosphate activities that are greater than that supported by brushite, an unstable calcium phosphate). Increasing the CO_2 partial pressure from 0.01 to 0.1 shifts the cerussite (and hydrocerussite), apatite, and brushite stability lines, but the phases predicted to control Pb^{2+} activity (cerussite, chloropyromorphite, and plumbogummite) remain unchanged.

The chemical conditions in the soil surrounding the lead pipe at Horace's Villa must be consistent with the stability of cerussite, as this mineral is present. Thus, soluble phosphate levels must be below those controlled by apatite dissolution. Further, the presence of apatite will always result in the formation of chloropyromorphite, irrespective of CO_2 partial pressure and unless $(pH + pCl^-)$ levels are greater than 15.93 (Cl^- activity less than $10^{-7.93}$ at pH 8), a condition that is atypical of natural systems. The fact that chloropyromorphite is not identified in the stable soil environment near the soil-pipe interface may indicate that phosphate levels are depressed relative to those supported by apatite, the stable mineral phosphate phase for the calcareous soil conditions.

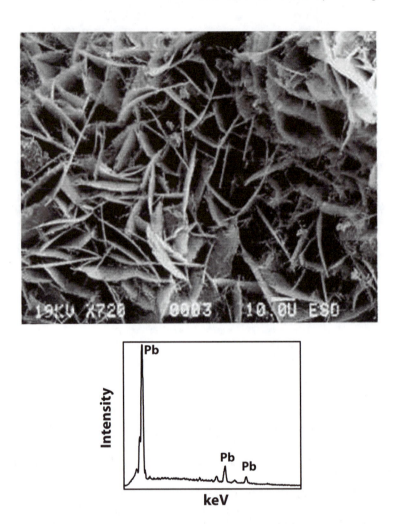

FIGURE 6.11 Scanning electron micrograph and energy dispersive x-ray analysis of material on the surface of a lead pipe buried for more than 2000 years at Horace's Villa near Rome, Italy (Courtesy of Dr. Yul Roh.)

Conversely, a pyromorphite may be present, but phosphorus concentrations are too low to allow for detection and predominance of the mineral.

It is conceivable that the 25°C stability diagrams may not represent the temperature condition of the soil environment at the archaeological site. Although the air temperatures at the site range from an annual average low of 10.6°C to an average high of 20.4°C, the mean air temperatures may not represent the range in soil temperatures. Using the enthalpy of formation data in Table 6.8, activity ratio diagrams representing potential low (10°C) and high (40°C) temperature conditions at the site can be constructed. The activity ratio diagrams in Figure 6.13 illustrate the stability of lead-bearing minerals in a 10°C environment as a function of CO_2 partial pressure. The 25°C (Figure 6.12) and the 10°C activity ratio diagrams do not differ relative to the minerals predicted to control Pb^{2+} activities (cerussite or chloropyromorphite depending on (pH + $pH_2PO_4^-$)). The significant differences between the 25°C and the 40°C activity ratio diagrams (Figure 6.14) are (1) the displacement of cerussite by hydrocerussite as the stable lead-bearing carbonate phase, and (2) a greater range of (pH + $pH_2PO_4^-$) levels for which the lead carbonate phases are stable for the higher temperature condition. The fact that cerussite is the identified phase in the Villa soil indicates

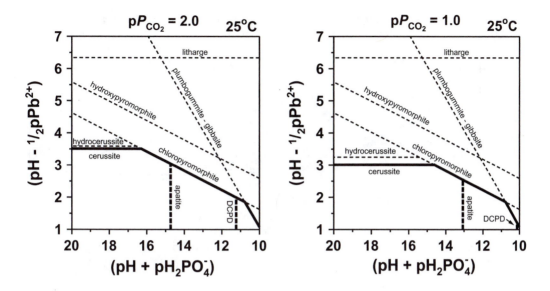

FIGURE 6.12 Activity ratio diagrams illustrating the stability of lead carbonates and lead phosphates at 25°C and 0.101 MPa as a function of solution pH, $H_2PO_4^-$ activity, and CO_2 partial pressure. Controlling conditions are $(pH + pCl^-) = 11.3$, $(pH - \frac{1}{2}pCa^{2+})$ controlled by calcite, and $(pH - \frac{1}{3}pAl^{3+})$ controlled by gibbsite. Mineral and liquid water activities are unity. Litharge, PbO; cerussite, $PbCO_3$; hydrocerussite, $Pb_3(CO_3)_2(OH)_2$; hydroxypyromorphite, $Pb_5(PO_4)_3OH$; chloropyromorphite, $Pb_5(PO_4)_3Cl$; plumbogummite, $PbAl_3(PO_4)_2(OH)_5 \cdot H_2O$; apatite, $Ca_5(PO_4)_3OH$; dicalcium phosphate dihydrate (DCPD), $CaHPO_4 \cdot 2H_2O$.

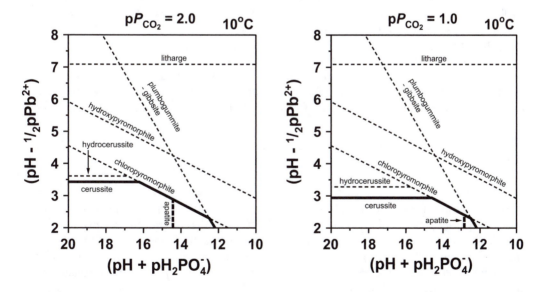

FIGURE 6.13 Activity ratio diagrams illustrating the stability of lead carbonates and lead phosphates at 10°C and 0.101 MPa as a function of solution pH, $H_2PO_4^-$ activity, and CO_2 partial pressure. Mineral and liquid water activities are unity. The controlling conditions and the chemical formula that correspond to the mineral names are given in the legend to Figure 6.12.

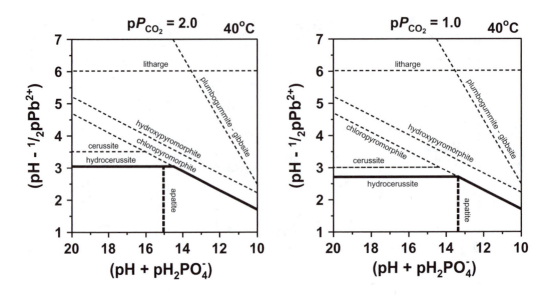

FIGURE 6.14 Activity ratio diagrams illustrating the stability of lead carbonates and lead phosphates at 40°C and 0.101 MPa as a function of solution pH, $H_2PO_4^-$ activity, and CO_2 partial pressure. Mineral and liquid water activities are unity. The controlling conditions and the chemical formula that correspond to the mineral names are given in the legend to Figure 6.12.

that soil temperatures may be more closely aligned with lower temperature conditions as indicated by the presence of cerussite. It is also evident that a lead carbonate phase (cerussite or hydrocerussite) will control Pb^{2+} activity in the calcareous environment, irrespective of temperature, given the restricted $H_2PO_4^-$ activities that are typical of calcareous soils, as well as the low total phosphorus levels in the soil.

Several important points relative to the potential long-term behavior of lead ammunition in alkaline soils can be obtained through the analysis of a soil system in contact with metallic lead for more than two centuries (Horace's Villa). Metallic Pb is a thermodynamically unstable phase in the presence of $O_2(g)$. The oxidation of metallic Pb to litharge is described by the reaction:

$$Pb^0(s) + \tfrac{1}{2}O_2(g) \rightarrow PbO(s)$$

$$\log K = 33.15 \tag{6.115}$$

Assuming the activities of metallic Pb and litharge are unity, it follows that litharge will predominate when the partial pressure of O_2 is greater than $10^{-16.6}$, which is insignificant relative to the atmospheric O_2 level ($P_{O_2} = 10^{-0.68}$). Despite the apparent thermodynamic instability of the metallic Pb, the pipe remains in the Villa soil two centuries after emplacement. Conceivably, spent lead ammunition, whole or in fragments, may also remain in affected calcareous soil for an extended period, limiting the effectiveness of *in situ* treatment options that require labile lead. It is also evident from the activity ratio diagrams that hydroxyapatite is an unstable phosphate mineral under the conditions necessary for cerussite precipitation. Since cerussite is directly observed, it follows that hydroxyapatite is not present or is not controlling soluble phosphate activity. The incorporation of apatite as a stabilization mechanism may result in the initial formation of a pyromorphite phase (as illustrated in laboratory evaluations); however, without continued management to counter the absorption of CO_2 and the nonlabile properties of the metallic Pb, the long-term stability of the pyromorphite will not be realized. Instead, it appears that lead metal in a calcareous or alkaline soil will ultimately oxidize to litharge, which in turn will weather to cerussite. The cerussite and litharge form an intimate association with the metallic lead and may serve to restrict O_2 diffusion, impeding the oxidation of Pb^0 and allowing for the long-term metastability of Pb^0 in an affected environment. Since cerussite is the stable lead-bearing mineral, *in situ* and long-term stabilization through the formation of a lead phosphate is not a feasible objective.

6.5.5 PREDOMINANCE DIAGRAMS

A predominance diagram is another type of activity diagram that illustrates the relative stability of minerals as a function of solution activity variables at equilibrium. A predominance diagram indicates mineral stability fields, or the ranges of solution variables over which a particular mineral is stable. The construction of a predominance diagram is performed in a manner similar to that used to construct an activity diagram. However, instead of congruent mineral dissolution, the reactions are written such that incongruent dissolution occurs. Consider the transition of kaolinite $[Al_2Si_2O_5(OH)_4]$ to the accessory mineral gibbsite $[Al(OH)_3]$:

$$Al_2Si_2O_5(OH)_4(s) + 5H_2O(l) \rightarrow 2Al(OH)_3(s) + 2H_4SiO_4^0(aq) \tag{6.116}$$

The K_{dis} for this reaction is $10^{-10.6}$, which is mathematically expressed as:

$$K_{dis} = \frac{f_{gibbsite}^2 [Al(OH)_3]^2 (H_4SiO_4^0)^2}{f_{kaolinite}[Al_2Si_2O_5(OH)_4](H_2O)^5} \tag{6.117}$$

where the brackets indicate concentration and the parentheses indicate activities. If the rational activity coefficients and the activity of water are assumed to be unity, Equation 6.117 can be written:

$$K_{dis} = \frac{[Al(OH)_3]^2 (H_4SiO_4^0)^2}{[Al_2Si_2O_5(OH)_4]} \tag{6.118}$$

Rearranging, substituting for K_{dis}, and taking the square root of both sides of the resulting expression:

$$\frac{10^{-5.3}}{(H_4SiO_4^0)} = \frac{[Al(OH)_3]}{[Al_2Si_2O_5(OH)_4]^{1/2}} \tag{6.119}$$

This expression states that when solution $H_4SiO_4^0$ activities are less than $10^{-5.3}$, the ratio of gibbsite concentration to kaolinite concentration will be greater than 1 and gibbsite will predominate over kaolinite (gibbsite concentrations will be greater than kaolinite concentrations). Conversely, when $H_4SiO_4^0$ activities in a soil solution are greater than $10^{-5.3}$, the ratio of gibbsite to kaolinite concentrations will be less than 1 and kaolinite will predominate over gibbsite. Equation 6.119 further indicates that the concentration ratio of gibbsite to kaolinite will be exactly 1 when the activity of $H_4SiO_4^0$ equals $10^{-5.3}$, or $[Al_2Si_2O_5(OH)_4] = [Al(OH)_3]$. Therefore, a two-dimensional diagram, with $\log(H_4SiO_4^0)$ as the dependent variable and pH as the independent variable (Figure 6.15), containing the line: $-5.3 = \log(H_4SiO_4^0)$, is a predominance diagram. The region described by $\log(H_4SiO_4^0) > -5.3$ is the stability field for kaolinite; whereas, the region described by $\log(H_4SiO_4^0) < -5.3$ is the stability field for gibbsite, irrespective of solution pH.

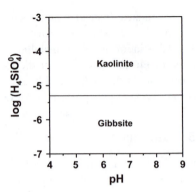

FIGURE 6.15 Predominance diagram illustrating the kaolinite [$Al_2Si_2O_5(OH)_2$] and gibbsite [$Al(OH)_3$] stability fields as a function of pH and $H_4SiO_4^0$ activity (unit activity of the minerals and liquid water, 25°C, and 0.101 MPa).

The assignment of minerals to their various regions is one of the more challenging aspects of predominance diagram construction. According to Equation 6.116, increasing the activity of $H_4SiO_4^0$ favors the stability of kaolinite relative to gibbsite. Conversely, depletion of $H_4SiO_4^0$ will favor kaolinite dissolution and gibbsite precipitation. Therefore, the high $H_4SiO_4^0$ activity side of the kaolinite-gibbsite stability line in Figure 6.15 must correspond to the kaolinite stability field, while the stability field consistent with low $H_4SiO_4^0$ activities must correspond to the gibbsite stability field.

The addition of a third mineral to the kaolinite-gibbsite diagram requires an evaluation of the stability of the mineral relative to both kaolinite and gibbsite. For example, the inclusion of muscovite is accomplished by considering the following incongruent transitions:

Muscovite to kaolinite

$$KAl_2(Si_3Al)O_{10}(OH)_2 + 1.5H_2O + H^+ \rightarrow 1.5Al_2Si_2O_5(OH)_4 + K^+$$

$$\log K_{sp} = 4.54 \tag{6.120}$$

Muscovite to gibbsite

$$KAl_2(Si_3Al)O_{10}(OH)_2 + 9H_2O + H^+ \rightarrow 3Al(OH)_3 + 3H_4SiO_4^0 + K^+$$

$$\log K_{sp} = -11.29 \tag{6.121}$$

The stability lines for the transitions are:

Muscovite to kaolinite

$$pH = 4.54 - \log K^+ \tag{6.122}$$

Muscovite to gibbsite

$$\log H_4SiO_4^0 = -3.76 - \tfrac{1}{3} pH - \tfrac{1}{3} \log K^+ \tag{6.123}$$

Equations 6.122 and 6.123 are plotted on the $\log(H_4SiO_4^0)$ vs. pH kaolinite-gibbsite predominance diagram, assuming the activity of K^+ is 10^{-2} (Figure 6.16a). Each line in the resulting diagram represents a phase transition. It has already been established that kaolinite predominates when $H_4SiO_4^0$ activity is greater than $10^{-5.3}$ and gibbsite predominates when $H_4SiO_4^0$ activity is less

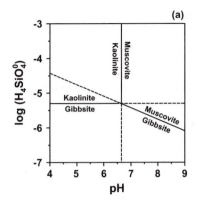
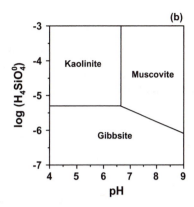

FIGURE 6.16 Predominance diagrams illustrating the muscovite [$KAl_2(Si_3Al)O_{10}(OH)_2$], kaolinite [$Al_2Si_2O_5(OH)_2$], and gibbsite [$Al(OH)_3$] stability fields as a function of pH and $H_4SiO_4^0$ activity ($\log K^+ = -2.0$, unit activity of the minerals and liquid water, 25°C, and 0.101 MPa): (a) stability lines representing the muscovite–kaolinite, muscovite–gibbsite, and kaolinite–gibbsite transitions; and (b) final predominance diagram with truncated stability lines.

than $10^{-5.3}$. The muscovite–kaolinite line (pH = 6.54) does not apply when $\log(H_4SiO_4^0)$ is less than -5.3 because kaolinite does not predominate. Similarly, the muscovite–gibbsite line ($\log(H_4SiO_4^0) = -\frac{1}{3}pH - 3.09$) does not apply when $\log(H_4SiO_4^0)$ is greater than -5.3 because gibbsite does not predominate. Finally, the original kaolinite–gibbsite line ($\log(H_4SiO_4^0) = -5.3$) does not apply when pH is greater than 6.54 because neither kaolinite nor gibbsite predominates. According to Equation 6.120, high pH (basic conditions, low H^+ activity) will favor muscovite stability relative to kaolinite, and high $H_4SiO_4^0$ activity will favor muscovite relative to gibbsite (Equation 6.121). Thus, the stability fields of muscovite, kaolinite, and gibbsite can be labeled accordingly resulting in the muscovite-kaolinite-gibbsite predominance diagram (Figure 6.16b).

The construction of the Figures 6.15 and 6.16 illustrate the mechanism employed to generate a predominance diagram. Typically however, predominance diagrams have much broader applicability relative to the number of minerals that may be considered and the resulting interpretations. Consider a system that contains the following minerals: K-feldspar, muscovite, montmorillonite, kaolinite, gibbsite, quartz, and amorphous silica. The following weathering reactions (in addition to Equations 6.116, 6.120, and 6.121) and associated $\log K_{sp}$ values are written such that Al^{3+} is conserved (all the Al^{3+} released during the dissolution process is precipitated into the mineral product):

K-feldspar to kaolinite

$$2KAlSi_3O_8 + 9H_2O + 2H^+ \rightarrow Al_2Si_2O_5(OH)_4 + 4H_4SiO_4^0 + 2K^+$$

$$\log K_{sp} = -3.20 \tag{6.124}$$

K-feldspar to muscovite

$$3KAlSi_3O_8 + 12H_2O + 2H^+ \rightarrow KAl_2(Si_3Al)O_{10}(OH)_2 + 6H_4SiO_4^0 + 2K^+$$

$$\log K_{sp} = -9.34 \tag{6.125}$$

K-feldspar to montmorillonite

$$1.71\text{KAlSi}_3\text{O}_8 + 0.11\text{Fe}_2\text{O}_3 + 0.485\text{Mg}^{2+} + 3.27\text{H}_2\text{O} + 0.74\text{H}^+$$

$$\rightarrow \text{Mg}_{0.195}(\text{Al}_{1.52}\text{Fe}^{\text{III}}_{0.22}\text{Mg}_{0.29})(\text{Al}_{0.19}\text{Si}_{3.81})\text{O}_{10}(\text{OH})_2 + 1.71\text{K}^+ + 1.32\text{H}_4\text{SiO}^0_4$$

$$\log K_{sp} = -1.65 \tag{6.126}$$

Muscovite to montmorillonite

$$0.57\text{KAl}_2(\text{Si}_3\text{Al})\text{O}_{10}(\text{OH})_2 + 0.11\text{Fe}_2\text{O}_3 + 0.485\text{Mg}^{2+} + 2.1\text{H}_4\text{SiO}^0_4$$

$$\rightarrow \text{Mg}_{0.195}(\text{Al}_{1.52}\text{Fe}^{\text{III}}_{0.22}\text{Mg}_{0.29})(\text{Al}_{0.19}\text{Si}_{3.81})\text{O}_{10}(\text{OH})_2 + 0.57\text{K}^+ + 3.57\text{H}_2\text{O} + 0.4\text{H}^+$$

$$\log K_{sp} = 3.68 \tag{6.127}$$

Montmorillonite to kaolinite

$$0.855\text{Al}_2\text{Si}_2\text{O}_5(\text{OH})_4 + 0.11\text{Fe}_2\text{O}_3 + 0.485\text{Mg}^{2+} + 2.1\text{H}_4\text{SiO}^0_4$$

$$\rightarrow \text{Mg}_{0.195}(\text{Al}_{1.52}\text{Fe}^{\text{III}}_{0.22}\text{Mg}_{0.29})(\text{Al}_{0.19}\text{Si}_{3.81})\text{O}_{10}(\text{OH})_2 + 4.425\text{H}_2\text{O} + 0.97\text{H}^+$$

$$\log K_{sp} = 1.17 \tag{6.128}$$

Quartz dissolution

$$\text{SiO}_2 + 2\text{H}_2\text{O} = \text{H}_4\text{SiO}^0_4$$

$$\log K_{sp} = 3.95 \tag{6.129}$$

Amorphous Silica dissolution

$$\text{SiO}_2 + 2\text{H}_2\text{O} = \text{H}_4\text{SiO}^0_4$$

$$\log K_{sp} = 2.66 \tag{6.130}$$

The independent variable for the activity diagram will be $(\text{pH} - \text{pK}^+)$ and the dependent variable will be the activity product $\text{pH}_4\text{SiO}^0_4$ (negative $\log(\text{H}_4\text{SiO}^0_4)$). The selection of an activity product as an axis variable allows for the evaluation of mineral stability as a function of three solution variables using a two-dimensional diagram. In this example, the influence of the pH and the activities of H_4SiO^0_4 and K^+ on aluminosilicate weathering can be examined. Two additional components must be controlled or fixed in order to illustrate mineral stability regions on the two-dimensional diagram: Fe^{3+} activity, which is controlled by hematite [Fe_2O_3] dissolution, and Mg^{2+} activity, which is fixed at 10^{-4} (required in transitions involving montmorillonite, Equations 6.126, 6.127, and 6.128). Due to the latter stipulation, the montmorillonite stability region will expand or contract as a function of pH. The linear functions that represent the various transitions from one stability field to another are:

K-feldspar to kaolinite	$pH_4SiO_4^0 = \frac{1}{2}(pH - pK^+) + 0.8$	(6.131)

K-feldspar to muscovite	$pH_4SiO_4^0 = \frac{1}{3}(pH - pK^+) + 1.56$	(6.132)

K-feldspar-montmorillonite

$pH = 4$ $pH_4SiO_4^0 = 1.295(pH - pK^+) - 0.224$

$pH = 8$ $pH_4SiO_4^0 = 1.295(pH - pK^+) - 3.164$ (6.133)

Muscovite to montmorillonite

$pH = 4$ $pH_4SiO_4^0 = -0.271(pH - pK^+) + 2.674$

$pH = 8$ $pH_4SiO_4^0 = 1.295(pH - pK^+) + 4.522$ (6.134)

Muscovite to kaolinite	$(pH - pK^+) = 4.54$	(6.135)

Muscovite to gibbsite	$pH_4SiO_4^0 = \frac{1}{3}(pH - pK^+) + 3.76$	(6.136)

Montmorillonite to kaolinite

$pH = 4$ $pH_4SiO_4^0 = 1.479$

$pH = 8$ $pH_4SiO_4^0 = 3.327$ (6.137)

Kaolinite to gibbsite	$pH_4SiO_4^0 = 5.3$	(6.138)

Quartz dissolution	$pH_4SiO_4^0 = 3.95$	(6.139)

Amorphous silica dissolution	$pH_4SiO_4^0 = 2.66$	(6.140)

Each of the above equations represents a phase change and identifies the solution properties where one mineral becomes more stable than another. The stability lines are plotted on the predominance diagram (Figure 6.17) and illustrate the stability fields for the aluminosilicate phases as a function of equilibrium solution activities. Recall that the transition from kaolinite to gibbsite was a function of only $H_4SiO_4^0$ activity. When the activity of $H_4SiO_4^0$ is reduced below $10^{-5.3}$ (e.g., through leaching), gibbsite is predicted to become the stable mineral phase, relative to kaolinite. In the muscovite to kaolinite transition, the reaction is written such that silica is conserved (all the silica in mica goes into kaolinite). As a result, the transition is a function of only $(pH - pK^+)$. When $(pH - pK^+)$ is greater than 4.54, either through high soluble K^+ or high solution pH, muscovite is stable relative to kaolinite. However, low soluble K^+ activity or low solution pH contribute to muscovite instability. The montmorillonite stability region is specific to the pH 8 condition, and is consistent with the observed stability of smectite in alkaline systems with high soluble silica levels. As solution pH is

FIGURE 6.17 Predominance diagram illustrating the stability fields for K-feldspar [KAlSi$_3$O$_8$], muscovite–mica [KAl$_2$(Si$_3$Al)O$_{10}$(OH)$_2$], montmorillonite–smectite [Mg$_{0.195}$(Al$_{1.52}$Fe$^{III}_{0.22}$Mg$_{0.29}$)(Al$_{0.19}$Si$_{3.81}$)O$_{10}$(OH)$_2$], kaolinite [Al$_2$Si$_2$O$_5$(OH)$_4$], gibbsite [Al(OH)$_3$], quartz (SiO$_2$), and amorphous silica [H$_4$SiO$_4$(am)] (unit activity of the minerals and liquid water, 25°C, and 0.101 MPa). The Fe^{3+} activity is controlled by hematite [Fe$_2$O$_3$] and Mg^{2+} activity is fixed at 10^{-4}. The smectite stability field is valid for pH 8 solutions. The smectite–kaolinite transition will move up the pH$_4$SiO$_4^0$ axis as solution pH decreases from 8 to 6.56. When solution pH is less than 6.56, kaolinite will transition directly to H$_4$SiO$_4$(am), and the dashed lines representing the transition of K-feldspar to kaolinite will supplant the smectite stability field. If quartz is assumed to control H$_4$SiO$_4^0$ (instead of H$_4$SiO$_4$(am)), the quartz stability field will supplant smectite and K-feldspar, and the stability line, pH$_4$SiO$_4^0$ = 3.95, will define the kaolinite–quartz and muscovite–quartz transitions.

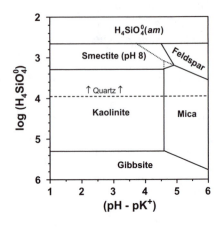

reduced, the smectite region will decrease in area (smectite to kaolinite transition will move up on the pH$_4$SiO$_4^0$ axis). At pH values less than 6.56 the smectite stability region is supplanted by that of kaolinite and, to a lesser degree feldspar and mica. According to Figure 6.17, muscovite can weather directly to gibbsite in alkaline and silica-poor environments; but montmorillonite will not weather directly to gibbsite. Further, K-feldspar may weather directly to kaolinite when solution pH decreases below 7.44 (a result of the shrinking smectite stability region). However, the diagram predicts that K-feldspar will not weather directly to gibbsite; rather, kaolinite would form first (or muscovite, depending on pH and K$^+$ activity). The predictions illustrated by the predominance diagram are also consistent with observed mineral weathering schemes:

- Minimum leaching, high base cation content, neutral to alkaline pH values, and high soluble silica content characterize relatively young soils. This condition is represented by the mid- to upper-right quadrant of the predominance diagram (Figure 6.17). In this region, the primary aluminosilicates, muscovite and K-feldspar, and the secondary phases, montmorillonite and amorphous silica, are predicted to predominate.
- As soils age, the chemical characteristics that are initially impacted are pH (which decreases) and base cation content (which decreases). The mineralogy of soils of intermediate age is consistent with that represented in the middle of the predominance diagram. In this region, kaolinite is the predominant aluminosilicate.
- Acidic conditions and effective leaching of bases and soluble silica characterize relatively old soils. This condition is represented in the lower-left region of the predominance diagram. The predicted stability of the accessory mineral gibbsite is consistent with the abundance of this mineral in highly weathered soils.

6.6 PREDICTING SOLUTION COMPOSITION

If a specific mineral phase is known to occur in a soil, the solution concentration of an element may be predicted. During the late 1970s and 1980s, the extraction of oil from marlstone (oil shale) was economically feasible. The extraction process, termed retorting, consisted of burning the rock at high temperatures to release the oil. In addition to the oil, a solid waste material was produced.

The spent oil shale was highly alkaline and contained many high-temperature mineral phases that were unstable and very soluble at terrestrial temperatures. Further, due to the alkalinity and high mineral solubility, there was the potential for the release of several deleterious substances to solutions that may come into contact with the solid waste, including selenate, arsenate, and molybdate. One such species was fluoride (F^-). Detailed examination of the solid waste indicated the occurrence of the mineral fluorite, CaF_2. Given this fact, the total concentration of fluoride in a spent oil shale leachate can be predicted. Predicting the total dissolved concentration of a substance in a soil solution requires a detailed understanding of the solution chemistry of the substance and the application of a geochemical (ion speciation) model. The total concentration of fluoride in a spent oil shale leachate is given by the mass balance expression:

$$F_T = [F^-] + [NaF^0] + [KF^0] + [CaF^+] + [MgF^+] \tag{6.141}$$

where F_T and the free and complex species are expressed in mol L^{-1} (under the chemical conditions prevalent in the waste, the indicated fluoride complexes are the significant fluoride species present in solution). The formation of each of the complexes is controlled by the following expressions:

$$[NaF^0] = {}^cK_{f,NaF}[Na^+][F^-] \tag{6.142}$$

$$[KF^0] = {}^cK_{f,KF}[K^+][F^-] \tag{6.143}$$

$$[CaF^+] = {}^cK_{f,CaF}[Ca^{2+}][F^-] \tag{6.144}$$

$$[MgF^+] = {}^cK_{f,MgF}[Mg^{2+}][F^-] \tag{6.145}$$

Substituting the ion association expressions (Equations 6.142 through 6.145) into the mass balance expression (Equation 6.141) yields:

$$F_T = [F^-] + {}^cK_{f,NaF}[Na^+][F^-] + {}^cK_{f,KF}[K^+][F^-] + {}^cK_{f,CaF}[Ca^{2+}][F^-] + {}^cK_{f,MgF}[Mg^{2+}][F^-] \tag{6.146}$$

Rearranging:

$$F_T = [F^-]\{1 + {}^cK_{f,NaF}[Na^+] + {}^cK_{f,KF}[K^+] + {}^cK_{f,CaF}[Ca^{2+}] + {}^cK_{f,MgF}[Mg^{2+}]\} \tag{6.147}$$

If fluorite is present and controlling the activity of fluoride in the solution, the dissolution reaction and associated K_{sp} are applicable:

$$CaF_2(s) \rightarrow Ca^{2+}(aq) + 2F^-(aq)$$

$$K_{sp} = (Ca^{2+})(F^-)^2$$

$$[F^-] = \{{}^cK_{sp}/[Ca^{2+}]\}^2 \tag{6.148}$$

Substituting Equation 6.148 into Equation 6.147 yields:

$$F_T = \{{}^cK_{sp}/[Ca^{2+}]\}^2\{1 + {}^cK_{f,NaF}[Na^+] + {}^cK_{f,KF}[K^+] + {}^cK_{f,CaF}[Ca^{2+}] + {}^cK_{f,MgF}[Mg^{2+}]\} \tag{6.149}$$

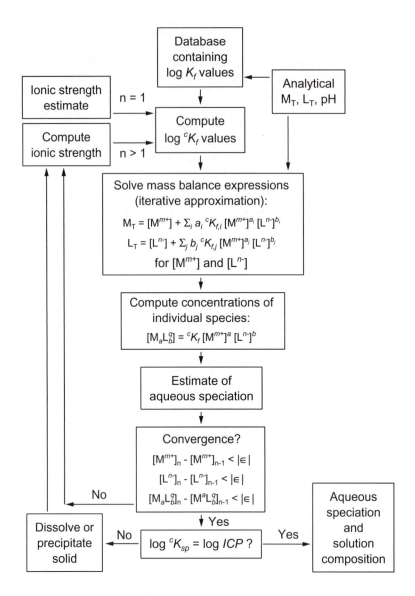

FIGURE 6.18 Schematic representation of an algorithm employed to compute ion speciation and mineral solubility in aqueous solutions.

In practice, computer codes that predict ion speciation using the total concentrations of substances in a solution (described in Chapter 5) are also employed to predict the total concentration of a substance in equilibrium with a specified mineral phase. The algorithm used to compute the total concentration of F_T in spent oil shale leachates is illustrated in Figure 6.18. The prediction of F_T is an iterative process. Initially, the concentrations of the major cations and anions, an estimate of F_T, and solution pH are used as input to a geochemical model to compute the free concentrations of F^-, Na^+, K^+, Ca^{2+}, and Mg^{2+}. The ion concentration product (ICP) of the solution with respect to fluorite ($ICP_{fluorite} = [Ca^{2+}][F^-]^2$) is then compared to the conditional solubility product constant ($^cK_{sp}$) of fluorite. If the estimated input value for F_T is too high and $ICP_{fluorite}$ is greater than $^cK_{sp}$, fluorite will precipitate such that $ICP_{fluorite} = ^cK_{sp}$. If the estimated input value for F_T is too low and $ICP_{fluorite}$ is less than $^cK_{sp}$, fluorite will dissolve such that $ICP_{fluorite} = ^cK_{sp}$. The precipitation or dissolution of fluorite alters the F^- content of the solution, necessitating the reevaluation of aqueous

TABLE 6.10
**Experimental and Predicted Total Fluoride Concentrations
in Spent Oil Shale Leachates Controlled by Fluorite
$[CaF_2(s)]$ Dissolution**

Waste type	Experimental	Predicted	
		$SI_{fluorite} = 0$	$SI_{fluorite} = -1$
		mg F L^{-1} ± 2σ	
Indirectly retorted oil shale	69 ± 8	99	31
Directly retorted oil shale A	9 ± 4	20	6
Directly retorted oil shale B	10 ± 4	9	3
Combusted oil shale	6 ± 4	15	5

speciation. The process of computing the aqueous speciation, checking for saturation with respect to fluorite, and adjusting the F$^-$ concentration continues until $ICP_{fluorite} = {}^cK_{sp}$. Once completed, the algorithm yields F_T and the aqueous speciation of the solution.

This approach was applied to four different highly alkaline waste materials thought to contain fluorite. A comparison of predicted solution fluoride concentrations to those actually measured in the spent oil shale leachates is shown in Table 6.10. The measured fluoride concentrations fall within the range of predicted values, assuming the condition $0.1K_{sp} < IAP_{fluorite} < K_{sp}$ brackets the expected $IAP_{fluorite}$ values supported by fluorite. In this example, the total soluble fluoride concentration in a waste leachate will depend on the free concentrations of calcium, magnesium, sodium, and potassium, which in turn depend upon the concentrations of complexing ligands. Essentially, one must have a thorough chemical analysis of the solution of interest in order to provide an accurate prediction of solution composition. This being the case, in all likelihood the concentration of the substance of interest will also be determined, and the need for model predictions will be academic. A modeling approach to predict and provide insight into the potential behavior of an element in a chemically complex environment is a tool. The application of this tool allows for a better understanding of the fate and behavior of an element, indicating how an element may potentially behave under a given set of environmental conditions.

6.7 EXERCISES

1. $\log K_{sp}$ for the reaction $AlOOH(s) + 3H^+ \rightarrow Al^{3+} + 2H_2O$ is 7.92. What is the activity of Al^{3+} in a pH 4 solution in equilibrium with $AlOOH(s)$?

2. Mineralogical characterization of a waste-affected soil has shown that lead occurs in the mineral hydroxypyromorphite $(Pb_5(PO_4)_3OH)$. If the phosphate activity in this soil is controlled by the dissolution of octacalcium phosphate $(Ca_8H_2(PO_4)_6 \cdot 5H_2O)$, what is the activity of Pb^{2+} in the soil solution if the pH is 6.5 and the free Ca^{2+} ion activity is 10^{-3}? Use the following dissolution reactions and associated solubility product constants in your computations:

$$Pb_5(PO_4)_3OH(s) + 7H^+ \rightarrow 5Pb^{2+} + 3H_2PO_4^- + H_2O; \quad \log K_{sp} = -4.14$$

$$Ca_8H_2(PO_4)_6 \cdot 5H_2O(s) + 10H^+ \rightarrow 8Ca^{2+} + 6H_2PO_4^- + 5H_2O; \quad \log K_{sp} = 23.52$$

If the EC of the soil solution is 0.8 dS m^{-1}, what is the concentration of Pb^{2+}?

3. Quartz (SiO_2) is a ubiquitous mineral in soils and the stable Si-bearing phase relative to amorphous silicic acid $(am\text{-}H_4SiO_4)$. However, when evaluating the stability of soil

silicates and aluminosilicates, the aqueous activity of $H_4SiO_4^0$ is often assumed to be controlled by the solubility of am-H_4SiO_4, rather than by quartz. Explain why this assumption is often employed.

4. The equilibrium solubility of $Pb_3(PO_4)_2(s)$ is $\log K_{sp} = -43.53$. A soil solution is analyzed and the activities of Pb^{2+} and PO_4^{3-} are computed to be $10^{-8.46}$ and $10^{-10.3}$. Compute the ion activity product (IAP) for the solution with respect to $Pb_3(PO_4)_2(s)$ and the saturation index (SI). What is the saturation status of this solution (undersaturated, saturated, or supersaturated) with respect to $Pb_3(PO_4)_2(s)$? Is $Pb_3(PO_4)_2(s)$ predicted to precipitate or dissolve (if present) in the system, or is the solid in equilibrium with the soil solution?

5. For the reaction $ZnCO_3(s) \rightarrow Zn^{2+} + CO_3^{2-}$, $\log K_{sp} = -10$. Construct a stability diagram with $\log(CO_3^{2-})$ as the independent variable (x-axis) and $\log(Zn^{2+})$ as the dependent variable (y-axis). Indicate on your diagram the undersaturated, saturated, and supersaturated regions.

6. A solution containing $10^{-1.27} M$ KNO_3 is equilibrated at 25°C with $Ba_3(AsO_4)_2(s)$. After 28 d the equilibrium solution is found to contain $10^{-3.59} M$ Ba, $10^{-3.69} M$ AsO_4, and a pH of 11.07. Assuming minimal complexation of the aqueous components and complete dissociation of arsenic acid, compute the solubility product constant for the dissolution of $Ba_3(AsO_4)_2(s)$ according to the reaction: $Ba_3(AsO_4)_2(s) \rightarrow 3Ba^{2+} + 2AsO_4^{3-}$.

7. Figure 6.9a illustrates the relative stability of various Zn-bearing minerals at 25°C and 0.101 MPa and under the stated controlling conditions. The diagram indicates that smithsonite and franklinite control approximately the same levels of Zn^{2+}, irrespective of solution pH.

     a. How might the controlling conditions be changed in this system such that smithsonite would be predicted to control a lower Zn^{2+} activity relative to franklinite?

     b. How might the controlling conditions be changed in this system such that franklinite would be predicted to control a lower Zn^{2+} activity relative to smithsonite?

8. The dissolution reaction, $Al(OH)_3(s) + 3H^+ \rightarrow Al^{3+} + 3H_2O$ is described by $\log K_{sp}$ values of 8.05 for macrocrystalline gibbsite, 8.77 for "soil gibbsite," 9.35 for microcrystalline gibbsite, and 10.8 for poorly crystalline gibbsite.

     a. Construct a stability diagram ($\log(Al^{3+})$ vs. pH) to describe the stability of these four phases in the pH 4 to 7 range. Employ the GLO step rule and predict which of these three minerals will form first when Al is released by mineral weathering reactions.

     b. With the exception of macrocrystalline gibbsite, which has an activity of unity, assume that each of these $Al(OH)_3(s)$ solids represents a gibbsite phase that has an activity that is not equal to unity. Compute the activity of each of these phases.

9. A leachate from a solid waste is analyzed and determined to have the following activity ratio, activity product characteristics: $(pH - \frac{1}{2}pSr^{2+}) = 9.34$, $(pH + \frac{1}{2}pCO_3^{2-}) = 13.98$, and $(pH + \frac{1}{2}pSO_4^{2-}) = 13.38$. Given that the dissolution reactions for strontianite ($SrCO_3$) and celestite ($SrSO_4$) are:

$$SrCO_3(s) \rightarrow Sr^{2+} + CO_3^{2-}; \quad \log K_{sp} = -9.28$$

$$SrSO_4(s) \rightarrow Sr^{2+} + SO_4^{2-}; \quad \log K_{sp} = -6.62$$

Identify the mineral that appears to be present and controlling the activity of Sr^{2+} in the waste leachate.

10. Comment on the following statements: "Soil solution composition strongly influences the total concentration of a soluble substance that is controlled by a mineral dissolution reaction. However, the activity of the substance is not influenced by solution composition if the solid and solution are in equilibrium."

11. Smectite and gibbsite have occasionally been found together in soils. Using the information below, construct a stability diagram with $pH_4SiO_4^0$ as the independent variable and pAl^{3+} as the dependent variable and determine if this mineral association is possible under equilibrium conditions. Minerals to consider, dissolution reactions, and $\log K_{sp}$ values are:

Gibbsite:

$$Al(OH)_3(s) + 3H^+(aq) \rightarrow Al^{3+}(aq) + 3H_2O(l): \quad \log K_{sp} = 8.05$$

Kaolinite:

$$Al_2Si_2O_5(OH)_4(s) + 6H^+ \rightarrow 2Al^{3+} + 2H_4SiO_4^0 + H_2O: \quad \log K_{sp} = 5.6$$

Montmorillonite:

$$Mg_{0.17}(Al_{1.5}Mg_{0.27}Fe_{0.23})(Si_{3.93}Al_{0.07})O_{10}(OH)_2(s) + 3.72H_2O + 6.28H^+$$
$$\rightarrow 0.44Mg^{2+} + 1.57Al^{3+} + 0.23Fe^{3+} + 3.93H_4SiO_4^0: \quad \log K_{sp} = 1.88$$

Also assume the following: $pH = 6$; $pMg^{2+} = 3.7$; Fe^{3+} activity controlled by hematite dissolution: $Fe_2O_3(s) + 6H^+ \rightarrow 2Fe^{3+} + 3H_2O$, $\log K_{sp} = -3.99$.

12. The dissolution of the lead and aluminum phosphate, plumbogummite, may be described by the reaction:

$$PbAl_3(PO_4)_2(OH)_5 \cdot H_2O(s) + 5H^+ \rightarrow Pb^{2+} + 3Al^{3+} + 2PO_4^{3-} + 6H_2O$$

which has a $\log K_{sp}$ of -32.79. The dissolution of this mineral may also be described by:

$$PbAl_3(PO_4)_2(OH)_5 \cdot H_2O(s) + 6H^+ \rightarrow Pb^{2+} + 3AlOH^{2+} + 2H_2PO_4^- + 3H_2O$$

Determine the solubility product constant for the latter plumbogummite dissolution reaction given the following:

$$Al^{3+} + H_2O \rightarrow AlOH^{2+}; \quad \log K_{a,1} = -5.00$$

$$H_2PO_4^- \rightarrow 2H^+ + PO_4^{3-}; \quad \log K_{a,2} = -19.55$$

13. The lead carbonate cerussite ($PbCO_3$) or hydrocerussite ($Pb_3(CO_3)_2(OH)_2$) are often found to occur in environments (acidic and basic) contaminated with metallic Pb.

   a. Develop a dissolution reaction for each of these minerals and use the standard free energy of formation (ΔG_f°) values provided below to compute the associated $\log K_{sp}$ values.

   b. The relative stability of these two minerals (the mineral that is predicted to control soluble Pb^{2+} activities) is a function of $CO_2(g)$ partial pressure, but not solution pH. What is the partial pressure of $CO_2(g)$ where both cerussite and hydrocerussite are predicted to control Pb^{2+} activities in solution (state all assumptions)?

   c. Which mineral is predicted to control Pb^{2+} activities in solutions when $CO_2(g)$ partial pressures are greater than the value determined in (b) above?

Species	ΔG_f°, kJ mol^{-1}	Species	ΔG_f°, kJ mol^{-1}
Pb^{2+}	-24.4	$PbCO_3(s)$	-625.34
$CO_2(g)$	-394.4	$Pb_3(CO_3)_2(OH)_2(s)$	-1711.59
H_2O	-237.2		

14. Determining the stability of a mineral in a soil system often requires a number of assumptions about the system. For example, the stability of a vermiculite $(K_{0.7}(Mg_{2.9}Fe_{0.1})(Si_{3.2}Al_{0.8})O_{10}(OH)_2)$ in a soil solution at 25°C and under 1 atm pressure as a function of pH and the activity of K^+ can only be determined with knowledge of the Mg^{2+}, Fe^{3+}, $H_4SiO_4^0$, and Al^{3+} activities, and the activity of the soil water and the vermiculite. It is convenient and not altogether inappropriate to impose the assumption of unit activity for water and the solid (although these properties may also be varied from unity). However, the assumed activities of the soluble components must be defined and they must be realistic for the environmental conditions. Two mechanisms have been employed in this chapter to control components and reduce the degrees of freedom: (1) assume an accessory solid is present and controlling the activity of a soluble species or (2) fix the activities of solutes at specified values. In the case of vermiculite, the activity of Fe^{3+} may be controlled by goethite as a function of pH, Al^{3+} by gibbsite as a function of pH, $H_4SiO_4^0$ by amorphous silicic acid, and Mg^{2+} at a fixed value of 10^{-4}. For each of the following minerals, associated activity diagram variables, and soil conditions: (1) identify the soluble components that must be controlled to obtain a stability line on a two-dimensional diagram and (2) select appropriate minerals or aqueous activities necessary to control each component (assume unit activities of water and all solids and standard state temperature and pressure conditions).

   a. Plumbogummite $(PbAl_3(PO_4)_2(OH)_5 \cdot H_2O)$ as a function of pH and Pb^{2+} activity in an alkaline soil that contains apatite $(Ca_5(PO_4)_3OH)$

   b. Hydrocerussite $(Pb_3(CO_3)_2(OH)_2)$ as a function of pH and Pb^{2+} activity in an alkaline system that contains calcite

   c. Leadhillite $(Pb_4SO_4(CO_3)_2(OH)_2)$ as a function of pH and Pb^{2+} activity in an alkaline system

   d. Hinsdalite $(PbAl_4PO_4SO_4(OH)_6)$ as a function of pH and Pb^{2+} activity in an acidic system

REFERENCES

Chen, C.H. A method of estimation of standard free energies of formation of silicate minerals at 298.15°K. *Am. J. Sci.* 275:801–817, 1975.

Essington, M.E. Solubility of barium arsenate. *Soil Sci. Soc. Am. J.* 52:1566–1570, 1988a.

Essington, M.E. Estimation of the standard free energy of formation of metal arsenates, selenates, and selenites. *Soil Sci. Soc. Am. J.* 52:1574–1579, 1988b.

Essington, M.E. Calcium molybdate solubility in spent oil shale and a preliminary evaluation of the association constants for the formation of $CaMoO_4^0(aq)$, $KMoO_4^-(aq)$, and $NaMoO_4^-(aq)$. *Environ. Sci. Technol.* 24:214–220, 1990.

Essington, M.E. and S.V. Mattigod. Trace element solid-phase associations in sewage sludge and sludge-amended soil. *Soil Sci. Soc. Am. J.* 55:350–356, 1991.

Holden, W.L. The Solubilization of Phosphates in the Presence of Organic Acids. M.S. thesis, The University of Tennessee, Knoxville, 1996.

Isaure, M.P., A. Laboudigue, A. Manceau, G. Sarret, C. Tiffreau, P. Trocellier, G. Lamble, J.L. Hazemann, and D. Chateigner. Quantitative Zn speciation in a contaminated dredge sediment by μ-PIXE, μ-SXRF, EXAFS spectroscopy and principal component analysis. *Geochim. Cosmochim. Acta* 66:1549–1567, 2002.

Lindsay, W.L. *Chemical Equilibria in Soils.* John Wiley & Sons, New York, 1979.

Longnecker, D.E. and P.J. Lyerly. Chemical characteristics of soils of West Texas as affected by irrigation water quality. *Soil Sci.* 87:207–216, 1959.

Mattigod, S.V., A.L. Page, and I. Thornton. Identification of some trace metal minerals in a mine-waste contaminated soi. *Soil Sci. Soc. Am. J.* 50:254–258, 1986.

Naumov, G.B., R.N. Ryzhenko, and I.L. Khodakovsky. *Handbook of Thermodynamic Data.* Translated by G.J. Soleimani. PB-226 722. National Technology Information Service, Springfield, VA, 1974.

Nriagu, J.O. Lead orthophosphates. IV. Formation and stability in the environment. *Geochim. Cosmochim. Acta* 38:887–898, 1974.

Parker, D.R., W.A. Norvell, and R.L. Chaney. GEOCHEM-PC—A chemical speciation program for IBM and compatible personal computers. In *Chemical Equilibrium and Reaction Models*. R.L. Loeppert et al. (Eds.) SSSA Spec. Publ. 42. SSSA Madison, WI, 1995, pp. 253–269.

Robie, R.A., B.S. Hemingway, and J.R. Fisher. Thermodynamic properties of minerals and related substances at 298.15 K and 1 bar (10^5 pascals) pressure and at higher temperature. U.S. Geological Survey Bull. 1452. U.S. Government Printing Office, Washington, D.C., 1978.

Sadiq, M. and W.L. Lindsay. Selection of standard free energies of formation for use in soil chemistry, Colorado State Univ. Tech. Bull. 134 and Supplements. Colorado State University Experimental Station, Fort Collins, CO, 1979.

Sangameshwar, S.R. and H.L. Barnes. Supergene processes in zinc-lead-silver sulfide ores in carbonates. *Econ. Geol.* 78:1379–1397, 1983.

Smith, R.M. and A.E. *Martell. Critical Stability Constants. Volume 4: Inorganic Complexes.* Plenum Press, New York, 1976.

Sposito, G. *The Thermodynamics of Soil Solutions.* Oxford Clarendon Press, New York, 1981.

Sposito, G. *The Chemistry of Soils.* Oxford University Press, New York, 1989.

Sposito, G. *Chemical Equilibria and Kinetics in Soils.* Oxford University Press, New York, 1994.

Tardy, Y. and R.M. Garrels. Prediction of Gibbs energies of formation. I. Relationships among Gibbs energies of formation of hydroxides, oxides and aqueous ions. *Geochim. Cosmochim. Acta* 40:1051–1056, 1976.

Tardy, Y. and R.M. Garrels. Prediction of Gibbs energies of formation of compounds from the elements. II. Monovalent and divalent metal silicates. *Geochim. Cosmochim. Acta* 41:87–92, 1977.

Tardy, Y. and L. Gartner. Relationships among Gibbs energies of formation of sulfates, nitrates, carbonates, oxides and aqueous ions. *Contrib. Mineral. Petrol.* 63:89–102, 1977.

Tardy, Y. and P. Vieillard. Relationships among Gibbs free energies and enthalpies of formation of phosphates, oxides and aqueous ions. *Contrib. Mineral. Petrol.* 63:75–88, 1977.

Tesoriero, A.J. and J.F. Pankow. Solid solution partitioning of Sr^{2+}, Ba^{2+}, and Cd^{2+} to calcite. *Geochim. Cosmochim. Acta* 60:1053–1063, 1996.

Wagman, D.D., W.H. Evans, V.B. Parker, R.H. Schumm, I. Harlow, S.M. Bailey, K.L. Churney, and R.L. Nutall. Selected values for inorganic and C_1 and C_2 organic substances in SI units. *J. Phys. Chem. Ref. Data* 11, Suppl. 2, 1982.

7 Surface Chemistry and Adsorption Reactions

Adsorption is arguably the most important of the physical-chemical processes responsible for the retention of inorganic and organic substances in the soil environment. The soil solution composition of inorganic and organic substances is controlled by surface or near-surface processes. With respect to trace elements, the activities of aqueous metal ions in equilibrium with an adsorbed phase are controlled to levels that can be orders of magnitude lower than levels controlled by even the most stable mineral phases. Strictly defined, adsorption is a surface process that results in the accumulation of a dissolved substance (an adsorbate) at the interface of a solid (the adsorbent) and the solution phase (Figure 7.1a). This interfacial region incorporates the volume of the soil solution that is under the direct influence of the surface, and is commonly referred to as the solid-solution interface. The process of adsorption can be contrasted with that of precipitation, in which the crystal structure of a mineral increases in volume as a result of the three-dimensional growth of the structure (Figure 7.1b). Inorganic and organic substances can also be retained through the process of absorption. In this process, a substance diffuses into the three-dimensional framework of a solid structure. Partitioning (absorption) is a mechanism that is frequently responsible for the retention of organic compounds by soil organic matter.

The particular retention mechanism operating in a soil system for a given substance can be quite difficult to ascertain. Indeed, the removal of a soluble compound from a solution phase is often identified simply as a *sorption* process, particularly when only the mass disappearance of a substance is determined. In this way, no mechanism (adsorption, partitioning, or precipitation) is implied, because the specific retention process is not known. It is also important to note that all adsorption processes are also exchange processes, and that these processes occur at specific locations at the solid-solution interface. Surfaces bear electrical charge that is either (1) formal charge, such as created through isomorphic substitution or protonation and deprotonation reactions in inorganic and organic functional groups; or (2) partial charge, as expressed through the polarity (natural or induced) of atoms at a crystal surface or in neutral organic functional groups. Since in nature there are no unsatisfied formal or partial charges, there always exist counter ions or polar substances that reside in close proximity to surfaces, which serve to neutralize the surface charge. Therefore, in order for the adsorption of a substance to occur, the resident substance on the surface must be displaced.

7.1 SURFACE FUNCTIONAL GROUPS AND COMPLEXES

Soil particle surfaces, both mineral and organic, are highly reactive. Surfaces are responsible for the retention of all classes of substances: ionic, nonionic, polar (hydrophilic), and nonpolar (hydrophobic). Particle surface characteristics also influence the flocculation and dispersion of soil particles, and thus the water infiltration or percolation rate. The reactivity of the soil is primarily confined to the clay-sized particles and results from the combined influence of reactive surface functional groups and a large specific surface area. The surface functional groups may be charged (positive or negative) or neutral, their reactivity determined by the specific type of mineral to which they are attached or, in the case of organic materials, the type of organic functional group and *R*-group (molecular framework) to which they are attached.

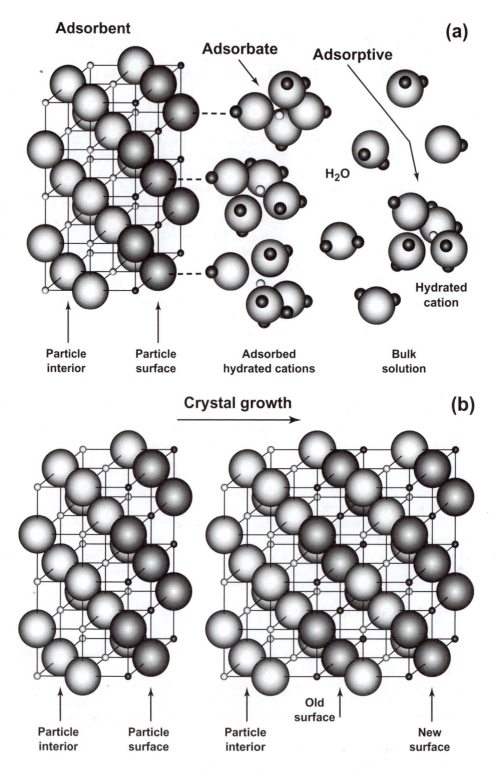

FIGURE 7.1 Adsorption processes (a) occur at mineral surfaces. This diagram illustrates the adsorption of hydrated metal cations, where the adsorbent is the mineral surface, the adsorbate is the adsorbed hydrated metal, and the adsorptive is the hydrated metal in the bulk solution. Precipitation processes (b) result in the three-dimensional growth of a crystal structure.

7.1.1 INORGANIC SOIL PARTICLE SURFACES AND THE SOURCE OF CHARGE

There are two basic types of charge that may develop on mineral surfaces: permanent and pH dependent. Permanent charge on a mineral is developed at the time the mineral crystallizes from liquid magma or precipitates from a supersaturated solution. It is a property of the mineral that cannot be altered by the chemistry of the environment in which the mineral resides once the mineral is formed. Permanent charge results from the process of isomorphic substitution. Further, permanent charge development is specific to the phyllosilicates. In general, isomorphic substitution in the phyllosilicates results in the mineral bearing a net negative charge. For example, the isomorphic substitution of Al^{3+} for Si^{4+} in the tetrahedral layer or the isomorphic substitution of Mg^{2+} for Al^{3+} or Li^+ for Mg^{2+} in the octahedral layer of the phyllosilicates will result in a deficit of the positive charge required to neutralize the coordinating anion charge. Thus, negative charge in the mineral structure is created and must be neutralized by an interlayer cation. Positive charge may also result from isomorphic substitution. For example, the isomorphic substitution of Fe^{3+} for Mg^{2+} in the octahedral layer results in more positive charge than can be neutralized by the coordinating anions. As we have already seen, the extent of isomorphic substitution (and the development of net negative charge in the mineral structure) is a defining characteristic for the classification of the layer silicates and in the expansiveness of the clay layers. This property also influences the ease of displacement of interlayer cations.

The functional site that arises from isomorphic substitution is located on the interstitial siloxane surface (in the interlayer region) of 2:1 layer silicates. This functional site is the siloxane ditrigonal cavity, which is the distorted hexagonal cavity formed by the six corner-sharing Si tetrahedral (Figure 7.2). Because of the distorted nature of the ditrigonal cavity, only three of the oxygen atoms in the cavity coordinate adsorbed metals. The diameter of the siloxane ditrigonal cavity is ~0.26 nm. The reactivity of the siloxane cavity depends on the electronic charge distribution in the layer silicate (i.e., the location and extent of isomorphic substitution). If there is no isomorphic substitution in close proximity to the siloxane cavity, the surface oxygen atoms carry a partial negative charge (as a result of the polarity of oxygen) and the functional site is only mildly reactive. The cavity acts as a relatively soft Lewis base (does not complex protons) and tends to weakly attract only charge-neutral dipolar molecules (such as H_2O). The cavity also displays hydrophobic character, lending the interlayers of low-charge-density and expansive phyllosilicates (e.g., smectites) a capacity to retain hydrophobic organic compounds.

If the siloxane cavity is in close proximity to a source of permanent negative charge, the reactivity of the functional site will depend on the extent and location of the isomorphic substitution. If the substitution occurs in the octahedral layer, such as in montmorillonite, the negative charge (normally of magnitude -1) is distributed over approximately 18 surface oxygen atoms on the two interlayer surfaces of the 2:1 structure (Figure 7.3a). A -0.5 charge will be spread over nine oxygen atoms on one surface, and a -0.5 charge will be spread over nine oxygen atoms on the opposite side of the layer structure. If the isomorphic substitution occurs in the tetrahedral layer, such as in the vermiculites and beidellite, the negative charge is localized and distributed over three surface oxygen atoms (Figure 7.3b). The higher charge density associated with the surface oxygen atoms of minerals having tetrahedral substitution will result in stronger surface complexes than those formed on octahedrally substituted layer silicates. The octahedral occupation will also influence the reactivity of the siloxane cavity. In dioctahedral minerals, the proton on the structural OH of the octahedral layer is shifted to the empty octahedral location, pointing away from the interlayer and an adsorbed cation (see Chapter 3). In tetrahedral minerals, the proton on the structural OH points directly toward the interlayer surface, resulting in a repulsive force between an adsorbed cation and the structural proton.

The development of pH-dependent charge results from the combined influence of the mineral surface and the environment in which the mineral resides. This type of charge development is termed pH dependent because it results from the protonation and deprotonation (dissociation of protons) of

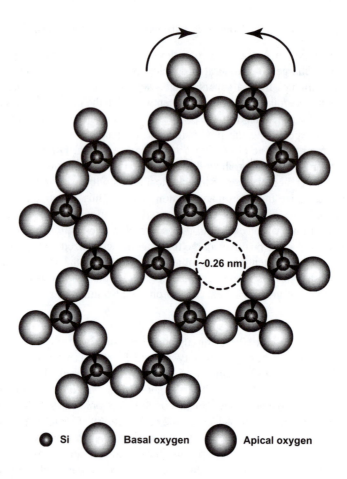

FIGURE 7.2 Schematic of the siloxane ditrigonal cavity on the planar surface of a phyllosilicate. The rotation of the silica tetrahedral about the Si-apical oxygen axis (in the directions of the arrows) distorts the ideal hexagonal configuration, resulting in the ditrigonal arrangement of the silica tetrahdra and a cavity diameter of ~0.26 nm.

surface hydroxyl groups ($\equiv SOH$ groups on the surface, where $\equiv S$ represents a metal bound in the crystal structure). The pH-dependent charge associated with a surface functional group can be negative, positive, or neutral. These types of surface functional groups are commonly found on phyllosilicates and crystalline and amorphous metal oxides, hydroxides, and oxyhydroxides. The reactivity of the surface hydroxyl group to protons is a function of the number of structural metal atoms bound to the surface hydroxyl, and the valency and coordination of the structural metal atoms. Consider the mineral goethite (Figure 7.4). There are three types of surface hydroxyl groups evident in the goethite structure, each differing in their reactivity with protons. Type A groups consist of a surface hydroxyl coordinated with one Fe^{3+} cation, $\equiv FeOH^{-0.5}$ (also called a terminal hydroxyl). Each type B surface hydroxyl is coordinated with three Fe^{3+} cations, $\equiv Fe_3OH^{+0.5}$. Type C surface hydroxyls are coordinated with two Fe^{3+} cations, $\equiv Fe_2OH^0$. Type B and C functional groups can only deprotonate, or, in the case of type B, may not be protonated at normal soil solution pH. This is due to the large amount of cation charge (2 or 3 Fe^{3+} ions) associated with the hydroxyl and the extent to which the Fe^{3+} ions polarize the surface oxygen atoms.

    There are two ways in which to view the protonation and deprotonation of inorganic surface functional groups. The first and most common is the 2–pK approach. This approach assumes that positive surface charge (protonation) and negative surface charge (deprotonation) develop as a result

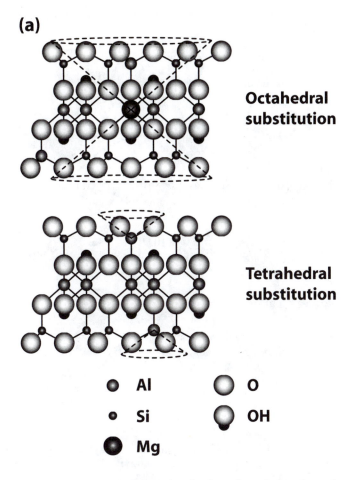

(a)

Octahedral
substitution

Tetrahedral
substitution

Al   O

Si   OH

Mg

FIGURE 7.3 The reactivity of the siloxane ditrigonal cavity depends on the location and extent of isomorphic substitution in the 2:1 phyllosilicate structure: side view (a) of the 2:1 structure and the radiation of permanent charge to the surface; planar view (b) of the surface and the approximate extent of surface charge distribution. Substitution of Al^{3+} for Si^{4+} in the tetrahedral layer results in negative surface charge that is localized to three basal oxygen atoms. Substitution of Mg^{2+} for Al^{3+} in the octahedral layer results in negative charge that is dispersed over approximately nine basal oxygen atoms.

of two protolysis reactions. This approach arbitrarily assigns +1 bond strength to the $\equiv S$–O bond, irrespective of the structural cation and the actual charge distribution (bond strength obtained by applying Pauling's electrostatic valency principle). For example, the goethite type A surface hydroxyls protonate and deprotonate in the 2–pK approach according to the reactions:

$$\equiv FeOH^0 + H^+ \rightarrow \equiv FeOH_2^+ \tag{7.1}$$

$$\equiv FeOH^0 \rightarrow \equiv FeO^- + H^+ \tag{7.2}$$

The protonation reaction and the development of positive surface charge is described by a stability constant, K_+, which is equal to $10^{6.2}$ for goethite (Equation 7.1). This value is also equal to $1/K_{s,1}$, the acid dissociation constant (note that p$K_{s,1} = 6.2$, the pH value at which $[\equiv FeOH^0] = [\equiv FeOH_2^+]$). The development of negative charge on goethite is described by K_- (also denoted by $K_{s,2}$) which is equal to $10^{-11.8}$ (p$K_{s,2} = 11.8$, the pH value at which $[\equiv FeOH^0] = [\equiv FeO^-]$).

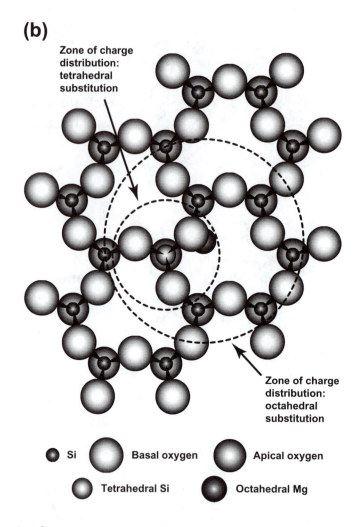

FIGURE 7.3 (*continued*).

The less common method used to describe surface acidity, but employed in this text, is the 1–pK approach. This approach assigns surface charge by imposing the electrostatic valency principle (Pauling Rule 2, Chapter 2). In a stable structure, the total strength (s) of each bond that radiates from a coordinated cation is equal to the cation charge (Z) divided by the coordination number (CN), or $s = Z/CN$. In the goethite structure, Fe^{3+} is in octahedral ($CN = 6$) coordination, and the strength of each Fe—O bond is $+^3/_6 = +0.5$. It then follows that a type A surface OH is only partially neutralized by the structural Fe^{3+}, resulting in a $\equiv FeOH^{-0.5}$ functional group. The equilibrium between $\equiv FeOH_2^{+0.5}$ and $\equiv FeOH^{-0.5}$ is described by the dissociation reaction:

$$\equiv FeOH_2^{+0.5} \rightarrow \equiv FeOH^{-0.5} + H^+ \tag{7.3}$$

where $K_{s,H}$ is $10^{-8.5}$ (or p$K_{s,H} = 8.5$, the pH value at which $[\equiv FeOH_2^{+0.5}] = [\equiv FeOH^{-0.5}]$).

The ability of the surface hydroxyls to protonate and deprotonate is also a function of the metal cation present in the structure. More specifically, surface site acidity is a function of metal coordination and valence (M—O bond strength), bond strength to M—OH bond length ratio (s/r_{M-OH}), and electronegativity (Table 7.1). For example, type A groups on α-Al_2O_3 ($\equiv AlOH_2^{+0.5}$ and $\equiv AlOH^{-0.5}$)

TABLE 7.1
Comparison of the Metal-Oxygen Bond Characteristics That Influence the Reactivity of Surface Functional Groups on Various Metal Oxides[a]

Mineral [formula]	Valence	CN_M	s_{M-O}	s/r_{M-OH}	EN_M	η_M	$pK_{s,1}$[b]	$pK_{s,2}$[b]	$pK_{s,H}$[c]
Quartz [α-SiO$_2$]	IV	4	+1	3.818	1.90	3.38	−1.2	7.2	3.0
Vernadite [δ-MnO$_2$]	IV	6	+$\frac{2}{3}$	2.300	1.55	3.72	0.16	7.36	3.76
Rutile [TiO$_2$]	IV	6	+$\frac{2}{3}$	2.248	1.54	3.37	2.6	9.0	5.8
Corundum [α-Al$_2$O$_3$]	III	6	+$\frac{1}{2}$	1.711	1.61	2.77	6.1	11.8	8.95
Hematite [α-Fe$_2$O$_3$]	III	6	+$\frac{1}{2}$	1.645	1.83	3.81	5.7	11.3	8.5

[a] Data obtained from Sahai (2002), unless noted otherwise. Valence is specific to the metal in the metal oxide, such as Si^{4+} in quartz or Fe^{3+} in hematite; CN_M is the coordination number of the metal; s_{M-O} is the Pauling bond strength of the M—O bond; s/r_{M-OH} is the ratio of bond strength to radius of M—OH bond (in nm); and η_M is the absolute acid hardness (Mulliken scale) of the metal. Higher η_M values represent greater acid hardness.

[b] $pK_{s,1}$ and $pK_{s,2}$ represent the surface deprotonation reactions, $\equiv SOH_2^+ \rightarrow \equiv SOH^0 + H^+$ and $\equiv SOH^0 \rightarrow \equiv SOH^- + H^+$, respectively. Data obtained from Sahai and Sverjensky (1997).

[c] $pK_{s,H} = \frac{pK_{s,1} + pK_{s,2}}{2}$.

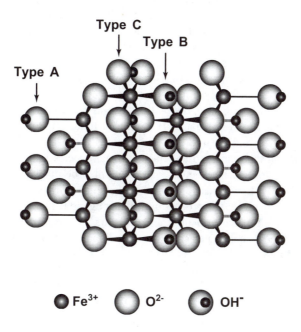

FIGURE 7.4 Goethite (FeOOH) has three types of surface functional groups: type A hydroxyls are singly coordinated by Fe^{3+}; type C hydroxyls and oxygen atoms are doubly coordinated by Fe^{3+} atoms; and type B hydroxyls and oxygen atoms are triply coordinated by Fe^{3+}.

and α-SiO$_2$ ($\equiv$SiOH^0 and $\equiv$SiO$^-$) have very different acid-base character. According to the 1–pK approach, the acidity of the α-Al$_2$O$_3$ and α-SiO$_2$ type A functional groups is illustrated in the reactions:

$$\equiv AlOH_2^{+0.5} \rightarrow \equiv AlOH^{-0.5} + H^+ \tag{7.4}$$

$$\equiv SiOH^0 \rightarrow \equiv SiO^- + H^+ \tag{7.5}$$

From Table 7.1, the strength of the Al—O bond is $s = +0.5$, the s/r_{M-OH} ratio is 1.711 nm^{-1}, and the p$K_{s,H}$ for the reaction in Eq. 7.4 is 8.95 (p$K_{s,H} = \frac{pK_{s,1} + pK_{s,2}}{2}$). The type A silica group, as found on the quartz (α-SiO$_2$) surface, is bound to structural Si^{4+} in tetrahedral coordination. The bond strength of the Si—O bond is $s = +1$, the s/r_{M-O} ratio is 3.818 nm^{-1}, and the p$K_{s,2}$ for the reaction in Equation 7.5 is approximately 7.2. Based on the p$K_{s,H}$ values, the $\equiv$SiOH^0 group is a stronger Lowry-Brønsted acid relative to the $\equiv$AlOH$_2^{+0.5}$ group. The higher bond strength (+1 for Si—O vs. +0.5 for Al—O), higher s/r_{M-OH} ratio (3.812 nm^{-1} for Si—O vs. 1.711 nm^{-1} for Al—O), the greater electronegativity (1.90 for Si vs. 1.61 for Al, and the greater covalent character of the Si—O bond relative to the Al—O bond), and the greater absolute hardness (3.38 for Si^{4+} vs. 2.77 for Al^{3+}) are characteristics that allow the surface Si atom to more strongly attract electrons, relative to Al, and polarize the surface O. As a result, $\equiv$SiOH^0 is more acidic and the proton is held less tightly, relative to the   AlOH$_2^{+0.5}$ group. It should be noted that this comparison is between a monoprotic ($\equiv$SiOH^0) and a diprotic ($\equiv$AlOH$_2^{+0.5}$) group; however, the diprotic $\equiv$SiOH$_2^+$, perhaps more appropriately comparable to $\equiv$AlOH$_2^{+0.5}$, has a p$K_{s,1}$ of −1.2. Thus, $\equiv$SiOH$_2^+$ is a strong acid and does not exist in aqueous soil environments, a result of the strong polarizing effect of Si^{4+} and the resulting lack of sufficient electron charge to effectively bond a second proton.

The complexation chemistry of inorganic surface functional groups, or the ability to adsorb ligands and metals, is a function of the Lewis acidity or basicity of the surface sites (the ability to accept or donate electrons). Inorganic surface functional groups in an aqueous environment (with the exception of the $\equiv$SiOH^0 group) usually exist as diprotic Lewis acid species, such as the $\equiv$AlOH$_2^{+0.5}$ and $\equiv$FeOH$_2^{+0.5}$ groups on gibbsite and goethite. Indeed, these diprotic species predominate when solution pH is below p$K_{s,H}$ (8.95 for gibbsite and 8.5 for hematite). The structural cation is essentially hydrated, directly bound to a water molecule that acts as a ligand in the coordination sphere of the cation. Conceptually, this is similar to a water molecule in the primary hydration sphere of a metal cation in solution (Chapter 5). Indeed, the protonated, type A surface functional groups on gibbsite and goethite may be depicted as $\equiv$Al$^{+0.5}$—OH$_2$ and $\equiv$Fe$^{+0.5}$—OH$_2$. Just as waters of hydration that surround an aqueous metal species may be displaced by another ligand to form a complex ion, surface-bound water may also be displaced by a ligand from the solution phase, assuming the Lewis base character of the soluble ligand and the acid character of the structural metal are compatible. The process by which a ligand from the aqueous phase displaces surface-bound water is called ligand exchange, and it results in the formation of a surface complex that is more stable than the bound water complex. The ligand exchange process is illustrated by the reaction (Figure 7.5a):

$$\equiv AlO'H_2^{+0.5} + H_2PO_4^- \rightarrow \equiv AlOPO_3H_2^{-0.5} + H_2O' \tag{7.6}$$

In this example, a phosphate ligand displaces the charged surface water, resulting in the formation of an inner-sphere surface complex and a charge-neutral water molecule (Figure 7.5b).

When singly protonated (e.g., $\equiv$SiOH^0, $\equiv$FeOH$^{-0.5}$, and $\equiv$AlOH$^{-0.5}$) or deprotonated (e.g., $\equiv$SiO$^-$), hydrous metal oxide surface functional groups display Lewis base character. The basic

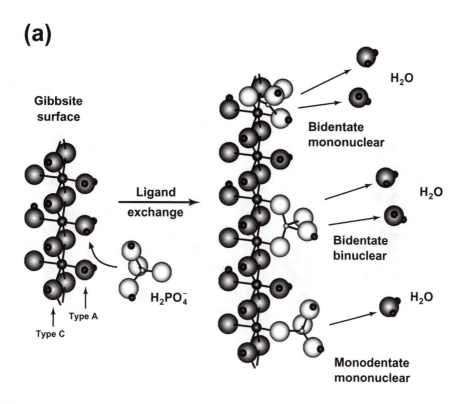

(a)

Gibbsite
surface

Ligand
exchange

$H_2PO_4^-$

Type A

Type C

H₂O

Bidentate
mononuclear

H₂O

Bidentate
binuclear

H₂O

Monodentate
mononuclear

FIGURE 7.5 The ligand exchange process at the gibbsite surface (a) may result in the formation of bidentate–mononuclear, bidentate–binuclear, or monodentate–mononuclear surface complexes. Note that one or two doubly protonated surface oxygen atoms (water molecules) are displaced in the ligand exchange process by the orthophosphate ligand. The complexation of a metal cation (b), such as Pb^{2+}, is a proton exchange process that may lead to the formation of bidentate–mononuclear, bidentate–binuclear, or monodentate–mononuclear inner-sphere surface complexes. The specific adsorption of Pb^{2+} is accompanied by the release of protons.

character of these groups is evident in the proton exchange reactions on quartz and goethite:

$$\equiv SiO'H^0 + Ni^{2+} + H_2O \rightarrow \equiv SiO'NiOH^0 + 2H^+ \tag{7.7}$$

$$\equiv FeO'H^{-0.5} + Ni^{2+} + H_2O \rightarrow \equiv FeO'NiOH^{-0.5} + 2H^+ \tag{7.8}$$

The surface oxygen remains bound to the structural cation (Si^{4+} and Fe^{3+} in Equations 7.7 and 7.8), employing a doubly occupied electronic orbital to complex the adsorbed metal. The nature of the Lewis base is determined by the same properties that influence surface Lowry-Brønsted acidity (Table 7.1). The Si—O bond is approximately 50% ionic; whereas, the Fe—O, Al—O and Ti—O bonds have greater ionic character (>60%). The greater covalent character of the Si—O bond implies that the oxygen is polarized to a greater degree and the electrons required for effective metal complexation are less available. Thus, the $\equiv SiOH^0$ group is a softer Lewis base relative to the surface groups of other metal oxides (based on the relative abilities of surface groups to complex H^+). Furthermore, the affinity of metals for the silica surface is generally less than for other surfaces.

Like the hydrous metal oxides, the layer silicate minerals also develop pH-dependent charge. This occurs at the edges of the layer structure (Figure 7.6). Two essentially different functional groups are evident: the silanol and the aluminol groups. The silanol group is terminal (type A) and consists of a surface hydroxyl bound to a single Si^{4+} cation. The reactivity of the silanol group is confined to protonation-deprotonation (Equation 7.5) and proton exchange reactions (Equation 7.7),

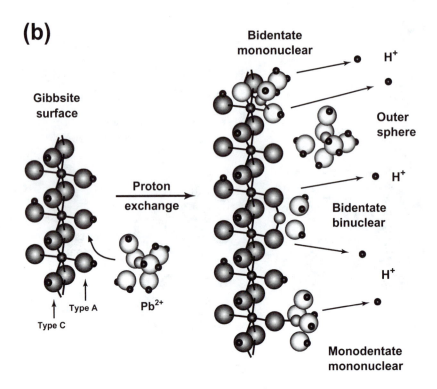

(b)

FIGURE 7.5 (*continued*).

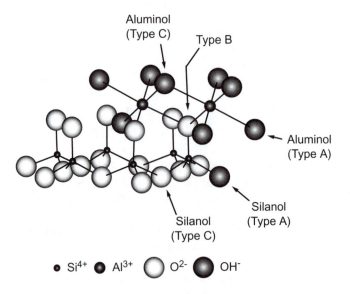

FIGURE 7.6 Aluminol and silanol groups at the layer silicate edge occur in both type A (singly coordinated oxygen) and type C (doubly coordinated oxygen) configurations. In addition, a type B (triply coordinated) oxygen atom is also evident and bound by two Al^{3+} atoms and a single Si^{4+} atom.

as described above for the $\equiv SiOH^0$ group on quartz. The type A aluminol functional group is also terminal and similar in character and reactivity to the $\equiv AlOH^{-0.5}$ group on hydrous aluminum oxides. The aluminol group participates in protonation and deprotonation (Equation 7.4), proton exchange (Equation 7.8), and ligand exchange (Equation 7.6) reactions, as described above. In addition to the type A functional groups, the edges of layer silicates also possess type C and type B surface oxygens and hydroxyls. The doubly coordinated $\equiv Al_2OH^0$ surface groups (type C, which are also prevalent on hydrous aluminum oxides) dissociate only at solution pH values greater than ~10 (pK_a ~12). These groups are considered to be inert in natural systems, as are the $\equiv Si_2O^0$ type C and the $\begin{smallmatrix}\equiv Si \\ \equiv Al_2\end{smallmatrix}>O^0$ type B surface groups.

7.1.2 SURFACE COMPLEXES

When a surface functional group reacts with an ion or molecule dissolved in the soil solution, a surface complex is formed. There are two types of surface complexes: inner-sphere and outer-sphere. These are defined in a manner similar to their counterparts in aqueous solutions (Chapter 5). An inner-sphere surface complex is formed when there is no water molecule between the adsorbed ion or molecule and a surface ligand (Figure 7.5a). In contrast, an outer-sphere complex is formed when there is at least one water molecule between a charged surface functional group and the adsorbed ion (Figure 7.5b). Outer-sphere surface complexes are maintained by electrostatics, as interceding water molecules prevent electron sharing (required for covalent bonding character). Inorganic and organic ionic species retained by the outer-sphere mechanism are said to be nonspecifically retained (or adsorbed), and may be easily displaced from the surface by ions from the solution phase. This process, termed ion exchange, is indiscriminate relative to the types of charged substances that participate; it is a significant process in the retention of all cationic and anionic species (Chapter 8 discussed cation exchange phenomena). However, there are groups of substances that interact with surfaces almost exclusively by outer-sphere complexation, with some exceptions. Metal cations that are retained only by outer-sphere complexation are those that do not hydrolyze in the pH range typical of the natural environment (pH < 9). Classically, these Lewis acid metal cations are identified as the base cations: Na^+, K^+ (and NH_4^+), Ca^{2+}, and Mg^{2+}, although all alkali and alkaline earth metals (periodic groups IA and IIA) are included. Ligands that are exchangeable and that are retained almost exclusively by outer-sphere complexation are the strong acid anions, such as Cl^-, Br^-, NO_3^-, and ClO_4^-.

Ions or molecules retained by the inner-sphere mechanism are said to be specifically retained or specifically adsorbed, indicating that they are tightly bound by the surface through bonds with a relatively high degree of structural configuration and covalent character (as might be observed in a metal–oxygen bond in a mineral structure). The type of inner-sphere surface complex formed by an adsorbed species may be characterized by the combination of two descriptive terms, such as monodentate–mononuclear. The first term identifies the number of positions in the coordination sphere of the adsorbed species that are involved in the bonding (and occupied by surface functional groups). The surface complex formed is monodentate if one position in the coordination sphere of the adsorbate is occupied by a surface functional group, or bidentate if two positions are occupied (Figure 7.5). Bidentate surface complexes are also called bridging complexes. The second term identifies the number of structural metal atoms involved in the bonding. The complex is mono-nuclear or binuclear if one or two structural metals are bound to the adsorbate, respectively. For example, a ligand exchange reaction involving $H_2PO_4^-$ may result in the formation of monodentate–mononuclear or bidentate-binuclear surface complexes (Figure 7.5a). Similarly, the inner sphere surface complexation of a metal cation, such as Pb^{2+}, can be monodentate–mononuclear or bidentate–binuclear, as depicted in Figure 7.5b. Metal cations that hydrolyze in the pH range typical of the natural environment (pH < 9) can interact with hydrous oxides and the surface sites on the edges of layer silicates by forming monodentate and bidentate inner-sphere structures (Figure 7.5b). At very low surface coverage, metals tend to form mondentate complexes. As surface coverage increases, bidentate complexes become prevalent. Metal cations that reportedly form inner-sphere surface complexes include the first row transition metals: VO^{2+}, Cr^{3+}, Ni^{2+}, Cu^{2+}, Zn^{2+}; and the heavy metals: Cd^{2+}, Hg^{2+}, and Pb^{2+}. Ligands that are derived from weak acids form inner-sphere surface complexes. A specifically adsorbed ligand, such as the fluoride ion (F^-), forms a monodentate surface complex. Polynuclear ligands, such as phosphate (HPO_4^{2-} and $H_2PO_4^-$), arsenate ($HAsO_4^{2-}$ and $H_2AsO_4^{2-}$), borate ($H_4BO_4^-$), selenite (SeO_3^{2-}), chromate ($HCrO_4^-$ and CrO_4^{2-}), molybdate (MoO_4^{2-}), carbonate (HCO_3^- and CO_3^{2-}), and silicate ($H_4SiO_4^0$), form monodentate-mononuclear and bidentate-binuclear surface complexes.

Unlike outer-sphere surface complexes, inner-sphere complexes are quite stable. The kinetics of formation appears to be a function of the prevailing type of surface complex formed. In general, monodentate surface complex formation is rapid; whereas, bidentate complexation displays a rapid initial uptake, followed by a period of relatively slow kinetics. Irrespective of the type of inner-sphere complex formed, desorption kinetics are exceedingly slow, indicating that a high activation energy of dissociation is required.

The siloxane ditrigonal cavity of layer silicates can form both inner-sphere and outer-sphere surface complexes. For example, the adsorption of K^+ (also NH_4^+) by vermiculite results in an inner-sphere complex where the K^+ ion is directly bonded to six interstitial surface oxygen molecules: three oxygen atoms on the lower interlayer surface and three on the upper (Figure 7.7a). The ability of K^+ to form an inner-sphere surface complex is primarily based on the following three factors. First, the highly localized negative charge that results from the isomorphic substitution of Al^{3+} for Si^{4+} in the tetrahedral layer of vermiculite lends the surface oxygen atoms relatively hard base character, although not as hard as water. These surface oxygen atoms have sufficient base strength to displace the waters of hydration that surround the K^+ ion. Second, the waters of hydration around the K^+ ion are less tightly held relative to those of other mono- and divalent cations (e.g., Li^+, Na^+, Mg^{2+}, and Ca^{2+}), as indicated by the less negative heat of hydration (ΔH_h) for K^+ (Table 5.2). Finally, the diameter of the K^+ ion is 0.27 nm, which is nearly identical to that of the siloxane ditrigonal cavity (0.26 nm). In other words, K^+ fits.

The adsorption of Ca^{2+} by montmorillonite results in an outer-sphere surface complex (Figure 7.7b). In this example, Ca^{2+} retains the waters of hydration. The negative charge on the siloxane functional site is relatively low and diffused over numerous surface oxygen atoms. Thus, the Lewis base strength of the surface oxygen atoms on montmorillonite is not of sufficient strength to displace waters of hydration from almost all cations with the exception of the Cs^+ ion, for which a hydration sphere may not even exist.

7.2 THE SOLID-SOLUTION INTERFACE: A MICROSCOPIC VIEW

The solid-solution interface is a transitional zone that is neither mineral nor bulk solution. The interfacial region exists because soil minerals bear electronic charge and metal and ligand complexation capabilities. The interfacial region of the soil consists of adsorbed protons, metals, and ligands (as described above), and an interfacial solution phase. The composition of the interfacial phase is directly influenced by the structural, chemical, and electrochemical characteristics of the soil minerals and the chemical characteristics of the bulk soil solution. In the preceding sections, it was established that essentially two types of surface charge exist on soil minerals: structural charge and pH-dependent charge. While these two categories of charge may account for the preponderance of surface charge, they are by no means the only sources of charge on mineral surfaces.

7.2.1 SURFACE CHARGE DENSITY

The interfacial region between the solid and solution phases, the region that exists between structural solid and bulk solution, conceptually consists of laminated layers that differ in chemical and electrochemical characteristics. Charge development on a mineral surface and the neutralization of

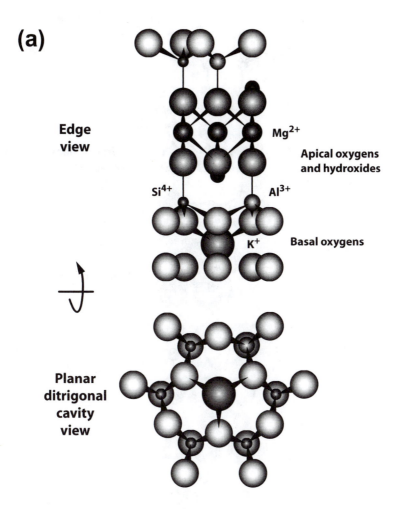

(a)

Edge view

Mg^{2+}

Apical oxygens and hydroxides

Si^{4+} Al^{3+}

K$^+$ Basal oxygens

Planar ditrigonal cavity view

FIGURE 7.7 Example of inner-sphere complexation of K$^+$ by trioctahedral vermiculite (a) as seen from side and planar views of the siloxane ditrigonal cavity. The K$^+$ ion is bound in octahedral coordination by three basal oxygen atoms of one silica sheet and by three in the opposing sheet. An example of an outer-sphere complexation of a hydrated cation, such as Ca^{2+}, is shown with montmorillonite (b).

the surface charge by dissolved ions and molecules under the influence of the surface occur in the interfacial layers. The solid-solution interface and the types of surface complexes that form on soil minerals are illustrated in Figure 7.8. Within the mineral structure, isomorphic substitution results in a permanent structural charge. This type of charge is generally confined to the 2:1 layer silicates and is almost always negative. The charge density associated with the extent of isomorphic substitution is defined as the permanent structural charge density, σ_o, and is expressed in units of Coulombs per square meter (C m^{-2}). Surfaces that display this type of charge are also called constant charge surfaces, as this type of charge cannot be modified by the environment. A value for σ_o can be computed for any layer silicate if the cation exchange capacity (CEC) and the specific surface area (S) of the mineral are known:

$$\sigma_o = \frac{F(-CEC)(0.001 \text{ kg g}^{-1})(0.01 \text{ mol}_c\text{cmol}^{-1})}{S} \tag{7.9}$$

where F is the Faraday constant (96,485 C mol^{-1}) and CEC and S have units of cmol$_c$ kg^{-1} and m^2 g^{-1}, respectively. Consider the reference smectite SWy-1, 2 (Wyoming bentonite), for which

(b)

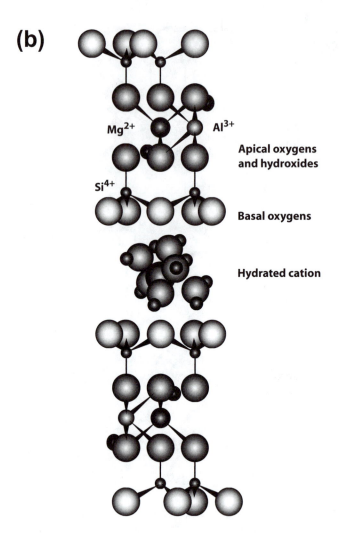

Mg^{2+}

Al^{3+}

Apical oxygens
and hydroxides

Si^{4+}

Basal oxygens

Hydrated cation

FIGURE 7.7 (*continued*).

$CEC = 85$ cmol$_c$ kg^{-1} and $S = 785$ m^2 g^{-1} (internal specific surface area). The structural charge density is:

$$\sigma_o = \frac{(96,485 \text{ C mol}^{-1})(-85 \text{ cmol}_c\text{ kg}^{-1})(0.001 \text{ kg g}^{-1})(0.01 \text{ mol}_c\text{cmol}^{-1})}{785 \text{ m}^2 \text{ g}^{-1}} = -0.104 \text{ C m}^{-2} \quad (7.10)$$

The first layer of charge, proceeding out from the surface, is the charge that results from the adsorption of the potential-determining ions, H$^+$ and OH$^-$. This type of charge, development of which occurs in the s-plane of the mineral surface, is called the net proton charge density, σ_H. This is essentially pH-dependent charge and can also be defined as the moles of protons (q_H) minus the moles of hydroxide (q_{OH}), per unit mass, complexed by surface functional groups:

$$\sigma_H = \frac{F(q_H - q_{OH})}{S} \quad (7.11)$$

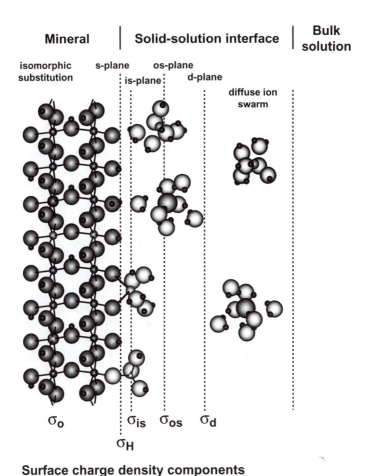

Surface charge density components

FIGURE 7.8 Schematic representation of the solid-solution interface illustrating the adsorption planes (s, is, os, and d) and the associated charge densities (σ_H, σ_{is}, σ_{os}, and σ_d). Isomorphic substitution may result in structural charge density, σ_o, which is not influenced by the composition of the solid-solution interface or the bulk solution. In the is-plane, a metal cation is retained in a bidentate–binuclear complex and a ligand is adsorbed in a monodentate–mononuclear complex. The nonspecific adsorption of hydrated cations (protons on the waters of hydration are directed away from the cation) and hydrated anions (protons on the waters of hydration directed toward the larger anion) occurs in the os- and d-planes.

The net proton charge density can be positive, zero, or negative depending on soil solution pH and mineral-specific characteristics (e.g., the number of bonds each surface site receives from structural cations—type A, B, or C; the coordination and valency of the structural metal; the number of sites available for protonation). Surfaces with this type of charge are found on hydrous metal oxides and edges of layer silicates. Surfaces that bear net proton charge are called constant potential surfaces (variable charge surfaces), as the surface will have a fixed or constant number of functional groups that may potentially protonate (or deprotonate). The variation in σ_H, as a function of pH, is commonly determined by potentiometric acid-base titration of a suspension containing the mineral of interest at a controlled ionic strength. For any point on a titration curve,

$$(q_H - q_{OH}) = \frac{V}{c_s}\{C_A - C_B - [H^+] + [OH^-]\}$$

(7.12)

where c_s is the mass (in kg) of the solid suspended in volume V of titrand, C_A is the molar concentration of added acid, C_B is the molar concentration of added base, [H$^+$] is the molar solution proton concentration, and [OH$^-$] is the molar solution hydroxide concentration. The values for V, C_A, C_B, and [H$^+$] are unique to each titration point. The concentration of the hydroxide ion ([OH$^-$]) is computed from [H$^+$]:

$$[OH^-] = \frac{^c K_w}{[H^+]} \tag{7.13}$$

where $^c K_w$ is the conditional ionization constant for water at the controlled ionic strength. In practical application, Equation 7.12 is:

$$(q_H - q_{OH}) = \frac{V}{c_s \gamma}\{[10^{-pH_B} - 10^{-pH_S}] - [10^{-(14-pH_B)} - 10^{-(14-pH_S)}]\} \tag{7.14}$$

where γ is the single-ion activity coefficient for a monovalent ion, pH_B is the pH of a blank solution titration point (titration in the absence of the solid), and pH_S is the pH of the suspension titration point. The σ_H value associated with each titration point is computed using Equation 7.11. However, the computed values are generally not the true σ_H values, because the speciation of the surface functional groups (protonated vs. deprotonated) is unknown when the titrations are initiated. However, this deficiency is rectified by employing the following information: (1) σ_H is a function of ionic strength (as well as pH) and (2) the pH at which $\sigma_H = 0$ is a unique value that is independent of ionic strength. Therefore, titration curves obtained under differing ionic strength conditions will cross at the pH value where $\sigma_H = 0$ (Figure 7.9a). The σ_H data are then adjusted to establish the true variation in σ_H as a function of pH (Figure 7.9b).

The permanent structural charge plus the net proton charge density is defined as the intrinsic surface charge density, σ_{in} ($= \sigma_o + \sigma_H$). The intrinsic surface charge density is a function of the

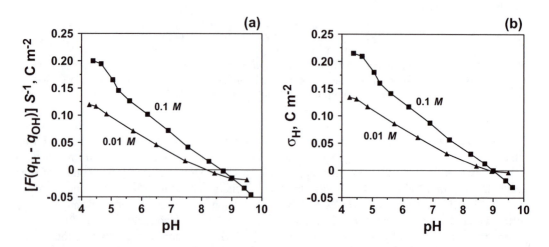

FIGURE 7.9 The charge density characteristics of gibbsite: (a) the difference between adsorbed proton concentration (q_H) and adsorbed hydroxide concentration (q_{OH}) as a function of ionic strength and pH (the pH at which titration curves, obtained at different ionic strengths, cross is the point of zero net proton charge (pH_{pznc})) and (b) the $F(q_H - q_{OH})S^{-1}$ data are adjusted to yield σ_H (the net proton charge density) as a function of ionic strength and pH.

mineral itself, representing the charge density that results from isomorphic substitution and the charge density that is a function of the number of surface functional groups that can potentially complex H^+ or OH^-. Mathematically, the intrinsic surface charge density of a mineral is given as:

$$\sigma_{in} = \frac{F(q_+ - q_-)}{S} \tag{7.15}$$

where q_+ and q_- represent moles per kilogram of adsorbed cation and anion charge at fixed pH. Inherent to this definition is that the cation and anion used in the determination must be nonspecifically adsorbed; the absence of specific adsorption is required.

Specifically adsorbed ions and molecules (other than H^+ and OH^-) reside in the is-plane on the mineral surface. These are species that form inner-sphere surface complexes. The charge density that results from the adsorption of these species is termed the inner-sphere complex charge density, σ_{is}. For example, the specific adsorption of $PbOH^+$ to a metal oxide results in the surface charge becoming more positive. Conversely, the specific adsorption of the phosphate ion (HPO_4^{2-}) to a metal oxide results in a surface that bears greater negative charge. The σ_{is} can be positive, negative, or zero.

Ions and molecules that are held at the surface via an outer-sphere surface complex reside in the os-plane, resulting in a charge density termed σ_{os}, the outer-sphere surface complex charge. The nonspecific adsorption of Ca^{2+} to a mineral surface results in the surface charge becoming more positive. Conversely, the nonspecific adsorption of SO_4^{2-} results in the surface becoming more negative. Note that both the σ_{is} and σ_{os} charge densities arise solely from the adsorption of charged constituents from the soil solution. Because of this, these surface charge densities are not intrinsic to the mineral, but rather to the chemical characteristics of the soil solution; their presence on the surface is in response to the intrinsic surface characteristics but is not an intrinsic surface charge characteristic. The σ_{os} can be positive, negative, or zero.

The net total particle charge, σ_p, is the surface charge density that arises from all the surface charge components: $\sigma_p = \sigma_o + \sigma_H + \sigma_{is} + \sigma_{os}$. As previously indicated, σ_o is virtually always negative; whereas, σ_p, σ_H, σ_{is}, and σ_{os} can be positive, negative, or zero depending on the soil solution chemical characteristics. The fact that σ_p is a measurable characteristic of mineral surfaces indicates that soil particles can bear electrical charge. However, soils are electronically neutral. Therefore, σ_p must be balanced by another kind of surface charge. This is accomplished by ions that are not bound as surface complexes, but are still associated with the surface. Ions that balance σ_p are said to reside in the diffuse ion swarm, or d-plane. These ions are freely mobile but remain near enough to the particle surface to balance σ_p. The distinction between ions that are adsorbed by outer-sphere complexation (reside in the os-plane), and those that reside in the diffuse ion swarm (reside in the d-plane), is one of distance. Essentially, ions in the os-plane are separated from the surface by approximately the diameter of a single water molecule (~ 0.3 nm) (Figure 7.8). Ions in the diffuse layer are separated from the surface by two or more water molecules (>0.6 nm), the maximum distance determined by the extent of the diffuse layer. The charge density that arises from ions in the d-plane is termed σ_d, the diffuse ion swarm charge. Because the particle charge is exactly balanced by the diffuse ion swarm charge, the following relationship is valid: $\sigma_p + \sigma_d = 0$.

7.2.2 POINTS OF ZERO CHARGE

The points of zero charge of a mineral surface are defined as the pH values at which one or more of the surface charge densities vanish (described in Section 7.2.1). One of the points of zero charge of a mineral surface, pH_{pzc}, is defined as the solution pH value when σ_d, the diffuse layer charge density, equals zero. This is also the pH at which the total net particle charge density, σ_p, equals zero (because σ_p is balanced by σ_d, when σ_d disappears, so must σ_p). When the solution pH is

TABLE 7.2
Point of Zero Charge (pH_{pzc}) Values for Selected Soil Minerals[a]

Mineral	pH_{pzc}
Quartz [α-SiO_2]	2.9
Amorphous silica [$SiO_2 \cdot 2H_2O$]	3.5
Birnessite [δ-MnO_2]	3.76
Kaolinite [$Al_2Si_2O_5(OH)_4$]	4.7
Rutile [TiO_2]	5.8
Anatase [TiO_2]	6.0
Magnetite [Fe_3O_4]	6.9
Muscovite [$KAl_2(Si_3Al)O_{10}(OH)_2$]	7.5
γ-Alumina [γ-Al_2O_3]	8.5
Hematite [α-Fe_2O_3]	8.5
Gibbsite [$Al(OH)_3$]	8.9
Corundum [α-Al_2O_3]	8.9
Goethite [α-$FeOOH$]	9.0

[a] Values obtained from Sverjensky and Sahai (1996) and Sahai and Sverjensky (1997).

below the pH_{pzc}, the mineral surface will have a net positive charge. Correspondingly, when solution pH is greater than the pH_{pzc}, the mineral surface will have a net negative charge. The pH_{pzc} values for selected soil minerals appear in Table 7.2. Soil minerals that have a low pH_{pzc} (e.g., SiO_2) have a greater ability to attract and retain cations over a broad soil pH range than those with a higher pH_{pzc}, because the surface is predominately negative for normal soil pH values. Conversely, minerals with a high pH_{pzc} (e.g., α-Al_2O_3) have a greater ability to attract and retain anions over a broad soil pH range.

The solution pH at which $\sigma_H = 0$ is termed the point of zero net proton charge (pH_{pznpc}). The pH_{pznpc} is defined by the condition $(q_H - q_{OH}) = 0$, where q_H is the surface concentration of adsorbed H^+ and q_{OH} is the surface concentration of adsorbed OH^- (or dissociated H^+). For example, the pH_{pznpc} for gibbsite, as determined by the potentiometric titration (Figure 7.9 b), is 9.0. Further, pH_{pznpc} can be obtained from the intrinsic surface acidity constants:

$$pH_{pznpc} = \tfrac{1}{2}(p\,K_{a1}^{int} + p\,K_{a2}^{int}) \tag{7.16}$$

where $p\,K_{a1}^{int}$ and $p\,K_{a2}^{int}$ are the equilibrium constants that apply to the surface dissociation reactions:

$$\equiv SOH_2^+ \rightarrow \equiv SOH^0 + H^+ \tag{7.17}$$

$$\equiv SOH^0 \rightarrow \equiv SO^- + H^+ \tag{7.18}$$

Intrinsic surface acidity constants for selected soil minerals are shown in Table 7.3. In the absence of metals and ligands that form inner–sphere surface complexes, $pH_{pzc} = pH_{pznpc}$ (Table 7.2). The pH value at which $\sigma_H = 0$ and $\sigma_o = \sigma_{is} = \sigma_{os} = 0$ is termed the isoelectric point (IEP). This is the pH at which structural charge and surface complexes are assumed absent (σ_d also equals zero at the IEP).

TABLE 7.3
Intrinsic Surface Acidity Constants for Selected Soil Minerals[a]

Mineral	pK_{a1}^{intb}	pK_{a2}^{intb}
Quartz [α-SiO$_2$]	−1.2	7.2
Amorphous silica [SiO$_2$ · 2H$_2$O]	−0.7	7.7
Birnessite [δ-MnO$_2$]	0.16	7.36
Rutile [TiO$_2$] O$_z^z$	2.6	9.0
Hematite [α-Fe$_2$O$_3$]	5.7	11.3
Goethite [α-FeOOH]	6.1	11.7
Corundum [α-Al$_2$O$_3$]	8.50	9.70

[a] Data obtained from Sahai and Sverjensky (1997).
[b] Values for pK_{a1}^{int} and pK_{a2}^{int} are specific to the reactions in Equations 7.17 and 7.18.

The point of zero net charge (pH_{pznc}) is the pH at which the cation exchange capacity (*CEC*) of the surface is equal to the anion exchange capacity (*AEC*), or $CEC - AEC = 0$. In terms of surface charge density, pH_{pznc} is met when $\sigma_{os} + \sigma_d = 0$, only if non–specifically adsorbed ions (those that are retained by outer-sphere and diffuse ion swarm mechanisms) are present. However, if ions that may form inner-sphere complexes are also present, pH_{pznc} is met when $\sigma_{is} + \sigma_{os} + \sigma_d = 0$. The pH_{pzc} will equal pH_{pznc} when $\sigma_{os} = 0$.

The point of zero salt effect (pH_{pzse}) is not actually a point of zero charge. The pH_{pzse} occurs when a surface has a common σ_H value when determined at different ionic strengths. This condition is met when ionic strength makes no difference to σ_H, i.e., the pH at which two or more potentiometric titration curves, each obtained at a unique ionic strength, intersect (Figure 7.9a). The pH_{pzse} will equal pH_{pzc} if σ_{is} remains constant when ionic strength is varied (or when $\sigma_{is} = 0$). Consider the two examples that follow. At pH_{pzse}, $\sigma_{H,I1} = \sigma_{H,I2}$, and at the pH_{pzc}, $\sigma_{d,I1} = \sigma_{d,I2}$, or $(\sigma_o + \sigma_H + \sigma_{os} + \sigma_{is})_{I1} = (\sigma_o + \sigma_H + \sigma_{os} + \sigma_{is})_{I2}$. Since σ_o does not depend on ionic strength, $(\sigma_H + \sigma_{os} + \sigma_{is})_{I1} = (\sigma_H + \sigma_{os} + \sigma_{is})_{I2}$. Since $\sigma_{H,I1} = \sigma_{H,I2}$, $pH_{pzc} = pH_{pzse}$ when $(\sigma_{os} + \sigma_{is})_{I1} = (\sigma_{os} + \sigma_{is})_{I2}$. This condition may be met if the activity of a specifically adsorbed ion is held constant while the ionic strength is varied and $\sigma_{os} = 0$. Conversely, if a salt containing a specifically adsorbed cation is used to control the ionic strength, σ_H will decrease with increasing ionic strength at any pH value, and the pH_{pzse} will shift downward relative to the pH_{pzc} value obtained in the absence of specifically adsorbed species (when $\sigma_{is} = 0$). In this case, $pH_{pzse} \neq pH_{pzc}$.

The pH_{pzc} will also be influenced by the specific adsorption (formation of inner-sphere complexes) of metals and ligands. At the pH_{pzc}, $\sigma_d = 0$ or $\sigma_H = -(\sigma_o + \sigma_{is} + \sigma_{os})$. If it assumed (for convenience) that $\sigma_o = \sigma_{os} = 0$, then $\sigma_H = -\sigma_{is}$. This leads to the following statements: (1) if cations are specifically adsorbed, σ_H will be smaller than when $\sigma_{is} = 0$ and the pH_{pzc} will shift upward; and (2) if anions are specifically adsorbed, σ_H will be larger than when $\sigma_{is} = 0$ and the pH_{pzc} will shift upward. Essentially, specific adsorption shifts the pH_{pzc} in the same direction as the sign of the valence of the specifically adsorbed ion. For example, the specific adsorption of phosphate, HPO_4^{2-}, will shift pH_{pzc} to a lower value because the surface becomes more negative:

$$\equiv SOH^0 + HPO_4^{2-} \rightarrow \equiv SOPO_3^{2-} + H_2O \tag{7.19}$$

$$\equiv SOPO_3^{2-} + 2H^+ \rightarrow \equiv SOPO_3H_2^0 \tag{7.20}$$

Neutralization of the more negative surface charge that results from the ligand exchange reaction (Equation 7.19) requires additional proton charge (Equation 7.20). Conversely, the specific adsorption of lead, Pb^{2+}, will shift pH_{pzc} to a higher value because the surface becomes more positive:

$$\equiv SOH^0 + Pb^{2+} \rightarrow \equiv SOPb^+ + H^+ \qquad (7.21)$$

$$\equiv SOPb^+ + OH^- \rightarrow \equiv SOPbOH^0 \qquad (7.22)$$

The neutralization of the positive surface charge generated through the exchange of a surface proton for Pb^{2+} (Equation 7.21) requires additional base (Equation 7.22).

7.2.3 THE ELECTRIC DOUBLE-LAYER

When applied to the oxide surface ($\sigma_o = 0$), the conceptual model of a particle solid-solution interface (Figure 7.8) defines the location of adsorbed species and the nature of the surface charge densities. This model of the particle surface is a combination of many previously developed and more simplistic surface models. All surface models, however, have a number of key assumptions in common. First, the proton and the hydroxide ion are potential determining ions; they are responsible for the generation of surface charge. Adsorption of a proton produces positive surface charge through the formation of $\equiv SOH_2^+$. Conversely, the adsorption of hydroxide, which is equivalent to the dissociation of a proton, produces negative surface charge, $\equiv SO^-$. Because the surface is charged and attracts ions of opposite charge, an electric potential develops at the solid-solution interface:

$$\psi_0 = \frac{RT \ln 10}{F} (pH_{zpc} - pH) \qquad (7.23a)$$

or

$$\psi_0 = 0.059155 (pH_{zpc} - pH) \qquad (7.23b)$$

where ψ_0 is the surface potential at 25°C with units of volts, pH_{zpc} is the solution pH at which the surface has no net charge ($\sigma_p = 0$), and pH is the pH of the bulk solution in which the particle resides. At the particle surface an electric double layer develops. The first layer is formed by the charge on the surface created by the adsorption of potential determining ions. A second layer is in the solution and is formed by the charge on the counter ions that balance the surface charge.

The simplest double-layer model is the von Helmoltz model (Figure 7.10a). In the Helmholtz model, the solid-solution interface consists of two layers of charge, one layer on the surface and one in solution. The negative surface charge is assumed to be evenly distributed over the surface with a charge density of σ_{H0}. The counter charge in the solution layer is concentrated in a plane that is parallel to the surface at some distance x. The charge density in the solution layer is σ_{isH}. The surface potential as a function of the distance from the surface, $\psi_{(x)}$, is at a maximum at the surface ($\psi_{(0)}$) and drops off linearly with distance from the surface. The concentration of anions ($[A^-]_H$) in the Helmholtz surface layer is zero, and the excess of cation charge ($[C^+]_H$) at the surface is determined by the charge density of the solid.

If the counter ions in the solution layer of the double layer are exposed to a balance between electrostatic and diffusion forces, the Gouy-Chapman double-layer theory, or diffuse double layer theory, is obtained (Figure 7.10b). In this model, the surface is assumed to be planar and infinite in extent, the negative surface charge density (σ_{GC0}) is evenly distributed over the colloid surface,

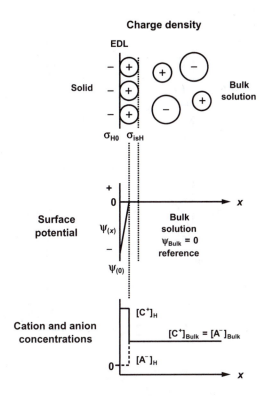

(a) Helmholtz double layer

FIGURE 7.10 Electric double layer (EDL) models that describe the solid-solution interface. The Helmholtz double layer (a) consists of the solid with surface charge density σ_{H0} that is exactly balanced by a layer of cations (counterions) with a charge density of σ_{isH}. The electrical potential in the Helmholtz model decreases from a value of $\psi_{(0)}$ at the surface to zero at a distance of one counterion radius. Counterion concentration in the Helmholtz layer is dictated by the magnitude of the surface charge, and anions are completely excluded from the double layer. The surface charge density, σ_{GC0}, in the Gouy-Chapman model (b) is balanced by a diffuse ion swarm of counterion charge, with charge density σ_d. The electrical potential, ψ_d, and the concentrations of cations and anions in the diffuse layer are modeled as a function of distanced from the surface by a Boltzman distribution. The Stern EDL model (c) incorporates properties of both the Helmholtz and Gouy-Chapman models. This model assumes that the surface charge is balanced by a layer of cations that are directly associated with the surface (Helmholtz component) and only partially satisfy the surface charge. The remaining surface charge is satisfied by a diffuse layer of cations and anions (Gouy-Chapman component).

and the cation charge is dispersed in the solution surface layer. The diffuse distribution of dissolved cations and anions at the surface is the result of a force balance. The attractive electrostatic force pulls cations toward the surface, resulting in higher cation concentrations in this zone than in the bulk solution. Conversely, the negatively charged surface repels, or excludes, anions from the surface region, resulting in lower anion concentrations than in the bulk solution. The chemical potentials of cations in this region are higher than in the bulk solution. Since ions move from a condition of high chemical potential to low chemical potential, a diffusive force pulls cations away from the surface (just as a diffusive force also pulls anions into the surface region). When equilibrium is attained (i.e., when the electrostatic and diffusive forces balance), the solution surface layer will extend out away from the surface, with cation concentration and surface potential decreasing with

(b) Gouy-Chapman double layer

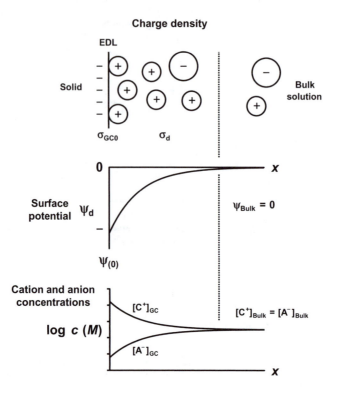

FIGURE 7.10 (*continued*).

distance, and the anion concentration increasing with distance. The surface potential decreases from the soil surface in a nonlinear fashion:

$$\psi_{(x)} = \psi_d = \psi_{(0)} \exp(-\kappa x) \tag{7.24}$$

where $\psi_{(0)}$ is the surface potential, ψ_d is the potential in the diffuse layer as a function of distance (x in meters) from the surface, and κ is a parameter that is related to the reciprocal of the double layer thickness (units of m^{-1}):

$$\kappa = ZF \left(\frac{2I(10^3)}{\varepsilon\varepsilon_0 RT} \right)^{0.5} \tag{7.25}$$

where Z is the counter ion charge, F is the Faraday constant (96,487 C mol^{-1}), I is the ionic strength of the bulk solution in moles per liter (or moles per cubic decimeter) counter ion valence, ε is the dielectric constant of water (78.54 at 298.15 K), ε_0 is the permittivity of a vacuum (8.854 × 10^{-12} C^2 J^{-1} m^{-1}), R is the gas constant (8.314 J mol^{-1} K^{-1}), and T is the solution temperature (K). The center of charge of the diffuse layer may also be computed as a function of ionic strength (I) and counter ion charge:

$$\kappa^{-1} = \frac{3.042(10^{-10})}{ZI^{0.5}} \tag{7.26}$$

(c) Stern triple layer

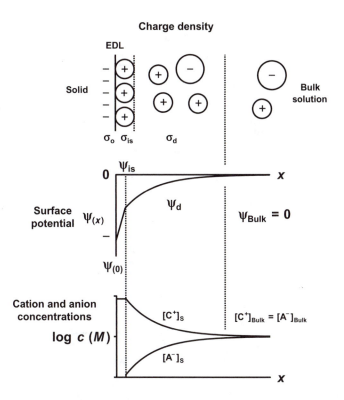

FIGURE 7.10 (*continued*).

where κ^{-1}, in m, is often characterized as the "thickness" of the diffuse layer. The concentration distribution in the diffuse layer follows the Boltzman distribution:

$$c_x = c_B \exp\left(\frac{-ZF\psi_d}{RT}\right)$$ (7.27a)

where c_x is the concentration of an ion at distance x from the surface, c_B is the bulk solution concentration of the ion, and all other parameters are previously defined.

The distribution of cations and anions in the diffuse layer may be computed by combining Equations 7.24 and 7.26 (or 7.25) into Equation 7.27a (for a solution at 25°C):

$$c_x = c_B \exp\left(\frac{-ZF\psi_0 \exp(ZI^{0.5}3.287(10^9)x)}{2.4788(10^3)}\right)$$ (7.27b)

The impacts of ionic strength and counter ion valence on diffuse layer thickness are illustrated in Figure 7.11. The solution factors that influence the thickness (κ^{-1}) and the extent of the diffuse layer are ionic strength and counter ion charge (valence). The diffuse layer thickness decreases linearly with increasing ionic strength, as indicated in Equation 7.26. A tenfold increase in I will decrease the double layer thickness by a factor of $\sqrt{10}$, or approximately 3. For example, in Figure 7.11a the thickness of the diffuse layer (κ^{-1}) in a 0.01 I solution controlled by a monovalent

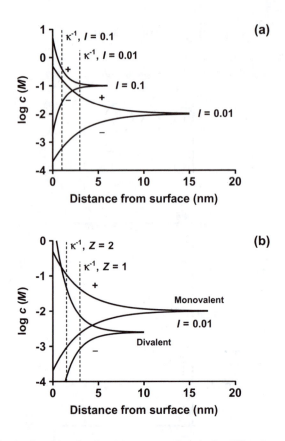

FIGURE 7.11 Monovalent cation (+) and anion (−) concentrations in the diffuse layer as a function of distance from a surface bearing a potential ($\psi_{(0)}$) of −0.1 V as a function of ionic strength (a) and counterion charge (b). The thickness of the double layer parameter, κ^{-1}, represents the center of counterion charge and is representative of the distance from the surface the double layer extends (although the actual extent of the double layer is greater than κ^{-1}).

electrolyte is 3.04 nm; whereas, in a 0.1 I solution, κ^{-1} is 0.96 nm. A higher electrolyte concentration in the bulk solution decreases the diffusive force acting on the adsorbed cations, resulting in a more compact diffuse layer. The thickness of the double layer also decreases directly (linearly) with increasing counter ion valence. In the 0.01 I solution controlled by a divalent electrolyte (c_B = 0.0025 M), the thickness of the double layer is 1.52 nm, compared to the 3.04 nm thickness of the monovalent electrolyte (c_B = 0.01 M) system (Figure 7.11b).

Stern theory (Figure 7.10c) divides the solution region near the surface into two layers by combining the Helmholtz and Gouy-Chapman double layer models. In the layer closest to the surface, termed the Stern layer or the inner Helmholtz layer, ions are specifically adsorbed and form a compact layer of counter ion charge that is similar to that depicted in the Helmholtz model. The second solution layer is described by the diffuse layer model developed in the Gouy-Chapman theory.

7.3 QUANTITATIVE DESCRIPTION OF ADSORPTION

Adsorption is the mechanism most commonly responsible for the retention of ions and molecules by soils, particularly trace element cations (e.g., Ni^{2+}, Cu^{2+}, Zn^{2+}, Pb^{2+}, Cd^{2+}), anions (e.g., CrO_4^{2-}, SeO_3^{2-}, SeO_4^{2-}, AsO_3^{3-}, AsO_4^{3-}, MoO_4^{2-}), and organic compounds (e.g., pesticides and industrial solvents). In addition to adsorbing, organic compounds can partition, or *absorb*, into soil organic matter.

During absorption, ions or compounds migrate into a structure, thus this retention process is not strictly a surface process. It is very difficult to distinguish between adsorption and absorption mechanisms. Thus, it is common to see the term *sorption* used to indicate that the exact retention mechanism is unknown. Because the adsorption (or sorption) process often restricts compound mobility and bioavailability, a mechanism is needed to quantitatively indicate the intensity and extent of the sorption process. The quantitative measures of adsorption could then provide information on the potential mobility of a compound relative to other compounds as a function of the soil and the chemical conditions or potential chemical conditions acting in the soil.

Two general approaches, nonmechanistic and mechanistic, have been employed to provide a quantitative measure of substance adsorption by soil and soil minerals. Nonmechanistic techniques rely on determining the mass distribution of a substance between the solid and solution phases at equilibrium. Typically, solutions containing known quantities of the substance of interest (the adsorptive or adsorbate) are equilibrated with a soil (the adsorbent). At equilibrium, the amount of substance that has disappeared from the solution phase is assumed to be adsorbed by the solid. This solid-solution distribution is then characterized by a distribution coefficient:

$$K_d = \frac{q}{c_{eq}} \tag{7.28}$$

where q is the equilibrium mass of adsorbed substance per unit mass of adsorbent, c_{eq} is the equilibrium mass of the substance in solution per unit volume of solution, and K_d has units of volume per mass. The variation in the distribution coefficient with surface coverage (q) or solution concentration (c_{eq}) is then mathematically described to elucidate the adsorption behavior of a substance when reacted with the particular adsorbent (or soil), and under the chemical conditions imposed in the adsorption study. An example of a nonmechanistic technique is the adsorption isotherm. Information obtained from this type of study is not transferable from one soil to another or even from one chemical environment to another for the same soil, which somewhat limits the applicability of mass distribution information. However, nonmechanistic evaluations of substance behavior are popular, easy to perform, and the information gained is useful in assessing the relative affinity of different substances for the same soil, or the relative affinity of a single substance for several soils.

Mechanistic techniques for evaluating substance adsorption are actually quite similar to non-mechanistic techniques, except in that the amount of substance adsorbed by a solid is determined as a function of a solution property other than substance concentration. Typically, this involves the evaluation of adsorption as a function of solution pH, as the proton is a master species that directly impacts the reactivity of surface functional groups, and ionic strength at constant adsorbate concentration. Adsorption is then described by chemical reactions involving specific chemical forms of the adsorbent and specific surface functional groups. Each reaction is further characterized by an equilibrium constant that can be applied to any environment, irrespective of the characteristics of the environment in which the adsorption reaction occurs.

7.3.1 THE ADSORPTION ISOTHERM

A common nonmechanistic technique used to assess substance adsorption is the adsorption isotherm. An adsorption isotherm is a graph of the equilibrium surface excess or amount of a compound adsorbed (e.g., in units of mmol kg^{-1} or mg kg^{-1}), designated by q, plotted against the equilibrium solution concentration of the compound (e.g., in units of mmol L^{-1} or mg L^{-1}), designated by c_{eq}, at fixed temperature (thus the term *isotherm*), pressure, and solution chemistry (e.g., pH and ionic strength). To obtain the data necessary to construct an isotherm, the following batch equilibration procedure may be employed:

- Place a mass of soil (adsorbent), m_s, in contact with a volume of solution, V_l, which contains an initial concentration, c_{in}, of the compound of interest (adsorbate) in a reaction vessel. The ionic strength of the solution is typically selected to mimic that of the soil solution and is normally controlled by a salt, such as $CaCl_2$, which is also common to the environment under study. Typically, solution pH is allowed to be controlled by the soil.
- Repeat the above procedure using several different reaction vessels, each identical but differing in c_{in}.
- Equilibrate the contents of reaction vessels, while agitating, under constant temperature conditions.
- Following equilibration, separate the solution from the soil (via centrifugation and filtration) and analyze the solution for the compound of interest. The concentration of the compound in this solution is c_{eq}.
- Compute q by difference:

$$q = \frac{V_l(c_{in} - c_{eq})}{m_s} \tag{7.29}$$

- Plot q (the y-axis variable) against c_{eq} (the x-axis variable).

The application of this procedure is illustrated in the following example. Ten-gram samples of a loess-derived soil are placed in contact with 10 mL of a 10 mM $CaCl_2$ solution containing varying concentrations of the herbicide fluometuron (Figure 7.12a) and equilibrated, with constant shaking, for 24 h. The initial solutions and the equilibrated solutions are then analyzed by high-pressure liquid chromatography (HPLC) to determine fluometuron concentrations. Consider the data for system 3 in Table 7.4, in which the initial concentration of fluometuron, c_{in}, is 143 µg L^{-1}. Following

TABLE 7.4
Adsorption Data for the Retention of Fluometuron by a Lexington Silt Loam (Ultic Hapludalf)[a]

| System | c_{in} | c_{eq} | q |
	µg L^{-1}		µg kg^{-1}
1	51.9	11.5	40.4
2	96.4	20.4	76.0
3	143	31.7	111
4	192	44.2	148
5	288	65.1	223
6	487	125	362

[a] These data are used to construct the adsorption isotherm in Figure 7.12b. Each batch system consisted of 10 g soil equilibrated with 10 mL of a 10 mM $CaCl_2$ solution containing an initial fluometuron concentration, c_{in}. Following the 24-h equilibration period at 25°C, the concentration of fluometuron in solution is c_{eq}. The concentration of adsorbed fluometuron, q, is computed using Equation 7.29. The data represent the means of triplicate systems.

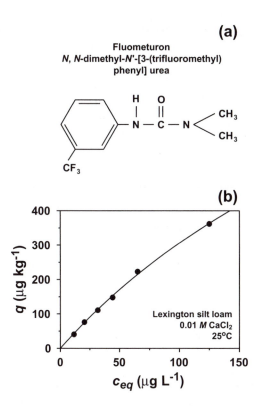

FIGURE 7.12 The structure (a) and adsorption isotherm (b) characterizing the retention of the herbicide fluometuron by soil from the Ap horizon of a Lexington silt loam (Ultic Hapludalf). The adsorption data were obtained by equilibrating 10-g samples of soil and 10 mL of 10 mM CaCl$_2$ solutions containing varied concentrations of the herbicide. The suspensions were equilibrated for 24 h at 25°C.

the 24-h equilibration period, the equilibrium concentration of fluometuron, c_{eq}, is 31.7 µg L^{-1}. Using Equation 7.29, the concentration of fluometuron adsorbed by the soil is computed:

$$q = \frac{(0.01\ \text{L}) \times (143\ \mu\text{g L}^{-1} - 31.7\ \mu\text{g L}^{-1})}{0.01\ \text{kg}} = 111\ \mu\text{g kg}^{-1} \tag{7.30}$$

The results for each initial fluometuron concentration are summarized in Table 7.4 and the adsorption isotherm is shown in Figure 7.12b. When examining the isotherm closely, notice that a straight line cannot be used to connect all the data points. However, the data can be fit to a curved line, whose slope is greatest at low surface coverage (low values of q) and decreases with increasing surface coverage. In other words, the slope of the isotherm does not either increase or remain constant with increasing surface coverage. This is an example of an L-curve (or L-type) isotherm (Figure 7.13a). The L-type isotherm indicates that the adsorbate has a relatively high affinity for the soil surface at low surface coverage. However, as coverage increases, the affinity of the adsorbate for the soil surface decreases. The L-type isotherm is the most commonly encountered type of isotherm in soil chemistry.

In addition to the L-curve isotherm, there are three other major types of isotherms. An adsorbate that has a very high affinity for the soil surface will display an H-curve isotherm (Figure 7.13b). The H-type isotherm is an extreme version of the L-type isotherm and is usually interpreted to indicate the formation of inner-sphere surface complexes.

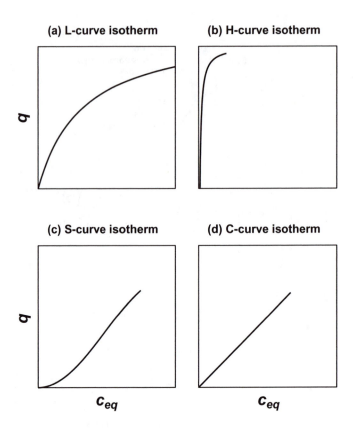

FIGURE 7.13 Isotherm types that are commonly observed in the environmental sciences: (a) L-type; (b) H-type; (c) S-type; (d) C-type.

The S-curve isotherm is characterized by a small slope at low surface coverage that increases with adsorbate concentration (Figure 7.13c). The adsorption of trace elements by soil is often described by an S-type isotherm, particularly in soils with high dissolved organic carbon concentrations. This type of isotherm suggests that the affinity of the soil for the adsorbate is less than that of the aqueous solution when the solution concentration of the adsorbate is low. For example, consider the adsorption of Cu by soil. At low solution Cu concentrations, the surface is in competition with dissolved organic carbon (DOC) for Cu^{2+}. The DOC is very effective at complexing Cu^{2+}, due to the abundance of N-bearing moieties. As Cu concentration increases and exceeds the complexation capacity of the DOC, the surface gains preference and L-type behavior is evident. The adsorption of Ni in the presence of the synthetic chelate EDTA is also S-type (Figure 7.14). At low Ni solution concentrations, EDTA chelates the Ni, so that Ni^{2+} is not available to participate in adsorption reactions. However, as the concentration of Ni increases and exceeds the capacity of EDTA to complex it (saturates the EDTA), the adsorption reverts to L-type. S-curves are also associated with the adsorption of neutral organic compounds. Initially, adsorption is low because the ionic (charged) environment of the soil surface is not compatible with the nonionic character of the organic compound. As adsorbate concentration increases and greater amounts of the organic compound are forced onto the soil surface, the surface becomes stabilized (essentially becoming nonionic). The organic compound is then attracted to the surface where it self-polymerizes.

The C-curve isotherm is also commonly associated with the adsorption of nonionic and hydrophobic organic compounds (Figure 7.13d). In this type of isotherm the initial slope remains independent of the surface coverage until some maximum adsorption capacity is achieved. This type of

FIGURE 7.14 Adsorption isotherms characterizing the retention of nickel (Ni) by surface horizon (0 to 15 cm) samples of a Glendale soil (Typic Torrifluvent) from 5 m*M* CaCl$_2$ (H-type isotherm) and 5m*M* CaCl$_2$ + 0.17 m*M* EDTA (S-type isotherm) solutions. The adsorption data were obtained by equilibrating 11-g samples of soil and 11 mL of 5 m*M* CaCl$_2$ solutions containing varied concentrations of NiCl$_2$. The suspensions were equilibrated for 24 h at 20 to 25°C. (Modified from Bowman, R.S., M.E. Essington, and G.A. O'Connor. Soil sorption of nickel: influence of solution composition. *Soil Sci. Soc. Am. J.* 45:860–865, 1981.)

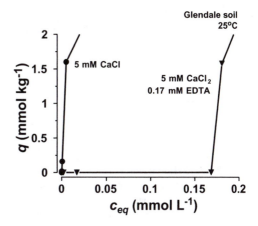

adsorption behavior is also termed constant partitioning, as the partitioning of an organic compound between water (a hydrophilic phase) and a phase that is immiscible with water (such as *n*-octanol, a hydrophobic phase) will display a C-type isotherm.

The objective of performing an adsorption isotherm study is to obtain compound-specific adsorption parameters that quantitatively describe adsorption in a specific environment. Therefore, the adsorption isotherm must lend itself to a mathematical description. Several mathematical models have been used to describe adsorption isotherms. The three most commonly employed isotherm equations are the Langmuir equation, the Freundlich equation, and the linear partition model.

Of the three isotherm models, only the Langmuir equation is theoretically derived; however, the assumptions employed in the derivation of the Langmuir equation are generally not applicable to a soil system. The Langmuir equation was originally derived to describe the adsorption of gas molecules by a solid surface, and assumes the following:

- Adsorption occurs at specific locations (or sites) on a surface.
- All adsorption sites are identical in character; the surface is homogeneous.
- A monolayer of adsorbed molecules is formed on the surface and an adsorption maximum is achieved as the monolayer becomes filled by the adsorbate.
- The heat or energy of adsorption is constant over the entire surface.
- Adsorbed species do not interact.
- The volume of the monolayer and the energy of adsorption are independent of temperature.
- Equilibrium is attained.

When these assumptions are met in a physical system, the adjustable parameters that are empirically determined from an adsorption isotherm have physical meaning. The Langmuir equation is:

$$q = \frac{bK_L c_{eq}}{(1 + K_L c_{eq})} \tag{7.31}$$

where b and K_L are adjustable parameters. As classically defined, b is the adsorption maxima, having the units of q. It is the value of q that is approached asymptotically as c_{eq} becomes infinitely large (Figure 7.15). The adsorption constant, K_L, is a measure of the intensity of the adsorption isotherm (in units of L kg^{-1} of adsorbent). The adsorption maxima times the adsorption constant (bK_L) is the initial slope (slope of isotherm as c_{eq} approaches zero) of the adsorption isotherm.

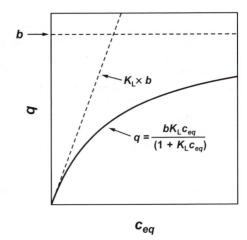

FIGURE 7.15 Empirical constants of adsorption obtained from the Langmuir isotherm equation, $K_L \times b$, is the slope the isotherm as q approaches zero, and b is the maximum value that q approaches asymptotically as c_{eq} becomes arbitrarily large.

The Langmuir equation can also be derived for the adsorption of a solute in an aqueous system. Consider the exchange of a substance for adsorbed water (recalling that all adsorption reactions are also exchange reactions):

$$\equiv S - H_2O(ad) + M(aq) \rightarrow \equiv S - M(ad) + H_2O(l) \tag{7.32}$$

where $\equiv S$ represents a specific surface site, M is the adsorbate, and $\equiv S - H_2O$ and $\equiv S - M$ are adsorbed water and compound M phases. The exchange reaction in Equation 7.32 can be described by the equilibrium constant:

$$K_{ex} = \frac{(\equiv S - M)(H_2O)}{(\equiv S - H_2O)(M)} \tag{7.33}$$

The adsorption of M is described by a single exchange reaction and a single equilibrium constant (and a single energy of adsorption). Therefore, there is only one type of surface site; the surface is homogeneous. Recall that the activity of a solid in a solid solution is the mole fraction multiplied by a rational activity coefficient, fN. This treatment is also valid for adsorbed phases; the activity of $\equiv S - M$ is equal to $f_{\equiv S-M} N_{\equiv S-M}$. If it is further assumed that the surface solid solution is ideal, the activities of adsorbed water and M are equal to their mole fractions at the surface. Similarly, if it is assumed that the solution is dilute in M, the activity of M can be approximated by the concentration of M(C_M). Employing these assumptions in Equation 7.33 yields:

$$K_{ex} = \frac{N_{\equiv S-M}(H_2O)}{N_{\equiv S-H_2O} C_M} \tag{7.34}$$

Since there are only two adsorbed species present on the surface, the mole fractions must sum to $1(N_{\equiv S-H_2O} + N_{\equiv S-M} = 1)$. Equation 7.34 can then be written in terms of $N_{\equiv S-M}$:

$$K_{ex} = \frac{N_{\equiv S-M}(H_2O)}{(1 - N_{\equiv S-M}) C_M} \tag{7.35}$$

Rearranging Equation 7.35 leads to:

$$N_{\equiv S-M} = \frac{\left[\dfrac{K_{ex}C_M}{(H_2O)}\right]}{\left[1 + \dfrac{K_{ex}C_M}{(H_2O)}\right]} \qquad (7.36)$$

The activity of the liquid water is a constant. Therefore, the exchange equilibrium constant divided by the activity of water is also a constant: $K'_{ex} = K_{ex}/(H_2O)$. Substituting K'_{ex} into Equation 7.36 yields:

$$N_{\equiv S-M} = \frac{K'_{ex}C_M}{(K'_{ex}C_M + 1)} \qquad (7.37)$$

An identical expression is obtained if the activity of liquid water is assumed to be unity. The mole fraction of adsorbed M is equal to the moles of adsorbed M (n_M) divided by the moles of adsorption sites ($n_{\equiv S}$): $N_{\equiv S-M} = n_M/n_{\equiv S}$. Substituting this expression in 7.37 and rearranging:

$$n_M = \frac{n_{\equiv S}K'_{ex}C_M}{(K'_{ex}C_M + 1)} \qquad (7.38)$$

The total number of adsorption sites, n_M, is the maximum number of moles of M that can be adsorbed. According to Equation 7.32, n_M is also a monolayer, as each surface site can only accept a single M. Dividing both sides of Equation 7.38 by the mass of the adsorbent, m, yields:

$$q = \frac{bK'_{ex}C_M}{(K'_{ex}C_M + 1)} \qquad (7.39)$$

where b is the adsorption maximum ($n_{\equiv S}/m$) and q is the concentration of adsorbed M at equilibrium n_M/m. Of course, Equation 7.39 is identical to Equation 7.31, the Langmuir equation; however, Equation 7.39 is developed for the adsorption of a solute from an aqueous environment. The assumptions used to develop Equation 7.39 are similar to those employed to develop Equation 7.31 (gas adsorption). Solute adsorption is an exchange process that occurs at specific sites on a homogeneous surface. A monolayer of adsorbed solute is formed on the surface and an adsorption maximum is achieved as the monolayer becomes saturated by the adsorbate. Adsorbed species do not interact, although adsorbates can interact with solutes. The energy of adsorption (ΔG_r°) is identical for each adsorption site (as implied in the definition of a homogeneous surface), equilibrium is attained, and the temperature is constant. Again, when these conditions are met, the Langmuir equation yields the adsorbate concentration that is a monolayer (the b parameter) and a parameter that is related to the energy of adsorption and the exchange equilibrium constant (the K_L parameter). Otherwise, as is the case in soil, the Langmuir model parameters are merely empirical, describing the shape of the adsorption isotherm.

The Langmuir adsorption constants are obtained by regressing c_{eq} on q using linear or nonlinear least-squares regression. In order to employ linear regression analysis, the Langmuir equation must first be transformed into an expression that has the form: $y = mx + b$ (m is the slope, b is the y-intercept, and x and y are the independent and dependent variables, respectively). Beginning with the Langmuir equation (Equation 7.31), multiply both sides by $(1 + K_L c_{eq})$:

$$q(1 + K_L c_{eq}) = bK_L c_{eq}$$

or

$$q + qK_L c_{eq} = bK_L c_{eq} \tag{7.40}$$

Dividing both sides of this expression by c_{eq} yields

$$\frac{q}{c_{eq}} + qK_L = bK_L \tag{7.41}$$

Rearranging,

$$\frac{q}{c_{eq}} = bK_L - qK_L \tag{7.42}$$

For convenience, q/c_{eq} is defined as the distribution coefficient, K_d (from Equation 7.28). Equation 7.42 then becomes

$$K_d = bK_L - qK_L \tag{7.43}$$

Therefore, a plot of K_d (y-axis variable) against q (x-axis variable) will yield a straight line with a slope equal to $-K_L$, a y-intercept of bK_L, and an x-intercept of b, **if the Langmuir model describes the adsorption data**.

Another way to transform the Langmuir equation so that K_L and b may be obtained by linear least-squares regression begins by dividing Equation 7.31 through by q:

$$1 = \frac{bK_L c_{eq}}{[q(1 + K_L c_{eq})]} \tag{7.44}$$

Multiplying through by $(1 + K_L c_{eq})$:

$$(1 + K_L c_{eq}) = \frac{bK_L c_{eq}}{q} \tag{7.45}$$

Dividing both sides of the above expression by $bK_L c_{eq}$:

$$\frac{(1 + K_L c_{eq})}{bK_L c_{eq}} = \frac{1}{q} \tag{7.46}$$

which leads to:

$$\frac{1}{bK_L c_{eq}} + \frac{1}{b} = \frac{1}{q} \tag{7.47}$$

Finally, multiplying through by c_{eq} yields:

$$\frac{1}{bK_L} + \frac{c_{eq}}{b} = \frac{c_{eq}}{q} \tag{7.48}$$

For this expression, a plot of c_{eq}/q (y-axis variable) against c_{eq} (x-axis variable) will yield a straight line with a slope of $1/b$ and a y-intercept intercept of $1/bK_L$, **if the Langmuir model describes the adsorption data**.

The Freundlich isotherm model has broad application to L-, H-, and C-type isotherms. Unlike the Langmuir isotherm model, the Freundlich equation is not theoretically based, although the equation can be derived by assuming a specific distribution of adsorption site ΔG_r° values (heterogeneous surface). However, in practical use, the Freundlich model is merely an equation that can account for the variation in q as a function of c_{eq}. The Freundlich equation is:

$$q = K_F c_{eq}^N \tag{7.49}$$

where K_F and N are positive-valued adjustable parameters and N is constrained to lie between 0 and 1. The adjustable parameters of the Freundlich model are not normally interpreted as having physical meaning. However, it is evident that K_F is numerically equal to q when c_{eq} is unity, and N is a measure of inflection in the curve that fits the isotherm data. It has also been shown mathematically that N is a measure of the heterogeneity of adsorption sites on the adsorbent surface (Sposito, 1980). As N approaches 0, surface site heterogeneity increases, indicating that there is a broad distribution of adsorption site types. Conversely, as N approaches unity, surface site homogeneity increases, indicating that there is a narrow distribution of adsorption site types. The Freundlich adsorption constants can be obtained using linear or nonlinear least-squares regression. In order to employ linear regression analysis, the Freundlich equation must first be transformed into an expression that has the form $y = \mathrm{m}x + b$ (m is the slope, b is the y-intercept, and x and y are the independent and dependent variables, respectively). Taking the common logarithm of both sides of Equation 7.49 transforms the Freundlich equation into the linear expression:

$$\log q = \log K_F + N \log c_{eq} \tag{7.50}$$

Therefore, a plot of $\log q$ (y-axis variable) against $\log c_{eq}$ (x-axis variable) will yield a straight line with slope N and a y-intercept $\log K_F$, *if the Freundlich model describes the adsorption data*.

The application of adsorption isotherm models to adsorption data can be viewed by examining the adsorption of the aquatic herbicide fluridone (Figure 7.16a) by a Ca^{2+}-saturated montmorillonite. The background electrolyte in the experiment is 10 mM $CaCl_2$. The adsorption results are presented in Table 7.5. A q vs. c_{eq} graph of the fluridone adsorption data (Figure 7.16b) indicates an L-type adsorption isotherm. At this point, the data may be analyzed using nonlinear regression analysis

---

TABLE 7.5
Adsorption Data for the Retention of Fluridone by Ca^{2+}-Saturated STx-1 Montmorillonite[a]

c_{eq} (µg L⁻¹)	q (µg kg⁻¹)	c_{eq} (µg L⁻¹)	q (µg kg⁻¹)
0.94	3,118	24.8	16,105
2.11	4,585	50.1	21,820
4.73	6,244	91.1	29,020
6.21	7,850	133.6	34,130
9.9	9,510	199.9	37,635
13.9	11,665		

[a] Each batch system consisted of 0.3 g Ca^{2+}-saturated clay equilibrated at 25°C with 30 mL of a 10 mM $CaCl_2$ solution containing varied initial fluridone concentrations. The concentration of adsorbed fluridone, q, is computed using Equation 7.29. The data represent the means of duplicate systems.

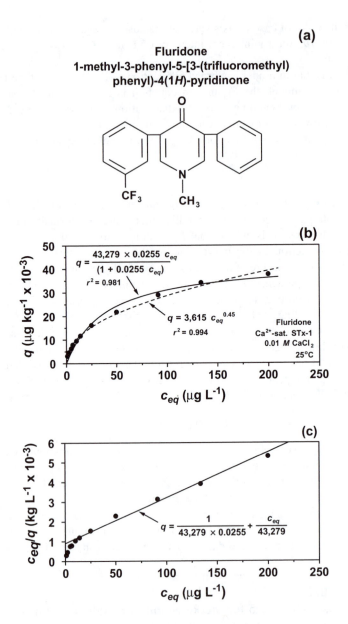

FIGURE 7.16 The structure (a) and adsorption isotherm (b) characterizing the retention of the herbicide fluridone by Ca^{2+}-saturated STx-1 montmorillonite. The experimental data are indicated by the closed circles. The Langmuir (solid line) and Freundlich (dashed line) adsorption isotherm models were generated by nonlinear least-squares regression analysis. Application of a transformed version (linearized) of the Langmuir model to a c_{eq}/q vs. c_{eq} plot of the adsorption data is illustrated in (c). Log-log transformation of the adsorption data and the application of the linearlized Freundlich model is shown in (d). The adsorption data were obtained by equilibrating 0.3-g samples of Ca^{2+}-saturated clay and 30 mL of 10 mM $CaCl_2$ solutions containing varied concentrations of the herbicide. The suspensions were equilibrated for 24 h at 25°C.

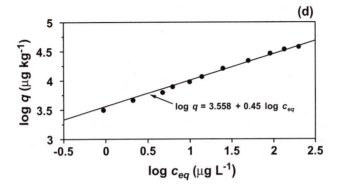

FIGURE 7.16 (continued).

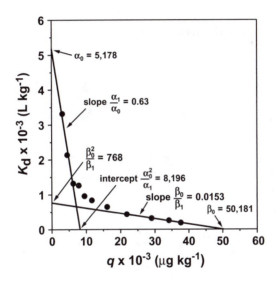

FIGURE 7.17 Application of a transformed version (linearized) of the Langmuir model to a K_d vs. q plot of the fluridone adsorption isotherm data described in Figure 7.16. In this example, the adsorption of fluridone by Ca^{2+}-saturated STx-1 montmorillonite is described by two separate Langmuir isotherm equations. The parameters, α_0, α_1, β_0, and β_1, are defined in Equations 7.53a through 7.53d.

to fit the data to the Langmuir and the Freundlich adsorption models. The results of this regression analysis are also plotted on the graph. As is indicated by the r^2 values (0.981 for Langmuir and 0.994 for Freundlich), both models provide a highly significant fit to the data (both models describe the variability of q as a function of c_{eq}). In Figures 7.16c and 7.16d, the data are plotted according to the Equation 7.43 and Equation 7.50 versions of each model. The Freundlich model adequately describes the adsorption data over the entire range of c_{eq} studied; whereas, the Langmuir model adequately describes fluridone adsorption as c_{eq} becomes large.

The observation that a single Langmuir model does not describe fluridone adsorption over the entire range of c_{eq} does not preclude its use. It has been recognized that the description of compound adsorption over broad ranges of surface coverage may require more than one adsorption model. Adsorption isotherms that display differences in adsorption behavior as a function of surface coverage have been interpreted to indicate multisite or multilayer adsorption. For example, the fluridone adsorption isotherm can be divided into two distinct regions, as illustrated by the K_d vs. q plot in Figure 7.17. Because K_d decreases as a function of q, and goes to zero at a finite value of q, the fluridone adsorption data can be described mathematically by a two-site Langmuir model of the form:

$$q = \frac{b_1 K_1 c_{eq}}{(1 + K_1 c_{eq})} + \frac{b_2 K_2 c_{eq}}{(1 + K_2 c_{eq})} \tag{7.51}$$

Evaluation of the four adsorption constants in Equation 7.51 (b_1, b_2, K_1, and K_2) is aided by the information in Figure 7.17 and by further mathematical manipulation of Equation 7.51. With respect to the latter, it can be shown that as q goes to 0, the expression that relates K_d to q is:

$$\lim_{q \to 0} K_d = (b_1 K_1 + b_2 K_2) - q \left[\frac{(b_1 K_1^2 + b_2 K_2^2)}{(b_1 K_1 + b_2 K_2)} \right] \tag{7.52a}$$

As K_d goes to 0, the expression that relates K_d to q is:

$$\lim_{K_d \to 0} K_d = \frac{(b_1 + b_2)^2}{\left(\dfrac{b_1}{K_1} + \dfrac{b_2}{K_2} \right)} - q \left[\frac{(b_1 + b_2)}{\left(\dfrac{b_1}{K_1} + \dfrac{b_2}{K_2} \right)} \right] \tag{7.52b}$$

Both Equations 7.52a and 7.52b are straight-line functions that model the behavior of K_d at the limits of the K_d vs. q diagram (Figure 7.17). Employing the following substitutions:

$$\alpha_0 = b_1 K_1 + b_2 K_2 \tag{7.53a}$$

$$\alpha_1 = -(b_1 K_1^2 + b_2 K_2^2) \tag{7.53b}$$

$$\beta_0 = b_1 + b_2 \tag{7.53c}$$

$$\beta_1 = \frac{b_1}{K_1} + \frac{b_2}{K_2} \tag{7.53d}$$

Equations 7.52a and 7.52b become:

$$K_d = \alpha_0 + \left(\frac{\alpha_1}{\alpha_2} \right) q \tag{7.54a}$$

and

$$K_d = \frac{\beta_0^2}{\beta_1} - \left(\frac{\beta_0}{\beta_1} \right) q \tag{7.54b}$$

Numerical values for α_0, $\frac{\alpha_1}{\alpha_0}$, $\frac{\beta_0}{\beta_1}$, and $\frac{\beta_0^2}{\beta_1}$ can be obtained by evaluating K_d as a linear function of q as q goes to 0 and as K_d goes to 0. Using the fluridone adsorption data as an example (Figure 7.17), $\alpha_0 = 5,178$, $\frac{\alpha_1}{\alpha_0} = 0.63$, $\frac{\beta_0}{\beta_1} = 0.0153$, and $\frac{\beta_0^2}{\beta_1} = 768$. It follows that $\alpha_1 = 3,271$, $\beta_0 = 50,181$, and $\beta_1 = 3,279,823$. Equations 7.53a through 7.53d then become:

$$5,178 = b_1K_1 + b_2K_2 \tag{7.55a}$$

$$3,271 = -(b_1K_1^2 + b_2K_2^2) \tag{7.55b}$$

$$50,181 = b_1 + b_2 \tag{7.55c}$$

$$3,279,823 = \frac{b_1}{K_1} + \frac{b_2}{K_2} \tag{7.55d}$$

These four equations can then be solved explicitly for the four constants of adsorption in the two-site Langmuir model. For fluridone adsorption, the two-site model is:

$$q \ (\text{in } \mu g \ kg^{-1}) = 44,812 \times \frac{0.0117 \times c_{eq}}{(1 + 0.0117 \times c_{eq})} + 5369 \times \frac{0.8633 \times c_{eq}}{(1 + 0.8633 \times c_{eq})} \tag{7.56}$$

This function is plotted with the adsorption data in Figure 7.18. Continuing with the two-site theme, Equation 7.56 indicates fluridone is adsorbed by a limited number ($b = 5.4$ mg kg^{-1} or 16.3 μmol kg^{-1}) of high-energy sites ($K_L = 0.8633$ L μg^{-1}) and a large number ($b = 44.8$ mg kg^{-1} or 136 μmol kg^{-1}) of low energy sites ($K_L = 0.0117$ L μg^{-1}) that are present on the smectite surface, assuming that the constraints implicit to the development of the Langmuir equation are met.

The ability to describe the adsorption of a compound by a multisite or multilayer adsorption model is not, in and of itself, evidence that compound adsorption occurs at multiple sites or in multiple layers at the soil surface. Such an interpretation implies that adsorption isotherms provide information on the mechanism of retention, rather than just a description of the nonmechanistic mass distribution of a substance between two phases. One must remember the governing rule of adsorption isotherms: adsorption isotherms are nonmechanistic (macroscopic) mass distribution

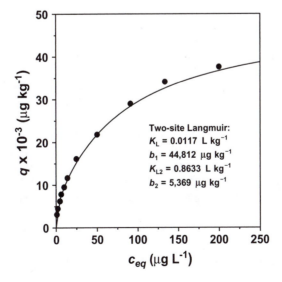

FIGURE 7.18 Two-site Langmuir model that describes the adsorption of fluridone by Ca^{2+}-saturated STx-1 montmorillonite (described in Figure 7.16 and using Equation 7.56).

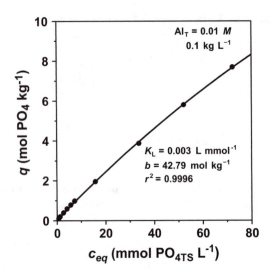

FIGURE 7.19 The solubility of variscite [$AlPO_4 \cdot 2H_2O$] described by the Langmuir adsorption isotherm model. The total soluble concentration of aluminum (Al_T) is 0.01 M and the initial concentration of varascite is 0.1 kg L^{-1}.

characterizations, and adsorption isotherm equations are *empirical* and are used explicitly for the purpose of describing solute adsorption by soil. They cannot be used to infer any particular adsorption mechanism, because the mechanism is a microscopic characteristic. Further, they cannot be used to indicate which retention mechanisms are operating (adsorption or precipitation). Consider the precipitation of phosphate to form the mineral variscite ($AlPO_4 \cdot 2H_2O$). If a phosphate adsorption-like experiment is conducted, such that the adsorbent is variscite and the adsorbate is phosphate, and the initial solutions are supersaturated with respect to variscite, an adsorption isotherm similar to that in Figure 7.19 should be obtained. The data can be described by a Langmuir model, with an adsorption maximum (b) of 42.79 mol kg^{-1}, a Langmuir adsorption constant (K_L) of 0.003 L mol^{-1}, and an r^2 value of 0.9996. The Langmuir adsorption model provides an excellent description of the data, which may lead one to conclude that the mechanism of phosphate retention is adsorption. However, this is not the case. The mechanism of phosphate retention, or the removal of phosphate from solution, is the precipitation of variscite. In this example, the precipitation of variscite has been described by an adsorption isotherm model.

7.3.2 LINEAR PARTITION THEORY

The adsorption behavior of nonpolar or hydrophobic organic compounds is most commonly characterized by the C-type isotherm. This type of behavior, also known as linear partitioning, constant partitioning or hydrophobic sorption, is quantified using the expression $q = K_p c_{eq}$, where K_p is the partition coefficient. This expression is identical to the Freundlich equation with $N = 1$. The sorption of hydrophobic organic compounds is associated with the following characteristics:

- Equilibrium sorption isotherms are linear (C-type) up to relatively high adsorbate concentrations.
- The extent of compound sorption is highly correlated to the soil organic carbon (SOC) content, or the soil organic matter (SOM) content: K_p increases with increasing SOC or SOM content.
- Compound sorption is not influenced by temperature.

- The soil sorption of a compound is not influenced by the presence of a second compound (no apparent competition between the organic compounds for the soil surface, which implies that specific adsorption sites are not involved in compound retention).
- Organic compounds with lower water solubility tend to show higher K_p values.
- Soil properties, such as type and amount of clay, soil pH, and hydrous oxide content, have little effect on the sorption of nonpolar compounds (except in low SOC soils).

The above observations are not generally consistent with a surface process. Thus, the partition theory was developed.

The partition theory states that hydrophobic organic compounds dissolve into the SOM. In this theory, the sorbed organic compound permeates into the network of SOM and is held by weak, physical, van der Waals forces. This retention process is analogous to the extraction of an organic compound from water into an immiscible organic phase (this process is called partitioning); however, in the case of soil partitioning the immiscible organic phase is SOM. Compounds that partition in the soil would not be adsorbed, because adsorption is a surface process. Instead, these compounds would be *absorbed* and homogeneously distributed throughout the entire volume of the SOM; they would be dissolved into the SOM. Again, since the partition theory is just a theory, and the exact mechanism (adsorption vs. absorption) is not known, the term *sorption* is used to denote compound retention.

Organic compounds vary widely in their hydrophobic-hydrophilic character. Even within a single compound, an ionic moiety (e.g., a carboxyl or an amine group) can coexist with nonpolar alkyl chains and/or aromatic rings. There is also no dividing line that separates hydrophobic compounds from hydrophilic compounds, although an ionic compound generally meets the definition of a hydrophilic compound. Nonionic organic compounds make up a continuum of hydrophobic-hydrophilic character. Two compound-specific properties that have been employed to indicate the hydrophobic-hydrophilic character of a compound are water solubility (S_w, in mol L^{-1}) and octanol-water partition coefficient (K_{ow}). Water solubility is a direct measure of the hydrophilic character of a compound, while K_{ow} is a measure of the hydrophobic character of a compound. The K_{ow} is determined by measuring the partitioning of the compound between two immiscible phases that are in contact with one another: *n*-octanol (the hydrophobic phase) and water (the hydrophilic phase). The K_{ow} is computed as the equilibrium concentration of the substance in the *n*-octanol phase, $c_{octanol}$, divided by the concentration in the aqueous phase, c_{aq}:

$$K_{ow} = \frac{c_{octanol}}{c_{aq}} \tag{7.57}$$

The greater the K_{ow} for a compound, the greater is the hydrophobic character.

Because S_w and K_{ow} are describing contrasting characteristics of nonpolar compounds, they are inversely related. Figure 7.20a illustrates the relationship between S_w and K_{ow} for a large number ($n = 98$) of nonpolar organic compounds. The relationship is log-log linear:

$$\log K_{ow} = 1.396 - 0.634 \, (\pm 0.015) \times \log S_w \tag{7.58}$$

where S_w is the water solubility in mol L^{-1}. The correlation coefficient of $r^2 = 0.949$ is highly significant and the standard error of the slope of $s_{yx} = 0.015$ translates to a relative standard error (rse_{yx}) of 2.37% ($rse_{yx} = \frac{s_{yx}}{b} \times 100$, where b is the slope of the regression line). Both the r^2 and s_{yx} values indicate that water solubility is an excellent predictor of the octanol-water partition coefficient for nonpolar compounds, and vice verse. The relationship between S_w and K_{ow} is generally improved by considering specific classes of compounds separately. For example, a separate evaluation of the

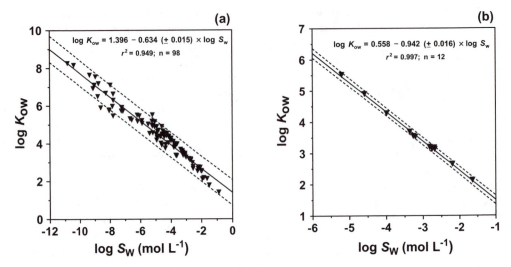

FIGURE 7.20 The relationship between the octanol-water partition coefficient (log K_{ow}) and water solubility (log S_w) for 98 nonpolar organic compounds (a), and 12 halogenated aromatic compounds (b). The solid line was obtained by regressing log K_{ow} on log S_w; the dashed lines represent the 95% confidence interval about the regression line. (Data from Miller, M.M., S.P. Wasik, G.L. Huang, W.Y. Shiu, and D. Mackay. Relationships between octanol–water partition coefficient and aqueous solubility. *Environ. Sci. Technol.* 19:522–529, 1985.)

FIGURE 7.21 The relationship between the octanol-water partition coefficient (log K_{ow}) and water solubility (log S_w) for 187 organic pesticides. The solid line was obtained by regressing log K_{ow} on log S_w; the dashed lines represent the 95% confidence interval about the regression line. The data were obtained from numerous compilations and studies of individual compounds, and no attempt was made to group compounds according to similarities in physical–chemical properties.

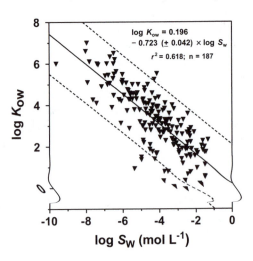

log S_w–log K_{ow} relationship for halogenated aromatic compounds (Figure 7.20b) results in the regression equation:

$$\log K_{ow} = 0.558 - 0.942\ (\pm 0.016) \times \log S_w \tag{7.59}$$

with a correlation coefficient of $r^2 = 0.997$ and an rse_{yx} of 1.70%.

While the correlation between log S_w and log K_{ow} for nonpolar compounds is highly significant, the correlation between log S_w and log K_{ow} for polar, nonionic compounds is somewhat less satisfying. Figure 7.21 illustrates the relationship between S_w and K_{ow} for a large number ($n = 187$) of polar organic pesticides. Although considerable scatter in the data exists, the general inverse relationship between log S_w and log K_{ow} is evident. The regression equation:

$$\log K_{ow} = 0.196 - 0.723\ (\pm 0.042) \times \log S_w \tag{7.60}$$

with a correlation coefficient of $r^2 = 0.618$ and an $rse_{yx} = 5.81\%$, is a statistically significant model for the description of log K_{ow} as a function of log S_w.

As was previously indicated, a characteristic of hydrophobic compound sorption is the dependence of K_p on the organic carbon content of the soil. Indeed, the dependence is linear and is quantified by the expression:

$$K_p = K_{oc} f_{oc} \tag{7.61}$$

where K_{oc} is the organic carbon–normalized partition coefficient and f_{oc} is the fractional organic carbon content of the soil for which the K_p value is valid. A similar expression:

$$K_p = K_{om} f_{om} \tag{7.62}$$

relates K_p to the organic matter–normalized partition coefficient, K_{om}, and the fractional organic matter content of the soil, f_{om}. The K_{oc} and K_{om} for nonpolar substances are considered to be compound-specific properties, as are S_w and K_{ow}. Indeed, both can be predicted from S_w or K_{ow}. The water solubility, S_w, and the K_{ow} of nonpolar organic compounds (aromatic and chlorinated hydrocarbons) are related to K_{oc} values by log-log linear relationships (Figure 7.22, data from Karickhoff et al. 1979):

$$\log K_{oc} = 1.381 - 0.541 \, (\pm 0.049) \times \log S_w \tag{7.63}$$

$$\log K_{oc} = -0.223 + 1.004 \, (\pm 0.017) \times \log K_{ow} \tag{7.64}$$

Log K_{oc} is inversely and highly correlated to log S_w ($r^2 = 0.938$) and positively correlated to log K_{ow} ($r^2 = 0.998$), with the rse_{yx} values of 9.06 and 1.69%, respectively. Similar expressions have been developed for predicting the log K_{om} of chlorobenzenes and polychlorinated biphenols (data from Chiou et al., 1983):

$$\log K_{om} = -0.030 - 0.736 \, (\pm 0.017) \times \log S_w \tag{7.65}$$

$$\log K_{om} = -0.856 + 0.921 \, (\pm 0.032) \times \log K_{ow} \tag{7.66}$$

Log K_{om} is inversely and highly correlated to log S_w ($r^2 = 0.996$) and positively correlated to log K_{ow} ($r^2 = 0.989$), with the rse_{yx} values of 2.31% and 3.47%, respectively.

For nonionic, polar pesticides (Figure 7.23), the log K_{oc}–log K_{ow} and log K_{oc}–log S_w relationships are also significantly correlated:

$$\log K_{oc} = 0.760 - 0.510 \, (\pm 0.025) \times \log S_w \tag{7.67}$$

$$\log K_{oc} = 1.188 + 0.532 \, (\pm 0.030) \times \log K_{ow} \tag{7.68}$$

The correlation coefficients for the above equations are $r^2 = 0.688$ and 0.626, respectively. As was noted for the log K_{ow} and log S_w relationship for these polar compounds (Equation 7.60), there is considerable scatter in the data. The poorer predictability of these pesticide K_{oc} values from either S_w or K_{ow} can be attributed to the fact that adsorption mechanisms other than partitioning are predominantly responsible for compound retention (discussed in Section 7.6). Thus, Equations 7.61 and 7.62 and the partition theory will not apply.

In addition to the expressions developed above for predicting K_{oc} from S_w and K_{ow}, several other expressions have been developed for specific groups of organic compounds. A sampling of these expressions is provided in Table 7.6. Once the K_{oc} is computed for the specific hydrophobic organic compound, the partition coefficient, K_p, for the compound in any soil can be determined, if the

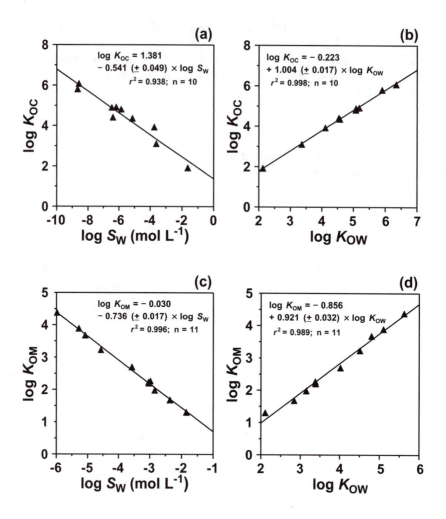

FIGURE 7.22 Relationships between soil organic carbon-normalized partition coefficient (log K_{oc}) and water solubility (log S_w) (a); log K_{oc} and octanol-water partition coefficient (log K_{ow}) (b); soil organic matter–normalize partition coefficient (log K_{om}) and log S_w (c); and log K_{om} and log K_{ow} for aromatic and chlorinated hydrocarbons (d) (data from Karickhoff et al., 1979). The solid lines represent the regression equation that appears on each diagram.

organic carbon content of the soil is known. For example, a soil with an SOC content of 24.5 g kg^{-1} (2.45%) is inadvertently contaminated with a PCB (assume a model compound: 2,4,4´-PCB, even though industrial PCBs are mixtures and are called Aroclors). The water solubility of 2,4,4´-PCB is 1.047 µmol L^{-1}, and log S_w = −5.98 (mol L^{-1}). Using the expression above (Equation 7.63) that relates log K_{oc} to log S_w, the log K_{oc} for 2,4,4´-PCB is 4.616, and K_{oc} = 41,325. The SOC content of the soil is 24.5 g kg^{-1}; therefore, f_{oc} is 0.0245. Recall that $K_{oc} f_{oc} = K_p$. Then, K_p = 41,325 × 0.0245, or K_p = 1012 L kg^{-1} is the predicted linear partition coefficient for 2,4,4´-PCB in this soil and q = 1012 × c_{eq} would describe the adsorption isotherm. However, employing the equation of Hassett et al. (1980) (Table 7.6) leads to a K_p value of 452 L kg^{-1} from a computed log K_{oc} of 4.265. Note that the log K_{oc} values computed using Equation 7.63 and the equation of Hassett et al. (1980) differ by only 0.351 units (as do the log K_p values). Although the two models predict comparable log K_{oc} and log K_p values, the resulting predicted K_p values differ by a factor of greater than two: 1012 L kg^{-1} vs. 452 L kg^{-1}, which is a consequence of transforming from logarithmic to nonlogarithmic values.

TABLE 7.6
Statistical Relationships for Estimating the Organic Carbon-Normalized Partition Coefficient (log K_{oc}) for Several Compound Classes Determined Using Water Solubility (log S_w) or Octanol-Water Partition Coefficient (log K_{ow})

Regression Equation	r^2	n	Compound Classes[a]	Reference
$\log K_{oc} = -0.68 \log S_w$ (mg L^{-1}) $+ 4.273$	0.99	15	CH	Green and Karickoff (1990)
$\log K_{oc} = -0.557 \log S_w$ (µmol L^{-1}) $+ 4.277$	0.93	23	PAH, AA, CH	Chiou et al. (1979)
$\log K_{oc} = -0.736 \log S_w$ (mol L^{-1}) $+ 0.211$	0.996	12	PAH	Chiou et al. (1983)
$\log K_{oc} = -0.62 \log S_w$ (mg L^{-1}) $+ 3.95$	0.86	107	PAH, AA, CH	Hassett et al. (1983)
$\log K_{oc} = \log K_{ow} - 0.317$	0.98	23	PAH, AA, CH	Green and Karickoff (1990)
$\log K_{oc} = 0.72 \log K_{ow} + 0.49$	0.95	13	pesticides	Schwarzenbach and Westall (1981)
$\log K_{oc} = 0.909 \log K_{ow} + 0.088$	0.93	34	PAH, AA, CH	Hassett et al. (1983)
$\log K_{oc} = 1.029 \log K_{ow} - 0.18$	0.91	13	pesticides	Rao and Davidson (1982)
$\log K_{oc} = 0.921 \log K_{ow} - 0.615$	0.989	12	PAH, CH	Chiou et al. (1983)

[a] CH, chlorinated hydrocarbons; PAH, poly- and hetero-nuclear aromatic hydrocarbons; AA, aromatic amines.

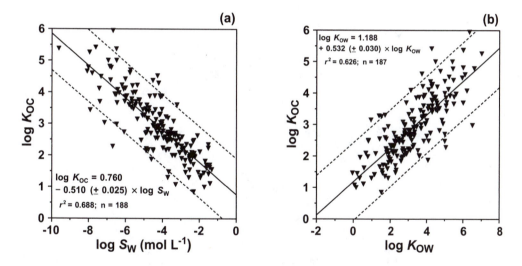

FIGURE 7.23 Relationship between soil organic carbon-normalized partition coefficient (log K_{oc}) and water solubility (log S_w) (a); and log K_{oc} and octanol-water partition coefficient (log K_{ow}) for organic pesticides (b). The solid lines represent the regression equation that appears on each diagram. The dashed lines represent the 95% confidence interval about the regression line. The data were obtained from numerous compilations and studies of individual compounds, and no attempt was made to group compounds according to similarities in physical–chemical properties.

Partition theory offers a mechanism to predict the retention of organic compounds in a soil environment. However, there are a number of limitations and potential abuses associated with the application of K_{oc}–S_w and K_{oc}–K_{ow} relationships. For nonpolar compounds, the relationships between K_{oc}–S_w and K_{oc}–K_{ow} are log–log linear, with a 95% confidence interval of approximately 1 log unit. Therefore, the error associated with the predicted K_{oc} value can be one order of magnitude (a factor of 10). This means that the use of partition theory to predict K_{oc} and subsequently K_p values is valuable only as a first order approximation (i.e., measurements alone can provide accurate K_p values).

Partition theory strictly applies to compounds that display hydrophobic character. Organic compounds that are polar, hydrophilic, ionic, and have a high water solubility are retained by soils by mechanisms other than, or in addition to, partitioning, and their adsorption isotherms are not linear over broad concentration ranges (those isotherms are usually L-type). Partition theory also does not account for the potential contribution of soil minerals to organic compound adsorption. The interlayer surfaces of clay minerals (basal oxygen atoms of silica tetrahedral) have hydrophobic character in the absence of isomorphic substitution and can contribute to hydrophobic compound retention. It is known that soil minerals are particularly important in organic compound retention in low organic carbon soils. For example, atrazine sorption is strictly correlated to SOC content in soils containing greater than 60 to 80 g kg^{-1} SOC. However, in soils containing less than 60 to 80 g kg^{-1} SOC, atrazine sorption is correlated to both SOC content and clay content.

Partition theory has also been employed to describe the association of nonionic organic compounds with dissolved organic carbon (DOC) in soil solutions. Dissolved fulvate substances, as well as co-contaminants, can effectively compete with SOC (humates) and clay materials for an organic adsorbate, resulting in greater adsorbate concentration in a soil solution relative to those which would be present in the absence of the DOC. Figure 7.24 illustrates the partitioning of an organic solute into mobile and immobile components. The mobile component might be composed of the "free" solute and DOC-solute complexes. The immobile component might include the

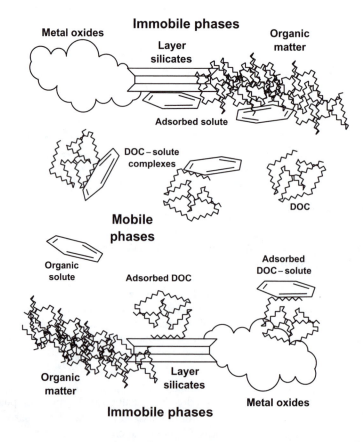

FIGURE 7.24 Naturally occurring dissolved organic carbon (DOC) may impact the fate and behavior of an organic solute of anthropogenic origin. The organic solute may remain free in the mobile soil solution phase, or participate in adsorption (or partition) reactions with immobile soil solids. The organic solute may also associate with DOC, which may enhance solute mobility or adsorption, depending on the hydrophobic–hydrophilic character of the solute and the affinity of the DOC for the soil surfaces.

FIGURE 7.25 The adsorption of fluridone by DOC- and Ca^{2+}-saturated STx-1 montmorillonite in the presence and absence of 100 mg L^{-1} DOC. The experimental data are indicated by the closed triangles. The Freundlich adsorption isotherm models, represented by the lines and corresponding equations, were generated by nonlinear least-squares regression analysis. The adsorption data were obtained by equilibrating 0.3-g samples of DOC- and Ca^{2+}-saturated clay and 30 mL of 10 mM CaCl$_2$ containing varied concentrations of the herbicide. The DOC was extracted from a Lexington silt loam (Ultic Hapludalf) surface soil. The suspensions were equilibrated for 24 h at 20 to 22°C.

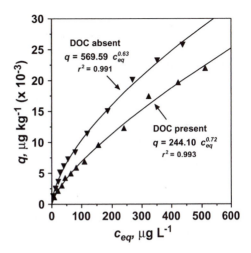

adsorbed (or absorbed) solute, or adsorbed DOC-solute complexes. An example of this effect, termed the solubility enhancement effect, is illustrated in Figure 7.25. The retention of the aquatic herbicide fluridone (Figure 7.16a; S_w = 36.4 μmol L^{-1} and log K_{ow} = 1.87) by Ca^{2+}- and DOC-saturated montmorillonite is decreased in the presence of DOC (obtained by extracting the surface 2 cm of a Lexington silt loam (Ultic Hapludalf) with a 10 mM CaCl$_2$ solution). In the absence of solution DOC, the adsorption of fluridone to the DOC-saturated smectite is characterized by the Freundlich model:

$$q = 569.6 c_{eq}^{0.63} \qquad (7.69)$$

Fluridone adsorption in the presence of approximately 100 mg L^{-1} DOC is also modeled using the Freundlich equation:

$$q = 244.1 c_{eq}^{0.72} \qquad (7.70)$$

DOC is chemically similarly to immobile solid organic carbon and may bind contaminant organic compounds, promoting greater mobility. In order for this process to occur, the DOC molecules must be sufficiently large and possess a sizeable intramolecular nonpolar environment with which to promote compound binding. If the partition mechanism is assumed, the compound must exhibit very low water solubility and significant compatibility with DOC. Additionally, polar and ionic compounds would be expected to interact with the functional moieties on DOC. Numerous methods have been used to measure the intensity of compound-DOC interactions, including: equilibrium dialysis, ultrafiltration, size-exclusion chromatography, reverse-phase separation, fluorescence quenching, and water solubility enhancement. For example, the difference between fluridone adsorption isotherms for Ca^{2+}- and DOC-saturated clay obtained in the presence and absence of DOC might reflect the formation of soluble DOC-bound fluridone (Figures 7.24 and 7.25).

Organic compound–DOC associations in soil solutions are typically described by linear distribution coefficients, conceptually similar to those used to describe compound partitioning between solid-phase SOC and the solution phase. Linear partitioning implies that the DOC does not have a maximum capacity for the compound, but only has constant interaction intensity denoted by the K_{DOC} value:

$$K_{DOC} = \frac{q_{DOC}}{c_{f,eq}} \qquad (7.71)$$

where K_{DOC} is the distribution coefficient (L kg^{-1}), q_{DOC} is the concentration of a compound partitioned with DOC in an equilibrium solution (μg kg^{-1}), and $c_{f,eq}$ is the equilibrium concentration of the free, unbound compound (μg L^{-1}). The concentration of the DOC-bound compound may also be expressed as:

$$c_{DOC} = q_{DOC}[DOC] \tag{7.72}$$

where c_{DOC} is the volume-based concentration of the DOC-bound compound (μg L^{-1}) and [DOC] is the kg L^{-1} concentration of DOC. Combining Equation 7.71 and Equation 7.72 yields an expression for c_{DOC} as a function of $c_{f,eq}$:

$$c_{DOC} = c_{f,eq}K_{DOC}[DOC] \tag{7.73}$$

The mass balance expression for the organic compound in solution is:

$$c_T = c_{DOC} + c_{f,eq} \tag{7.74}$$

where c_T is the total soluble concentration of the compound, in μg L^{-1}. Substituting Equation 7.73 into Equation 7.74 for c_{DOC} and rearranging yields an expression for $c_{f,eq}$ as a function of the total compound concentration (a directly measurable quantity), the concentration of DOC (also directly measurable), and K_{DOC}:

$$c_{f,eq} = \frac{c_T}{(1 + K_{DOC}[DOC])} \tag{7.75}$$

The adsorption of an organic compound by soil or sediment in the absence of appreciable quantities of DOC (negligible complexation of the organic compound) may be conveniently described using a general adsorption isotherm model:

$$q = Ac_{f,eq}^N \tag{7.76}$$

where A is the partition coefficient (K_p) if $N = 1$, or the Freundlich constant (K_F) if N lies between 0 and 1. Equation 7.76 describes adsorption relative to the free concentration of an organic compound of interest. Incorporating Equation 7.75 into the general adsorption isotherm model (Equation 7.76) yields:

$$q = A\left[\frac{c_T}{(1 + K_{DOC}[DOC])}\right]^N \tag{7.77}$$

or

$$q = A^{app}c_T^N \tag{7.78}$$

where A^{app} is an apparent soil adsorption constant that describes compound adsorption in a system containing sufficient DOC to impact adsorption. Comparison of Equations 7.77 and 7.78 leads to the following equality:

$$A^{app} = \frac{A}{(1 + K_{DOC}[DOC])^N} \tag{7.79}$$

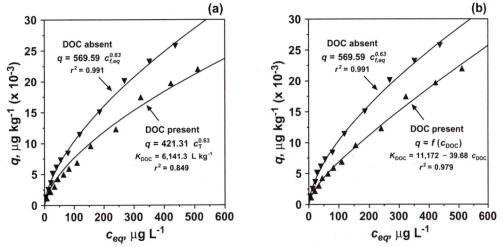

FIGURE 7.26 The adsorption of fluridone by DOC- and Ca^{2+}-saturated STx-1 montmorillonite in the presence and absence of 100 mg L^{-1} DOC. The experimental design is described in Figure 7.25. The adsorption of fluridone (a) in the presence of DOC is described by the solid line that represents Equation 7.80 and assumes that the fluridone and DOC interact with constant intensity, with $K_{DOC} = 6{,}141$ L kg^{-1}. Fluridone adsorption (b) in the presence of DOC is described by the solid line that represents Equation 7.86 and assumes that the fluridone and DOC interact with variable intensity and constant capacity, where K_{DOC} is a function of the concentration of DOC-bound fluridone concentration ($K_{DOC} = 11{,}172 - 39.68\ c_{DOC}$).

If it is assumed that the decreased adsorption of fluridone (Figure 7.25) by DOC-saturated montmorillonite is due to aqueous fluridone-DOC partitioning, Equation 7.77 can be employed to establish K_{DOC}. Employing Equations 7.69 and 7.77 leads to:

$$q = 569.6 \left[\frac{c_T}{(1 + K_{DOC}[1 \times 10^{-4}\ \text{kg DOC L}^{-1}])} \right]^{0.63} \tag{7.80}$$

where 569.6 is A, and 0.63 is N (from Equation 7.69). Nonlinear least-squares regression of Equation 7.80, using the fluridone adsorption isotherm data collected in the presence of DOC, leads to $K_{DOC} = 6141.3$ L kg^{-1} and $A^{app} = 421.3$ (Figure 7.26a).

Constant partitioning has been the assumption of choice to characterize DOC-organic compound interactions in the aqueous phase, particularly in assessing the impact of DOC on adsorption isotherms. In soils, however, DOC represents a minor portion of the total organic carbon content and the hydrophobic character of DOC is much less than that of solid-phase organic carbon. Due to the hydrophilic character of DOC (fulvates), the *intensity* of the hydrophobic compound-DOC association is small compared to the intensity of partitioning by SOC (humates). Furthermore, the low concentrations of DOC may lead to a limited *capacity* to bind organic compounds. Thus, linear partitioning behavior for the herbicide-DOC interaction, as described above (Equation 7.71), is not anticipated over wide ranges of compound concentrations and fixed DOC concentrations. Instead, the aqueous association of an organic compound with DOC can be described by the Langmuir model (Equation 7.31), which incorporates both intensity and capacity parameters:

$$q_{DOC} = \frac{^m q_{DOC} c_{f,eq}}{(K_m + c_{f,eq})} \tag{7.81}$$

where q_{DOC} and $c_{f,eq}$ are previously defined, $^mq_{\text{DOC}}$ is the maximum amount of compound that can be retained by a unit mass of DOC, and K_m is the value of $c_{f,eq}$ at which $q_{\text{DOC}} = \frac{1}{2}\,^mq_{\text{DOC}}$. Furthermore,

$$K_{\text{DOC}} = B\,^mq_{\text{DOC}} - Bq_{\text{DOC}} \tag{7.82a}$$

and

$$K_{\text{DOC}} = B\frac{^mc_{\text{DOC}}}{[\text{DOC}]} - B\frac{c_{\text{DOC}}}{[\text{DOC}]} \tag{7.82b}$$

where $B = K_m^{-1}$ is the Langmuir adsorption constant, and $^mc_{\text{DOC}}$ is the volume-based maximum concentration of a compound that may be bound by DOC in the aqueous phase. Applying the mass balance expression (Equation 7.74) and solving for $c_{f,eq}$ yields:

$$c_{f,eq} = \frac{1}{2}\{[(K_m + \,^mc_{\text{DOC}} - c_{\text{T}})^2 + 4K_mc_{\text{T}}]^{\frac{1}{2}} - (K_m + \,^mc_{\text{DOC}} - c_{\text{T}})\} \tag{7.83}$$

Again, using the data for fluridone adsorption on DOC-saturated clay (Figure 7.25; $A = 569.6$ L kg^{-1} and $N = 0.63$) and Equation 7.83, the adsorption isotherm in the presence of 0.0001 kg L^{-1} DOC is:

$$q = 569.6\{\frac{1}{2}([(K_m + \,^mc_{\text{DOC}} - c_{\text{T}})^2 + 4K_mc_{\text{T}}]^{\frac{1}{2}} - (K_m + \,^mc_{\text{DOC}} - c_{\text{T}}))\}^{0.63} \tag{7.84}$$

Nonlinear regression analysis results in $K_m = 251.99$ µg L^{-1}, $^mc_{\text{DOC}} = 281.53$ µg L^{-1}, $K_{\text{DOC}} = 11{,}172 - 39.68\,c_{\text{DOC}}$ (from Equation 7.82b), and

$$q_{\text{DOC}} = \frac{281.53c_{f,eq}}{(251.99 + c_{f,eq})} \tag{7.85}$$

The Freundlich model that best fits the soil adsorption data, assuming Langmuir fluridone-DOC binding behavior is (Figure 7.26b):

$$q = 569.6\{\frac{1}{2}([(533.52 - c_{\text{T}})^2 + 1007.96]^{\frac{1}{2}} - (533.52 - c_{\text{T}}))\}^{0.63} \tag{7.86}$$

Equation 7.86 appears to be a better predictor of q relative to that offered by Equation 7.80 for explaining fluridone adsorption in the presence of DOC, although both models offer statistically significant predictions (r^2 values of 0.979 and 0.849, respectively). This observation does not necessarily confirm the hypothesis that DOC has a limited capacity to bind the fluridone, nor does it imply that the aqueous binding of DOC is the mechanism responsible for reduced adsorption. A mathematical model that employs two independent variable regression parameters (such as Equation 7.86 with K_m and $^mc_{\text{DOC}}$) will always provide improved predictability over a model with only one regression parameter (such as Equation 7.80 with K_{DOC}).

7.3.3 ADSORPTION KINETICS AND DESORPTION HYSTERESIS

Implicit to the development of adsorption isotherms is the condition of equilibrium. Adsorption is characterized as having two distinct kinetic phases (biphasic kinetics): a rapid and reversible initial stage followed by a much slower, nonreversible stage. For organic compounds, the rapid phase is characterized by the retention of the compound in a labile form that is easily desorbed. For metals and inorganic ligands, this phase is also characterized by easily desorbed, exchangeable forms. For all compound types, inorganic and organic, the rapid phase is thought to include the retention of compounds by easily accessible sites on macroparticles (agglomerates of smaller inorganic particles that are held together by various inorganic and organic cementing agents), and on the edges of layer silicates. The slower reaction phase that follows the initial phase generally involves the entrapment of organic compounds in a nonlabile form that is difficult to desorb. Also occurring during the slow phase is the formation of inner-sphere surface complexes and bonds that have

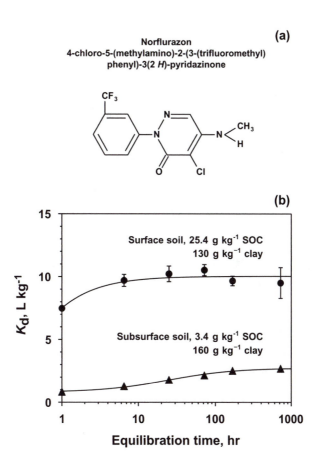

FIGURE 7.27 The structure of norflurazon (a), and the influence of equilibration time and soil organic carbon (SOC) and clay content (b) on norflurazon K_d values (c_{in} = 170 µg L^{-1}) and adsorption isotherms (c). The adsorption data were obtained by equilibrating 10-g samples of Lexington silt loam (Ultic Hapludalf) surface (0 to 8 cm) or subsurface (30 to 45 cm) soil, and 20 mL of 10 mM CaCl$_2$ solutions containing varied concentrations of the herbicide. The suspensions were equilibrated at 20 to 22°C. The error bars in (b) represent the range of K_d values obtained from replicate systems. When error bars are absent, the variation in K_d is smaller than the dimensions of the data point. The lines through the data in (c) represent Freundlich adsorption models obtained through nonlinear least-squares regression analysis. (*continued*)

covalent character. In this phase, diffusion of compounds into the micropores of the SOC matrix and inorganic soil components is also a rate-limiting step.

Figure 7.27 illustrates the sorption kinetics of the nonionic herbicide norflurazon. This compound is relatively hydrophilic, S_w = 112 µmol L^{-1} (34 mg L^{-1}), K_{ow} = 282 (log K_{ow} = 2.45), and K_{oc} = 353 (log K_{oc} = 2.55). In Figure 7.27b, the variation in the distribution coefficient, K_d, as a function of reaction time is shown for 0 to 8 cm surface (25.4 g kg^{-1} SOC, 130 g kg^{-1} clay) and a 30- to 40-cm subsurface (3.4 g kg^{-1} SOC, 160 g kg^{-1} clay) sample of a loess-derived soil. The diagram illustrates several points. First, the extent of compound sorption is related to the SOC content. The K_d for norflurazon sorption by the surface soil is approximately 10.2 L kg^{-1}; whereas, the K_d for the subsurface soil samples is approximately 2.9 L kg^{-1}, at equilibrium. This type of response, increasing sorption with increasing SOC, is a common characteristic of organic compound sorption. Second, sorption equilibrium is essentially attained after a 6-h equilibration period in the surface soil and the kinetics do not appear to be biphasic. However, even after a 200-h equilibration, the norflurazon K_D values had not stabilized for the relatively low SOC subsurface soil. Apparently, norflurazon rapidly partitions and equilibrates with SOC, while in the low SOC subsurface soil the

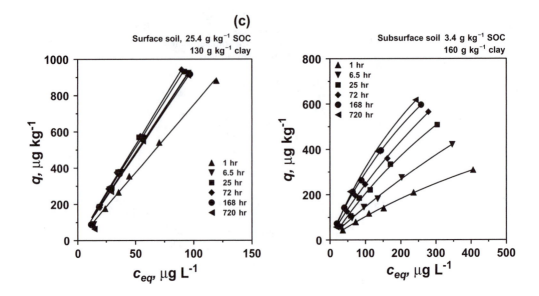

FIGURE 7.27 (*continued*).

diffusion of the organic compound to adsorption sites (e.g., clay interlayers) is a rate-limiting step. The impact on sorption kinetics on the norflurazon sorption isotherms is also evident in Figure 7.27c. Note that the adsorption of norflurazon by the surface soil is characterized by partition (C-type) isotherms; whereas, adsorption by subsurface soil is characterized by Langmuir (L-type) isotherms, illustrating that sorption processes other than partitioning apply in low SOC soils. Other examples of how sorption K_d values are impacted by the slow kinetics of sorption include the following:

- The K_d values for the herbicides atrazine and metolachlor in soils sampled 1 to 5 years after treatment are 2.3 to 42 times greater than the Kd values determined after a 24-h equilibration.
- The K_d value for the fumigant ethylene dibromide is 1.49 L kg^{-1} after a 24-h equilibration period and 230 L kg^{-1} after an 11-month equilibration.
- The K_d value for the industrial solvent trichloroethane is 6.5 L kg^{-1} after a 24-h equilibration and 190 L kg^{-1} after a 20-d equilibration.

As was previously indicated, the entrapment of organic compounds, the diffusion of dissolved ions to adsorption sites within micropores and clay interlayers, and the formation of covalent bonds can result in the apparent irreversible retention of compounds and ions. These processes lead to a commonly observed characteristic of sorption isotherms: desorption hysteresis. Defined, desorption hysteresis is the apparent increase in the sorption constant (e.g., K_d, K_L, K_F, or K_p) when equilibrium is approached from the desorption direction. The nonhysteretic behavior of the herbicide norflurazon, and the hysteretic behavior of the herbicide fluometuron are illustrated in Figure 7.28. Under equilibrium conditions, the dilution of a solution with respect to an adsorbate, and the reestablishment of equilibrium should result in a system whose solution (c_{eq}) and sorption (q) composition lies on the sorption isotherm. Desorption of norflurazon is nonhysteretic because desorption c_{eq} and q data are consistent with the sorption isotherm (Figure 7.28a). However, fluometuron displays hysteretic desorption behavior, as the desorption data points do not fall on the sorption isotherm (Figure 7.28b). As the solution is diluted and reequilibrated, the sorbed fluometuron is not desorbed and the measured K_d values for each desorption point are elevated relative to the expected K_d value of 2.90 (Table 7.7). However, the dilution of the equilibrium solution does lead to some fluometuron

TABLE 7.7
The Hysteretic Behavior of the Herbicide Fluometuron in a Lexington Silt Loam (Ultic Hapludalf), as Qualified by Measured Distribution Coefficients (K_d Values)

System[a]	c_{eq}, µg L^{-1}	Measured K_d, L kg^{-1}	K_{dH}, L kg^{-1}[b]
Adsorption	124.8	2.90	—
Desorption step 1	114.6	3.14	3.22
Desorption step 2	103.8	3.46	3.58
Desorption step 3	87.9	4.03	4.48
Desorption step 4	76.2	4.57	5.60

[a] 10 g soil samples were equilibrated with 10 mL volumes of a 0.01 M CaCl$_2$ solution initially containing approximately 5 µg fluometuron. All systems, adsorption and desorption, were equilibrated for 48 h at 25°C. The results represent the means of triplicate systems. Desorption steps were performed in a sequential manner. Desorption steps 1 and 2 were initiated by removing 10% of the equilibrating solution from adsorption system (for desorption step 1) or the previous desorption step (for desorption step 2) and replacing with an equal volume of 0.01 M CaCl$_2$ solution. Desorption steps 3 and 4 were initiated by removing 20% of the equilibrating solution from the previous desorption steps, and replacing with an equal volume of 0.01 M CaCl$_2$ solution.
[b] K_{dH} is the distribution coefficient that represents complete hysteresis; desorption does not occur. These values are computed by assuming q, the adsorbed concentration of fluometuron at adsorption equilibrium, remains constant throughout the four desorption steps (only soluble fluometuron concentrations are impacted by the dilutions).

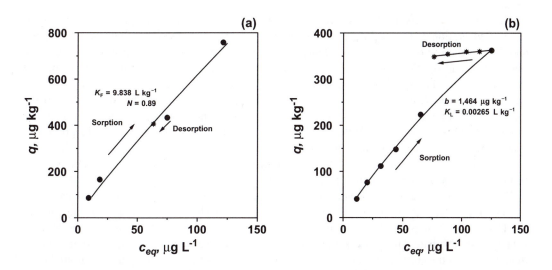

FIGURE 7.28 Nonhysteretic desorption (a) of an adsorbed substance is indicated when desorption data (closed stars) lie on the adsorption isotherm (closed circles), as illustrated for norflurazon. Desorption hysteresis (b) is evidenced when the desorption isotherm does not mimic the adsorption isotherm, as shown for fluometuron. The adsorption data for diagrams (a) and (b) were generated by equilibrating 10-g samples of a Lexington silt loam (Ultic Hapludalf) surface (0 to 8 cm depth) soil with 20 mL of 10 mM CaCl$_2$ containing varied concentrations of herbicide, and equilibrated for 72 h (a) or 48 h (b). The desorption data point for norflurazon (a) was obtained by diluting replicate equilibrium adsorption solutions by removing 5 mL and reequilibrating for 14 d. The fluometuron desorption isotherm (b) was generated by successive dilution (1 to 2 mL) equilibration (48 h) steps that were started from the equilibrated adsorption systems.

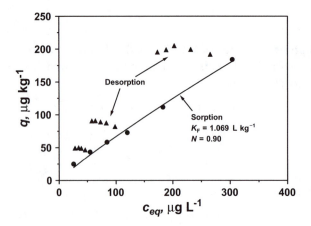

FIGURE 7.29 Adsorption and desorption isotherms for fluometuron in contact with a Lexington silt loam (Ultic Hapludalf) subsurface soil (30 to 45 cm depth). The experimental design is described in Figure 7.28, with an adsorption equilibration time of 24 h. The continued removal of the herbicide from solution during the desorption steps suggests nonequilibrium of the adsorption process.

desorption, as the measured K_d values for each desorption step are less than those expected for complete irreversibility, indicated by K_{dH}. In Table 7.7, K_{dH} is the distribution coefficient that represents complete hysteresis; desorption does not occur. The K_{dH} values for each desorption step are computed by assuming q, the adsorbed concentration of fluometuron at adsorption equilibrium, remains constant throughout the four desorption steps, and that only dilution of soluble fluometuron occurs.

The apparent irreversibility of adsorption processes is a confounding complication in the application of adsorption constants for predicting compound fate and behavior. Equilibrium is a defined condition for the use of adsorption constants. Accordingly, the distribution of a substance between the adsorbed and solution phases should respond to any perturbation in the chemistry of the soil environment, such that equilibrium is maintained. The hysteretic behavior that is commonly observed in laboratory desorption studies may arise from: (1) nonequilibrium of the adsorption process (continued adsorption of a substance even after the equilibrating solution is diluted; described in Figure 7.29); (2) nonequilibrium of desorption processes; (3) experimental artifacts or inappropriate experimental design (an experimental design that does not allow for the characterization of a reversible chemical process); and (4) true irreversibility of the adsorption process. Probably the two most common reasons for desorption hysteresis are engendered in items (2) and (3).

Desorption is typically examined as part of an adsorption isotherm study. Following adsorption equilibration, an aliquot of the solution phase is removed for chemical analysis (determination of c_{eq}). The aliquot is replaced by a solution having identical ionic character to that of the aliquot, excluding the adsorbate. This process dilutes the adsorbate concentration in the equilibrium solution, while preserving the controlling chemical characteristics (e.g., pH, ionic strength, background electrolyte). The system is then reequilibrated, and the solution analyzed to establish the new, equilibrium distribution of the adsorbate between the solid and solution phases. Of course, the key to establishing the reversibility or irreversibility of the adsorption process lies in the expression: equilibrium distribution. If equilibrium is not established, desorption behavior will appear to be hysteretic. Desorption K_d values illustrate the reversibility of the fluometuron adsorption reaction (Figure 7.30). In a surface sample of a Lexington silt loam (Ultic Hapludalf), which contains a relatively high SOC content, the desorption process requires approximately 1 week to achieve equilibrium and return to the adsorption K_d value. However, in the low SOC subsurface soil sample, the desorption reaction requires approximately 1 month to achieve equilibrium. The rate of the fluometuron desorption

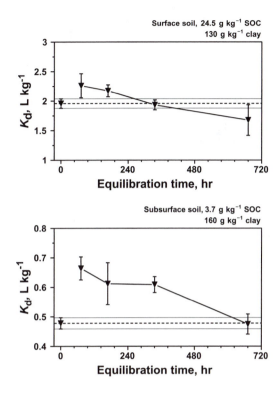

FIGURE 7.30 Fluometuron desorption K_d values as a function of equilibration time and soil organic carbon (SOC) and clay content. The dashed line represents the mean adsorption K_d; the solid lines and error bars represent the 95% confidence intervals. (Modified from Suba, J.D. and M.E. Essington. Adsorption of fluometuron and norflurazon: effect of tillage and dissolved organic carbon. *Soil Sci.* 164:145–155, 1999.)

reaction is apparently a function of SOC content, a soil characteristic that also influences the rate of substance adsorption (shown for norflurazon in Figure 7.27b). Desorption K_d values for norflurazon indicate that the adsorption process is either nonreversible (truly hysteretic), or that equilibrium is not achieved within the 1-month reaction period, irrespective of SOC content (Figure 7.31). These conclusions are evidenced by the fact that the adsorption K_d value is not reestablished.

Adsorption-desorption nonhysteresis and adsorption reaction reversibility are terms that are synonymous in common practice; they are used interchangeably. In fact, however, the determination of reaction reversibility (or irreversibility) requires a much higher standard of proof than does nonhysteresis (or hysteresis). As described above, nonhysteresis of desorption is established by examining the equilibrium between sorbed and solution phases that results when an initial adsorption equilibrium system is perturbed (diluted). There is no standard for the degree of perturbation that must be imposed on an equilibrium system to initiate desorption, as is evidenced by the numerous techniques employed in the scientific literature. However, for the assessment of reversibility there exists a very strict standard. If a system is in an equilibrium state, and a small change occurs in the activity of the adsorbate (in physical chemistry this is called an infinitesimally small change), a natural process will result in a new equilibrium state. If the small change in the adsorbate concentration is reversed back to its original value, the system may revert back to its original equilibrium state. If the original state of the system is attained when equilibrium is reestablished, a reversible process has occurred. It is a requirement that during reversible processes the system must always proceed through states of equilibrium. By definition, in order for a reaction in a soil to undergo reversible processes, the soil must always be close to equilibrium. Further, this requires

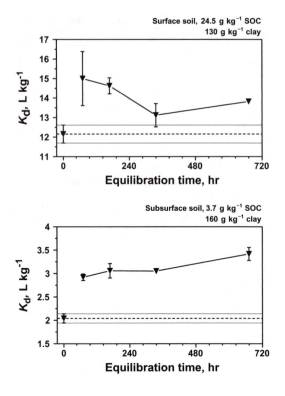

FIGURE 7.31 Norflurazon desorption K_d values as a function of equilibration time and soil organic carbon (SOC) and clay content (from Suba and Essington, 1999). The dashed line represents the mean adsorption K_d; the solid lines and error bars represent the 95% confidence intervals. (Modified from Suba, J.D. and M.E. Essington. Adsorption of fluometuron and norflurazon: effect of tillage and dissolved organic carbon. *Soil Sci.* 164:145–155, 1999.)

an infinitesimal process, as only an infinitesimal process will maintain equilibrium (maximum entropy). It is important to recognize that proceeding through states of equilibrium may not produce reversibility, only that proceeding through states of equilibrium is a necessary condition for reversibility to occur. Therefore, a reversible process cannot result if an equilibrium system is perturbed to a large degree; only a new equilibrium state will be attained.

As the preceding discussion implies, a reversible adsorption process can only be expected to occur when a system at equilibrium is perturbed by a small amount. Adherence to this strict standard is required when investigating the desorption behavior of a substance. In a practical sense, a small change must be large enough to obtain an analytically detectable change in the concentration of the substance of interest. In many instances found in the literature, the extent of dilution in desorption studies is excessive, approaching an extraction process. In such systems, disequilibrium has been imposed and reversibility cannot occur, and adsorption hysteresis is the conclusion. Even when equilibrated adsorption systems are diluted or otherwise modified to a minor degree (in an effort to maintain conditions that are near equilibrium), no effort is made to reestablish the initial adsorption equilibrium. Thus, true reversibility of the adsorption reaction is not established.

7.4 SPECIFIC RETENTION OF METALS AND LIGANDS

The type of chemical bond formed at the solid-solution interface defines the mechanism responsible for the retention of metals and ligands. Metal cations that bond directly to surface oxygen atoms are said to be specifically adsorbed, forming inner-sphere surface complexes. The metal cations that are commonly associated with inner-sphere surface complex formation are the divalent first-row transition metals (e.g., Co^{2+}, Ni^{2+}, Cu^{2+}, and Zn^{2+}) and the heavy metals (e.g., Cd^{2+}, Pb^{2+}, and Hg^{2+}). As a group, these metal cations are often described as trace elements, as they are found in trace concentrations in uncontaminated environments. The monodentate and bidentate specific binding mechanisms that have been identified for these metal cations are illustrated in Figure 7.5b.

The specific retention of trace metal cations is distinguished from nonspecific adsorption (cation exchange) through a number of key experimental observations. Nonspecific, or outer-sphere adsorption is a direct result of negative charge formation on the mineral surface. For example, the adsorption of an electrolyte cation, such as Na^+, by a constant potential mineral surface, such as gibbsite, will increase with increasing pH in a manner consistent with the increasing concentration of negatively charged surface sites ($[\equiv AlOH^{-0.5}]$, Figure.7.32). Conversely, the adsorption of an electrolyte anion, such as NO_3^-, is a function of positively charged surface site concentration ($[\equiv AlOH_2^{+0.5}]$). The outer-sphere retention of a substance is also easily reversible and decreases with increasing ionic strength. The adsorption characteristics of p-toluidine, an aromatic amine ($Ar-NH_3^+$), illustrate that the increased ionic strength, from 0.03 M to 0.3 M KCl, decreases the adsorption of the amine by montmorillonite (Figure 7.33). This occurs because both p-toluidine and K^+ are retained on montmorillonite by the outer-sphere adsorption mechanism; they are both

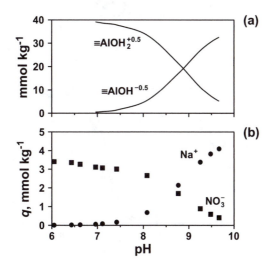

FIGURE 7.32 (a) The concentrations of positively and negatively charged surface sites on gibbsite and (b) the corresponding adsorption behavior of Na^+ and NO_3^- as a function of solution pH.

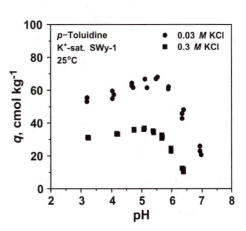

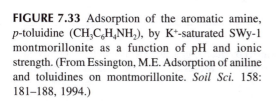

FIGURE 7.33 Adsorption of the aromatic amine, p-toluidine ($CH_3C_6H_4NH_2$), by K^+-saturated SWy-1 montmorillonite as a function of pH and ionic strength. (From Essington, M.E. Adsorption of aniline and toluidines on montmorillonite. *Soil Sci.* 158: 181–188, 1994.)

exchangeable ions. The replacement of the amine on the surface by K^+ that results from increasing the ionic strength of the solution (increasing the aqueous concentration of K^+) is a mass action response and an intrinsic characteristic of cation exchange reactions (discussed in Chapter 8).

In contrast, inner-sphere metal complexation by a constant potential mineral surface is generally not tied to the formation of negative surface charge. Instead, metal adsorption increases rapidly from a minimum to a maximum value in a narrow pH range (approximately 2 pH units or less), as the adsorption envelopes for Pb, Ni, and Hg indicate (Figure 7.34). Metal adsorption reactions

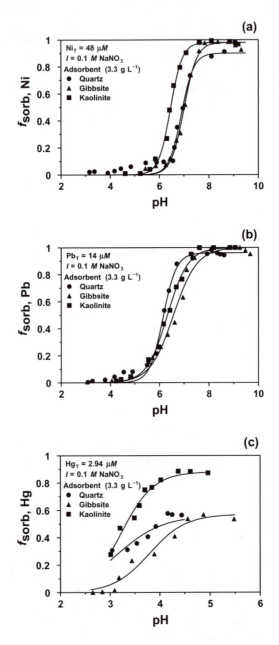

FIGURE 7.34 Adsorption of (a) Ni, (b) Pb, and (c) Hg(II) by quartz, gibbsite, and kaolinite as a function of pH (from Sarkar, 1997; Sarkar et al. 1999, 2000). Adsorption is expressed as the fraction of the total metal concentration, f_{sorb}.

tend to be irreversible (highly hysteretic), suggesting that a high degree of covalent character is associated with the surface complexes. The desorption behavior of Ni, illustrated in Figure 7.35, is indicative of irreversible adsorption behavior. Nickel adsorption by a Harvey soil (Ustic Haplocalcid) is described by the Freundlich model over an extensive (several orders of magnitude) range of surface coverage. When equilibrium solutions are diluted and reequilibrated, the surface concentration (q) of Ni does not decrease, even through several desorption (dilution) steps. Metal adsorption also tends to be unaffected by ionic strength, indicating a high degree of specific bonding (nonelectrostatic) in the surface complexes. The influence of ionic strength on the adsorption of Hg(II) by gibbsite from 0.01 M, 0.1 M, and 0.5 M NaNO$_3$ solutions is illustrated in Figure 7.36. In this study, the initial concentration of Hg(II) was 2.94×10^{-6} M. Sodium concentrations in the background electrolyte, which range from approximately 4 to 5 orders of magnitude greater than

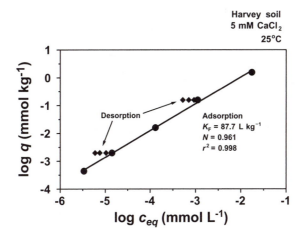

FIGURE 7.35 Adsorption isotherm characterizing the adsorption and desorption of nickel by surface (0 to 15 cm) samples of a Harvey soil (coarse-loamy, mixed, superactive, mesic Ustic Haplocalcid) from 5 mM CaCl$_2$ solutions. The adsorption data were obtained by equilibrating 11-g samples of soil and 11 mL of 5 mM CaCl$_2$ containing varied concentrations of NiCl$_2$ for 24 h at 20 to 25°C. Nickel desorption was evaluated by removing a 5-mL volume of the equilibrating solution and replacing with an equal volume of Ni-free 5 mM CaCl$_2$, and reequilibrating for 1 h (preliminary evaluations showed that a 1 h desorption equilibration achieved the same state as a 3 month desorption equilibration). This desorption sequence was repeated two additional times to generate the desorption isotherms. (Modified from Bowman, R.S., M.E. Essington, and G.A. O'Connor. Soil sorption of nickel: influence of solution composition. *Soil Sci. Soc. Am. J.* 45:860–865, 1981.)

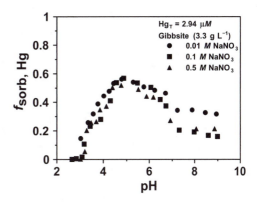

FIGURE 7.36 Adsorption of Hg(II), expressed as a fractional amount of the total Hg(II) concentration (f_{sorb}), by gibbsite as a function of pH and ionic strength. (Modified from Sarkar, D., M.E. Essington, and K.C. Misra. Adsorption of mercury(II) by variable charge surfaces of quartz and gibbsite. *Soil Sci. Soc. Am. J.* 63:1626–1636, 1999.)

TABLE 7.8
Equation 7.46 Parameters for Metal Adsorption Illustrated in Figure 7.34[a]

Surface	Hg		Pb		Ni	
	b	pH_{50}	b	pH_{50}	b	pH_{50}
Quartz [α-SiO$_2$]	2.767	3.80	3.653	6.15	4.043	6.90
Gibbsite [Al(OH)$_3$]	2.202	3.15	2.101	6.57	3.893	7.01
Kaolinite [Al$_2$Si$_2$O$_5$(OH)$_4$]	3.106	3.24	2.367	6.33	4.145	6.46
$pK_{a,1}$[b]		3.40		7.71		9.86

[a] pH_{50} is the pH at which 50% of the metal is adsorbed, and b indicates the steepness of the adsorption edge at pH_{50}.

[b] $pK_{a,1}$ is the negative common logarithm of the equilibrium constant for the metal hydrolysis reaction: $M^{2+}(aq) + H_2O(l) \rightarrow MOH^+(aq) + H^+(aq)$.

the initial Hg(II) concentration, did not significantly influence Hg(II) adsorption. These results can be compared with those in Figure 7.33, where p-toluidine adsorption is strongly influenced by K$^+$ concentrations that are up to approximately 1 order of magnitude greater than that of the organic amine.

The pH region where metal adsorption increases is termed the adsorption edge. The experimental data in Figure 7.34 are described mathematically by the equation:

$$f_{sorb} = \frac{f_{max}}{\{1 + \exp[-b(pH - pH_{50})]\}} \tag{7.87}$$

where f_{sorb} is the fraction of the total metal concentration that is adsorbed in the system, f_{max} is the adsorption maximum (as a fraction of the total metal), pH_{50} is the pH at which 50% of f_{max} is obtained, and b indicates the steepness of the adsorption edge at pH_{50}. The pH_{50} values as a function of selected metals and surfaces are presented in Table 7.8. The surface type does not appear to influence Hg, Pb, or Ni adsorption, even though the pH_{pzc} varies with surface (2.0 for SiO$_2$, 4.7 for kaolinite, and 8.8 for gibbsite). In general, the pH_{50} values of divalent metal cations are positively correlated with the first hydrolysis constant of the metal (irrespective of surface type), essentially indicating that those metals that are more easily hydrolyzed (lower pK_a values) are more strongly adsorbed. Further, pH_{50} is usually lower than pK_a, suggesting that the surface may actually promote metal hydrolysis. These observations, pH_{50} correlated to the pK_a and usually lower than the pK_a, are seen to some degree in the adsorption data presented in Figure 7.34 and Table 7.8. Surface-enhanced metal hydrolysis is described by the reaction:

$$\equiv SO'H^0 + M^{2+} + H_2O \rightarrow \equiv SO'MOH^0 + 2H^+ \tag{7.88}$$

It has been suggested that this specific adsorption process is analogous to metal cation hydrolysis, as the coordination of a metal cation by surface oxygen is similar to the formation of aqueous complexes with OH$^-$:

$$HO'H^0 + M^{2+} + H_2O \rightarrow HO'MOH^0 + 2H^+ \tag{7.89}$$

The proton exchange process is confirmed by the observation that protons are released to solution during metal adsorption. Experimental evidence indicates that the number of protons released per metal ion adsorbed usually ranges between 1 and 2, with an average stoichiometry of >1.5. If it is assumed that some of the displaced protons are readsorbed, or otherwise neutralized via aqueous complexation reactions (and thus, not measured), the proton stoichiometry is approximately 2 and may be explained by surface hydrolysis (Equation 7.88) or a bidentate surface complexation process:

$$\begin{aligned} \equiv SOH^0 \\ \equiv SOH^0 \end{aligned} + M^{m+} \rightarrow \begin{aligned} \equiv SO \\ \equiv SO \end{aligned} \rangle M^{m-2} + 2H^+ \tag{7.90}$$

Specific anion adsorption occurs through the ligand exchange process, with H_2O as the displaced ligand:

$$\equiv SOH_2^+ + L^- \rightarrow \equiv SL^0 + H_2O \tag{7.91a}$$

$$\equiv SO'H_2^+ + MO_x^{l-} \rightarrow \equiv SOMO_{x-1}^{1-l} + H_2O' \tag{7.91b}$$

or with OH^- as the displaced ligand:

$$\equiv SOH^0 + L^- \rightarrow \equiv SL^0 + OH^- \tag{7.92a}$$

$$\equiv SO'H^0 + MO_x^{l-} \rightarrow \equiv SOMO_{x-1}^{1-l} + O'H^- \tag{7.92b}$$

Ligands that participate in ligand exchange reactions are the weak acid anions, many of which are the partially and completely dissociated oxyanions (such as $B(OH)_4^-$; $H_4SiO_4^0$; $H_2PO_4^-$ and HPO_4^{2-}; $H_2AsO_4^-$ and $HAsO_4^{2-}$; and $HCrO_4^-$ and CrO_4^{2-}). Ligand exchange is also specific to constant potential surfaces, such as hydrous metal oxides and the edges of layer silicates. The specific retention of ligands is distinguished from nonspecific adsorption (anion exchange) through a number of key experimental observations. As shown in Figure 7.32 for NO_3^- adsorption, nonspecific retention of anions is a direct result of the development of positive charge on a mineral surface. Ligand exchange, however, occurs over a wide range of solution pH values and the reactions tend to be irreversible (hysteretic), suggesting that a high degree of covalent character is associated with the surface complexes.

Representative adsorption envelopes for selected weak acid anions are illustrated in Figure 7.37. For ligands that are derived from monoprotic acids (e.g., $B(OH)_4^-$), and ligands that are derived from polyprotic acids but have high pK_a values (e.g., $H_4SiO_4^0 \rightarrow H_3SiO_4^- + H^+$, $pK_{a1} = 9.86$), the adsorption gradually increases with increasing pH, reaching a maximum value when solution pH approximately equals the pK_a. Adsorption rapidly decreases with further increases in pH due to competition with OH^- for surface functional groups (this process is identical to the loss of protonated surface functional groups and exchangeable ligand —H_2O and OH^- — on the surface). The adsorption of polyprotic acid anions (e.g., phosphate) is at a maximum at low pH, and decreases with increasing pH. When pH is approximately equal to the pK_a of a deprotonation reaction (such as, $H_2PO_4^- \rightarrow HPO_4^{2-} + H^+$, $pK_{a2} = 7.20$), there is an inflection in the adsorption envelope such that decreasing adsorption with increasing pH is more pronounced.

The adsorption envelope that is observed for ligands is a combined result of surface chemistry (concentration of $\equiv SOH_2^+$) and acid strength (pK_a value). Weak acid dissociation and the formation of the anion appear to be a necessary step in the adsorption process, as the anion is the specifically retained species. For monoprotic, weak acid anions that have a relatively low pK_a (e.g., $HF^0 \rightarrow F^- + H^+$, $pK_a = 3.17$), there is a rapid rise in adsorption as pH increases to pK_a. This is a direct result of L^- formation. At pH values above the pK_a, the concentration of L^- does not change; however, there is a decrease in $\equiv SOH_2^+$ concentration (loss of the exchangeable ligand —H_2O— from the surface). Thus, adsorption decreases with further increases in pH.

For monoprotic, weak acid anions that have a relatively high pK_{a1} (e.g., $H_4SiO_4^0$), there is a slow rise in adsorption as pH increases to pK_{a1}. This also is a result of L^- formation. Adsorption is lower for these anions (relative to F^-), as there will be less exchangeable ligand present on the surface (principally H_2O) when pH = pK_{a1} and greater competition from the hydroxide ion for the adsorption

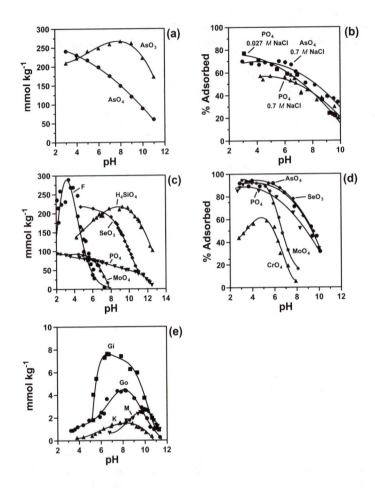

FIGURE 7.37 The adsorption envelopes of various anions: (a) 2.5 g L^{-1} goethite suspension containing 1 mM AsO$_4$ or 1 mM AsO$_3$ in a 0.01 M NaNO$_3$ solution (from Grafe et al., 2001); (b) 0.234 g L^{-1} goethite suspension containing 24.2 μM PO$_4$ or a 22.9 μM AsO$_4$ in a 0.7 M NaCl solution, or 25.0 μM PO$_4$ in a 0.027 M NaCl solution (from Gao and Mucci, 2001); (c) 4 g L^{-1} goethite suspension containing 4 mM F, 1 mM PO$_4$, 1.3 mM SeO$_3$, 1.2 mM H$_4$SiO$_4$, or 2 mM MoO$_4$ in a 0.1 M NaCl solution (from Hingston et al., 1972); (d) 6 g L^{-1} goethite suspension containing 0.3 mM PO$_4$, AsO$_4$, SeO$_3$, MoO$_4$, or CrO$_4$ in a 0.1 M NaClO$_4$ solution (from Okazaki et al., 1989); and (e) 25 g L^{-1} goethite (Go), 25 g L^{-1} gibbsite (Gi), 30 g L^{-1} KGa-2 kaolinite (K), or 30 g L^{-1} SWy-1 montmorillonite (M) suspensions containing 54 mM B(OH)$_3$ in a 0.1 M NaCl solution. (From Goldberg et al., 1993.)

sites. At pH values above the pK_{a1}, the concentration of L$^-$ reaches a maximum and does not change; however, there is a decrease in $\equiv SOH_2^+$ and $\equiv SOH^0$ concentrations, and an increase in $\equiv SO^-$. Thus, adsorption decreases abruptly with further increases in pH. For the polyprotic acids that are already deprotonated (i.e., strong acid pK_{a1} and weak acid pK_{a2}), the anionic form already exists at low pH values (e.g., H$_2$PO$_4^-$). Decreasing adsorption with increasing pH results from loss of the exchangeable ligand on the surface (decreasing concentrations of $\equiv SOH_2^+$ and $\equiv SOH^0$).

7.5 LIGAND EFFECTS ON METAL ADSORPTION

A ligand (other than hydroxide) can influence the adsorption of a metal by forming soluble complexes that may be more or less attractive to a surface than the free metal cation, or by modifying the surface characteristics of the mineral through specific, ligand exchange interactions. The potential

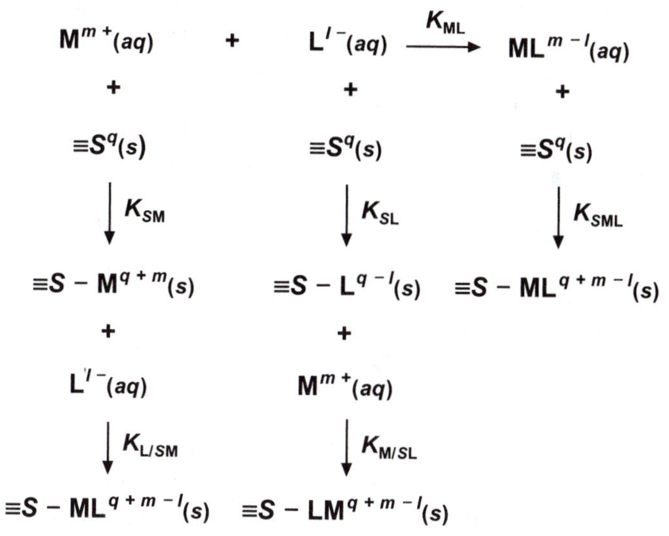

$$\mathbf{M}^{m+}(aq) \quad + \quad \mathbf{L}^{l-}(aq) \xrightarrow{\quad K_{\mathrm{ML}} \quad} \mathbf{ML}^{m-l}(aq)$$

$$+ \qquad\qquad + \qquad\qquad +$$

$$\equiv\mathbf{S}^q(s) \qquad\qquad \equiv\mathbf{S}^q(s) \qquad\qquad \equiv\mathbf{S}^q(s)$$

$$\Big\downarrow K_{\mathrm{SM}} \qquad\qquad \Big\downarrow K_{\mathrm{SL}} \qquad\qquad \Big\downarrow K_{\mathrm{SML}}$$

$$\equiv\mathbf{S}-\mathbf{M}^{q+m}(s) \qquad \equiv\mathbf{S}-\mathbf{L}^{q-l}(s) \quad \equiv\mathbf{S}-\mathbf{ML}^{q+m-l}(s)$$

$$+ \qquad\qquad\qquad +$$

$$\mathbf{L}^{l-}(aq) \qquad\qquad \mathbf{M}^{m+}(aq)$$

$$\Big\downarrow K_{\mathrm{L/SM}} \qquad\qquad \Big\downarrow K_{\mathrm{M/SL}}$$

$$\equiv\mathbf{S}-\mathbf{ML}^{q+m-l}(s) \quad \equiv\mathbf{S}-\mathbf{LM}^{q+m-l}(s)$$

FIGURE 7.38 The potential metal-ligand interactions that influence metal cation adsorption. $M^{m+}(aq)$ represents a metal cation, L^{l-} is a ligand, and $\equiv S^q(s)$ is a surface functional group. The depicted equilibrium constants (denoted by K and defined in Equations 7.93a through 7.93d) describe the aqueous complexation or surface complexation reactions of the metal species.

metal-ligand interactions that influence metal cation adsorption are diagrammed in Figure 7.38. The equilibrium constants depicted in Figure 7.38 represent the following reactions:

$$M^{m+}(aq) + L^{l-}(aq) \rightarrow ML^{m-l}(aq)$$

$$K_{\mathrm{ML}} = \frac{(ML^{m-l})}{(M^{m+})(L^{l-})} \tag{7.93a}$$

$$\equiv S^q(s) + ML^{m-l}(aq) \rightarrow \ \equiv S-ML^{q+m-l}(s)$$

$$K_{\mathrm{SML}} = \frac{(\equiv S-ML^{q+m-l})}{(ML^{m-l})(\equiv S^q)} \tag{7.93b}$$

$$\equiv S^q(s) + L^{l-}(aq) \rightarrow \ \equiv S-L^{q-l}(s)$$

$$K_{\mathrm{SL}} = \frac{(\equiv S-L^{q-l})}{(L^{l-})(\equiv S^q)} \tag{7.93c}$$

$$\equiv S-L^{q-l}(s) + M^{m+}(aq) \rightarrow \ \equiv S-LM^{q+m-l}(s)$$

$$K_{\mathrm{M/SL}} = \frac{(\equiv S-LM^{q+m-l})}{(M^{m+})(\equiv S-L^{q-l})} \tag{7.93d}$$

In general, ligand effects on metal adsorption can be classified into five categories (after Sposito, 1984):

1. The metal and the ligand have a high affinity for each other (K_{ML} is large), forming a soluble complex that has a high affinity for the surface (K_{SML} is large).
2. The ligand has a high affinity for the surface (K_{SL} is large) and the metal has a high affinity for the adsorbed ligand ($K_{\mathrm{M/SL}}$ is large).

3. The metal and the ligand have a high affinity for each other (K_{ML} is large), forming a soluble complex that has a low affinity for the surface (K_{SML} is small).
4. The ligand has a high affinity for the surface (K_{SL} is large) and the metal has a low affinity for the adsorbed ligand ($K_{M/SL}$ is small).
5. The ligand has a low affinity for both the metal (K_{ML} is small) and the surface (K_{SL} is small), and has little or no impact on metal adsorption.

The influence of several ligands (OH^-, NO_3^-, Cl^-, SO_4^{2-}, and HPO_4^{2-} and $H_2PO_4^-$) on the adsorption of Hg(II) by gibbsite is illustrated in Figure 7.39. In this example, Hg(II) adsorption from a $NaNO_3$ solution as a function of pH (OH^- concentration) is the reference system. Hydroxide forms strong complexes with Hg^{2+} (K_{ML} is large for both $HgOH^+$ and $Hg(OH)_2^0$ formation) and the hydrolysis products are specifically retained by the gibbsite surface (K_{SML} is large). Therefore, OH^- is a Category 1 ligand. Nitrate does not complex Hg^{2+} or any of the Hg^{2+} hydrolysis products [$HgOH^+$ or $Hg(OH)_2^0$] in solution. Further, NO_3^- is not specifically adsorbed and does not influence gibbsite surface functionality. Thus, NO_3^- is a Category 5 ligand. In the NO_3^- system, Hg(II) retention increases with increasing pH to an adsorption maximum of 57% of the total Hg(II) present at a pH value of approximately 5. Mercury(II) retention then decreases with increasing pH. The inclusion of Cl^- in the system drastically reduces the amount of Hg(II) adsorbed at pH 5 to approximately 1% of the total Hg(II) in the system. Similar to NO_3^-, Cl^- does not influence gibbsite surface functionality, but significantly influences the aqueous speciation of Hg(II). Chloride is a Category 3 ligand for Hg(II), as the chloride complexes $HgCl^+$ and $HgCl_2^0$ account for approximately 100% of the total soluble Hg(II) at pH values below 7 (K_{ML} is large). The gibbsite surface has little affinity for Hg(II)-chloride complexes, as Hg(II) adsorption in the presence of Cl^- is at a minimum (K_{SML} is small). The inclusion of SO_4 reduces the amount of Hg(II) adsorption to approximately 40% at pH 5. Similar to NO_3^- and Cl^-, SO_4^{2-} is primarily nonspecifically retained by the surface and would not be expected to significantly impact surface chemistry (although the specific interaction of SO_4^{2-} and constant potential surfaces is known to occur). Thus, SO_4^{2-} is a Category 3 ligand for Hg(II). The soluble $Hg(OH)SO_4^{2-}$ species, which is predicted to form in the SO_4 system, apparently has

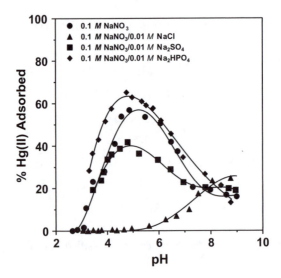

FIGURE 7.39 The influence of several ligands on the adsorption of Hg(II) (total concentration equals 2.94 μM) by gibbsite (3.3 g L^{-1}) as a function of pH in 10 mM $NaNO_3$. (From Sarkar, D. M.E. Essington, and K.C. Misra. Adsorption of mercury(II) by variable charge surfaces of quartz and gibbsite. *Soil Sci. Soc. Am. J.* 63:1626–1636, 1999.)

low affinity for the gibbsite surface. The PO_4 ligands (HPO_4^{2-} and $H_2PO_4^-$) are classified as Category 1 or 2 ligands for Hg(II). Mercury(II) retention is enhanced to approximately 65% at pH 5 when PO_4 is present in the system (relative to 57% when PO_4 is absent). This enhanced Hg(II) adsorption may be due to the formation of the soluble $Hg(OH)_2H_2PO_4^-$ and $Hg(OH)_2HPO_4^{2-}$ complexes which may have an affinity for the surface, or by the high affinity of Hg^{2+}, $HgOH^+$, or $Hg(OH)_2^0$ for adsorbed PO_4.

7.6 ORGANIC SURFACE FUNCTIONAL GROUPS AND ORGANIC MOLECULAR RETENTION MECHANISMS

The pH-dependency of soil surface charge characteristics is not limited to hydrous metal oxides or the edges of layer silicates. The charging characteristics of SOC play an important role in the pH dependency of the adsorption capacity of soil. Although SOC is structurally and chemically complex and structurally labile, surface reactivity due to SOC can be ascribed to a relatively small number of different functional groups. In addition to SOC, there is a vast array of other organic compounds whose presence in the soil environment may be natural or anthropogenic. The reactivity of these non SOC compounds with the soil surface depends upon the functional moieties they possess.

7.6.1 ORGANIC SURFACE FUNCTIONAL GROUPS

Soil organic matter contains a variety of functional groups that may reside at the solid-solution interface. Many of these groups may develop pH-dependent charge, resulting in the retention of cations or anions, depending on solution pH. The most common functional groups on SOM are the carboxyl, phenolic-OH, and carbonyl groups. Other prominent surface functional groups include the amino, imide, sulfhydryl, and sulfonic groups. These organic functional groups vary in their ability to develop charge (ionize), which occurs either through protonation or deprotonation reactions. Further, for any given functional group, the ability to ionize is also a function of the R-groups to which they are attached. For example, the pK_a for the acid dissociation of the phenolic-OH is 9.98; the molecule is not negatively ionized in the pH range of soil solutions. However, substitution on the phenolic benzene ring may lead to pK_a values that are within the pH range of soil solutions. The substituted phenol, 4-hydroxyacetophenone, has an ethanone group ($-CHOCH_3$) in the *para* position and a pK_a for the phenolic-OH of 8.05. Similarly, the substituted phenol, 4-hydroxybenzaldehyde, contains methylaldehyde ($-CHO$) in the *para* position and a pK_a for the phenolic-OH of 7.62. Thus, both of these substituted phenols develop negative charge in the soil pH range.

Carboxyl and sulfonic groups generally develop only very weak complexes with the proton. Thus, these groups tend to develop negative charge and the ability to retain cations. The amine, imide, phenolic-OH, and the sulfhydryl groups are proton-selective, potentially developing positive charge and the ability to retain anions. The carbonyl group does not ionize, but is polarized and may only interact weakly with soluble species. The structures of important organic surface functional groups are displayed in Table 7.9. Also indicated in this table are the configurations of the functional groups in acidic and alkaline environments.

7.6.2 MOLECULAR RETENTION MECHANISMS

Several mechanisms or combinations of mechanisms can be responsible for the retention of organic compounds by soils. The adsorption mechanism for an organic solute is governed, most importantly, by its functional groups. Adsorbate compounds that bear functional groups that can develop positive charge in the pH range of soils may displace exchangeable monovalent and divalent cations. The principle organic functional groups that are involved in cation exchange reactions are the protonated

TABLE 7.9
Functional Groups that are Responsible for the Surface Charge of Soil Organic Matter

Functional Group	Structure	
	Acid Soil	Alkaline Soil

Carboxyl — Acid Soil: $R—C(=O)—OH$; Alkaline Soil: $R—C(=O)—O^-$

Phenolic-OH — Acid Soil: benzene ring with R_1, R_2, R_3, R_4, R_5 and $—OH_2^+$; Alkaline Soil: benzene ring with R_1, R_2, R_3, R_4, R_5 and $—OH^0$

Carbonyl — Acid Soil: $R—C(=O)—R'$; Alkaline Soil: $R—C(=O)—R'$

Sulfonic — Acid Soil: $R—S(=O)(=O)—O^-$; Alkaline Soil: $R—S(=O)(=O)—O^-$

Sulfhydryl — Acid Soil: $R—SH_2^+$; Alkaline Soil: $R—SH^0$

Amine — Acid Soil: $R—NH_3^+$; Alkaline Soil: $R—NH_2^0$

Imide — Acid Soil: ring with NH_2^+; Alkaline Soil: ring with NH^0

Heterocyclic N — Acid Soil: ring with R_1, R_2, R_3, R_4, R_5 and NH^+; Alkaline Soil: ring with R_1, R_2, R_3, R_4, R_5 and N^0

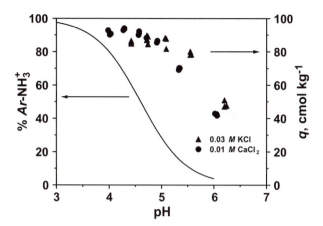

FIGURE 7.40 The adsorption of aniline ($C_6H_5NH_2$) as a function of pH and background electrolyte type (KCl or $CaCl_2$) by K^+- or Ca^{2+}-saturated SWy-1 montmorillonite (right axis). The percentage of the total concentration of aniline that exists as the protonated amine is indicated by the solid line (left axis). (Modified from Essington, M.E. Adsorption of aniline and toluidines on montmorillonite. *Soil Sci.* 158:181–188, 1994.)

amines, imides, and heterocyclic N groups. The cation exchange mechanism for organic molecular retention is generally restricted to exchange with monovalent, exchangeable cations:

$$R\text{-}NH_3^+(aq) + Na^+ - {}^-X(s) \rightarrow R\text{-}NH_3^+ - {}^-X(s) + Na^+(aq) \qquad (7.94)$$

where ^-X is a mole of exchanger phase charge. Divalent cations on the exchange complex tend to collapse the 2:1 structures of clay minerals, although not completely, limiting the access of large organic cations to exchange sites. Retention in the case of divalent cation exchange with a large organic cation would then be restricted to the external surfaces of the clays. However, relatively small organic cations can effectively compete with divalent cations for exchange sites. Aniline, which is an aromatic amine, effectively competes with both K^+ and Ca^{2+} for surface sites on a montmorillonite surface (Figure 7.40). Aniline displays an adsorption envelope that is similar to that displayed by specifically adsorbed ligands (illustrated in Figure 7.37). However, aniline is not a ligand and montmorillonite is not a constant potential surface. The observed adsorption envelope for aniline is a direct result of anilinium ion ($Ar\text{-}NH_3^+$) formation. As pH decreases from a value of 7, the proportion of the total soluble amine that is protonated increases according to the reaction: $Ar\text{-}NH_2^0 + H^+ \rightarrow Ar\text{-}NH_3^+$ (log $K = 4.6$ for aniline). Consistent with the cation exchange process, the adsorption of aniline increases with decreasing pH until an adsorption maximum is reached. The concentration of adsorbed aniline at the adsorption maxima (~80 to 90 cmol kg^{-1} for both the K^+ and Ca^{2+} systems) is approximately equal to the cation exchange capacity of the montmorillonite (85 cmol kg^{-1}); further evidence that a cation exchange mechanism is responsible for aniline adsorption and that both mono- and divalent cations can be displaced.

Organic functional groups that develop negative charge can participate in anion exchange reactions. However, this mechanism is limited to molecules that contain a carboxyl group with a pK_a for the dissociation reaction: $R\text{-}COOH \rightarrow R\text{-}COO^- + H^+$, which is in the pH range of soil solutions. As with the cation exchange mechanism, the anion exchange mechanism is primarily specific to carboxyl-monovalent anion exchange reactions:

$$R\text{-}COO^-(aq) + Cl^- - {}^+X(s) \rightarrow R\text{-}COO^- - {}^+X(s) + Cl^-(aq) \qquad (7.95)$$

where ^+X is a mole of exchanger phase charge. The anion exchange mechanism is not commonly observed due to the weakness of the surface bonds and the net negative surface charge of soils (anions are repulsed from surfaces). However, in acidic soils dominated by clay-sized hydrous metal oxides, the anion exchange mechanism may be more commonly encountered.

At low pH, the surface functional groups of hydrous metal oxides are protonated: $\equiv SOH_2^+$. The waters of hydration of adsorbed metals (outer-sphere) may also be protonated: $\equiv SO^- - H_2O - M^{2+}(H_2O)_n OH_2^+$. Both types of surface sites can attract proton-selective organic functional groups, resulting in adsorption through a protonation reaction. The organic functional groups that are most susceptible to this retention mechanism are the proton selective neutral amine and heterocyclic N groups:

$$R\text{-}NH_2^0(aq) + \equiv SOH_2^+(s) \rightarrow \equiv SOH_2^+ - NH_2\text{-}R(s) \tag{7.96a}$$

$$R\text{-}NH_2^0(aq) + \equiv SO^- - H_2O - M^{2+}(H_2O)_n OH_2^+(s) \rightarrow$$
$$\equiv SO^- - (H_2O) - M^{2+}(H_2O)_n OH_2 - NH_2 - R^+(s) \tag{7.96b}$$

the neutral carbonyl group, bearing a partial negative charge ($\delta-$):

$$R_2\text{-}C{=}O^{\delta-}(aq) + \equiv SOH_2^+(s) \rightarrow \equiv SOH_2^+ - {}^{-\delta}O{=}C\text{-}R_2(s) \tag{7.96c}$$

and the dissociated carboxyl group:

$$R\text{-}COO^-(aq) + \equiv SOH_2^+(s) \rightarrow \equiv SOH_2^+ - {}^-OOC - R(s) \tag{7.96d}$$

A retention mechanism that is similar to the protonation mechanism is hydrogen bonding. This mechanism is thought to be of only limited significance. The un-ionized (neutral charge) amine, carbonyl, carboxyl, and phenolic-OH groups may form a weak hydrogen bond with neutral surface oxygens or hydroxides. While this mechanism is minor for organic solute retention by soil minerals, it is thought to be an important mechanism by which organic solutes are retained by organic surfaces.

Amines, carboxyl, carbonyl, and alcoholic-OH (aliphatic R-OH) groups may also be retained by complex formation with an adsorbed, exchangeable cation:

$$R\text{-}COO^-(aq) + \equiv SO^- - (H_2O) - M^{2+}(H_2O)_n(s) \rightarrow$$
$$\equiv SO^- - H_2O - M^{2+}(H_2O)_n - {}^-OOC - R(s) \tag{7.97a}$$

This outer-sphere complex formation reaction is termed water bridging. If the metal is a relatively hard Lewis acid and the organic compound is of sufficient base character, the organic compound may displace a water of hydration, forming an inner-sphere surface complex:

$$R\text{-}COO^-(aq) + \equiv SO^- - (H_2O) - M^{2+}(H_2O)_n(s) \rightarrow$$
$$\equiv SO^- - (H_2O) - M^{2+} - {}^-OOC - R(H_2O)_{n-1}(s) \tag{7.97b}$$

The formation of an inner-sphere surface complex is termed cation bridging.

Carboxyl groups that have a pK_a value within the soil pH range can form inner-sphere surface complexes with Al and Fe oxide surfaces and with the aluminol group at the edges of layer silicates. The specific retention of these weak organic acids occurs through the ligand exchange mechanism:

$$R\text{-}COO^-(aq) + \equiv SOH_2^+(s) \rightarrow \equiv SOOC - R(s) + H_2O(l) \tag{7.98}$$

This mechanism is identical to that discussed for the inorganic weak acid anions and commonly results in multidentate surface complexes. This is indicative of a very strong chemical bond between the metal oxide surface and the carboxyl group. Low-molecular-mass organic acid anions (e.g., citrate and 2-ketogluconate) have been found to inhibit the precipitation of metal oxyhydroxides

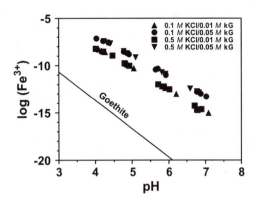

FIGURE 7.41 An activity diagram illustrating the stability of goethite (solid line) and the activity of Fe^{3+} as a function of pH, ionic strength, and 2-keto-gluconate (kG) in solutions equilibrated with goethite. The activity of Fe^{3+} was determined using GEOCHEM-PC and did not include Fe^{3+}–kG complexation reactions in the computations (ion pair formation constants for Fe^{3+}–kG complexes have not been determined).

and phosphates, and have caused instead the formation of amorphous phases, rather than crystalline phases. The specific retention of the organic acids by the surface is thought to be of sufficient strength to disrupt crystal growth. For example, the elevated solubility of gibbsite (Chapter 6) and goethite (Figure 7.41) may be ascribed, in part, to the ability of 2-ketogluconate to inhibit mineral precipitation.

The adsorption of uncharged, nonpolar molecules occurs through van der Waals interactions. This mechanism is a result of dipole–dipole interactions produced by correlations between the fluctuating dipoles of two molecules (or a neutral surface and a molecule) in close proximity to each other. Relative to the previously discussed organic compound adsorption mechanisms, van der Waals interactions are very weak. However, the intensity of the van der Waals dispersive force (the electron correlation effect) is proportional to the number of atoms in the two structures that are interacting (the effect is additive). Thus, for large organic compounds the interaction between a surface and an organic solute can be very strong.

7.7 SURFACE COMPLEXATION MODELS

Consider a very simple natural system consisting of two phases: a solid phase and an aqueous phase. Further, consider the solid phase (e.g., Al_2O_3) is composed of a homogeneous surface with one type of surface functional group, the inorganic hydroxyl group ($\equiv AlO^{-1.5}$). The aqueous phase contains a swamping background electrolyte ($NaNO_3$) that completely dissociates into Na^+ and NO_3^- and controls the ionic strength. If a trace element (e.g., lead) is introduced into this system, in the form of the common salt $[Pb(NO_3)_2]$, such that the lead nitrate concentration is exceedingly minor compared to that of the swamping sodium nitrate, there are a number of reactions that are specific to Pb that may occur. The $Pb(NO_3)_2$ will dissolve and dissociate, forming the free species $Pb^{2+}(aq)$ and $NO_3^-(aq)$. The free $Pb^{2+}(aq)$ may form a soluble complex $[PbNO_3^+(aq)]$ and a hydrolysis product $[PbOH^+(aq)]$. Further, $Pb^{2+}(aq)$ may precipitate to form $PbO(s)$ or $Pb(OH)_2(s)$, depending on the degree of saturation relative to the K_{sp} values of the two solids. As we have seen, the complexation chemistry of Pb^{2+} and the potential for various Pb-bearing solids to precipitate can be determined through the application of the ion association model and equilibrium solubility concepts (Chapters 5 and 6). The $Pb^{2+}(aq)$, $PbNO_3^+(aq)$, and $PbOH^+(aq)$ species (which, when summed, equal the total concentration of Pb in the aqueous phase, Pb_T) may be adsorbed by the $Al_2O_3(s)$ surface. Total Pb adsorption may be quantified by computing the distribution coefficient, K_d. Further, if several of these natural systems are constructed, each with a different initial Pb_T concentration, an adsorption isotherm can be generated, yielding Langmuir or Freundlich equations that quantify Pb adsorption as a function of Pb_T at equilibrium.

It may appear from the above discussion that we have a complete understanding of our simple natural system: aqueous chemistry, mineral solubility, and adsorption. However, there are still a number of unknowns that must be addressed in order to have a complete description of the natural system; unknowns that are specific to the adsorbed phase. An adsorption isotherm will not indicate exactly which Pb species—$Pb^{2+}(aq)$, $PbNO_3^+(aq)$, or $PbOH^+(aq)$—is adsorbed by the $Al_2O_3(s)$ surface. The isotherm will not identify the mechanism of retention: inner-sphere, outer-sphere, or diffuse ion swarm. Finally, the isotherm cannot even be employed to establish that the mechanism of retention is truly adsorption (recall that an adsorption isotherm could also describe a precipitation process). In short, while we have discussed the tools to predict to a high degree of confidence aqueous and solid-phase speciation, we have yet to critically address compound adsorption behavior. Such a critical evaluation requires that we have a clear understanding of the adsorption mechanisms; those processes that can be quantitatively described by chemical reactions and associated equilibrium constants.

A surface complexation model (SCM) is a tool used to indirectly identify the chemical reactions and associated equilibrium constants responsible for compound adsorption. Further, if a mechanism is known, as might be identified through a spectroscopic technique, an SCM can be employed to determine the specific equilibrium constant. However, the true utility of an SCM is associated with its ability to predict the distribution of a substance between the adsorbed and aqueous phases, assuming the validity of the mechanism(s) of retention and the associated equilibrium constant(s). When an SCM is used in concert with models that predict aqueous speciation and mineral solubility (as described in Chapters 5 and 6), the resulting product of the computation, the distribution of a substance within a phase and between phases, may closely resemble the actual speciation of the substance (or at a minimum identify the relative importance of the processes that affect fate and behavior).

There are several unique models that describe surface complexation; the more popular models will be described below. These models vary according to how they conceptualize the distribution of charge and the electrical potentials, as well as the location of adsorbed species, in the solid-solution interface. Despite these differences, there exists a set of assumptions common to all models that describe the solid-solution interface. First, all mineral surfaces in an aquatic environment are assumed to contain functional groups (surface sites) that have well-defined coordinative properties. In essence, an adsorbed metal cation is complexed (inner sphere or outer sphere) by one or more ligands (typically oxygen atoms) that are bound to the structural metal (recall the definitions of type A, C, and B surface functional groups). Second, for each type of surface site, a total concentration (S_T) can be defined. For example, the gibbsite surface bears two types of surface sites (Figure 7.42):

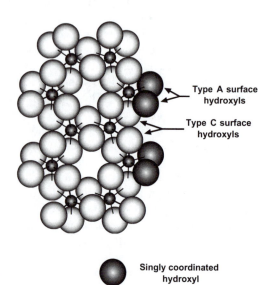

FIGURE 7.42 Representation of a gibbsite surface showing an equal number of type A and C functional groups on the edge surface and a planar surface that bears only type C functional groups.

type A (singly coordinated) and type C (doubly coordinated). Like the layer silicates, gibbsite consists of a layer structure and exhibits both planar and edge surfaces. On the edge surface there are equal numbers of type A ($\equiv$AlOH) and type C ($\equiv$Al$_2$OH) surface OH ions; whereas, the planar surface consists of only type C surface OH ions. Based on well-established crystallographic parameters, the density of the two types of functional groups on the gibbsite surface is known. The site density for type A ($n_{S,A}$) and type C ($n_{S,C}$) groups on the edge surface are each equal to 8.15 sites nm^{-2}, and the site density of the type C group on the planar surface is $n_{S,C} = 13.8$ sites nm^{-2}. The total concentration of the two types of sites can be computed using the site densities, planar and edge surface areas, and the concentration of gibbsite in the aqueous environment. The total concentration of sites (S_T) in an aqueous suspension is computed using the expression:

$$S_T = \frac{n_S \times 10^{18} \times a \times S_A}{A_N} \tag{7.99}$$

where a is the concentration of the solid in the suspension (g L^{-1}), A_N is the Avogadro constant (6.022×10^{23} mol^{-1}), S_A is the specific surface of the solid (m^2 g^{-1}), 10^{18} (nm^2 m^{-2}) is a conversion factor, and n_S (sites nm^{-2}) and S_T (mol L^{-1}) are defined above. For a suspension containing 3.33 g L^{-1} gibbsite with a specific edge surface of 3.5 m^2 g^{-1}, the total concentration of type A functional groups is:

$$S_{T,A} = \frac{8 \text{ sites nm}^{-2} \times 10^{18} \text{ nm}^2 \text{ m}^{-2} \times 3.33 \text{ g L}^{-1} \times 3.5 \text{ m}^2 \text{ g}^{-1}}{6.022 \times 10^{23} \text{ sites mol}^{-1}} = 1.55 \times 10^{-4} \text{ mol L}^{-1} \tag{7.100}$$

The total concentration of type C functional groups in the suspension (assuming a specific planar surface of 20 m^2 g^{-1}) is the summation of the concentration of sites on the edge and planar surfaces:

$$S_{T,C} = \frac{8 \text{ sites nm}^{-2} \times 10^{18} \text{ nm}^2 \text{ m}^{-2} \times 3.33 \text{ g L}^{-1} \times 3.5 \text{ m}^2 \text{ g}^{-1}}{6.022 \times 10^{23} \text{ sites mol}^{-1}}$$
$$+ \frac{14 \text{ sites nm}^{-2} \times 10^{18} \text{ nm}^2 \text{ m}^{-2} \times 3.33 \text{ g L}^{-1} \times 20 \text{ m}^2 \text{ g}^{-1}}{6.022 \times 10^{23} \text{ sites mol}^{-1}} = 1.70 \times 10^{-3} \text{ mol L}^{-1} \tag{7.101}$$

The final assumption that is common to all surface complexation models deals with the energetics of the adsorption reactions, and that is that a free energy of adsorption (ΔG°_{ads}) can be assigned for each adsorption reaction. Mathematically, ΔG°_{ads} is the sum of intrinsic and coulombic free energy terms:

$$\Delta G^\circ_{ads} = \Delta G^\circ_{int} + \Delta G^\circ_{coul} \tag{7.102}$$

where ΔG°_{int} is the portion of the total free energy of adsorption that is intrinsic, or specific to the surface. It is also the chemical portion of ΔG°_{ads}. The coulombic free energy portion (ΔG°_{coul}) of ΔG°_{ads} is the electrostatic term. It is variable with distance from a surface, influenced by the electrical double-layer, and accounts for the electrostatic work necessary to move ions through the surface potential gradient. Correspondingly, the adsorption constant, K_{ads}, is a product of the intrinsic (K^{int}) and coulombic (K_{coul}) constants:

$$K_{ads} = K^{int} K_{coul} \tag{7.103}$$

with each K related to ΔG_r° by:

$$K = \exp\left(\frac{-\Delta G^\circ}{RT}\right) \tag{7.104}$$

where R is the universal gas constant (8.3143 J K^{-1} mol^{-1}) and T is the temperature expressed in Kelvin. For the surface complexation-adsorption reaction,

$$\equiv SOH^0(s) + M^{m+}(aq) \rightarrow \equiv SOM^{m-1}(ads) + H^+(aq) \tag{7.105}$$

the equilibrium constant is:

$$K_{ads} = \frac{N_{\equiv SOM^{m-1}}(H^+)}{N_{\equiv SOH^0}(M^{m+})} = K^{int}K_{coul} \tag{7.106}$$

where parentheses ( ) represent activities,

$$N_{\equiv SOH^0} = \frac{[\equiv SOH^0]}{[\equiv SOH^0]+[\equiv SOM^{m-1}]} \tag{7.107a}$$

$$N_{\equiv SOM^{m-1}} = \frac{[\equiv SOM^{m-1}]}{[\equiv SOH^0]+[\equiv SOM^{m-1}]} \tag{7.107b}$$

are the mole fractions, and brackets [ ] represent the concentrations of the surface species in mol kg^{-1}. Equation 7.106 simplifies to:

$$K_{ads} = \frac{[\equiv SOM^{m-1}](H^+)}{[\equiv SOH^0](M^{m+})} = K^{int}K_{coul} \tag{7.108}$$

The adsorption constant, K_{ads}, is a conditional equilibrium constant, and varies with the pH and ionic strength of the media, and the composition of the adsorbed phase. The intrinsic equilibrium constant, K^{int}, is a true constant that is independent of the composition of the adsorbed phase, but only at a fixed ionic strength. Therefore, K^{int} is also a conditional equilibrium constant (as it is a function of ionic strength), applicable only to those aqueous environments or systems that have an ionic strength identical to that used to determine K^{int}. Further, K^{int} is defined as the equilibrium constant ($K^{int} = K_{ads}$) when the mole fractions of the adsorbed species are unity ($N_{\equiv SOH^0} = N_{\equiv SOM^{m-1}} = 1$) and the surface charge density is zero, $\sigma_p = 0$. In order to describe this adsorption process on the basis of a surface complexation model, it is necessary to define the stoichiometry of the adsorption reaction, obtain the intrinsic constants for the formation of the various surface species, and to provide an expression for the coulombic term.

7.7.1 THE NONELECTROSTATIC MODEL

The simplest surface complexation model is the nonelectrostatic model (NEM). In this model, the variable electrostatic interaction is ignored and the intrinsic constant alone is used to describe the adsorption process: $K_{ads} = K^{int}$ and $\Delta G_{ads}^\circ = \Delta G_{int}^\circ$. This approach is equivalent to studying the adsorption process under conditions where the surface charge does not change when adsorption occurs. An example of this is seen in the specific adsorption of a hydrolyzed metal cation: $\equiv SOH^0$

+ $NiOH^+ \rightarrow \; \equiv SONiOH^0 + H^+$. This model can also be used when the coulombic term is constant and can be simply included in the adsorption constant (K_{ads}). The NEM approach is similar to that used to determine the aqueous speciation of ions in a soil solution using the ion association model (Figure 5.14).

As an example of the application of NEM, consider the adsorption of Pb^{2+} by $Al(OH)_3$ (gibbsite) in a 0.1 M $NaNO_3$ background electrolyte and a pH range of 4 to 7. The experimentally determined adsorption behavior of Pb is shown in Figure 7.42. The adsorption of Pb^{2+} occurs at the type A sites on the gibbsite surface. Because the gibbsite surface is well characterized, the density of available type A adsorption sites and the specific surface area are known (8 sites nm^{-2} with an edge surface area of 3.5 $m^2\ g^{-1}$). Further, the concentration of gibbsite is fixed in the experimental design (0.1 g in 30 mL or 3.33 g L^{-1}); thus, the total number of available adsorption sites is known (Equation 7.99). Also by experimental design, the total concentration of Pb (Pb_T) in the system (14 μM) is known. Finally, for computational ease, it can be assumed that the surface will not be close to saturation with respect to Pb, as the total concentration of $Pb(1.4 \times 10^{-5}\ mol\ L^{-1})$ is small relative to the total concentration of surface sites ($1.55 \times 10^{-4}\ mol\ L^{-1}$), or Pb_T is small relative to S_T.

The chemical model for the gibbsite system consists of the surface protolysis reaction (proton dissociation reaction), the hypothesized Pb^{2+} surface complexation reaction, and the reaction that describes the hydrolysis of Pb^{2+} in the aqueous solution. Equilibrium constants for all but the Pb^{2+} surface complexation reaction are known. The type A surface functional group on gibbsite is essentially a weak Lowry-Brønsted acid. At pH values less than approximately 8, the functional group is doubly protonated and the site bears a positive charge. The dissociation of protons from the site is described by the protolysis reaction and the associated acid dissociation constant:

$$\equiv AlOH_2^{+0.5} \rightarrow \; \equiv AlOH^{-0.5} + H^+ \tag{7.109a}$$

$$K_{1,2} = 10^{-8.87} = \frac{[\equiv AlOH^{-0.5}](H^+)}{[\equiv AlOH_2^{+0.5}]} \tag{7.109b}$$

Thus, at solution pH values below 8.87, the surface displays a predominance of positive charge. It is also evident that the charge associated with the surface groups are not whole numbers. This is a consequence of Pauling Rule 2. The doubly protonated surface oxygen atom in $\equiv AlOH_2^{+0.5}$ receives three cation bonds, two +1 bonds from the surface protons and a +0.5 bond from the octahedrally coordinated Al^{3+}. Thus, the total cation charge received by the oxygen anion is +2.5, which overcompensates for the −2 charge of the oxygen and results in a +0.5 functional group charge. A hypothesized surface complexation reaction that may describe the mechanism of Pb^{2+} adsorption is:

$$\equiv AlOH^{-0.5} + Pb^{2+} + H_2O \rightarrow \; \equiv AlOPbOH^{-0.5} + 2H^+ \tag{7.110a}$$

$$K_{Pb} = \frac{[\equiv AlOPbOH^{-0.5}](H^+)^2}{[\equiv AlOH^{-0.5}](Pb^{2+})} \tag{7.110b}$$

where K_{Pb} is an unknown quantity that will be determined when the predicted $[\equiv AlOPbOH^{-0.5}]$ values are made to match, as closely as possible, the experimentally determined $[\equiv AlOPbOH^{-0.5}]$ values, as a function of pH (assuming the adsorption mechanism displayed in Equation 7.110a is appropriate). Finally, the hydrolysis of Pb^{2+} to form $PbOH^+$ is described by the reaction:

$$Pb^{2+} + H_2O \rightarrow PbOH^+ + H^+ \tag{7.111a}$$

$$K_{PbOH} = 10^{-7.7} = \frac{(PbOH^+)(H^+)}{(Pb^{2+})} \tag{7.111b}$$

The salient mole balance expressions for the system are:

$$Pb_T = [Pb^{2+}] + [PbOH^+] + [\equiv AlOPbOH^{-0.5}] = 1.4 \times 10^{-5}\ M \tag{7.112}$$

$$S_T = [\equiv AlOH_2^{+0.5}] + [\equiv AlOH^{-0.5}] + [\equiv AlOPbOH^{-0.5}] = 1.55 \times 10^{-4}\ M \tag{7.113}$$

With the above expressions, the adsorption behavior of Pb^{2+} by the gibbsite surface becomes an algebraic exercise and can be determined as a function of pH by obtaining an expression for $\frac{[\equiv AlOPbOH^{-0.5}]}{Pb_T}$.

An expression for $\frac{[\equiv AlOPbOH^{-0.5}]}{Pb_T}$ as a function of pH can be obtained by first modifying the equilibrium constant expressions in Equations 7.109b, 7.110b, and 7.111b. Using Equations 7.109b and 7.110b, the concentrations of $\equiv AlOH_2^{+0.5}$ and $\equiv AlOPbOH^{-0.5}$ are expressed in terms of $[\equiv AlOH^{-0.5}]$:

$$[\equiv AlOH_2^{+0.5}] = 10^{8.87}[\equiv AlOH^{-0.5}](H^+) \tag{7.114}$$

$$[\equiv AlOPbOH^{-0.5}] = \frac{K_{Pb}[\equiv AlOH^{-0.5}](Pb^{2+})}{(H^+)^2} \tag{7.115}$$

The activity of $PbOH^+$ is expressed in terms of (Pb^{2+}) using Equation 7.111b:

$$(PbOH^+) = \frac{10^{-7.7}(Pb^{2+})}{(H^+)} \tag{7.116}$$

Equations 7.115 and 7.116 are substituted into the mass balance expression for Pb_T (Equation 7.112), assuming $[Pb^{2+}]$ equals (Pb^{2+}) and $[PbOH^+]$ equals $(PbOH^+)$ (brackets denote molar concentrations):

$$Pb_T = [Pb^{2+}] + \frac{10^{-7.7}[Pb^{2+}]}{(H^+)} + \frac{K_{Pb}[\equiv AlOH^{-0.5}][Pb^{2+}]}{(H^+)^2} \tag{7.117}$$

The mass balance expression for the total concentration of surface functional groups (Equation 7.113) is also simplified by employing the observation that Pb_T is minor relative to S_T. Because the total concentration of Pb in the system is much less that the total concentration of surface sites ($1.4 \times 10^{-5}\ M\ Pb_T$ vs. $1.55 \times 10^{-4}\ M\ S_T$), if follows that the concentration of adsorbed Pb^{2+}, $[\equiv AlOPbOH^{-0.5}]$, can be neglected in Equation 7.113. Thus, Equation 7.113 becomes:

$$S_T = [\equiv AlOH_2^{+0.5}] + [\equiv AlOH^{-0.5}] \tag{7.118}$$

Substituting Equation 7.114 for $[\equiv AlOH_2^{+0.5}]$ yields:

$$S_T = 10^{8.87}[\equiv AlOH^{-0.5}](H^+) + [\equiv AlOH^{-0.5}] \tag{7.119}$$

or

$$[\equiv AlOH^{-0.5}] = \frac{S_T}{[1 + 10^{8.87}(H^+)]} \tag{7.120}$$

Substituting Equations 7.120 for $[\equiv AlOH^{-0.5}]$ in Equation 7.117:

$$Pb_T = [Pb^{2+}] + \frac{10^{-7.7}[Pb^{2+}]}{(H^+)} + \frac{K_{Pb}[Pb^{2+}]S_T}{(H^+)^2[1 + 10^{8.87}(H^+)]} \tag{7.121}$$

Substituting Equations 7.120 into Equation 7.115:

$$[\equiv\text{AlOPbOH}^{-0.5}] = \frac{K_{\text{Pb}}[\text{Pb}^{2+}]S_{\text{T}}}{(\text{H}^+)^2[1+10^{8.87}(\text{H}^+)]} \tag{7.122}$$

Finally, dividing Equation 7.122 by Equation 7.121 yields an expression for $\frac{[\equiv\text{AlOPbOH}^{-0.5}]}{\text{Pb}_{\text{T}}}$:

$$\frac{[\equiv\text{AlOPbOH}^{-0.5}]}{\text{Pb}_{\text{T}}} = \frac{K_{\text{Pb}}S_{\text{T}}}{\{(\text{H}^+)^2[1+10^{8.87}(\text{H}^+)]+10^{-7.7}(\text{H}^+)[1+10^{8.87}(\text{H}^+)]+K_{\text{Pb}}S_{\text{T}}\}} \tag{7.123}$$

If $\equiv\text{AlOPbOH}^{-0.5}$ is the surface complex formed, and all other assumptions are appropriate, Equation 7.123 will describe the experimental adsorption data (plotted as $\frac{[\equiv\text{AlOPbOH}^{-0.5}]}{\text{Pb}_{\text{T}}}$ vs. pH in Figure 7.43a). The adsorption equilibrium constant, K_{Pb}, is an adjustable parameter in Equation 7.123 and is determined by nonlinear least-squares regression analysis. The best fit of Equation 7.123 to the adsorption is obtained when $K_{\text{Pb}} = 10^{-7.2}$. The determination of the suitability of the chemical model (the surface complexation reaction) will depend on how well the predicted adsorption behavior matches the experimental behavior (the goodness-of-fit), particularly in the pH region where adsorption increases from nearly zero to almost complete adsorption (the adsorption edge). If a reasonably good fit is obtained, the conclusion that the surface complexation model is appropriate is reasonable. In this case, Pb adsorption by gibbsite, the NEM consisted of one reaction, the adsorption of Pb^{2+} to form $\equiv\text{AlOPbOH}^{-0.5}$. As Figure 7.43a indicates, this chemical model appears to be appropriate, except for the discrepancy at pH values below the pH_{50}. Although highly subjective, one may also conclude, in this instance,

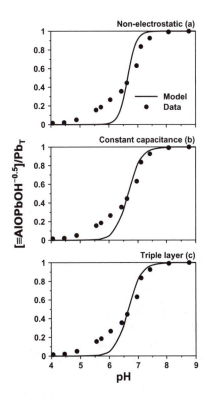

FIGURE 7.43 Comparison of different surface complexation models used to describe Pb adsorption by gibbsite as a function of pH. The closed circles represent the experimental adsorption data (3.33 g L^{-1} gibbsite, 0.1 M NaNO_3 background electrolyte, and a total Pb concentration of 14 μM). The solid lines represent the predicted Pb adsorption behavior by employing the (a) nonelectrostatic model, (b) constant capacitance model, and (c) triple layer model.

that the NEM model provides a satisfactory description of Pb adsorption by gibbsite (variable electrostatic interaction is ignored).

In Section 7.3.1 it was stated that L- and H-type isotherms were usually indicative of inner-sphere surface complex formation. Returning to the Pb adsorption experiment described above, the following mass balance expression was developed:

$$S_T = [\equiv AlOH_2^{+0.5}] + [\equiv AlOH^{-0.5}] + [\equiv AlOPbOH^{-0.5}] \tag{7.124}$$

If solution pH is maintained at a value below $pK_{1,2}$ (8.87, surface protolysis constant) and pK_{PbOH} (7.7, first Pb^{2+} hydrolysis constant), Equation 7.124 becomes:

$$S_T = [\equiv AlOH_2^{+0.5}] + [\equiv AlOPbOH^{-0.5}] \tag{7.125}$$

Using this expression and Equation 7.115, the total concentration of surface functional groups can be expressed as a function of the adsorbed Pb and Pb^{2+} concentrations:

$$S_T = \frac{[\equiv AlOPbOH^{-0.5}](H^+)^2}{K_{Pb}[Pb^{2+}]} + [\equiv AlOPbOH^{-0.5}] \tag{7.126}$$

Rearranging leads to:

$$S_T = [\equiv AlOPbOH^{-0.5}]\left\{ \frac{(H^+)^2}{K_{Pb}[Pb^{2+}]} + 1 \right\} \tag{7.127}$$

or

$$[\equiv AlOPbOH^{-0.5}] = \frac{S_T}{\left\{ \dfrac{(H^+)^2}{K_{Pb}[Pb^{2+}]} + 1 \right\}} \tag{7.128}$$

Finally,

$$[\equiv AlOPbOH^{-0.5}] = \frac{S_T \dfrac{K_{Pb}}{(H^+)^2}[Pb^{2+}]}{\left\{ 1 + \dfrac{K_{Pb}}{(H^+)^2}[Pb^{2+}] \right\}} \tag{7.129}$$

This is the Langmuir expression with the adsorption maximum (b) equal to S_T, and the Langmuir adsorption constant (K_L) equal to $\frac{K_{Pb}}{(H^+)^2}$. Equation 7.129 is plotted in Figure 7.44 as a function of pH using the K_{Pb} derived from the NEM ($10^{-7.2}$). The three adsorption isotherms represent Pb^{2+} adsorption when (1) pH < pH_{50} (pH = 6) and retention is at a minimum; (2) pH ≈ pH_{50} (pH = 6.6) and approximately 50% of the total Pb in the system is adsorbed; and (3) pH > pH_{50} (pH = 7) and approximately 100% of the Pb is adsorbed. When pH = 6, the adsorption isotherm is linear (C-type) in the $[Pb^{2+}]$ range examined. However, the isotherms display L-type then H-type character as pH is increased to 6.6 then 7. In this example, the increase in K_L with each incremental increase in pH is two orders of magnitude (from $10^{4.78}$ L mol^{-1} at pH 6 to $10^{6.78}$ L mol^{-1} at pH 7).

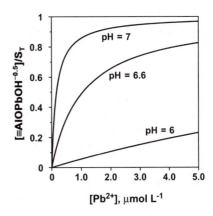

FIGURE 7.44 Predicted adsorption isotherms are shown for Pb adsorption by gibbsite as a function of pH using Equation 7.129 and the nonelectrostatic for systems containing 3.33 g L^{-1} gibbsite in 0.1 M NaNO$_3$ background electrolyte.

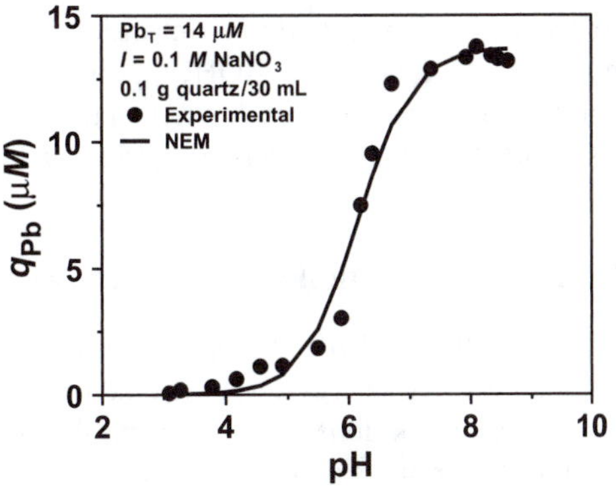

FIGURE 7.45 Lead adsorption by quartz as a function of pH (3.33 g L^{-1} quartz, 0.1 M NaNO$_3$, and a total Pb concentration of 14 μM). The solid line represents the predicted Pb adsorption behavior by employing the nonelectrostatic model.

Metal adsorption by constant potential surfaces, or more specifically the pH of the adsorption edge, is not particularly influenced by surface type (as illustrated in Figure 7.34). Rather, the pH of the adsorption edge is correlated (although loosely) with pK_{MOH}. This is a particularly interesting observation, considering that surface site characteristics, such as pH$_{pzc}$ and surface acidity (pK_a values), vary according to the mineral. In the above example, the adsorption of Pb by gibbsite is described by a pH$_{50}$ of 6.6, and the formation of $\equiv$AlOPbOH$^{-0.5}$ according to the reaction in Equation 7.110a with a K_{Pb} of 10$^{-7.2}$. The pH$_{pzc}$ of the gibbsite is 8.87, which coincides with the pH where [$\equiv$AlOH$_2^{+0.5}$] = [$\equiv$AlOH$^{-0.5}$]. Using the methodology described above, the K_{Pb} that describes the adsorption of Pb by quartz according to the following reaction can be determined:

$$\equiv SiOH^0 + Pb^{2+} + H_2O \rightarrow \equiv SiOPbOH^0 + 2H^+ \tag{7.130}$$

The quartz surface has little in common with the gibbsite surface, with the exception of the pH$_{50}$, which for Pb adsorption by quartz is 6.2 (Table 7.8). The pH$_{pzc}$ of quartz is ~2, and as previously discussed, the $\equiv$SiOH^0 functional group does not protonate (to form $\equiv$SiOH$_2^+$) to any relevant degree in soils. The $\equiv$SiOH^0 deprotonates according to the reaction:

$$\equiv SiOH^0 \rightarrow \equiv SiO^- + H^+ \tag{7.131}$$

with K_{s2} = 10$^{-7.2}$. The adsorption of Pb by quartz is described by Equation 7.130 using the NEM to generate a K_{Pb} of 10$^{-4.1}$ (Figure 7.45).

The Pb adsorption process depicted in Equation 7.130 is actually the sum of two separate chemical reactions, Equation 7.131 and:

$$\equiv SiO^- + Pb^{2+} + H_2O \rightarrow \equiv SiOPbOH^0 + H^+ \tag{7.132}$$

The equilibrium constant for the reaction in Equation 7.132 is $10^{3.1}$. A similar analysis can be performed for Pb adsorption by the gibbsite surface (Equation 7.110a). The Pb adsorption step is:

$$\equiv AlO^{-1.5} + Pb^{2+} + H_2O \rightarrow \equiv AlOPbOH^{-0.5} + H^+ \tag{7.133}$$

However, the K value for the complete dissociation of the $\equiv AlOH^{-0.5}$ group to form $\equiv AlO^{-1.5}$ is unknown, although this value is probably less than 10^{-12}. Using this K value as a maximum, and the K_{Pb} for Equation 7.110a ($10^{-7.2}$), the equilibrium constant for Equation 7.133 will be greater than approximately 10^5. Thus, K for Pb adsorption to quartz (Equation 7.132, $K = 10^{3.1}$) is less than that for gibbsite (Equation 7.133, $\log K > 5$). Surface functional groups that display stronger Lowry-Brønsted acidity (lower intrinsic K values), such as the quartz $\equiv SiOH^0$ group ($K_- = 10^{-7.2}$) relative to the gibbsite $\equiv AlOH_2^{+0.5}$ group ($K_{s2} < 10^{-12}$), also display relatively soft base character, which tends to be offset by lower metal adsorption (lower K_M values). This offset results in metal adsorption that is not particularly influenced by surface type (as illustrated in Figure 7.34).

7.7.2 Electrostatic Effects on Adsorption

Recall that the total free energy of adsorption is the sum of chemical (intrinsic) and electrostatic (coulombic) free energy terms (Equation 7.102). The coulombic term (ΔG°_{coul}) results from the electrostatic work done in transporting the adsorbate from the bulk solution to the surface and the adsorbate from the surface to the bulk solution. The coulombic term is described by:

$$\Delta G^\circ_{coul} = F \Delta Z \psi_0 \tag{7.134}$$

where F is the Faraday constant (96,485 C mol^{-1}), ΔZ is the net change in surface charge due to adsorption, and ψ_0 is the surface potential relative to the reference potential of zero in the bulk solution ($\psi_b = 0$). The coulombic portion of the adsorption equilibrium constant is:

$$K_{coul} = \exp\left(\frac{-\Delta G^\circ_{coul}}{RT}\right) \tag{7.135}$$

Substituting Equation 7.134 for ΔG°_{coul}:

$$K_{coul} = \exp\left(\frac{-F \Delta Z \psi_0}{RT}\right) \tag{7.136}$$

Consider the protolysis reaction: $\equiv SOH^0 + H^+ \rightarrow \equiv SOH_2^+$. The adsorption constant for this reaction is:

$$K_+ = \frac{[\equiv SOH_2^+]}{[\equiv SOH^0](H^+)} = K_+^{int} K_{coul} \tag{7.137}$$

$K_+ = K_+^{int} K_{coul}$. The ΔZ for this reaction is +1. Substituting Equation 7.136 into the expression for K_+ (Equation 7.137) yields:

$$K_+ = K_+^{int} \exp\left(\frac{-F\psi_0}{RT}\right) \tag{7.138}$$

The protolysis (adsorption) constant, K_+, is a conditional equilibrium constant and varies with solution and adsorbed phase compositions. The intrinsic constant, K_+^{int}, is a true constant that is independent of the composition of the adsorbed phase at a fixed ionic strength. Therefore, K_+^{int} resembles a true equilibrium constant and is valid for describing an adsorption process in any environment having an ionic strength similar to the one in which K_+^{int} was determined. Rearranging Equation 7.138 leads to an expression for K_+^{int}:

$$K_+^{int} = K_+ \exp\left(\frac{F\psi_0}{RT}\right) \tag{7.139}$$

Substituting Equation 7.137 for K_+:

$$K_+^{int} = \frac{[\equiv SOH_2^+]}{[\equiv SOH^0](H^+)} \exp\left(\frac{F\psi_0}{RT}\right) \tag{7.140}$$

If there is no net change in the surface charge as a result of an adsorption process, the adsorption constant, K_{ads}, and the intrinsic adsorption constant, K_+^{int}, will be equal, as ΔZ for the reaction is zero. This is the case for the reaction that describes the adsorption of Pb^{2+} by gibbsite in Equation 7.110a.

In the NEM, the determination of an adsorption constant that describes a surface complexation reaction simply reduces to an exercise in solving mass balance expressions. Alternatively, the determination of the distribution of a species between the adsorbed and solution phases reduces to the application of the ion association model if the adsorption constant is known. However, further refinement of an adsorption constant and the determination of an adsorption constant that can be translated from one environment to another requires knowledge of the behavior of ions in the electrified interfacial region. Specifically, the planes of adsorption at the solid-solution interface must be assigned, and the variation of ψ in each plane must be known. The surface complexation models that consider electrostatics vary according to the number of planes allowed for complexation and the mathematical description of ψ for each surface plane.

7.7.2.1 The Constant Capacitance Model

In the constant capacitance model (CCM), adsorption is assumed to occur in a single surface plane that is a combination of the s-plane (for proton and hydroxide adsorption) and the is-plane (for metal and ligand adsorption) (Figure 7.46). Thus, the inorganic surface hydroxyl groups are constrained to form only inner-sphere surface complexes with adsorbed species. Outer-sphere surface complexes are not considered by the CCM. The surface charge density on the particle is given by $\sigma_0 = \sigma_H + \sigma_{is}$, so that both the proton and the specifically adsorbed species contribute to surface charge density. The CCM also does not require charge balance at the surface, i.e., the surface charge created through protonation-deprotonation and metal or ligand adsorption, σ_p, is not balanced by counterions, such as those that would reside in the d-plane and contribute to σ_d. In the CCM, the surface charge density is proportional to the potential at the particle surface:

$$\sigma_p = C\psi_{(0)} \tag{7.141}$$

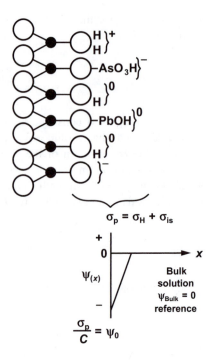

FIGURE 7.46 The solid-solution interface of a hydrous metal oxide surface as described by the constant capacitance model.

where σ_p is expressed in C m^{-2} (Coulombs per square meter), C is the capacitance in F m^{-2} (farads per square meter), and $\psi_{(0)}$ is the surface potential in volts.

In general, the chemical reactions that describe surface complexation of a proton, hydroxide, metal, and ligand are:

$$\equiv SOH^0 + H^+ \rightarrow \equiv SOH_2^+ \tag{7.142a}$$

$$\equiv SOH^0 \rightarrow \equiv SO^- + H^+ \tag{7.142b}$$

$$\equiv SOH^0 + M^{m+} \rightarrow \equiv SOM^{m-1} + H^+ \tag{7.142c}$$

$$\equiv SOH^0 + L^{n-} \rightarrow \equiv SL^{1-n} + OH^- \tag{7.142d}$$

The formation of bidentate surface complexes may also occur, particularly for oxyanions (e.g., $HAsO_4^{2-}$ and HPO_4^{2-}) and trace metals (e.g., Pb^{2+} and Cu^{2+}):

$$2\equiv SOH^0 + M^{m+} \rightarrow (\equiv SO)_2 M^{m-2} + 2H^+ \tag{7.142e}$$

$$2\equiv SOH^0 + L^{n-} \rightarrow \equiv S_2 L^{2-n} + 2OH^- \tag{7.142f}$$

Each surface complexation reaction is described by an intrinsic equilibrium constant:

$$K_+^{int} = \frac{[\equiv SOH_2^+]}{[\equiv SOH^0](H^+)} \exp\left(\frac{F\psi_{(0)}}{RT}\right) \tag{7.143a}$$

$$K_-^{int} = \frac{[\equiv SO^-](H^+)}{[\equiv SOH^0]} \exp\left(\frac{-F\psi_{(0)}}{RT}\right) \tag{7.143b}$$

$$K_M^{int} = \frac{[\equiv SOM^{m-1}](H^+)}{[\equiv SOH^0](M^{m+})} \exp\left(\frac{(m-1)F\psi_{(0)}}{RT}\right) \tag{7.143c}$$

$$K_L^{\text{int}} = \frac{[\equiv SL^{1-n}](OH^-)}{[\equiv SOH^0](L^{n-})} \exp\left(\frac{(1-n)F\psi_{(0)}}{RT}\right) \tag{7.143d}$$

$$^2K_M^{\text{int}} = \frac{[(\equiv SO)_2 M^{m-2}](H^+)^2}{[\equiv SOH^0]^2(M^{m+})} \exp\left(\frac{(m-2)F\psi_{(0)}}{RT}\right) \tag{7.143e}$$

$$^2K_L^{\text{int}} = \frac{[\equiv S_2 L^{2-n}](OH^-)^2}{[\equiv SOH^0]^2(L^{n-})} \exp\left(\frac{(2-n)F\psi_{(0)}}{RT}\right) \tag{7.143f}$$

The mass and charge balance expressions for the surface are:

$$S_T = [\equiv SOH^0] + [\equiv SOH_2^+] + [\equiv SO^-] + [\equiv SOM^{m-1}] + [\equiv SL^{1-n}]$$
$$+ 2[(\equiv SO)_2 M^{m-2}] + 2[\equiv S_2 L^{2-n}] \tag{7.144}$$

$$\sigma_0 = \frac{F}{aS_A}\{[\equiv SOH_2^+] - [\equiv SO^-] + (m-1)[\equiv SOM^+] + (1-n)[\equiv SL^{1-n}]$$
$$+ (m-2)[(\equiv SO)_2 M^{m-2}] + (2-n)[\equiv S)_2 L^{2-n}]\} \tag{7.145}$$

where a is the concentration of the solid in the suspension (g L^{-1}), and S_A is the specific surface area (m^2 g^{-1}).

In the previous section, it was shown that the NEM could be used to describe the adsorption of Pb by gibbsite as a function of solution pH. Specifically, the chemical model that considered $\equiv AlOPbOH^{-0.5}$ formation was found to adequately predict Pb adsorption behavior (Figure 7.43a) and particularly the adsorption edge. The CCM may also be employed to describe and predict the adsorption of Pb by gibbsite. The chemical model is again described by the following reactions:

$$\equiv AlOH_2^{+0.5} \rightarrow \equiv AlOH^{-0.5} + H^+ \tag{7.146a}$$

$$\equiv AlOH^{-0.5} + Pb^{2+} + H_2O \rightarrow \equiv AlOPbOH^{-0.5} + 2H^+ \tag{7.146b}$$

$$Pb^{2+} + H_2O \rightarrow PbOH^+ + H^+ \tag{7.146c}$$

The associated equilibrium relationships are:

$$K_+^{\text{int}} = 10^{-8.87} = \frac{[\equiv AlOH^{-0.5}](H^+)}{[\equiv AlOH_2^{+0.5}]} \exp\left(\frac{-F\psi_{(0)}}{RT}\right) \tag{7.147a}$$

$$K_{Pb}^{\text{int}} = \frac{[\equiv AlOPbOH^{-0.5}](H^+)^2}{[\equiv AlOH^{-0.5}](Pb^{2+})} \tag{7.147b}$$

$$K_{PbOH} = 10^{-7.7} = \frac{(PbOH^+)(H^+)}{(Pb^{2+})} \tag{7.147c}$$

The mass balance expressions are:

$$S_T = [\equiv AlOH^{-0.5}] + [\equiv AlOH_2^{+0.5}] + [\equiv AlOPbOH^{-0.5}] \tag{7.148}$$

$$Pb_T = [Pb^{2+}] + [PbOH^+] + [\equiv AlOPbOH^{-0.5}] \tag{7.149}$$

The charge balance expression for σ_p is a function of the expressions that describe adsorbed proton and hydroxide charge balance, σ_H, and surface charge that results from adsorbed Pb, σ_{is}. The determination of σ_H is not a straightforward task. Consider the hypothetical, completely dissociated gibbsite surface functional group: $\equiv AlO^{-1.5}$. When all functional groups are completely dissociated, σ_H (in mol_c L^{-1}) will equal $-1.5[\equiv AlO^{-1.5}]$, or $-1.5S_T$. The adsorption of a proton will add charge to the surface, such that σ_H may be expressed:

$$\sigma_H = -1.5S_T + [\equiv AlOH^{-0.5}] + 2[\equiv AlOH_2^{+0.5}] \tag{7.150}$$

where $\equiv AlOH^{-0.5}$ results from the adsorption of one mole of protons and $\equiv AlOH_2^{+0.5}$ from two moles. In reality, $\equiv AlO^{-1.5}$ does not exist; therefore, S_T is the summation of $[\equiv AlOH^{-0.5}]$ and $[\equiv AlOH_2^{+0.5}]$, and Equation 7.150 can be expressed as:

$$\sigma_H = -1.5\{[\equiv AlOH^{-0.5}] + [\equiv AlOH_2^{+0.5}]\} + [\equiv AlOH^{-0.5}] + 2[\equiv AlOH_2^{+0.5}] \tag{7.151}$$

Further, the adsorption of $PbOH^+$ removes a proton from the s-plane:

$$\sigma_H = -1.5\{[\equiv AlOH^{-0.5}] + [\equiv AlOH_2^{+0.5}]\} + [\equiv AlOH^{-0.5}]$$
$$+ 2[\equiv AlOH_2^{+0.5}] - [\equiv AlOPbOH^{-0.5}] \tag{7.152}$$

Simplifying,

$$\sigma_H = -0.5[\equiv AlOH^{-0.5}] + 0.5[\equiv AlOH_2^{+0.5}] - [\equiv AlOPbOH^{-0.5}] \tag{7.153}$$

Converting the units of σ_H from mol_c L^{-1} to C m^{-2}:

$$\sigma_H = \frac{F}{aS_A}\{-0.5[\equiv AlOH^{-0.5}] + 0.5[\equiv AlOH_2^{+0.5}] - [\equiv AlOPbOH^{-0.5}] \tag{7.154}$$

The inner-sphere surface charge results strictly from the adsorption of $PbOH^+$:

$$\sigma_{is} = \frac{F}{aS_A}\{[\equiv AlOPbOH^{-0.5}]\} \tag{7.155}$$

Combining the expressions for σ_H and σ_{is} (Equations 7.154 and 7.155) leads to the charge balance expression:

$$\sigma_0 = \sigma_H + \sigma_{is} = \frac{F}{aS_A}\{-0.5[\equiv AlOH^{-0.5}] + 0.5[\equiv AlOH_2^{+0.5}]\} \tag{7.156}$$

Recalling Equation 7.141 and employing Equation 7.156, the surface potential is:

$$\psi_0 = \frac{F\sigma_0}{C} = \frac{F^2}{aCS_A}\{-0.5[\equiv AlOH^{-0.5}] + 0.5[\equiv AlOH_2^{+0.5}]\} \tag{7.157}$$

Application of the CCM to the Pb adsorption envelope, with $C = 1.25$ F m^{-2} (a reasonable value for gibbsite), leads to a K_{Pb}^{int} value of $10^{-9.4}$ for the adsorption reaction in Equation 7.146b (Figure 7.42b). In this instance, K_{Pb} also equals $10^{-9.4}$ because ΔZ for the adsorption reaction is zero. The predicted adsorption offered by the CCM is somewhat improved relative to that offered

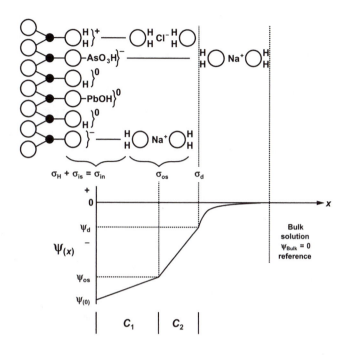

FIGURE 7.47 The solid-solution interface of a hydrous metal oxide surface as described by the modified triple layer model.

by the NEM, particularly as solution pH approaches the adsorption maximum. However, both the CCM and the NEM fail to adequately predict Pb adsorption when solution pH is below the pH_{50} (the pH at which adsorption is 50% of maximum).

7.7.2.2 The Modified Triple Layer Model

In the original triple layer model (TLM), the inorganic surface hydroxyl groups were assumed to form inner-sphere surface complexes only with the proton and the hydroxide ion. All other metals and ligands were restricted to forming outer-sphere complexes with the surface. The modified version of the TLM allows metal and ligand adsorption in the s-plane (the formation of inner-sphere complexes) while maintaining the ability to model the formation of outer-sphere complexes (Figure 7.47). In the modified version, the surface charge density is $\sigma_{in} + \sigma_{os} = -\sigma_d$, where $\sigma_{in} = \sigma_H + \sigma_{is}$ (as defined in the CCM), σ_{os} is the surface charge associated with the outer-sphere cation and anion complexation, and σ_d defines the charge density associated with the counterions.

In addition to the reactions considered by the CCM (Equations 7.142a through 7.142f), the following outer-sphere surface complexation reactions are considered:

$$\equiv SOH^0 + M^{m+} \rightarrow \equiv SO^- \!-\! M^{m+} + H^+ \tag{7.158a}$$

$$\equiv SOH^0 + H^+ + L^{n-} \rightarrow \equiv SOH_2^+ \!-\! L^{n-} \tag{7.158b}$$

$$\equiv SOH^0 + C^+ \rightarrow \equiv SO^- \!-\! C^+ + H^+ \tag{7.158c}$$

$$\equiv SOH^0 + H^+ + A^- \rightarrow \equiv SOH_2^+ \!-\! A^- \tag{7.158d}$$

where C^+ and A^- are the cation and anion of a background electrolyte. The intrinsic equilibrium constants for protonation and deprotonation, and the inner-sphere complexation of M^{m+} and L^{n-}

(Equations 7.142a through 7.142f) are identical to those defined for the CCM (Equations 7.143a through 7.143f). The intrinsic equilibrium constants for the outer-sphere complexation reactions in Equations 7.158a through 7.158d are:

$$K_{M,os}^{int} = \frac{[\equiv SO^- - M^{m+}](H^+)}{[\equiv SOH^0](M^{m+})} \exp\left(\frac{F(m\psi_{os} - \psi_{(0)})}{RT}\right) \tag{7.159a}$$

$$K_{L,os}^{int} = \frac{[\equiv SOH_2^+ - L^{n-}]}{[\equiv SOH^0](L^{n-})(H^+)} \exp\left(\frac{F(\psi_{(0)} - n\psi_{os})}{RT}\right) \tag{7.159b}$$

$$K_C^{int} = \frac{[\equiv SO^- - C^+](H^+)}{[\equiv SOH^0](C^+)} \exp\left(\frac{F(\psi_{os} - \psi_{(0)})}{RT}\right) \tag{7.159c}$$

$$K_A^{int} = \frac{[\equiv SOH_2^+ - A^-]}{[\equiv SOH^0](A^-)(H^+)} \exp\left(\frac{F(\psi_{(0)} - \psi_{os})}{RT}\right) \tag{7.159d}$$

The surface potential of each adsorption plane is given by the following expressions (where C_1 is the inner layer capacitance and C_2 is the outer layer capacitance):

$$\sigma_{in} = C_1(\psi_{(0)} - \psi_{os}) \tag{7.160}$$

$$\sigma_d = C_2(\psi_d - \psi_{os}) = -(\sigma_{in} + \sigma_{os}) \tag{7.161}$$

The diffuse layer charge is also described by a Boltzman distribution according to the Gouy-Chapman double layer theory:

$$\sigma_d = -(8RTI\varepsilon_0\varepsilon)^{0.5} \sinh\left(\frac{F\psi_d}{2RT}\right) \tag{7.162}$$

where I is the ionic strength of a 1:1 background electrolyte solution, ε_0 is the permittivity of vacuum, and ε is the dielectric constant of water. At 25°C, the expression for σ_d simplifies to:

$$\sigma_d = -0.1174 \times I^{0.5} \sinh\left(\frac{F\psi_d}{2RT}\right) \tag{7.163}$$

The mass and charge balance expressions are:

$$S_T = [\equiv SOH^0] + [\equiv SOH_2^+] + [\equiv SO^-] + [\equiv SOM^{m-1}] + [\equiv SL^{1-n}]$$

$$+ 2[(\equiv SO)_2M^{m-2}] + 2[\equiv S_2L^{2-n}] + [\equiv SO^- - M^{m+}] + [\equiv SOH_2^+ - L^{n-}]$$

$$+ [\equiv SO^- - C^+] + [\equiv SOH_2^+ - A^-] \tag{7.164}$$

$$\sigma_{in} = \frac{F}{aS_A}\{[\equiv SOH_2^+] - [\equiv SO^-] + (m-1)[\equiv SOM^+] + (1-n)[\equiv SL^{1-n}]$$

$$+ (m-2)[(\equiv SO)_2M^{m-2}] + (2-n)[(\equiv S)_2L^{2-n}] - [\equiv SO^- - C^+]$$

$$- [\equiv SO^- - M^{m+}] + [\equiv SOH_2^+ - A^-] + [\equiv SOH_2^+ - L^{n-}]\} \tag{7.165}$$

$$\sigma_{os} = \frac{F}{aS_A}\{m[\equiv SO^- - M^{m+}] - n[\equiv SOH_2^+ - L^{n-}]$$

$$+ [\equiv SO^- - C^+] - [\equiv SOH_2^+ - A^-]\} \tag{7.166}$$

The TLM can be applied to the Pb adsorption data described by the NEM and CCM in the previous sections. The chemical model that describes Pb adsorption is qualified in Equations 7.146a through 7.146c. The associated intrinsic and hydrolysis constants are given in Equations 7.147a through 7.147c. The following outer-sphere complexation reactions will also be considered:

$$\equiv\text{AlOH}^{-0.5} + \text{Na}^+ \rightarrow \equiv\text{AlOH}^{-0.5}\text{—Na}^+ \tag{7.167a}$$

$$\equiv\text{AlOH}^{-0.5} + \text{H}^+ + \text{NO}_3^- \rightarrow \equiv\text{AlOH}_2^{+0.5}\text{—NO}_3^- \tag{7.167b}$$

The intrinsic constants are:

$$K_{\text{Na}}^{\text{int}} = 10^{0.1} = \frac{[\equiv\text{AlOH}^{-0.5} - \text{Na}^+]}{[\equiv\text{AlOH}^{-0.5}](\text{Na}^+)}\exp\left(\frac{F(\psi_{\text{is}} - \psi_{(0)})}{RT}\right) \tag{7.168a}$$

$$K_{\text{NO}_3^-}^{\text{int}} = 10^{8.77} = \frac{[\equiv\text{AlOH}_2^+ - \text{NO}_3^-]}{[\equiv\text{AlOH}^{-0.5}](\text{NO}_3^-)(\text{H}^+)}\exp\left(\frac{F(\psi_{(0)} - \psi_{\text{os}})}{RT}\right) \tag{7.168b}$$

The mass balance expressions are:

$$S_T = [\equiv\text{AlOH}^{-0.5}] + [\equiv\text{AlOH}_2^{+0.5}] + [\equiv\text{AlOPbOH}^{-0.5}]$$

$$+ [\equiv\text{AlOH}^{-0.5}\text{—Na}^+] + [\equiv\text{AlOH}_2^{+0.5}\text{—NO}_3^-] \tag{7.169}$$

$$\text{Pb}_T = [\text{Pb}^{2+}] + [\text{PbOH}^+] + [\equiv\text{AlOPbOH}^{-0.5}] \tag{7.170}$$

and the charge balance expressions are:

$$\sigma_{\text{in}} = \frac{0.5F}{aS_A}\{-[\equiv\text{AlOH}^{-0.5}] + [\equiv\text{AlOH}_2^{+0.5}] - [\equiv\text{AlOH}^{-0.5} - \text{Na}^+]$$

$$+ [\equiv\text{AlOH}_2^{+0.5} - \text{NO}_3^-]\} \tag{7.171}$$

$$\sigma_{\text{os}} = \frac{F}{aS_A}\{[\equiv\text{AlOH}^{-0.5} - \text{Na}^+] - [\equiv\text{AlOH}_2^{+0.5} - \text{NO}_3^-]\} \tag{7.172}$$

In addition to the inner layer capacitance defined in the CCM ($C_1 = 1.25$ F m^{-2}), the outer layer capacitance is set at $C_2 = 0.2$ F m^{-2}. The results of the TLM predictions are illustrated in Figure 7.43c. Application of the TLM yields a $K_{\text{Pb}}^{\text{int}}$ of $10^{-9.4}$, which is identical to the intrinsic constant generated by the CCM. The fact that the CCM and the TLM predict identical $K_{\text{Pb}}^{\text{int}}$ values for the reaction $\equiv\text{AlOH}^{-0.5} + \text{Pb}^{2+} + \text{H}_2\text{O} = \equiv\text{AlOPbOH}^{-0.5} + 2\text{H}^+$ should not come as a surprise. In both models, the inner-sphere complexation of Pb is restricted to the is-plane. The TLM considers the outer-sphere complexation of the background electrolytes; however, the retention of these species in the os-plane will not impact Pb adsorption.

The chemical model used to describe Pb adsorption by gibbsite (the formation of $\equiv\text{AlOPbOH}^{-0.5}$) predicts Pb retention at pH values above the pH$_{50}$, but underpredicts retention at pH values below the pH$_{50}$. This suggests that an additional adsorption reaction is needed to predict Pb retention when solution pH is less than pH$_{50}$. Since the pH range in question is acidic (<6) and less than the pK_{PbOH}, Pb occurs predominantly as the divalent cation, Pb^{2+}. Further, when solution pH is less than the pH$_{50}$, the gibbsite surface sites predominantly consists of $\equiv\text{AlOH}_2^{+0.5}$ sites. Thus, the outer-sphere

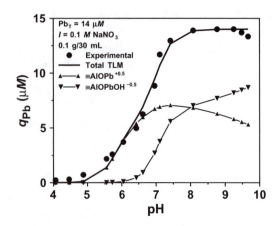

FIGURE 7.48 The adsorption of Pb by gibbsite as a function of pH (3.33 g L^{-1} gibbsite, 0.1 M NaNO$_3$, and a total Pb concentration of 14 μM). The solid lines represent the predicted Pb adsorption behavior by employing the modified triple layer model and by assuming the Pb adsorption occurs through the formation of two distinct surface complexes: $\equiv$AlOPb$^{+0.5}$ and $\equiv$AlOPbOH$^{-0.5}$.

complexation of Pb^{2+}, as described in Equation 7.158a, is not plausible. However, the inner-sphere complexation of Pb^{2+} to form $\equiv$AlOPb$^{+0.5}$, according to Equation 7.142c, may be used in conjunction with Equation 7.146b ($\equiv$AlOPbOH$^{-0.5}$ formation) to describe the Pb adsorption envelope. Application of the TLM as described above, but with the additional surface complexation reaction, predicts an adsorption envelope that closely approximates the experimental Pb retention (Figure 7.48). The TLM-predicted K_{Pb}^{int} values are $10^{-10.4}$ for $\equiv$AlOPbOH$^{-0.5}$ formation (Equation 7.146b), and $10^{2.4}$ for $\equiv$AlOPb$^{+0.5}$ formation (Equation 7.142c).

7.8   EXERCISES

1. The surfaces of soil minerals can be highly reactive. Identify and describe the various surface functional groups that exist on soil minerals. Include in your discussion the factors that influence the reactivity of the fuctional groups.

2. In order to assess the factors that impact the fate and behavior of the antibiotic chlortetracycline in soil, an adsorption isotherm study was conducted. Five-gram samples of high-organic-matter surface soil or low-organic-matter subsurface soil were equilibrated with 30 mL volumes of solution containing varying initial concentrations (c_{in}) of the antibiotic. The suspensions were equilibrated for 24 h and the solid and solution phases separated by centrifugation. The solution was then analyzed by HPLC for chlortetracycline to yield equilibrium concentrations (c_{eq}). The results are presented in the Table below. Answer the following:

   a. On a single graph, plot the chlortetracycline adsorption isotherms for both soils.

   b. For each isotherm answer the following:

      i. What are the isotherm types?

      ii. What adsorption model(s) should be employed to offer a quantitative characterization of the isotherms?

      iii. Fit the adsorption data with the appropriate adsorption isotherm model(s) and generate the adsorption constants.

iv. Why might the isotherms for the surface and subsoils samples differ (which soil adsorbs the greatest amount of the antibiotic and why)?

Initial (c_{in}) and Equilibrium Concentrations (c_{eq}) (μmol L^{-1}) of Chlortetracycline in Solutions in Contact with Surface and Subsurface Soil Samples

Initial Concentration	Equilibrium Concentration	
	Surface	Subsurface
9.86	0.031	0.010
19.89	0.09	0.013
38.58	0.066	0.018
86.84	0.162	0.032
170.97	0.358	0.076

3. Adsorption and precipitation are two mechanisms responsible for the loss of inorganic substances from a soil solution.
   a. Compare and contrast the two mechanisms.
   b. What experimental evidence is sufficient to determine which mechanism is responsible for the loss of an inorganic substance from the soil solution?
   c. Indicate why the adherence of experimental data to an adsorption isotherm equation provides no evidence as to the mechanism of inorganic substance retention.
4. Soil organic matter is a highly reactive material and can be responsible for a significant portion of a soil's capacity to retain inorganic and organic substances.
   a. Identify and describe the various surface-reactive functional groups that exist on soil organic matter.
   b. Indicate the soil solution composition variables that influence the reactivity of the organic fuctional groups and the manner in which these variables impact reactivity.
   c. Describe the mechanisms responsible for the retention of cations and anions by soil organic matter.
5. Compare and contrast the characteristics of surface functional groups on quartz, birnessite, and gibbsite.
6. Surface charge generated on hydrous metal oxide minerals is commonly referred to a pH-dependent charge. Some would consider this moniker to be a misnomer. Explain why this could be the case.
7. The structural formula of a 2:1 layer silicate is $Ca_{0.13}Na_{0.03}(Al_{1.21}Mg_{0.29})Si_4O_{10}(OH)_2$. The unit cell parameters for this mineral are $a = 0.517$ nm and $b = 0.894$ nm. Answer the following:
   Name the mineral.
   Compute a value for the cation exchange capacity of the mineral.
   Compute a value for the internal surface area of the mineral.
   Compute the permanent structural charge density (σ_o) of the mineral.
8. The water solubility of 1,1,1-trichloroethane is 5.40 mol m^{-3}. Calculate the K_{oc} for this compound. If a solution containing this compound is equilibrated with a soil containing 12 g kg^{-1} SOC, what is the predicted linear partition coefficient (K_p) that describes 1,1,1-trichloroethane adsorption by the soil?

9. Five-gram samples of a low-organic-matter subsurface soil were equilibrated with 30-mL volumes of solution containing varying background electrolytes and initial concentrations (c_{in}) of the antibiotic. The suspensions were equilibrated for 24 h and the solid and solution phases separated by centrifugation. The solutions were then analyzed by HPLC for chlortetracycline to yield equilibrium concentrations (c_{eq}). The results are presented in the table below. Answer the following:
   a. On a single graph, plot the chlortetracycline adsorption isotherms for both the 10 mM CaCl$_2$ and 10 mM KCl systems.
   b. For each isotherm answer the following:
      i. Fit the adsorption data with an appropriate adsorption isotherm model and generate the adsorption constants.
      ii. Examine the isotherms and the adsorption isotherm equations. The isotherms should indicate that chlortetracycline adsorption is a function of the background electrolyte. Provide plausible explanations as to how the background electrolyte influences chlortetracycline adsorption.

Initial (c_{in}) and Equilibrium Concentrations (c_{eq}) (in mg L^{-1}) of Chlortetracycline in Solutions in Contact with Subsurface Soil Samples as a function of Background Electrolyte

Initial Concentration	Equilibrium Concentration	
	10 mM CaCl$_2$	10 mM KCl
0.806	0.0058	0.0046
1.905	0.0069	0.0063
2.890	0.0130	0.0084
8.277	0.0306	0.0153
16.380	0.0692	0.0364

10. An adsorption isotherm examines the retention of a substance in a soil system that is vastly different from a soil as it exists in the environment.
   a. Discuss the positive and negative aspects associated with the application of adsorption isotherm constants to describe and predict the adsorption behavior of substances in the soil environment. Include in your discussion an assessment of the the value of adsorption constents when desorption hysteresis is significant.
   b. Assume that laboratory adsorption studies are inadequate for assessing the retention characteristics of a substance in the soil environment. How might adsorption studies be designed and performed to provide information on the adsorption and desorption characteristics of a substance in a soil as it exists in the environment?
11. Plot the adsorption edge for Ca adsorption by kaolinite using the data given in the table below. Assume that the adsorption maximum is 50 mmol kg^{-1} and calculate a value for the pH$_{50}$.

pH	q_{Ca} (mmol kg^{-1})	pH	q_{Ca} (mmol kg^{-1})
4.68	0.6	7.12	19.8
5.63	0.6	7.95	28.2
5.99	2.4	8.47	42.8
6.38	8.4	8.77	46.8

12. Data describing the adsorption of Hg(II) by kaolinite in 0.1 M $NaNO_3$ as a function of pH and in the absence or presence of 0.01 M $Ca(NO_3)_2$ is given in the table below. Plot the Hg(II) adsorption edge for both systems and answer the following:

   a. Determine the pH_{50} values for Hg(II) adsorption in the absence and presence of Ca. If the error associated with any given pH determination is ±0.05, are the two pH_{50} values significantly different?

   b. Offer plausible explanations as to why the adsorption behavior of Hg(II) is different when Ca is present, relative to when Ca is absent.

Hg(II) Adsorption, no $Ca(NO_3)_2$				Hg(II) Adsorption, 0.01 M $Ca(NO_3)_2$			
pH	q_{Hg}	pH	q_{Hg}	pH	q_{Hg}	pH	q_{Hg}
3.01	50.40	4.96	158.5	3.21	70.85	4.79	143.7
3.20	85.40	5.74	150.1	3.34	85.65	5.58	142.7
3.49	99.04	6.14	148.7	3.57	94.80	6.32	133.2
3.59	115.5	6.74	148.5	3.76	118.7	6.81	125.3
3.70	136.2	7.34	145.8	3.95	125.9	7.45	111.4
3.84	139.5	7.95	144.4	4.26	144.1		
4.02	149.3	8.13	142.3				
4.37	160.8	8.9	140.8				
4.61	160.4						

13. Surface complexation models are tools that are used to predict metal or ligand adsorption by constant potential mineral surfaces. For the (a) nonelectrostatic, (b) constant capacitance, and (c) modified triple layer models, describe how the model views the solid-solution interface (i.e., where does adsorption occur and what types of adsorption processes are considered) and identify the assumptions employed.

14. An average value for the site density of singly-coordinated (type A) surface functional groups on the gibbsite surface is 4 sites nm^{-2}. Compute the concentration of (a) positive and (b) negative surface charge that is present at the pH_{pzc} given that the specific surface area of gibbsite is 40.8 m^2 g^{-1}. Express your results in units of $cmol_c$ kg^{-1}.

15. Montmorillonite and birnessite share common chemical and structural characteristics that result in points of zero charge values in the 2 to 4 range. Examine the chemical and structural characteristics of birnessite and indicate why the pH_{pzc} of this hydrous oxide is similar to that of montmorillonite ($pH_{pzc} \approx 2.5$), rather than to the pH_{pzc} values of anatase and rutile (information presented in Chapter 2 may be relevant).

REFERENCES

Bowman, R.S., M.E. Essington, and G.A. O'Connor. Soil sorption of nickel: influence of solution composition. *Soil Sci. Soc. Am. J.* 45:860–865, 1981.

Chiou, C.T., L.J. Peters, and V.H. Freed. A physical concept of soil–water equilibria for nonionic organic compounds. *Science* 206:831–832, 1979.

Chiou, C.T., P.E. Porter, and D.W. Schmedding. Partition equilibria of nonionic organic compounds between soil organic matter and water. *Environ. Sci. Technol.* 17:227–231, 1983.

Essington, M.E. Adsorption of aniline and toluidines on montmorillonite. *Soil Sci.* 158:181–188, 1994.

Gao, Y. and A. Mucci. Acid base reactions, phosphate and arsenate complexation, and their competitive adsorption at the surface of goethite in 0.7 M NaCl solution. *Geochim. Cosmochim. Acta* 65:2361–2378, 2001.

Goldberg, S., H.S. Forster, and E.L. Heick. Boron adsorption mechanisms on oxides, clay minerals, and soils inferred from ionic strength effects. *Soil Sci. Soc. Am. J.* 57:704–708, 1993.

Grafe, M., M.J. Eick, and P.R. Grossl. Adsorption of arsenate (V) and Arsenite (III) on goethite in the presence and absence of dissolved organic carbon. *Soil Sci. Soc. Am. J.* 65:1680–1687, 2001.

Green, R.E. and S.W. Karickhoff. Sorption estimates for modeling. In *Pesticides in the Soil Environment*. H.H. Cheng (Ed.) SSSA Book Series, no. 2, SSSA, Madison, WI, 1990, pp. 79–101.

Hassett, J.J., J.C. Means, W.L. Banwart, and S.G. Wood. Sorption Properties of Sediments and Energy Related Pollutants. EPA-600/3080-041, 1980.

Hassett, J.J., W.L. Banwart, and R.A. Griffin. Correlation of compound properties with sorption characteristics of nonpolar compounds by soils and sediments: concept and limitations. In *Environment and Solid Wastes*. C.W. Francis and S.I. Auerbach (Ed.) Butterworths, Boston, 1983, pp. 161–178.

Hingston, F.J., A.M. Posner, and J.P. Quirk. Anion adsorption by goethite and gibbsite. I. The role of the proton in determining adsorption envelopes. *J. Soil Sci.* 23:177–192, 1972.

Karickhoff, S.W., D.S. Brown, and T.A. Scott. Sorption of hydrophobic pollutants on natural sediments. *Water Res.* 13:241–248, 1979.

Miller, M.M., S.P. Wasik, G.L. Huang, W.Y. Shiu, and D. Mackay. Relationships between octanol–water partition coefficient and aqueous solubility. *Environ. Sci. Technol.* 19:522–529, 1985.

Okazaki, M., K. Sakaidani, T. Saigusa, and N. Sakaida. Ligand exchange of oxyanions on synthetic hydrated oxides of iron and aluminum. *Soil Sci. Plant Nutr.* 35:337–346, 1989.

Rao, P.S.C. and J.M. Davidson. Estimation of pesticide retention and transformation parameters required for non–point source pollution models. In *Environmental Impact of Nonpoint Source Pollution*. M.R. Overcash and J.M. Davidson (Ed.) Ann Arbor Science Publisher, Ann Arbor, MI, 1980, pp. 23–67.

Sahai, N. Is silica really an anomalous oxide? Surface acidity and aqueous hydrolysis revisited. *Environ. Sci. Technol.* 36:445–452, 2002.

Sahai, N. and D.A. Sverjensky. Evaluation of internally consistent parameters for the triple-layer model by the systematic analysis of oxide surface titration data. *Geochim. Cosmochim. Acta* 61:2801–2826, 1997.

Sarkar, D., Adsorption of Mercury onto Variable Charge Surfaces. Ph.D. dissertation. The University of Tennessee, Knoxville, 1997.

Sarkar, D., M.E. Essington, and K.C. Misra. Adsorption of mercury(II) by variable charge surfaces of quartz and gibbsite. *Soil Sci. Soc. Am. J.* 63:1626–1636, 1999.

Sarkar, D., M.E. Essington, and K.C. Misra. Adsorption of mercury(II) by kaolinite. *Soil Sci. Soc. Am. J.* 64:1968–1975, 2000.

Schwarzenbach, R.P. and J. Westall. Transport of nonpolar organic compounds from surface water to ground water. Laboratory sorption studies. *Environ. Sci. Technol.* 15:1360–1367, 1981.

Sposito, G. Derivation of the Freundlich equation for ion exchange reactions in soils. *Soil Sci. Soc. Am. J.* 44:652–654, 1980.

Sposito, G. *The Surface Chemistry of Soils*. Oxford University Press, New York, 1984.

Suba, J.D. and M.E. Essington. Adsorption of fluometuron and norflurazon: effect of tillage and dissolved organic carbon. *Soil Sci.* 164:145–155, 1999.

Sverjensky, D.A. and N. Sahai. Theoretical prediction of single-site surface-protonation equilibrium constants for oxides and silicates in water. *Geochim. Cosmochim. Acta* 60:3773–3797, 1996.

8 Cation Exchange

Perhaps one of the most studied phenomena in soil chemistry has been the process of cation exchange. Cation exchange is an adsorption process, and like all other adsorption processes it involves the displacement of an adsorbate from the soil surface by an adsorptive. Cation exchange is distinguished from other retention mechanisms by the nature of the interaction between the surface functional group and the adsorbed ion. Exchangeable ions are held by soil surfaces strictly through a relatively weak electrostatic (Coulombic) and nonspecific interaction. All aqueous species, inorganic and organic, that exist as cations in soil solutions may participate in cation exchange reactions. Cations that participate in exchange reactions remain hydrated and form outer-sphere surface complexes or reside in the diffuse ion swarm of the solid-solution interface. Outer-sphere complexation occurs because the Lewis base surface functional group does not have the base strength necessary to dislodge waters of hydration (which are also Lewis bases) from the primary hydration sphere of the metal cation or ionized organic moiety. Most commonly, the surface functional groups associated with cation exchange are the ditrigonal siloxane cavity, found in the interstitial regions of phyllosilicates, and the deprotonated carboxylic and phenolic moieties of soil organic matter. The inorganic functional groups on hydrous metal oxides may also participate in ion exchange reactions, particularly in highly weathered soils. However, the contribution of hydrous metal oxides to the soil's capacity to retain exchangeable ions is relatively minor.

8.1 CATION EXCHANGE: A BEGINNING FOR SOIL CHEMISTRY

Before the year 1850, the belief that soil was an inert material was prevalent among agricultural scientists. Even Justus von Liebig, a revered agricultural chemist of the period, the first to demonstrate that plants obtain mineral nutrients from soil and the first to recognize that the addition of a single essential nutrient would increase crop yield only if all other nutrients were present in sufficient levels (Liebig's "Law of the Minimum"), espoused the belief that soil was merely a nonreactive filter and support media for plants. However, all this changed in 1845 when an agriculturalist and Yorkshire farmer named Harris Stephan Thompson performed a small series of experiments (Thompson, 1850). To glass columns containing a "*light sandy loam of good quality*," a "*black soil (from the bottom of an old stick heap)*," and a "*strong clay soil*," Thompson added a $(NH_4)_2SO_4$ solution. Upon leaching each column with a volume of water that exceeded "*in amount the heaviest continuous fall of rain which is ordinarily experienced in this country*," a leachate was obtained which, when dried, yielded a precipitate that was found to be principally gypsum ($CaSO_4 \cdot 2H_2O$), as "*the whole of the ammonia was retained by the soil.*"

In 1848, H.S. Thompson related to John Thomas Way, a consulting chemist to the Royal Agricultural Society, the results described above. In addition, a Dorset farmer named Huxtable related to Way that, "*he made an experiment in the filtration of the liquid manure in his tanks through a bed of an ordinary loamy soil; and that after its passage through the filter-bed, the urine was found to be deprived of colour and smell—in fact, that it went in manure and came out water,*" (Way, 1850). The findings of Huxtable, and particularly the quantitative findings of Thompson, lead J.T. Way (Way, 1850; 1852) to perform the first comprehensive studies of the process later termed base exchange. Way's studies were principally confined to two soils, a "red" soil from Berkshire and a loam from Dorsetshire Downs, although white pottery clay, washed sand, red brick dust, powdered tobacco pipe (composed of clay), and aluminosilicate precipitates were also used. In general, Way's experiments consisted of passing solutions of common salts (including NH_4, K, Na, Mg, and Ca salts of OH, SO_4, Cl, NO_3, and CO_3) through columns or filter beds of the various materials. In addition, experiments were conducted using soil particle size separates, dried soil (50 to 60°C), combusted soil, and HNO_3- and HCl-digested materials. Soil leachates of sodium phosphate solution, guano extract, human urine, flax water, sewer water, and Thames River water were also examined. The results and conclusions of J.T. Way were revolutionary (from Way, 1852; the salient statements are in bold):

…it was found that ordinary soils possessed the power of separating from solution in water the different earthy and alkaline substances presented to them in manure; **thus, when solutions of salts of ammonia, of potash, magnesia, &c ., were made to filter slowly through a bed of dry soil, 5 or 6 inches deep, arranged in a flower-pot or other suitable vessel, it was observed that the liquid which first ran through no longer containing any of the ammonia or other salt employed**.

But further, this power of the soil was found not to extend to the whole salt of ammonia or potash, but only to the alkali itself. **If, for instance, sulphate of ammonia were the compound used in the experiments, the ammonia would be removed from solution, but the filtered liquid would contain sulfuric acid in abundance – not in the free or uncombined form, but united with lime; … and this result was obtained whatever the acid of the salt experimented on might be**.

It may be mentioned, also, in this place, that, at a later period of the investigation, it was satisfactorily proven that **the quantity of lime acquired by the solution corresponded exactly to that of the ammonia removed from it – the action was therefore a true chemical decomposition**.

Again, it was found **that the combination between the soil and the alkaline substance was rapid, if not instantaneous**, partaking therefore of the nature of the ordinary union between an acid and alkali. In the course of these experiments several different soils were operated upon, and it was found that all soils capable of profitable cultivation possessed the property in question in a greater or lesser degree. **It was shown that the power to absorb alkaline substances did not exist in sand; that the organic matters of the soil had nothing to do with it**; that the addition of carbonate of lime to a soil did not increase its absorptive power for these salts; and indeed that a soil in which carbonate of lime did not occur, might still possess in a high degree the power of removing ammonia or potash from solution, and **it was evident that the active ingredient in all these cases was clay**.

In addition to the above findings, Way also observed the complete soil removal of phosphorus from guano extract and sodium phosphate leachates (a result that Way attributed to the formation of calcium phosphate precipitates, rather than "absorption"), that the acid digestion of soil did not destroy the absorptive power, and that the combustion of soil or clay diminished the ability of the material to absorb. An initial conclusion of J.T. Way—that organic matter is unimportant in cation retention—was later refuted through the work of Samuel W. Johnson (1859), who noted that swamp-muck was "capable of absorbing 1.3 per cent of ammonia, while ordinary soil absorbs but 0.5 to 1 per cent." The experiments of Way were limited to exchange reactions involving the displacement of native Ca^{2+} by the common cations, Na^+, K^+, NH_4^+, and Mg^{2+}. Indeed, until the work of J.M. van Bemmelen in the 1880s, it was generally thought that only Ca^{2+} was displaced from soils by exchange reactions, although Way himself noted the appearance of more Na^+ in white clay leachate than in the influent flax water: "This soda can only have been derived from the clay, which we find from the analysis contains this alkali in considerable quantity. It would seem, therefore, that in the present instance soda, and not lime, had acted the part of the substituting base" (Way, 1852).

8.2 QUALITATIVE ASPECTS OF CATION EXCHANGE

With the exception of a small number of misinterpretations, the conclusions of J.T. Way have stood as the defining characteristics of cation exchange reactions. Ion exchange reactions are reversible, rapid (rate is controlled by site accessibility and is said to be diffusion controlled), and stoichiometric with respect to charge. The following exchange reactions illustrate the stoichiometric and equivalent replacement of an ion on the exchange complex by an ion from the aqueous phase:

$$0.5CaCl_2(aq) + Mg_{0.5}X(ex) \rightarrow Ca_{0.5}X(ex) + 0.5MgCl_2(aq) \tag{8.1a}$$

$$NH_4^+(aq) + Ca_{0.5}X(ex) \rightarrow NH_4X(ex) + 0.5Ca^{2+}(aq) \tag{8.1b}$$

$$K^+(aq) + Al_{0.33}X(ex) \rightarrow KX(ex) + 0.33Al^{3+}(aq) \tag{8.1c}$$

where X^- represents an equivalent of exchange phase charge. In the context of exchange reactions, the equivalent quantity of a cation may be described as the moles of cation that replace a mole of hydrogen ions. Since a mole of proton is identical to a mole of charge (mol_c), an equivalent quantity of a cation may also be described as the moles of cation that are equivalent to a mole of charge. For example, ⅓ mol of Al^{3+} is equivalent to 1 mol charge (a mole of Al^{3+} is equivalent to 3 mol of charge), a mole of Ca^{2+} or Mg^{2+} is equivalent to 2 mol of charge, and a mole of NH_4^+ or K^+ is equivalent to 1 mol of charge. In Equations 8.1a to 8.1c, one equivalent of exchange phase charge (the same as one mole of exchange phase charge) is satisfied by one equivalent of exchangeable cation. Thus, one mole of NH_4^+ or K^+, ½ mol of Ca^{2+} or Mg^{2+}, or ⅓ mol of Al^{3+} satisfies a mole of exchange charge. The cation exchange reactions also illustrate the exchange stoichiometry. For example, displacement of one equivalent of Al^{3+} from the exchange phase requires one equivalent of K^+, or 3 mol of K^+ are required to displace 1 mol of Al^{3+}.

Equations 8.1a to 8.1c correctly display cation exchange equilibria for Mg^{2+}-Ca^{2+}, Ca^{2+}-NH_4^+, and Al^{3+}-K^+ exchange, as the requirements of charge and mass balance are satisfied. Ion exchange reactions may also be expressed, for example, as indicated for Ca^{2+}-NH_4^+ exchange:

$$2NH_4^+(aq) + CaX_2(ex) \rightarrow 2NH_4X(ex) + Ca^{2+}(aq) \tag{8.2a}$$

$$2NH_4^+(aq) + CaX(ex) \rightarrow (NH_4)_2X(ex) + Ca^{2+}(aq) \tag{8.2b}$$

where X^- represents a mole of exchange phase charge in Equation 8.2a (as in Equation 8.1b), and X^{2-} represents two moles of exchange phase charge in Equation 8.2b. Both reactions in Equation 8.2 satisfy mass and charge balance requirements, and are equally valid mechanisms for describing the Ca^{2+}-NH_4^+ exchange process (along with Equation 8.1b). Finally, exchange reactions are also qualified with respect to the nature of the ions involved in the process. Equation 8.1a is an example of a symmetrical (homovalent) cation exchange reaction, as the ions involved have the same charge. Equations 8.1b and c and 8.2a and b are examples of nonsymmetrical (heterovalent) cation exchange reactions, as the ions involved have unequal charges. Equations 8.1b and 8.2a and b are uni-bivalent exchange reactions; Equation 8.1c is a uni-trivalent exchange reaction.

The use of equivalents to describe chemical behavior (as in ion exchange) or ion concentrations is no longer an acceptable practice. Prior to the favored use of SI units in the soil sciences, it was common to express the concentrations of ions in soil solutions in terms of normality (N), eq L^{-1} or meq L^{-1}, and the cation exchange capacity of a soil in meq 100 g^{-1} (milliequivalents per 100 grams). The composition of a soil solution that contains 10 meq L^{-1} Na as Na^+ and 10 meq L^{-1} Ca as Ca^{2+} contains 10 mmol L^{-1} Na and 5 mmol L^{-1} Ca, assuming that an equivalent is defined as the moles of a substance that are identical to a mole of charge. One of the ambiguities associated with the use of equivalent units is demonstrated in the above example. Expressing the total concentration of a dissolved substance in equivalent units requires an assumption relative to ion speciation. Implicit in the 10 meq L^{-1} Ca concentration is the assumption that Ca exists and reacts as the divalent ion, even though a significant proportion of the total soluble Ca may exist in ion pairs (e.g., $CaCl^+$ and $CaSO_4^0$) or organic complexes (which are species that will not replace two equivalents of proton charge).

An additional ambiguity in the use of equivalents is that the unit is specific to the type of chemical reaction that occurs. For example, in the exchange reactions:

$$FeCl_2(aq) + 2KX(ex) \rightarrow FeX_2(ex) + 2KCl(aq) \tag{8.3a}$$

$$FeCl_3(aq) + 3KX(ex) \rightarrow FeX_3(ex) + 3KCl(aq) \tag{8.3b}$$

1 mol of Fe^{2+} in Equation 8.3a will replace 2 mol of K^+. Similarly, 1 mol of Fe^{3+} in Equation 8.3b will replace 3 mol of K^+. Thus, the equivalent mass of Fe^{2+} (g eq^{-1}) is one half the molar mass (g mol^{-1}) of Fe, and the equivalent mass of Fe^{3+} is one third the molar mass of Fe (there are 2 eq mol^{-1} of Fe^{2+} and 3 eq mol^{-1} of Fe^{3+}). However, in redox reactions, the equivalent quantity of a cation may be described as the moles of cation that are equivalent to a mole of electrons transferred. For example, in the reaction:

$$FeCl_2(aq) + 0.25O_2(g) + HCl(aq) \rightarrow FeCl_3(aq) + 0.5H_2O(l) \tag{8.4}$$

both Fe^{2+} and Fe^{3+} are capable of donating and accepting only one electron; thus, the equivalent masses of Fe^{2+} and Fe^{3+} are identical to the molar mass of Fe (1 eq mol^{-1} of Fe^{2+} and 1 eq mol^{-1} of Fe^{3+}). Therefore, a solution that contains 10 meq L^{-1} $FeCl_3$ may contain 3.3 mmol L^{-1} $FeCl_3$ if Equation 8.3b is the reaction of interest, or 10 mmol L^{-1} if Equation 8.4 is the reaction of interest. Because there is no ambiguity in the use of molar concentration units, their use is favored.

Ion exchange reactions abide by the Law of Mass Action and respond to perturbations of the soil chemical environment as described by Le Châtelier's principle. For the reaction,

$$2Na^+(aq) + CaX_2(ex) \rightarrow 2NaX(ex) + Ca^{2+}(aq) \tag{8.5}$$

the retention of Ca^{2+} is favored if the system initially contains equal concentrations of soluble Ca^{2+} and Na^+ (retention of higher valence species is favored). Le Châtelier's principle states that when a stress is brought upon a system at equilibrium, a change will occur such that the equilibrium is displaced in a direction that tends to undo the effect of the stress. Although the retention of the Na^+ ion is not favored over Ca^{2+}, it can be forced onto the exchange complex by any stress that disturbs the equilibrium in favor of the formation of NaX. Thus, any process that increases Na^+ in the solution or removes soluble Ca^{2+} will increase NaX and decrease CaX_2 concentrations. For example, loading the system with Na^+, forming a Ca precipitate, or forming Ca^{2+} ion pairs (e.g., $CaSO_4^0$) will favor the formation of NaX by forcing Equation 8.5 to proceed further to the right in response to the stress placed on the equilibrium of the system. The creation of homoionic clays, as required for effective particle size separation and x-ray diffraction analysis (Chapter 2), is an example of Le Châtelier's principle in action. A soil can be Na^+ saturated, irrespective of the native exchange phase composition, by a sequential centrifuge washing technique. In this technique, the soil is reacted with a concentrated (1 M) NaCl solution in a centrifuge tube. The suspension is centrifuged, the supernatant liquid is removed and replaced by another aliquot of NaCl solution, and then the process is repeated (a minimum of three times). The high concentration of Na in the equilibrating solution stresses the system and favors the displacement of Ca^{2+} and other native cations, which are removed in the supernatant after centrifugation, leaving the soil exchange complex Na^+ saturated.

A characteristic of nonsymmetrical ion exchange reactions is that the dilution of a solution in equilibrium with the exchanger phase favors the retention of more highly charged exchangeable ion. This characteristic is known as the valence dilution effect, and has been useful in the reclamation of sodic environments. Correspondingly, increasing the salt content of the equilibrating solution (e.g., through evaporation) will result in the preferential retention of the lower valence ion. For the reaction:

$$2Na^+(aq) + CaX(ex) \rightarrow Na_2X(ex) + Ca^{2+}(aq) \qquad (8.6a)$$

where X^{2-} represents 2 mol of surface charge, the Law of Mass Action states that an exchange selectivity coefficient (reaction quotient), K_S, may be written:

$$K_S = \frac{\{Na_2X\}[Ca^{2+}]}{\{CaX\}[Na^+]^2} \qquad (8.6b)$$

where { } and [ ] denote effective concentration variables of the exchange and aqueous phase species. Rearranging Equation 8.6b yields:

$$K_S \frac{[Na^+]^2}{[Ca^{2+}]} = \frac{\{Na_2X\}}{\{CaX\}} \qquad (8.6c)$$

If $[Na^+] = [Ca^{2+}] = 1.0$ in the equilibrating solution, Equation 8.6c reduces to:

$$K_S = \frac{\{Na_2X\}}{\{CaX\}} \qquad (8.6d)$$

If this solution is subjected to a tenfold dilution, as might occur during a rainfall event, then $[Na^+] = [Ca^{2+}] = 0.1$. When the exchange and solution phases have reequilibrated, the new equilibrium condition, according to Equation 8.6c is:

$$K_S \frac{[0.1]^2}{[0.1]} = 0.1K_S = \frac{\{Na_2X\}}{\{CaX\}} \qquad (8.6e)$$

In this example, a tenfold decrease in concentration in the equilibrating solution will result in a tenfold decrease in the ratio of Na to Ca on the exchange complex, if K_S is assumed to remain constant; or a tenfold increase in the ratio of Ca to Na on the exchange complex (ten times as much Ca^{2+} as Na^+ on the exchange phase relative to the original system). Thus, dilution of a solution in equilibrium with an exchanger phase will favor the retention of the more highly charged exchangeable ion.

Exchangeable cations are highly hydrated and do not actually form chemical bonds with the adsorbing surface. Instead, exchangeable cations are held at the surface through an electrostatic interaction and their retention is influenced by the variables described in Coulomb's law:

$$F = \frac{q_+ q_-}{\varepsilon r^2} \qquad (8.7)$$

As this equation states, the force of attraction (F) between opposing charges is directly related to the magnitude of the charges q_+ and q_- and inversely related to the square of the separation distance r in a uniform medium having a dielectric constant ε. Therefore, the effective size and valence of an exchangeable ion determine its exchangeability. The ease with which adsorbed ions can be displaced from the surface by completing ions can be predicted from size and valence parameters. As the valence of an exchangeable cation increases, so does the force of attraction to a charged surface. The greater the valence, the greater is the selectivity of the surface for the cation. Further, for a given valence, the hydrated radius of an ion determines exchangeability. As the hydrated radius of an exchangeable cation decreases (r decreases), the force of cation attraction to a charged surface increases.

The relative replaceability of exchangeable cations (ease of removal) is described by a lyotropic series. Beginning with the most easily removed cation, the lyotropic series for the monovalent and divalent exchangeable cations are (hydrated radii in nanometers are displayed in parentheses):

$$Li^+ (0.382) \approx Na^+ (0.358) > K^+ (0.331) \approx NH_4^+ (0.331) > Rb^+ (0.329) > Cs^+ (0.329)$$

and

$$Mg^{2+} (0.428) > Ca^{2+} (0.412) > Sr^{2+} (0.412) \approx Ba^{2+} (0.404)$$

All cations may participate in cation exchange reactions, even though some may also participate in specific retention processes with surface functional groups. Thus, the lyotropic series can be expanded to include trace metal cations:

$$Li^+ (0.382) > Na^+ (0.358) > NH_4^+ (0.331) > K^+ (0.331) > Rb^+ (0.329) >$$

$$Cs^+ (0.329) > Ag^+ (0.341)$$

and

$$Mg^{2+} (0.428) > Zn^{2+} (0.430) > Co^{2+} (0.423) > Cu^{2+} (0.419) > Cd^{2+} (0.426) > Ni^{2+} (0.404) >$$

$$Ca^{2+} (0.412) > Sr^{2+} (0.412) > Pb^{2+} (0.401) > Ba^{2+} (0.404)$$

8.3 CATION EXCHANGE CAPACITY AND EXCHANGE PHASE COMPOSITION

There are two distinctly different definitions of an exchangeable ion; one operational and the other mechanistic. An exchangeable cation is operationally defined as an ion that is removed from the soil by a solution containing a neutral salt. The operational definition of an exchangeable cation must not be confused with that of a soluble cation, which is a cation that can be removed by the action of water alone. An exchangeable cation is mechanistically defined as an ion in the fully hydrated state that is bound at the soil surface solely by outer-sphere and diffuse-ion swarm mechanisms. If the operation to obtain a measure of exchangeable ions is performed correctly, then only those cations that are nonspecifically adsorbed, residing in the os- and d-planes of the solid-solution interface (Chapter 7), will be displaced and the two definitions will be equivalent.

8.3.1 EXCHANGE PHASE COMPOSITION

The readily exchangeable cations that are present in the soil can generally be predicted on the basis of the soil solution pH. In alkaline soils, the cations that dominate the exchange complex are collectively known as the base cations: Ca^{2+}, Mg^{2+}, Na^+, and K^+. Although labeled as such, these cations are not actually basic. They do not donate hydroxide to a solution, nor do they consume protons. They are completely dissociated from the hydroxide anion throughout the range of soil solution pH values. The base cations are correctly termed nonacidic, as they do not hydrolyze in soil solutions (although 'base cations' remains in common usage). Sodium and K hydroxide are both strong bases,

TABLE 8.1
Average Composition (Mean ± Standard Deviation) of the
Exchange Phase of Soils from Around the World[a]

	cmol$_c$ kg^{-1}	
Exchangeable Ion	pH$_w$ < 6, n = 1027[b]	pH$_w$ > 7, n = 249
Ca^{2+}	3.80 ± 5.65	25.18 ± 16.28
Mg^{2+}	1.65 ± 2.49	10.06 ± 8.49
Na^+	0.249 ± 1.487	1.21 ± 4.31
K^+	0.234 ± 0.324	0.737 ± 0.684
Al^{3+c}	8.76 ± 11.71	0

[a] Data were obtained from the International Soil Reference and Information Centre, Wageningen, The Netherlands (Batjes, 1995).

[b] n represents the number of soil samples, collected from the A and B horizons, included in the mean.

[c] 1 mol kg^{-1} exchangeable Al is assumed to satisfy 3 mol$_c$ kg^{-1} of exchange phase charge.

with pK_b values for NaOH and KOH dissociation of –0.2 and –0.5. Calcium and Mg hydroxides ($Ca(OH)_2$ and $Mg(OH)_2$) are also strong bases. Further, the pK_{b2} values for $MgOH^+$ and $CaOH^+$ dissociation are 2.58 and 1.3. Alternatively, the pK_{a1} values for Mg^{2+} and Ca^{2+} hydrolysis are 11.42 and 12.7 (where pK_{a1} = 14 – pK_{b2}). Thus, these cations are nonacidic because they do not hydrolyze in the pH range of normal soil environments. On average, the exchange phase charge satisfied by the readily exchangeable cations in alkaline soils (pH > 7) is dominated by Ca^{2+}, followed by Mg^{2+} (~40% of exchangeable Ca^{2+}), Na (~4.8% of Ca^{2+}), and K^+ (~2.9% of Ca^{2+}) (Table 8.1). In acid soils, with solution pH values that are generally less than 6, Al^{3+} and the associated hydrolysis products $AlOH^{2+}$ and $Al(OH)_2^+$ dominate exchange phase charge, followed by Ca^{2+} (43% of exchangeable Al^{3+}), Mg^{2+} (19% of Al^{3+}), Na^+ (2.9% of Al^{3+}), and K^+ (2.7% of Al^{3+}). Acid soils may also contain an appreciable amount of exchangeable Mn^{2+}. For example, the average exchangeable Mn^{2+} concentrations (0.27 cmol$_c$ kg^{-1} or 0.135 cmol kg^{-1}) in loessial east Tennessee surface soils (pH$_w$ < 6) exceed exchangeable Al^{3+} (0.18 cmol$_c$ kg^{-1} or 0.06 cmol kg^{-1}).

The cation exchange capacity (*CEC*) of a soil can be defined as the moles of adsorbed cation charge that can be displaced by an index ion per unit mass of soil. More critically, *CEC* is defined as the total charge excess of cations over anions in the os- and d-planes, under stated (controlled) conditions of temperature, pressure, soil solution composition, soil-to-solution mass ratio, and other method dependent variables. The *CEC* is expressed in units of cmol$_c$ kg^{-1}, which are equivalent to the units of meq 100 g^{-1} common in the early literature (i.e., 1 cmol$_c$ kg^{-1} = 1 meq 100 g^{-1}). In general, the reported *CEC* of a soil often refers to the maximum negative surface charge and indicates the potential *CEC* of the soil. The potential *CEC* of a soil indicates surface charge arising from constant charge surfaces (from isomorphic substitution) and deprotonated inorganic surface functional groups and organic functional groups.

In mineral soils, and with the exception of highly weathered soils, the *CEC* is primarily dictated by the abundance and types of phyllosilicates that are present, although hydrous metal oxides may in some situations also significantly contribute to the *CEC* of a soil (*CEC* values for the various phyllosilicate and hydrous metal oxide minerals are given in Chapter 2). Although highly variable, the representative *CEC* values of 10 mineral soil orders are generally characteristic of the types of clay materials common to the respective order (Table 8.2). On average, Vertisols display the greatest *CEC* values owing to their high smectitic clay content. Similarly, Andisols display a relatively high *CEC* due to the abundance of allophane and smectite. Highly weathered soils, as found in the

TABLE 8.2
The Potential Cation Exchange Capacity
(Mean CEC ± Standard Deviation) of
Surface Mineral Soils (A and B
Horizons) from Around the World[a]

Soil Order (*n*)[b]	CEC, cmol$_c$ kg^{-1}
Alfisols (295)	15.4 ± 11.1
Andisols (35)	30.9 ± 18.4
Aridisols (35)	17.8 ± 11.3
Entisols (53)	19.9 ± 13.7
Inceptisols (364)	21.1 ± 15.7
Mollisols (238)	24.0 ± 11.8
Oxisols (240)	7.6 ± 5.9
Spodosols (23)	26.7 ± 30.1
Ultisols (223)	8.9 ± 6.3
Vertisols (80)	50.1 ± 16.6

[a] Data were obtained from the International Soil Reference and Information Centre, Wageningen, The Netherlands (Batjes, 1995).

[b] *n* represents the number of soil samples included in the mean.

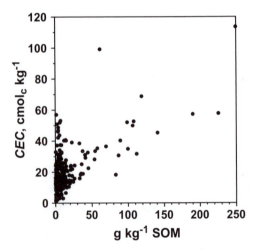

FIGURE 8.1 The cation exchange capacity (*CEC*) of 249 surface soil samples collected from around the U.S. as a function of soil organic matter content (SOM). Data were obtained from Batjes (1995).

Ultisol and Oxisol orders, are characterized by relatively low *CEC* values, due the absence of constant charge surfaces and the presence of constant potential mineral surfaces that have relatively low *CEC* values. The *CEC* of Histosols (organic soils) and the O horizons of mineral soils can be considerable, due to the abundance of dissociated carboxylic and phenolic functional groups (discussed in Chapter 4). The representative *CEC* of Histosols is 140 ± 30 cmol$_c$ kg^{-1} (Sposito, 2000), a value that far exceeds that of Vertisols. On average, the *CEC* associated with the O horizons of mineral soils is 133.2 ± 58.2 cmol$_c$ kg^{-1} (average soil organic matter content of 426.5 g kg^{-1}).

The variation of *CEC* as a function of organic matter content and clay content is illustrated in Figures 8.1 and 8.2. The data in Figure 8.1 indicate that soil *CEC* is directly influenced by the soil organic matter (SOM) content. In general, as SOM increases so does *CEC*, especially when SOM is greater than approximately 20 g kg^{-1}. However, when SOM is less than 20 g kg^{-1} there is

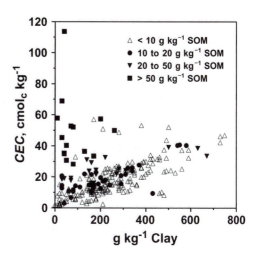

FIGURE 8.2 The cation exchange capacity (*CEC*) of 249 surface soil samples collected from around the U.S. as a function of soil clay content. The marker used to represent a particular data point indicates the soil organic matter (SOM) content of the soil sample. Data were obtained from Batjes (1995).

considerable variability in *CEC* and no apparent relationship between *CEC* and SOM. The same *CEC* data, plotted as a function of soil clay content, with the SOM content ranges superimposed on the data, illustrate a strong correlation between *CEC* and clay content (Figure 8.2). Again this strong relationship is especially evident when SOM is less than 20 g kg^{-1}. When the concentration of SOM exceeds this value, the *CEC* is no longer correlated to clay content, but to the SOM content, as illustrated in Figure 8.1. Thus, the *CEC* is directly correlated to clay content when SOM content is low; otherwise *CEC* is directly correlated to the SOM content.

8.3.2 Methods of Measuring the Cation Exchange Capacity

Several methods are available to determine the *CEC* of a soil. However, all methods rely on the ability of an index cation to replace the native, readily exchangeable ions. The index cation is also termed an indifferent cation, as it must only displace cations that are nonspecifically retained in the os- and d-planes of the soil surface. Two general categories of *CEC* methodologies are employed. The first involves the displacement of native cations and the saturation of the exchange complex with an index cation, such as Na^+. The adsorbed Na^+ is then displaced by a second index cation, such as NH_4^+. The moles of Na^+ displaced by the NH_4^+ are equivalent to the moles of exchangeable surface charge per mass of soil. The second technique involves the displacement of native, readily exchangeable cations by an index cation that is normally not native, such as NH_4^+, and the subsequent determination of displaced Ca^{2+}, Mg^{2+}, K^+, and Na^+ (and Al^{3+} and Mn^{2+} if the soil is acidic). A summation of the moles of displaced charge is then a measure of the effective *CEC* (*ECEC*). This second method also allows for a determination of the exchangeable base cations present on the exchange complex (sum of the cmol$_c$ kg^{-1} of Ca^{2+}, Mg^{2+}, K^+, and Na^+ adsorbed).

8.3.2.1 Ammonium Acetate (pH 7 or 8.2)

The ammonium acetate method employs two index ions, Na^+ and NH_4^+, in pH-buffered acetate solutions to determine the potential *CEC* of a soil. Since the pH of the index cation solutions are buffered, the measured *CEC* can either over- or underestimate the field (actual) *CEC* of a soil, depending on the field pH of the soil. If the field pH is greater than the buffer pH, the ammonium acetate *CEC* will underestimate *CEC*. Conversely, if soil pH is less than the buffer pH, the *CEC* will be overestimated. In general, the buffer pH is selected with knowledge of the soil pH. For the *CEC* of acid soils, a pH 7 buffer is used; while the *CEC* of alkaline soils is determined at pH 8.2 (in an effort to minimize calcite dissolution). Thus, the ammonium acetate procedure provides a measure of the potential *CEC* of the soil: the *CEC* when all surface functional groups are deprotonated.

The pH 8.2 $C_2H_3O_2Na$-$C_2H_3O_2NH_4$ (sodium acetate–ammonium acetate) method was developed primarily for application to alkaline soils. Initially, the soil is centrifuge-washed a number of times with a pH 8.2, 1 M $C_2H_3O_2Na$ solution. The successive washings remove soluble salts and saturate the exchange complex with the Na^+ index ion. The soil is then centrifuge-washed with a 95% ethanol solution (minimizing dispersion) to remove occluded Na^+ and minimize the displacement of Na^+ from the exchange complex. Once the electrical conductivity of the ethanol supernatant is below 0.04 dS m^{-1}, the soil is repeatedly washed with a pH 7, 1 M $C_2H_3O_2NH_4$ solution to displace Na^+ for analysis. The cmol$_c$ of displaced Na^+ divided by the sample mass, in kg, is the *CEC*. Consider the determination of *CEC* for a soil sample collected from an Alfisol. A 2-g soil sample is treated with a pH 8.2, 1 M $C_2H_3O_2Na$ solution followed by the removal of entrained Na^+. The Na^+-saturated soil is then treated by sequential washings with a pH 7, 1 M $C_2H_3O_2NH_4$ solution. The supernatant solutions from the $C_2H_3O_2NH_4$ washings are collected and combined into a 50-mL volumetric flask, which is brought to volume with distilled-deionized water. This solution is analyzed by atomic emission spectrometry (described in Chapter 5), and the total Na concentration is determined to be 119.4 mg L^{-1}. The potential *CEC* of the soil is computed:

$$CEC = \frac{119.4 \text{ mg}}{\text{L}} \times \frac{0.05 \text{ L}}{0.002 \text{ kg}} \times \frac{\text{mmol}}{22.9898 \text{ mg}} \times \frac{\text{cmol}_c}{10 \text{ mmol}} = 12.98 \text{ cmol}_c\text{kg}^{-1} \qquad (8.8a)$$

This *CEC* computation for the ammonium acetate method is generalized in the expression:

$$CEC, \text{ in cmol}_c \text{ kg}^{-1} = C_{Na} \times \frac{V_L}{m_{kg} \times MW_{Na} \times 10} \qquad (8.8b)$$

where C_{Na} is the mg L^{-1} concentration of Na in the combined wash solutions having volume V_L in liters, m_{kg} is the mass of the soil sample in kg, and MW_{Na} is the molecular mass of Na.

Another *CEC* method is the $C_2H_3O_2NH_4$ pH 7 leach method. Soil is placed in a column that is attached to a vacuum extractor. The soil is leached with an initial volume of pH 7, 1 M $C_2H_3O_2NH_4$ to remove soluble ions and the leachate discarded. An additional overnight equilibration of the soil with $C_2H_3O_2NH_4$ results in the displacement of exchangeable ions. Analysis of the leachate following this equilibration may be performed to determine exchangeable cations. The soil is then leached with ethanol to remove occluded NH_4^+. Once the leachate is free of NH_4^+, the soil is leached with 1 M KCl. The cmol$_c$ of displaced NH_4^+ in the leachate divided by the mass of soil in the column, in kg, is the *CEC*.

8.3.2.2 Cation Exchange Capacity of Arid Land Soils

Arid-zone soils often contain a number of relatively soluble minerals (carbonates, sulfates, zeolites) that are difficult to remove during the initial washing step in the *CEC* procedures. Thus, the *CEC* procedure employed for these soils must distinguish between the soluble cations that are not removed during the initial soil washing step and those that are truly exchangeable. Initially, the soil is repeatedly washed with a pH 7, 0.4 M $C_2H_3O_2Na$-0.1 M NaCl-60% ethanol solution to maximize soluble salt removal and Na^+-saturate the exchange complex. The soil is then immediately (without a washing step to remove occluded ions) equilibrated with a 0.25 M $Mg(NO_3)_2$ extracting solution (this step is repeated a number of times). The solutions from the $Mg(NO_3)_2$ washings are combined and analyzed for both Na and Cl. Chloride is determined so that any soluble Na carried over from the saturation step to the extraction step can be subtracted from the total Na in the extracting solution to obtain exchangeable Na^+. Again, the cmol$_c$ of exchangeable Na^+ divided by the mass of sample is the *CEC*.

8.3.2.3 Cation Exchange Capacity of Acid Soils

Unlike alkaline soils, the surface charge characteristics of acid soils are strongly influenced by pH and ionic strength (due to the abundance of metal oxide surfaces). Acid soils may also bear a substantial anion exchange capacity (*AEC*). Further, the exchange complex may contain adsorbed Al^{3+} and associated hydrolysis products ($AlOH^{2+}$ and $Al(OH)_2^+$), and Mn^{2+} which required a more competitive index cation than Na^+ or NH_4^+ for displacement. Acid soils are initially washed with a 0.1 M $BaCl_2$ solution to remove soluble salts and to initially saturate the exchange complex with Ba^{2+}. The pH of the 1 M $BaCl_2$ solution can be adjusted to pH 7 or any other desirable value. The soil is then repeatedly centrifuge-washed with a 0.002 M $BaCl_2$ solution. An aliquot of 0.005 M $MgSO_4$ solution is added and the suspension equilibrated for 1 h. The reactant suspension is then diluted to achieve 0.0015 M $MgSO_4$, and the suspension is equilibrated overnight. The suspension is centrifuged and supernatant retained for pH, Mg, and Cl determinations. The *CEC* is computed by difference ($cmol_c$ of Mg^{2+} initially added minus $cmol_c$ of Mg^{2+} in the final supernatant liquid, divided by the mass of soil, in kg). The *AEC* may be computed from the $cmol_c$ Cl^- displaced by SO_4^{2-}.

8.3.2.4 Unbuffered Salt Extraction

The unbuffered salt extraction method is employed to measure the field or actual *CEC* of a soil. A soil sample is repeatedly centrifuge washed with a 0.2 M NH_4Cl solution. If the soil is nonsaline and noncalcareous, the supernatant solutions can be saved and analyzed to determine exchangeable Na, K, Ca, Mg, and Al and provide a measure of the effective *CEC*. The soil is then repeatedly washed with a 0.04 M NH_4Cl solution to provide an ionic strength that is representative of the soil environment. A 0.2 M KNO_3 solution is then employed to displace NH_4^+ from the exchange complex. The $cmol_c$ of NH_4^+ in the combined KNO_3 extracts per kg of soil is the *CEC*. The $cmol_c$ of Cl^- in the combined KNO_3 extracts per kg of soil is the *AEC*.

8.4 QUANTITATIVE DESCRIPTION OF CATION EXCHANGE

Since the groundbreaking work of Thompson and Way, considerable effort has been expended to develop a mechanism to quantitatively describe ion exchange. Much of this effort has been driven by a need to either characterize and manage problem soils, such as those that are sodic (excessive levels of exchangeable Na^+) or acidic (contain exchangeable Al species), or to understand the fate and behavior of fertilizer nutrients, such as NH_4^+ and K^+. As a result of these efforts, several models have been proposed to characterize and predict the equilibrium distribution of an exchangeable ion between the adsorbed and aqueous phases. These models employ mass action principles and describe ion partitioning using an exchange selectivity coefficient, a parameter that is not a true constant but nevertheless a utilitarian one and one that can be employed to approximate the true exchange equilibrium constant.

The exchange behavior of cations is generally evaluated in binary exchange systems, although ternary systems have elicited more recent interest. For example, consider a batch study to examine the soil exchange behavior of K^+ and Mg^{2+} in a binary system beginning with K^+-saturated soil. Initially, soil samples are placed in several centrifuge tubes and saturated with K^+ by repeatedly washing the soil samples with a 1 M KCl solution in a centrifuge-washing procedure. After saturation, additional washings with a 70% ethanol solution are used to remove entrained K^+. The soil samples are now K^+-saturated, or homoionic with respect to K. To each centrifuge tube a volume of solution is then added, each containing a different ratio of KCl to $MgCl_2$, such that the total normality (or ionic strength, depending on the experimental design) of the solution in each tube is identical. The systems are then equilibrated, centrifuged to separate the solid and solution phases, and the solution phase removed for K and Mg analysis. The composition of the soil exchange phase is determined by first washing the soil remaining in the centrifuge tubes with a 70% ethanol solution and then displacing adsorbed K^+ and Mg^{2+} with sodium acetate washings. The supernatant solutions are collected, combined, brought to volume in a volumetric flask, and analyzed for K and Mg. As a result of the study, solution composition as a function of exchange phase composition is known, and the data provide the necessary information to evaluate cation preference and various thermodynamic parameters.

The following sections examine the exchange selectivity coefficients that are commonly employed to describe ion exchange behavior. Data sets describing CaX_2-KX, MgX_2-KX, and CaX_2-MgX_2 exchange in a Loring silt loam (Oxyaquic Fragiudalf) (from the dissertation of Smith, 2000) will be examined to provide a basis for comparisons among the various coefficients. The Loring silt loam is a loessial soil that has a *CEC* of 9.7 $cmol_c$ kg^{-1}, and contains approximately 20% clay that is primarily composed of hydroxy-interlayered vermiculite (HIV) and vermiculite, with lesser amounts of kaolinite and mica.

8.4.1 Cation Exchange Selectivity

Cation exchange selectivity, and the preference of the soil exchange phase for one cation over another, can be described by a mass action coefficient, termed the selectivity coefficient. Selectivity coefficients (there are several types) are similar in form to a true equilibrium constant. In fact, selectivity coefficients represent the efforts of early researchers to develop a pseudo-constant that could be employed to describe the equilibrium status of an exchange system throughout a range of exchange phase compositions. However, they are not true constants, as their values for a particular exchange reaction vary with exchange phase composition. Selectivity coefficients may actually be described as conditional exchange equilibrium constants, in a manner similar to that employed in Chapter 5 for aqueous speciation and mineral solubility reactions. Consider the nonsymmetrical uni-bivalent ion exchange of Ca^{2+} for K^+ by a soil exchange phase denoted by X^-:

$$Ca^{2+}(aq) + 2KX(ex) \rightarrow CaX_2(ex) + 2K^+(aq) \tag{8.9}$$

The true equilibrium constant for this reaction is:

$$K_{ex} = \frac{(CaX_2)(K^+)^2}{(KX)^2(Ca^{2+})} \tag{8.10}$$

where the parentheses () denote exchange and aqueous phase activities. A selectivity coefficient for this reaction may take the form:

$$K_S = \frac{\{CaX_2\}(K^+)^2}{\{KX\}^2(Ca^{2+})} \tag{8.11}$$

Again, the parentheses denote aqueous activities and the braces { } denote concentration variables or functions that model the activities of adsorbed cations. The various types of selectivity coefficients differ with regard to how adsorbed cation activities are defined. A summary of commonly employed selectivity coefficients for symmetrical and nonsymmetrical (uni-bivalent) cation exchange, as well as the assumptions employed in modeling the exchange phase activities, are provided in Table 8.3.

8.4.1.1 Kerr Selectivity Coefficient

H.W. Kerr (1928) investigated CaX_2-MgX_2 exchange on Miami (Oxyaquic Hapludalf) and Colby silt loam (Aridic Ustorthent) soils in an effort to establish the physical–chemical character of the soil exchange phase. Specifically, Kerr considered two hypotheses: (1) in a binary CaX_2-MgX_2

TABLE 8.3
Cation Exchange Selectivity Coefficients Applied to Symmetrical (KX-NaX) and Nonsymmetrical (KX-CaX$_2$) Exchange

Model	Uni-Univalent Exchange[a]	Uni-Bivalent Exchange[b]
Kerr	$K_K = \dfrac{N_{Na}(K^+)}{N_K(Na^+)}$ [c]	$K_K = \dfrac{N_{Ca}(K^+)^2}{N_K^2\{[CaX_2]+[KX]\}(Ca^{2+})}$ [c]
Vanselow	$K_V = K_K = \dfrac{N_{Na}(K^+)}{N_K(Na^+)}$	$K_V = \dfrac{N_{Ca}(K^+)^2}{N_K^2(Ca^{2+})}$
Rothmund-Kornfeld	$K_{RK} = \dfrac{N_{Na}^{1/\beta}(K^+)}{N_K^{1/\beta}(Na^+)}$ [d]	$K_{RK} = \dfrac{E_{Ca}^{1/\beta}(K^+)^2}{(E_K^{1/\beta})^2(Ca^{2+})}$ [c,d]
		$(KX) = E_K^{1/\beta}$ and $(CaX_2) = E_{Ca}^{1/\beta}$
		or
		$K_{RK} = \dfrac{E_{Ca}^{2/\beta}(K^+)^2}{(E_K^{1/\beta})^2(Ca^{2+})}$;
		$(KX) = E_K^{1/\beta}$ and $(CaX_2) = E_{Ca}^{2/\beta}$
Gapon	$K_G = K_K = \dfrac{N_{Na}(K^+)}{N_K(Na^+)}$	$K_G = \dfrac{E_{Ca}(K^+)}{E_K(Ca^{2+})^{1/2}}$ [e]
Gaines-Thomas	$K_{GT} = K_K = \dfrac{N_{Na}(K^+)}{N_K(Na^+)}$	$K_{GT} = \dfrac{E_{Ca}(K^+)^2}{E_K^2(Ca^{2+})}$
Davies	$K_D = K_K = \dfrac{N_{Na}(K^+)}{N_K(Na^+)}$	$K_D = \dfrac{N_{Ca}\left(N_K + \dfrac{2(n-1)N_{Ca}}{n}\right)}{N_K^2}\dfrac{(K^+)^2}{(Ca^{2+})}$ [f]
Regular solution	$K_{RS} = \dfrac{\{\exp(QN_K^2)\}N_{Na}(K^+)}{\{\exp(QN_{Na}^2)\}N_K(Na^{2+})}$ [g]	$K_{RS} = \dfrac{\{\exp(QN_K^2)\}N_{Ca}(K^+)^2}{\{\exp(QN_{Ca}^2)\}^2 N_K^2(Ca^{2+})}$ [g]

[a] For the symmetrical (KX-NaX) exchange reaction: $Na^+(aq) + KX(ex) \rightarrow NaX(ex) + K^+(aq)$.

[b] For the nonsymmetrical (KX-CaX$_2$) exchange reaction: $Ca^{2+}(aq) + 2KX(ex) \rightarrow CaX_2(ex) + 2K^+(aq)$.

[c] N_{Na}, N_K, and N_{Ca} represent the mole fractions of Na^+, K^+, and Ca^{2+} that occupy the exchange phase (Equation 8.82); parentheses ( ) denote aqueous activities; brackets [ ] denote adsorbed concentrations in mol kg^{-1}.

[d] β is an adjustable parameter related to surface homogeneity; E_K, and E_{Ca} represent the equivalent fractions K^+ and Ca^{2+} that occupy the exchange phase (Equations 8.27 and 8.28).

[e] K_G describes the reaction, $\frac{1}{2}Ca^{2+}(aq) + 2KX(ex) \rightarrow Ca_{1/2}X(ex) + K^+(aq)$.

[f] n is a parameter that indicates the number of nearest neighbors associated with an adsorbed cation.

[g] Q is defined by the expressions: $Q_K = \lim_{N_K \to 0} f_K$ and $Q_{Ca} = \lim_{N_{Ca} \to 0} f_{Ca}$, where $Q = Q_K = Q_{Ca}$.

exchange system, the exchange phase is an assemblage consisting of two discrete solid phases, CaX and MgX (where X^{2-} represents the soil exchanger); or (2) the exchange phase consists of a homogeneous mixture of exchangeable cations that is described as a solid solution composed of (Ca, Mg)X. In the case of the discrete-phase hypothesis, the exchange phases, CaX and MgX, are assumed to exist as independent and pure solids with unit activities (the activity of a pure solid is discussed in Chapter 6). Therefore, for the reaction,

$$Mg^{2+}(aq) + CaX(s) \rightarrow MgX(s) + Ca^{2+}(aq) \qquad (8.12)$$

the equilibrium constant reduces to:

$$K_{discrete} = \frac{(Ca^{2+})}{(Mg^{2+})} \tag{8.13}$$

where the parentheses () denote activities (although Kerr employed milliequivalents). In the case of the homogeneous mixture hypothesis, the activity of CaX and MgX in the single, mixed exchange phase, $(Ca, Mg)X$, is assumed to be a function of composition; the concentrations of Ca^{2+} and Mg^{2+} in $(Ca, Mg)X$. In this case, the equilibrium condition proposed by Kerr for Equation 8.12 is:

$$K_{homogeneous} = \frac{[MgX](Ca^{2+})}{[CaX](Mg^{2+})} \tag{8.14}$$

where the brackets represent the activities, or "*active masses*" as described by Kerr, of Ca^{2+} and Mg^{2+} on the exchange phase. Kerr employed milliequivalents to describe solution composition, as well as to describe the active masses of the adsorbed species.

Although Kerr did not determine the free aqueous concentrations of the divalent ions in his studies, he reasoned that the equilibrium concentrations of total dissolved Ca and Mg in the exchange systems were proportional to that of the corresponding free species, Ca^{2+} and Mg^{2+}; an assumption he based on the similar aqueous complexation behavior of Ca and Mg in a chloride matrix. Further, since only the ratio of Mg^{2+} to Ca^{2+} was needed, any error associated with the assumption on the computed $K_{discrete}$ and $K_{homogeneous}$ values would be minimal. Moreover, but not explicitly stated by Kerr, it was assumed that Mg^{2+} and Ca^{2+} have the same activity coefficients. Therefore, the concentration ratio of Mg^{2+} to Ca^{2+} is essentially the activity ratio, as denoted in Equations 8.13 and 8.14.

Kerr employed a similar line of thought to relate the CaX to MgX concentration ratio to the active mass (or activity) ratio in Equation 8.14. Specifically, he tacitly assumed that CaX and MgX behave as if they were in a true solution with identical activity coefficients. Thus, the determination of CaX and MgX concentrations in an exchange system at equilibrium was presumably sufficient to generate an activity ratio for determining values of $K_{homogeneous}$. Data obtained by Kerr for CaX$_2$-MgX$_2$ exchange by the Miami and Colby soils illustrate the influence of exchange phase composition on computed $K_{discrete}$ and $K_{homogeneous}$ values (Figure 8.3). As the composition of the exchange phase was varied from 20 to 70% coverage by Ca^{2+}, $K_{discrete}$ values varied considerably; whereas, $K_{homogeneous}$ values were relatively constant. The constancy of $K_{homogeneous}$ throughout a range of exchange phase compositions led Kerr to conclude that the exchangeable cations indeed form a solid solution. Actually, it was fortuitous for Kerr that he examined symmetrical exchange, as well as the binary CaX$_2$-MgX$_2$ exchange system, as it is a general rule that these ions form a nearly ideal solid solution on exchangers. If Kerr had chosen to study a nonsymmetrical exchange system, or a symmetrical system that was nonideal, $K_{homogeneous}$ could have varied with exchange phase composition, as seen for the CaX$_2$-KX exchange system illustrated in Figure 8.4a, perhaps resulting in a set of very different conclusions. Finally, Kerr was not the first to employ mass action principles to describe exchange reactions. According to Sposito (1981), R. Gansses in 1913 was the first to describe binary exchange using a mass action coefficient. However, the selectivity coefficient introduced by Gansses and employed by Kerr ($K_{homogeneous}$ in Equation 8.14) is formally defined as the Kerr selectivity coefficient, K_K. With few exceptions, selectivity coefficients developed since the pivotal work of Kerr are equivalent to K_K for describing symmetrical ion exchange (Table 8.3).

8.3.1.2 Vanselow Selectivity Coefficient

Vanselow (1932) noted that the study of symmetrical exchange equilibria was not sufficient to distinguish between Kerr's two hypotheses of exchange phase character (discrete phases versus a solid solution). However, he proposed that the study of nonsymmetrical exchange could generate

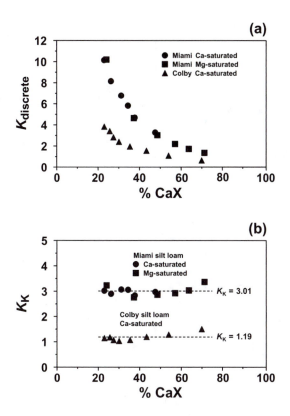

FIGURE 8.3 Data from Kerr (1928) illustrating the dependence of (a) $K_{discrete}$ (Equation 8.13) or (b) $K_{homogeneous}$ (defined as K_K, Equation 8.14) on the composition of the soil exchange phase (the percentage occupied by Ca^{2+}). Both $K_{discrete}$ and K_K are for the $Mg^{2+} + CaX \rightarrow MgX + Ca^{2+}$ exchange reaction initiated from either Ca^{2+}- or Mg^{2+}-saturation in Colby (Aridic Ustorthent) and Miami (Oxyaquic Hapludalf) soils. Kerr offered the K_K versus % CaX results in diagram (b) as proof that the soil exchange phase consisted of a single solid solution phase [(Ca,Mg)X] rather than discrete CaX and MgX phases.

data for distinguishing between the two hypotheses. Additionally, Vanselow found Kerr's definition of the "active masses" as the concentrations of CaX and MgX to be *"difficult to subscribe to."* Instead, Vanselow preferred to employ the thermodynamic approach, defining the activity of a component in a solid solution by its mole fraction as developed by G.N. Lewis in 1913 (discussed in Chapter 6). Thus, for the nonsymmetrical uni-bivalent exchange reaction examined by Vanselow,

$$Ca^{2+}(aq) + 2NH_4X(ex) \rightarrow CaX_2(ex) + 2NH_4^+(aq) \qquad (8.15)$$

the equilibrium constant is a function of activities,

$$K_{ex} = \frac{(CaX_2)(NH_4^+)^2}{(NH_4X)^2(Ca^{2+})} \qquad (8.16)$$

and can be represented by

$$K_V = \frac{N_{Ca}(NH_4^+)^2}{N_{NH_4}^2(Ca^{2+})} \qquad (8.17)$$

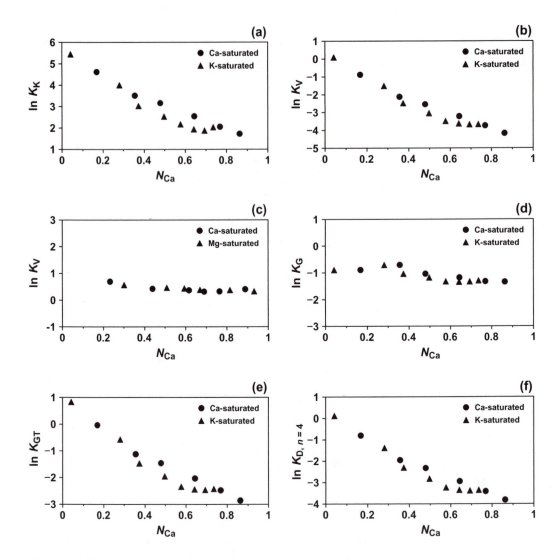

FIGURE 8.4 The dependence of exchange selectivity coefficients on the composition of the exchange phase, as defined by the mole fraction (N_{Ca}) of adsorbed Ca^{2+}, for the KX-CaX$_2$ exchange reaction: $Ca^{2+} + 2KX \rightarrow CaX_2 + 2K^+$. The data were obtained for Loring silt loam (Oxyaquic Fragiudalf) and the exchange reactions were initiated from either Ca^{2+} or K^+ saturation with an ionic strength of 0.15. The selectivity coefficients are (a) Kerr (K_K), (b) Vanselow (K_V), (c) Vanselow for MgX$_2$-CaX$_2$ exchange (which is identical to K_K, K_G, K_{GT}, and $K_{D,n=4}$), (d) Gapon (K_G), (e) Gaines-Thomas (K_{GT}), and (f) Davies with $n = 4$ ($K_{D,n=4}$).

where K_V is the Vanselow selectivity coefficient. In the binary NH$_4$X-CaX$_2$ exchange system, the mole fractions, N_{Ca} and N_{NH_4}, are defined by:

$$N_{Ca} = \frac{[CaX_2]}{[NH_4X]+[CaX_2]} \tag{8.18a}$$

$$N_{NH_4} = \frac{[NH_4X]}{[NH_4X]+[CaX_2]} \tag{8.18b}$$

where the brackets [ ] represent the concentrations of adsorbed cations in mol kg^{-1}.

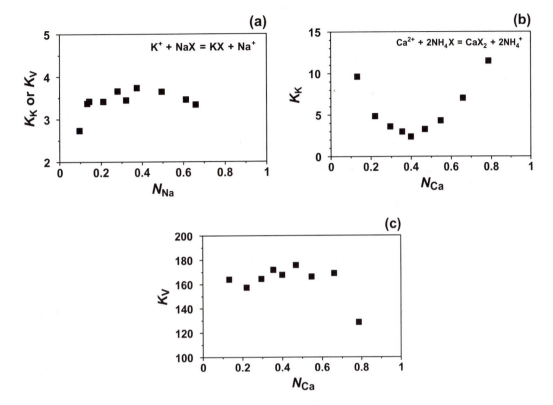

FIGURE 8.5 Data from Vanselow (1932) illustrate the dependence of K_K or K_V on exchange phase composition. The diagram in (a) shows that for symmetric exchange, K_V (identical to K_K) is relatively invariant with the mole fraction of adsorbed Na^+ (N_{Na}) for NaX-KX exchange. However, for nonsymmetric uni-bivalent NH_4X-CaX_2 exchange, K_K varies with exchange phase composition (b), while K_V is relatively invariant (c).

Among the numerous studies Vanselow performed, both nonsymmetrical exchange, as described in Equation 8.15, and symmetrical NaX-KX exchange on bentonite clay were examined. Selected results of his evaluations are displayed in Figure 8.5. Like Kerr, Vanselow found K_K (and K_V) values for symmetrical exchange (NaX-KX) to be relatively constant with exchange phase composition (Figure 8.5a). On average, K_K and K_V equal 3.42 throughout the range of N_{Na}. Also indicated in Figure 8.5a is that K_K and K_V are synonymous for symmetrical exchange reactions (this is also indicated in Table 8.3). For example, the Vanselow selectivity coefficient for the exchange reaction

$$K^+(aq) + NaX(ex) \rightarrow KX(ex) + Na^+(aq) \tag{8.19}$$

is

$$K_V = \frac{N_K(Na^+)}{N_{Na}(K^+)} \tag{8.20}$$

Substituting the exchange phase concentrations, [NaX] and [KX], for the mole fractions, N_{Na} and N_K, yields:

$$K_V = \frac{\dfrac{[KX]}{[NaX]+[KX]}(Na^+)}{\dfrac{[NaX]}{[NaX]+[KX]}(K^+)} \tag{8.21}$$

Simplifying,

$$K_V = \frac{[KX](Na^+)}{[NaX](K^+)} \tag{8.22}$$

which is also the K_K expression for Na^+-K^+ exchange (Equation 8.19).

For nonsymmetrical NH_4X-CaX_2 exchange, Vanselow found computed K_K values to vary with exchange phase composition, from a minimum value of 2.38 to a maximum of 11.49 (Figure 8.5b). When mole fractions were employed to describe exchange phase compositions, values for K_V were relatively invariant throughout a wide range of N_{Ca} values (Figure 8.5c). This finding further substantiated the salient conclusion of Kerr, as stated by Vanselow:

> The results of the experimental investigation of the calcium-ammonium exchange reactions of a bentonite and a soil colloid can be best interpreted by the hypothesis of the formation of a mixed crystal when one replaceable cation is partially replaced by another.

However, it was the introduction of mole fractions to describe the activities of adsorbed ions for which Vanselow is best known. Implicit to the selection of mole fractions to represent exchange phase activities is the assumption that adsorbed cations form an ideal mixture (an ideal solid solution) on the soil exchanger. Recalling Equation 8.10, which defines K_{ex} for KX-CaX_2 exchange according to Equation 8.9, and introducing the following expressions that define the activities of solids in a solid solution (from Chapter 6):

$$(KX) \equiv f_K N_K \text{ and } (CaX_2) \equiv f_{Ca} N_{Ca} \tag{8.23}$$

where the f values are rational activity coefficients, Equation 8.10 becomes:

$$K_{ex} = \frac{f_{Ca} N_{Ca} (K^+)^2}{f_K^2 N_K^2 (Ca^{2+})} \tag{8.24}$$

Vanselow's selectivity coefficient will be identical to K_{ex} when $f_K = f_{Ca} = 1.0$. This condition only occurs when a solid is in the standard state (pure, macrocrystalline, 298.15 K, and 0.101 MPa) or, in the case of a solid solution, when the mixture forms an ideal solid solution. Thus, for K_V to approximate K_{ex}, the exchangeable cations must be nearly identical in size and charge, the distribution of the exchangeable cations on the surface must be random, and steric and electronic structural effects must be absent. A plot of $\ln K_V$ vs. N_{Ca} for KX-CaX_2 and MgX_2-CaX_2 exchange by the Loring silt loam (Oxyaquic Fragiudalf) illustrates both nonideal exchange and nearly ideal behavior (Figure 8.4b and 8.4c). For KX-CaX_2 exchange, $\ln K_V$ varies with N_{Ca}, similar to the variation of the Kerr selectivity coefficient (K_K) with N_{Ca}. However, the $\ln K_V$ for the MgX_2-CaX_2 system is relatively invariant with N_{Ca}, with a mean of 0.44 ($K_V = 1.56 \pm 0.01$) (Figure 8.4a).

8.4.1.3 Rothmund and Kornfeld Selectivity Coefficient

Rothmund and Kornfeld (1918, 1919) observed that the partitioning of exchangeable ions between the aqueous and exchanger phases could be described by the van Bemmelen adsorption model (also known as the Freundlich adsorption isotherm model). Applying the van Bemmelen model to each cation independently in a uni-bivalent binary exchange system, and taking the ratio of the two expressions lead to (for example, using KX-CaX_2 exchange in Equation 8.9)

$$k = \frac{\{2CaX_2\}}{\{KX\}^2} \left[\frac{\{K\}^2}{\{2Ca\}} \right]^\beta \tag{8.25}$$

where the braces { } denote charge concentrations, and k and β $(0 < \beta \leq 1)$ are empirical constants. Since the introduction of the Rothmund-Kornfeld equation, there have been two principal and uniquely different modifications to the expression. Both modifications use solution phase activities, but one version uses exchange phase equivalent fractions raised to the power of $\frac{1}{\beta}$ to model exchange phase activities:

$$K_{RK} = \frac{(K^+)^2}{(Ca^{2+})} \frac{E_{Ca}^{\frac{1}{\beta}}}{(E_K^{\frac{1}{\beta}})^2} \tag{8.26}$$

where, K_{RK} is the Rothmund-Kornfeld selectivity coefficient. The equivalent fractions of adsorbed K^+ and Ca^{2+} (E_K and E_{Ca}) are defined by:

$$E_K = \frac{n_K}{n_{Ca} + n_K} = \frac{[KX]}{2[CaX_2] + [KX]} \tag{8.27a}$$

$$E_{Ca} = \frac{n_{Ca}}{n_{Ca} + n_K} = \frac{2[CaX_2]}{2[CaX_2] + [KX]} \tag{8.27b}$$

where n_K and n_{Ca} are the equivalents (moles of charge) of adsorbed K^+ and Ca^{2+} and brackets [ ] denote mol kg^{-1}. The equivalent fractions are related to mole fractions (defined by Equations 8.18a and 8.18b) by:

$$E_K = \frac{N_K}{2N_{Ca} + N_K} \tag{8.28a}$$

$$E_{Ca} = \frac{2N_{Ca}}{2N_{Ca} + N_K} \tag{8.28b}$$

In Equation 8.26, the exchange phase activities are given by:

$$(KX) = E_K^{\frac{1}{\beta}} \quad \text{and} \quad (CaX_2) = E_{Ca}^{\frac{1}{\beta}} \tag{8.29}$$

A second version of the Rothmund-Kornfeld equation describes the activity of each adsorbed cation as the equivalent fraction raised to the power of their valence multiplied by $\frac{1}{\beta}$:

$$K_{RK} = \frac{(K^+)^2}{(Ca^{2+})} \frac{E_{Ca}^{\frac{2}{\beta}}}{(E_K^{\frac{1}{\beta}})^2} \tag{8.30}$$

In this form of the Rothmund-Kornfeld equation, the activities of the exchangeable ions are given by:

$$(KX) = E_K^{\frac{1}{\beta}} \quad \text{and} \quad (CaX_2) = E_{Ca}^{\frac{2}{\beta}} \tag{8.31}$$

Evaluation of the second version of the Rothmund-Kornfeld equation may be performed in a manner similar to that for determining Freundlich K_F and n values (Chapter 7). For example, rearranging Equation 8.30 yields:

$$\left[\frac{(Ca^{2+})}{(K^+)^2}\right]^{\beta} K_{RK}^{\beta} = \frac{E_{Ca}^2}{E_K^2} \tag{8.32}$$

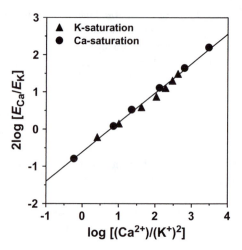

FIGURE 8.6 A plot of KX-CaX$_2$ binary exchange data (ionic strength = 0.15) for Loring (Oxyaquic Fragiudalf) soil according to the Rothmund-Kornfeld model. The equation of the line through the data, obtained by least-squares linear regression analysis, is $2\log[E_{Ca}/E_K] = 0.791 \times \log[(Ca^{2+})/(K^+)^2] - 0.608$, $r^2 = 0.994$.

The logarithmic transformation of this expressions leads to:

$$2\log\left[\frac{E_{Ca}}{E_K}\right] = \beta\log K_{RK} + \beta\log\left[\frac{(Ca^{2+})}{(K^+)^2}\right] \tag{8.33}$$

Thus, the Rothmund-Kornfeld model is obeyed if a plot of $2\log[E_{Ca}/E_K]$ vs. $\log[(Ca^{2+})/(K^+)^2]$ is a straight line with slope β and intercept $\beta\log K_{RK}$. This is the case for the K$^+$-Ca^{2+} exchange data plotted in Figure 8.6. Least-squares linear regression analysis leads to an expression that describes the variation in $2\log[E_{Ca}/E_K]$ as a function of $\log[(Ca^{2+})/(K^+)^2]$ with a highly significant r^2 value (=0.994). The slope of the regression line is $\beta = 0.791$, and the intercept is $\beta\log K_{RK} = -0.608$. Therefore, $\log K_{RK} = -0.769$, or $K_{RK} = 0.170$. Although the constant parameters K_{RK} and β are empirical, the variation in KX-CaX$_2$ exchange behavior is accurately described throughout the range of exchange phase composition. Therefore, assuming the validity of Equation 8.31, K_{RK} estimates the true equilibrium constant (K_{ex}, Equation 8.10) and indicates that combination of Ca^{2+}(aq) + 2KX(ex) is stable relative to CaX$_2$(ex) + 2K$^+$(aq) (K$^+$ is thermodynamically preferred by the soil exchanger phase).

8.4.1.4 Gapon Selectivity Coefficient

Y.N. Gapon (1933) developed an empirical expression to describe the nonsymmetric NH$_4$X-CaX$_2$ exchange reaction and elected to write the exchange reaction in terms of equivalents. For example, the KX-CaX$_2$ exchange reaction written in the Gapon convention is:

$$\tfrac{1}{2}Ca^{2+}(aq) + KX(ex) \rightarrow Ca_{\frac{1}{2}}X(ex) + K^+(aq) \tag{8.34}$$

The Gapon selectivity coefficient is defined as:

$$K_G = \frac{[Ca_{\frac{1}{2}}X][K^+]}{[KX][Ca^{2+}]^{\frac{1}{2}}} \tag{8.35}$$

where the exchangeable cation concentrations are expressed in mmol$_c$ g^{-1} (or meq g^{-1}) and the soluble cation concentrations are expressed in mmol L^{-1}. A contemporary version of the Gapon selectivity coefficient employs aqueous ion activities and equivalent fractions of the adsorbed cations to model exchanger activities:

$$K_G = \frac{E_{Ca}}{E_K} \frac{(K^+)}{(Ca^{2+})^{\frac{1}{2}}}$$

(8.36)

The Gapon equation may also be derived from the Rothmund-Kornfeld equation. In the Rothmund-Kornfeld equation, β is interpreted as a measure of the heterogeneity of exchange sites on the soil surface, relative to the affinities of sites for the exchangeable cations (the same interpretation given to the Freundlich n value). Large values of β (as β approaches unity) indicate a narrow distribution of surface site affinities (homogeneous surface); whereas, small values of β suggest a broad distribution of surface site affinities (heterogeneous surface). Employing the limiting condition of $\beta = 1$, which implies that the exchanger phase is composed of a very narrow distribution of surface site affinities (essentially one type of surface site), Equation 8.30 becomes:

$$K_{RK,\beta=1} = \frac{(K^+)^2}{(Ca^{2+})} \frac{E_{Ca}^2}{E_K^2}$$

(8.37)

A comparison of this expression to Gapon selectivity coefficient (Equation 8.36) indicates that $K_G^2 = K_{RK,\beta=1}$. It can also be shown that the Rothmund-Kornfeld equation collapses to the Kerr equation for symmetrical cation exchange when $\beta = 1$. Compared to the Vanselow K_V values for K^+-Ca^{2+} exchange, which vary from 1.1 to 0.02 as N_{Ca} values vary from 0.08 to 0.9, the K_G values are relatively stable, varying from 0.5 to 0.23 (Figure 8.4d). Indeed, the minimal variation of the Gapon selectivity coefficient with change in exchange phase composition, particularly over a wide range of N values, has led to its popular application in soil chemistry.

The importance of the Gapon expression is not in its ability to describe exchange equilibria *per se*, independent of surface composition, but in its utility and popularity for the diagnosis and management of sodic soils (discussed in detail in Chapter 11). The genesis of a sodic soil can be described by the CaX_2-NaX exchange reaction and results from the saturation of the exchange complex by Na^+ and the concomitant displacement of divalent cations, Ca^{2+} and Mg^{2+}. Thus, it is common to illustrate this process by the reaction,

$$Na^+(aq) + (Ca + Mg) X_{\frac{1}{2}}(ex) \rightarrow NaX(ex) + \tfrac{1}{2}(Ca^{2+} + Mg^{2+})(aq)$$

(8.38)

where the Gapon selectivity coefficient (Equation 8.35) is expressed:

$$K_G = \frac{[NaX][Ca^{2+} + Mg^{2+}]^{\frac{1}{2}}}{[(Ca+Mg)_{\frac{1}{2}}X][Na^+]}$$

(8.39)

(the actual impetus for including both Ca^{2+} and Mg^{2+} was due to the inability of early analytical techniques to differentiate between the two cations). The ratio of [NaX] to $[(Ca^{2+} + Mg^{2+}) \tfrac{1}{2}X]$ on the soil exchange complex is defined as the exchangeable sodium ratio (*ESR*):

$$ESR = \frac{[NaX]}{[(Ca+Mg)_{\frac{1}{2}}X]}$$

(8.40)

where the concentration units are cmol$_c$ kg^{-1} (selected for convenience, as they are the units of CEC). Further, because Na$^+$, Ca^{2+}, and Mg^{2+} are the dominant exchangeable cations in sodic soils, or soils that may potentially become sodic, the ESR may also be described by:

$$ESR = \frac{[NaX]}{[CEC - NaX]} \qquad (8.41)$$

The ratio of Na$^+$ to divalent ion concentrations in the soil solutions or irrigation waters is defined as the sodium adsorption ratio (SAR):

$$SAR = \frac{[Na^+]}{[Ca^{2+} + Mg^{2+}]^{\frac{1}{2}}} \qquad (8.42)$$

where the concentration units are mmol L^{-1} or mol L^{-1}. Combining Equations 8.39, 8.40 (or 8.41), and 8.42 yields:

$$K_G = \frac{ESR}{SAR} \qquad (8.43)$$

As this expression indicates, the exchangeable Na$^+$ component of the soils exchange capacity can be predicted given the Na$^+$ and divalent cation composition of the soil solution and a value for K_G. Correspondingly, the impact of irrigation waters or other leachates on the sodicity of a soil can also be established.

8.4.1.5 Gaines and Thomas Selectivity Coefficient

Gaines and Thomas (1953), two Yale University chemists, proposed an ion exchange selectivity coefficient that mirrored Vanselow's expression, but employed equivalent fractions to model exchange phase activities. For the KX-CaX$_2$ exchange reaction (Equation 8.9), the Gaines-Thomas selectivity coefficient is defined as:

$$K_{GT} = \frac{E_{Ca}(K^+)^2}{E_K^2(Ca^{2+})} \qquad (8.44)$$

The comparison of K_{GT} and K_V values for KX-CaX$_2$ exchange (Figure 8.4c and 8.4e) illustrates that the use of equivalent fractions instead of mole fractions to describe exchange phase activities does not decrease the variability of the selectivity coefficient with changes in N_{Ca}. Indeed, K_{GT} and K_V mirror one another.

The relevant contribution of Gaines and Thomas principally lies in their definition of the reference state of an ion in the exchange phase. As will be shown in a later section, their definition of the reference state allows for the unambiguous determination of the exchange equilibrium constant (K_{ex}). According to Gaines and Thomas, the reference state is defined as follows:

> For the solid phase we choose the reference state as the mono-ion solid in equilibrium with an infinitely dilute solution of that ion, i.e., so that for $E_A = 1$ and $a_s = 1$, $f_A = 1$ and for $E_B = 1$ and $a_s = 1$, $f_b = 1$.

(In the original text the equivalent fractions of exchangeable ions A and B are represented by N_A and N_B. I have replaced N_A and N_B with E_A and E_B in the quotation to minimize confusion with my notation for mole fractions.)

Thus, the reference state is a homoionic exchange phase in equilibrium with a solution that is infinitely dilute in the saturating cation. Therefore, the activity of the solution (a_s) is equal to 1 and the rational activity coefficient (f) is equal to 1.

8.4.1.6 Davies Selectivity Coefficient

Davies (1950) and Davies and Rible (1950) employed a statistical mechanical approach to derive an expression for an exchange selectivity coefficient by assuming the exchanger surface to be composed of a regular array of negative point charges. For the generalized uni-bivalent exchange reaction:

$$B^{2+}(aq) + 2AX(ex) \rightarrow BX_2(ex) + 2A^+(aq) \tag{8.45}$$

Davies showed that the distribution of cations on a surface could be described by:

$$K_D = \left\{ \frac{N_B(q_A N_A + q_B N_B)}{N_A^2} \right\} \frac{(A^+)^2}{(B^{2+})} \tag{8.46}$$

where K_D is the Davies selectivity coefficient, values for q_A and q_B depend on the number of nearest neighbors associated with each adsorbed ion, and all other parameters are previously defined. The q values are computed according to the expression:

$$q_i = \frac{(n_i z_i - 2z_i + 2)}{n_i} \tag{8.47}$$

where z_i is the valence of ion i and n_i is the number of nearest neighbors. For the uni-bivalent exchange reaction (Equation 8.45), $q_A = 1$ and $q_B = 2(n-1)/n$. Substituting for q_A and q_B, Equation 8.46 becomes:

$$K_D = \frac{N_B\left(N_A + \dfrac{2(n-1)N_B}{n}\right)}{N_A^2} \frac{(A^+)^2}{(B^{2+})} \tag{8.48}$$

which describes the Davies selectivity coefficient as a function of n for uni-bivalent exchange.

There are several versions of the Davies distribution equation, each assuming a different value for n. Krishnamoorthy and Overstreet (1950) selected $n = 4$, which represents a square array of surface charges. Each adsorbed cation sits inside an imaginary square box, the corners of which are the four nearest neighbor-adsorbed cations. The Krishnamoorthy-Overstreet selectivity coefficient ($K_{D,n=4}$) is defined as:

$$K_{D,n=4} = \frac{N_B(N_A + 1.5N_B)}{N_A^2} \frac{(A^+)^2}{(B^{2+})} \tag{8.49}$$

However, using the approach of Davies to describe cation exchange, as applied to the KX-CaX$_2$ exchange data using the Krishnamoorthy and Overstreet model (Figure 8.4f), results in a selectivity coefficient that varies with N_{Ca} in much the same way K_V and K_{GT} vary (Figures 8.4c and 8.4e). The similar trends noted for $K_{D,n=4}$, K_V, and K_{GT} as a function of N_{Ca} could be predicted, as K_D

collapses to K_V and K_{GT} at the limits of n. If n is assumed to be very large, as might be conceptualized for an exchangeable ion residing in a diffuse ion swarm, Equation 8.48 is written:

$$K_{D,n\uparrow\infty} = \frac{N_B(N_A + 2N_B)}{N_A^2} \frac{(A^+)^2}{(B^{2+})}$$

(8.50)

Replacing the equivalent fractions with mole fractions in the Gaines-Thomas model (Equation 8.44) yields:

$$K_{GT} = \left\{ \frac{\dfrac{2N_B}{(N_A + 2N_B)}}{\dfrac{N_A^2}{(N_A + 2N_B)^2}} \right\} \frac{(A^+)^2}{(B^{2+})}$$

(8.51)

Simplifying,

$$K_{GT} = \frac{2N_B(N_A + 2N_B)}{N_A^2} \frac{(A^+)^2}{(B^{2+})}$$

(8.52)

Comparing Equation 8.52 with Equation 8.50 indicates that $2K_{D,n\uparrow\infty} = K_{GT}$. Finally, if a small value for n is selected ($n = 2$, a linear array of surface cations), Equation 8.48 becomes:

$$K_{D,n=2} = \frac{N_B(N_A + N_B)}{N_A^2} \frac{(A^+)^2}{(B^{2+})}$$

(8.53)

which reduces to $K_{D,n=2} = K_V$ (defined in Equation 8.17), because $N_A + N_B = 1$. Therefore, if exchange data cannot be described by the Vanselow or the Gaines-Thomas selectivity coefficients (which effectively represent the lower and upper limits of n), it is improbable that the Davies distribution function will do so at any value of n.

8.4.1.7 Regular Solution Model

The regular solution theory, which is often employed to characterize mineral phase solid solutions, is a byproduct of the van Laar model of the rational activity coefficients. The van Laar model is derived from the Margules equations of the rational activity coefficients (Sposito, 1994). When applied to the rational activity coefficients of ions involved in the AX-BX$_2$ exchange reaction (Equation 8.45), f_A and f_B are described by:

$$\ln f_A = a_2 N_B^2 + a_3 N_B^3$$

(8.56a)

$$\ln f_B = \left(a_2 + \tfrac{3}{2} a_3\right) N_A^2 - a_3 N_A^3$$

(8.56b)

Additionally, the coefficients a_2 and a_3 are defined by:

$$a_2 = 2Q_B - Q_A$$

(8.57a)

$$a_3 = 2(Q_A - Q_B)$$

(8.57b)

The parameters Q_A and Q_B are independent of exchanger phase composition and are defined as:

$$Q_A = \lim_{N_A \to 0} f_A \quad \text{and} \quad Q_B = \lim_{N_B \to 0} f_B \tag{8.58}$$

The regular solution model assumes $Q_A = Q_B$. Thus, employing Equations 8.57a and 8.57b leads to $a_2 = Q_A = Q_B \ (= Q)$, and $a_3 = 0$. Substituting for a_2 and a_3 in Equations 8.56a and 8.56b results in:

$$\ln f_A = Q N_B^2 \quad \text{and} \quad \ln f_B = Q N_A^2 \tag{8.59}$$

The van Laar model leads to the Vanselow model when $Q_A = Q_B = 1$. In this case, the solution to Equations 8.59 is $\ln f_A = \ln f_B = 0$, or $f_A = f_B = 1$.

For the AX-BX$_2$ exchange reaction (Equation 8.45), the regular solution exchange selectivity coefficient is given by:

$$K_{RS} = \frac{\{\exp(Q N_A^2)\} N_B (A^+)^2}{\{\exp(Q N_B^2)\}^2 N_A^2 (B^{2+})} \tag{8.60}$$

or

$$K_{RS} = \frac{\{\exp(Q N_A^2)\}}{\{\exp(Q N_B^2)\}^2} K_V \tag{8.61}$$

Simplifying by taking the natural logarithm of both sides and rearranging such that $\ln K_V$ is the dependent variable:

$$\ln K_V = \ln K_{RS} - Q N_A^2 + 2 Q N_B^2 \tag{8.62}$$

Substituting $(1 - N_B)$ for N_A, with additional algebraic manipulation leads to:

$$\ln K_V = \ln K_{RS} + Q[N_B(N_B + 2) - 1] \tag{8.63}$$

Therefore, the regular solution model is obeyed if a plot of $\ln K_V$ vs. $[N_{Ca}(N_{Ca} + 2) - 1]$ is a straight line with slope Q and intercept $\ln K_{RS}$. This appears to be the case for the KX-CaX$_2$ exchange data plotted in Figure 8.7. Least-squares linear regression analysis leads to an expression that describes

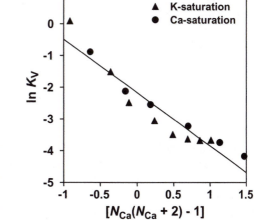

FIGURE 8.7 A plot of KX-CaX$_2$ binary exchange data (ionic strength = 0.15) for Loring (Oxyaquic Fragiudalf) soil according to the regular solution model. The equation of the line through the data, obtained by least-squares linear regression analysis, is $\ln K_V = -1.68 \times [N_{Ca}(N_{Ca} + 2) - 1] - 2.17$, $r^2 = 0.908$.

the variation in $\ln K_V$ as a function of $[N_{Ca}(N_{Ca} + 2) - 1]$ with a significant r^2 value (= 0.908). The slope of the regression line is $Q = -1.68$, and the intercept is $\ln K_{RS} = -2.17$ ($K_{RS} = 0.114$). Therefore, assuming the validity of Equation 8.59, K_{RS} estimates the true equilibrium constant (K_{ex}, Equation 8.10) and indicates that combination of $Ca^{2+}(aq) + 2KX(ex)$ is stable relative to $CaX_2(ex) + 2K^+(aq)$ (K^+ is thermodynamically preferred by the soil exchanger phase). This conclusion is identical to that derived from the application of the Rothmund-Kornfeld model (indeed, $K_{RS} = 0.114$ is similar to $K_{RK} = 0.179$).

8.4.2 EXCHANGE ISOTHERMS AND PREFERENCE

The Vanselow equation offers a true thermodynamic description of cation exchange if the exchangeable ions on the surface form an ideal mixture. However, if this is not the case, the Vanselow selectivity coefficient is a conditional equilibrium constant that can be employed to establish the surface preference for one ion over another. An additional aid to establishing cation preference is the exchange isotherm. An exchange isotherm is a plot of the equivalent fraction of an ion in the exchange phase (E) against the equivalent fraction of the ion in the aqueous phase ($\tilde{E}$). For a binary symmetrical exchange system, the equivalent fractions of ions in the aqueous and exchanger phases are identical to the corresponding mole fractions. However, in nonsymmetrical exchange systems the equivalent fractions differ from mole fractions and are defined in Equations 8.27 and 8.28. Similarly, the equivalent fractions of the ions in the aqueous phase are:

$$\tilde{E}_{Ca} = \frac{2[Ca^{2+}]}{[K^+] + 2[Ca^{2+}]} \tag{8.64a}$$

$$\tilde{E}_{K} = \frac{[K^+]}{[K^+] + 2[Ca^{2+}]} \tag{8.64b}$$

where the brackets denote molar concentrations.

Exchange isotherms are classified in a manner similar to adsorption isotherms (Figure 8.8). The L- and H-class isotherms are characterized as having an initial slope that decreases as values of E increase. Conversely, an S-class isotherm is characterized as having an initial slope that increases as E increases. Finally, a C-class isotherm can be described as having a constant slope throughout the range of E. The hydrated radius and the valence of the cation determine the selectivity of a surface for an exchangeable cation. Qualitatively, selectivity is described in terms of ease of replacement, as illustrated by the lyotropic series for cation exchange in soils. However, quantitatively the preference of a surface for one exchangeable ion over another can be defined relative to the Vanselow selectivity coefficient. For the completely general exchange reaction:

$$aB^{b+}(aq) + bAX_a(ex) \rightarrow aBX_b(ex) + bA^{a+}(aq) \tag{8.65}$$

$$K_V = \frac{N_B^a (A^{a+})^b}{N_A^b (B^{b+})^a} \tag{8.66}$$

If K_V is greater than 1 for the reaction in Equation 8.65, B^{b+} is preferred by the exchange complex (the formation of BX_b is preferred relative to AX_a). If K_V is less than 1, A^{a+} is preferred by the exchange complex (the formation of AX_a is preferred). If K_V is approximately equal to 1, the surface shows no particular preference for either A^{a+} or B^{b+}.

For symmetric ion exchange (e.g., CaX_2-MgX_2 exchange), preference and the magnitude of K_V can be determined from the appearance of the exchange isotherm. For example, an exchange

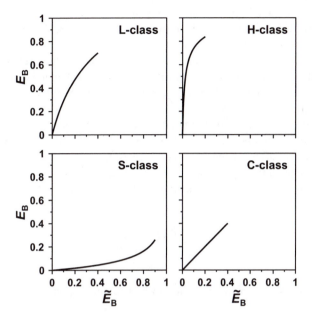

FIGURE 8.8 The classification of exchange isotherms: $\tilde{E}_B$ is the equivalent fraction of B^{b+} in solution and E_B is the equivalent fraction of B^{b+} on the exchange complex. Both $\tilde{E}_B$ and E_B range from 0 to 1.

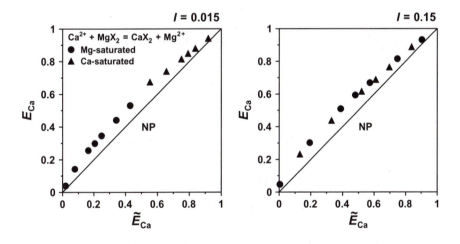

FIGURE 8.9 Exchange isotherms for binary MgX_2-CaX_2 exchange in Loring (Oxyaquic Fragiudalf) soil as a function of ionic strength (0.015 and 0.15). The nonpreference isotherm (NP) is the line represented by the equation $E_{Ca} = \tilde{E}_{Ca}$.

isotherm (E_{Ca} vs. $\tilde{E}_{Ca}$) for CaX_2-MgX_2 exchange by a Loring silt loam (Figure 8.9) is class L and indicates that Ca^{2+} is preferred by the soil exchange complex. The preference of the soil exchanger for Ca^{2+} is also indicated by the location of the exchange isotherm relative to the nonpreference isotherm. The nonpreference isotherm is a theoretically derived line or curve on the E vs. $\tilde{E}$ diagram which represents the condition where K_V equals 1 (the exchanger has no preference for either exchangeable ion). The data in Figure 8.9 indicate that there is more Ca^{2+} on the exchange complex relative to the nonpreference isotherm; therefore, Ca^{2+} is preferred by the soil exchange phase relative to Mg^{2+}. For symmetrical exchange, the nonpreference isotherm is C-class and is represented by the equation $E = \tilde{E}$. However, for nonsymmetric ion exchange, the preference of a surface for one cation over another cannot be inferred by the shape of an exchange isotherm. Such a determination requires that the exchange isotherm be compared to the nonpreference isotherm ($K_V = 1$), for which E is a nonlinear function of $\tilde{E}$.

The generation of an equation for the nonpreference isotherm for uni-bivalent exchange is a rather straightforward, but algebraically cumbersome exercise. For the AX-BX_2 exchange reaction (Equation 8.65), the derivation of the nonpreference isotherm for A^+, as a function of the total normality (N_T) of cations in the equilibrium solution, is initiated by setting $K_V = 1$ and substituting aqueous concentrations for activities in Equation 8.66 ($a = 1$ and $b = 2$):

$$1 = \frac{N_B \gamma_A^2 [A^+]^2}{N_A^2 \gamma_B [B^{2+}]} \tag{8.67}$$

Equations 8.28a and 8.28b can be manipulated to find N_B and N_A as a function of E_B and E_A:

$$N_B = \frac{E_B}{2 - E_B} \tag{8.68a}$$

$$N_A = \frac{2E_A}{1 + E_A} \tag{8.68b}$$

Equation 8.68a is modified further to express N_B in terms of E_A:

$$N_B = \frac{1 - E_A}{1 + E_A} \tag{8.68c}$$

Substituting Equations 8.68b and 8.68c into Equation 8.67 and rearranging:

$$\frac{4E_A^2}{1 - E_A^2} = \Gamma \frac{[A^+]^2}{[B^{2+}]} \tag{8.69}$$

where $\Gamma = \frac{\gamma_A^2}{\gamma_B}$. Additional algebraic manipulation to isolate E_A yields:

$$\frac{1}{E_A^2} = 1 + \frac{4[B^{2+}]}{\Gamma[A^+]^2} \tag{8.70}$$

which is equal to:

$$\frac{1}{E_A^2} = 1 + \frac{4[B^{2+}]}{\Gamma[A^+]^2} \frac{(2[B^{2+}] + [A^+])^2}{(2[B^{2+}] + [A^+])^2} \tag{8.71}$$

Applying Equations 8.64a and 8.64b, which define the equivalent fractions of the aqueous species:

$$\frac{1}{E_A^2} = 1 + \frac{4\tilde{E}_B}{2\Gamma\tilde{E}_A^2(2[B^{2+}]+[A^+])} = 1 + \frac{2(1-\tilde{E}_A)}{\Gamma\tilde{E}_A^2(2[B^{2+}]+[A^+])} \qquad (8.72)$$

The total normality of cations in the aqueous solution of a binary, uni-bivalent exchange system is defined as $N_T = 2[B^{2+}] + [A^+]$. Equation 8.21j becomes:

$$\frac{1}{E_A^2} = 1 + \frac{2(1-\tilde{E}_A)}{\Gamma\tilde{E}_A^2 N_T} \qquad (8.73)$$

which is also written:

$$E_A = \left\{ 1 + \frac{2}{\Gamma N_T} \left[\frac{1}{\tilde{E}_A^2} - \frac{1}{\tilde{E}_A} \right] \right\}^{-0.5} \qquad (8.74)$$

The nonpreference isotherm for A^+ as a function of the ionic strength of the aqueous phase (I) is similarly derived:

$$E_A = \left\{ 1 + \frac{1}{\Gamma I} \left[\frac{3}{\tilde{E}_A^2} - \frac{4}{\tilde{E}_A} + 1 \right] \right\}^{-0.5} \qquad (8.75)$$

For binary ion exchange, there are only two cations in the system (e.g., B^{2+} and A^+), thus, the total normality is computed from the concentrations of B^{2+} and A^+ ($N_T = 2[B^{2+}] + [A^+]$) and the ionic strength computing from the concentrations of B^{2+}, A^+, and associated anion ($I = 3[B^{2+}] + [A^+]$, if the anion is monovalent). The impact of ionic strength on the nonpreference isotherm is illustrated in Figure 8.10. As the salt concentration increases, the surface becomes more selective for the lower valence cation, A^+; while dilution favors the retention of the higher valence cation, B^{2+} (valence dilution effect). Note that while the nonpreference isotherm lines indicate that the surface is selective for the divalent ion (the equivalent fraction of AX is less than that of BX_2 throughout the entire range of $\tilde{E}$), there is no preference, in the sense that $K_V = 1$.

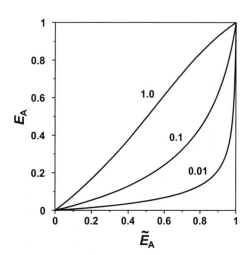

FIGURE 8.10 The influence of ionic strength (0.01, 0.1, and 1) on the nonpreference isotherm for binary uni-bivalent exchange.

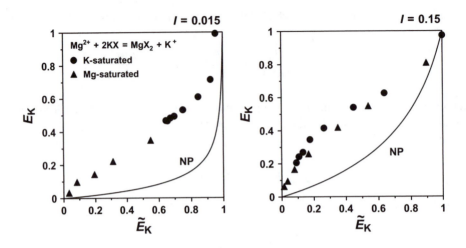

FIGURE 8.11 Exchange isotherms for binary KX-MgX$_2$ exchange in Loring (Oxyaquic Fragiudalf) soil as a function of ionic strength (0.015 and 0.15). The nonpreference isotherms (NP) were computed using Equation 8.75.

The exchange isotherm for KX-MgX$_2$ exchange by Loring silt loam soil for two ionic strength conditions is illustrated in Figure 8.11. The clay fraction of this soil is composed of HIV, vermiculite, mica, and kaolinite. Also shown are the nonpreference isotherms. The apparent selectivity of the soil for K$^+$ depends on ionic strength. High concentrations of Mg and K ($I = 0.15$) result in a class L isotherm, indicating a soil preference for K$^+$ ($E_K > \tilde{E}_K$ for $0 < \tilde{E}_K < 0.6$). Reducing the concentrations of K and Mg (from $I = 0.15$ to 0.015) results in an S-class isotherm, suggesting that the soil is selective for Mg^{2+} (there is a greater concentration of Mg^{2+} satisfying exchange phase charge than there is K$^+$; $E_K < \tilde{E}_K$ for all $\tilde{E}_K$). However, because there is more K$^+$ on the exchange complex relative to the nonpreference isotherm (for any given value of $\tilde{E}_K$), the exchange phase has a thermodynamic preference for K$^+$, irrespective of the ionic strength. Therefore, K_V is greater than 1 for the reaction $2K^+(aq) + MgX_2(ex) = 2KX(ex) + Mg^{2+}(aq)$. The valence dilution effect is evident in the K$^+$ exchange isotherms, as decreasing the ionic strength from 0.15 to 0.015 favors the greater retention of Mg^{2+} relative to K$^+$. For symmetrical CaX$_2$-MgX$_2$ exchange, as shown for a Loring (Oxyaquic Fragiudalf) soil in Figure 8.9, decreasing the ionic strength from 0.15 to 0.015 does not influence the exchange isotherms, providing further evidence that the valence dilution effect is specific to nonsymmetrical ion exchange.

The computation of an exchange selectivity coefficient assumes a number of conditions. Among these is the stipulation that exchange reactions are stoichiometric and reversible. Reversible ion exchange implies that the identical equilibrium state will always be achieved, regardless of the initial condition of the system. For example, an equilibrium exchange isotherm generated by reacting Mg^{2+}-saturated soil with solutions having varied Mg^{2+} and Ca^{2+} compositions should be identical to the isotherm generated by reacting Ca^{2+}-saturated soil with the different solutions. If identical equilibrium isotherms are obtained, the exchange reaction is reversible. The exchange isotherm generated from the Mg^{2+}-saturated Loring soil (Figure 8.9) is identical to that generated from the Ca^{2+}-saturated soil. Thus, the symmetrical exchange reaction is reversible and also unaffected by ionic strength.

Hysteretic ion exchange may be evident if the exchange system is in disequilibrium. However, exchange equilibrium is relatively rapid (nearly instantaneous) due to the hydrated nature of an ion on the soil exchange phase, particularly on clay minerals bearing readily accessible surfaces (e.g., smectites and kaolinites). Relatively long exchange equilibration times may be required for vermiculitic soils, particularly when the exchange involves cations that can collapse the clay

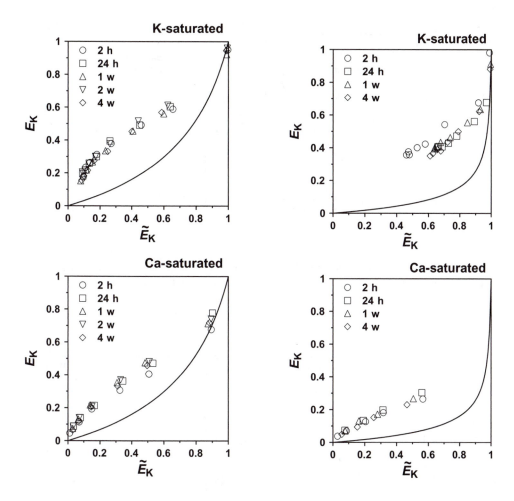

FIGURE 8.12 Exchange isotherms for the KX-CaX$_2$ binary system in Loring (Oxyaquic Fragiudalf) soil as a function of ionic strength: (a) 0.15 and (b) 0.015, equilibration time, and initial cation saturation. The line represents the nonpreference isotherm.

structure. For example, consider KX-CaX$_2$ exchange by a Loring soil as a function of equilibration time and initial cation saturation when ionic strength is 0.15 (Figure 8.12a). For both systems, exchange equilibrium is achieved in approximately 2 h, irrespective of the initial cation saturation. However, when the initially K$^+$- and Ca^{2+}-saturated equilibrium exchange isotherms (0.15 ionic strength) are viewed together (Figure 8.13a), the two data sets do not overlap. This is evidence of exchange hysteresis and indicates that the soil exchanger has greater preference for K$^+$ when initially K$^+$ saturated, relative to the preference for K$^+$ demonstrated when initially Ca^{2+} saturated. Likewise, the soil has greater relative preference for Ca^{2+} when initially Ca^{2+} saturated. Greater relative preference for the initially saturating cation is a known phenomenon. This fact is further illustrated by comparing the computed K_V values (Figure 8.14a). In this case, the hysteretic behavior of the exchange systems is not a function of disequilibrium, but may be attributed to the clay mineralogy of the Loring soil, which contains an abundance of vermiculite and HIV that may fix K$^+$. Decreasing the ionic strength of the exchange systems affects the equilibration rate for K$^+$-saturated soil, requiring a 24-h equilibration period for the 0.015 ionic strength systems, compared to 2 h for the 0.15 ionic strength systems (Figure 8.12a and b). Higher Ca^{2+} concentrations in the

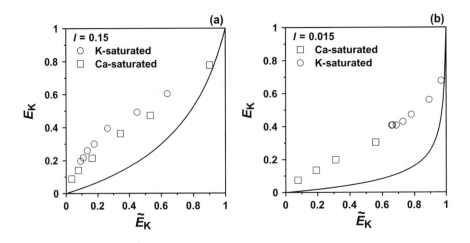

FIGURE 8.13 Exchange isotherms for the KX-CaX$_2$ binary system in Loring (Oxyaquic Fragiudalf) soil as a function of ionic strength, (a) 0.15 and (b) 0.015, that displays the influence of initial cation saturation on preference.

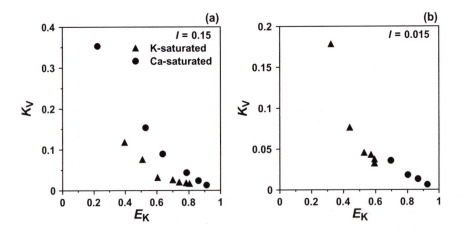

FIGURE 8.14 The Vanselow selectivity coefficient for KX-CaX$_2$ binary exchange in Loring (Oxyaquic Fragiudalf) soil as a function of E_K and ionic strength: (a) 0.15 and (b) 0.015.

higher ionic strength systems result in a more rapid exchange equilibration. Combined, the 0.015 ionic strength exchange isotherms initiated from K$^+$- and Ca^{2+}-saturated soil tend to suggest exchange reversibility, as a single curve may be drawn to describe both sets of data (Figure 8.13b). However, the computed K_V values demonstrate that exchange hysteresis occurs in the 0.015 ionic strength system, as there appears to be a discontinuity between the two datasets (Figure 8.14b).

8.4.3 EXCHANGE EQUILIBRIUM CONSTANT

Considerable effort has been afforded the development of a universally applicable expression that describes cation exchange behavior in terms of a thermodynamic equilibrium constant and composition variables. These efforts have generally resulted in the establishment of various selectivity coefficients, often termed conditional exchange constants (as described in the previous sections and

TABLE 8.4
Values for the Various Selectivity Coefficients Discussed in the Text for CaX$_2$-KX Binary Exchange on a Loring Silt Loam (Oxyaquic Fragiudalf) According to the Reaction: 2K$^+$(aq) + CaX$_2$(ex) → 2KX(ex) + Ca^{2+}(aq)[a]

N_K[b]	K_K	K_V	K_{RK}[c]	K_G	K_{GT}	$K_{D, n=4}$	K_{RS}[d]
			Initially K$^+$-Saturated				
0.264	0.131	39.12	5.36	3.65	11.27	28.60	5.14
0.307	0.153	39.55	6.43	3.78	11.68	29.37	6.44
0.359	0.145	37.71	7.49	3.84	11.49	28.55	7.98
0.423	0.114	32.78	8.35	3.77	10.39	25.44	9.58
0.503	0.079	21.10	7.40	3.26	7.05	16.90	9.23
0.627	0.048	11.95	7.28	2.83	4.35	10.07	9.74
0.719	0.018	4.53	4.60	2.03	1.77	3.97	5.91
			Initially Ca^{2+}-Saturated				
0.137	0.178	64.96	4.94	4.34	17.44	45.38	4.50
0.231	0.129	42.14	4.95	3.71	11.89	30.40	4.60
0.357	0.079	25.11	4.89	3.13	7.63	18.98	5.16
0.522	0.042	12.73	4.83	2.58	4.30	10.27	6.12
0.644	0.030	8.35	5.53	2.42	3.08	7.08	7.40

[a] The selectivity coefficients are defined in Table 8.3.
[b] N_K is the mole fraction of K$^+$ on the soil exchange complex.
[c] Computed using $\beta = 0.791$.
[d] Computed using $Q = -1.68$.

in Table 8.3). The computed values for selectivity coefficients generally vary as a function of exchange phase composition, or remain constant only over a limited range of exchange phase compositions. Table 8.4 illustrates the dependence of selectivity coefficients (described in the previous sections) on exchange phase composition for CaX$_2$-KX exchange in the Loring soil. The Kerr, Vanselow, Gaines-Thomas, and Davies selectivity coefficients are a function of N_K, the mole fraction of adsorbed K$^+$. The Rothmund-Kornfeld and the regular solution models of exchange behavior generate relatively stable selectivity coefficients, owing to the inclusion of the adjustable β and Q parameters in the respective models. The Gapon model, which does not incorporate an adjustable parameter, also generates a relatively invariant exchange selectivity coefficient. However, despite the apparent stability of the Gapon, Rothmund-Kornfeld, and regular solution coefficients for CaX$_2$-KX exchange, these selectivity coefficients do not provide a true thermodynamic description of the exchange reaction.

The ability to determine exchange equilibrium constants was first demonstrated in 1950 by Argersinger et al. (1950) and Hogfeldt (Ekedahl et al., 1950; Hogfeldt et al., 1950). A thermodynamic treatment of ion exchange requires the ability to describe the exchange system in terms of an equilibrium constant that is independent of exchange phase composition. A thermodynamic treatment requires the following assumptions: (1) cation and anion exchange occur separately and do not affect each other; (2) the exchange capacity of the adsorbent is constant and unaffected by exchangeable ion or solution composition: (3) the exchange reaction is stoichiometric and reversible; and (4) exchange equilibrium exists.

Consider the nonsymmetrical uni-bivalent exchange reaction:

$$\text{Ca}^{2+}(aq) + 2\text{KX}(ex) \rightarrow \text{CaX}_2(ex) + 2\text{K}^+(aq) \tag{8.76}$$

The thermodynamic equilibrium constant for the exchange reaction is:

$$K_{ex} = \frac{(CaX_2)(K^+)^2}{(KX)^2(Ca^{2+})} \tag{8.77}$$

where the parentheses ( ) denote activities. The activities of the soluble species can be related to composition (where molar concentrations are denoted by brackets [ ]) by employing the expressions:

$$(Ca^{2+}) = \gamma_{Ca^{2+}}[Ca^{2+}] \quad \text{and} \quad (K^+) = \gamma_{K^+}[K^+] \tag{8.78}$$

where the γ values represent the single-ion activity coefficients. Substituting these expressions for ion activities into Equation 8.77 yields:

$$K_{ex} = \frac{(CaX_2)\gamma_{K^+}^2[K^+]^2}{(KX)^2\gamma_{Ca^{2+}}[Ca^{2+}]} \tag{8.79}$$

If the exchange phase is assumed to be a solid solution, with the end-members KX and CaX$_2$, the mole fractions of K$^+$ and Ca^{2+} on the exchange phase are composition variables that can be related to exchange phase activities (solid solutions are discussed in Chapter 6):

$$(CaX_2) = f_{Ca}N_{Ca} \quad \text{and} \quad (KX) = f_K N_K \tag{8.80}$$

where N_{Ca} and N_K are the mole fractions and f_{Ca} and f_K are the rational activity coefficients, the values for which approach 1.0 as the corresponding mole fraction approaches 1.0. The rational activity coefficients will also be unity if the exchangeable cations form an ideal mixture on the surface. However, ideal mixing of ions on the exchange complex is the exception rather than the rule. In general, ideal mixing is observed for ions with similar properties (e.g., valence and hydration energy) that do not strongly interact with one another when adsorbed and which randomly occupy surface sites. For example, Mg^{2+} and Ca^{2+} display near-ideal exchange behavior on the Loring soil, with a relatively constant K_V (1.56 ± 0.01) that is independent of N_{Ca} (Figure 8.4c). Both divalent ions are highly hydrated ($\Delta H_h = -1922$ kJ mol^{-1} for Mg^{2+} and -1592 kJ mol^{-1} for Ca^{2+}), have similar hydrated radii (428 pm for Mg^{2+} and 412 pm for Ca^{2+}), are well-shielded due to their large hydration numbers (36 for Mg^{2+} and 29 for Ca^{2+}), and are nonspecifically adsorbed (outer-sphere complexation) by constant charge surface sites.

Substituting the expressions for (CaX$_2$) and (KX) into Equation 8.79 yields

$$K_{ex} = \frac{f_{Ca}N_{Ca}\gamma_{K^+}^2[K^+]^2}{f_K^2 N_K^2 \gamma_{Ca^{2+}}[Ca^{2+}]} \tag{8.81}$$

Values for N_K and N_{Ca} are computed with knowledge of the moles of adsorbed K$^+$ (m_K) and Ca^{2+} (m_{Ca}), or the mol kg^{-1} concentration of K$^+$ [KX] and Ca^{2+} [CaX$_2$]:

$$N_K = \frac{m_K}{m_K + m_{Ca}} = \frac{[KX]}{[KX]+[CaX_2]} \tag{8.82a}$$

and

$$N_{Ca} = \frac{m_{Ca}}{m_K + m_{Ca}} = \frac{[CaX_2]}{[KX]+[CaX_2]} \tag{8.82b}$$

Values for N_K and N_{Ca} (in a binary system) are also related through the expression,

$$N_K + N_{Ca} = 1 \tag{8.82c}$$

Since the activities of soluble species can be determined by well-known methods from measurable composition data (as illustrated in Chapter 5 using the ion association model), and the exchange phase composition can be directly measured by displacement with an index cation, the application of equilibrium concepts to ion exchange reactions reduces to a determination of the rational activity coefficients, f_K and f_{Ca}. As described above, Vanselow provided the first truly thermodynamic representation of ion exchange. Recall that the exchange coefficient for Equation 8.76, written in the Vanselow convention is:

$$K_V = \frac{N_{Ca}(K^+)^2}{N_K^2(Ca^{2+})} = \frac{N_{Ca}\gamma_{K^+}^2[K^+]^2}{N_K^2\gamma_{Ca^{2+}}[Ca^{2+}]} \tag{8.83}$$

The Vanselow selectivity coefficient (K_V) describes K_{ex} if the exchangeable ions form an ideal mixture at the surface ($f_K = f_{Ca} = 1$). Otherwise, K_V is related to K_{ex} by:

$$K_{ex} = \frac{f_{Ca}}{f_K^2} K_V \tag{8.84}$$

Again, the activities of the aqueous species (K^+) and (Ca^{2+}) can be readily obtained, and for binary systems, K_V is readily determined. Then, Equation 8.84 can be used to obtain f_K and f_{Ca}, assuming the following two constraints:

1. The dependence of K_V on the composition of the exchange phase results from the dependence of the rational activity coefficients on the composition of the exchange phase. This constraint is made evident in Equation 8.84. Since K_{ex} is a true constant, any variation in K_V will depend on the variation in the rational activity coefficients, f_K and f_{Ca}. The mathematical interpretation of this constraint is applied by first taking the natural logarithm of Equation 8.84:

$$\ln K_{ex} = \ln K_V + \ln f_{Ca} - 2\ln f_K \tag{8.85}$$

   Because K_{ex} is a true equilibrium constant, its value does not depend on composition. Therefore, the derivative of $\ln K_{ex}$ (the change in $\ln K_{ex}$ relative to any compositional change in the exchange phase) equals zero: $d \ln K_{ex} = 0$. Then:

$$d \ln K_{ex} = 0 = d \ln K_V + d \ln f_{Ca} - 2d \ln f_K \tag{8.86a}$$

   or

$$d \ln K_V = 2d \ln f_K - d \ln f_{Ca} \tag{8.86b}$$

2. The composition dependence of the activity of one exchange component is compensated for by the activity of the other such that mass is conserved on the exchange complex. This constraint is interpreted through the application of the Gibbs-Duhem equation: in a single phase consisting of two components, such as the binary exchanger phase, the chemical potentials of the components cannot be independently varied. Mathematically, this constraint is expressed:

$$m_K d\mu_K + m_{Ca} d\mu_{Ca} = 0 \tag{8.87}$$

where μ_K and μ_{Ca} are the chemical potentials of KX and CaX_2, and m_K and m_{Ca} are the moles of adsorbed K^+ and Ca^{2+}. Divided Equation 8.87 through by $(m_K + m_{Ca})$ and substituting the chemical potentials with the expressions:

$$\mu_K = \mu_K^\circ + RT \ln(KX) \tag{8.88a}$$

$$\mu_{Ca} = \mu_{Ca}^\circ + RT \ln(CaX_2) \tag{8.88b}$$

yields:

$$\frac{m_K}{(m_K + m_{Ca})} d[\mu_K^\circ + RT \ln(KX)] + \frac{m_{Ca}}{(m_K + m_{Ca})} d[\mu_{Ca}^\circ + RT \ln(CaX_2)] = 0 \tag{8.89}$$

Note that $\frac{m_K}{(m_K + m_{Ca})}$ and $\frac{m_{Ca}}{(m_K + m_{Ca})}$ are the mole fractions defined in Equations 8.82a and 8.82b (N_K and N_{Ca}), and $d\mu^\circ = 0$, as the standard-state chemical potential is not dependent on composition (and does not vary with changes in composition). Dividing Equation 8.89 through by RT yields:

$$N_K d \ln(KX) + N_{Ca} d \ln(CaX_2) = 0 \tag{8.90}$$

Since $(KX) = f_K N_K$ and $(CaX_2) = f_{Ca} N_{Ca}$, Equation 8.90 becomes:

$$N_K d \ln(f_K N_K) + N_{Ca} d \ln(f_{Ca} N_{Ca}) = 0 \tag{8.91}$$

This expression can be simplified further, because any change in the mole fraction of adsorbed K^+ must occur concurrently with an equal and opposite change in the mole fraction of adsorbed Ca^{2+}, or $dN_K = -dN_{Ca}$. Therefore,

$$N_K d \ln f_K + N_{Ca} d \ln f_{Ca} = 0 \tag{8.92}$$

or

$$d \ln f_{Ca} = -\frac{N_K}{N_{Ca}} d \ln f_K \tag{8.93}$$

Equations 8.86b and 8.93 represent the mathematical interpretations of the two constraints. They also represent two equations that can be solved explicitly for the two rational activity coefficients, f_K and f_{Ca}. Substituting Equation 8.93 into Equation 8.86b and rearranging results in an expression for f_K:

$$2d \ln f_K = d \ln K_V - \frac{N_K}{N_{Ca}} d \ln f_K \tag{8.94}$$

Combining $\ln f_K$ terms and rearranging,

$$2d \ln f_K = \left[\frac{2N_{Ca}}{2N_{Ca} + N_K} \right] d \ln K_V \tag{8.95}$$

where

$$E_{Ca} = \frac{2N_{Ca}}{2N_{Ca} + N_K}$$

is the equivalent fraction of Ca^{2+} in the exchange phase. Then, Equation 8.95 becomes:

$$2d \ln f_K = E_{Ca} d \ln K_V \qquad (8.96)$$

The K^+ ion on the exchanger phase is in the reference state when $E_K = 1.0$ and $f_K = 1.0$ (Gaines-Thomas reference state). Correspondingly, when $f_K = 1.0$ ($\ln f_K = 0$), $E_{Ca} = 0$ ($E_{Ca} = 1 - E_K$). Equation 8.96 is solved for $\ln f_K$ by integration from the reference state ($\ln f_K = 0$, $E_{Ca} = 0$) to a state of arbitrary exchange composition:

$$2 \int_0^{\ln f_K} d \ln f_K = \int_0^{E_{Ca}} E_{Ca} d \ln K_V \qquad (8.97)$$

Integration leads to

$$2 \ln f_K = E_{Ca} \ln K_V - \int_0^{E_{Ca}} \ln K_V dE_{Ca} \qquad (8.98)$$

Expressions for $\ln f_{Ca}$ and $\ln K_{ex}$ may be obtained in a similar manner:

$$\ln f_{Ca} = -(1 - E_{Ca}) \ln K_V + \int_{E_{Ca}}^{1} \ln K_V dE_{Ca} \qquad (8.99)$$

Substituting the above expressions for $\ln f_K$ and $\ln f_{Ca}$ into Equation 8.85 leads to:

$$\ln K_{ex} = \int_0^1 \ln K_V dE_{Ca} \qquad (8.100)$$

Analytical solutions for $\ln K_{ex}$, $\ln f_K$, and $\ln f_{Ca}$ at any value of E_{Ca} require an expression for $\ln K_V$ as a function of E_{Ca}. Such an expression may take any form, although it is common to employ a power series of the form:

$$\ln K_V = b + mE_{Ca} + nE_{Ca}^2 + pE_{Ca}^3 \qquad (8.101)$$

However, the variability of $\ln K_V$ can be satisfactorily described by employing a truncated form of Equation 8.101, which models $\ln K_V$ as a linear function of E_{Ca}:

$$\ln K_V = b + mE_{Ca} \qquad (8.102)$$

where m and b are the slope and y-intercept of an $\ln K_V$ versus E_{Ca} plot of the exchange data. Using Equation 8.102, a generalized expression for computing $\ln K_{ex}$ may be obtained by substituting Equation 8.102 into Equation 8.100:

$$\ln K_{ex} = \int_0^1 [b + mE_{Ca}] dE_{Ca} \qquad (8.103)$$

Integrating and evaluating the resulting expression between the limits of integration yields

$$\ln K_{ex} = \left[bE_{Ca} + \frac{m}{2} E_{Ca}^2 \right]_0^1 = b + \frac{m}{2} \tag{8.104}$$

Note that $\ln K_{ex}$ is equal to $\ln K_V$ when E_{Ca} equals 0.5, irrespective of the values obtained for b and m (compare Equations 8.104 and 8.102 for $E_{Ca} = 0.5$). Substitution of Equation 8.102 into the expressions for $\ln f_K$ and $\ln f_{Ca}$ (Equations 8.98 and 8.99), integrating, and evaluating accordingly yields generalized expressions for the rational activity coefficients:

$$\ln f_K = \frac{m}{4} E_{Ca}^2 \tag{8.105}$$

$$\ln f_{Ca} = \frac{m}{2}(1 - E_{Ca})^2 = \frac{m}{2} E_K^2 \tag{8.106}$$

A thermodynamic analysis of nonsymmetrical ion exchange may be performed by considering the data presented in Table 8.5 which describes binary KX-CaX$_2$ exchange on a Loring silt loam according to the reaction in Equation 8.76. The Vanselow selectivity coefficient (K_V) decreases with increasing Ca^{2+} on the exchange phase (Table 8.5), from a value of 1.10 when E_{Ca} is 0.08, to 0.015 when E_{Ca} is 0.93. Because K_V varies with E_{Ca}, the exchangeable ions apparently form a nonideal mixture at the surface. The decrease in $\ln K_V$ with increasing E_{Ca} may be described by a linear expression (Equation 8.102), as illustrated in Figure 8.15. The expression obtained by least-squares linear regression analysis is

$$\ln K_V = 0.526 - 5.109 E_{Ca} \tag{8.107}$$

TABLE 8.5
Values for the Vanselow Selectivity Coefficient as a Function of the Equivalent Fraction of Ca^{2+} on the Exchange Complex for KX-CaX$_2$ Binary Exchange on a Loring Silt Loam (Oxyaquic Fragiudalf) According to the Reaction: Ca^{2+}(aq) + 2KX(ex) → CaX$_2$(ex) + 2K$^+$(aq)[a]

E_{Ca}	K_V	$\ln K_V$	f_{Ca}	f_K	K_{ex}
0.078	1.098	1.10	0.115	0.992	0.128
0.287	0.414	−0.88	0.276	0.898	0.141
0.439	0.221	−1.51	0.449	0.781	0.162
0.525	0.120	−2.12	0.562	0.703	0.136
0.544	0.084	−2.48	0.582	0.689	0.103
0.647	0.079	−2.54	0.731	0.583	0.169
0.664	0.047	−3.05	0.744	0.573	0.107
0.732	0.031	−3.49	0.830	0.506	0.099
0.782	0.027	−3.63	0.884	0.460	0.111
0.819	0.025	−3.68	0.921	0.424	0.130
0.848	0.026	−3.67	0.944	0.397	0.153
0.869	0.024	−3.74	0.958	0.380	0.157
0.926	0.015	−4.17	0.988	0.331	0.138

[a] The values for each parameter were computed using data obtained from Smith (2000).

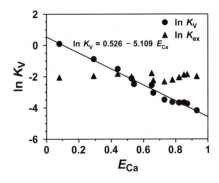

FIGURE 8.15 Plot of ln K_V vs. E_{Ca} for the KX-CaX$_2$ binary exchange in Loring (Oxyaquic Fragiudalf) soil with an ionic strength of 0.15. The line was obtained using least-squares linear regression analysis.

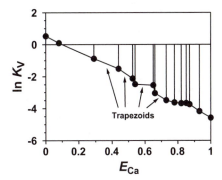

FIGURE 8.16 Plot of ln K_V vs. E_{Ca} for the KX-CaX$_2$ binary exchange in Loring (Oxyaquic Fragiudalf) soil with an ionic strength of 0.15. The area bounded by the ln K_V vs. E_{Ca} data and the ln $K_V = 0$ axis is numerically equal to ln K_{ex} and may be estimated using the trapezoid method (Table 8.6).

with $r^2 = 0.980$. Using Equation 8.104 and $m = -5.109$ and $b = 0.526$, the thermodynamic equilibrium constant for the exchange reaction is ln $K_{ex} = -2.028$ ($K_{ex} = 0.132$). Values for f_K and f_{Ca}, computed using Equations 8.105 and 8.106 for each value of E_{Ca}, are also shown in Table 8.5. The deviation of the rational activity coefficients from unity is further evidence that the exchangeable ions (K$^+$ and Ca^{2+}) form a nonideal mixture at the surface. The rational activity coefficients can also be employed, using Equation 8.84 or 8.85, to compute a value for ln K_{ex} at each value of E_{Ca}. The Table 8.5 K_{ex} values range from 0.099 to 0.169, with a mean of $K_{ex} = 0.133$ ($\pm$0.023 standard deviation), which is in good agreement with the K_{ex} computed using Equation 8.104.

The determination of K_{ex} may also be completed without the use of integral calculus, although the procedure is somewhat tedious. As stated in Equation 8.100, ln K_{ex} is equal to the integral of an expression that describes the variance of ln K_V as a function of E_{Ca} (such as found in Equation 8.107). Further, this function is integrated with respect to E_{Ca} throughout the entire range of E_{Ca} (0 to 1). Therefore, ln K_{ex} is mathematically equivalent to the area under the line or curve that describes ln K_V as a function of E_{Ca}, and is bounded by the ln $K_V = 0$ axis. The area may be estimated by using a series of trapezoids, as shown in Figure 8.16. Again, using the data from the binary KX-CaX$_2$ exchange system (Table 8.5), the estimated area encompassed by the ln K_V data is ln $K_{ex} = -2.017$ ($K_{ex} = 0.133$) (Table 8.6). This value is in good agreement with ln $K_{ex} = -2.028$ ($K_{ex} = 0.132$) obtained by integrating Equation 8.107.

In addition to the computation of the equilibrium constant (K_{ex}) for a binary exchange reaction, the free energy change for the reaction (ΔG_{ex}°) may also be computed. Assuming standard state conditions ($T = 298.15$ K and $P = 0.101$ MPa), the relationship between $\log K_{ex}$ and ΔG_{ex}° is (from Chapter 5):

$$\Delta G_{ex}^{\circ} = -5.708 \log K_{ex} \tag{8.108}$$

TABLE 8.6
Computation of ln Kex for the Reaction: $Ca^{2+}(aq) + 2KX(ex) \rightarrow$
$CaX_2(ex) + 2K^+(aq)$, Using the Trapezoid Technique Illustrated
in Figure 8.16[a]

E_{Ca}	ln K_V[b]	Area[c]
0	0.53	
0.078	0.09	0.024
0.287	*−0.88*	−0.083
0.439	*−1.51*	−0.182
0.525	−2.12	−0.156
0.544	−2.48	−0.044
0.647	−2.54	−0.259
0.664	−3.05	−0.048
0.732	−3.49	−0.222
0.782	−3.63	−0.178
0.819	−3.68	−0.135
0.848	−3.67	−0.107
0.869	−3.74	−0.078
0.926	−4.17	−0.225
1	−4.58	−0.324

$\Sigma(\text{Area}) = -2.017 \ (= \ln K_{ex})$

[a] The values for each parameter were computed using data obtained from Smith (2000).
[b] ln K_V values for $E_{Ca} = 0$ and 1 were estimated using Equation 8.107.
[c] The area of each trapezoid described by the ln K_V versus E_{Ca} data in Figure 8.17 is defined by the expression: Area = $\frac{1}{2}(\ln K_{V2} + \ln K_{V1}) \times (E_{Ca2} - E_{Ca1})$. For example, the area described by the trapezoid with corners (italicized data): $E_{Ca2} = 0.439$, $E_{Ca1} = 0.287$, ln $K_{V2} = -1.51$, and ln $K_{V1} = -0.88$, is $\frac{1}{2}(-1.51 - 0.88) \times (0.439 - 0.287) = -0.182$. Summation of all the trapezoid areas [$\Sigma(\text{Area})$] is the estimate of ln K_{ex}.

where ΔG_{ex}° has units of kJ mol^{-1}. Using the binary KX-CaX$_2$ exchange data (Table 8.5 and described above), with $K_{ex} = 0.132$ and $\log K_{ex} = -0.879$, the computed ΔG_{ex}° for the reaction described in Equation 8.76 is 5.02 kJ mol^{-1}. The values for both K_{ex} and ΔG_{ex}° indicate that the soil exchange phase is selective for K$^+$, as K_{ex} is less than 1 ($\log K_{ex}$ is negative) and ΔG_{ex}° is positive. Actually, the correct conclusion is that the exchange reaction products, the combination of CaX$_2$ + 2K$^+$, is less stable then the reactants, 2KX + Ca^{2+}. This particular result would appear to be an anomaly (relative to the lyotropic series and Coulombs Law, Equation 8.7), as one might predict soil exchange phase preference for the divalent Ca^{2+} ion over the monovalent K$^+$. However, the finding that K$^+$ is preferred by the Loring soil exchange phase relative to Ca^{2+} is consistent with both the clay mineralogy of this soil (dominance of HIV-vermiculite), as well as results from other binary K$^+$-Ca^{2+} exchange studies (Table 8.7). In general, soil exchanger phases prefer K$^+$ (and NH$_4^+$) relative to Ca^{2+}, Mg^{2+}, and Na$^+$, irrespective of the clay mineralogy. It is also evident from the data in Table 8.7 that the exchanger phase for a number of soils prefers Ca^{2+} over Mg^{2+} and Ca^{2+} over Na$^+$. Again, these findings are consistent with Coulomb's Law predictions. Exchange equilibrium constants and ΔG_{ex}° values for KX-CaX$_2$, KX-MgX$_2$, and MgX$_2$-CaX$_2$ for the Loring as a function of ionic strength are shown in Table 8.7. Also shown are the ΔG_{ex}° values for various exchange reactions in the Yolo (smectitic clay fraction) and Brucedale (illitic clay fraction) soils as a function of ionic strength or total molarity (Bond, 1995). The computed K_{ex} for a particular exchange reaction should not vary with ionic strength. In general,

TABLE 8.7

Thermodynamic Data for the Binary Exchange Systems KX-CaX$_2$, KX-MgX$_2$, MgX$_2$-CaX$_2$ in a Loring Soil (Oxyaquic Fragiudalf) that Contains Hydroxy Interlayered Vermiculite, Kaolinite, and Mica (Computed Using Data Obtained from Smith, 2000), and a Yolo Soil (Mollic Xerofluvent) from the Study of Jensen and Babcock (1973) (as Evaluated by Bond, 1995) as a Function of Ionic Strength*

Ionic Strength	K_{ex}[a]			ΔG_{ex}°, kJ mol^{-1}		
	KX-CaX$_2$	KX-MgX$_2$	MgX$_2$-CaX$_2$	KX-CaX$_2$	KX-MgX$_2$	MgX$_2$-CaX$_2$
			Loring Soil[b]			
0.015	0.057	0.029	1.47	7.11	8.79	−0.95
0.15	0.124	0.055	1.62	5.18	7.21	−1.19
			Yolo Soil[c]			
0.001	0.152	0.059	2.85	4.68	7.04	−2.60
0.01	0.114	0.053	2.86	5.38	7.26	−2.60

Reported ΔG_{ex}° (kJ mol^{-1}) Values for Other Soils

Dominate Clays[d]	NaX-KX	NaX-CaX$_2$	KX-CaX$_2$	MgX$_2$-CaX$_2$
I, K	−4.26 (0.05 M)[e]	−2.17 (0.05 M)[e]	5.52 (0.05 M)[e]	
	−4.69 (0.2 M)[e]	−4.22 (0.2 M)[e]	4.47 (0.2 M)[e]	
K			0.78 to 3.22[f]	
HIV, V, K		−3.50 to −6.08[g]	−1.93[h]	−1.09 to −2.81[g]
Sm	−4.13 (0.01)[c]		4.74 to 4.61[i]	
	−4.00 (0.1)[c]			

*Thermodynamic exchange data for other soils is included to illustrate the influence of clay mineralogy and ionic environment on ΔG_{ex}° values.

[a] K_{ex} and ΔG_{ex}° are computed using Equations 8.100 and 8.108 for KX-CaX$_2$ exchange: Ca^{2+} + 2KX → CaX$_2$ + K$^+$; KX-MgX$_2$ exchange: Mg^{2+} + 2KX → MgX$_2$ + K$^+$; and MgX$_2$-CaX$_2$ exchange: Ca^{2+} + MgX$_2$ → CaX$_2$ + Mg^{2+}.

[b] K_{ex} and ΔG_{ex}° are computed from the data of Smith (2000).

[c] Cation exchange in a Yolo loam containing smectite clay (Jensen and Babcock, 1973). Experiments were conducted in 0.01 and 0.1 ionic strength solution for the NaX-KX exchange studies.

[d] I, illite; K, kaolinite; HIV, hydroxyl interlayered vermiculite; V, vermiculite; Sm, smectite.

[e] From Bond (1995) for cation exchange in Brucedale subsoil containing >60% clay composed of illite and kaolinite. Exchange experiments were performed using 0.05 M and 0.2 M total cation concentrations.

[f] Levy et al. (1988) examined KX-MgX$_2$ exchange in kaolinitic soils.

[g] Ogwada and Sparks (1986) examined KX-MgX$_2$ exchange in Chester and Downer soils (both are Typic Hapludults) that contain hydroxy-interlayered vermiculite, kaolinite, and mica clay.

[h] From Jardine and Sparks (1984) for KX-CaX$_2$ exchange in the Ap horizon of an Evesboro soil (Typic Quartzipsamment) whose clay fraction is composed of hydroxy-interlayered vermiculite, kaolinite, and mica.

[i] Van Bladel and Gheyi (1980) examined NaX-CaX$_2$ and MgX$_2$-CaX$_2$ exchange in Mollisols, Inceptisols, and Vertisols containing smectitic clay.

this is the case for the exchange in dilute Yolo soil systems (0.001 and 0.01 ionic strengths). While generally not significant, K_{ex} (and ΔG_{ex}°) values vary as a function of ionic strength in all KX-CaX$_2$ exchange systems, and for all exchange reactions in the relatively high ionic strength Loring (0.015 and 0.15 ionic strength) and Brucedale (0.05 and 0.2 total normality) soils. The observation that some exchange data generate variations in the computed K_{ex} value with ionic strength while others do not is common and likely due to the heterogeneity of soil surfaces. High aqueous concentrations of K^+ may also lead to the collapse of the phyllosilicate structures (K fixation), leading to the apparent dependence of K^+ exchange reactions on ionic strength.

8.5 EXERCISES

1. Using the data in Table 8.1 compute the average *ECEC* values for soils with $pH_w < 6$ and for soils with $pH_w > 7$. Offer explanations as to why the *ECEC* values for acidic and alkaline soils differ.

2. In the table below are equilibrium data concerning the exchange reaction: $Ca^{2+}(aq) + 2KX(ex) \rightarrow CaX_2(ex) + 2K^+(aq)$.
   a. Complete the table.
   b. Assume $\ln K_V$ is a linear function of E_K. What is the equation of this line?
   c. What is K_{ex} for the KX-CaX$_2$ exchange reaction?
   d. What ion is preferred by the exchange phase?

E_K	(K^+)	(Ca^{2+})	N_K	N_{Ca}	K_V	$\ln K_V$
0.210	0.0041	0.0068				
0.900	0.0369	0.00075				

3. The *CEC* of a pH 5.22 surface soil sample is determined using the ammonium acetate pH 7 procedure. In this procedure a 2-g soil sample is repeatedly centrifuge-washed with pH 8.2, 1 M $C_2H_3O_2Na$ to remove soluble salts and to saturate the exchange complex with Na^+. After removing occluded Na^+, the soil is repeatedly washed with pH 7, 1 M $C_2H_3O_2NH_4$ to displace Na^+ from the exchange complex. The supernatant $C_2H_3O_2NH_4$ rinsates are combined in a 50-mL volumetric flask, which is brought to volume with deionized water. Additionally, a 5-mL volume of solution from the 50-mL volumetric flask is placed in a 100-mL volumetric flask, brought to volume, and analyzed for Na using atomic absorption spectrophotometry. The Na concentration in the diluted supernatant solution is 7.49 mg L^{-1}. Compute the *CEC* of the soil and express your result in cmol$_c$ kg^{-1}.

4. For a *CEC* procedure that uses Na^+ as the index cation, H_2O/ethanol as the wash solvent, and NH_4^+ as the displacing cation, discuss how each of the following influence the measured *CEC* value relative to the actual *CEC* of a soil:
   a. Hydrolysis due to excess washing.
   b. Incomplete removal of the index cation.
   c. Presence of large amounts of calcite and gypsum.
   d. Incomplete index cation saturation.
   e. Precipitation of an insoluble Na^+ salt in the ethanol.

5. The cation exchange data given in the table below are for KX-CaX$_2$ and NaX-CaX$_2$ exchange in a soil that has >60% illitic and kaolinitic clay and a *CEC* of 21 cmol$_c$ kg^{-1}.
   a. Construct the exchange isotherms, E_K vs. $\tilde{E}_K$ and E_{Na} vs. $\tilde{E}_{Na}$, for KX-CaX$_2$ and NaX-CaX$_2$ exchange and include the nonpreference isotherm (total normality is 50 meq L^{-1}) for the reactions:

   $Ca^{2+}(aq) + 2KX(ex) \rightarrow CaX_2(ex) + 2K^+(aq)$
   $Ca^{2+}(aq) + 2NaX(ex) \rightarrow CaX_2(ex) + 2Na^+(aq)$

b. Classify the exchange isotherms.

KX-CaX₂ Exchange				NaX-CaX₂ Exchange			
Solution Concentration		Exchange Concentration		Solution Concentration		Exchange Concentration	
Ca	K	Ca	K	Ca	Na	Ca	Na
mmol L⁻¹		cmol_c kg⁻¹		mmol L⁻¹		cmol_c kg⁻¹	
23.15	3.96	18.49	2.43	22.88	4.95	19.75	0.43
21.44	8.81	16.96	3.96	19.83	9.66	19.51	0.46
16.15	17.66	14.39	6.41	15.64	19.53	19.11	0.97
9.82	31.26	10.83	9.65	8.70	33.41	17.90	2.25
4.96	40.43	8.10	12.29	3.74	42.73	15.49	5.13
2.86	43.27	5.70	16.05	1.61	46.49	12.33	7.73

6. Use the data for NaX-CaX₂ exchange in Question 5 and determine the Vanselow selectivity coefficients (K_V) for the reaction: $Ca^{2+}(aq) + 2NaX(ex) \rightarrow CaX_2(ex) + 2Na^+(aq)$ (assume that the molar concentrations of Ca^{2+} and Na^+ are equivalent to the activities). Construct a plot of $\ln K_V$ vs. E_{Ca} and develop a linear relationship that predicts $\ln K_V$ as a function of E_{Ca}. Compute the exchange equilibrium constant (K_{ex}) and the free energy change for the reaction (ΔG_{ex}°). Are the reactants ($Ca^{2+}(aq) + 2NaX(ex)$) or the products ($CaX_2(ex) + 2Na^+(aq)$) more stable?

7. The findings and conclusions of H.S. Thompson and J.T. Way were revolutionary for their time. List and discuss the conclusions of J.T. Way and indicate how these conclusions relate to our current understanding of ion exchange phenomena.

8. Valence dilution is a technique that is employed in the reclamation of sodic soils. The technique involves the initial irrigation of a sodic soil with highly saline water. Thereafter, and with successive irrigations, the salinity of the irrigation water is reduced. Suppose a soil solution from a montmorillonitic soil is dominated by Ca^{2+} and Na^+, containing 0.1 M Na^+ and 0.025 M Ca^{2+}.

   a. If the Gapon selectivity coefficient is 0.02 for the reaction, $\frac{1}{2}Ca^{2+}(aq) + NaX(ex) \rightarrow Ca_{1/2}X(ex) + Na^+(aq)$, and the soil has a *CEC* of 20 cmol_c kg⁻¹, what is the composition of the exchange phase (assume that Ca^{2+} and Na^+ are the only ions present)?

   b. What is the value for the Vanselow selectivity coefficient for the reaction, $Ca^{2+}(aq) + 2NaX(ex) \rightarrow CaX_2(ex) + 2Na^+(aq)$?

   c. If rainfall results in a tenfold dilution in the composition of the soil solution, what will be the new composition of the exchange phase? How does this new composition differ from the original exchange phase composition determined in (a)?

9. Studies involving exchange equilibria on a Loring soil indicate that K^+ is preferred by the exchanger phase relative to Ca^{2+} and Mg^{2+}. It is also noted that the soil preference for K^+ is greater when the soil is initially K^+ saturated, relative to the preference observed when the soil is initially saturated by a divalent cation (exchange hysteresis). These findings appear to contradict the lyotropic series, Coulomb's Law (Equation 8.7), and the assumption of exchange reaction reversibility and nearly instantaneous equilibration. How can these data be explained, given that the clay fraction of this soil is composed of vermiculite-HIV, mica, and kaolinite?

10. The *CEC* value for a soil in the environment is not constant, but varies according to depth in a profile and the location on the landscape (as shown in Chapter 1). However, at any given position, soil *CEC* is also a function of the compositional characteristics of

the soil solution. Explain how the chemical characteristics of the soil solution can influence soil *CEC* and indicate the substances and processes that may increase soil *CEC* and those that may decrease soil *CEC*.

11. A 200-mg sample of K^+-saturated Cheto montmorillonite (CEC = 120 $cmol_c$ kg^{-1}) is equilibrated with a 45-mL volume of a pH 4 KCl solution that initially contains 12 mmol L^{-1} aniline, an aromatic amine (pK_a = 4.60 for the amine deprotonation reaction: $ArNH_3^+ \rightarrow ArNH_2^0 + H^+$). At equilibrium, the solution contains 30 mmol L^{-1} K^+ and 8 mmol L^{-1} aniline. Compute the Vanselow selectivity coefficient (K_V) for the reaction: $ArNH_3^+ (aq) + KX(ex) \rightarrow ArNH_3X(ex) + K^+(aq)$ (assume that molar concentrations and activities are equivalent).

12. Use the data for KX-CaX_2 exchange in Question 5 and determine the Rothmund-Kornfeld model parameters, K_{RK} and β, and the regular solution model parameters, K_{RS} and Q, for the reaction: $Ca^{2+}(aq) + 2KX(ex) \rightarrow CaX_2(ex) + 2K^+(aq)$ (assume that the molar concentrations of Ca^{2+} and Na^+ are equivalent to their activities). Compare the empirical K_{RS} and K_{RK} values to the exchange equilibrium constant (K_{ex} = 0.0985; $\ln K_{ex}$ = –2.318).

REFERENCES

Argersinger, W.J., A.W. Davidson, and O.D. Bonner. Thermodynamics and ion exchange phenomena. *Trans. Kansas Acad. Sci.* 53:404–410, 1950.

Batjes, N.H. (ed.) A homogenized soil data file for global environmental research: a subset of FAO, ISRIC and NRCS profiles (Version 1.0). Working Paper and Peprint 95/10b, International Soil Reference and Information Centre, Wageningen, the Netherlands, 1995.

Bond, W.J. On the Rothmund-Kornfeld description of cation exchange. *Soil Sci. Soc. Am. J.* 59:436–443, 1995.

Davies, L.E. Ionic exchange and statistical thermodynamics. I. Equilibria in simple exchange systems. *J. Coll. Sci.* 5:71–79, 1950.

Davies, L.E. and J.M. Rible. Monolayers containing polyvalent ions. *J. Coll. Sci.* 5:81–83, 1950.

Ekedahl, E., E. Högfeldt, and L.G. Sullén. Activities of the components in ion exchangers. *Acta Chem. Scan.* 4:556–558, 1950.

Gaines, G.L., Jr. and H.C. Thomas. Adsorption studies on clay minerals. II. A formulation of the thermodynamics of exchange adsorption. *J. Chem. Phys.* 21:714–718, 1953.

Gapon, Y.N. On the theory of exchange adsorption in soils. *J. Gen. Chem. USSR* (Engl. Transl.) 3:144–160, 1933.

Högfeldt, E., E. Ekedahl, and L.G. Sullén. Activities of the components in ion exchangers with multivalent ions. *Acta Chem. Scand.* 4:828–829, 1950.

Jardin, P.M. and D.L. Sparks. Potassium-calcium exchange in a multireactive soil system. II. Thermodynamics. *Soil Sci. Soc. Am. J.* 48:45–50, 1984.

Jensen, H.E. and K.L. Babcock. Cation exchange equilibria on a Yolo loam. *Higardia* 41:475–487, 1973.

Johnson, S.W. On some points of agricultural science. *Am. J. Sci. Arts. Ser.* 2. 28:71–85, 1859.

Kerr, H.W. The nature of base-exchange and soil acidity. *J. Am. Soc. Agron.* 20:309–335, 1928.

Krishnamoorthy, C. and R. Overstreet. An experimental evaluation of ion exchange relationships. *Soil Sci.* 69:41–53, 1950.

Levy, G.J., H.v.H. van der Watt, I. Shainberg, and H.M. du Plessis. Potassium-calcium and sodium-calcium exchange on kaolinite and kaolinitic soils. *Soil Sci. Soc. Am. J.* 52:1259–1264, 1988.

Ogeada, R.A. and D.L. Sparks. A critical evaluation on the use of kinetics for determining thermodynamics of ion exchange in soils. *Soil Sci. Soc. Am. J.* 50:300–305, 1986.

Rothmund, V. and G. Kornfeld. Die basenaustausch im permutit. I. *Z. Anorg. Allg. Chem.* 103:129–163, 1918.

Rothmund, V. and G. Kornfeld. Die basenaustausch im permutit. II. *Z. Anorg. Allg. Chem.* 108:215–225, 1919.

Smith, S.A. Potassium Dynamics and Exchange Equilibria in Loess-Derived Soils. Ph.D. dissertation (Diss. Abstr. AA19996392), The University of Tennessee, Knoxville, 2000.

Sposito, G. *The Thermodynamics of Soil Solutions*. Oxford Press, New York, 1981.

Sposito, G. *Chemical Equilibria and Kinetics in Soils*. Oxford Press, New York, 1994.

Sposito, G. Ion exchange phenomena. In *Handbook of Soil Science*. M.E. Sumner (Ed.) CRC Press, Boca Raton, FL, 2000, pp. B-241-B263.

Thompson, H.S. On the absorbent power of soils. *J. Royal Agric. Soc. Engl.* 11:68–74, 1850.

Van Bladel, R. and H.R. Gheyi. Thermodynamic study of calcium-sodium and calcium-magnesium exchange in calcareous soils. *Soil Sci. Soc. A. J.* 44:938–942, 1980.

Vanselow, A.P. Equilibria of the base-exchange reactions of bentonites, permutites, soil colloids, and zeolites. *Soil Sci.* 33:95–113, 1932.

Way, J.T. On the power of soils to absorb manure. *J. Royal Agric. Soc. Engl.* 11:313–379, 1850.

Way, J.T. On the power of soils to absorb manure. *J. Royal Agric. Soc. Engl.* 13:123–143, 1852.

9 Oxidation-Reduction Reactions in Soils

The fate and behavior of numerous elements in the soil environment are either directly or indirectly influenced by chemical reactions involving the transfer of electrons. The oxidation-reduction (redox) status of an element is a function of the abundance of electrons in a system. In a general sense, systems that are rich in electrons are termed reduced, while those that are depleted in electrons are oxidized. Major elements in the soil environment that are directly influenced by electron transfer include Fe (Fe^{II} [Fe^{2+}], Fe^{III} [Fe^{3+}]), Mn (Mn^{II} [Mn^{2+}], Mn^{VI} [MnO_4^{2-}]), N (N^{-III} [NH_4^+], N^V [NO_3^-]), S (S^{-II} [H_2S,] S^{VI} [SO_4^{2-}]), O (O^0 [$O_2(g)$], O^{-II} [H_2O]) and C ($C^{\pm IV}$ [organic C], C^{IV} [CO_3^{2-}]). Trace elements that are directly influenced by electron transfer include Cr (Cr^{III} [Cr^{3+}], Cr^{VI} [CrO_4^{2-}]), As (As^{III} [$As(OH)_3^0$], As^V [AsO_4^{3-}]), and Se (Se^{IV} [SeO_3^{2-}], Se^{VI} [SeO_4^{2-}]). The environmental behavior of elements that are not directly impacted by electron transfer may also be subject to the influence of redox sensitive elements. For example, many of the transition metal cations, such as Ni^{2+}, Zn^{2+}, Cd^{2+}, and Pb^{2+}, are moderately mobile in an oxidized environment. However, the relative mobility of these elements in a reduced environment is very low to immobile. In an oxidized system, the concentrations of Ni^{2+}, Zn^{2+}, Cd^{2+}, and Pb^{2+} in solutions are controlled by ion exchange and specific adsorption reactions, as well as the solubility of carbonate, hydroxycarbonate, and phosphate minerals. In a reducing environment, sulfur is found in the reduced form (S^{2-}) and metal concentrations in solution are controlled by the solubilities of sulfide minerals (e.g., NiS, ZnS, CdS, and PbS), which are much more stable than the minerals found in oxidized systems.

Electron transfer involves the coupling of two chemical reactions, or half reactions. In one half-reaction, a reduced element (an element in an electron rich state) is oxidized by donating electrons to produce the oxidized species (a relatively electron-poor form of the element) and free electrons. For example, the oxidation of Fe^{II} (which occurs in the Fe^{2+} species) to Fe^{III} (in the Fe^{3+} species) is a driving force in primary mineral weathering. The oxidation half-reaction is:

$$Fe^{2+}(aq) \rightarrow Fe^{3+}(aq) + e^- \tag{9.1a}$$

In a reduction half-reaction, the electron is a reactant and an oxidized species accepts the electron to become reduced. A common oxidant in soils is O_2:

$$\tfrac{1}{4}O_2(g) + e^- + H^+(aq) \rightarrow \tfrac{1}{2}H_2O(l) \tag{9.1b}$$

If the Fe^{2+} and the O_2 come into contact electron transfer will occur and the overall redox reaction will result (electrons may also be transferred through bridging atoms):

$$Fe^{2+}(aq) + \tfrac{1}{4}O_2(g) + H^+(aq) \rightarrow Fe^{3+}(aq) + \tfrac{1}{2}H_2O(l) \tag{9.1c}$$

where O_2 is the oxidant and Fe^{2+} is the reductant. A properly balanced redox reaction will show that the electrons are completely transferred (electrons do not appear as products or reactants).

Oxidation-reduction reactions also occur in the absence of oxygen. For example, the redox reaction primarily responsible for the generation of acidity from pyritic materials involves the oxidation of the disulfide (S_2^{2-}) in pyrite by ferric iron to produce sulfate, ferrous iron, and protons:

$$\text{FeS}_2(s) + 14\text{Fe}^{3+}(aq) + 8\text{H}_2\text{O}(l) \rightarrow 15\text{Fe}^{2+}(aq) + 2\text{SO}_4^{2-}(aq) + 16\text{H}^+(aq) \qquad (9.2a)$$

In this reaction Fe^{3+} is the oxidant and S_2^{2-} is the reductant. Equation 9.2a is the combination of the oxidation and reduction half-reactions:

$$\text{FeS}_2(s) + 8\text{H}_2\text{O}(l) \rightarrow \text{Fe}^{2+}(aq) + 2\text{SO}_4^{2-}(aq) + 16\text{H}^+(aq) + 14e^- \qquad (9.2b)$$

and

$$14\text{Fe}^{3+}(aq) + 14e^- \rightarrow 14\text{Fe}^{2+}(aq) \qquad (9.2c)$$

Individually, oxidation and reduction half-reactions offer a convenient mechanism to evaluate the number of electrons donated and accepted by each species in a redox couple. For example, the oxidative dissolution of 1 mol of pyrite produces 14 mol of electron (Equation 9.2b). Thus, there must also be a sufficient number of electron acceptor atoms available to complete the transfer. Since each Fe^{3+} atom can only consume one electron during the production of an Fe^{2+} atom, the reduction half-reaction must involve 14 mol of Fe^{3+} atoms (Equation 9.2c): *the number of electrons generated in an oxidation half-reaction must equal the number consumed in the reduction half-reaction.* However, half-reactions are not intended to imply that electrons exist as free aqueous species. Indeed, the free electron is only a useful conceptual construct.

9.1 THE ELECTRON ACTIVITY

The abundance of electrons in a soil environment can be quantitatively expressed in terms of an electrode potential (E) or the electron activity. Consider the generalized reduction half-reaction involving the transfer of a single electron:

$$m\text{A}_{\text{ox}} + n\text{H}^+(aq) + e^- \rightarrow p\text{A}_{\text{red}} + q\text{H}_2\text{O}(l) \qquad (9.3)$$

where A_{ox} is the oxidized species and A_{red} is the reduced species. The equilibrium constant for this reaction is

$$K_{\text{R}} = \frac{(\text{A}_{\text{red}})^p}{(\text{A}_{\text{ox}})^m(\text{H}^+)^n(e^-)} \qquad (9.4)$$

Rearranging,

$$(e^-) = \frac{(\text{A}_{\text{red}})^p}{(\text{A}_{\text{ox}})^m(\text{H}^+)^n K_{\text{R}}} \qquad (9.5)$$

Thus, the activity of e^- can be used to calculate, or is directly related to the distribution of an element between the oxidized and reduced species. The activity of an electron can range several orders of magnitude and is commonly expressed as $-\log(e^-)$ (or pe) in a manner similar to that used to express the proton activity in solutions (pH). Unlike pH, pe can have a negative value.

The electrode potential of the standard hydrogen electrode (E_H) is related to pe through the expression:

$$E_H = \frac{RT \ln 10}{F} pe \tag{9.6}$$

where F is the Faraday constant (96484.56 C mol^{-1}). Substituting for F, R (natural gas constant) = 8.314 JK^{-1} mol^{-1}, and $T = 298.15$ K in Equation 9.6 yields

$$E_H = 0.05916 \; pe \tag{9.7}$$

where E_H is expressed in volts. Therefore, if an electrochemical cell can provide an accurate measure of the chemical potential of an electron in a soil system (note that E_H is a function of the electrochemical potential of the electron, $E_H = -\tilde{\mu}_{e^-}/F$), the resulting E_H can be employed to establish the electron activity (Equation 9.7), and subsequently the ratio of oxidized to reduced species in the system (Equation 9.5). Further, for any half-reaction the corresponding equilibrium constant is related to the standard electrode potential ($E_H^o = -\tilde{\mu}_{e^-}^o/F$) by the expression:

$$\log K_R = \frac{n_e E_H^o}{0.05916} \tag{9.8}$$

where n_e is the number of electrons transferred in the reduction half-reaction. Equilibrium constants and E_H^o values for several reduction reactions are listed in Table 9.1.

In aquatic systems, the pe can range several orders of magnitude; however, the range of pe that is of environmental significance is determined, in part, by the stability of H_2O. A large positive value of pe results in an environment that is depleted of electrons, potentially resulting in the stability of $O_2(g)$ relative to that of $H_2O(l)$. Conversely, a large negative value of pe results in an environment that is enriched in electrons, potentially resulting in the stability of $H_2(g)$ relative to that of $H_2O(l)$. Using the $\log K_R$ data presented in Table 9.1, the region of pH and pe in which H_2O is stable can be determined. Water is oxidized to $O_2(g)$ according to the equation:

$$\tfrac{1}{2}H_2O(l) \rightarrow \tfrac{1}{4}O_2(g) + H^+(aq) + e^- \tag{9.9a}$$

and

$$\log K_{Ox} = -20.8 = \tfrac{1}{4}\log P_{O_2} - \text{pH} - pe \tag{9.9b}$$

where K_{Ox} is the equilibrium constant for the oxidation half-reaction. Rearranging Equation 9.9b and assuming $P_{O_2} = 0.21$ atm (activity is equal to partial pressure in atmospheres; $\log P_{O_2} = -0.68$):

$$pe = 20.61 - \text{pH} \tag{9.9c}$$

This expression indicates the pe at which water will decompose to form $O_2(g)$. There are two important characteristics of this expression that are worth noting. First, the decomposition of water is not a function of pe alone, but of pe and pH. As solutions become more acidic, higher pe values will be required to form $O_2(g)$. Second, the (pe + pH) activity product term is the true variable representing the redox status of the solution. In this case, water will remain stable when (pe + pH) is less than 20.61.

TABLE 9.1
Selected Reduction Half-Reactions and Associated log K_R and E_H^o Values ($T = 298.15$ K, $P = 0.101$ MPa)[a]

Reaction	Log K_R	E_H^o, V
$\frac{1}{2}Mn_3O_4(s) + 4H^+ + e^- \rightarrow \frac{3}{2}Mn^{2+} + 2H_2O$	30.7	1.816
$Co^{3+} + e^- \rightarrow Co^{2+}$	30.6	1.810
$\frac{1}{2}NiO_2 + 2H^+ + e^- \rightarrow \frac{1}{2}Ni^{2+} + H_2O$	29.8	1.763
$\frac{1}{2}Mn_2O_3(s) + 3H^+ + e^- \rightarrow Mn^{2+} + \frac{3}{2}H_2O$	25.7	1.520
$Mn^{3+} + e^- \rightarrow Mn^{2+}$	25.5	1.508
$\gamma - MnOOH(s) + 3H^+ + e^- \rightarrow Mn^{2+} + 2H_2O$	25.4	1.503
$\frac{1}{2}PbO_2 + 2H^+ + e^- \rightarrow \frac{1}{2}Pb^{2+} + H_2O$	24.8	1.467
$\frac{1}{2}NO_2^- + \frac{3}{2}H^+ + e^- \rightarrow \frac{1}{4}N_2O(g) + \frac{3}{4}H_2O$	23.6	1.396
$\frac{1}{5}NO_3^- + \frac{6}{5}H^+ + e^- \rightarrow \frac{1}{10}N_2(g) + \frac{3}{5}H_2O$	21.1	1.248
$\frac{1}{4}O_2(g) + H^+ + e^- \rightarrow \frac{1}{2}H_2O$	20.8	1.230
$\frac{1}{2}MnO_2(s) + 2H^+ + e^- \rightarrow \frac{1}{2}Mn^{2+} + H_2O$	20.8	1.230
$\frac{1}{2}Fe_3O_4(s) + 4H^+ + e^- \rightarrow \frac{3}{2}Fe^{2+} + 2H_2O$	17.8	1.053
$\frac{3}{4}MnO_2(s) + H^+ + e^- \rightarrow \frac{1}{4}Mn_3O_4(s) + \frac{1}{2}H_2O$	15.9	0.941
$Fe(OH)_3(s) + 3H^+ + e^- \rightarrow Fe^{2+} + 3H_2O$	15.8	0.935
$\frac{1}{6}NO_2^- + \frac{4}{3}H^+ + e^- \rightarrow \frac{1}{6}NH_4^+ + \frac{1}{3}H_2O$	15.1	0.893
$\frac{1}{8}NO_3^- + \frac{5}{4}H^+ + e^- \rightarrow \frac{1}{8}NH_4^+ + \frac{3}{8}H_2O$	14.9	0.881
$\frac{1}{2}NO_3^- + H^+ + e^- \rightarrow \frac{1}{2}NO_2^- + \frac{1}{2}H_2O$	14.1	0.834
$\frac{1}{2}Fe_2O_3(s) + 3H^+ + e^- \rightarrow Fe^{2+} + \frac{3}{2}H_2O$	13.4	0.793
$Fe^{3+} + e^- \rightarrow Fe^{2+}$	13.0	0.769
$FeOOH(s) + 3H^+ + e^- \rightarrow Fe^{2+} + 2H_2O$	13.0	0.769
$\frac{1}{2}O_2(g) + H^+ + e^- \rightarrow \frac{1}{2}H_2O_2$	11.6	0.686
$\frac{1}{2}CH_3OH + H^+ + e^- \rightarrow \frac{1}{2}CH_4(g) + \frac{1}{2}H_2O$	9.9	0.586
$\frac{1}{6}SO_4^{2-} + \frac{4}{3}H^+ + e^- \rightarrow \frac{1}{6}S(s) + \frac{2}{3}H_2O$	5.3	0.314
$\frac{1}{8}SO_4^{2-} + \frac{5}{4}H^+ + e^- \rightarrow \frac{1}{8}H_2S + \frac{1}{2}H_2O$	5.2	0.308
$\frac{1}{4}SO_4^{2-} + \frac{5}{4}H^+ + e^- \rightarrow \frac{1}{8}S_2O_3^{2-} + \frac{5}{8}H_2O$	4.9	0.290
$\frac{1}{6}N_2(g) + \frac{4}{3}H^+ + e^- \rightarrow \frac{1}{3}NH_4^+$	4.6	0.272
$\frac{1}{12}C_6H_{12}O_6 + H^+ + e^- \rightarrow \frac{1}{4}C_2H_5OH$	4.4	0.260
$\frac{1}{8}SO_4^{2-} + \frac{9}{8}H^+ + e^- \rightarrow \frac{1}{8}HS^- + \frac{1}{2}H_2O$	4.3	0.254
$\frac{1}{2}SO_4^{2-} + \frac{1}{2}H^+ + e^- \rightarrow \frac{1}{2}SO_2(g) + H_2O$	2.9	0.172
$\frac{1}{8}CO_2(g) + H^+ + e^- \rightarrow \frac{1}{8}CH_4(g) + H_2O$	2.9	0.172
$Cu^{2+} + e^- \rightarrow Cu^+$	2.6	0.154
$\frac{1}{2}CH_2O + H^+ + e^- \rightarrow \frac{1}{2}CH_3OH$	2.1	0.124
$\frac{1}{2}HCOOH + H^+ + e^- \rightarrow \frac{1}{2}CH_2O + \frac{1}{2}H_2O$	1.5	0.089
$H^+ + e^- \rightarrow \frac{1}{2}H_2(g)$	0.0	0.0
$\frac{1}{4}CO_2 + H^+ + e^- \rightarrow \frac{1}{24}C_6H_{12}O_6 + \frac{1}{4}H_2O$	−0.21	−0.012
$\frac{1}{4}CO_2 + H^+ + e^- \rightarrow \frac{1}{4}CH_2O + \frac{1}{4}H_2O$	−1.2	−0.071
$\frac{1}{2}CO_2 + H^+ + e^- \rightarrow \frac{1}{2}HCOOH$	−1.9	−0.112

[a] Data were obtained from the compilations of James and Bartlett (1999).

Water is reduced according to the reaction,

$$H^+(aq) + e^- \rightarrow \frac{1}{2}H_2(g) \tag{9.10a}$$

and

$$\log K_R = 0 = \frac{1}{2}\log P_{H_2} + pH + pe \tag{9.10b}$$

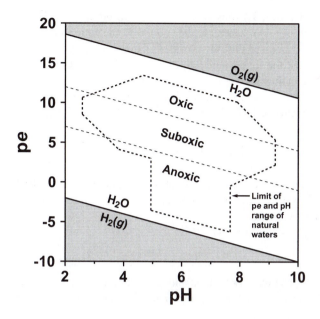

FIGURE 9.1 The pe–pH stability region of water ($P_{H_2} = 1.0$ atm, $P_{O_2} = 0.21$ atm, $T = 25°C$, $P = 1$ atm, and the activity of H_2O is unity) and the redox zones (oxic, suboxic, and anoxic) of Sposito (1989). The diagram also illustrates the measured limits of pe and pH in natural waters. (Modified from Baas Becking, L.G.M., L.R. Kaplan, and D. Moore. Limits of the natural environment in terms of pH and oxidation-reduction potentials. *J. Geology* 68:224–284, 1960.).

Rearranging and assuming $P_{H_2} = 1$ atm ($\log P_{H_2} = 0$):

$$pe = -pH \qquad\qquad (9.10c)$$

As was noted for the oxidation of water, the reduction of water is a function of both pe and pH, and ($pe + pH$) > 0 represents the redox conditions where water is stable relative to $H_2(g)$. The stability regions for water, $O_2(g)$, and $H_2(g)$ as a function of pH and pe, are illustrated in Figure 9.1. The line representing the transition from H_2O to $O_2(g)$ is Equation 9.9c, and that representing the transition from H_2O to $H_2(g)$ is Equation 9.10c. The measured pe and pH characteristics of natural aqueous environments occupy a smaller area of the pe–pH diagram than the restrictions placed on the limits of pe and pH by the stability of water (Figure 9.1).

9.2 REDOX POTENTIAL MEASUREMENTS

The Pt electrode system is a common method for determining the electrode potential (E_H) and the electron activity of a solution. This electrode system is illustrated for a simple solution containing Fe^{3+} and Fe^{2+} and using the calomel reference electrode:

Electrode L (reference)		Solution	Electrode R
$Hg(l)$; $Hg_2Cl_2(s)$ | KCl(aq, saturated)	|	$Fe^{3+}(aq) \rightarrow Fe^{2+}(aq)$	| Pt

In the electrode system depicted above, the Pt electrode (electrode **R**) is nonreactive (inert) and is only a source or sink for electrons. The Pt electrode responds to the transfer of electrons between the ferrous and ferric ions ($Fe^{3+} + e^- \rightarrow Fe^{2+}$) in the sample solution. The calomel reference electrode (electrode **L**) consists of a gel composed of liquid Hg and solid Hg_2Cl_2 bathed in a saturated KCl solution. An asbestos fiber connects the saturated KCl solution to the sample by allowing K^+ and Cl^- ions to diffuse from the reference electrode and into the sample solution. This connection is called the liquid junction, or salt bridge, and electrically connects the reference electrode to the sample solution.

In this example, the Fe^{3+}–Fe^{2+} couple controls the electron transfer process in the sample solution. In the left electrode, $Hg(l)$ is oxidized to $Hg_2Cl_2(s)$ according to the reaction:

$$Hg(l) + Cl_L^- \rightarrow \tfrac{1}{2} Hg_2Cl_2(s) + e_L^- \tag{9.11}$$

At equilibrium, the sum of the chemical potentials (μ, if a neutral salt or solid) or the electrochemical potentials ($\tilde{\mu}$, if a charged species) of the products must be equal to that of the reactants. Therefore, within the reference (**L**) electrode:

$$\mu_{Hg} + \tilde{\mu}_{Cl^-,L} = \tfrac{1}{2}\mu_{Hg_2Cl_2} + \tilde{\mu}_{e^-,L} \tag{9.12}$$

where μ represents the chemical potentials and $\tilde{\mu}$ represents the electrochemical potentials. Rearranging to obtain an expression for the electrochemical potential of the electron in the calomel electrode yields:

$$\tilde{\mu}_{e^-,L} = \mu_{Hg} + \tilde{\mu}_{Cl^-,L} - \tfrac{1}{2}\mu_{Hg_2Cl_2} \tag{9.13}$$

Since the Pt electrode is inert, the electrochemical potential of the electron at the Pt-solution interface ($\tilde{\mu}_e^{sol}$) is controlled by the oxidation or reduction of Fe in the solution. The potential difference between electrons in the two systems, the calomel electrode and the Pt electrode-solution, is related to the electrochemical potential difference:

$$-F\mathrm{E} = \tilde{\mu}_e^{sol} - \tilde{\mu}_{e^-,L} \tag{9.14}$$

The electrochemical potential of the electron in the calomel electrode ($\tilde{\mu}_{e^-,L}$) is dependent on the electrochemical potential of Cl^- (Equation 9.13). Further, $\tilde{\mu}_{Cl^-,L}$ is equal to $\tilde{\mu}_{Cl^-,KCl}$, the electrochemical potential of Cl^- in the saturated KCl solution. However, the K^+ and Cl^- components in the calomel electrode are not in equilibrium with the sample solution. Although the salt bridge allows for the transfer of Cl^- and K^+, and provides an electrical connection between the reference electrode and the sample, the chemical potentials of Cl^- and K^+ in the electrode and the sample solution do not equalize. Because a transport process and not an equilibrium process controls the chemical potentials of these ions, the electrical potential difference is separated into two components; a thermodynamic component (E_{cell}) and a part that depends on transport processes at the liquid junction (E_j, the liquid junction potential). The liquid junction potential is not a measurable quantity and can be a source of error in any electrochemical measurement; although in most cases this error is removed through electrode calibration procedures. Equation 9.14 then becomes:

$$-F(E_{cell} - E_j) = \tilde{\mu}_e^{sol} - \tilde{\mu}_{e^-,L} \tag{9.15}$$

The activity of an electron is related to the electrochemical potential through the expression:

$$\tilde{\mu}_{e^-} = \tilde{\mu}_{e^-}^{\circ} + RT\ln(e^-) \tag{9.16}$$

where $\tilde{\mu}^{\circ}_{e^-}$ is the standard state electrochemical potential of the electron. By definition, $\tilde{\mu}^{\circ}_{e^-} = 0$. Then,

$$\tilde{\mu}_{e^-} = RT\ln(e^-) \tag{9.17}$$

Substituting Equation 9.17 into Equation 9.15 yields:

$$-F(E_{cell} - E_j) = RT\ln(e^-_{sol}) - RT\ln(e^-_L) \tag{9.18}$$

Rearranging:

$$E_{cell} = -\frac{RT\ln(e^-_{sol})}{F} + \frac{RT\ln(e^-_L)}{F} + E_j \tag{9.19}$$

The latter part of Equation 9.19 can be combined into a single term that represents the potential of the reference electrode:

$$E_{ref} = -\frac{RT\ln(e^-_L)}{F} - E_j \tag{9.20}$$

Equation 9.19 may then be written:

$$E_{cell} = \frac{RT(\ln 10)pe_{sol}}{F} - E_{ref} \tag{9.21}$$

or

$$E_{cell} = 0.05916pe_{sol} - E_{ref} \tag{9.22}$$

Recalling Equation 9.6 and substituting:

$$E_{cell} = E_H - E_{ref} \tag{9.23}$$

This expression relates the measurement obtained from a Pt electrode system to that of the standard hydrogen electrode (E_H). For the calomel reference electrode and at 25°C, $E_{ref} = 244.4$ mV. The derivation also illustrates that the electrode measurement (E_{cell}) is directly related to the activity of the electron (pe) in the sample solution (Equation 9.22).

Recall that the electrode system illustrated above responds to electron transfer in a simple solution containing Fe^{3+} and Fe^{2+}:

$$Fe^{3+} + e^- \rightarrow Fe^{2+} \tag{9.24}$$

The equilibrium constant for this reaction is:

$$K_R = \frac{(Fe^{2+})}{(Fe^{3+})(e^-)} \tag{9.25}$$

A logarithmic transformation of this expression yields:

$$\log K_R = \log\left[\frac{(Fe^{2+})}{(Fe^{3+})}\right] + pe \tag{9.26}$$

Substituting Equation 9.8 for $\log K_R$ (with $n_e = 1$) leads to:

$$E_H^\circ = 0.05916\log\left[\frac{(Fe^{2+})}{(Fe^{3+})}\right] + 0.05916pe \tag{9.27}$$

where E_H° is the standard electrode potential for half-reaction in Equation 9.24 that describes the reduction of ferric iron (769 mV). Using Equation 9.22 and substituting for pe:

$$E_H^\circ = 0.05916\log\left[\frac{(Fe^{2+})}{(Fe^{3+})}\right] + E_{cell} + E_{ref} \tag{9.28}$$

or

$$E_H = E_H^\circ - 0.05916\log\left[\frac{(Fe^{2+})}{(Fe^{3+})}\right] \tag{9.29}$$

where $E_H = E_{cell} + E_{ref}$. Equation 9.29 is the general form of the Nernst equation for a system at 298.15 K (25°C) and indicates that the response of the Pt electrode (E_{cell}) is related to the activity ratio of the reduced species (Fe^{2+}) to the oxidized species (Fe^{3+}).

From the above derivations, it would appear that the Pt electrode is a viable mechanism for determining the E_H and pe of a solution, as well as the activity ratio of the electroactive species in the solution. Indeed, historically the Pt electrode has been a common mechanism for determining the redox status of soil solutions. However, there are several shortcomings associated with the use of the Pt electrode that perhaps preclude its ability to accurately determine the redox status of the chemically complex soil system.

Unlike pH measurements that depend on the activity of H^+ in a solution, E_H and pe determinations depend on the ability of the electron to be transferred (not on its activity) and on the inert nature of the Pt electrode. With respect to the latter requirement, when dissolved O_2 is present in a solution the Pt electrode can become fouled as a result of the reaction of O_2 with Pt on the electrode surface which can form $Pt(OH)_2$. Unlike Pt, $Pt(OH)_2$ is electroactive and develops a potential with Pt [$\frac{1}{2}Pt(OH)_2 + e^- + H^+ \rightarrow \frac{1}{2}Pt + H_2O$; $\log K_R = 16.6$, $E_H^\circ = 982$ mV] and masks the electrode response from electron transport between redox-sensitive solution species.

The Pt electrode is also insensitive to the O_2–H_2O couple (which dominates when dissolved oxygen is present). In a system of pure water at pH 7 in equilibrium with a gaseous phase having $P_{O_2} = 0.21$ (atmospheric level), the O_2–H_2O couple will poise pe at 13.63 (poise is to pe as buffer is to pH). A thousand-fold decrease in P_{O_2} (to 0.0021, the minimum for aerobic respiration) will result in a measured pe of 13.13, a difference of only 0.5 pe units. It is also recognized that the Pt electrode responds to pH in aerobic soils (dissolved oxygen present). Thus, in aerobic soils the Pt electrode is not considered a reliable indicator of redox status. Instead, a direct measure of O_2 partial pressure in the soil air is the preferred mechanism for determining the redox potential in aerated soils.

In a soil system, several redox couples may be present: Fe^{2+}–Fe^{3+}, Mn^{2+}–MnO_2, NO_3^-–NO_2^-–NH_4^+, H_2S–SO_4^{2-}, and CH_4–CO_2. If all of these redox couples are at equilibrium, each couple will have a different activity ratio of reduced to oxidized species [e.g., $(Fe^{2+})/(Fe^{3+})$]. In theory, the Pt electrode should respond to all of these couples simultaneously, resulting in a single and unique E_H value for the system. In reality, many couples do not accept or donate electrons at the Pt electrode easily. Indeed, the Pt electrode responds well to only a small number of couples, chiefly those involving soluble Fe and S species. In systems where Fe is present at activities that exceed 10^{-5}, the measured E_H values primarily reflect the $(Fe^{2+})/(Fe^{3+})$ activity ratio. Again, if equilibrium is achieved, the E_H determined from the Fe^{2+}–Fe^{3+} couple should be sufficient to compute the redox distribution of other redox-sensitive species, irrespective of the inability of the Pt electrode to respond to all the couples. Unfortunately, redox equilibrium does not generally exist in natural systems because (1) precipitation limits the electroactivity of some elements, (2) some elements form nonelectroactive gases or molecules, and (3) different redox couples react at different rates.

Iron (III) and Mn^{VI} form relatively stable oxide minerals in slightly acidic and alkaline soils limiting the electroactivity of these oxidized species with the Pt electrode and with soluble species in soil solutions. However, structural Fe^{III} and Mn^{VI} (and Mn^{III}) are electroactive and known to participate in soil redox processes. For example, arsenite ($As(OH)_3^0$), a mobile and relatively toxic form of As, is adsorbed by Mn oxides, presumably by the formation of an inner-sphere surface complex. The adsorbed As^{III} is readily oxidized by structural Mn^{IV} and Mn^{III} to form Mn^{2+} (Mn^{II}) and arsenate ($H_2AsO_4^-$, As^V), both of which are released to the soil solution. The surface mediated oxidation of As^{III} is a two-step process. Beginning with birnessite, Mn^{IV} is reduced to Mn^{III} (where $MnOOH^*(s)$ represents a reaction intermediary existing at the birnessite solid-solution interface):

$$2MnO_2(s) + As(OH)_3^0 \rightarrow 2MnOOH^*(s) + H_2AsO_4^- + H^+ \tag{9.30}$$

This reaction is followed by the reaction of additional As^{III} with Mn^{III} to form Mn^{II} and As^V species:

$$2MnOOH^*(s) + As(OH)_3^0 + 3H^+ \rightarrow 2Mn^{2+} + H_2AsO_4^- + 3H_2O \tag{9.31}$$

The majority of the arsenate is readsorbed, forming a strong bidentate-binuclear surface complex:

$$2\equiv MnOH^{+\frac{1}{6}} + H_2AsO_4^- \rightarrow (\equiv MnO)_2AsOOH^{-\frac{2}{3}} + 2H_2O \tag{9.32}$$

The surface-mediated oxidation of As^{III} by Mn oxides is a beneficial process for the stabilization of As-contaminated soils. Arsenic(V) is a substantially less toxic form of arsenic than As^{III}, and As^V is strongly retained by soil surfaces relative to As^{III}. Conversely, the oxidation of relatively innocuous and stable Cr^{3+} (Cr^{III}) to $HCrO_4^-$ (Cr^{VI}) at Mn^{IV}–Mn^{III} oxide surfaces also occurs. In this case, the surface-catalyzed process results in a mobile and carcinogenic form of chromium ($HCrO_4^-$ is the anion of a strong acid and is nonspecifically adsorbed).

The $As^{III} \rightarrow As^V$ and $Cr^{III} \rightarrow Cr^{VI}$ surface-catalyzed redox reactions are examples of reductive dissolution reactions. The reduction of structural Mn^{IV} and Mn^{III} atoms in MnO_2 results in the formation of Mn^{2+}, which is released to the aqueous phase. Another surface-mediated redox process that is specific to the Mn oxides can result in the incorporation of adsorbed metal cations into the oxide structure. Manganese oxide surfaces have a high affinity for Cu^{2+}, Ni^{2+}, Co^{2+}, and Pb^{2+}. Adsorption of these metal cations is accompanied by the release of protons, indicating the occurrence of specific adsorption processes and the formation of inner-sphere surface complexes (Chapter 7). However, in the case of Co^{2+}, adsorption is also accompanied by the expulsion of Mn^{2+} from the oxide structure. Spectroscopic studies indicate that adsorbed Co^{2+} is oxidized by structural Mn^{IV} and Mn^{III}, producing Co^{3+} and Mn^{2+}. The Mn^{2+} is expelled from the oxide structure and is replaced by the Co^{3+} resulting in a Co_2O_3 inclusion in the Mn^{IV}–Mn^{III} oxide. Incorporation of cobalt into the crystalline structure of the Mn oxide results in an environmentally stable form of Co, as minor structural inclusions are more stable (less soluble) than adsorbed species or pure mineral phases (see Chapter 6 for a discussion on the stabilities of solid-solutions).

Redox processes involving the "light" elements (e.g., C, O, N, H, and S) are irreversible, or nearly so, owing to the formation of nonelectroactive gases or molecules or because the reactions are kinetically constrained. For example, the reduction of O_2 to H_2O occurs at a lower E_H than predicted because a relative excess of electrons is required to overcome the dissociation energy of the $O{=}O$ covalent double bond. The inverse of this reaction does not occur, and the only mechanism to generate O_2 in soil solutions is through aeration. Similarly, the oxidation of N_2 to NO_3^- is thermodynamically favored in the presence of O_2, but does not occur due to the high dissociation energy of the $N{\equiv}N$ triple bond. However, the reverse reaction, NO_3^- to N_2, readily proceeds through the intermediaries, NO_2^- and N_2O, in the presence of microbes and under moderately reducing conditions. Thus, equilibrium between the N_2 and NO_3^- nitrogen forms is not achieved. As a result of nonequilibrium and the differing rates of redox reactions, the E_H value measured by the Pt electrode in a soil system is called a mixed potential, as it reflects the status of all the redox couples in a system at disequilibrium. Finally, and because a vast majority of the E_H measurements in soil systems represent mixed potentials, a quantitative interpretation of their meaning is virtually unattainable.

9.3 REDOX STATUS IN SOILS

As indicated above, measured redox potentials in chemically complex soil systems may be of limited value, particularly if the objective is to predict the distribution of an element between oxidized and reduced species. An alternative to predicting redox speciation based on questionable E_H measurements is to analytically determine the total concentrations of the oxidized and reduced species involved in the couple. Speciation of the solution using an ion association model (discussed in Chapter 5) would then lead to the activities of the species in the couple, from which the pe (or $pe + pH$) of the solution could be computed. If the analysis of another couple leads to a similar $pe + pH$ value, then redox equilibrium may be assumed to exist in the system. However, if the analysis of the two different couples results in dissimilar $pe + pH$ values, the system is at disequilibrium. For example, in soil solutions selenium may exist as selenite [SeO_3^{2-}] or selenate [SeO_4^{2-}]. The redox reaction governing Se distribution is:

$$\tfrac{1}{2}SeO_4^{2-} + e^- + H^+ \rightarrow \tfrac{1}{2}SeO_3^{2-} + \tfrac{1}{2}H_2O \tag{9.33a}$$

The equilibrium constant for this reaction is:

$$\log K_R = 14.9 = \tfrac{1}{2}\log\left[\frac{(SeO_3^{2-})}{(SeO_4^{2-})}\right] + pe + pH \tag{9.33b}$$

Rearranging,

$$pe + pH = 14.9 - \tfrac{1}{2}\log\left[\frac{(SeO_3^{2-})}{(SeO_4^{2-})}\right] \tag{9.33c}$$

Using Equation 9.33c, the $pe + pH$ can be computed from measured properties of the solution, namely, the activities of SeO_3^{2-} and SeO_4^{2-}.

The primary drawback of using solution properties to compute $pe + pH$ is associated with the sensitivity of analytical techniques. Using the equation for the Se^{IV}–Se^{VI} couple (Equation 9.33c) and assuming $pe = 7$, pH = 7 ($pe + $ pH = 14), and total soluble Se is 10^{-5} M (assuming a dilute solution and therefore that activities equal concentrations), the activity ratio of reduced to oxidize Se is:

$$\log\left[\frac{(SeO_3^{2-})}{(SeO_4^{2-})}\right] = 1.80 \qquad (9.34)$$

Since $[SeO_3^{2-}] + [SeO_4^{2-}] = 10^{-5}$, $[SeO_4^{2-}] = 1.56 \times 10^{-7}$ M (0.02 mg L^{-1}) and $[SeO_3^{2-}] = 9.84 \times 10^{-6}$ M (1.25 mg L^{-1}). In this example, both SeO_3^{2-} and SeO_4^{2-} concentrations are greater than analytical detection limits. Consider the same couple with $pe = 5$, pH = 7 ($pe + $ pH = 12), and total soluble Se is 10^{-5} M (again assuming a dilute solution such that activities equal concentrations), the activity ratio of reduced to oxidized Se is:

$$\log\left[\frac{(SeO_3^{2-})}{(SeO_4^{2-})}\right] = 5.80 \qquad (9.35)$$

Since $[SeO_3^{2-}] + [SeO_4^{2-}] = 10^{-5}$, $[SeO_4^{2-}] = 1.585 \times 10^{-11}$ M (2.3×10^{-6} mg L^{-1}) and $[SeO_3^{2-}] = 9.99 \times 10^{-6}$ M (1.27 mg L^{-1}). In this example, essentially all of the Se is reduced and found in the SeO_3^{2-} species and the concentration of SeO_4^{2-} is below detectable levels. In this situation, the redox distribution of Se would have to be inferred using the computed $pe + $ pH value from another analytically detectable couple, requiring the assumption of redox equilibria (again, a condition that does not necessarily occur in natural systems).

9.3.1 REDUCTION-OXIDATION SEQUENCES IN SOILS

Despite the uncertainty associated with the redox potential values obtained with the Pt-electrode, such measurements still have value, particularly when coupled with the observed behavior of redox-sensitive elements in the soil environment. Patrick and Jugsujinda (1992) examined the reduction and oxidation sequences of redox-sensitive elements in a soil under controlled E_H conditions (Figure 9.2). Beginning in a reduced soil system ($E_H = -100$ mV), Fe^{2+} concentrations continually decrease with increasing E_H. The concentrations of Fe^{2+} drop below detectable levels when E_H values are increased to between approximately 50 and 100 mV (pe between 0.8 and 1.7). Above these E_H values, essentially all of the Fe^{2+} has been oxidized to sparingly soluble Fe^{III} phases, such as goethite (FeOOH):

$$Fe^{2+} + 2H_2O \rightarrow FeOOH(s) + 3H^+ + e^- \qquad (9.36)$$

The equilibrium constant for this reaction is $\log K = -13.0$, and the equilibrium relationship is (assuming an activity of $Fe^{2+} = 10^{-7}$ denotes a limit of detection):

$$pe = 20 - 3pH \qquad (9.37)$$

Therefore, in the pH 6.5 solution, a pe of 0.5 ($E_H = 30$ mV) is predicted to indicate the redox status of the soil when Fe^{2+} drops to below detectable levels. In this case, the E_H condition at which Fe^{2+} is observed to be depleted (50 to 100 mV) is consistent with the predicted value (30 mV).

Ammonium concentrations also decrease with increasing E_H. The soil is depleted of NH_4^+ when E_H values are increased above approximately 100 to 150 mV (pe between 1.7 and 2.5). Nitrate, however, which is presumably formed via the oxidation of NH_4^+, is not detected until solution E_H values are greater than 200 mV ($pe > 3.4$). This discrepancy is perhaps due to the formation of the

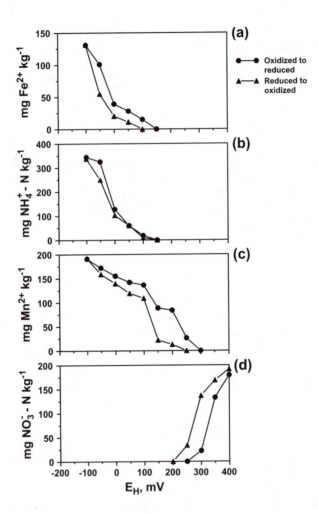

FIGURE 9.2 The concentrations of soluble constituents in soil solutions (expressed on a soil mass basis) as a function of E_H (redox potential) and initial redox state of the soil (oxidized to reduced or reduced to oxidized). (a) Iron(II); (b) ammonium-N; (c) manganese(II); and (d) nitrate-N. The data were obtained from Patrick and Jugsujinda (1996).

intermediary NO_2^-, which was not analytically characterized in the study. Theoretical consideration indicates that NH_4^+ will convert to NO_2^-, which is converted further to NO_3^- with increasing pe (Figure 9.3); however, in contrast to the data in Figure 9.2b and d the predicted (theoretical) conversion of NH_4^+ to NO_2^- begins at approximately $pe = 6.2$, while NO_3^- is not predicted to appear until pe values exceed approximately 6.5. The concentrations of soluble Mn^{2+}, like NH_4^+ and Fe^{2+}, also fall below detectable levels when E_H values are greater than 200 mV ($pe > 3.4$). Further, Mn^{2+} is not detected in the soil when detectable concentrations of NO_3^- are observed, indicating that the oxidation of Mn^{2+} and the nitrification of NH_4^+ are sequential, in that NO_3^- is not produced until Mn^{2+} is depleted. In this system the formation of Mn^{IV} from Mn^{2+} also occurs at a much lower pe value than predicted. According to the oxidation reaction ($\log K = -20.8$):

$$\tfrac{1}{2}Mn^{2+} + H_2O \rightarrow \tfrac{1}{2}MnO_2(s) + 2H^+ + e^- \tag{9.38}$$

the pe at which Mn^{2+} concentrations should fall below detectable levels is 11.3 ($E_H = 668$ mV) (assuming pH = 6.5 and Mn^{2+} activity is 10^{-7}), compared to a measured pe of 3.4.

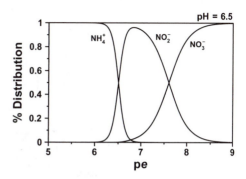

FIGURE 9.3 The predicted distribution of nitrogen as a function of pe in a pH 6.5 solution. The thermodynamic data in Table 9.1 were used to construct the figure.

When approached from an oxidized soil system ($E_H = 400$ mV), the $NO_3^- \rightarrow N_2$, $Fe^{III} \rightarrow Fe^{2+}$, and $MnO_2 \rightarrow Mn^{2+}$ transitions occurred at E_H values that were approximately 50 mV more positive than when the study was initiated from a reduced condition. Nitrate concentrations decrease with decreasing E_H until E_H values are below 300 and 250 mV ($5.1 > pe > 4.2$). The decrease in NO_3^- concentrations is presumed to be a result of NO_3^- reduction to N_2. The appearance of NH_4^+ (generated from organic-N, but not NO_3^- or N_2) does not occur until E_H is less than 150 mV ($pe < 2.5$). Once NO_3^- has been depleted, microorganisms begin to reduce Mn^{IV}, resulting in the appearance of Mn^{2+} when E_H is less than approximately 250 mV ($pe < 4.2$). Again, this observation is an indication that the reduction of Mn^{IV} will not occur until NO_3^- has been depleted. Ferrous iron also forms when NO_3^- has been depleted (as well as O_2), and first appears in the soil solution when E_H decreases below 100 mV ($pe < 1.7$).

Based on the findings of Patrick and Jugsujinda (1992) and others, a reduction and oxidation sequence for the major electroactive elements in soils can be constructed (Figure 9.4). Also illustrated are the theoretical redox potentials at which the various transformations are predicted to occur. As E_H decreases from 668 mV (the upper limit of E_H observed in pH 7 waters) in a closed environment (such as a flooded soil), the electron activities become sufficient to promote the reduction of O_2 to H_2O. This is a microbially mediated process, as O_2 is consumed during aerobic respiration. Below an E_H of approximately 350 mV, O_2 is depleted from soil solutions and microbes catalyze the reduction of NO_3^- to N_2 through a number of intermediaries, such as NO_2^- and N_2O. Once NO_3^- and O_2 have been depleted (E_H less than 275 to 250 mV), the soil becomes suboxic and a sufficient supply of electrons are available to support the reduction of Mn^{IV} (and Mn^{III}) and Fe^{III}. Both of these elements exist in stable oxides in oxic systems (such as MnO_2 and FeOOH and others as described in Chapter 2), as both Fe^{2+} and Mn^{2+} are readily oxidized by O_2, thus the reduction of Mn^{IV} and Fe^{III} is a function of the supply of electrons and the solubilities of the Mn and Fe solids. As the E_H decreases below approximately -150 mV, the soil becomes anoxic, and anaerobic bacteria catalyze the reduction of SO_4^{2-} and CO_2. Electron abundance is also sufficient in anoxic systems to reduce H_2O to H_2.

With the exception of the $Fe(OH)_3(s) \rightarrow Fe^{2+}$ and $SO_4^{2-} \rightarrow H_2S$ couples (Figure 9.4), the measured E_H values at which the redox transformations occur do not compare well with the theoretically predicted E_H values. For nitrogen and oxygen this discrepancy is due to kinetic constraints and the inert behavior of the O_2–H_2O and NO_3^-–N_2 couples at the Pt electrode. However, the disparity between the observed and predicted appearance of Mn^{2+} is probably associated with the nature of the Mn oxides found in soils. Manganese oxides are notoriously impure, poorly crystalline, and contain Mn in a number of oxidations states (Mn^{II}, Mn^{III}, and Mn^{IV}). For example, the chemical formula for birnessite cited in Chapter 2 is (Na, Ca, Mn^{II})(Mn^{III}, Mn^{IV})$_7O_{14} \cdot 2.8H_2O$. This formula is substantially different from that of synthetic birnessite, MnO_2, upon which the thermodynamic predictions are based. Thus, the actual form of oxidized Mn may be quite different from MnO_2, leading to the discrepancy between the predicted and actual E_H values at which the $Mn^{2+} \rightarrow MnO_2$ transition occurs. Manganese oxides also accumulate in soils through an autocatalytic process. This process

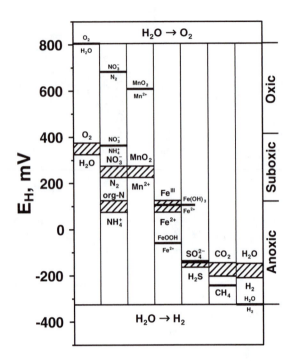

FIGURE 9.4 The reduction sequence observed in pH 6.5 to 7 soils (data obtained from Patrick and Jugsujinda, 1996; McBride, 1994; and Sposito, 1989). The observed transformation from oxidized to reduced species occurs in the E_H range indicated by the hatched region. The E_H values where the redox transformations are theoretically predicted to occur are indicated by horizontal lines. The oxic ($E_H > 414$ mV), suboxic ($120 < E_H < 414$ mV), and anoxic ($E_H < 120$ mV) for Sposito (1989) are also indicated.

involves the specific adsorption of Mn^{2+} by freshly formed or "seed" Mn oxide surfaces. The adsorbed Mn^{2+} is then readily oxidized by O_2, increasing the volume and the surface area of the Mn oxide precipitate. This is a self-perpetuating process and results in the further accumulation of Mn oxides where Mn oxides already exist. This is also the process by which Mn nodules form.

9.3.2 REDOX ZONES

Several investigators have defined "redox zones" to indicate the ranges of E_H or pe that are controlled by various redox couples. Sposito (1981) defined the following zones at pH 7: *oxidized* soils, $pe > 7$ ($E_H > 414$ mV; O_2 present); *moderately reduced* soils, $2 < pe < 7$ ($120 < E_H < 414$ mV; O_2 absent, $NO_3^- \leftrightarrow N_2$ and $MnO_2 \leftrightarrow Mn^{2+}$); *reduced* soils, $-2 < pe < 2$ ($-120 < E_H < 120$; $Fe^{3+} \leftrightarrow Fe^{2+}$); and *highly reduced* soils, $pe < -2$ ($E_H < -120$; $SO_4^{2-} \leftrightarrow S^{2-}$). Sposito (1989) also proposed the redox zones illustrated in Figure 9.1. In the *oxic* region of the pe–pH diagram ($pe +$ pH > 14; $E_H > 414$ mV at pH 7), redox status is primarily controlled by the redox reactions of oxygen, and then nitrogen when oxygen has been depleted. Manganese and Fe control the redox processes in the *suboxic* region ($9 < pe +$ pH < 14; 120 mV $< E_H < 414$ mV at pH 7). The system is called *anoxic* and the redox reactions of sulfur control redox chemistry when potentials are below 120 mV at pH 7 ($pe +$ pH < 9). Liu and Narasimhan (1989) defined the soil redox zones as: $100 < E_H < 250$ mV, oxygen-nitrogen range; $0 < E_H < 100$ mV, iron range; $-200 < E_H < 0$ mV, sulfate range; and $E_H < -200$ mV, methane-hydrogen range.

Bartlett (1999) argues that measured E_H or pe values are useless for the redox classification of a soil. He cites the nonelectroactive nature or limited solubility of the oxidized species of N, S, Mn, Fe, C, and H that preclude any meaningful interpretation of measured E_H values (using a Pt electrode). Instead, Bartlett (1999) proposes a qualitative soil redox classification scheme that is based on field observations (redoximorphic features, vegetative cover, and moisture and temperature regimes) and field test procedures (reactivity with various reagents, such as tetramethylbenzidine and Cr^{III}). Also assigned to each redox category in Bartlett's classification is the inferred electron lability (reactivity or availability) of the soil. *Superoxic* soils have very low electron lability and contain appreciable amounts of Mn-oxides and Mn^{3+}. *Manoxic* soils are highly oxidized and well drained, contain mature humus, are mineralizing organic matter and nitrifying when moist, and do not contain gley mottles in the profile. Manoxic soils have low electron lability. *Suboxic* soils have medium electron lability, are oxidized, and contain nitrates. However, these soils have significant potentials for reduction and contain easily oxidizable organic compounds. Soils that have medium high electron lability are *redoxic*. These soils are slightly acidic and contain balanced reducing and oxidizing tendencies, Fe^{II} is generally absent (except in highly acidic systems), and Fe^{III} is found in fresh (amorphous) precipitates. *Anoxic* soils have high electron lability, are pH neutral, and are identified by the presence of Fe^{II} and the absence of NO_3^-. Soils that have very high electron lability are *sulfidic*. These soils are very strongly reducing (oxygen is absent), and produce methane (CH_4) and hydrogen sulfide (H_2S). Sulfidic soils are also characteristically malodorous (H_2S).

9.4 pe–pH PREDOMINANCE DIAGRAMS

The electron activity (pe) and the pH have a pronounced influence on the chemistry of compounds in the soil environment. For this reason, they are often termed "master variables." In order to provide some insight into the speciation of elements as a function of these master variables, one can utilize predominance diagrams with pH and pe selected as the independent and dependent variables. Such diagrams are termed pe–pH or Pourbaix diagrams. Figure 9.1 is a pe–pH diagram for the system containing pure H_2O.

Critical to the construction of pe–pH diagrams, as well as to their validity, is the availability of thermodynamic data that can be employed to describe the relevant redox transformations. In many instances, $\log K_R$ values for a vast array of reduction reactions are readily available, as in Table 9.1. However, there is a mechanism to determine an equilibrium constant for almost any reaction of interest if a $\log K_R$ value is not available. Recall from Chapter 5 that an equilibrium constant for any chemical reaction can be computed, given that the standard free energies of formation (ΔG_f^o values) for the reaction products and reactants are known. For example, consider the reduction of $H_2AsO_4^-$ (As^V) to $As(OH)_3^0$ (As^{III}). The balanced chemical reaction is:

$$H_2AsO_4^- + 3H^+ + 2e^- \rightarrow As(OH)_3^0 + H_2O \tag{9.39}$$

The free energy change for the reduction reaction (ΔG_r^o) is given by:

$$\Delta G_r^o = \Delta G_{f,As(OH)_3^0}^o + \Delta G_{f,H_2O}^o - \Delta G_{f,H_2AsO_4^-}^o - 3\Delta G_{f,H^+}^o - 2\Delta G_{f,e^-}^o \tag{9.40}$$

The ΔG_f^o values for the proton and the electron are defined:

$$\Delta G_{f,H^+}^o \equiv 0.0 \tag{9.41}$$

and

$$\Delta G_{f,e^-}^o \equiv 0.0 \tag{9.42}$$

Therefore, Equation 9.40 simplifies to:

$$\Delta G_r^o = \Delta G_{f,As(OH)_3^0}^o + \Delta G_{f,H_2O}^o - \Delta G_{f,H_2AsO_4^-}^o \tag{9.43}$$

The ΔG_f^o values are obtained from Wagman et al. (1982): $\Delta G_{f,As(OH)_3^0}^o = -639.9$ kJ mol^{-1}; $\Delta G_{f,H_2AsO_4^-}^o = -753.29$ kJ mol^{-1}; and $\Delta G_{f,H_2O}^o = -237.18$ kJ mol^{-1}. Substituting these values into Equation 9.40 yields:

$$\Delta G_r^o = -639.9 + (-237.18) - (-753.29) = -123.79 \text{ kJ mol}^{-1} \tag{9.44}$$

It was also established in Chapter 5 that the logarithm of the equilibrium constant is related to ΔG_r^o (in kJ mol^{-1}, at 25°C and 0.101 MPa) by:

$$\log K = -\frac{\Delta G_r^o}{5.708} \tag{9.45}$$

Therefore, $\log K_R$ for the arsenate reduction reaction is:

$$\log K_R = -\frac{(-123.79)}{5.708} = 21.69 \tag{9.46}$$

The reaction in Equation 9.39 can be written such that one electron is transferred (as are the reactions in Table 9.1):

$$\tfrac{1}{2}H_2AsO_4^- + \tfrac{3}{2}H^+ + e^- \rightarrow \tfrac{1}{2}As(OH)_3^0 + \tfrac{1}{2}H_2O \tag{9.47}$$

with $\log K_R = 10.84$.

9.4.1 Construction of pe–pH Diagrams: The Cr–H$_2$O System

The mechanics of constructing a pe–pH diagram are similar to those used in constructing a predominance diagram (Chapter 6). Consider the Cr–H$_2$O system using the following chemical reactions and with the activity of dichromate controlled at $(Cr_2O_7^{2-}) = 10^{-6}$:

$$\tfrac{1}{3}CrO_4^{2-} + \tfrac{8}{3}H^+ + e^- \rightarrow \tfrac{1}{3}Cr^{3+} + \tfrac{4}{3}H_2O \qquad (\log K_R = 25.0) \tag{9.48}$$

$$\tfrac{1}{6}Cr_2O_7^{2-} + \tfrac{7}{3}H^+ + e^- \rightarrow \tfrac{1}{3}Cr^{3+} + \tfrac{7}{6}H_2O \qquad (\log K_R = 22.5) \tag{9.49}$$

$$\tfrac{1}{6}Cr_2O_7^{2-} + \tfrac{4}{3}H^+ + e^- \rightarrow \tfrac{1}{3}Cr(OH)_3(s) + \tfrac{1}{6}H_2O \quad (\log K_R = 18.6) \tag{9.50}$$

Each of the above reduction reactions can be used to establish the equation of a boundary line that predicts pe as a function of pH where the activity of the reduced chromium species is equal to the activity of the oxidized species. For example, consider the reaction in Equation 9.48 that describes the reduction of CrO_4^{2-} to Cr^{3+}. The equilibrium constant for the reduction reaction is given by the expression (assuming that the activity of H$_2$O is unity):

$$K_R = \frac{(Cr^{3+})^{\frac{1}{3}}}{(CrO_4^{2-})^{\frac{1}{3}}(H^+)^{\frac{8}{3}}(e^-)} \tag{9.51}$$

A logarithmic transformation yields:

$$\log K_R = \tfrac{1}{3}\log \mathrm{Cr}^{3+} - \tfrac{1}{3}\log \mathrm{CrO}_4^{2-} + \tfrac{8}{3}\mathrm{pH} + pe \tag{9.52}$$

At the boundary between the CrO_4^{2-} and Cr^{3+} predominance zones, $(\mathrm{CrO}_4^{2-}) = (\mathrm{Cr}^{3+})$, and Equation 9.52 becomes:

$$\log K_R = 25.0 = \tfrac{8}{3}\mathrm{pH} + pe \tag{9.53}$$

Rearranging, such that pe is the dependent variable and pH is the independent variable:

$$pe = 25.0 - \tfrac{8}{3}\mathrm{pH} \tag{9.54}$$

This expression is plotted in Figure 9.5a (line 1), along with the expressions that indicate the stability region of liquid water (Equations 9.9c and 9.10c). Equation 9.54 designates the boundary between the CrO_4^{2-} and the Cr^{3+} predominance regions. According to Equation 9.48, acidic (high H^+ activities or low pH values) and reducing (high e^- activities or low pe values) conditions favor the formation of Cr^{3+}. Therefore, Cr^{3+} predominates in the region of the diagram that lies below and to the left of the boundary line (Equation 9.54). Conversely, CrO_4^{2-} predominates in the region above and to the right of the boundary line (Figure 9.5a). Although the two regions of the pe–pH diagram are identified by a predominate species, such as Cr^{3+}, it does not imply that the Cr^{3+} region is devoid of CrO_4^{2-}. The Cr^{3+} species predominates because its activity is greater than that of CrO_4^{2-}, not because it is the only chromium species present. For example, the distribution of nitrogen as a function of pe in Figure 9.3 illustrates that more than one redox species may occur in significant proportions for a specified value of pe and pH.

The boundary line for the $\mathrm{Cr}_2\mathrm{O}_7^{2-} \rightarrow \mathrm{Cr}^{3+}$ transition is generated in a similar manner. The equilibrium constant for the reduction reaction in Equation 9.49 is:

$$K_R = \frac{(\mathrm{Cr}^{3+})^{\frac{1}{3}}}{(\mathrm{Cr}_2\mathrm{O}_7^{2-})^{\frac{1}{6}}(H^+)^{\frac{7}{3}}(e^-)} \tag{9.55}$$

and

$$\log K_R = \tfrac{1}{3}\log \mathrm{Cr}^{3+} - \tfrac{1}{6}\log \mathrm{Cr}_2\mathrm{O}_7^{2-} + \tfrac{7}{3}\mathrm{pH} + pe \tag{9.56}$$

At the boundary between the $\mathrm{Cr}_2\mathrm{O}_7^{2-}$ and Cr^{3+} predominance zones, $(\mathrm{Cr}^{3+}) = (\mathrm{Cr}_2\mathrm{O}_7^{2-}) = 10^{-6}$, and Equation 9.56 becomes:

$$\log K_R = 22.5 = \tfrac{1}{3}(-6) - \tfrac{1}{6}(-6) + \tfrac{7}{3}\mathrm{pH} + pe \tag{9.57}$$

Simplifying and rearranging yields the boundary line that distinguishes between the $\mathrm{Cr}_2\mathrm{O}_7^{2-}$ and Cr^{3+} predominance zones (line 2):

$$pe = 23.5 - \tfrac{7}{3}\mathrm{pH} \tag{9.58}$$

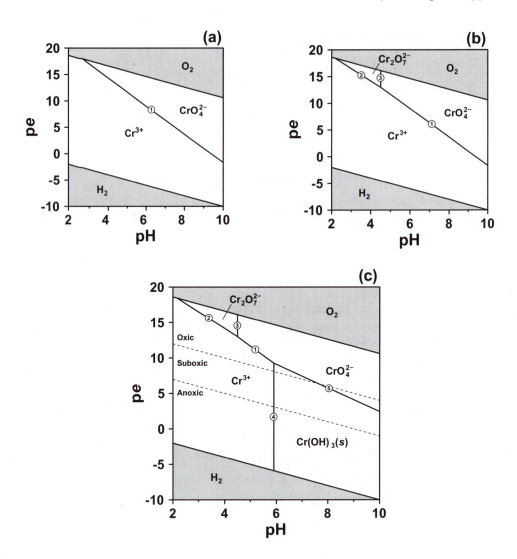

FIGURE 9.5 The redox speciation of Cr. Diagram (a) illustrates the predominance regions of Cr^{3+} and CrO_4^{2-} separated by boundary line 1 (Equation 9.54). In diagram (b), the $Cr_2O_7^{2-}$ species is delineated by boundary lines 2 and 3 (Equations 9.58 and 9.62). The completed diagram in (c) includes $Cr(OH)_3(s)$, which is delineated by lines 4 and 5 (Equations 9.66 and 9.70). The pe–pH diagrams are specific to the following conditions: activities of the soluble Cr species are 10^{-6}; activities of H_2O and $Cr(OH)_3(s)$ are unity; $T = 25°C$; and $P = 0.101$ MPa. The pe–pH regions consistent with oxic, suboxic, and anoxic conditions are also illustrated.

According to Equation 9.49, acidic (high H^+ activities or low pH values) and reducing (high e^- activities or low pe values) conditions favor the formation of Cr^{3+}. Therefore, Cr^{3+} predominates in the region of the diagram that lies below and to the left of the $Cr_2O_7^{2-} \rightarrow Cr^{3+}$ boundary line (Equation 9.58). Conversely, $Cr_2O_7^{2-}$ predominates in the region above and to the right of the boundary line (Figure 9.5b).

When pH = 4.5, the $CrO_4^{2-} \rightarrow Cr^{3+}$ and $Cr_2O_7^{2-} \rightarrow Cr^{3+}$ boundary lines (Equations 9.54 and 9.58) intersect. The equation, pH = 4.5, is also the boundary line that distinguishes between the $Cr_2O_7^{2-}$ and CrO_4^{2-} predominance regions. This may be confirmed by using Equation 9.48 (inverted to describe the oxidation of Cr^{3+}) and Equation 9.46 to generate the chemical reaction for $Cr_2O_7^{2-} \rightarrow CrO_4^{2-}$ and the associated equilibrium constant:

$$\tfrac{1}{3}Cr^{3+} + \tfrac{4}{3}H_2O \rightarrow \tfrac{1}{3}CrO_4^{2-} + \tfrac{8}{3}H^+ + e^- \qquad \log K_{Ox} = -25.0$$

$$+ \quad \tfrac{1}{6}Cr_2O_7^{2-} + \tfrac{7}{3}H^+ + e^- \rightarrow \tfrac{1}{3}Cr^{3+} + \tfrac{7}{6}H_2O \qquad \log K_R = 22.5$$

$$\overline{\tfrac{1}{6}Cr_2O_7^{2-} + \tfrac{1}{6}H_2O \rightarrow \tfrac{1}{3}CrO_4^{2-} + \tfrac{1}{3}H^+} \qquad \log K_R = -2.5 \tag{9.59}$$

The equilibrium constant for Equation 9.59 is:

$$K_R = \frac{(CrO_4^{2-})^{\frac{1}{3}}(H^+)^{\frac{1}{3}}}{(Cr_2O_7^{2-})^{\frac{1}{6}}} \tag{9.60}$$

or

$$\log K_R = \tfrac{1}{3}\log CrO_4^{2-} - \tfrac{1}{6}\log Cr_2O_7^{2-} - \tfrac{1}{3}pH \tag{9.61}$$

Applying the condition $(CrO_4^{2-}) = (Cr_2O_7^{2-}) = 10^{-6}$, and substituting for $\log K_R$ (line 3):

$$pH = 3\left[\tfrac{1}{3}(-6) - \tfrac{1}{6}(-6) - (-2.5)\right] = 4.5 \tag{9.62}$$

When solution pH values are greater than 4.5, CrO_4^{2-} will predominate; whereas, when pH is less than 4.5, $Cr_2O_7^{2-}$ will predominate.

The next boundary line to develop is the $Cr^{3+} \rightarrow Cr(OH)_3(s)$ line. The chemical reaction and associated equilibrium constant may be obtained by combining Equations 9.49 and 9.50:

$$\tfrac{1}{3}Cr^{3+} + \tfrac{7}{6}H_2O \rightarrow \tfrac{1}{6}Cr_2O_7^{2-} + \tfrac{7}{3}H^+ + e^- \qquad \log K_{Ox} = -22.5$$

$$+ \quad \tfrac{1}{6}Cr_2O_7^{2-} + \tfrac{4}{3}H^+ + e^- \rightarrow \tfrac{1}{3}Cr(OH)_3(s) + \tfrac{1}{6}H_2O \qquad \log K_R = 18.6$$

$$\overline{\tfrac{1}{3}Cr^{3+} + H_2O \rightarrow \tfrac{1}{3}Cr(OH)_3(s) + H^+} \qquad \log K_R = -3.9 \tag{9.63}$$

The equilibrium constant for Equation 9.63 is (assuming the solid has unit activity):

$$K_R = \frac{(H^+)}{(Cr^{3+})^{\frac{1}{3}}} \tag{9.64}$$

or

$$\log K_R = -\tfrac{1}{3}\log Cr^{3+} - pH \tag{9.65}$$

Applying the condition $(Cr^{3+}) = 10^{-6}$, and substituting for $\log K_R$ (line 4):

$$pH = -\tfrac{1}{3}(-6) - (-3.9) = 5.9 \tag{9.66}$$

When solution pH values are greater than 5.9, $Cr(OH)_3(s)$ will predominate; whereas, when pH is less than 5.9, Cr^{3+} will predominate (Figure 9.5c).

With the inclusion of the $Cr^{3+} \rightarrow Cr(OH)_3(s)$ boundary line, the $CrO_4^{2-} \rightarrow Cr^{3+}$ line at pH values above 5.9 becomes irrelevant, as Cr^{3+} is no longer predicted to predominate in this region under the stated conditions. Instead, a boundary line describing the $CrO_4^{2-} \rightarrow Cr(OH)_3(s)$ transition must be developed. This may be accomplished by employing Equations 9.50 and 9.59:

$$\frac{1}{3}CrO_4^{2-} + \frac{1}{3}H^+ \rightarrow \frac{1}{6}Cr_2O_7^{2-} + \frac{1}{6}H_2O \qquad \log K_{Ox} = 2.5$$

$$+ \quad \underline{\frac{1}{6}Cr_2O_7^{2-} + \frac{4}{3}H^+ + e^- \rightarrow \frac{1}{3}Cr(OH)_3(s) + \frac{1}{6}H_2O \qquad \log K_R = 18.6}$$

$$\frac{1}{3}CrO_4^{2-} + \frac{5}{3}H^+ + e^- \rightarrow \frac{1}{3}Cr(OH)_3(s) + \frac{1}{3}H_2O \qquad \log K_R = 21.1 \tag{9.67}$$

The equilibrium constant for this reaction is:

$$K_R = \frac{1}{(CrO_4^{2-})^{1/3}(H^+)^{5/3}(e^-)} \tag{9.68}$$

or

$$\log K_R = -\frac{1}{3}\log(CrO_4^{2-}) + \frac{5}{3}pH + pe \tag{9.69}$$

Applying the condition $(CrO_4^{2-}) = 10^{-6}$, and substituting for $\log K_R$ (line 5):

$$pe = 19.1 - \frac{5}{3}pH \tag{9.70}$$

Figure 9.5c illustrates the completed $pe - pH$ diagram for the $Cr^{3+}-H_2O$ system. The diagram is valid for the conditions (parentheses denote activities): $(Cr_2O_7^{2-}) = 10^{-6}$, $(Cr(OH)_3(s)) = (H_2O) = 1.0$, $T = 25°C$, and $P = 0.101$ MPa. Absent from the diagram is a boundary line representing the transition from $Cr_2O_7^{2-} \rightarrow Cr(OH)_3(s)$:

$$pe = 17.6 - \frac{4}{3}pH \tag{9.71}$$

As Figure 9.5c indicates, the formation of $Cr(OH)_3(s)$ from $Cr_2O_7^{2-}$ is superceded by the predominance regions of Cr^{3+} and CrO_4^{2-}. Thus, the predominance regions of $Cr_2O_7^{2-}$ and $Cr(OH)_3(s)$ are not connected. Each of the above boundary equations (Equations 9.54, 9.58, 9.62, 9.64, and 9.70) represents the pH and pe characteristics of a solution where one species becomes more stable than another. In this example, the redox speciation of chromium is dominated by the Cr^{III} state, occurring as Cr^{3+} in acidic environments (pH < 6) and as $Cr(OH)_3(s)$ in alkaline systems. Further, these reduced species dominate in suboxic and anoxic conditions where redox processes are controlled by the $Mn^{II}-Mn^{III}-Mn^{IV}$ and $Fe^{II}-Fe^{III}$ couples (suboxic), and $S^{-II}-S^{VI}$ couples (anoxic) (Figure 9.4). In oxic systems, where redox processes are controlled by the reactions of O_2, Cr is predicted to exist in the Cr^{VI} state when solution pH values are greater than 6, principally as CrO_4^{2-}. Therefore, it would appear that the absence of O_2 is a condition that is necessary for maintaining Cr in the relatively innocuous or sparingly soluble reduced forms in neutral to alkaline systems. However, in acidic systems the Cr^{III} state may be maintained in the presence of O_2.

9.4.2 EXAMPLES OF pe–pH DIAGRAMS FOR REDOX-SENSITIVE ELEMENTS

The interpretations obtained through the examination of pe–pH diagrams are significantly less concrete than the mechanisms employed in their creation. Figure 9.6 illustrates pe–pH diagrams for the $Fe-SO_4-CO_2-H_2O$ system. The chemical reactions, associated $\log K_R$ and $\log K_a$ values, and

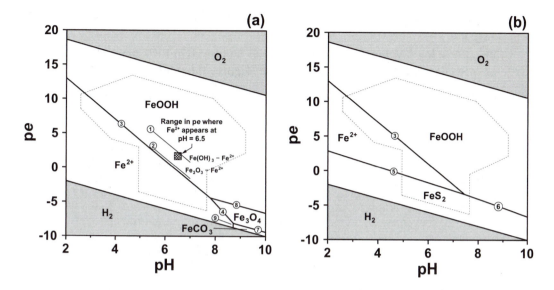

FIGURE 9.6 The redox speciation of Fe. The $Fe-CO_2-H_2O$ system is illustrated in diagram (a) and the $Fe-S-CO_2-H_2O$ system in (b). In both diagrams the anvil-shaped outline identifies the pe–pH region of natural waters. The range in pe at pH 6.5 where Fe^{2+} is first detected in soil solutions is shown in (a). The boundary lines for metastable $Fe(OH)_3 \leftrightarrow Fe^{2+}$ (line 1) and $Fe_2O_3 \leftrightarrow Fe^{2+}$ (line 2) transitions are also shown. In diagram (b), pyrite $(FeS_2(s))$ supercedes Fe^{2+}, $Fe_3O_4(s)$, and $FeCO_3(s)$ in anoxic environments. The pe–pH diagrams are specific to the following conditions: activities of the soluble Fe species are 10^{-6}; $CO_2(g)$ activity is $10^{-3.52}$; SO_4^{2-} activity is 10^{-4}; activities of H_2O and all solids are unity; $T = 25°C$; and $P = 0.101$ MPa. The boundary lines are identified by number in Table 9.2.

boundary equations used to create the diagrams are shown in Tables 9.1 and 9.2. The diagrams are specific to the following conditions: the activities of liquid water and the solids (FeOOH, $Fe(OH)_3$, Fe_2O_3, Fe_3O_4, $FeCO_3$, and FeS_2) are unity; the activities of all soluble species at a boundary are 10^{-6} (unless specified otherwise); the partial pressure of H_2 is 1.0 atm, O_2 is 0.21 atm, and CO_2 is $10^{-3.52}$ atm; and $T = 25°$ and $P = 1$ atm (0.101 MPa). The diagrams are not valid for systems that do not adhere to these conditions, although changes in any one or more of the imposed conditions may not significantly impact the appearance of the diagrams or the interpretations. The diagrams also assume chemical equilibrium and it is assumed that the thermodynamic data employed in their construction are accurate. Thus, chemical transformations that are not kinetically favored or do not couple well (electron transfer is hindered, in the case of redox reactions) are not indicated, as only the final equilibrium state is predicted. Finally, the diagrams indicate the regions of pe and pH where a particular species predominates, but not to the exclusion of other species. Indeed, the pH-distribution diagrams presented in Chapter 5 illustrate that several species of a particular element may exist at any pH value, and that the change in speciation with a change in solution pH is gradual. Speciation is similarly influenced by pe, as illustrated for N in Figure 9.3 for a pH 7 solution.

9.4.2.1 Iron and Manganese

Despite all the caveats listed above, pe–pH diagrams are useful in predicting the redox speciation of an element, as well as the redox reactions that might occur. Conversely, the diagrams may be employed to indicate redox speciation and processes that are not possible, under the stated conditions. The pe–pH diagrams in Figure 9.6 examine the redox speciation of Fe under differing environmental conditions. In Figure 9.6a, ferric iron (Fe^{III}), in the form of goethite (FeOOH), is predicted to predominate throughout a wide range of pe and pH conditions (Fe^{2+} activity is 10^{-6}). According to the findings of Patrick and Jugsujinda (1996), ferrous Fe appears in pH 6.5 soil

TABLE 9.2

Reactions, log K Values, and Boundary Lines Employed in the Construction of pe–pH Diagrams for Iron and Manganese (Figures 9.6a, 9.6b, and 9.7)[a]

Reaction	log K	Boundary Line[b]
Iron		
$Fe(OH)_3(s) \rightarrow Fe^{2+}$ [c]		$pe = 21.8 - 3pH$ [1]
$Fe_2O_3(s) \rightarrow Fe^{2+}$ [c]		$pe = 19.4 - 3pH$ [2]
$FeOOH(s) \rightarrow Fe^{2+}$ [c]		$pe = 19 - 3pH$ [3]
$Fe_3O_4(s) \rightarrow Fe^{2+}$ [c]		$pe = 26.8 - 4pH$ [4]
$\frac{1}{14}Fe^{2+} + \frac{1}{7}SO_4^{2-} + \frac{8}{7}H^+ + e^- \rightarrow \frac{1}{14}FeS_2(s) + \frac{4}{7}H_2O$	6.0	$pe = 5.14 - \frac{8}{7}pH$ [5][d]
$\frac{1}{15}FeOOH(s) + \frac{2}{15}SO_4^{2-} + \frac{19}{15}H^+ + e^- \rightarrow \frac{1}{16}FeS_2(s) + \frac{1}{3}H_2O$	6.5	$pe = 6.07 - \frac{19}{15}pH$ [6][d]
$\frac{1}{2}Fe_3O_4(s) + \frac{3}{2}CO_2(g) + H^+ + e^- \rightarrow FeCO_3(s) + \frac{1}{2}H_2O$	5.92	$pe = 0.64 - pH$ [7][d]
$3FeOOH(s) + H^+ + e^- \rightarrow Fe_3O_4(s) + \frac{1}{2}H_2O$	3.4	$pe = 3.4 - pH$ [8]
$FeCO_3(s) + 2H^+ \rightarrow Fe^{2+} + CO_2 + H_2O$	7.92	$pH = 8.72$ [9][d]
Manganese		
$\gamma - MnOOH(s) \rightarrow Mn^{2+}$ [c]		$pe = 31.4 - 3pH$ [1]
$MnO_2(s) \rightarrow Mn^{2+}$ [c]		$pe = 23.8 - 2pH$ [2]
$Mn_3O_4(s) \rightarrow Mn^{2+}$ [c]		$pe = 39.7 - 4pH$ [3]
$MnO_2(s) + H^+ + e^- \rightarrow \gamma - MnOOH(s)$	16.2	$pe = 16.2 - pH$ [4]
$3MnOOH(s) + H^+ + e^- \rightarrow Mn_3O_4(s) + 2H_2O$	14.8	$pe = 14.8 - pH$ [5]
$MnCO_3(s) + 2H^+ \rightarrow Mn^{2+} + CO_2(g) + H_2O$	8.87	$pH = 9.20$ [6][d]
$\frac{1}{2}Mn_3O_4(s) + \frac{3}{2}CO_2(g) + H^+ + e^- \rightarrow \frac{3}{2}MnCO_3(s) + \frac{1}{2}H_2O$	17.4	$pe = 12.12 - pH$ [7][d]

[a] Equilibrium constants where obtained from the compilations of James and Bartlett (1999), Lindsay (1979), and Smith and Martell (1976).

[b] The assumed conditions are: $(Fe^{2+}) = (Mn^{2+}) = 10^{-6}$; activities of H_2O and solids are unity; $T = 25°C$ and $P = 0.101$ MPa. The number in brackets indicates the line number in Figures 9.6 and 9.7.

[c] Chemical reaction and log K_R values are presented in Table 9.1.

[d] $(CO_2) = 10^{-3.52}$; $(SO_4^{2-}) = 10^{-3}$.

---

solutions when E_H drops below approximately 125 to 75 mV ($pe = 2.11$ to 1.26). These observed values are greater than the predicted pe of –0.5 ($E_H = -30$ mV), assuming the activity of Fe^{2+} is 10^{-6} and FeOOH (goethite) is the controlling Fe^{III} phase. One reason for this discrepancy may be that the Fe^{III} solid that is actually present and controlling the Fe activities in the soil solutions may not be the most stable phase (which is goethite). When unstable Fe^{III} phases (hematite and amorphous $Fe(OH)_3$) are also included in the diagram, it is evident that the $Fe(OH)_3$–Fe^{2+} boundary line models the experimental E_H range in which Fe^{2+} is initially observed. Indeed, the $Fe(OH)_3$–Fe^{2+} boundary is predicted to occur at $pe = 2.3$ ($E_H = 136$ mV) when the solution pH is 6.5. As indicated in Chapter 2, the prediction in Figure 9.6a that Fe chemistry may be controlled by a fast-forming metastable and amorphous mineral is supported by direct observations, particularly when Fe^{2+}-rich soil solutions are rapidly oxidized. In alkaline, anoxic conditions, and in the absence of appreciable sulfur concentrations, magnetite (Fe_3O_4) and siderite ($FeCO_3$) are the predicted stable iron phases ($P_{CO_2} = 10^{-3.52}$). However, these phases are superceded by pyrite when sulfur is included as a component in the pe–pH diagram (SO_4^{2-} activity is 10^{-3}) (Figure 9.6b).

According to the pe–pH diagram for Mn (Figure 9.7), Mn^{2+} is predicted to predominate in all but alkaline and oxic environments. The diagram also indicates that the transition from MnO_2 to Mn^{2+} is predicted to occur when pe is 10.8 ($E_H = 639$ mV) and pH is 6.5. However, measurable concentrations of Mn^{2+} do not appear in pH 6.5 solutions until E_H drops below 275 to 225 mV ($pe = 4.65$ to 3.80). Again, the inconsistency between the measured and predicted E_H values at which Mn^{2+} first appears

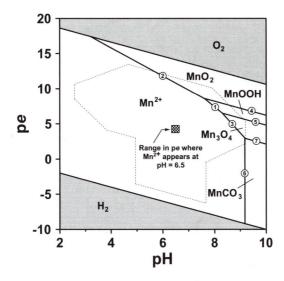

FIGURE 9.7 The redox speciation of Mn. The anvil-shaped outline identifies the pe–pH region of natural waters. The range in pe at pH 6.5 where Mn^{2+} is first detected in soil solutions is shown, and the anvil-shaped outline identifies the pe–pH region of natural waters. The pe–pH diagram is specific to the following conditions: activities of the soluble Fe species are 10^{-6}; $P_{CO_2} = 10^{-3.52}$; SO_4^{2-} activity is 10^{-4}; activities of H_2O and all solids are unity; $T = 25°C$; and $P = 0.101$ MPa. The boundary lines are identified by number in Table 9.2.

in pH 6.5 soil solutions may be due to the nature of Mn oxide phases found in soils (impure, noncrystalline, and containing Mn^{II}, Mn^{III}, and Mn^{IV}) relative to the pure MnO_2 phase employed in theoretical predictions. Indeed, it is because Mn solid phases are notoriously disordered that the ability of pe–pH, or any other stability diagrams, to accurately predict the distribution of Mn between the soil solid and solution phase is virtually nonexistent.

9.4.2.2 Selenium

Selenium is a redox-sensitive trace element of environmental concern and may be particularly problematic in agricultural drainage from arid regions, such as observed in the western San Joaquin Valley of California (discussed in Chapter 11). Selenium may also accumulate in vegetation grown on high-Se soils (seleniferous soils). Typically, soils that form on Cretaceous shale, such as those found in the mountain west region of the U.S., contain sufficient plant-available Se to produce seleniferous-vegetation. Selenium is a required mineral nutrient for animals and humans. However, there is a narrow gap between sufficient and toxic Se concentrations. The consumption of seleniferous vegetation (Se-accumulators) by grazing animals for extended periods leads to two chronic Se toxicosis syndromes: alkali disease and blind staggers. Selenium is also widely dispersed into the environment by aerosols and particulates from coal-fired power generation plants, and via various mining activities (most notably coal, bentonite, and uranium mining) that impact soils, surface waters, and groundwater.

The environmental behavior of Se is very similar to that of sulfur. Like SO_4^{2-}, the selenate species (SeO_4^{2-}) is a mobile form of the element that is highly soluble, forming only simple salts and other evaporite minerals (often substituting for SO_4^{2-}). Ligand exchange processes do not play a significant role in the soil retention of SeO_4^{2-} because the species is an exchangeable anion that requires positive surface charge for retention. The pe–pH diagrams presented in Figures 9.8a and 9.8b show that selenate is stable in oxic environments (when O_2 is present), a conclusion that is consistent with the high mobility and accumulation of soluble Se in agricultural drainage waters. The chemical reactions, associated $\log K_R$ and $\log K_a$ values, and boundary equations used to create

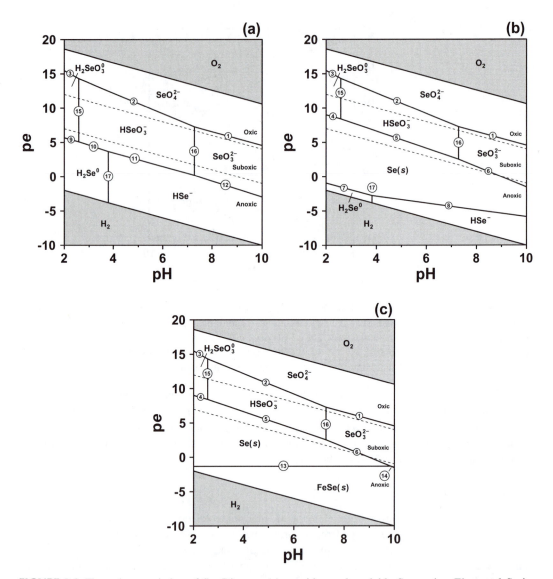

FIGURE 9.8 The redox speciation of Se. Diagram (a) considers only soluble Se species. Elemental Se is included in (b), and both elemental Se and FeSe(s) are included in (c). The pe–pH diagrams are specific for the following conditions: activities of the soluble Se species are 10^{-6}; Fe^{2+} activity is 10^{-6}; activities of H_2O and all solids are unity; $T = 25°C$; and $P = 0.101$ MPa. The boundary lines are identified by number in Table 9.3, and the oxic, suboxic, and anoxic pe–pH regions are illustrated.

the diagram are given in Table 9.3. The selenite species ($HSeO_3^-$ and SeO_3^{2-}) are predicted to predominate in suboxic environments, and in oxic-acidic (pH < 6) soil. The adsorption of selenite by soils, hydrous metal oxides, and aluminosilicates is pH dependent, indicating that ligand exchange processes are involved. The adsorption maximum occurs below pH 6, with selenite adsorption decreasing with increasing pH above 6 (see Chapter 7). Selenite minerals are also relatively soluble. Equilibrium solubility computations indicate that the $Fe_2(SeO_3)_3$ and $MnSeO_3$ phases may potentially control selenite activities in soil solutions. However, these minerals require unrealistically high activities of Fe^{3+} and Mn^{2+} in order to predominate in any region of the pe–pH diagrams in Figure 9.8. In general, distinct selenite phases do not exist in soils, even in contaminated environments. In anoxic systems, the Se^{-II} species, H_2Se^0 and HSe^-, are predicted to predominate.

TABLE 9.3

Redox and Acid-Base Reactions, K_R and K_a Values, and Boundary Lines Used to Construct the pe–pH Diagrams for Selenium (Figure 9.8a and b)[a]

Reaction	$\log K_R$	Boundary Line[b]
$\frac{1}{2}SeO_4^{2-} + H^+ + e^- \rightarrow \frac{1}{2}SeO_3^{2-} + \frac{1}{2}H_2O$	14.55	$pe = 14.55 - pH$ [1]
$\frac{1}{2}SeO_4^{2-} + \frac{3}{2}H^+ + e^- \rightarrow \frac{1}{2}HSeO_3^- + \frac{1}{2}H_2O$	18.19	$pe = 18.19 - \frac{3}{2}pH$ [2]
$\frac{1}{2}SeO_4^{2-} + 2H^+ + e^- \rightarrow \frac{1}{2}H_2SeO_3^0 + \frac{1}{2}H_2O$	19.48	$pe = 19.48 - 2pH$ [3]
$\frac{1}{4}H_2SeO_3^0 + e^- + H^+ \rightarrow \frac{1}{4}Se(s) + \frac{3}{4}H_2O$	12.5	$pe = 11.0 - pH$ [4]
$\frac{1}{4}HSeO_3^- + e^- + \frac{5}{4}H^+ \rightarrow \frac{1}{4}Se(s) + \frac{3}{4}H_2O$	13.14	$pe = 11.64 - \frac{5}{4}pH$ [5]
$\frac{1}{4}SeO_3^{2-} + e^- + \frac{3}{2}H^+ \rightarrow \frac{1}{4}Se(s) + \frac{3}{4}H_2O$	14.96	$pe = 13.46 - \frac{3}{2}pH$ [6]
$\frac{1}{2}Se(s) + e^- + H^+ \rightarrow \frac{1}{2}H_2Se^0$	−1.94	$pe = 1.06 - pH$ [7]
$\frac{1}{2}Se(s) + e^- + \frac{1}{2}H^+ \rightarrow \frac{1}{2}HSe^-$	−3.85	$pe = -0.85 - \frac{1}{2}pH$ [8]
$\frac{1}{6}H_2SeO_3^0 + e^- + H^+ \rightarrow \frac{1}{6}H_2Se^0 + \frac{1}{2}H_2O$	7.68	$pe = 7.68 - pH$ [9]
$\frac{1}{6}HSeO_3^- + e^- + \frac{7}{6}H^+ \rightarrow \frac{1}{6}H_2Se^0 + \frac{1}{2}H_2O$	8.11	$pe = 8.11 - \frac{7}{6}pH$ [10]
$\frac{1}{6}HSeO_3^- + e^- + H^+ \rightarrow \frac{1}{6}HSe^- + \frac{1}{2}H_2O$	7.48	$pe = 7.48 - pH$ [11]
$\frac{1}{6}SeO_3^{2-} + e^- + \frac{7}{6}H^+ \rightarrow \frac{1}{6}HSe^- + \frac{1}{2}H_2O$	8.69	$pe = 8.69 - \frac{7}{6}pH$ [12]
$Se(s) + Fe^{2+} + 2e^- \rightarrow FeSe(s)$	−3.35	$pe = -1.33$ [13]
$\frac{1}{6}SeO_3^{2-} + \frac{1}{6}Fe^{2+} + e^- + H^+ \rightarrow \frac{1}{6}FeSe(s) + \frac{1}{2}H_2O$	−10.53	$pe = 8.53 - pH$ [14]
	$\log K_a$	
$H_2SeO_3^0 \rightarrow HSeO_3^- + H^+$	−2.58	$pH = 2.58$ [15]
$HSeO_3^- \rightarrow SeO_3^{2-} + H^+$	−7.29	$pH = 7.29$ [16]
$H_2Se(g) \rightarrow HSe^- + H^+$	−3.81	$pH = 3.81$ [17]

[a] $\log K_R$ and $\log K_a$ values from Smith and Martell (1976) or computed using the standard free energies of formation values tabulated by Wagman et al. (1982).

[b] The activities of soluble species are 10^{-6}. The activities of H_2O and solids are unity, and $T = 25°C$ and $P = 0.0101$ MPa. The number in brackets indicates the line number in Figure 9.8.

However, when elemental Se and FeSe are included in the pe–pH diagram (Figure 9.8b and 9.8c) they are predicted to supplant the soluble selenides and predominate throughout a wide pH range in anoxic environments (poised by the $SO_4^{2-} \leftrightarrow S^{2-}$ couple) and in the lower pe range of suboxic systems (controlled by $Fe^{2+} \leftrightarrow Fe^{3+}$ couple). These reduced forms of Se restrict the mobility and bioavailablity of Se.

9.4.2.3 Arsenic

Arsenic is a trace element that has been widely dispersed in the environment as a result of human activities. Like Se, As is dispersed in aerosols and particulates from coal-fired power plants, via various mining and smelting activities (primarily those associated with copper production), and through the extensive and historic use of As-containing pesticides. Arsenic has no known biological function in animals, nor is there evidence to support its essentiality for plants. However, As is a poultry feed additive that is used for disease control and to promote chick growth. In general, plants are a barrier that restrict the movement of As in the soil–plant–animal continuum. This is particularly the case in food crops that exhibit As phytotoxicity symptoms and reduced yields before edible plant parts can accumulate sufficient As levels to cause human toxicity. The primary route of As exposure to humans is in drinking water obtained from wells drilled in high As aquifer materials.

The mobility and toxicity of As is highly dependent on redox state. There are essentially two As oxidation states of consequence in soils: As^{III} in arsenite (principally $As(OH)_3^0$) and As^V in arsenate (such as $H_2AsO_4^-$). The As^{III} forms are substantially more mobile in the environment and more toxic than the As^V forms. Figure 9.9 illustrates the pe–pH diagram for the As–H_2O system.

TABLE 9.4
Redox and Acid–Base Reactions, K_R and K_a Values, and Boundary Lines Used to Construct the pe–pH Diagrams for Arsenic (Figure 9.9)[a]

Reaction	$\log K_R$	Boundary line[b]
$\frac{1}{2}H_2AsO_4^- + e^- + \frac{3}{2}H^+ \rightarrow \frac{1}{2}As(OH)_3^0 + \frac{1}{2}H_2O$	10.84	$pe = 10.84 - \frac{3}{2}pH$ [1]
$\frac{1}{2}HAsO_4^{2-} + e^- + 2H^+ \rightarrow \frac{1}{2}As(OH)_3^0 + H_2O$	14.22	$pe = 14.22 - 2pH$ [2]
$\frac{1}{2}H_3AsO_4^0 + e^- + H^+ \rightarrow \frac{1}{2}As(OH)_3^0 + \frac{1}{2}H_2O$	9.72	$pe = 9.72 - pH$ [3]
$\frac{1}{2}HAsO_4^{2-} + e^- + \frac{3}{2}H^+ \rightarrow \frac{1}{2}As(OH)_4^- + H_2O$	9.61	$pe = 9.61 - \frac{3}{2}pH$ [4]
	$\log K_a$	
$H_3AsO_4^0 \rightarrow H_2AsO_4^- + H^+$	-2.24	$pH = 2.24$ [5]
$H_2AsO_4^- \rightarrow HAsO_4^{2-} + H^+$	-6.96	$pH = 6.96$ [6]
$As(OH)_3^0 + H_2O \rightarrow As(OH)_4^- + H^+$	-9.29	$pH = 9.29$ [7]

[a] $\log K_R$ and $\log K_a$ values computed using the standard free energies of formation values tabulated by Wagman et al. (1982).

[b] The activities of soluble species are 10^{-6}. The activity of H_2O is unity, $T = 25°C$, and $P = 0.0101$ MPa. The number in brackets indicates the line number in Figure 9.9.

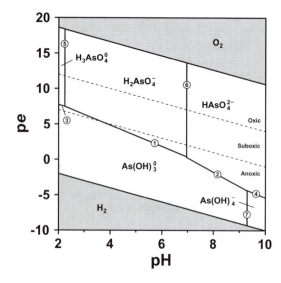

FIGURE 9.9 The redox speciation of As. The pe–pH diagram is specific to the following conditions: activities of the soluble As species are 10^{-6}; activity of H_2O is unity; $T = 25°C$; and $P = 0.101$ MPa. The boundary lines are identified by number in Table 9.4 and the oxic, suboxic, and anoxic pe–pH regions are illustrated.

The chemical reactions, associated $\log K_R$ and $\log K_a$ values, and boundary equations used to create the diagram are given in Table 9.4. The arsenate species dominate in oxic and suboxic systems, while the arsenite species are limited to anoxic systems. The environmental behavior and chemistry of arsenate is very similar to that of phosphate. Both phosphate and arsenate are polyprotic acids, and their pK_a values are comparable:

$$H_3AsO_4^0 \xrightarrow{pK_a=2.24} H_2AsO_4^- \xrightarrow{pK_a=6.96} HAsO_4^{2-} \xrightarrow{pK_a=11.50} AsO_4^{3-} \tag{9.72}$$

$$H_3PO_4^0 \xrightarrow{pK_a=2.15} H_2PO_4^- \xrightarrow{pK_a=7.2} HPO_4^{2-} \xrightarrow{pK_a=12.35} PO_4^{3-} \tag{9.73}$$

Like phosphate, arsenate has a high affinity for hydrous metal oxide surfaces, particularly in acidic systems, and forms stable bidentate-binuclear surface complexes throughout a wide range of soil pH values (recall from Chapter 7 that ligand adsorption is greatest when pH is equal to the pK_a). However, arsenate does not occur in soils in discrete mineral phases, although several have been proposed based on thermodynamic considerations. The predicted stable phases are similar to those established for phosphate: Ca arsenates in neutral to alkaline soils; and Fe^{III} and Al arsenates in neutral to acidic soils. Usually, AsO_4^{3-} is a minor substituent of phosphate minerals, even in relatively contaminated soils. The more toxic and mobile As^{III} species are predicted to predominate in only reducing systems. Because $As(OH)_3^0$ has a high pK_a value (9.29), adsorption increases with increasing solution pH, and it may be more strongly adsorbed than arsenate in alkaline environments (see Chapter 7 for a direct comparison of arsenate and arsenite adsorption behavior). The adsorption of $As(OH)_3^0$ by Fe^{III}, and particularly Mn^{III} and Mn^{IV} oxide minerals results in surface-mediated oxidation to the less toxic arsenate forms. This process is described in Equations 9.30 through 9.32.

9.5 EERCISES

1. Write balanced oxidation-reduction reactions for the following transformations:
   a. The oxidation of VO^{2+} by $O_2(g)$ to form $VO_2(OH)_2^-$
   b. The oxidation of elemental mercury by soil organic matter (CH_2O) to form dimethylmercury, $(CH_3)_2Hg(g)$
   c. The oxidation of realgar $(AsS(s))$ by $O_2(g)$ to form $H_2AsO_4^-$ and SO_4^{2-}
2. The Pt electrode is perhaps the most common method for quantifying electron activity in soils and natural waters. However, potentiometric measurements by the Pt electrode are generally considered unreliable for accurately assessing the redox status of chemically complex systems, particularly aerobic soil systems. Discuss the reasons for the unreliability of the Pt electrode for determining the redox status of soil systems.
3. Write balanced oxidation-reduction reactions and determine the equilibrium constants for the following transitions:
   a. The oxidation of $H_2(g)$ by formic acid to produce $CO_2(g)$
   b. The oxidation of arsenite $(As(OH)_3^0)$ by birnessite to produce arsenate $(H_2AsO_4^-)$ and soluble manganese
   c. The oxidation of HS^- by NO_3^- to form SO_4^{2-} and $N_2(g)$
   d. The oxidation of SeO_3^{2-} by nitrate to form SeO_4^{2-} and $N_2(g)$
4. Figure 9.5c illustrates the speciation of chromium as a function of pe and pH. Construct a new pe–pH diagram for chromium that includes the $HCrO_4^-$ species by modifying Figure 9.5c. The dissociation of $HCrO_4^-$ is described by: $HCrO_4^- \rightarrow CrO_4^- + H^+$, $\log K_a = -6.51$. Under what conditions of pe and pH will $Cr_2O_7^-$ predominate over $HCrO_4^-$?
5. In a pH 4 soil solution in equilibrium with goethite and a soil atmosphere containing an $O_2(g)$ partial pressure of 0.21, determine the activities of Fe^{3+} and Fe^{2+}.
6. Many of the redox transformations that occur in soils are microbially mediated. Can the oxidation of NH_4^+ by SeO_4^{2-} to form NO_3^- be mediated by microorganisms in a pH 7 soil?
7. Construct a pe–pH predominance diagram for sulfur that incorporates the species: SO_4^{2-}, HS^-, H_2S^0, and $S(s)$. Limit your diagram to the pH 3 to 9 range; include the stability range for water. Assume unit activity of H_2O and $S(s)$, the activities of all soluble species are 10^{-3}, and standard state T and P conditions.
8. A pH 6 soil system is poised at a pe of 6. Compute the following concentration ratios (assuming concentrations equal activities) and determine if the concentrations of the oxidized and reduced species can be analytically measured if the detection limit for each species is 0.01 mg L^{-1}:

    a. $H_2AsO_4^-$ and $As(OH)_3^0$, total As is 2 mg L^{-1}

    b. SeO_4^{2-} and $HSeO_3^-$, total Se is 40 mg L^{-1}

    c. Mn^{2+} and $MnO_2(s)$

9. A sodium thiosulfate solution is used to solubilize and remove iron oxides from soils in a procedure that cleans clay minerals and prepares them for x-ray diffraction analysis. Write a balanced chemical reaction for the solubilization process. Indicate the experimental conditions that may be imposed during the treatment process to facilitate the removal of iron oxides.

10. Figure 9.2a through d illustrates the concentrations of dissolved Fe^{2+}, NH_4^+, Mn^{2+}, and NO_3^- that occur in pH 6.5 soil suspensions that are poised at various E_H values. Determine the theoretical E_H values where the concentrations of Fe^{2+}, NH_4^+, Mn^{2+}, and NO_3^- in pH 6.5 solutions become detectable, assuming that the detection limit for each species is 10^{-5} M and that concentrations equal activities. Assume Fe^{2+} is controlled by FeOOH; NH_4^+ is controlled by alanine decomposition and a CO_2 partial pressure of $10^{-3.52}$ atm ($\frac{1}{12}C_3H_4O_2NH_3 + \frac{1}{3}H_2O = \frac{1}{4}CO_2(g) + \frac{1}{12}NH_4^+ + \frac{11}{12}H^+ + e^-$, log $K_{ox} = -0.8$); Mn^{2+} is controlled by birnessite; and NO_3^- decomposes to $N_2(g)$ with a partial pressure of 0.79atm. Compare your results to the measured E_H values in Figure 9.2 where Fe^{2+}, NH_4^+, Mn^{2+}, and NO_3^- are first detected, and explain any differences.

REFERENCES

Baas Becking, L.G.M., L.R. Kaplan, and D. Moore. Limits of the natural environment in terms of pH and oxidation-reduction potentials. *J. Geol.* 68:224–284, 1960.

Bartlett, R.J. Characterizing soil redox behavior. In *Soil Physical Chemistry*, 2nd ed. D.L. Sparks (Ed.) CRC Press, Boca Raton, FL, 1999, pp. 371–397.

James, B.R. and R.J. Bartlett. Redox phenomena. In *Handbook of Soil Science*. M.E. Sumner (Ed.) CRC Press, Boca Raton, FL, 1999, pp. B169–B194.

Lindsay, W.L. *Chemical Equilibria in Soils*. John Wiley & Sons, New York, 1979.

McBride, M.B. *Environmental Chemistry of Soils*. Oxford University Press, New York, 1994.

Patrick, W.H., Jr. and A. Jugsujinda. Sequential reduction and oxidation of inorganic nitrogen, manganese, and iron in flooded soil. *Soil Sci. Soc. Am. J.* 56:1071–1073, 1992.

Smith, R.M. and A.E. Martell. *Critical Stability Constants. Volume 4: Inorganic Complexes*. Plenum Press, New York, 1976.

Sposito, G. *The Thermodynamics of Soil Solutions*. Oxford University Press, New York, 1981.

Sposito, G. *The Chemistry of Soils*. Oxford University Press, New York, 1989.

Wagman, D.D., W.H. Evans, V.B. Parker, R.H. Schumm, I. Harlow, S.M. Bailey, K.L. Churney, and R.L. Nutall. Selected values for inorganic and C_1 and C_2 organic substances in SI units. *J. Phys. Chem.* Ref. Data 11, Suppl. 2, 1982.

10 Acidity in Soil Materials

A vast array of chemical reactions occurring in the soil environment involves either the consumption or the release of protons. Because proton activity is a controlling factor in many chemical processes, the pH of a soil solution is considered a master chemical variable. The pH of a solution is defined as the negative common logarithm (denoted by "p") of the hydrogen ion (H^+) activity. A common misconception is that the pH of a solution is the negative logarithm of the molar concentration of the hydrogen ion. As will be shown in a later section, the pH electrode, which is an ion selective electrode that senses protons, responds to H^+ activity. Therefore, the pH electrode is providing a measure of the true activity of the proton in solution.

The pH of soil solutions varies widely, from pH < 3 in pyritic- and other metal sulfide-bearing soils, to pH > 9 in sodium-affected or black-alkali soils. Although it is difficult to fix a range of soil pH values that encompasses all environments, soils that have pH values less than 4.0 to 4.5 or greater than 8.5 have usually been impacted by human activities. Excessive soil acidity, which is generally indicated by solution pH values less than 5.0 to 5.5, is a concern from both an environmental and an agronomic perspective. In excessively acidic systems (such as pyritic spoils), macronutrient availability and microbial activity is restricted, phytotoxic levels of soluble Al and Mn are observed, and the solubility and mobility of many potentially deleterious trace elements is relatively high.

Despite the ever-present temptation to ascribe soil acidity to human activities, it is a natural consequence of weathering. During the weathering process, carbonic and organic acids promote the hydrolysis (dissolution) of soil minerals. As a result, in regions where precipitation exceeds evapotranspiration, base cations (Ca^{2+}, Mg^{2+}, K^+, and Na^+, also called nonacidic cations) and silica are solubilized and leached from the soil. Other weathering products, such as soluble Al^{3+} and Fe^{3+}, hydrolyze and release protons as they form hydrolysis products (principally $AlOH^+$ and $Al(OH)_2^+$) and accessory minerals (e.g., gibbsite and goethite). These hydrolysis reactions, coupled with the displacement of nonacidic cations from the exchange complex by Al^{3+}, $AlOH^+$, and $Al(OH)_2^+$, and the subsequent leaching of the nonacidic cations which do not hydrolyze to produce protons, are processes that lead to the development of soil acidity. However, it is the occupation of the soil exchange complex by Al species, which are labile and available for further hydrolysis and proton production that is the root cause of soil acidity. Many soils worldwide, such as those found in the humid tropics and forested areas of temperate zones, are acidic. Strictly speaking, an acid soil is defined as the pH of the soil solution being less than 7.0; however, the proton activity in the soil solution (pH) is not equivalent to soil acidity. It is an indicator of soil acidity, just as running a fever is an indicator of the flu. The virus is the problem and the fever is a consequence—an indicator that a virus is present.

The genesis of an acid soil is a complex process, resulting from the interrelated actions of physical, chemical, and biological processes. The physical processes that impact soil acidity are the transport processes associated with the movement of acidic and alkaline substances into, out of, and within the soil profile. Acid-producing materials may enter the soil through precipitation (rain, snow, fog, drizzle), which carries carbonic acid and protons from the reaction of water with industrial byproducts. Examples of this are sulfuric acid from SO_2 from coal combustion for electric power generation, and nitric acid from NO_2 from automotive exhaust. Precipitation is also the source of fresh water that is responsible for the leaching of base cations through the soil profile. Natural and anthropogenic, inorganic and organic particulates are deposited on the soil surface as dryfall. Nitrogen and phosphorus fertilizers are examples of anthropogenic dryfall that can result in soil acidification. Dust storms in the western U.S. can redistribute soil alkalinity (base cation–bearing and carbonate minerals) as natural dryfall.

10.1 THE MEASUREMENT OF SOIL SOLUTION pH

The pH of the soil solution provides a plethora of information concerning the chemical characteristics of the soil and the chemical processes operating in the soil. However, as with all other soil solution chemical properties, the measured soil solution pH is a function of the soil solution isolation technique and the analytical method employed.

10.1.1 THE pH ELECTRODE SYSTEM

The glass membrane electrode, coupled with a suitable reference electrode, is the standard method for determining the pH of a soil solution. An example of this electrode system is diagrammed for the glass membrane–calomel reference electrode system:

| **Electrode L (reference)** | | | | **Electrode R (glass membrane)** | |
| Hg(l); Hg$_2$Cl$_2$(s) | KCl(aq, saturated) | solution | glass | 0.1 M HCl(aq), AgCl(aq, saturated) | Ag(s) |

In the pH electrode system, the glass membrane electrode (electrode **R**) consists of Ag wire suspended in a 0.1 M HCl solution that is saturated with AgCl. This solution has a constant H$^+$ activity (and electrochemical potential), and it is enclosed in a thin glass membrane (Na- or Li-doped) that separates the internal electrode solution from the soil solution. The surface of the glass electrode is selective for protons, and when protons exchange with ions that are adsorbed at the glass surface, a potential difference between the constant H$^+$ electrochemical potential inside the electrode and that of the adsorbed proton develops. The calomel reference electrode (electrode **L**) consists of a gel composed of liquid Hg and solid Hg$_2$Cl$_2$ bathed in a saturated KCl solution. An asbestos fiber acts as the liquid junction (or salt bridge) and provides an electrical connection by allowing the minimal diffusion of a saturated KCl solution into the soil solution. The difference between the electrochemical potentials of the electrons in the calomel ($\tilde{\mu}_{e^-,L}$) and Ag/AgCl ($\tilde{\mu}_{e^-,R}$) electrode is related to the electric potential difference (E, in volts) between the reference and glass membrane electrodes:

$$-F\text{E} = \tilde{\mu}_{e^-,\text{L}} - \tilde{\mu}_{e^-,\text{R}} \tag{10.1}$$

where F is the Faraday constant. The components in the calomel electrode are not in equilibrium with the sample solution. Although the salt bridge allows for the transfer of Cl$^-$ and K$^+$, their chemical potentials in the electrode and the sample solution do not equalize. Because a transport process, and not an equilibrium process, controls the chemical potentials of Cl$^-$ and K$^+$, the electrical potential difference is separated into two components: a thermodynamic component, E$_{cell}$, and a part that depends on transport processes at the liquid junction, E$_j$, the liquid junction potential. The liquid junction potential is not a measurable quantity and can be a source of error in any electrochemical measurement, although in most cases this error is removed through electrode calibration procedures. In addition, the glass membrane does not allow for the equilibration of H$^+$ between the 0.1 M HCl solution in the Ag/AgCl electrode and the sample solution. This results in a proton electrochemical potential difference, E$_{H^+}$. Therefore, the electric potential difference between the two electrodes is given by:

$$\text{E} = \text{E}_{cell} - \text{E}_j - \text{E}_{H^+} \tag{10.2}$$

Equation 10.1 then becomes:

$$-F(E_{cell} - E_j - E_{H^+}) = \tilde{\mu}_{e^-,L} - \tilde{\mu}_{e^-,R} \tag{10.3}$$

Rearranging yields:

$$(E_{cell} - E_j - E_{H^+}) = -\frac{\tilde{\mu}_{e^-,L}}{F} + \frac{\tilde{\mu}_{e^-,R}}{F} \tag{10.4}$$

and

$$E_{cell} = -\frac{\tilde{\mu}_{e^-,L}}{F} + \frac{\tilde{\mu}_{e^-,R}}{F} + E_j + E_{H^+} \tag{10.5}$$

The electrochemical potential difference between the protons in the solution and inside the glass ion selective electrode is:

$$E_{H^+} = \frac{\tilde{\mu}_{H^+,sol}}{F} - \frac{\tilde{\mu}_{H^+,R}}{F} \tag{10.6}$$

The activity of a proton, denoted by parentheses ( ), is related to the electrochemical potential through the expression:

$$\tilde{\mu}_{H^+} = \tilde{\mu}^o_{H^+} + RT \ln(H^+) \tag{10.7}$$

where $\tilde{\mu}^o_{H^+}$ is the standard-state electrochemical potential of the proton. By definition, $\tilde{\mu}^o_{H^+} = 0$; therefore:

$$\tilde{\mu}_{H^+} = RT \ln(H^+) \tag{10.8}$$

Substituting Equation 10.8 into Equation 10.6 yields:

$$E_{H^+} = \frac{RT \ln(H^+)_{sol}}{F} - \frac{RT \ln(H^+)_R}{\cdot \quad F} \tag{10.9}$$

Equation 10.5 becomes:

$$E_{cell} = -\frac{\tilde{\mu}_{e^-,L}}{F} + \frac{\tilde{\mu}_{e^-,R}}{F} + E_j + \frac{RT \ln(H^+)_{sol}}{F} - \frac{RT \ln(H^+)_R}{F} \tag{10.10}$$

The electrode potential is defined by $E_{Electrode} = -\frac{\tilde{\mu}_{e^-}}{F}$. Equation 10.10 becomes:

$$E_{cell} = E_{Hg/Hg_2Cl_2} - E_{Ag/AgCl} + E_j - \frac{RT \ln(H^+)_R}{F} + \frac{RT \ln(H^+)_{sol}}{F} \tag{10.11}$$

where E_{Hg/Hg_2Cl_2} is the potential generated by the $Hg(l)–Hg_2Cl_2(s)$ redox couple and $E_{Ag/AgCl}$ is the potential generated by the $Ag(s)–AgCl(aq)$ redox couple within the respective electrodes.

The electrode potentials, E_{Hg/Hg_2Cl_2} and $E_{Ag/AgCl}$, are constant and the proton activity in the proton selective glass electrode, $(H^+)_R$, is fixed by the composition of the solution in the glass electrode. Therefore, Equation 10.11 may be written:

$$E_{cell} = C - \frac{RT \ln(10)}{F} pH_{sol}$$

(10.12)

where

$$C = E_{Hg/Hg_2Cl_2} - E_{Ag/AgCl} + E_j - \frac{RT \ln(H^+)_R}{F}$$

(10.13)

Note that in Equation 10.12, pH_{sol} refers to the negative common logarithm of $(H^+)_{sol}$, the activity of the proton in the sample solution. At 25°C, Equation 10.12 becomes:

$$E_{cell} = C - 0.05916 pH_{sol}$$

(10.14)

where E_{cell} is in volts. Calibration of the electrode system against standard buffer solutions is the mechanism used to account for any variability in C that is caused by variability in E_j (which varies with the chemical characteristics of the solutions whose pH values are being measured). However, irrespective of the behavior of E_j, electrode response is logarithmically related to the activity of the hydrogen ion in a solution.

10.1.2 SOIL SOLUTION PH

Soil pH measurements are standard for any type of soil characterization. Yet, the pH value obtained for a particular soil solution is influenced by the methodology used to isolate the soil solution (solid-to-solution ratio, soil equilibration with water or salt solution) and the placement of the electrode in a soil suspension (electrodes in supernatant liquid or in soil suspension). For this reason, soil pH values are reported with reference to the methodology employed. For example, the pH of several loessial west Tennessee soils was determined during 1996 and 1998 (Figure 10.1). Soil solution pH was determined by reacting soil samples with either 10 mM CaCl$_2$ (pH$_s$) or deionized water (pH$_w$). On average, pH$_s$ was 0.36 pH units lower than pH$_w$ when using a 1:1

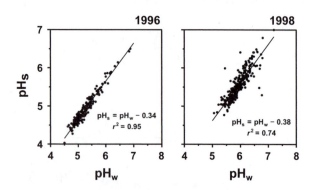

FIGURE 10.1 The relationship between the water pH and salt pH of loessial west Tennessee surface soil samples collected during 1996 and 1998. The soil series included in this comparison are the Calloway (Aquic Fraglassudalf), Grenada (Oxyaquic Fraglassudalf), Henry (Typic Fragiaqualf), Loring (Oxyaquic Fragiudalf), Memphis (Typic Hapludalf), and Routon (Typic Epiaqualf). Soil solution pH was determined by reaction soil samples with either a 10 mM CaCl$_2$ solution (pH$_s$) or deionized water (pH$_w$) using a 1:1 solid-to-solution ratio (10 g dry soil reacted with 10 mL solution).

solid-to-solution ratio (10 g soil reacted with 10 mL solution). Further, pH_s and pH_w were highly correlated, indicating that either method provides a measure of soil reactivity. In this example, the observation that pH_s is lower than pH_w may reflect the composition of the soil exchange phase. Because these soils are acidic, the native exchange complex contains the base (nonacidic) cations (Ca^{2+}, Mg^{2+}, K^+, and Na^+) and Al and Mn species (Al^{3+}, $AlOH^+$, $Al(OH)_2^+$, and Mn^{2+}). The introduction of Ca^{2+} in the 10 mM $CaCl_2$ solution results in the displacement of exchangeable Al species, as well as Mn^{2+}, Mg^{2+}, K^+, and Na^+. Once in solution, Al hydrolyzes, releasing protons and resulting in a lower pH_s relative to pH_w.

In the above example, the pH of the soil was determined by first reacting the soil with either a salt solution or with deionized water, then separating the soil solution from the solids by filtration, and finally measuring the pH of the extracts. An equally viable method is to determine the pH of the soil suspension (solution plus solids). It is in this type of system where anomalous electrode behavior may lead to erroneous pH measurements. In the electrode system, the liquid junction potential, E_j, arises because the KCl solution in the reference electrode is not in equilibrium with the sample solution. One of the reasons KCl is used in the salt bridge is that K^+ and Cl^- have approximately the same mobility in water and move through the liquid junction at approximately the same rate. It is a well-established characteristic of soil pH measurements that the pH of a clear supernatant solution sitting on top of soil suspension will differ from the pH measured in the sediment. This characteristic is termed the suspension effect.

Prior to pH determinations, the electrode system is calibrated against standard buffer solutions. This process accounts for the liquid junction potential. If the calibrated electrode system is then inserted into a soil suspension, the mobility of K^+ and Cl^- through the liquid junction may not be the same (as it was in the standard buffer solutions). If the soil has a high *CEC*, K^+ may have a greater mobility than Cl^- through the liquid junction, as the K^+ will be attracted to the negatively charged soil colloids. This results in an anomalously low pH reading (pH of suspension less than pH of supernatant). Conversely, if the soil contains an appreciable amount of hydrous Fe and Al oxides and is acidic, the mobility of Cl^- through the liquid junction may be greater than that of K^+. This results in an anomalously high pH reading (pH of suspension greater than pH of supernatant). However, even though the suspension effect is known to occur, there are standard methods in soil science that require the determination of a suspension pH (e.g., saturation paste pH).

10.2 CHEMICAL AND BIOCHEMICAL PROCESSES THAT INFLUENCE SOIL SOLUTION pH

The processes that govern the pH of a soil cannot be easily separated into strictly chemical or biochemical compartments, because soil pH is a result of the combined activities of soil organisms and abiotic soil chemical processes. These processes have been discussed in the previous chapters. An ever-present contributor to soil acidity is carbonic acid. Carbon dioxide (CO_2) in the soil atmosphere readily dissolves in the soil solution and hydrates to yield $H_2CO_3^*$ (a combination of $CO_2 \cdot H_2O$ and $H_2CO_3^0$ as described in Chapter 5). The distribution of CO_2 between the gaseous and aqueous phase is described by the Henry's Law constant:

$$k_H = \frac{(H_2CO_3^*)}{P_{CO_2}} \tag{10.15}$$

where the parentheses ( ) denote activity and P_{CO_2} is the partial pressure (or activity) of $CO_2(g)$ in the atmosphere. Carbonic acid is polyprotic, with the first deprotonation reaction characterized by K_{a1}:

$$K_{a1} = \frac{(H^+)(HCO_3^-)}{(H_2CO_3^*)} \tag{10.16}$$

and the second deprotonation reaction characterized by K_{a2}:

$$K_{a2} = \frac{(H^+)(CO_3^{2-})}{(HCO_3^-)} \tag{10.17}$$

The pH of a solution in equilibrium with a gaseous phase can be computed for any level of P_{CO_2}. This computation was performed in Chapter 5. The pH of water in equilibrium with atmospheric CO_2 ($P_{CO_2} = 0.0003$ atm) is approximately 5.70. In a soil, CO_2 can range from $P_{CO_2} = 0.01$ to 0.1 atm, with the higher levels found in the rhizosphere (the volume of soil directly impacted by plant roots) where there is a greater concentration of respiring microorganisms feeding on root exudates. If CO_2 alone controlled soil pH, the computed pH for a $P_{CO_2} = 0.01$ atm system would be 4.91, and for a $P_{CO_2} = 0.1$ atm system the pH would be 4.41.

Carbonate chemistry can also be responsible for controlling soil solution pH in alkaline soils, such that soil pH is generally never greater than approximately 8.5. In alkaline soils there is an abundance of base cations, particularly Ca (a result of limited leaching). The dominance of Ca in alkaline soils, coupled with the propensity for alkaline solutions to absorb $CO_2(g)$, results in the precipitation of $CaCO_3$ (calcite). Indeed, soils in arid regions are typically alkaline and contain calcite. The influence of calcite and P_{CO_2} on soil solution pH in the Ca-CO_2-H_2O system may be examined in a manner similar to that used to predict the influence of P_{CO_2} on the pH of standing water (Chapter 5). The dissolution of calcite:

$$CaCO_3(s) + 2H^+(aq) \rightarrow Ca^{2+}(aq) + CO_2(g) + H_2O(l) \tag{10.18}$$

is described by the solubility product constant:

$$K_{sp} = \frac{(Ca^{2+})P_{CO_2}}{(H^+)^2} \tag{10.19}$$

Calcite solubility may also be expressed as a function of concentrations in the form of a conditional equilibrium constant:

$$^cK_{sp} = \frac{[Ca^{2+}]P_{CO_2}}{[H^+]^2} \tag{10.20}$$

where the brackets [ ] denote molar concentrations. The pH of a solution in equilibrium with calcite and a gaseous phase can be computed for any level of P_{CO_2}. For a Ca-CO_2-H_2O system, a controlling equality is the charge balance expression:

$$[H^+] = [OH^-] + [HCO_3^-] + 2[CO_3^{2-}] - 2[Ca^{2+}] - [CaHCO_3^+] \tag{10.21}$$

Each term in this charge balance expression may be described as a function of conditional equilibrium constants, [H$^+$], and P_{CO_2}. The formation of the $CaHCO_3^+$ ion pair is described by:

$$Ca^{2+}(aq) + CO_2(g) + H_2O(l) \rightarrow CaHCO_3^+(aq) + H^+(aq) \tag{10.22}$$

$$^cK_f = \frac{[CaHCO_3^+][H^+]}{[Ca^{2+}]P_{CO_2}} \tag{10.23}$$

Since $[Ca^{2+}]$ is controlled by calcite dissolution, from Equation 10.19:

$$[Ca^{2+}] = \frac{{}^cK_{sp}[H^+]^2}{P_{CO_2}} \tag{10.24}$$

Substituting Equation 10.24 into Equation 10.23 yields:

$$[CaHCO_3^+] = {}^cK_f{}^cK_{sp}[H^+] \tag{10.25}$$

The concentrations of HCO_3^- and CO_3^{2-} may also be expressed as a function of conditional constants, P_{CO_2}, and $[H^+]$. Beginning with Equation 10.17, replacing activities with molar concentrations, K values with ${}^cK$ values, and rearranging:

$$[CO_3^{2-}] = \frac{{}^cK_{a2}[HCO_3^-]}{[H^+]} \tag{10.26}$$

Similarly, Equation 10.16 may be rearranged:

$$[HCO_3^-] = \frac{{}^cK_{a1}[H_2CO_3^*]}{[H^+]} \tag{10.27}$$

Finally, Equation 10.15 is solved for $[H_2CO_3^*]$:

$$[H_2CO_3^*] = {}^ck_H P_{CO_2} \tag{10.28}$$

Substituting for $[H_2CO_3^*]$ in Equation 10.27 (from Equation 10.28) leads to an expression for HCO_3^- as a function of conditional equilibrium constants, $[H^+]$, and P_{CO_2}:

$$[HCO_3^-] = \frac{{}^cK_{a1}{}^ck_H P_{CO_2}}{[H^+]} \tag{10.29}$$

Substituting Equation 10.29 in Equation 10.26 for $[HCO_3^-]$ yields an expression for $[CO_3^{2-}]$:

$$[CO_3^{2-}] = \frac{{}^cK_{a2}{}^cK_{a1}{}^ck_H P_{CO_2}}{[H^+]^2} \tag{10.30}$$

Finally, $[OH^-]$ is given by:

$$[OH^-] = \frac{{}^cK_w}{[H^+]} \tag{10.31}$$

Substituting Equation 10.24 for $[Ca^{2+}]$, Equation 10.25 for $[CaHCO_3^+]$, Equation 10.29 for $[HCO_3^-]$, Equation 10.30 for $[CO_3^{2-}]$, and Equation 10.31 for $[OH^-]$ into the charge balance expression (Equation 10.21) yields:

$$[H^+] = \frac{{}^cK_w}{[H^+]} + \frac{{}^cK_{a1}{}^ck_H P_{CO_2}}{[H^+]} + \frac{2{}^cK_{a2}{}^cK_{a1}{}^ck_H P_{CO_2}}{[H^+]^2} - \frac{2{}^cK_{sp}[H^+]^2}{P_{CO_2}} - {}^cK_f{}^cK_{sp}[H^+] \tag{10.32}$$

If the solution is dilute in Ca and CO_3, which is a valid assumption in this example (i.e., $[Ca^{2+}] = 10^{-5.26}$ and $[CO_3^{2-}] = 10^{-3.145}$ using Equations 10.24 and 10.30 with pH = 8 and $P_{CO_2} = 0.1$ atm), the true equilibrium constants may be substituted for the conditional constants. The equilibrium constants are: $K_w = 10^{-14}$, $k_H = 10^{-1.464}$, $K_{a1} = 10^{-6.352}$, $K_{a2} = 10^{-10.329}$, $K_{sp} = 10^{9.74}$, and $K_f = 10^{-6.70}$. Substituting these values for the conditional constants in Equation 10.32 yields:

$$[H^+] = \frac{10^{-14}}{[H^+]} + \frac{10^{-7.816} P_{CO_2}}{[H^+]} + \frac{2 \times 10^{-18.145} P_{CO_2}}{[H^+]^2} - \frac{2 \times 10^{9.74}[H^+]^2}{P_{CO_2}} - 10^{3.04}[H^+] \quad (10.33)$$

Rearranging leads to a fourth-order polynomial:

$$0 = -\frac{2 \times 10^{9.74}[H^+]^4}{P_{CO_2}} - 10^{3.04}[H^+]^3 + (10^{-14} + 10^{-7.816} P_{CO_2})[H^+] + 2 \times 10^{-18.145} P_{CO_2} \quad (10.34)$$

An exact solution to Equation 10.34 with atmospheric CO_2 levels, $P_{CO_2} = 10^{-3.52}$ atm, results in a computed pH of 8.30 (compared to a pH of 5.70 in the absence of calcite). Higher partial pressures of CO_2 lead to lower predicted pH values. When $P_{CO_2} = 0.01$ atm, the computed pH is 7.29. If the solution phase is in equilibrium with a CO_2 level representative of the rhizosphere, e.g., $P_{CO_2} = 0.1$ atm, the computed pH would be in the acidic range and equal to 6.62. Although in general, the presence of calcite in a system will support an alkaline soil solution, greatly elevated P_{CO_2} can lead to acidic soil solutions.

While the production of carbonic acid through the microbial oxidation of soil organic matter and the absorption of $CO_2(g)$ from the soil atmosphere could result in acidic soil solutions if $H_2CO_3^*$ were the only component controlling pH, numerous other acid–base reactions also affect soil solution pH. Both mineral weathering and the acid–base reactions of organic functional groups serve to buffer the pH of soils. These reactions have been discussed in previous chapters (specifically in Chapters 3, 4, and 5) and will briefly be reviewed here. Soil organic matter contains numerous functional groups that are weak Lowry-Brønsted acids (e.g., carboxyl, phenolic, and amino). Each proton-selective organic functional group is characterized by a specific pK_a value that is a function of the configuration of the molecular framework to which the functional group is attached. Protons released through metal hydrolysis reactions, or other natural or anthropogenic sources, can be consumed through the protonation of proton-selective organic functional groups (e.g., $R-COO^- + H^+ \rightarrow R-COOH^0$ and $R-NH_2^0 + H^+ \rightarrow R-NH_3^+$). Conversely, increasing the basicity of a soil solution can result in the dissociation of the functional groups (e.g., $R-COOH^0 \rightarrow R-COO^- + H^+$ and $R-NH_3^+ \rightarrow R-NH_2^0 + H^+$).

In general, primary mineral weathering consumes protons while secondary mineral precipitation releases protons. The ability of the soil mineral phase to consume or release protons may be qualitatively expressed by examining the changes in the acid-neutralization capacity (ANC) of the solid phase. When mineral dissolution consumes protons, as is typical of hydrolysis weathering reactions, the ANC of the mineral phase is reduced; minerals that can neutralize protons are lost. For example, the dissolution of a K-feldspar reduces the ANC of the soil solids by consuming protons at the expense of the feldspar:

$$KAlSi_3O_8(s) + 4H_2O(l) + 4H^+(aq) \rightarrow Al^{3+}(aq) + K^+(aq) + 3H_4SiO_4^0(aq) \quad (10.35)$$

Conversely, the precipitation of a metal hydrous oxide, such as gibbsite, produces protons and increases the ANC of the soil solids by creating a mineral that can neutralize protons:

$$Al^{3+}(aq) + 3H_2O(l) \rightarrow Al(OH)_3(s) + 3H^+(aq) \quad (10.36)$$

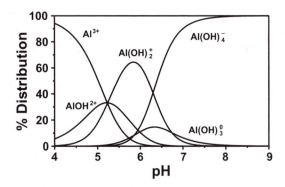

FIGURE 10.2 Distribution diagram illustrating Al speciation as a function of solution pH at 25°C. The percent distribution is the concentration of a specified species divided by the total concentration of the Al in the solution. The diagram was constructed using a total Al concentration of 10^{-5} M in a 10^{-2} M $NaClO_4$ background electrolyte. The pK_a values employed to construct the diagrams are compiled in Chapter 5.

The most detrimental natural chemical reactions, with respect to the genesis of acid soils, are those that involve the dissolution of Al-bearing minerals (Equation 10.35). It is these reactions, coupled with the physical movement of water through a soil, that lead to acidic soil environments. An Al-bearing primary mineral, such as anorthite, dissolves according to the reaction:

$$CaAl_2Si_2O_8(s) + 8H^+(aq) \rightarrow Ca^{2+}(aq) + 2Al^{3+}(aq) + 2H_4SiO_4^0(aq) \tag{10.37}$$

Again, loss of anorthite through the consumption of protons reduces the *ANC* of the soil mineral phase. The released Al^{3+} hydrolyzes, the extent to which depends on the soil solution pH (Figure 10.2):

$$Al^{3+}(aq) + H_2O(l) \rightarrow AlOH^{2+}(aq) + H^+(aq) \tag{10.38}$$

$$Al^{3+}(aq) + 2H_2O(l) \rightarrow Al(OH)_2^+(aq) + 2H^+(aq) \tag{10.39}$$

$$Al^{3+}(aq) + 3H_2O(l) \rightarrow Al(OH)_3^0(aq) + 3H^+(aq) \tag{10.40}$$

$$Al^{3+}(aq) + 4H_2O(l) \rightarrow Al(OH)_4^-(aq) + 4H^+(aq) \tag{10.41}$$

In slightly acidic to highly acidic solutions, Al^{3+} and $Al(OH)_2^+$ are dominant cationic Al species (Figure 10.2). Further, the solubility of hydrous Al oxides increases with decreasing pH (for every unit pH decrease the activity of Al^{3+} controlled by gibbsite increases by 3 orders of magnitude, Equation 10.36), allowing for high concentrations of soluble Al^{3+} and $AlOH^{2+}$. The soil exchanger phase is highly selective for free Al^{3+}, as well as the Al hydrolysis products, principally $Al(OH)_2^+$. These Al species readily displace the exchangeable base cations (nonacidic cations), reducing the *ANC* of the exchanger phase:

$$\{Ca^{2+}, Mg^{2+}, K^+, Na^+\}X(s) + \{Al^{3+}, AlOH^{2+}, Al(OH)_2^+\}(aq) \rightarrow \{Al^{3+}, AlOH^{2+}, Al(OH)_2^+\}X(s)$$

$$+ \{Ca^{2+}, Mg^{2+}, K^+, Na^+\}(aq) \tag{10.42}$$

If the displaced base cations are not leached, the *ANC* of the soil solution will increase, as the acidic Al species are consumed, leaving the nonacidic cations. However, as the flow of water through the soil system preferentially mobilizes the released base cations and soluble silica, relative to the adsorbed Al species, the *ANC* of the whole soil will decrease and soil acidity will increase. A similar exchange process occurs at organic functional groups, which may result in the direct release of protons:

$$R-COOH^0 + Al(OH)_2^+ \rightarrow R-COO^- -Al(OH)_2^+ + H^+ \tag{10.43}$$

or the release of base cations:

$$R-[COO]_2^{2-} -Ca^{2+} + 2Al(OH)_2^+ \rightarrow 2R-COOAl(OH)_2^0 + Ca^{2+} \tag{10.44}$$

Similar to the exchange behavior of Al species on the mineral exchange complex (Equation 10.42), the organic-bound Al is not easily displaced by a base cation (and may actually be chelated by the organic moieties), resulting in the preferential mobilization and loss of base cations.

Most reduction reactions consume protons. Conversely, most oxidation reactions produce protons. However, the impact of microbially catalyzed redox reactions on soil acidity cannot be fully appreciated by viewing only reduction or oxidation half-reactions. Instead, the full redox reaction must be examined, as well as the fate of the oxidized and reduced species in the soil (holistic approach). The oxidation of naturally occurring or anthropogenic ammonium can potentially produce soil acidity through the nitrification process:

$$NH_4^+(aq) + 2O_2(g) \rightarrow NO_3^-(aq) + 2H^+(aq) + H_2O(l) \tag{10.45}$$

However, the nitrification process will only result in soil acidification if NO_3^- and base cations are lost from the soil by leaching. In this case, protons produced by the nitrification process will remain in the soil, occupying organic exchange sites and participating in mineral weathering reactions. The Al solubilized during mineral weathering will hydrolyze (producing additional acidity) and displace base cations from mineral and organic exchange sites (reducing *ANC* of the solution and exchange phase). If NO_3^- does not leach, but instead is absorbed by plant roots, there may be no net acidification, since the plant uptake of an anionic nutrient generates alkalinity (OH^- and HCO_3^- are exuded):

{equivalents of anion plant uptake} – {equivalents of cation plant uptake}

$$= \{equivalents\ OH^-\ and\ HCO_3^-\ exuded\} \tag{10.46}$$

Conversely, cation uptake by plant roots generates acidity:

{equivalents of cation plant uptake} – {equivalents of anion plant uptake}

$$= \{equivalents\ H^+\ exuded\} \tag{10.47}$$

Since cation uptake by plant roots generally exceeds anion uptake, the exudation of protons generally exceeds the production of alkalinity by the roots. In addition, plant roots exude low-molecular-mass organic acids that are transient in the soil and easily oxidized to CO_2 by soil microbes:

$$C_6H_{12}O_6(aq) + 6O_2(g) \rightarrow 6CO_2(g) + 6H_2O(l) \tag{10.48}$$

The dissolved CO_2 hydrates to carbonic acid ($H_2CO_3^*$), which dissociates (Equations 10.16 and 10.17) to generate protons. Due to the processes described in Equations 10.47 and 10.48, the pH of the rhizosphere soil (the volume of soil directly influenced by plant roots) can be up to 2 pH units lower than that of the bulk soil.

10.3 ACID-NEUTRALIZING CAPACITY AND THE QUANTIFICATION OF SOIL ACIDITY

The soil solution, minerals, and exchange complex all have the capacity to neutralize soil acidity. Soil minerals neutralize soil acidity by consuming protons during chemical weathering reactions (Equations 10.35 and 10.37). However, relative to acid-base reactions in the soil solution and on the soil exchange complex, mineral weathering reactions are kinetically slow and are generally not considered when describing or managing soil acidity. The reactions of the soil exchange complex with solution components are rapid (not kinetically restricted) and generally reversible (Chapter 8). Thus, the exchange complex composition responds rapidly to changes in the soil solution composition and can be manipulated. The exchange complex is also characterized by a clearly defined *ANC*.

The *ANC* of the soil exchange phase is the moles of surface charge per unit mass of soil that are balanced by the readily exchangeable bases (metals that do not hydrolyze in the normal pH range of soils): Na^+, K^+, Mg^{2+}, and Ca^{2+}. The *ANC* of the soil exchange phase is also known as the base saturation. The definition of the exchange phase *ANC* leads directly to the definition of the total acidity (*TA*) of a soil. The potential (maximum) *TA* of a soil is the potential *CEC* of the soil (e.g., *CEC* determined using a pH-buffered CEC method, described in Chapter 8) minus the *ANC*:

$$TA = CEC - ANC \qquad (10.49)$$

The actual *TA* of a soil is defined as the moles of titratable protons per unit mass of soil that are displaced by an unbuffered, 1.0 *M* KCl solution. The actual *TA* is also termed *salt-replaceable acidity* or *exchangeable acidity*, and principally represents the Al-species (Al^{3+} and $Al(OH)_2^+$) retained at clay exchange sites, and the Al-species and protons that are complexed by organic functional groups. The variation of actual *TA*, relative to the potential *CEC*, as a function of solution pH for 572 surface soil samples collected from around the world and from all soil orders is illustrated in Figure 10.3a. In general, the *TA* of soils increases precipitously as soil solution pH decreases below a value of 6.5 to 6. Other defined forms of soil acidity include: *reserve acidity*, all titratable acidity in the soil that is associated with the solid phase; *nonexchangeable acidity*, surface-bound Al-species and protons that are not displaced, or are only slowly displaced, by an unbuffered 1.0 *M* KCl solution (reserve acidity minus exchangeable acidity); and *active acidity*, all titratable acidity in the soil solution (reserve acidity minus exchangeable acidity).

The *ANC* of the soil solution is defined as the moles of protons per unit volume (or mass) of solution required to change the pH of the soil solution to the pH at which the net charge from ions that do not react with OH^- or H^+ is zero. In this case, ions that do not react with OH^- or H^+ include those that do not protonate (Cl^-, SO_4^{2-}, NO_3^-) or hydrolyze (Na^+, K^+, Mg^{2+}, Ca^{2+}) in the normal pH range of acidic soil solutions ($3.5 < pH < 7.0$). In equation form:

$$ANC = [Na^+] + [K^+] + 2[Ca^{2+}] + 2[Mg^{2+}] - [Cl^-] - [NO_3^-] - 2[SO_4^{2-}] \qquad (10.50)$$

where brackets [ ] denote molar concentrations. For a soil solution in equilibrium with the soil atmosphere, dissolved CO_2 will be present, and the charge balance for the solution is:

$$0 = [Na^+] + [K^+] + 2[Ca^{2+}] + 2[Mg^{2+}] - [Cl^-] - [NO_3^-] - 2[SO_4^{2-}]$$
$$- [OH^-] - [HCO_3^-] - 2[CO_3^{2-}] + [H^+] \qquad (10.51)$$

Combining the charge balance expression (Equation 10.51) and the *ANC* expression (Equation 10.50) yields:

$$ANC = [OH^-] + [HCO_3^-] + 2[CO_3^{2-}] - [H^+] \qquad (10.52)$$

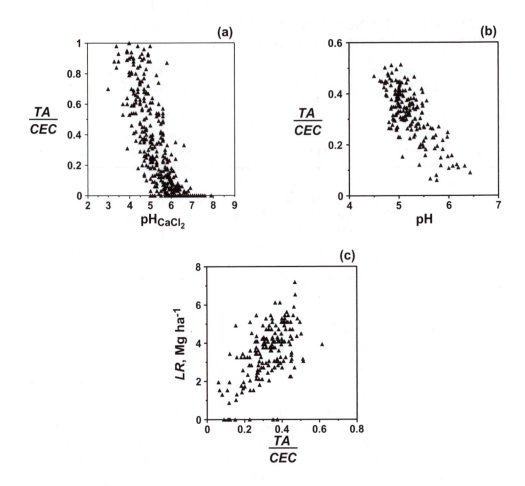

FIGURE 10.3 (a) The variation of total acidity (*TA*) relative to potential *CEC* as a function of solution pH for 572 surface soil samples collected from around the world and from all soil orders (data obtained from Batjes, 1995). (b) The variation of *TA* relative to potential *CEC* as a function of solution pH for soil samples collected from a Milan, TN cotton field. (c) The lime requirement (*LR*), determined by the Adams-Evans buffer test, as a function of the *TA* for loess-derived west Tennessee surface soils.

Equation 10.52 is defined as the alkalinity of a solution. This expression is also the reference system for evaluating the impact of solutes on the *ANC* of a solution. For example, when Al is released to the soil solution through mineral weathering reactions, the *ANC* of the solution is:

$$ANC = [OH^-] + [HCO_3^-] + 2[CO_3^{2-}] - [H^+] - 3[Al^{3+}] - 2[AlOH^{2+}] - [Al(OH)_2^+] \quad (10.53)$$

This equation illustrates that the capacity of the soil solution to neutralize acidity is reduced through the solubilization of Al-bearing minerals.

10.4 NEUTRALIZATION OF SOIL ACIDITY

A soil becomes acidic because the processes that neutralize soil acidity (e.g., mineral weathering, anion uptake by plants, anion or surface functional group protonation, base cation retention on the exchange complex, reduction reactions) neutralize less H^+ than is produced by the acid-producing processes. Recall that by strict definition an acid soil is one in which the pH of the soil solution is less than 7.0. The decision to mitigate soil acidity, however, is not based upon soil pH. Instead, acid soil management practices are based on *TA*. When the *TA* exceeds approximately 15% of the *CEC*, a variety of problems that inhibit plant growth arise, including Al and Mn toxicities and Ca, Mg, and Mo deficiencies. When this situation occurs, soil amendments are required to decrease *TA*.

By definition, the *TA* of a soil is that portion of the *CEC* that is not occupied by exchangeable base cations. Thus, in order to reduce *TA* to an acceptable level, base cations must displace acidic species from the exchange complex. The moles of base cation (most commonly Ca^{2+} or Mg^{2+}) charge per kilogram of soil required to decrease *TA* to a value deemed acceptable for the desired use of the soil defines the lime requirement of the soil. Note that the lime requirement is not based on the pH of the soil, but on the amount of base cation that must be added to a soil to decrease *TA*. Typically, the *TA* value that is acceptable is *TA* = 0. As illustrated in Figure 10.3b for soil samples collected from a Milan, TN cotton field, when *TA* = 0 the soil pH will be approximately 6 to 6.5 (a similar conclusion was noted for world soils from Figure 10.3a).

In practice, the lime requirement of a soil is determined by rapidly equilibrating a buffer solution with a soil sample. In this method, a buffer solution is equilibrated for a 15- to 40-min period with a soil sample, and the pH of the soil and buffer mixture determined. The buffer solution is a mixture of a weak acid and a salt of the weak acid anion, whose pH has been adjusted to a test-specific level. A buffer resists drastic changes in pH by neutralizing both acids and bases produced in the soil. The decrease in buffer pH during the rapid soil equilibration, coupled with the soil pH increase desired in the field, are measures of the soil acidity that must be neutralized by liming. Several buffer tests are used in the U.S, including the Shoemaker-McLean-Pratt (SMP), Mehlich, and Adams-Evans buffer methods. The SMP buffer is a pH 7.5 solution containing *p*-nitrophenol, K_2CrO_4, $CaCl_2$, Ca-acetate, and triethanolamine; whereas, the Mehlich buffer is a pH 6.6 solution containing acetic acid, triethanolamine, NH_4Cl, $BaCl_2$, and sodium glycerophosphate. Detailed descriptions of the SMP and Mehlich buffer tests are given by Sims (1996).

The Adams-Evans buffer solution consists of 1 *M* KCl, 0.2 *M* KOH, 0.15 *M* *p*-nitrophenol, and 0.25 *M* boric acid, adjusted to pH 8.0 with KOH or HCl. In the procedure, a 10-g (dry weight) soil sample is equilibrated with 10 mL of deionized water for 10 min and the pH is then determined (water pH). A 10-mL volume of the Adams-Evans buffer is then added to the soil-water suspension and equilibrated for an additional 40 min and the pH is again determined (buffer pH). The lime requirement can be computed using regression equations, which are generally state or region specific and based on a database developed by the respective state agricultural experiment stations. Such regression equations estimate the acidity of a soil at the water pH and at a producer-defined target pH, and predict the tons of ground agricultural limestone per acre required to achieve the target soil pH. Lime recommendations are typically presented in tabular form. Portions of the Tennessee lime recommendation tables are presented in Table 10.1. The lime requirement, as a function of the *TA* for some loess-derived west Tennessee soils is shown in Figure 10.3c. This diagram illustrates the close relationship between *TA* and lime requirement.

One should recognize, however, that the lime requirement determined by a rapid equilibration technique, such as the Adams-Evans buffer test only offers a gross estimation of the actual lime requirement of a soil. Indeed, buffer test results themselves do not indicate the lime requirement of a soil. Instead, the soil water pH and buffer pH results are only correlated to soil pH after adding lime and for the production of a specific crop. Techniques that involve longer equilibration times for reacting soil with a liming material (anywhere from days to months) provide a better estimation of the lime requirement. However, the price for improved accuracy is the loss of time. Further, and irrespective of the lime test performed, there may be chemical conditions in the soil that bring the lime test results and the resulting recommendations into question. For example, high concentrations of soil organic carbon (and elevated concentrations of dissolved soil organic carbon) minimize the phytotoxic effects of Al and Mn in acidic soils, such that a lime test overestimates the lime

TABLE 10.1
Lime Recommendations, Expressed in t A^{-1} (Mg ha^{-1} in Parentheses), from The University of Tennessee Agricultural Extension Service (Lime is Defined as Ground Agricultural Limestone That Is Greater Than 75% CaCO$_3$ Equivalents)

Soil pH in Water	Soil pH in Adams-Evans Buffer Solution				
	7.8	7.6	7.4	7.2	7.0
	Target pH in Soil Water = 6.0[a]				
5.4	1.0 (2.2)	2.0 (4.5)	2.0 (4.5)	2.0 (4.5)	2.5 (5.6)
5.2	1.5 (3.4)	2.0 (4.5)	2.0 (4.5)	2.5 (5.6)	3.0 (6.7)
5.0	1.5 (3.4)	2.0 (4.5)	2.0 (4.5)	3.0 (6.7)	3.5 (7.8)
4.8	1.5 (3.4)	2.0 (4.5)	2.5 (5.6)	3.0 (6.7)	4.0 (9.0)
4.6	2.0 (4.5)	2.0 (4.5)	2.5 (5.6)	3.5 (7.8)	4.0 (9.0)
	Target pH in Soil Water = 6.5[b]				
6.0	1.0 (2.2)	2.0 (4.5)	2.0 (4.5)	2.0 (4.5)	2.5 (5.6)
5.8	1.5 (3.4)	2.0 (4.5)	2.0 (4.5)	2.5 (5.6)	3.0 (6.7)
5.6	1.5 (3.4)	2.0 (4.5)	2.5 (5.6)	3.0 (6.7)	3.5 (7.8)
5.4	1.5 (3.4)	2.0 (4.5)	2.5 (5.6)	3.0 (6.7)	4.0 (9.0)
5.2	2.0 (4.5)	2.0 (4.5)	2.5 (5.6)	3.5 (7.8)	4.0 (9.0)
	Target pH in Soil Water = 7.0[c]				
6.4	1.5 (3.4)	2.0 (4.5)	2.5 (5.6)	3.0 (6.7)	3.5 (7.8)
6.2	1.5 (3.4)	2.0 (4.5)	2.5 (5.6)	3.0 (6.7)	4.0 (9.0)
6.0	2.0 (4.5)	2.0 (4.5)	3.0 (6.7)	3.5 (7.8)	4.0 (9.0)
5.8	2.0 (4.5)	2.0 (4.5)	3.0 (6.7)	4.0 (9.0)	4.5 (10.1)
5.6	2.0 (4.5)	2.5 (5.6)	3.0 (6.7)	4.0 (9.0)	5.0 (11.2)

[a] Crops include: beans, brambles, peppers, sweet potatoes, tobacco (dark), and strawberries.
[b] Crops include: bermudagrass, broccoli, cauliflower, corn, cotton, pasture-hay-silage, sorghum, soybeans, tobacco (burly), and tomatoes.
[c] Crops include: alfalfa, apples, pears, and peaches.

recommendations. Under such conditions, crops can tolerate higher *TA* because Al^{3+} (and associated hydrolysis products) and Mn^{2+} are complexed by the organic component.

Limestone [CaCO$_3$] or dolomitic limestone [CaMg(CO$_3$)$_2$] are the materials commonly used to neutralize soil acidity; their use is based on economics (easily obtained and inexpensive) and they provide a mild alkalizing effect on the soil solution pH. Since the purpose of liming is to reduce the total acidity of the soil to a value deemed acceptable for maximizing crop production, any base cation source should suffice. Indeed, while the carbonate supplied by CaCO$_3$ does consume protons [CO$_3^{2-}$(aq) + 2H$^+$(aq) → CO$_2$(g) + H$_2$O(l)], the overall soil acidity neutralization process conserves protons when CaCO$_3$ is employed. In addition to the displacement of Al species from the exchange phase (by base cations), an equally important objective of liming is to precipitate the displaced Al. The neutralization of an acid soil using CaCO$_3$ is illustrated in the following series of reactions:

$$3CaCO_3(s) + 6H^+(aq) \rightarrow 3Ca^{2+}(aq) + 3CO_2(g) + 3H_2O(l) \qquad (10.54)$$

This initial reaction results in a rapid increase in soil solution pH due to neutralization of active acidity (or due to the inefficiency of soil mixing). The increased pH and elevated Ca^{2+} levels also favor the retention of the Ca^{2+} on the exchange complex, limiting base mobility:

$$3Ca^{2+}(aq) + 2AlX_3(ex) \rightarrow 3CaX_2(ex) + 2Al^{3+}(aq) \qquad (10.55)$$

The elevated pH also promotes the precipitation of Al:

$$2Al^{3+}(aq) + 3H_2O(l) \rightarrow 2Al(OH)_3(s) + 6H^+(aq) \tag{10.56}$$

The summation of Equations 10.54 through 10.56 yields the overall soil acidity neutralization reaction:

$$3CaCO_3(s) + 2AlX_3(ex) + 3H_2O(aq) \rightarrow 3CaX_2(ex) + 2Al(OH)_3(s) + 3CO_2(g) \tag{10.57}$$

Note that protons consumed during the initial dissolution of $CaCO_3$ (Equation 10.54) are released during the precipitation of gibbsite ($Al(OH)_3$, Equation 10.56) and there is no net consumption of H^+.

Neutral salts, such as gypsum ($CaSO_4 \cdot 2H_2O$) and $CaCl_2$ are not considered liming materials because they do not neutralize the active acidity of the soil to create an environment conducive to the precipitation of an Al solid. Theoretically, it would appear that gypsum could be employed as a liming agent. The series of reactions that leads to the reduction in soil acidity begins with the dissolution of gypsum, followed by cation exchange and the precipitation of an Al solid:

$$3CaSO_4 \cdot 2H_2O(s) \rightarrow 3Ca^{2+}(aq) + 3SO_4^{2-}(aq) + 6H_2O(l) \tag{10.58}$$

$$3Ca^{2+}(aq) + 2AlX_3(ex) \rightarrow 3CaX_2(ex) + 2Al^{3+}(aq) \tag{10.59}$$

$$2Al^{3+}(aq) + 2SO_4^{2-}(aq) + 2H_2O(l) \rightarrow 2AlOHSO_4(s) + 2H^+(aq) \tag{10.60}$$

The increase in soluble Ca^{2+} hastens the displacement of Al^{3+}, and associated hydrolysis products from the soil exchange phase, resulting in a reduction in *TA*. The Al species that are released must precipitate to reduce the *TA*. In this case, the basic Al sulfate (Al-jurbanite) is formed. Again, the desired objective, the reduction or elimination of *TA* is met, even though the overall process results in the production of protons (active acidity, but not total acidity):

$$3CaSO_4 \cdot 2H_2O(s) + 2AlX_3(ex) \rightarrow 3CaX_2(ex) + 2AlOHSO_4(s)$$

$$+ SO_4^{2-}(aq) + 4H_2O(l) + 2H^+(aq) \tag{10.61}$$

Again, the objective association with the application of a liming agent is to decrease the *TA* of the soil to an acceptable level and to precipitate Al. Although protons are generated in Equation 10.61, the desired objectives appear to be met. However, there are two problems with the process described in Equation 10.61 and with the use of neutral Ca salts in general. First, the basic Al sulfate minerals are only stable in environments that contain very high concentrations (activities) of SO_4^{2-}. These minerals are common to the acid mine spoil systems that contain high soluble Al and sulfate concentrations (described in the next section). Thus, Equation 10.61 will not occur in soils that are commonly subjected to liming (e.g., crop production). Second, soluble Al species are present on the exchange complex because soil solution pH values are low enough to drive aluminosilicate dissolution reactions and prevent the precipitation of gibbsite and other Al minerals. If protons are not consumed during the initial dissolution of a liming agent, solution pH values will not be adjusted to levels necessary for gibbsite precipitation. The net effect is that *TA* will not be reduced, as Al^{3+} and its associated hydrolysis products will remain soluble and continue to effectively compete with Ca^{2+} (and Mg^{2+}) for sites on the exchange complex.

TABLE 10.2
Total Soluble Concentrations of Inorganic Substances in an Aqueous Extract of a Pyritic Mine Spoil[a]

Substance	mg L^{-1}	Substance	mg L^{-1}
Al	201	Mn	66.1
B	<0.01	Ni	3.02
Ca	447	NO$_3$	3.63
Cl	0.53	P	0.01
Cu	0.82	SO$_4$	4783
Fe	37.6	Zn	8.56
F	11.9		
K	0.32	pH	2.90 (units)
Mg	178	EC	3.96 (dS m^{-1})

[a] The spoil was obtained from an abandoned strip mine site in the Cumberland Mountains of east Tennessee. The extract was obtained by reacting the spoil with water for 2 h at a 1:1 solid-to-solution ratio.

10.5 ACID GENERATION AND MANAGEMENT IN MINE SPOILS: THE OXIDATION OF PYRITE

Mining activities, including those conducted in the Appalachian region of the U.S., can expose geologic materials that contain pyrite (FeS$_2$) to surface water and O$_2$. As a result, acid drainage (or acid mine drainage) is produced. Unlike an acid soil environment, the drainage from pyritic materials is extremely acidic (pH as low as 2) and contains very high concentrations of Fe, Mn, Al, SO$_4$, and trace elements (e.g., Cu, Ni, and Zn) (Table 10.2). The off-site movement of this acid drainage can diminish surface and groundwater quality. Further, the acidity of pyritic materials, coupled with the phytotoxic levels of numerous elements and the deficiency of several plant nutrients, can hinder efforts to revegetate disturbed mined lands. As a result, abandoned strip mine land can remain devoid of vegetation and highly acidic for decades without the implementation of reclamation efforts (Figure 10.4a).

10.5.1 ACID GENERATION AND NEUTRALIZATION IN PYRITIC WASTES

Acid drainage is produced through the oxidation of pyrite, which occurs in the presence of O$_2$ and Fe^{3+}, and in both the presence and absence of sulfur-oxidizing bacteria (which serve to accelerate oxidation rates). Several chemical reactions result in pyrite oxidation and acid production (Figure 10.5). The initial oxidation of FeS$_2$ by O$_2$ produces Fe^{2+}, SO$_4^{2-}$, and protons:

$$FeS_2(s) + \tfrac{7}{2}O_2(g) + H_2O(l) \rightarrow Fe^{2+}(aq) + 2SO_4^{2-}(aq) + 2H^+(aq) \quad (10.62)$$

This reaction can occur in the absence of microbial activity, but can also be mediated by the elemental sulfur and sulfide-oxidizing bacterium, *Thiobacillus thiooxidans*, has been isolated from acidic mine wastes. *Thiobacillus ferrooxidans*, also present in acidic mine wastes, is able to catalyze the oxidation of Fe^{2+}, elemental sulfur, and sulfides:

$$Fe^{2+}(aq) + \tfrac{1}{4}O_2(g) + H^+(aq) \rightarrow Fe^{3+}(aq) + \tfrac{1}{2}H_2O(l) \quad (10.63)$$

Although this reaction does not produce acidity (it actually consumes protons), it is an important step in the overall production of acidic spoil leachates. The Fe^{3+} produced by *T. ferrooxidans* can

FIGURE 10.4 An abandoned coal mine located on a contour-cut mountain top in the Western Appalachian Plateau/Cumberland Mountains physiographic region of northeastern Tennessee, near Carryville in Campbell County. The disturbed, 1.6-ha site resulted from the strip-mining of the Cold Gap Coal Bed of the Cross Mountain Formation in the Middle Pennsylvanian. (a) Abandoned around 1968, the site remained devoid of vegetation and highly acidic into the early 1990s. (b) Approximately 1 year after municipal sewage sludge application and overseeding with a quail forage mixture, the treated area supports a vegetative cover dominated by tall fescue. (c) Seven years after the initial sewage sludge treatment the mine spoil site supports a lush and diverse stand of vegetation.

also oxidize pyritic-S:

$$FeS_2(s) + 14Fe^{3+}(aq) + 8H_2O(l) \rightarrow 15Fe^{2+}(aq) + 2SO_4^{2-}(aq) + 16H^+(aq) \qquad (10.64)$$

Indeed, in highly acidic systems (pH < 3 to 4), Fe^{3+} is a more effective oxidizer of pyritic-S than O_2, while O_2 is the preferred oxidizer of Fe^{2+}. Further, the oxidation of pyritic-S by Fe^{3+} is more rapid than the oxidation of Fe^{2+} by O_2, thus, the rate-limiting step in pyrite oxidation is the

(c)

FIGURE 10.4 (*continued*).

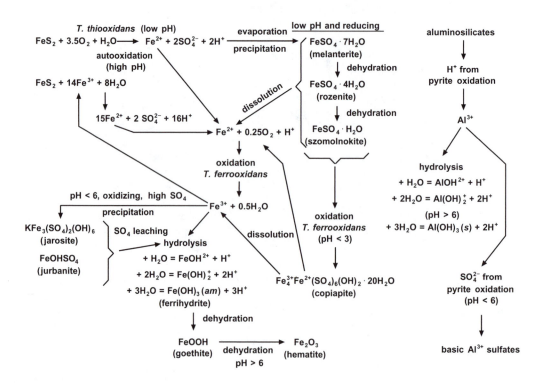

FIGURE 10.5 Iron and aluminum chemistry and the processes that control the generation of acidity in pyritic mine spoils (modified from Nordstrom [1982] and Sullivan and Essington [1987]).

oxidation of Fe^{2+} by O_2. The oxidation of Fe^{2+} by O_2 is hastened significantly by *T. ferrooxidans*, which may increase the rate of pyrite oxidation by 60 to 95%. The precipitation and dissolution of Fe(III) minerals, such as the basic sulfates (jarosite, jurbanite, and copiapite) control the flow of Fe^{3+} to and from solution and serve to regulate pyrite oxidation. In addition, the hydrolysis reactions of Fe^{3+} (and Al^{3+}) generate protons and serve to buffer the low pH of spoil leachates.

As a result of pyritic-S oxidation, spoil leachates also contain very high concentrations of Fe^{2+} and SO_4^{2-}, a condition that favors the formation of hydrous Fe(II) sulfates. These minerals and the Fe(II) carbonate, siderite ($FeCO_3$), also serve to regulate pyritic-S oxidation by controlling the availability of Fe^{2+} and production of Fe^{3+}.

The management of pyritic materials and the remediation of acid drainage require an ability to predict the extent of acid production. The potential for a geologic material to produce acidity is dependent on the amount of pyrite present and the amount of acid-neutralizing (alkaline) minerals present. A material will be acid-producing if the pyrite content exceeds the neutralization potential of the material. Conversely, if the alkalinity present in a material balances or exceeds the potential acidity from pyrite, the drainage will not be acidic. The method for characterizing geologic materials with respect to potential acidity and basicity is termed acid–base accounting. The acid–base account (*ABA*) of a material can be positive, negative, or zero, and is computed using the expression:

$$ABA = NP - AP \qquad (10.65)$$

where *NP* is the neutralization potential and *AP* is acid-producing potential. Typically, *ABA* is expressed in t $CaCO_3$ equivalent per 1000 t material, but may also be expressed in any other parts per thousand units, such as kg Mg^{-1}. A negative *ABA* indicates that a material does not contain sufficient neutralization power to counter the potential acid production from the oxidation of pyrite. Thus, the material may potentially produce acidic leachates. The *AP* of a material is determined by measuring the pyritic-S content of the material. Typically, this is accomplished by determining total-S following the removal of SO_4-S and organic-S. The complete oxidation and neutralization of the acidity from each mole of pyrite requires 2 mol of $CaCO_3$:

$$FeS_2(s) + 2CaCO_3(s) + \tfrac{15}{4}O_2(g) + \tfrac{1}{2}H_2O(l) \rightarrow FeOOH(s) + 2CO_2(g)$$
$$+ Ca^{2+}(aq) + 2SO_4^{2-}(aq) \qquad (10.66)$$

Since the molecular weight of S is 32 g mol^{-1}, and that of $CaCO_3$ is 100 g mol^{-1}, the acidity generated through the oxidation of each gram of pyritic-S will require 3.125 g of $CaCO_3$ (100 g $CaCO_3$ mol^{-1} divided by 32 g S mol^{-1}). Therefore, a material containing 1 g kg^{-1} pyritic-S will have an *AP* of 3.125 t $CaCO_3$ equiv./1000 t (or 3.125 kg $CaCO_3$ equiv. Mg^{-1}).

The *NP* of a material is the total amount of alkalinity present. The *NP* is determined by digestion (boiling for 1 min) of the material in standardized 0.1 *M* HCl. The quantity of HCl that is consumed is a direct measure of *NP*. Since each mole of $CaCO_3$ will consume 2 mol of H^+, a material that consumes 1 mol of protons per kg (or 1 g protons kg^{-1}) will have a *NP* of 50 t $CaCO_3$ equiv./1000 t (or 50 kg $CaCO_3$ equiv. Mg^{-1}). Consider the *ABA* of New Albany shale. The shale contains 42 kg Mg^{-1} (4.2%) pyritic-S. The *AP* is 131.3 kg $CaCO_3$ equiv. Mg^{-1} (42 kg S Mg^{-1} × 3.125). This shale consumes 9.8 cmol of protons kg^{-1}. The *NP* is 4.9 kg $CaCO_3$ equiv. Mg^{-1} (0.098 kg Mg^{-1} × 50). The *ABA* of this shale is: 4.9 − 131.3 = −126.4 kg $CaCO_3$ equiv. Mg^{-1} (or −126.4 t $CaCO_3$ equiv./1000 t). Because the *ABA* of this material is less than zero, the shale has the potential to generate acidity. In reality, the *ABA* is not an exact mechanism for determining the potential of materials to produce acidity. In general, long-term field studies indicate that if the *ABA* is less than −5 kg $CaCO_3$ equiv. Mg^{-1} the spoil material has the potential to produce acidity, while spoil material that has an *ABA* of greater than +33 kg $CaCO_3$ equiv. Mg^{-1} does not produce acid drainage.

One technique to manage and ameliorate the generation of acidity in pyritic materials is to apply an alkaline amendment. The primary objective is to neutralize the potential acidity (quantified by a negative *ABA*) of the pyritic spoil to the extent that the acidity produced through the oxidation of pyrite is neutralized by the basicity (quantified by a positive *ABA*) of the amendment. The New

Albany Shale has an *ABA* of -126.4 kg $CaCO_3$ equiv. Mg^{-1}. A by-product of an industrial combustion process (retorted oil shale) is found to be a suitable amendment and has an *ABA* of $+224.7$ kg $CaCO_3$ equiv. Mg^{-1}. The application rate of the amendment is:

$$\frac{126.4 \text{ kg Mg}^{-1}}{224.7 \text{ kg Mg}^{-1}} = 0.563 \; \frac{\text{Mg amendment}}{\text{Mg spoil}} = 563 \frac{\text{kg amendment}}{\text{Mg spoil}} \qquad (10.67)$$

or 563 t amendment per 1000 t spoil. If one assumes that an acre-6 in. of shale is equivalent to 1000 t, the application rate necessary to neutralize the acidity generated by the shale in the 6-in. surface layer is 563 t A^{-1} (in SI units the application rate is 1533 Mg ha^{-1} to neutralize to a depth of 15 cm).

Acid–base accounting is one of the most common techniques used for acid drainage remediation and mine spoil reclamation. Since the primary objective is to neutralize all the acidity that can be generated by pyrite, it is assumed that the applied alkalinity will remain until all the pyrite is oxidized. Herein lies one shortcoming associated with the *ABA* approach. Pyrite and alkaline minerals, such as $CaCO_3$, have different solubilization rates. Relative to pyrite, basic minerals have a rapid dissolution rate, which facilitates the rapid disappearance of the neutralization potential (due to leaching) from an amended acid mine spoil. While the *ABA* approach provides an estimate of the potential acidity of a mine spoil and the required rate of application of a neutralizing agent, there is no mechanism in the *ABA* procedure to account for the impact of the disposal environment on the long-term neutralization potential of an amendment. Attempts to account for this shortcoming in *ABA* evaluations are generally addressed by adjusting the amendment rates to higher levels than those determined by the *ABA*.

Another mechanism for evaluating the acid-producing potential of a pyritic material, and for assessing the ability of an amendment to neutralize acidity produced over the long term, is to simulate spoil weathering in the laboratory. These laboratory-based methods subject the spoil materials to repetitive wetting-drying and leaching cycles to accelerate weathering and compress into a matter of months the time required to observe the environmental chemistry that would require a number of years to observe in the field. Although the relative weathering rates of pyrite and of alkaline materials is considered, the specific correlation of laboratory weathering time to the field has not been established.

10.5.2 Pyritic Mine Spoil Reclamation: Case Study

The simulated weathering of a mine spoil material obtained from an abandoned coal strip mine was performed to determine the feasibility of using lime-stabilized sewage sludge for long-term remediation and revegetation. Use of municipal sewage sludge has been shown to aid in the sustainable revegetation and reclamation of mined lands when applied with or without a suitable liming agent. Sewage sludge is a source of organic matter and a pool of slow-release essential nutrients (N and P). If the pathogen reduction process during sewage sludge stabilization involves the addition of hydrated lime, resulting in a highly alkaline material, a built-in source of alkalinity is also realized. Upon initial sampling at the municipal wastewater treatment facility the sewage sludge may have an aqueous saturation extract pH of approximately 12. However, after exposure to atmospheric CO_2 levels the pH will decrease to approximately 8.5 as a result of calcite precipitation.

The mine spoil site in this case study is an abandoned coal mine located on a contour-cut mountain top (Grissel Knob) in the Western Appalachian Plateau/Cumberland Mountains physiographic region of northeastern Tennessee (Campbell County). The disturbed, 1.6-ha site resulted from the strip-mining of the Cold Gap Coal Bed of the Cross Mountain Formation in the Middle Pennsylvanian. The mine spoil is subsurface clay; the clay fraction composed of vermiculite, kaolinite, and mica. Although the exact age is unknown, the site was abandoned around 1968 (Figure 10.4a). The mine spoil has a water pH of 2.8. The sewage sludge was obtained from the Knoxville Utility Board's Kuwahee Waste Water Treatment Plant located in Knoxville, TN. This sewage sludge is a Class A, low metal sewage sludge as per the U.S. EPA Part 503 sludge management regulations (Chapter 40 Code of Federal Regulations, Part 503). The sludge, when it leaves the treatment plant, has a water pH of 12.4. The results of the acid–base accounting of the mine spoil and the sewage sludge were −13.3 and +151 kg $CaCO_3$ Mg^{-1} air-dry material. Thus, the sewage sludge amendment rate, computed on a dry-weight basis, was 88 kg Mg^{-1} (or 197 Mg ha^{-1}).

Prior to field application, the *ABA* determined rate was verified using a modified humidity cell technique for nonequilibrium laboratory weathering. In this method spoil samples are placed in polypropylene containers and the *ABA*-computed sewage sludge amendment rate is applied. The unamended and amended spoil samples are thoroughly mixed and a volume of deionized water (simulating natural precipitation) that is equal to the mass of the solid is then added and the samples are allowed to equilibrate for 2 h. The spoil solution is then extracted from the solids by vacuum filtration and the leachates subjected to chemical analysis. The solid cake that remains on the filter paper is returned to the corresponding plastic container and distributed evenly to facilitate drying. The solids were allowed to air-dry for 1 week, after which time the material was extracted again with a mass of water equal to the mass of the amended spoil material. This weathering technique of repeated wetting, extraction, and drying is continued for 18 cycles.

The acidity generated by the pyritic mine spoil was effectively neutralized by the alkaline sewage sludge amendment throughout the duration of the simulated weathering study (Figure 10.6a). The pH of leachates from the amended material was slightly greater than 7. The pH of the unamended spoil leachate never exceeded 3.5. The electrical conductivity (*EC*, a measure of leachate salinity) of the leachates was not influenced by treatment during the first two weathering cycles (Figure 10.6b). However, with continued weathering, the salinity of the unamended spoil decreased in a smooth and gradually fashion, while that of the amended material remained elevated until later weathering cycles. Elevated salinity is a common problem encountered during mine spoil reclamation, and has the effect of defeating revegetation efforts. Near the completion of the weathering study, the *EC* of the unamended and amended spoil leachates were similar. This illustrates that the salinity is mobile and can be mitigated by suitable leaching prior to the initiation of revegetation efforts (although leaching may also prematurely reduce the neutralization capacity of the amendment). The salinity of the sludge amended and unamended mine spoil leachates is primarily controlled by elevated concentrations of Ca and SO_4 (Figure 10.6c and d). The concentrations of other common salts, such as Mg and Cl, are one to several orders of magnitude lower than Ca and SO_4. Interestingly, the sewage sludge is the principal source of these salts (Ca, Mg, SO_4, and Cl) in the amended spoil. The alkaline sewage sludge amendment significantly depresses leachate concentrations of Al, Fe, Mn, Cu, Ni, and Zn relative to the unamended spoil leachates (Figure 10.6e and f). Indeed, following the initial weathering cycle, leachate concentrations of the metals were reduced to near or below detectable levels as a result of sludge amendment for the duration of the study. Mine spoil and amended mine spoil leachate concentrations of both Al and Mn are generally greater then Fe concentrations. This observation establishes the importance of Al and associated hydrolysis reactions in the production of acidity and buffering of acid conditions (Figure 10.5). Further, the high concentrations Al and Mn illustrate the toxic nature of the mine spoil environment with regards to the establishment and maintenance of vegetative cover. Geochemical analysis of the spoil leachates using an ion association model (described in Chapter 5) indicates that the aqueous chemistry of the metal cations is dominated by free metal cations and complexes with sulfate (Table 10.3). As a result of the elevated sulfate levels and the abundance of charge-neutral metal complexes, it is expected that the metals would be relatively mobile in untreated spoil.

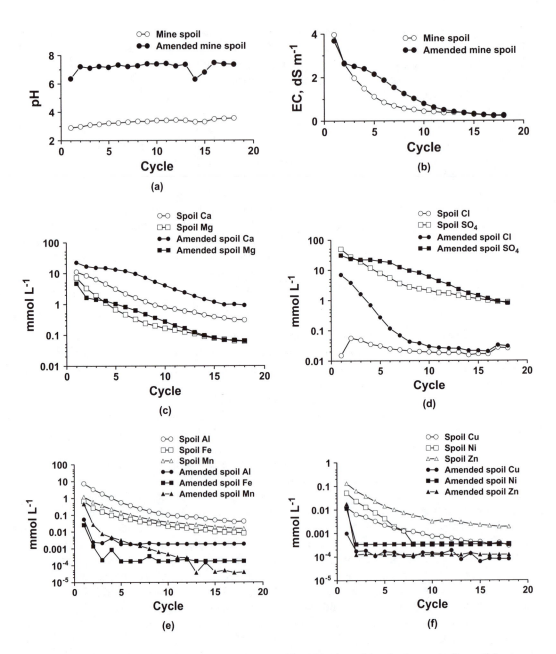

FIGURE 10.6 The chemical characteristics of leachates as a function of weathering cycle from a laboratory study of simulated weathering used to evaluate the potential long-term effectiveness of using lime-stabilized sewage sludge to neutralize acidity generated by a pyritic mine spoil: (a) pH, (b) electrical conductivity (*EC*), (c) Ca and Mg, (d) Cl and SO_4, (e) Al, Fe, and Mn, and (f) Cu, Ni, and Zn. (Modified from Abbott, D.E., M.E. Essington, M.D. Mullen, and J.T. Ammons. Fly ash and lime-stabilized sewage sludge mixtures in mine spoil reclamation: simulated weathering. *J. Environ. Qual.* 30:608–616, 2001.)

Results from the simulated weathering experiments indicate that the rate of sewage sludge application determined by the *ABA* method should provide adequate neutralization of the spoil acidity. Field application of the sewage sludge to the Grissel Knob abandoned mine site consisted of a surface application at the *ABA*-determined rate followed by disking to a maximum depth of 15 cm. The sludge application was augmented with a wildlife forage/legume seed mixture.

TABLE 10.3
Aqueous Speciation of Metals in the Mine Spoil Extract Described in Table 10.2[a]

Metal	Concentration	Aqueous Speciation (% Distribution)
pH	2.90 (units)	
	mmol L^{-1}	
Ca	11.14	Ca^{2+} (54.3), $CaSO_4^0$ (45.7)
Mg	7.311	Mg^{2+} (59.9), $MgSO_4^0$ (40.1)
Fe(III)	0.675	Fe^{3+} (3.0), $FeSO_4^+$ (77.9), $Fe(SO_4)_2^-$ (13.7), FeF^{2+} (0.9), $FeOH^{2+}$(4.5), $Fe(OH)_2^+$ (0.5)
Mn	1.202	Mn^{2+} (54.3), $MnSO_4^0$ (45.7)
Cu	0.014	Cu^{2+} (48.5), $CuSO_4^0$ (51.5)
Zn	0.131	Zn^{2+} (54.3), $ZnSO_4^0$ (45.7)
Ni	0.052	Ni^{2+} (54.0), $NiSO_4^0$ (45.6), $NiHSO_4^+$ (0.4)
Al	7.464	Al^{3+} (8.7), $AlSO_4^+$ (45.4), $Al(SO_4)_2^-$ (40.1), AlF^{2+} (5.4), AlF_2^+ (0.3)

[a] The ion speciation computations were performed using the GEOCHEM computer code.

Approximately 1 year after sludge application, the treated and untreated areas were readily discernable and the vegetative cover was dominated by an apparent monoculture of tall fescue (Figure 10.4b). Seven years after treatment the mine spoil site supports a lush and diverse stand of vegetation (Figure 10.4c).

10.6 EXERCISES

1. The measured salt pH values of acid soils are generally lower than the water pH values. Why is this result observed? The same results are true for alkaline-calcareous soils. For example, the pH of 1 M KCl extract of a calcareous soil will be less than the pH of a water extract. Why might this be the case?

2. Coal combustion for electric power generation and automotive exhaust produce $SO_2(g)$ and $NO_2(g)$ that dissolve in rainwater, potentially contributing to soil acidification. Write the balanced chemical reactions that lead to acid production when $SO_2(g)$ and $NO_2(g)$ dissolve in water.

3. Root mass of coffee plants is related to the activity of the Al^{3+} species in nutrient solutions according to the equation:

$$\text{mg plant}^{-1} = \exp[1.43 - 1.34 \times 10^6 \, (Al^{3+})], \quad r^2 = 0.99$$

   a. Compute the activity of Al^{3+} that results in a 90% reduction in the root mass of coffee plants.

   b. Compute the pH value of a soil solution in equilibrium with gibbsite that is required to maintain the Al^{3+} activity determined in part (a). Assume log K_{sp} = 8.05 for the gibbsite dissolution reaction: $Al(OH)_3(s) + 3H^+(aq) \rightarrow Al^{3+}(aq) + H_2O(l)$.

   c. Compute the pH value of a soil solution in equilibrium with Al-jurbanite and gypsum that is required to maintain the Al^{3+} activity determined in part (a). Assume log K_{sp} = −3.8 for the Al-jurbanite dissolution reaction: $AlOHSO_4(s) + H^+(aq) \rightarrow Al^{3+}(aq) + SO_4^{2-}(aq) + H_2O(l)$. Assume log K_{sp} = −4.64 for the gypsum dissolution reaction: $CaSO_4 \cdot 2H_2O(s) \rightarrow Ca^{2+}(aq) + SO_4^{2-}(aq) + 2H_2O(l)$, and that the activity of Ca^{2+} is 10^{-2}.

   d. Based on your answers in parts (b) and (c), which solid, gibbsite or Al-jurbanite, is likely to form in an acid soil treated with gypsum?

   e. Rewrite Equation 10.61 and develop a balance chemical reaction that assumes gibbsite is formed as a result of gypsum application to an acid soil, rather than Al-jurbanite.

4. The soil solid, solution, and exchange phases all have the inherent capacity to neutralize acidity. For each phase, discuss the mechanisms of acid neutralization, and for each mechanism write a balanced chemical reaction that illustrates acid neutralization.

5. The amount of lime needed to increase the pH of high organic matter soils to target pH values in the 6.0 to 6.5 range is substantially higher than that required for low-organic-matter soils. Generally, however, high-organic-matter soils are not limed to pH values greater than approximately 5.5. In answering the following, identify specific chemical processes or soil chemical characteristics to support your responses.

    a. Why is the lime requirement of high-organic-matter soils generally greater than low-organic-matter soils?

    b. Why are high-organic-matter soils only limed to achieve a pH of 5.5, instead of the 6.0 to 6.5 range that is standard for mineral soils?

6. A mechanism employed to dispose of fly ash, a product of coal combustion for electric power generation, is to return these materials to coal strip mine sites to be used as an amendment to aid in the reclamation and revegetation of pyritic coal spoil and overburden piles. A pyritic coal refuse with a saturation paste pH of 2.4 contains 75 g kg^{-1} total S, 10 g kg^{-1} sulfate S, and 18 g kg^{-1} organic S. The refuse also has a neutralization potential of 0.36 mol H$^+$ kg^{-1}. This spoil material is to be amended with fly ash, which has a neutralization potential of 1.8 mol H$^+$ kg^{-1} and insignificant acid potential. Determine the fly ash application rate necessary to neutralize the *ABA* of the coal refuse to a depth of 15 cm. Express your result in units of t A^{-1} and in Mg ha^{-1}.

7. Indicate and explain how each of the following salts will influence the *ANC* of a soil solution (increase, decrease, or remain the same) when added to a soil:

    a. NH_4NO_3
    b. $(NH_2)_2CO$
    c. KH_2PO_4
    d. H_3BO_3
    e. Na_3PO_4
    f. $Al(Cl)_3$
    g. $(NH_4)_2SO_4$
    h. CaO

8. The cation exchange data given in the table below are for CaX$_2$-AlX$_3$ and KX-AlX$_3$ exchange in 0.001 ionic strength systems using a soil that has a CEC of 7.62 cmol$_c$ kg^{-1}.

    a. Plot the exchange isotherms, E_{Ca} vs. $\tilde{E}_{Ca}$ and E_K vs. $\tilde{E}_K$, for CaX$_2$-AlX$_3$ and KX-AlX$_3$ exchange for the reactions:

$$2Al^{3+}(aq) + 3CaX_2(ex) \rightarrow 2AlX_3(ex) + 3Ca^{2+}(aq)$$

$$Al^{3+}(aq) + 3KX(ex) \rightarrow AlX_3(ex) + 3K^+(aq)$$

    b. Determine the Vanselow selectivity coefficient (K_V) as a function of E_{Ca} and E_K for CaX$_2$-AlX$_3$ and KX-AlX$_3$ exchange (assume the solution concentrations of K$^+$, Ca^{2+}, and Al^{3+} equal their activities).

    c. Plot ln K_V vs. E_{Ca} and ln K_V vs. E_K for the CaX$_2$-AlX$_3$ and KX-AlX$_3$ exchange systems and find linear relationships that describe ln K_V as a function of E_{Ca} and E_K. Determine the thermodynamic equilibrium constants for the CaX$_2$-AlX$_3$ and KX-AlX$_3$ exchange reactions.

    d. Based on your analyses, which ions in the CaX$_2$-AlX$_3$ and KX-AlX$_3$ exchange systems are preferred by the exchange complex of this soil? Are these findings in line with the expected metal ion exchange preference on soils: Al^{3+} > Ca^{2+} > K$^+$?

| CaX$_2$-AlX$_3$ Exchange | | | KX-AlX$_3$ Exchange | | |
| mmol L^{-1} | | | mmol L^{-1} | | |
Al	Ca	E_{Ca}	Al	K	E_K
0.02	0.29	0.180	0.030	0.82	0.202
0.04	0.25	0.086	0.056	0.67	0.163
0.08	0.18	0.029	0.095	0.43	0.128
0.11	0.11	0.014	0.125	0.25	0.105
			0.148	0.11	0.077

Data were obtained from Chung, J.B., R.J. Zasoski, and R.G. Burau. Aluminum-potassium and aluminum-calcium exchange equilibria in bulk and rhizosphere soil. *Soil Sci. Soc. Am. J.* 58:1376–1382, 1994.

9. In pyritic mine spoil reclamation, the pH of the spoil material is not considered when determining the application rate of an acid-neutralizing amendment. Using the data for the pyritic coal refuse given in Question 6, show that the acidity in the spoil solution (active acidity) is a very minor percentage of the potential acidity of the refuse material. Assume that the spoil pH refers to a saturation extract and that the spoil has a bulk density of 1.7 Mg m^{-3} and a porosity of 50%.

10. Figure 10.5 illustrates the processes that control the generation of acidity in pyritic mine spoils. Develop a similar type of diagram that describes the natural genesis of acid soils. Include in your diagram the physical, chemical, and biochemical processes that produce protons, consume base-reacting substances, or otherwise remove basic substances for the soil. Further, be sure to consider the interactions that occur between all three phases in a soil (gaseous, aqueous, and solid), and between the soil solution and plant roots. Use balance chemical reactions to describe the generation of acidity where applicable.

REFERENCES

Abbott, D.E., M.E. Essington, M.D. Mullen, and J.T. Ammons. Fly ash and lime-stabilized sewage sludge mixtures in mine spoil reclamation: simulated weathering. *J. Environ. Qual.* 30:608–616, 2001.

Batjes, N.H. (Ed.) A homogenized soil data file for global environmental research: a subset of FAO, ISRIC and NRCS profiles (Version 1.0). Working Paper and Peprint 95/10b, International Soil Reference and Information Centre, Wageningen, the Netherlands, 1995.

Nordstrom, D.K. Aqueous pyrite oxidation and the consequent formation of secondary iron minerals. In *Acid Sulfate Weathering*, J.A. Kittrick, D.S. Fanning, and L.R. Hossner (Ed.) SSSA Special Publication 10, SSSA, Madison, WI, 1982.

Sims, J.T. Lime requirement. In *Methods of Soil Analysis. Part 3. Chemical Methods*. D.L. Sparks et al. (Ed.) SSSA Book Series no. 5, SSSA, Madison, WI, 1996, pp. 491–515.

Sullivan, P.J. and M.E. Essington. Acid forming material characterization for planning and reclamation. In Acid Forming Materials Symposium, F.F. Munshower (Ed.) Reclamation Research Publication No. 9202, Montana State University, Bozeman, MT, 1987.

11 Soil Salinity and Sodicity

Salts are the bane of both irrigated agriculture and of civilizations that are based on irrigated agriculture, particularly if irrigation water and soil drainage are improperly managed. During the course of human history, thriving civilizations whose existence was based on irrigated agriculture have declined or disappeared, in part due to poor irrigation water management practices. For example, the Harappa civilization in the Indus Plain region of India and Pakistan, the inhabitants of the lower Viru Valley in Peru, and the Hohokam Indians in the Salt River region of Arizona have all succumbed to the degradative effects of soil salinization. However, nowhere in the recorded history of man is the influence of poor water management more graphically illustrated than in the desert region of the Middle East occupied by present day Iraq. The land between the Tigris and Euphrates Rivers in southern Iraq, known as the "cradle of civilization," ancient Mesopotamia ("the land between the rivers"), or Sumer in ancient times, is now desolate and barren, consisting of salt-encrusted soils. At one time (beginning over 6000 years ago) this region, a desert then as it is now, consisted of lush and productive fields of cereal grains, palm groves, and forage for livestock. The Sumerians colonized and transformed the desert by diverting water from the Euphrates River through a series of canals. They introduced irrigated agriculture to the region. The irrigation practices that were begun by the Sumerians continued under subsequent dynasties, such as the Akkadians and the Assyrians.

The center of civilization and power in Mesopotamia gradually shifted northwards as dynasties changed. Hillel (1992) points out that the decline of civilization in Sumer, and the northward migration of civilization, could be related to the decline of agriculture in the region. The salinization and sterilization of the soil in southern Mesopotamia resulted from a lack of adequate drainage and the introduction of salts in irrigation waters. The Euphrates was a silt-laden river during ancient times. As the river neared its lower reaches, the sediment settled onto the riverbed and banks, elevating the riverbed relative to the surrounding plains. As a result, water table levels in the region rose. Irrigation further contributed to the elevated level of the water table. It is estimated that the Euphrates loses approximately half its volume through evaporation (which concentrates salts) and seepage between its source and the Mesopotamian plain. Thus, the irrigation waters contained dissolved salts, which added to the salts released from the soil solids by mineral weathering. Although irrigation would move salts into the groundwater, the salts remained in close proximity to the surface, as the water table was near the soil surface and the groundwater did not have adequate natural flow out of the region. Further, when the water table became shallow, capillary rise moved salts up to the soil surface. Because drainage was inadequate and salts were added with continued irrigation, they accumulated at the surface soil and in the groundwater. With time, this process degraded the soil and destroyed the region's irrigation-based agriculture.

Excess salts in surface soils is a condition common in arid and semiarid regions where evapotranspiration exceeds precipitation. Poor irrigation water management exacerbates the problem of soil salinity. Indeed, irrigation will inevitably lead to the salinization of soils if proper water management practices are not employed. The impact that excess salts have on soil's physical and chemical characteristics depends on the type of salt present in soil or irrigation water. Excessive concentrations of Na (sodicity) can promote high soil pH, slaking of aggregates, and swelling and dispersion of soil clays. These physical conditions degrade soil structure and impede water and root penetration. Current data indicate that poor irrigation practices result in the loss of an estimated

10 million hectares of arable land every year as a result of soil salinization or sodification. It is estimated that approximately 7×10^9 ha of the Earth's land surface is arable, with 1.5×10^9 ha cultivated (Massoud, 1981). Szabolcs (1989) estimates that 351.2×10^6 ha of the Earth's cultivated land surface is saline and 581.0×10^6 ha is sodic. Thus, 5% of arable land and 23% of the world's cultivated lands are saline; 8% of arable and 39% of cultivated are sodic.

11.1 SOURCES OF SALTS

The common salts present in the solution phase of salt-affected soils consist of the nonhydrolyzing cations (also called the base cations), Ca^{2+}, Mg^{2+}, Na^+, and K^+; and the anions, Cl^-, SO_4^{2-}, HCO_3^- (and CO_3^{2-}), and NO_3^-. These salts are derived from a number of sources, including *in situ* weathering, saline water bodies (cyclic salts), atmospheric deposition, sedimentary rocks (fossil salts), and anthropogenic activities (secondary salinization sources). In addition to the common salts, there are also a small number of substances that are deleterious in low concentrations to plant growth, grazing animals, or human health and the environment, which may be readily solubilized in soils of certain regions where irrigated agriculture is practiced. Chief among these are boron (B^{III} in boric acid, $H_3BO_3^0$, and borate, $H_4BO_4^-$), selenium (Se^{IV} in selenite ($HSeO_3^-$ and SeO_3^{2-}) and Se^{VI} in selenate (SeO_4^{2-})), arsenic (As^{III} in arsenite ($HAsO_3^{2-}$) and As^V in arsenate ($H_2AsO_4^-$ and $HAsO_4^{2-}$)), and molybdenum (Mo^{VI} in molybdate, MoO_4^{2-})

Until the early 1980s, it was generally assumed that boron was the only trace element of concern in irrigated agriculture, as boron phytotoxicity is a rather common occurrence in arid regions. The optimal boron concentration range in soil (between deficiency and phytotoxicity) is narrow. Nutrient solutions must contain greater than 0.05 mg B L^{-1} to prevent boron deficiency; however, crops sensitive to boron will show toxic effects when solution concentrations exceed approximately 0.5 mg B L^{-1}. Crops that are very sensitive to boron will show injury symptoms and reduced yields when soluble levels exceed 0.3 mg B L^{-1}, while tolerant crops will not be affected until soil solution concentrations exceed 4 mg B L^{-1}.

The perception that boron was the only trace element of concern in arid-zone agricultural regions changed in 1983. During the 1950s and 1960s large tracts of land were brought under irrigation on the west side of the San Joaquin Valley in California. Irrigation water was imported to the region via various federal and state water projects; however, a drainage export facility (the partially completed San Luis Drain) was not utilized until 1981. Without adequate drainage, salinity levels in the soils of the region, and in the shallow groundwater, gradually increased during the 1960s and 1970s. Increasing soil salinity prompted even more intense irrigation management practices to maintain productivity, resulting in the additional build-up of salinity, water-logging, and finally the abandonment and loss of arable land. Beginning in 1981, approximately 8.6×10^6 m^3 (7000 acre-feet) of subsurface drainage water was discharged per year by the San Luis Drain into the Kesterson National Wildlife Refuge (Kesterson Reservoir). The reservoir was intended to control drainage discharge into the San Joaquin River. However, in 1983 it was discovered that high levels of selenium in the reservoir were causing a high incidence of deformity and mortality in waterfowl hatchlings at the Wildlife Refuge. Selenium, which exists predominately as the selenate oxyanion (SeO_4^{2-}) in alkaline soil environments, is quite mobile and bioavailable (its behavior in soil is similar to that of sulfate, SO_4^{2-}). The introduction of irrigated agriculture into the valley led to the solubilization of Se from the seleniferous soils that formed on alluvium derived from the sedimentary rocks that border the valley (particularly marine shales). Selenium concentrations and salinity in the shallow groundwater and soils continued to build during the two decades prior to the opening of the San Luis Drain. When the drain was finally opened, this poor quality water was transported to the wildlife refuge. The average Se concentration in the drainage water entering Kesterson Reservoir between 1984 and 1986 was approximately 300 µg L^{-1}, with the highest concentrations in excess of 4000 µg L^{-1}. The concentrations of selenium in these drainage waters far exceeded the average observed in freshwaters ($0.2\ \mu g$ Se L^{-1}), as well as the water quality criteria for irrigation waters ($10\ \mu g$ Se L^{-1}) and the U.S. EPA drinking water standard ($10\ \mu g$ Se L^{-1}).

Some minerals in the soil environment are inherently unstable. Mineral dissolution is facilitated through the action of carbonic acid (proton source) and soluble organic compounds (proton sources and complexing agents), as discussed in Chapter 3. As a result of primary mineral weathering, secondary minerals form and nonhydrolyzing (base) cations, HCO_3^- and CO_3^{2-} (weathering reaction products), and Cl^- and SO_4^{2-} (components of many minerals) are solubilized and tend to remain in solution. In the limited leaching environments of arid and semiarid regions, the pH can vary between 8 and 10 (depending on the mineral composition and soluble Na concentrations), and calcite is a common soil mineral. Typically, rainfall is sufficient to carry the salts released during *in situ* weathering through the soil profile (particularly Na), and the concentrations of salts in soil leachates generally do not exceed 5 $mmol_c\ L^{-1}$. However, when leaching is restricted and water-logging occurs such as in closed basins common in many arid regions, *in situ* weathering can result in the development of sodic systems. In these environments, elevated HCO_3^- and CO_3^{2-} levels coupled with alkaline pH promote the precipitation of Ca (and Mg) carbonates, leaving Na^+ as a significant component on the soil exchange complex. While *in situ* mineral weathering is a process common to all soils, the formation of saline and sodic soils by this mechanism is uncommon.

Soils adjacent to the ocean or other bodies of brackish water can receive cyclic salts. These soils may become salinized through tidal action (surface flow) or seepage (subsurface flow), where the degree of salinization depends on climate, tide character, hydrology, and soil properties. Soils adjacent to saline waters in arid regions tend to be extremely saline. Soils in humid regions tend to be saline only during high tide periods. Salt spray can contribute to the salinization of soils that are downwind of saline water bodies. Again, climate and soil drainage play a key role in salt accumulation from atmospheric sources. Salinization has also been observed in soils downwind from coal-fired power generation stations. The aerosols from the cooling towers, which are saline, drift over the landscape and contribute salts to soils in the fallout zone. Other wind-born salt sources include particulates from dry lakes (Playas) and arid-zone soils that are redeposited during dust storms and thunderstorms (calcite, gypsum, and other evaporate minerals); volcanic ash (a source of Cl^-); biological activities (HCO_3^-, CO_3^{2-}, NO_3^-, and H_2S); and anthropogenic activities (NO_3^- and SO_4^{2-} from automobile emissions and coal-fired power plants; common salts and B, Se, As, and Mo from fly ash).

Common salts, as well as the more problematic elements (e.g., Se and B), enter ground and surface waters through the dissolution of sedimentary rock. These salts, termed *fossil salts*, most commonly enter the soil in irrigation waters. Fossil salts are a primary salt source; however, soil salinization with fossil salts is a secondary and anthropogenic process. Although the application of irrigation water results in the addition of soluble salts to soils, it is poor water management that results in the accumulation of salts. Ineffective leaching of salts or the inadequate removal of drainage water are the processes that facilitate the buildup of salts in soil.

11.2 DIAGNOSTIC CHARACTERISTICS OF SALINE AND SODIC SOILS

A saline soil is one that contains an excess of soluble salts. This condition exists when salt concentrations in the root zone are too high for the normal growth and development of plants. Further, a soil is saline when the reduction in plant yield is proportional to the soil salinity and there are no other confounding factors impacting plant yields, such as a phytotoxic response to high concentrations of a particular element. In other words, the yield decrement is proportional to the osmotic potential of the soil water, and not to the concentrations of individual ions in the soil solution. Because plants vary greatly in their tolerance to salinity, it is difficult to assign a quantitative value of salinity that identifies a saline system.

Extraction of a soil sample that is saturated with deionized-distilled water (saturation extract) is the standard method used to obtain a soil solution sample for assessing soil salinity and sodicity. A saturation extract is obtained by first preparing a saturated paste. A saturated paste is achieved at the point when all voids in a soil sample are filled with solution. Making the perfect "mud pie" is more art than science; but uniformity, the ability to reproducibly achieve a paste, is a necessity. In general, the procedure for making a saturated paste is started by placing a mass of moist or air-dry soil (~200 g) in a plastic beaker. Deionized-distilled water is then mixed into the soil with a stainless-steel spatula just until the soil is uniformly moist. The mixture is allowed to sit, covered and undisturbed, for approximately 2 h. Small volumes of water are added thereafter, and the suspension is stirred after each addition until the paste achieves some rather nonscientific charac-teristics. After incorporation of each additional volume of water, the beaker is tapped on the lab bench to flatten-out the soil surface. A saturated paste is reached when the soil surface glistens and jiggles like Jell-O®. When the beaker is tipped, the soil will "tongue" (appear to want to flow). Further, the saturated soil will slide cleanly off the stainless-steel spatula. Finally, a trench formed in the paste with the spatula will disappear when the beaker is lightly tapped on the lab bench. If, after sitting for another 2-h period, water begins to puddle on the soil surface, the soil is oversat-urated, and additional soil must be added. Once a paste is prepared, it is allowed to sit covered and undisturbed for several hours to allow the soil and water to react. The soil paste is then transferred onto filter paper in a Büchner funnel and the soil solution extracted by vacuum filtration.

Once obtained, a saturation extract may be analyzed for any number of chemical characteristics. However, the most important extract properties required for assessing soil salinity and sodicity are the electrical conductivity (EC), and the concentrations of calcium, magnesium, and sodium. The ability of a solution to conduct electricity is directly related to the concentration of dissolved salts. The EC is determined by placing two flat, rectangular electrodes that are separated by a constant distance into a solution. These electrodes are contained in a conductivity cell. A constant potential is applied to the electrode and a current flows through the solution between the electrodes. Pure water is a perfect insulator; however, water conducts electricity when dissolved salts are present. Conductance and salt content are directly related. The current that flows through a solution is inversely proportional to the resistance of the solution, which is measured by a resistance bridge. Conductance is the reciprocal of resistance and has units of reciprocal ohms, or mhos (in cgs units). Each conductivity cell has a cell constant which is expressed in units of reciprocal distance (cm^{-1} or m^{-1}). The cell constant is determined by measuring the EC of a standard 0.01 M KCl reference solution. This solution has an EC of 1.4118 dS m^{-1} at 25°C (dS is decisiemens, where siemen is the SI unit of conductance). The cell constant is calculated as 1.4118 dS m^{-1}/dS$_{KCl}$ m^{-1}, where dS$_{KCl}$ m^{-1} is the measured conductivity of the 0.01 M KCl reference solution. The conductivity of an unknown solution multiplied by the cell constant yields the sample EC. The cgs unit for the EC of soil solutions and irrigation waters is mmhos cm^{-1}; whereas, the SI unit is dS m^{-1}. The units of mmhos cm^{-1} are comparable to dS m^{-1}, such that 1 mmhos cm^{-1} = 1 dS m^{-1}.

Historically, a soil is saline if the EC of a saturation extract (EC_e) exceeds 4 dS m^{-1} (United States Salinity Laboratory Staff, 1954). By the early 1970s it was recognized that the influence of soil salinity levels on crop response was crop-specific (see Bohn et al., 1985). It was during this time that the distinction between a nonsaline and a saline soil became entwined with crop type. For salt-sensitive plants, $EC_e > 2$ dS m^{-1} would denote a saline soil; whereas, for salt-tolerant plants, a saturation extract $EC_e > 8$ dS m^{-1} would denote a saline soil. More recently (see Maas, 1990), threshold EC_e levels have been defined to indicate the types of crops that can tolerate various levels of soil salinity. Saline soils are defined as $EC_e > 1.5$ dS m^{-1} for sensitive crops; $EC_e > 3.0$ dS m^{-1} for moderately sensitive crops; $EC_e > 6$ dS m^{-1} for moderately tolerant crops; and $EC_e > 10$ dS m^{-1} for tolerant crops (the crops specific to each category are described later in this chapter).

A soil is sodic if the exchange complex contains sufficient Na^+ to adversely impact soil structure and crop production. Whereas high salinity impacts the ability of plant roots to extract soil water, high sodium adversely impacts the soil physical structure and the movement of air and water into the soil and to plant roots. Excess sodium on the exchange complex causes clay dispersion and swelling (two separate processes) which subsequently clog soil pores and reduce soil hydraulic conductivity. The level of exchangeable Na necessary to encounter sodicity problems is dependent on a number of factors, including soil texture, clay mineralogy, organic matter content, EC_e, and the concentration of electrolytes (other than Na^+) in the soil solution.

Traditionally, a soil is by definition sodic if the exchangeable sodium percentage (*ESP*) is greater than 15 (United States Salinity Laboratory Staff, 1954):

$$ESP = 100 \times \frac{ESR}{(1 + ESR)} \tag{11.1}$$

where *ESR* is the exchangeable sodium ratio. The *ESR* is a property that is derived from the Gapon expression (discussed in Chapter 8). The genesis of a sodic soil can be described by the $(Ca + Mg)X_2$-NaX exchange reaction, and results from the displacement of divalent cations, Ca^{2+} and Mg^{2+}, from the exchange complex (denoted by X^-) by Na^+. Using the Gapon convention, this process is illustrated by the reaction:

$$Na^+(aq) + (Ca + Mg)_{\frac{1}{2}}X(ex) \rightarrow NaX(ex) + \frac{1}{2}(Ca^{2+} + Mg^{2+})(aq) \tag{11.2}$$

The Gapon selectivity coefficient for this reaction is:

$$K_G = \frac{[NaX][Ca^{2+} + Mg^{2+}]^{\frac{1}{2}}}{[(Ca + Mg)_{\frac{1}{2}}X][Na^+]} \tag{11.3}$$

The ratio of [NaX] to $[(Ca^{2+} + Mg^{2+})_{\frac{1}{2}}X]$ on the soil exchange complex is the *ESR*:

$$ESR = \frac{[NaX]}{[(Ca + Mg)_{\frac{1}{2}}X]} \tag{11.4}$$

where the brackets denote concentration in units of $cmol_c \ kg^{-1}$ (selected for convenience, as they are the units of *CEC*). Further, because Na^+, Ca^{2+}, and Mg^{2+} are the dominant exchangeable cations in sodic soils, or soils that may potentially become sodic, the *ESR* may also be described by:

$$ESR = \frac{[NaX]}{[CEC - NaX]} \tag{11.5}$$

The ratio of Na^+ to divalent ion concentrations in soil solutions or irrigation waters is defined as the sodium adsorption ratio (*SAR*):

$$SAR = \frac{[Na^+]}{[Ca^{2+} + Mg^{2+}]^{\frac{1}{2}}} \tag{11.6}$$

where the brackets denote concentration units of $mmol \ L^{-1}$. The units of *SAR* are $mmol^{\frac{1}{2}} \ L^{-\frac{1}{2}}$. Combining Equations 11.3, 11.4, and 11.6 yields:

$$K_G = \frac{ESR}{SAR} \tag{11.7}$$

As Equation 11.7 indicates, the exchangeable Na^+ component of the soil exchange capacity can be predicted given the Na^+ and divalent cation composition of the soil solution and a value for K_G. Correspondingly, the impact of irrigation waters or other leachates on the sodicity of a soil can also be established. When ESP is less than 25 to 30%, the Gapon selectivity coefficient for many irrigated soils is between 0.010 and 0.015. Since the values of $ESP/100$ and ESR are approximately equal (Equation 11.1), ESP and SAR are also approximately equal:

$$SAR = \frac{ESR}{K_G} \approx \frac{ESR}{0.01} \approx ESP \qquad (11.8a)$$

The United States Salinity Laboratory Staff (1954) examined 59 soil samples representing nine Western states and obtained the following statistical relationship:

$$ESP = \frac{100(-0.0126 + 0.01475 SAR_e)}{1 + (-0.0126 + 0.01475 SAR_e)} \qquad (11.8b)$$

where SAR_e is the SAR of the saturation extract. Using Equation 11.8b and an SAR_e of 10 mmol$^{\frac{1}{2}}$ L$^{-\frac{1}{2}}$, the predicted ESP is 11.9; while an estimate of $ESP = 10$ is obtained from Equation 11.8a. The SAR is a property of the soil solution concentrations of Na^+, Ca^{2+}, and Mg^{2+} (Equation 11.6), and is easily measured. Although the sodicity of a soil is a property of the exchange complex composition (ESP), the defining soil property that is currently employed to establish sodicity is the SAR, simply because it is easier to determine. In general, an SAR_e of greater than 13 to 15 mmol$^{\frac{1}{2}}$ L$^{-\frac{1}{2}}$ indicates a sodic soil, when the EC_e is less than 4 dS m^{-1}.

11.3 IRRIGATION WATER QUALITY PARAMETERS AND RELATIONSHIPS

Poor irrigation management practices can be responsible for the development of saline and sodic soils. Poor irrigation water quality can exacerbate the impact of these practices on soil salinity and sodicity. The key irrigation water quality parameters are the electrical conductivity (EC_{iw}) and the sodium adsorption ratio (SAR_{iw}). In addition to EC_{iw}, the total dissolved salts (TDS, in mg L^{-1}) in a solution are also a measure of salinity; however, the determination of TDS is fraught with error. The TDS are directly determined by evaporating a known volume of water to dryness. The mass of the solids is determined and reported on a water volume basis. The amount of hygroscopic water in the evaporite minerals (solids that precipitate when water is evaporated) obtained is highly dependent on the drying conditions. For example, the air-dry mass of salts precipitated from a known volume of water will be greater than that obtained when dried over a desiccant. The mass will be reduced further if the water sample is dried under elevated temperature conditions. In addition, water will be absorbed by evaporite minerals from the atmosphere as the mass is being determined and thus the mass obtained will be dependent on the relative humidity.

The EC of a solution is one of the easiest chemical characteristics to determine. When EC is in the 0.1 to 5 dS m^{-1} range, several useful empirical relationships are valid:

$$EC(\text{dS m}^{-1}) \times 640 \approx TDS \ (\text{mg L}^{-1}) \qquad (11.9)$$

$$EC(\text{dS m}^{-1}) \times 10 \approx C \ (\text{mmol}_c \ \text{L}^{-1}) \ [\text{sum of dissolved cation (or anion) charge}] \quad (11.10)$$

$$[\log EC \ (\text{dS m}^{-1}) \times 1.009] + 1.159 \approx \log I \ (\text{ionic strength in mmol L}^{-1}) \qquad (11.11)$$

$$EC \ (\text{dS m}^{-1}) \times 0.40 \approx \tau_o \ (\text{bars}) \ [\text{osmotic pressure at 25°C}] \qquad (11.12)$$

Table 11.1
Irrigation Water Quality Criteria for Salinity

Class	EC_{iw}, dS m^{-1}	Comments
Low salinity (no problem)	<0.75	No detrimental effects will usually be observed
Medium salinity (increasing problem)	0.75–3.00	May have detrimental effects on sensitive crops and will require careful management
High salinity (severe problem)	>3.00	To be used only for salt-tolerant crops on permeable soils with careful management

Several chemical characteristics of irrigation water are used to describe quality. Water quality criteria are based on broad generalizations about the crops to be grown, soil properties, water management, and climate. The potential impact of water quality on soil salinity and crops is based on the EC_{iw} (Table 11.1). In general, the classes also correspond to the salt tolerance of the crops that can be grown without impacting yield. Low salinity waters (EC_{iw} < 0.75 dS m^{-1}) can be used for production of relatively salt-sensitive crops; whereas, high salinity waters (EC_{iw} > 3.00 dS m^{-1}) are suitable for production of salt-tolerant crops (although only on permeable soils). Because irrigation water quality criteria are established at the state experiment station or local water management district levels, state-specific class designations and associated EC_{iw} values will vary slightly from those presented in Table 11.1.

Excessive sodium in irrigation water can produce soils with high exchangeable sodium levels. The sodium hazard of irrigation water is not a simple function of sodium concentration, but additionally of the concentrations of divalent cations (Ca^{2+} and Mg^{2+}), salinity (EC_{iw}), and alkalinity (CO_3^{2-} and HCO_3^-), the dominant soil clay type, and the soil texture. The basic characteristic that defines the sodicity hazard of irrigation water is the SAR_{iw} (Equation 11.6); however, there are two ways in which to compute SAR_{iw}. The true SAR of a solution (SAR_t) is computed using the free concentrations of Na^+, Ca^{2+}, and Mg^{2+} (in mmol L^{-1}). The practical SAR of a solution (SAR_p) is computed using the total concentrations of Na, Ca, and Mg (in mmol L^{-1}). The values obtained for SAR_t and SAR$_p$ will differ for a single solution because greater ion pair formation is observed for Ca and Mg than for Na. Thus, SAR_t values are greater than the corresponding SAR_p values. The regression equation that relates the two SAR parameters is (Sposito and Mattigod, 1977):

$$SAR_t = 0.08 + 1.115 \times SAR_p \tag{11.13}$$

In general, however, SAR_t and SAR_p do not differ significantly when they are less than 10 mmol$^{1/2}$L$^{-1/2}$.

An important factor that increases the sodium hazard of irrigation water is its alkalinity (CO_3^{2-} + HCO_3^-). In the soil solutions, calcium readily reacts with carbonate species to form calcite, which usually contains minor isomorphic substitution of Mg^{2+} for Ca^{2+} [$(Ca, Mg)CO_3(s)$]. Calcite precipitation removes Ca^{2+} (and to a lesser degree Mg^{2+}) from solution, favoring their replacement of on the exchange complex by Na^+ (Equation 11.2). This process increases the ESP of the soil. One measurement that has been used to describe the bicarbonate hazard of irrigation water is the residual sodium carbonate (RSC) parameter:

$$RSC = [\tfrac{1}{2}CO_3^{2-} + HCO_3^-] - \tfrac{1}{2}[Ca^{2+} + Mg^{2+}] \tag{11.14}$$

where the brackets represent mmol L^{-1}. If [$\tfrac{1}{2}CO_3^{2-}$ + HCO_3^-] is greater than $\tfrac{1}{2}[Ca^{2+} + Mg^{2+}]$, the soluble Ca^{2+} and Mg^{2+} in a soil will potentially precipitate as carbonates, leaving soluble Na^+ in

TABLE 11.2
Irrigation Water Quality Criteria for Permeability

Parameter	Dominate Clay	No Problem	Increasing Problem	Severe Problem
EC_{iw}, dS m^{-1}	—	>0.5	0.5–0.2	<0.2
adj. SAR_{iw}	Montmorillonite	<6	6–9[a]	>9
	Illite-vermiculite	<8	8–16[a]	>16
	Kaolinite-hydrous oxides	<16	16–4[a]	>24

[a] Actual adj. SAR criteria depends on EC_{iw}. If EC_{iw} < 0.4 dS m^{-1}, use values in the lower range of the criteria. If EC_{iw} < 0.4 to 1.6 dS m^{-1}, use values in the intermediate range of the criteria. If EC_{iw} > 1.6 dS m^{-1}, use values in the upper range of the criteria.

solution and increasing the sodium hazard. When RSC is less than 1.25, irrigation water is characterized as safe with respect to bicarbonate hazard. An RSC between 1.25 and 2.5 indicates that the water is potentially harmful; while an RSC greater than 2.5 indicates harm.

In practice, the use of RSC values to characterize the impact of bicarbonate on the sodium hazard of irrigation water is common. However, RSC values do not quantify calcite precipitation to any degree, and thus are not considered a reliable indicator on which to base water management decisions. The impact of the bicarbonate hazard on the SAR can be empirically determined by applying the Langelier index ($LI = pH_a - pH_c$):

$$\text{adj. } SAR_{iw} = SAR_{iw} \times [1 + (pH_a - pH_c)] \tag{11.15}$$

where adj. SAR_{iw} is an adjusted SAR, pH_a is the measured pH of an irrigation water or soil, and pH_c is the calculated pH of the irrigation water if equilibrated with calcite:

$$pH_c = (p^cK_2 - p^cK_{sp}) + p[Ca^{2+}] + p[HCO_3^-] \tag{11.16}$$

where p^cK_2 is the conditional second dissociation constant for carbonic acid ($HCO_3^- \rightarrow H^+ + CO_3^{2-}$), p^cK_{sp} is the conditional solubility product constant for calcite ($CaCO_3 \rightarrow Ca^{2+} + CO_3^{2-}$), and the brackets represent molar concentrations (conditional equilibrium constants are discussed in Chapter 5). Adjusted SAR_{iw} values are highly correlated with the SAR values of soil saturation extracts and drainage waters. Further, numerous modifications of the Langelier index have been proposed to better predict the SAR of soil drainage waters (SAR_{dw}) (see Suarez, 1981), because SAR_{dw} values are considered to be better predictors of irrigation water suitability.

The sodium hazard associated with irrigation waters is dependent on both SAR_{iw} and EC_{iw}, as well as the dominate clay minerals present in the soil. Tables 11.2 and 11.3 indicate the criteria used to characterize irrigation water quality and the potential impact of irrigation waters on soil permeability. Notice that problems may arise when using irrigation waters that are low in soluble salts. Irrigating with high quality water (from a salinity standpoint) can be just as problematic as irrigating with high SAR water (explained below).

11.4 GENESIS, MANAGEMENT, AND RECLAMATION OF SALT-AFFECTED SOILS

Soil salinity and sodicity are conditions that (1) limit the ability of plants to absorb water from the soil solution, (2) may cause phytotoxic effects from elevated concentrations of specific ions, and (3) alter the physical and chemical properties of soil such that long-term detrimental effects hinder

TABLE 11.3
Potential Sodium Hazard Associated with Irrigation Water
Quality as a Function of SAR_{iw} and EC_{iw}[a]

SAR_{iw} mmoL$^{1/2}$L$^{-1/2}$	None	Slight to Moderate EC_{iw}, dS m^{-1}	Severe
0–3	>0.7	0.7–0.2	<0.2
3–6	>1.2	1.2–0.3	<0.3
6–12	>1.9	1.9–0.5	<0.5
12–20	>2.9	2.9–1.3	<1.3

[a] From Ayers, R.S. and D.W. Westcot. Water quality for agriculture. FAO Irrig. Drain. Paper 29, 1985.

either productivity or the establishment of vegetative cover. Maintaining high levels of agricultural production, conserving the soil resource, conserving water (a very limited commodity in arid and semiarid regions), and minimizing the impact of agricultural drainage waters on aquatic environments requires sound irrigation and drainage water management. Many of the management practices necessary to achieve soil and water conservation and agricultural productivity were developed by the early 1950s and reported by the United States Salinity Laboratory Staff (1954) (USDA Handbook 60, also known as the "Green Bible"). An addendum to Handbook 60 was published by the American Society of Chemical Engineers (1990). Both publications should be referred to for detailed information on the management of saline and sodic soils, irrigation waters, and drainage waters.

11.4.1 SALINE SOILS

Saline soils result when leaching is not sufficient to move salts through the rooting zone and the progressive and harmful accumulation of salts in the rooting zone occurs. Saline soils are generally limited to arid environments where water is limiting and a precious commodity. Thus, irrigation water management in these regions must strike a balance between conservation and the needs of production agriculture. The long-term management of water in irrigated agriculture requires that salts entering a soil be eventually removed in drainage water. This requirement is termed salt balance (SB). The SB of a soil under steady-state conditions can be described as:

$$SB = V_{dw}C_{dw} + S_p + S_c - V_{iw}C_{iw} - S_f \qquad (11.17)$$

The volumes of drainage and irrigation water are represented by V_{dw} and V_{iw}, and the concentrations of salts in the drainage and irrigation water are represented by C_{dw} and C_{iw}. Salts entering the soil in fertilizers, as atmospheric deposition, and from mineral dissolution are contained in the S_f term. Salts removed from the soil through crop uptake are represented by S_c, and salts that precipitate are represented by S_p. In general, salts removed from the soil solution by precipitation and crop uptake should be roughly equal to the salts added through fertilization, atmospheric deposition, and mineral dissolution: $S_f = S_p + S_c$, except in cases of overfertilization or in natric soils with buried saline horizons. The salt balance becomes: $SB = V_{dw}C_{dw} - V_{iw}C_{iw}$. Since the goal of water management is to achieve $SB = 0$, $V_{dw}C_{dw} = V_{iw}C_{iw}$, or:

$$LF = \frac{V_{dw}}{V_{iw}} = \frac{C_{iw}}{C_{dw}} \qquad (11.18)$$

where LF is the leaching fraction. Since EC is a direct measure of the concentrations of salts in a solution (as described in Equations 11.9 and 11.10), LF is also given by:

$$LF = \frac{EC_{iw}}{EC_{dw}} \qquad (11.19)$$

The LF value that defines the fraction of applied irrigation water that *must* pass beyond the root zone to prevent the accumulation of salts to a level that adversely affects the production of a specific crop is defined as the leaching requirement (LR). Typically, target LR values range between 0.15 and 0.2. Therefore, the concentration of salts in the irrigation water must be between 6.7 ($LR = 0.15$) and 5 ($LR = 0.2$) times lower than the concentration of salts removed from the root zone.

Crop tolerance to soil salinity depends on LR, EC_{iw}, and the average salinity of the soil solution in the root zone ($\overline{EC_e}$). A threshold $\overline{EC_e}$ value describes crop tolerance to salts in the root zone. Saturation extract $\overline{EC_e}$ values below the threshold level for a given crop will not adversely impact yields. However, $\overline{EC_e}$ values above the threshold value for the crop will result in yield reductions. Threshold salinity levels ($\overline{EC_e}$ values) for crop tolerance to salinity are: sensitive (1.5 dS m^{-1}); moderately sensitive (3 dS m^{-1}); moderately tolerant (6 dS m^{-1}); and tolerant (10 dS m^{-1}). The relationship that describes the relative reduction in crop yield as soil salinity increases above a threshold value is:

$$Y = 100 - b \times (\overline{EC_e} - a) \qquad (11.20)$$

where Y is the relative crop yield expressed as a percentage of the maximum yield, a is the threshold $\overline{EC_e}$, and b is the percent yield reduction per unit increase in $\overline{EC_e}$ above the threshold value. The classification of various crops according to their salinity tolerance is presented in Table 11.4. Consider the production of tomatoes, which are moderately sensitive to salinity. If $\overline{EC_e}$ in the soil profile is 4 dS m^{-1}, the yield should be approximately 85% (85.15%) of the maximum yield [100 − 9.9 × (4.0 − 2.5)].

The LR value needed to achieve the maximum yield of a specific crop will depend on the quality of the irrigation water and the average EC_e of the root zone. The average EC_e of the root zone may be given by the expression (Rhoades, 1974):

$$\overline{EC_e} = k \frac{EC_t + EC_b}{2} \qquad (11.21)$$

where k is an empirical constant equal to 0.8, EC_t is the saturation extract EC value at the top of the root zone, and EC_b is the saturation extract EC at the bottom of the root zone. The EC_e of a soil is approximately ½ the EC of *in situ* soil water at field capacity. Therefore, $EC_t = \frac{1}{2} EC_{iw}$ and $EC_b = \frac{1}{2} EC_{dw}$. Substituting these expressions into Equation 11.21, with $k = 0.8$, yields:

$$\overline{EC_e} = 0.8 \left[\frac{\frac{1}{2} EC_{iw} + \frac{1}{2} EC_{dw}}{2} \right] \qquad (11.22a)$$

$$\overline{EC_e} = 0.4 \left[\frac{1}{2} EC_{iw} + \frac{1}{2} EC_{dw} \right] \qquad (11.22b)$$

$$\overline{EC_e} = 0.2 (EC_{iw} + EC_{dw}) \qquad (11.22c)$$

TABLE 11.4
Salt Tolerance of Selected Crops[a]

Crop	a	b	Crop	a	b
Sensitive			**Moderately Tolerant**		
Bean, *Phaseolus vulgaris*	1.0	19	Barley (forage), *Hordeum vulgare*	6.0	7.1
Blackberry, *Rubus* ssp.	1.5	22	Beet, *Beta vulgaris*	4.0	9.0
Carrot, *Daucus Carota*	1.0	14	Fescue, *Festuca elatior*	3.9	5.3
Onion, *Allium Cepa*	1.2	16	Soybean, *Glycine Max*	5.0	20
Peach, *Prunus Persica*	1.7	21	Wheat, *Triticum aestivum*	6.0	7.1
Strawberry, *Fragaria* ssp.	1.0	33			
Moderately Sensitive			**Tolerant**		
Alfalfa, *Medicago sativa*	2.0	7.3	Barley (grain), *Hordeum vulgare*	8.0	5.0
Corn, *Zea Mays*	1.7	12	Bermudagrass, *Cynodon Dactylon*	6.9	6.4
Peanut, *Arachis hypogaea*	3.2	29	Cotton, *Gossypium hirsutum*	7.7	5.2
Pepper, *Capsicum annuum*	1.5	14	Sugarbeet, *Beta vulgaris*	7.0	5.9
Rice, *Oryza sativa*	3.0	12			
Tomato, *Lycopersicon Lyco.*	2.5	9.9			

[a] a: threshold soil saturation extract electrical conductivity ($\overline{EC}_e$ in dS m^{-1}) at initial yield decline; b: percent yield decrease per unit increase in $\overline{EC}_e$ beyond threshold.

From Maas, E.V. Crop salt tolerance. In *Agricultural Salinity Assessment and Management*. K.K. Tanji (Ed.) American Society of Civil Engineers Manuals and Reports on Engineering Practices No. 71. American Society of Civil Engineers, New York, 1990, pp. 262–304.

Rearranging,

$$EC_{dw} = 5\overline{EC}_e - EC_{iw} \tag{11.23}$$

Substituting this expression into Equation 11.19:

$$LR = \frac{EC_{iw}}{5\overline{EC}_e - EC_{iw}} \tag{11.24}$$

For the maximum yield of tomatoes, the average EC_e of the soil profile ($\overline{EC}_e$) must not exceed 2.5 dS m^{-1} (a value for tomatoes, Table 11.4). If EC_{iw} is 3.21 dS m^{-1} (salinity of the Pecos River near Carlsbad, NM), the LR is:

$$LR = \frac{3.21}{5(2.5) - 3.21} = 0.35 \tag{11.25}$$

Irrigation of tomatoes using Pecos River water and employing an LR that is less than 0.35 would result in an $\overline{EC}_e$ value in the soil that exceeds the threshold. For example, an LR of 0.2 would result in an $\overline{EC}_e$ of 3.85 and a yield reduction of approximately 13% (computed using Equation 11.20).

Threshold $\overline{EC}_e$ values (a values) given in Table 11.4 are lower than those actually observed in field soils. The Table 11.4 values were determined from hydroponics studies with the roots submerged in circulating solutions. Thus, the total root volume is submerged in a solution whose salinity does not vary with location. However, in field soils the soil solution EC_e values increase

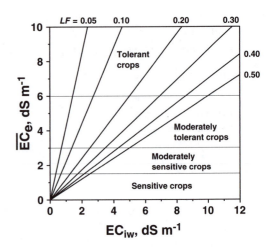

FIGURE 11.1 The relationship between irrigation water electrical conductivity (EC_{iw}) and the average electrical conductivity of the soil saturation paste extract ($\overline{EC}_e$) obtained using various values of leaching fraction (LF). Crop tolerance categories are also represented.

with depth, and low salinity soil solution is accessible to roots at the top of the root zone. Research has shown that plant root systems are capable of withstanding excessive soil salinity levels if the root system has access to water of low salinity. Therefore, for crop management it is common to compute LR values that are based on $\overline{EC}_e$ values that are predicted to produce 10 to 50% yield reductions (when using Table 11.4 data). For example, the $\overline{EC}_e$ value that produces a 10% reduction in tomato production is 3.51 dS m^{-1} (computed using Equation 11.20). Again, using Equation 11.24 and irrigation water with an $EC_{iw} = 3.21$, the LR needed to achieve maximum tomato yields is 0.22. Equation 11.24 was used to construct Figure 11.1, illustrating the relationships between EC_{iw}, $\overline{EC}_e$, LR, and crop sensitivity to soil salinity.

The reclamation of saline soils requires the movement of salts out of the profile to a depth that is well below the root zone. Reclamation is accomplished by leaching. Critical to this process is adequate drainage, either through the installation of an artificial drainage system or by the existence of a permeable soil and subsoil. The reclamation specifics employed are dependent on a number of factors, including: soil texture and structure, irrigation water quality, geology, topography, crop, and time limitations. Most irrigation waters are suitable for reclaiming saline soils; however, the amount of water required to achieve reclamation will be a function of the initial soil salinity, the soil type (texture), the depth of reclamation, and the technique for applying the water. The most common technique for applying water during reclamation is by flood irrigation (continuous ponding). The amount of irrigation water required to reclaim a saline soil is given by:

$$d_{iw} = Kd_s \frac{(\overline{EC}_e - EC_{iw})}{(a - \frac{1}{2}EC_{iw})} \tag{11.26}$$

where d_{iw} is the depth (cm) of ponded irrigation water needed to effect reclamation, d_s is the depth (cm) of soil to be reclaimed, K is a parameter based on the saturated volumetric water content and texture of the soil (0.45 for organic soils, 0.3 for clay loams, and 0.1 for sandy loams), a is the threshold or desired EC_e of the profile (crop specific and commonly obtained from Table 11.5), and $\overline{EC}_e$ and EC_{iw} are previously defined. For example, consider a saline soil with $\overline{EC}_e = 20$ dS m^{-1}. The depth of irrigation water ($EC_{iw} = 1.7$ dS m^{-1}) that must be ponded to permit the growing of alfalfa ($a = 2.0$ dS m^{-1}) on a clay loam without salinity stress (assuming a 60 cm active rooting depth) is:

$$d_{iw} = (0.3)(60)\frac{(20 - \frac{1}{2}(1.7))}{(2.0 - \frac{1}{2}(1.7))} = 300 \text{ cm} \tag{11.27}$$

The continuous ponding technique requires a level surface and is relatively inefficient compared with other techniques, such as intermittent ponding or sprinkler irrigation (which does not require a level surface). Flood irrigation results in saturated flow, and the water movement principally occurs in the macropores and at a relatively high velocity. The macropores account for only a small portion of the total porosity of a soil (and only a small portion of the salt burden contained therein), and salts must diffuse from the smaller micropores to the macropores before they can be flushed from the soil. This diffusion process is slow, and water movement through the profile under saturated flow is too rapid for efficient salt removal. Under unsaturated flow conditions, like those found under sprinkler irrigation, the macropores are still the principal conduit for water flow; however, water in the macropores is in contact with the micropore water for a greater period of time. Thus, there is greater diffusion of salts from the micropores and into the macropores, and greater efficiency of salt removal.

Since the bulk of the soil's salt burden is found in the micropores, sprinkler irrigation is a much more efficient technique for the reclamation of saline soils. The amount of irrigation water required to reclaim a saline soil using sprinkler irrigation is also given by Equation 11.26, but with $K = 0.1$. Using sprinkler irrigation, the amount of irrigation water required to reclaim the saline soil for alfalfa production, described in the above example, is 100 cm. In this example, the preparation of a clay loam soil for alfalfa production using sprinkler irrigation requires one third as much water as continuous ponding.

11.4.2 SODIC SOILS

The swelling and dispersion of clay minerals are the processes most commonly associated with sodic soils. These processes directly impact the stability of aggregates and the hydraulic properties of soils (e.g., infiltration rates and hydraulic conductivity). Although it is common to use the two terms interchangeably, swelling and dispersion are actually separate processes. Swelling involves the incorporation of water into the interlayers of clays minerals, principally the smectites. Swelling is a reversible process, in that the original particle associations and orientations are retained upon dewatering (or shrinking). Dispersion is an irreversible process, in that flocculation does not result in the same particle associations and orientations. However, in a different sense, dispersion is a reversible process in that flocculation can be initiated by increasing the electrolyte concentration of a solution. Both processes are dependent on the *ESP* and the total concentration of electrolytes in the soil solution. Figure 11.2 illustrates the dispersion and swelling of an expandable clay mineral. Individual clay particles (platelets) in soils are structured into tactoids or quasicrystals, which contain from four to nine platelets. The effective surface area of the tactoids is essentially limited to the external surfaces. Increasing sodicity acts on these external surfaces and results in dispersion, i.e., the mutual repulsion of individual tactoids; however, the internal surfaces remain saturated with Ca^{2+}.

Dispersion is a condition that is exacerbated by low electrolyte concentrations (discussed below). Dispersion is also the dominant mechanism responsible for the physical degradation of soils when *ESP* < 15. As the sodicity increases, Ca^{2+}-Na^+ exchange in the interlayers becomes significant and the influence of the diffuse layers extends (a clays swell). When *ESP* > 15, the physical characteristics of a soil are dictated by the swelling of clay minerals, particularly when $EC < 1$ dS m^{-1}. The tactoid breaks down completely when the *ESP* approaches 50.

11.4.2.1 Classification of Sodic Soils

It was indicated in previous sections that a sodic soil is generally described as one having an SAR_e of greater than 13 to 15 mmol$^{1/2}$ L$^{-1/2}$ and an EC_e of less than 4 dS m^{-1}. However, the disruptive influence of sodicity on the structural stability of soils is not solely a function of exchangeable Na$^+$ concentrations (as denoted by *ESP* and *SAR* values). It has long been recognized that the combination

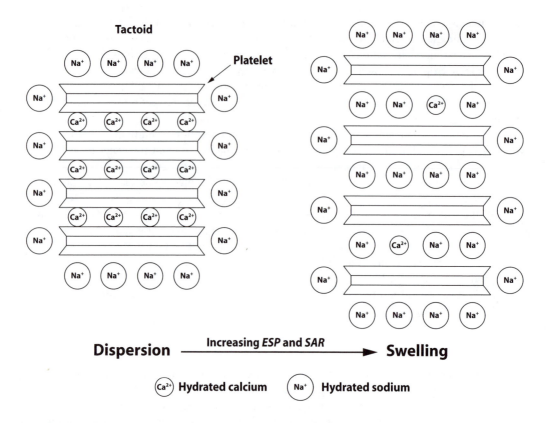

FIGURE 11.2 Clay tactoids are composed of four to nine clay platelets. A dispersed condition exists when tactoids are separated by Na^+ and associated waters of hydration ($ESP < 15$ and low salinity). When the ESP exceeds 15, swelling occurs as Na^+ and waters of hydration expand the clay interlayers.

of sodicity and the salinity of the ambient soil solution are important factors that influence soil structural stability (as does the clay mineralogy) (Table 11.2). The combined influences of irrigation water SAR and EC on sodicity hazard have already been addressed (Table 11.3).

Rengasamy et al. (1984) and Rengasamy and Olsson (1991) (reviewed by Rengasamy and Sumner, 1998) proposed a classification scheme for sodic soils based on potential dispersivity, and the SAR and EC of 1:5 soil-to-solution extracts ($SAR_{1:5}$ and $EC_{1:5}$) (Figure 11.3a). In their classification scheme, soils are dispersive, potentially dispersive, or flocculated. Dispersive soils are those that disperse spontaneously, without mechanical stress. These soils may be classically defined as sodic soils, based upon their physical state. Potentially dispersive soils cover a large area of the arable land and require mechanical energy to disperse, such as rainfall impact or tillage. Flocculated soils contain an abundance of soluble salts and may be typically classified as saline. The line that differentiates between the dispersive and potentially dispersive categories is a function of soil properties, particularly the soil properties that influence negative surface charge (e.g., clay mineralogy). The slope of this line will increase with increasing negative surface charge, and decrease with decreasing negative surface charge.

Sumner et al. (1998) proposed a scheme for classifying Na-affected soils in terms of soil dispersability, $SAR_{1:5}$, and $EC_{1:5}$. In their scheme, a nonsodic soil is described as having an $SAR_{1:5}$ of less than approximately 3 (which corresponds to $ESP < 6$). A soil having an $SAR_{1:5}$ between 3 and 10 (ESP between 6 and 15) is sodic, and a soil with an $SAR_{1:5} > 10$ ($ESP > 15$) is very sodic (Figure 11.3b). The $SAR_{1:5}$ values that place a soil into specific sodicity categories are not fixed, but shift as a function of $EC_{1:5}$ (salinity), pH, organic matter content, and clay type (e.g., swelling vs. nonswelling clays). The classification schemes illustrated in Figures 11.3a and b indicate that

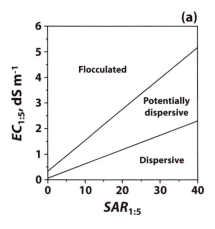
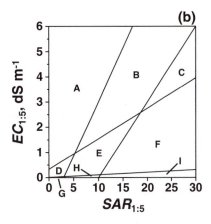

FIGURE 11.3 Sodicity and salinity influence the dispersivity of soil. In (a), Rengasemy et al. (1984) classify Alfisols containing swelling 2:1 clays according to $SAR_{1:5}$ and $EC_{1:5}$ (1:5 solid-to-solutions extracts). The categories are flocculated, potentially dispersive, and dispersive. In (b), Sumner et al. (1998) propose a nine category classification scheme based upon soil salinity ($EC_{1:5}$) and sodicity ($SAR_{1:5}$). The diagram illustrates the classification applied to soils containing nonswelling 2:1 (illitic) clays. The class designations for floccu-lated soils are: A, nonsodic–very saline; B, sodic–very saline; and C, very sodic–very saline. Mechanically dispersive soils are: D, nonsodic–saline; E, sodic–saline; and F, very sodic–saline. Spontaneously dispersive soils are : G, nonsolid–nonsaline; H, sodic–nonsaline; and I, very sodic–nonsaline. Notice that the $SAR_{1:5}$ and $EC_{1:5}$ region that defines dispersive soils is considerably smaller for illitic (nonswelling) soils in (b) than for soils containing swelling clays in (a).

even soils with very low sodicity levels ($ESP < 6$, nonsodic) can display sodic behavior (dispersed, restrictive permeability) when the electrolyte concentrations are not sufficient to maintain a floc-culated state.

11.4.2.2 Flocculation and Dispersion in Colloidal Suspensions

The impact of monovalent cations, such as Na^+, and electrolyte concentration on the extent of the diffuse layer is illustrated in Chapter 7. Classical diffuse layer theory (Gouy-Chapman) states that the thickness of the diffuse layer is inversely proportional to the valence of the exchangeable ions (counterions) and inversely proportional to the square root of the bulk electrolyte concentration:

$$\kappa^{-1} = \frac{3.042(10^{-10})}{ZI^{0.5}}$$

(11.28)

where κ^{-1} (in m) is the thickness of the diffuse layer, Z is the valence of an indifferent counterion, and I is the ionic strength. Thus, a diffuse layer is twice as thick when the saturating cation is monovalent, relative to divalent, and increases by a factor of 1.414 ($\sqrt{2}$) when the concentration of electrolytes in the bathing solution is halved. The thickness of the double layer is a factor that influences the dispersivity and swelling of soil clay. If the extent of the double layer is large, as is the case when the counterion is monovalent and/or the electrolyte solution is dilute, the double layers of adjacent particles will overlap, setting up an electrostatic repulsive force between the particles. Conversely, when the counterion is di- or trivalent and/or the electrolyte solution is concentrated, the double layers of adjacent particles are collapsed, allowing adjacent particles to closely approach one another and for attractive forces to operate. In this case, the attractive forces are van der Waals forces. As adjacent particles collapse there is also a force required to remove the waters of hydration from counterions. This force is called the salvation force. The total potential energy between adjacent particles (assumed to be planar) is the sum of the potential energies associated with each force:

$$\varphi_T = \varphi_R + \varphi_S - \varphi_A \tag{11.29}$$

where φ_T is the total potential energy, φ_R is the electrostatic repulsive potential energy, φ_S is the salvation potential energy, and φ_A is the potential energy associated with van der Waals attractive forces (by convention, positive potentials indicate repulsion and negative potentials attraction).

Colloidal suspensions flocculate because the electrostatic repulsive and solvation forces are balanced by the van der Waals attractive forces that operate between adjacent particles. This can only occur when the double layers are sufficiently compressed to allow the attractive forces to overcome the repulsive forces. An important point to recognize here is that van der Waals attractive forces alone do not cause flocculation. It is when clay particles are brought together by the compression of the double layer that attractive forces can begin to operate. The total potential energy at a particle surface is given by:

$$\varphi_T = \frac{64a^2}{Z\kappa} cRT \exp(-2\kappa d) - \frac{A_H}{12\pi d^2} + \frac{\alpha}{2\pi} \exp\left(-\frac{d}{\delta}\right) \tag{11.30a}$$

where

$$\varphi_R = \frac{64a^2}{Z\kappa} cRT \exp(-2\kappa d) \tag{11.30b}$$

$$\varphi_A = -\frac{A_H}{12\pi d^2} \tag{11.30c}$$

$$\varphi_S = \frac{\alpha}{2\pi} \exp\left(-\frac{d}{\delta}\right) \tag{11.30d}$$

In the above expressions, c is the concentration of a 1:1 electrolyte that contains the indifferent counterion (in mol m^{-3}), $a = \tanh(ZF\psi_0/4RT)$ (where Z is the counterion charge, F is the Faraday constant [96,487 C mol^{-1}], and ψ_0 is the potential at the particle surface in V), κ is the reciprocal of the double layer thickness (in nm^{-1}) (see Chapter 7 for computation), A_H is the Hamaker constant (2.2×10^{-20} J), α and δ are empirical constants ($\alpha \approx 0.03$ to 0.05 N m^{-1}, $\delta \approx 0.3$ to 1.0 nm), and d is the half-distance between two planar surfaces (in m).

The influence of electrolyte concentrations on the potential energy terms operating between adjacent planar particles is illustrated in Figures 11.4a through d for a divalent counterion (neglecting φ_S which is only significant when $d \ll 1$ nm). At a low electrolyte concentration (1 mol m^{-3}) (Figure 11.4a), repulsive forces (φ_R) dominate the total potential energy (φ_T) in the double layer at distances greater than the distance of closest approach (<1 nm). In this case, adjacent particles are predicted to remain dispersed, as the collisions between particles that result from their Brownian motion (thermal energy) will not generate the energy required to overcome the large positive potential energy barrier (the activation energy). In this case, flocculation will be retarded by the long-range repulsion and a stable colloidal suspension will result. As the electrolyte concentration is increased to 50 mol m^{-3} (Figure 11.4b), the double layer compresses, but φ_R continues to dominate φ_T. A large activation energy is still required to achieve flocculation; however, slow flocculation may occur. Increasing the electrolyte concentration to 250 mol m^{-3} (Figure 11.4c) sufficiently compressed the double layer to allow the attractive van der Waals forces (φ_A) to overcome φ_R and

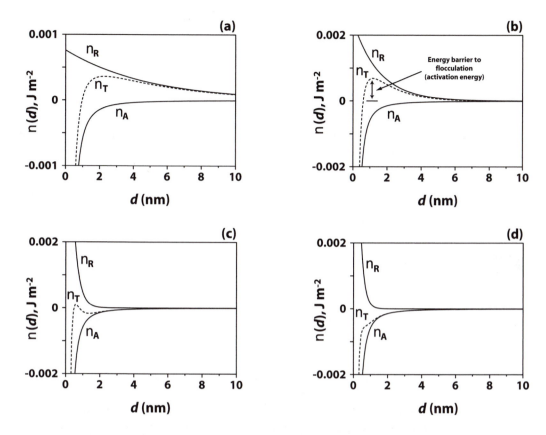

FIGURE 11.4 Total potential energy (φ_T), electrostatic repulsion potential energy (φ_R), and van der Waals attractive potential energy (φ_A) as a function of separation distance (d) between planar particles and divalent ion ($Z = 2$) electrolyte concentrations of (a) 1 mol m^{-3}, (b) 50 mol m^{-3}, (c) 250 mol m^{-3}, and (d) 500 mol m^{-3}. The potential energy curves were constructed using Equations 11.30a, 11.30b, and 11.30c and the following parameters: $\psi_0 = 0.203$ V, $A_H = 2.2 \times 10^{-20}$ J, $\varepsilon = 78.5$, $\varepsilon_0 = 8.854 \times 10^{-12}$ C^2 J^{-1} m^{-1}, $F = 96{,}487$ C mol^{-1}, $R = 8.314$ L mol^{-1} K^{-1}, and $T = 298$ K. The potential energy barrier that colliding particles must overcome to achieve flocculation is identified in (b), but also present in (a) and (c).

produce a local minimum (called a secondary minimum) and negative φ_T at approximately 1.2 nm. At this electrolyte concentration there is only a very small energy barrier which colliding particles may overcome as a result of their Brownian motion, potentially leading to slow flocculate. When electrolyte concentrations are sufficiently high, as illustrated for the 500 mol m^{-3} system in Figure 11.4d, the van der Waals attractive forces dominate φ_T and there is no repulsion predicted at any distance from the surface (and no potential energy barrier to overcome). In this high electrolyte system particle flocculation occurs spontaneously.

The electrolyte concentration necessary to cause flocculation can be computed using Equation 11.30a. A critical assumption is that the condition for flocculation is $\varphi_T = 0$ at the local maximum on a plane that is equidistant between adjacent planar particles (φ_T is never positive). The electrolyte concentration at and below which dispersion occurs is described by a critical coagulation concentration (CCC) (also called the critical flocculation concentration). Strictly defined, the CCC is the smallest concentration of electrolyte at which a colloidal suspension begins to undergo rapid coagulation. The CCC is dependent on the nature of the minerals in a colloidal suspension and on the composition of the electrolyte solution (counterion valence). Manipulation of Equation 11.30a and imposing the assumed condition for flocculation indicated above (i.e., $\varphi_R = \varphi_A$) leads to the expression:

$$CCC = \left(\frac{3072\pi a^2 RT}{e^2 \beta^{3/2} A_H} \right)^2 Z^{-6} \tag{11.31}$$

where β is a constant related to the inverse thickness of the double layer (κ) at the CCC. Equation 11.31 is known as the Schulze-Hardy rule, which can be simplified to:

$$CCC = \frac{k_{SH}}{Z^6} \tag{11.32}$$

where k_{SH} is an empirical constant and Z is the valence of the ion that produces flocculation. Equations 11.31 and 11.32 show that counterion charge strongly influences the CCC. For example, the ratio of $CCC_{Z=1}$ to $CCC_{Z=2}$ is:

$$\frac{CCC_{(Z=1)}}{CCC_{(Z=2)}} = \frac{\dfrac{k_{SH}}{1^6}}{\dfrac{k_{SH}}{2^6}} = 2^6 = 64 \tag{11.33}$$

This result indicates that the concentration of a monovalent electrolyte suspension must be 64 times that of a divalent electrolyte suspension to cause flocculation.

The CCC values for clay minerals are generally described by the Schulze-Hardy rule, although considerable variability in the experimental data exists. Consider a colloidal suspension containing kaolinite. For Na^+ systems, an average $CCC_{(z=1)}$ value is 8.5 mM (pH 7 – 8.3) (Sposito, 1984; Goldberg et al. 2000). When kaolinite suspensions contain Ca^{2+}, the average $CCC_{(z=2)}$ is 0.2 mM. Using these average CCC values, the ratio of $CCC_{(z=1)}$ to $CCCc_{(z=2)}$ for kaolinite is 43 (Equation 11.33). For colloidal montmorillonite systems, an average $CCC_{(z=1)}$ is 8 mM and an average $CCC_{(z=2)}$ is 0.12 mM (Sposito, 1984). The computed ratio of $CCC_{(z=1)}$ to $CCC_{(z=2)}$ for montmorillonite is 67. The ratios of $CCC_{(z=1)}$ to $CCC_{(z=2)}$ for kaolinite and montmorillonite (43 and 67) compare favorably to the theoretical ratio of 64 (2^6) predicted by the Schulze-Hardy rule.

11.4.2.3 Genesis and Management of Sodic Soils

Sodic soils most commonly occur as a result of natural processes (primary sodification). The nature of the parent material and the subsequent pedogenic processes dictate sodic soil development. Conditions that promote the formation of sodic soils include the presence of shallow saline ground-water or a perched water table (within 1.5 m of the surface), soil water rich in bicarbonate, impeded drainage, low slope, and textural discontinuities. Sodic, or black alkali, soils are associated with the presence of Na_2CO_3, high ESP values, and pH values greater than 9.0 (a consequence of the high Na_2CO_3 content). Secondary sodification is a process that occurs as a result of anthropogenic activities, where the development of a soil with sodic properties results from poor irrigation water management and poor drainage (waterlogging).

The natural formation of a sodic soil is closely tied to sulfate reduction in water-saturated systems. The necessary ingredients for primary sodic soil formation are impeded drainage, a shallow groundwater that is high in sulfate, high soil organic matter, and a reducing environment at the ground water interface. The reduction of sulfate produces alkalinity (consumes protons):

$$SO_4^{2-}(aq) + 2H^+(aq) \rightarrow H_2S(g) + 2O_2(g) \tag{11.34}$$

The decomposition of organic matter produces CO_2, which is absorbed by the alkaline waters to form HCO_3^-. The displacement of Ca^{2+} and Mg^{2+} from the exchange complex by Na^+ is favored as a result of calcite precipitation:

$$(Ca^{2+} + Mg^{2+})X_2(ex) + 2Na^+(aq) \rightarrow 2NaX(ex) + (Ca^{2+} + Mg^{2+})(aq) \qquad (11.35a)$$

$$(Ca^{2+} + Mg^{2+})(aq) + HCO_3^- (aq) \rightarrow (Ca, Mg)CO_3(s) + H^+(aq) \qquad (11.35b)$$

The high sulfate levels in the shallow groundwater can also provide favorable conditions for gypsum [$CaSO_4 \cdot 2H_2O$] precipitation, again favoring the retention of Na^+ on the exchange complex by removing dissolved Ca^{2+} from the soil solution:

$$Ca^{2+}(aq) + SO_4^{2-}(aq) + 2H_2O(l) \rightarrow CaSO_4 \cdot 2H_2O(s) \qquad (11.35c)$$

Excess bicarbonate coupled with elevated Na^+ levels result in the production of $NaHCO_3$. Finally, dehydration favors Na_2CO_3 formation: $2NaHCO_3 \rightarrow Na_2CO_3 + CO_2 + H_2O$. As a result, a high *ESP* characterizes the exchange complex, high *SAR* and high pH characterize the soil solution (controlled by Na_2CO_3), and calcite is present in the solid phase.

The impact of excessive sodicity on the permeability of soils is intimately associated with the salinity of the soil solution. As indicated in the paragraphs above, the degradative effects of sodium on soil structural properties can be overcome by two principle mechanisms, acting alone or in concert: (1) displacement of Na^+ from the soils exchange complex by divalent cations, or (2) increasing the salinity of the soil solution. These two mechanisms are complementary with regard to their beneficial effects on sodic soils. Irrigation with saline irrigation water may be initially employed to stabilize soil structure and increase soil permeability. However, increasing soil salinity is less effective in clayey, smectitic soils (which tend to remain relatively impermeable irrespective of *EC*) (Figure 11.5). Increasing the salinity of a soil solution also increases the threshold *ESP*, the *ESP* below which flocculation occurs. Increased permeability allows for the removal of soluble components that might interfere with reclamation (such as soluble sulfate when gypsum is the amendment) and the effective infusion of divalent cations necessary for sodic soil reclamation.

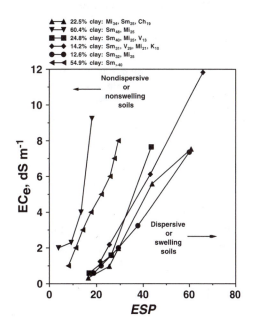

FIGURE 11.5 The saturation extract electrical conductivity (EC_e) and exchangeable sodium percentage (*ESP*) values needed to maintain soil hydraulic conductivity are a function of soil clay content and type. Each line represents the threshold *ESP-EC_e* values that result in a 20 to 25% reduction in soil permeability. Combinations of *ESP-EC_e* values that lie to the right of a line for a given soil result in a dispersive or swelling soil, while values to the left of the line result in a flocculated and permeable soil. Each soil is identified by the clay content and the dominate clay minerals: Mi, mica; Sm, smectite; Ch, chlorite; V, vermiculite; and K, kaolinite. Only clay minerals that account for greater than 10% of the total clay are indicated. Subscripts identify the approximate percentage of each clay mineral in the clay fraction. The data were obtained from McNeal and Coleman (1966) and Naghshineh-Pour et al. (1970).

The reclamation of a sodic soil requires adequate drainage, leaching, and a source of Ca^{2+}. Calcium, which is required to displace Na^+ from the exchange complex, can be derived from a number of sources. Chemical amendments that supply Ca^{2+} include gypsum and $CaCl_2$. If calcite is present in the soil, H_2SO_4 can be used to dissolve calcite. Highly saline water may also be employed to hasten Na^+ displacement by relying on the valence dilution effect (Chapter 8). In this procedure, a sodic soil is initially leached with highly saline, low SAR water, which also increases permeability (Figure 11.5). With each successive treatment, the irrigation water is diluted with high-quality (less saline) water. The dilution favors the displacement of Na^+ by divalent cations, and leaching removes the soluble Na^+. Irrespective of the mechanism used to reduce SAR and ESP, leaching to remove Na^+ from the soil profile is a necessity.

11.4.3 Sodic Mine Spoil Reclamation

Sodicity problems may also be encountered during the reclamation of mine spoil materials. Spoil materials from the strip mining of coal in the Four-Corners region of northern New Mexico (San Juan Basin) are predominantely saline and sodic. As a result, the very low permeability of the spoil materials complicates revegetation and management efforts. The spoil materials contain greater than 30% smectitic clay and have SAR_e values of approximately 40 $mmol^{1/2} L^{-1/2}$ and EC_e values around 14 dS m^{-1}. The available irrigation waters in the region have SAR values that range from approximately 10 to greater than 30 $mmol^{1/2} L^{-1/2}$, EC values that range from 4 to 8 dS m^{-1}, and elevated sulfate levels (24 to 50 mmol $SO_4 L^{-1}$).

Weber et al. (1979) examined the influence of irrigation water quality and the timing of gypsum ($CaSO_4 \cdot 2H_2O$) and sulfuric acid (H_2SO_4) amendments on the permeability of sodic mine spoil materials (Figure 11.6). Gypsum is a readily available and inexpensive source of Ca^{2+}, and H_2SO_4 is used to hasten the dissolution of calcite (4.8% of the spoil), releasing Ca^{2+} to the spoil solution. Their findings indicate that infiltration rates may be increased without chemical amendments by irrigating with a relatively low SAR, low sulfate water (irrigation water A) prior to rainwater applications (an example of the valence dilution effect). However, incorporation of either gypsum or H_2SO_4 resulted in additional increases in infiltration rates. Two treatments: the application of H_2SO_4 before irrigation (early acid treatment), or the incorporation of gypsum after irrigation (late gypsum treatment) and before rainwater application, increased infiltration rates to the greatest extent. The use of irrigation water with relatively high SAR and sulfate levels (irrigation water B) did not effectively increase infiltration rates. Further, the poor quality of irrigation water B was found to minimize the impact of the chemical amendments.

The observed treatment effects on spoil infiltration rates are due to the differing chemical characteristics of the two irrigation waters. The EC_e and SAR_e of the spoil material are 13.9 dS m^{-1} and 37.7 $mmol^{1/2} L^{-1/2}$. Irrigation water B has an EC of 7.8 dS m^{-1} and SAR of 32.5 $mmol^{1/2} L^{-1/2}$, and a sulfate concentration of 49.7 mmol L^{-1}. Using the conversion, $EC_e = \frac{1}{2} EC_{iw}$, the irrigation water would generate spoil EC_e and SAR_e values of 3.9 dS m^{-1} and 23 $mmol^{1/2} L^{-1/2}$ ($ESP \approx 25$, Equation 11.8b). Irrigation water B provides some reduction in EC_e (13.9 to 3.9 dS m^{-1}) and SAR_e (37.7 to 23 $mmol^{1/2} L^{-1/2}$). However, the ESP of the spoil is significantly greater than 15 (the ESP level that defines a sodic soil) and above the threshold level necessary to maintain stability in a smectitic system with an EC_e of 3.9 dS m^{-1} (Figure 11.5). The relative ineffectiveness of the gypsum and H_2SO_4 treatments can be attributed to the high sulfate concentration in irrigation water B (49.7 mmol L^{-1}) and in the spoil saturation extract (89.4 mmol L^{-1}). The dissolution of gypsum is retarded until sufficient rainwater has passed through the spoil to decrease soluble sulfate levels (Figure 11.6). Application of H_2SO_4 prior to irrigation resulted in a substantial increase in spoil permeability. However, the effect was short-lived as the solubilized Ca^{2+} was precipitated as gypsum (Equation 11.30c).

Spoil materials receiving irrigation water A responded to both the chemical treatments and to the valence dilution effect (Figure 11.6). The irrigation water would generate spoil EC_e and SAR_e

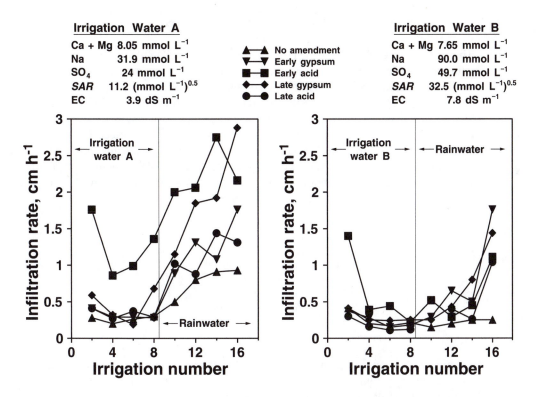

FIGURE 11.6 Weber et al. (1979) examined the influence of irrigation water quality and chemical amendments on the permeability of sodic mine spoils. Irrigation waters A and B (chemical composition shown) were applied for eight irrigation cycles, followed by eight cycles of irrigation with rainwater. The chemical treatments, gypsum or sulfuric acid, were applied at the beginning of the study (early amendment) or after irrigation number eight (late amendment). Amendment rates of 1.7 kg gypsum ha^{-1} and 1.0 kg H$_2$SO$_4$ ha^{-1} were computed to decrease *ESP* in the surface 30 cm of spoil by 10%.

values of 2 dS m^{-1} and 8 mmol$^{1/2}$ L$^{-1/2}$ (*ESP* ≈ 8). Irrigation water A provides significant reduction of spoil *EC*$_e$ (13.9 to 2 dS m^{-1}) and *SAR*$_e$ (37.7 to 8 mmol$^{1/2}$ L$^{-1/2}$). Although the *ESP* of the spoil irrigated with water A is less than 15, the spoil *ESP* is not clearly below the threshold level necessary to maintain stability in a smectitic system with an *EC*$_e$ of 2 dS m^{-1} (Figure 11.5). However, dilution of the spoil solution by rainwater increases the infiltration rates, indicating that sufficient concentrations of Ca^{2+} are present to produce the valence dilution effect. The effectiveness of the chemical treatments can be related to the relatively low sulfate concentrations in irrigation water A (24 mmol L^{-1}), relative to water B. Thus, Ca^{2+} released from calcite by H$_2$SO$_4$ and through the solubilization of gypsum remains soluble and displaces Na$^+$ from the exchange complex.

11.5 EXERCISES

1. An irrigation water has an EC_{iw} of 1.2 dS m^{-1} and a Na concentration of 45 mg L^{-1}. What is (a) the SAR_{iw} of the water, and (b) the *ESR* and *ESP* for a soil that is in equilibrium with this water if the Gapon selectivity coefficient is 0.015?

2. Well water from Roswell, NM contains 12.7 mmol$_c$ L^{-1} Ca, 6.37 mmol$_c$ L^{-1} Mg, and 27.1 mmol$_c$ L^{-1} Na. Compute the following: (a) *EC*, (b) *TDS*, and (c) *SAR*. Is this water suitable for irrigation for crops grown on soils of mixed mineralogy (assume that the

bicarbonate hazard is negligible)? Determine the *LR* required to produce peanuts using this irrigation water. What is the average EC_{dw}?

3. An irrigation water contains 1000 mg L^{-1} total dissolved solids. If used at an average leaching fraction of 0.2, what will be the average EC_{dw} of the drainage water leaving the soil?

4. The chemical characteristics of saturation extracts of three soils are provided in the table below. (a) Classify the soils using the criteria established in Figure 11.3a. (b) Indicate which soils may be described as saline. (c) Assume soil C is a clay loam soil and that the reported EC_e value represents $\overline{EC_e}$, the average salinity of the surface 45 cm. What is the depth of ponded irrigation water that must be applied to prepare this soil for alfalfa production using an irrigation water with $EC_{iw} = 0.74$ (assume a 45-cm active root zone)? (d) If sprinkler irrigation is employed instead of ponded irrigation, how much water is necessary to prepare the soil for alfalfa production?

Soil	pH	EC_e dS m^{-1}	Ca	Mg mmol L^{-1}	Na
A	7.41	0.70	1.00	0.50	2.80
B	8.12	3.36	5.85	1.85	9.20
C	6.99	8.90	41.6	8.30	7.40

5. A solution with a concentration of Na equal to 0.1 *M* and Ca equal to 0.05 *M* is leached through a soil until equilibrium is attained. The Na concentration of the leachate is then diluted to 0.05 *M*. What must the concentration of Ca in the leachate be in order to maintain the same ratio of Na to Ca on the exchange complex?

6. You are approached by a farmer with a problem. He farms approximately 100 acres of land in Arkansas, a considerable portion of which has naturally restricted permeability. A consultant told him that his soils also had relatively high sodium levels and you surmise that his infiltration problem is associated with sodicity. What is your recommendation to this farmer?

7. Equation 11.26 computes the depth of water necessary to reclaim a saline soil for crop production. In this expression, the *K* value for sprinkler irrigation is the same as that for flood irrigation of sandy loam soils ($K = 0.1$). Why is this the case?

8. Sumner et al. (1998) promote a saline and sodic soil classification scheme that is based on the *EC* and *SAR* of 1:5 soil-to-solution extracts ($EC_{1:5}$ and $SAR_{1:5}$) (Figure 11.3b) that relates the dispersability of soils. The lines that differentiate among the various categories in Figure 11.3b are not static, but move depending on the dominant clay mineralogy of a soil, as well as other soil-specific factors. Beginning with Figure 11.3b, qualitatively construct a diagram that may be used to classify soils that are predominantly composed of smetitic (expansive) clays, and indicate your reasoning behind any adjustments to Figure 11.3b.

9. Is irrigated agriculture a sustainable agronomic practice? Defend your opinion.

10. The *CCC* of Cs$^+$ for a montmorillonite suspension is 0.79 m*M*. The *CCC* of Na$^+$ for the same suspension is 2.1 m*M*. Explain why these *CCC* values differ (Hint: examine the Lewis acid character of each cation, the types of surface complexes formed by each cation in the smectite interlayers, and the resulting particle surface charge, σ_p).

11. Spoils resulting from strip mining in the Four-Corners region of New Mexico are predominantly sodic and contain calcite and a high smectitic clay content. Describe three possible reclamation strategies that could be employed to improve the hydraulic conductivity of the material.

12. Calcite ($CaCO_3$) is commonly present in sodic soils. Explain why this is observed.

13. Construct a potential energy diagram that illustrates the φ_T, φ_R, and φ_A as a function of separation distance between planar surfaces for a monovalent cation ($Z = 1$). Use Equations 11.30a, 11.30b, and 11.30c and the parameters employed in the construction of Figure 11.4. Compare your diagram to Figure 11.4b, which illustrates φ_T, φ_R, and φ_A as a function of separation distance for a divalent electrolyte system ($Z = 2$). Discuss your findings.

REFERENCES

American Society of Civil Engineers. *Agricultural Salinity Assessment and Management.* Tanji, K.K. (Ed.) ASCE Manuals and Reports on Engineering Practices No. 71. American Society of Civil Engineers, New York, 1990.

Ayers, R.S. and D.W. Westcot. Water quality for agriculture. FAO Irrig. Drain. Paper 29, 1985.

Bohn, H.L., B.L. McNeal, and G.A. O'Connor. *Soil Chemistry.* 2nd ed. John Wiley & Sons, New York, 1985.

Goldberg, S., I. Labron, and D.L. Suarez. Soil colloidal behavior. In *Handbook of Soil Science.* M.E. Sumner (Ed.) CRC Press, Boca Raton, FL, 2000, pp. B195–B240.

Hillel, D. *Out of the Earth: Civilization and the Life of the Soil.* University of California Press, Berkeley, 1992.

Maas, E.V. Crop salt tolerance. In *Agricultural Salinity Assessment and Management.* K.K. Tanji (Ed.) American Society of Civil Engineers Manuals and Reports on Engineering Practices No. 71. American Society of Civil Engineers, New York, 1990, pp. 262–304.

Massoud, F.J. Salt affected soils at a global scale and concepts for control. Tech. Paper. FAO Land and Water Development Div., Rome, 1981.

McNeal, B.L., W.A. Norvell, and N.T. Coleman. Effect of solution composition on the swelling of extracted soil clays. *Soil Sci. Soc. Am. Proc.* 30:313–317, 1966.

Naghshineh-Pour, B., G.W. Kunze, and C.D. Carson. The effect of electrolyte composition on hydraulic conductivity of certain Texas soils. *Soil Sci.* 110:124–127, 1970.

Rengasamy, P., R.S.B. Greene, G.W. Ford, and A.H. Mehanni. Identification of dispersive behavior and management of Red-Brown Earths. *Aust. J. Soil Res.* 24:229–237, 1984.

Rengasamy, P. and K.A. Olsson. Sodicity and soil structure. *Aust. J. Soil Res.* 29:935–952, 1991.

Rengasamy, P. and M.E. Sumner. Processes involved in sodic behavior. In *Sodic Soils: Distribution, Properties, Management, and Environmental Consequences.* M.E. Sumner and R. Naidfu (Eds.) Oxford University Press, New York, 1998, pp. 35–50.

Rhoades, J.D. Drainage for salinity control, *Agronomy* 17:433–461, 1974.

Sposito, G. *The Surface Chemistry of Soils.* Oxford University Press, New York, 1984.

Sposito, G. and S.V. Mattigod. On the chemical formation of the sodium adsorption ratio. *Soil Sci. Soc. Am. J.* 41:310–315, 1977.

Suarez, D.L. Relation between pH$_c$ and sodium adsorption ratio (SAR) and an alternate method of estimating SAR of soil or drainage waters. *Soil Sci. Soc. Am. J.* 45:469–475, 1981.

Sumner, M.E., P. Rengasamy, and R. Naidu. Sodic soils: a reappraisal. In *Sodic Soils: Distribution, Properties, Management, and Environmental Consequences.* M.E. Sumner and R. Naidfu (Eds.) Oxford University Press, New York, 1998, pp. 3–17.

Szabolcs, I. *Salt-Affected Soils.* CRC Press, Boca Raton, FL, 1989.

United States Salinity Laboratory Staff. Diagnosis and improvement of saline and sodic soils. L.A. Richards (Ed.). USDA Agric. Handbook No. 60. Washington, D.C., 1954.

Weber, S.J., M.E. Essington, G.A. O'Connor, and W.L. Gould. Infiltration studies with sodic mine spoil material. *Soil Sci.* 128:312–318, 1979.

Index